Snow and Ice-Related Hazards, Risks, and Disasters

T0305730

Snow and Ice-Related Hazards, Risks, and Disasters

Volume Editors

Wilfried Haeberli
University of Zurich, Switzerland

Colin Whiteman
University of Brighton, UK

Series Editor

John F. Shroder
University of Nebraska at Omaha, US

ELSEVIER

AMSTERDAM • BOSTON • HEIDELBERG • LONDON • NEW YORK • OXFORD
PARIS • SAN DIEGO • SAN FRANCISCO • SINGAPORE • SYDNEY • TOKYO

Elsevier
Radarweg 29, PO Box 211, 1000 AE Amsterdam, Netherlands
225 Wyman Street, Waltham, MA 02451, USA
The Boulevard, Langford Lane, Kidlington, Oxford, OX5 1GB, UK

Notices
Knowledge and best practice in this field are constantly changing. As new research and experience broaden our understanding, changes in research methods, professional practices, or medical treatment may become necessary.

Practitioners and researchers must always rely on their own experience and knowledge in evaluating and using any information, methods, compounds, or experiments described herein. In using such information or methods they should be mindful of their own safety and the safety of others, including parties for whom they have a professional responsibility.

To the fullest extent of the law, neither the Publisher nor the authors, contributors, or editors, assume any liability for any injury and/or damage to persons or property as a matter of products liability, negligence or otherwise, or from any use or operation of any methods, products, instructions, or ideas contained in the material herein.

Library of Congress Cataloging-in-Publication Data
Application submitted

British Library Cataloguing in Publication Data
A catalogue record for this book is available from the British Library

ISBN: 978-0-12-394849-6

For information on all Elsevier publications visit our
web site at http://store.elsevier.com/

This book has been manufactured using Print On Demand technology.

Contents

1.   Snow and Ice-Related Hazards, Risks, and Disasters: A General Framework

Wilfried Haeberli and Colin Whiteman

2.   Physical, Thermal, and Mechanical Properties of Snow, Ice, and Permafrost

Lukas U. Arenson, William Colgan and Hans Peter Marshall

3.   Snow and Ice in the Climate System

Atsumu Ohmura

11. Radioactive Waste Under Conditions
     of Future Ice Ages
     Urs H. Fischer, Anke Bebiolka, Jenny Brandefelt, Sven Follin,
     Sarah Hirschorn, Mark Jensen, Siegfried Keller, Laura Kennell,
     Jens-Ove Näslund, Stefano Normani, Jan-Olof Selroos
     and Patrik Vidstrand

12. Snow Avalanches
     Jürg Schweizer, Perry Bartelt and Alec van Herwijnen

13. Glacier Surges
     William D. Harrison, Galina B. Osipova, Gennady A. Nosenko,
     Lydia Espizua, Andreas Kääb, Luzia Fischer, Christian Huggel,
     Patty A. Craw Burns, Martin Truffer and Alexandre W. Lai

17.   Hazards at Ice-Clad Volcanoes: Phenomena, Processes, and Examples From Mexico, Colombia, Ecuador, and Chile

Hugo Delgado Granados, Patricia Julio Miranda,
Gerardo Carrasco Núñez, Bernardo Pulgarín Alzate,
Patricia Mothes, Hugo Moreno Roa, Bolívar E. Cáceres Correa
and Jorge Cortés Ramos

18.   Floating Ice and Ice Pressure Challenge to Ships

Ivana Kubat, Captain David Fowler and Mohamed Sayed

19.   Retreat Instability of Tidewater Glaciers and Marine Ice Sheets

Andreas Vieli

20.   Ice Sheets, Glaciers, and Sea Level

Ian Allison, William Colgan, Matt King and Frank Paul

Contributors

Ian Allison, Antarctic Climate and Ecosystems Cooperative Research Centre, Hobart, Australia

Bernardo Pulgarín Alzate, Servicio Geológico Colombiano, Observatorio Vulcanológico y Sismológico de Popayán, Barrio Loma de Cartagena, Popayán, Colombia

Oleg Anisimov, State Hydrological Institute, St. Petersburg, Russia

Lukas U. Arenson, BGC Engineering Inc., Vancouver, BC, Canada

Perry Bartelt, WSL Institute for Snow and Avalanche Research SLF, Davos, Switzerland

Anke Bebiolka, Bundesanstalt für Geowissenschaften und Rohstoffe (BGR), Stilleweg, Hannover, Germany

Tobias Bolch, Department of Geography, University of Zurich, Switzerland; Institute for Cartography; Technische Universität Dresden, Dresden, Germany

Jenny Brandefelt, Svensk Kärnbränslehantering AB (SKB), Blekholmstorget, Stockholm, Sweden

Michael Bründl, WSL Institute for Snow and Avalanche Research SLF, Davos, Switzerland

Bolívar E. Cáceres Correa, Instituto Nacional de Meteorología (INAMHI), Corea, Quito, Ecuador

Terry V. Callaghan, Royal Swedish Academy of Sciences, Lilla Frescativägen, Stockholm, Sweden; Department of Animal and Plant Sciences, University of Sheffield, Sheffield, UK; Department of Botany, National Research Tomsk State University, Tomsk, Siberia, Russia

Mark Carey, Robert D. Clark Honors College, University of Oregon, USA

Hanne H. Christiansen, Arctic Geology Department, The University Centre in Svalbard, UNIS, Longyearbyen, Norway; Center for Permafrost, CENPERM; Department of Geoscience and Natural Resource Management, University of Copenhagen, Denmark

John J. Clague, Centre for Natural Hazard Research, Simon Fraser University, Burnaby, B C, Canada

William Colgan, Geological Survey of Denmark and Greenland, Copenhagen, Denmark

Simon Cook, School of Science and the Environment, Manchester Metropolitan University, Manchester, United Kingdom

Patty A. Craw Burns, Department of Natural Resources, Division of Mining, Land & Water, Lands Section, Fairbanks, AK, USA

Reynald Delaloye, Department of Geosciences, Geography, University of Fribourg, Fribourg, Switzerland

Keith B. Delaney, Natural Disaster Systems, Department of Earth and Environmental Sciences, University of Waterloo, Waterloo, Ontario, Canada

Philip Deline, EDYTEM Lab, Université de Savoie, CNRS, Le Bourget-du-Lac Cedex, France

Lydia Espizua, Instituto Argentino de Nivología, Glaciología y Ciencias Ambientales (IANIGLA), Mendoza, Argentina

Stephen G. Evans, Natural Disaster Systems, Department of Earth and Environmental Sciences, University of Waterloo, Waterloo, Ontario, Canada

Tracy Ewen, Department of Geography, University of Zurich, Switzerland

Urs H. Fischer, Nationale Genossenschaft für die Lagerung radioactiver Abfälle (Nagra), Wettingen, Switzerland

Luzia Fischer, Norwegian Geological Survey, Trondheim, Norway

Sven Follin, SF GeoLogic AB, Täby, Sweden

Captain David Fowler, Retired Canadian Coast Guard Captain, McDougall, ON, Canada

Isabelle Gärtner-Roer, Department of Geography, University of Zürich, Switzerland

Marten Geertsema, Ministry of Forests, Lands, and Natural Resource Operations, Prince George, BC, Canada

Marco Giardino, GeoSitLab, Dipartimento di Scienze della Terra, Università di Torino, Italy

Hugo Delgado Granados, Departamento de Vulcanología, Instituto de Geofísica, Universidad Nacional Autónoma de México, México

Stephan Gruber, Department of Geography and Environmental Studies, Carleton University, Ottawa, Canada

Wilfried Haeberli, Department of Geography, University of Zurich, Switzerland

William D. Harrison, Geophysical Institute, University of Alaska, Fairbanks, AK, USA

Andreas Hasler, Department of Geography, University of Zurich, Switzerland

Tobias Heckmann, Department of Physical Geography, Catholic University of Eichstätt-Ingolstadt, Germany

Sarah Hirschorn, Nuclear Waste Management Organization (NWMO), Toronto, ON, Canada

Christian Huggel, Department of Geography, University of Zurich, Switzerland

Matthias Huss, Laboratory of Hydraulics, Hydrology and Glaciology (VAW), ETH, Zurich, Switzerland; Department of Geosciences, University of Fribourg, Switzerland

Jerrilynn Jackson, Department of Geography, University of Oregon, USA

Michal Jenicek, Department of Geography, University of Zurich, Switzerland; Department of Physical Geography and Geoecology, Faculty of Science, Charles University in Prague, Czech Republic

Mark Jensen, Nuclear Waste Management Organization (NWMO), Toronto, ON, Canada

Margareta Johansson, Department of Physical Geography and Ecosystem Science, Lund University, Lund, Sweden

Andreas Kääb, Department of Geosciences, University of Oslo, Norway

Siegfried Keller, Bundesanstalt für Geowissenschaften und Rohstoffe (BGR), Hannover, Germany

Laura Kennell, Nuclear Waste Management Organization (NWMO), Toronto, ON, Canada

Matt King, School of Geography and Environmental Studies, University of Tasmania, Hobart, Australia

Martin Kirkbride, Geography, School of the Environment, University of Dundee, United Kingdom

Oliver Korup, Institute of Earth and Environmental Science, University of Potsdam, Germany

Michael Krautblatter, Technische Universität München, Germany

Ivana Kubat, National Research Council of Canada, Coastal and River Engineering, Ottawa, Ontario, Canada

Alexandre W. Lai, Alyeska Pipeline Service Company, Integrity Management Department, Fairbanks, AK, USA

Florence Magnin, EDYTEM Lab, Université de Savoie, CNRS, Le Bourget-du-Lac Cedex, France

Stefan Margreth, WSL Institute for Snow and Avalanche Research SLF, Davos, Switzerland

Hans Peter Marshall, Department of Geosciences and Center for Geophysical Investigation of the Shallow Subsurface, Boise State University, ID, USA

Samuel McColl, Physical Geography Group, Institute of Agriculture and Environment, Massey University, Palmerston North, Australia

Graham McDowell, Department of Geography, McGill University, Montral, Canada

Patricia Julio Miranda, Escuela de Ciencias Sociales y Humanidades, Universidad Autónoma de San Luis Potosí, Frac. Talleres, México

Jeffrey Moore, Department of Geology and Geophysics, University of Utah, Salt Lake City, UT, United States

Patricia Mothes, Instituto Geofísico, Escuela Politécnica Nacional, Quito, Ecuador

Jens-Ove Näslund, Svensk Kärnbränslehantering AB (SKB), Blekholmstorget, Stockholm, Sweden

Stefano Normani, Civil and Environmental Engineering, University of Waterloo, ON, Canada

Gennady A. Nosenko, Institute of Geography, Russian Academy of Sciences, Moscow, Russia

Gerardo Carrasco Núñez, Centro de Geociencias, Campus UNAM Juriquilla, Querétaro, Qro

Jim E. O'Connor, U.S. Geological Survey, Oregon Water Science Center, Portland, Oregon, USA

Atsumu Ohmura, Institute for Atmospheric and Climate Science, Swiss Federal Institute of Technology (E.T.H), Zurich, Switzerland

Galina B. Osipova, Institute of Geography, Russian Academy of Sciences, Moscow, Russia

Frank Paul, Department of Geography, University of Zurich, Switzerland

César Portocarrero, Independent Consultant, Huaraz, Peru

Jorge Cortés Ramos, Departamento de Vulcanología, Instituto de Geofísica, Universidad Nacional Autónoma de México, México

Ludovic Ravanel, EDYTEM Lab, Université de Savoie, CNRS, Le Bourget-du-Lac Cedex, France

John M. Reynolds, Reynolds International Ltd, Mold, UK

Hugo Moreno Roa, Servicio Nacional de Geología y Minería (SERNAGEOMIN), Observatorio Volcanológico de los Andes del Sur (OVDAS), Rudecindo, Temuco, Chile

Mohamed Sayed, National Research Council of Canada, Coastal and River Engineering, Ottawa, Ontario, Canada

Philippe Schoeneich, Institut de Géographie Alpine, Université de Grenoble, CNRS, Grenoble, France

Jürg Schweizer, WSL Institute for Snow and Avalanche Research SLF, Davos, Switzerland

Jan Seibert, Department of Geography, University of Zurich, Switzerland; Department of Earth Sciences, Uppsala, University, Sweden

Jan-Olof Selroos, Svensk Kärnbränslehantering AB (SKB), Stockholm, Sweden

Dmitry Streletskiy, Department of Geography, George Washington University, Washington DC, USA

Darrel A. Swift, Department of Geography, University of Sheffield, United Kingdom

Martin Truffer, Geophysical Institute and the Department of Physics, University of Alaska, Fairbanks, AK, USA

Alec van Herwijnen, WSL Institute for Snow and Avalanche Research SLF, Davos, Switzerland

Alexander Vasiliev, Earth Cryosphere Institute RAS, Moscow, Russia

Luis Vicuña, Department of Geography, University of Zurich, Switzerland

Patrik Vidstrand, Svensk Kärnbränslehantering AB (SKB), Stockholm, Sweden

Andreas Vieli, Department of Geography, University of Zurich, Switzerland

Colin Whiteman, School of Environment and Technology, University of Brighton, UK

Hazards are processes that produce danger to human life and infrastructure. Risks are the potential or possibilities that something bad will happen because of the hazards. Disasters are that quite unpleasant result of the hazard occurrence that caused destruction of lives and infrastructure. Hazards, risks, and disasters have been coming under increasing strong scientific scrutiny in recent decades as a result of a combination of numerous unfortunate factors, many of which are quite out of control as a result of human actions. At the top of the list of exacerbating factors to any hazard, of course, is the tragic exponential population growth that is clearly not possible to maintain indefinitely on a finite Earth. As our planet is covered ever more with humans, any natural or human-caused (unnatural?) hazardous process is increasingly likely to adversely impact life and construction systems. The volumes on hazards, risks, and disasters that we present here are thus an attempt to increase understandings about how to best deal with these problems, even while we all recognize the inherent difficulties of even slowing down the rates of such processes as other compounding situations spiral on out of control, such as exploding population growth and rampant environmental degradation.

Some natural hazardous processes such as volcanos and earthquakes that emanate from deep within the Earth's interior are in no way affected by human actions, but a number of others are closely related to factors affected or controlled by humanity, even if however unwitting. Chief among these, of course, are climate-controlling factors, and no small measure of these can be exacerbated by the now obvious ongoing climate change at hand (Hay, 2013). Pervasive range and forest fires caused by human-enhanced or induced droughts and fuel loadings, mega-flooding into sprawling urban complexes on floodplains and coastal cities, biological threats from locust plagues and other ecological disasters gone awry; all of these and many others are but a small part of the potentials for catastrophic risk that loom at many different scales, from the local to planet girdling.

In fact, the denial of possible planet-wide catastrophic risk (Rees, 2013) as exaggerated jeremiads in media landscapes saturated with sensational science stories and end-of-the-world Hollywood productions is perhaps quite understandable, even if simplistically short-sighted. The "end-of-days" tropes promoted by the shaggy-minded prophets of doom have been with us for centuries, mainly because of Biblical verse written in the early Iron Age during remarkably pacific times of only limited environmental change. Nowadays, however, the Armageddon enthusiasts appear to want the worst to validate their death desires to validate their holy books. Unfortunately we are all entering times when just a few individuals could actually trigger societal breakdown by error or terror, if Mother Nature does not do it for us first. Thus we enter contemporaneous times of considerable peril that present needs for close attention.

These volumes we address here about hazards, risks, and disasters are not exhaustive dissertations about all the dangerous possibilities faced by the ever-burgeoning human populations, but they do address the more common natural perils that people face, even while we leave aside (for now) the thinking about higher-level existential threats from such things as bio- or cybertechnologies, artificial intelligence, ecological collapse, or runaway climate catastrophes. In contemplating existential risk (Rossbacher, 2013) we have lately come to realize that the new existentialist philosophy is no longer the old sense of disorientation or confusion at the apparently meaninglessness or hopelessly absurd worlds of the past, but instead an increasing realization that serious changes by humans appear to be afoot that even threaten all life on the planet (Kolbert, 2014; Newitz, 2013). In the geological times of the Late Cretaceous an asteroid collision with Earth wiped out the dinosaurs and much other life; at the present time by contrast, humanity itself appears to be the asteroid.

Misanthropic viewpoints aside, however, an increased understanding of all levels and types of the more common natural hazards would seem a useful endeavor to enhance knowledge accessibility, even while we attempt to figure out how to extract ourselves and other life from the perils produced by the strong climate change so obviously underway. Our intent in these volumes is to show the latest good thinking about the more common endogenetic and exo-genetic processes and their roles as threats to everyday human existence. In this fashion, the chapter authors and volume editors have undertaken to show you overviews and more focused assessments of many of the chief obvious threats at hand that have been repeatedly shown on screen and print media in recent years. As this century develops, we may come to wish that these ex-amples of hazards, risks, and disasters are not somehow eclipsed by truly existential threats of a more pervasive nature. The future always hangs in the balance of opposing forces; the ever-lurking, but mindless threats from an implacable nature, or heedless bureaucracies countered only sometimes in small ways by the clumsy and often febrile attempts by individual humans to improve our little lots in life. Only through improved education and under-standing will any of us have a chance against such strong odds; perhaps these volumes will add some small measure of assistance in this regard.

Specifically in this volume, the chapters presented herein show us the myriad hazards, risks, and disasters associated with the cryosphere that bedevil those who live in the higher latitudes and altitudes of the Earth where ice abounds. These regions are where the solid and liquid phases of the H_2O system are always problematic for humans to deal with anyway, and now undergoing the various regimes of climate change, are being subjected to various changes of state and location of the H_2O system that can cause further problems. Wilfried Haeberli and Colin Whiteman have had a many decades of essential personal experience studying, teaching, and writing about the dangerous phenomena associated with the cryosphere, with the result that the two were able to attract a suite of chapter authors who have given us almost all the essentials to better understand and avoid such risks. Thermokarstic degradation of permafrost, Arctic coastal erosion, damaging glacial surges, snow avalanches into populated regions, variable sea-level changes; these are just a few of the cryospheric hazards and risks that are ever more problematic of late as the magnitude of damaging climate change in the high latitudes is even larger than in mid- and tropical latitudes. Similarly in the cryospheric regions of high altitude, the changes are also profound as cliffs collapse through permafrost- and glacier-loss debuttressing, loss of meltwater supplies, ice collapse, and other related phenomena.

This volume offers a fresh new look at hazards, risks, and disasters associated with the snow and ice of the world, and as such is of considerable interest to those who live in places that are subject to freeze and thaw. As world climates continue to change, in fact, and as sea levels vary as a result of melting glaciers and shifts in gravitational attraction, the problems associated with changes in the cryosphere will reverberate outward around much of the world into countries close to sea level but far from the polar regions. Perhaps the changes of the far-away cryosphere will appear to them to be relatively unimportant, but for some lowland countries, their very survival into the future will be highly doubtful. Thus attention must be paid to the cryospheric phenomena discussed in this book, even by those apparently far from such conditions of the H_2O system; this volume does offer considerable insight and new observations.

John (Jack) Shroder
Editor-in-Chief

REFERENCES

Hay, W.W., 2013. Experimenting on a Small Planet: A Scholarly Entertainment. Springer-Verlag, Berlin, 983 p.

Kolbert, E., 2014. The Sixth Extinction: An Unnatural History. Henry Holt & Company, NY, 319 p.

Newitz, A., 2013. Scatter, Adapt, and Remember. Doubleday, NY, 305 p.

Rees, M., 2013. Denial of catastrophic risks. Science 339 (6124), 1123.

Rossbacher, L.A., October, 2013. Contemplating existential risk. Earth, Geologic Column 58 (10), 64.

Snow and ice occur widely at higher latitudes and altitudes as perennial, seasonal, or sporadic elements of the Earth's surface and near surface, depending on climatic conditions. Such snow and ice phenomena include ice sheets, glaciers, permafrost, snow fields—sea, lake, and river ice, and constitute the Earth's cryosphere. The cold regions of the Earth are characteristically inherently hazardous, due to their topography, the dynamics of large ice masses, and the distinctive cryogenic geological processes associated with phase changes from ice to water. Such hazards are diverse and their associated risks to people, infrastructure, and economic activity are undoubtedly becoming greater as population densities increase.

In addition, however, if we are to accurately forecast the potential scale and distribution of future hazards associated with, or arising from, the Earth's cryosphere, it is critically important that the response of these thermally sensitive environments to changing global climate is understood and accurately modeled. Hence, the challenge that the present volume addresses is to present our current knowledge of snow- and ice-related hazards and their distribution, both spatially and temporally, together with our understanding of the impact that likely future global climate changes may have on the scale and distribution of these hazards. Site-specific, very short-term events such as localized landslides or snow avalanches lie at one extreme of spatial scale, impacts on the hydrological cycle and water resources of drainage basins provide an example of potential intermediate scale hazard and risk, whereas the truly global scale includes the effects of sea level changes on coastal environments and communities. Temporal scales likewise range from the virtually instantaneous, such as the release of a snow avalanche, through decades or centuries over which time the volume of ice sheets and glaciers may change significantly, to tens of millennia, when the Earth may experience further glacial periods with serious potential impacts on, for instance, deep nuclear repositories.

Scientific and technological advances in measuring snow and ice phenomena and monitoring their changes in space and time have greatly increased our knowledge and understanding of cryospheric phenomena; this knowledge is now increasingly applied to risk assessment and avoidance, and to informing strategies designed to increase our resilience to hazardous events. The present volume brings together contributions from an international group of scientists and covers a very wide spectrum of hazards that arise from the presence of, and changes in, snow, ice, and frozen ground, and shows how the risks these may pose to human activity may be assessed. The authorship is of necessity multidisciplinary, with contributions from physicists, engineers, geologists, geographers, glaciologists, hydrologists, biologists, and climatologists. The work builds upon the coordinating efforts of international organizations such as the United Nations Environment Programme and the reports of the Inter-governmental Panel on Climate Change, and many of the authors in the present volume have made major contributions to these international initiatives.

It is clear to me that this book should be required reading for professionals, academics, and politicians involved with developing strategies for risk avoidance and reduction, not only within the cold regions of the world, but beyond them too, where the potential impacts of cryospheric change may affect many more people and nations.

Charles Harris
Cardiff

Preface

An Earth with moving plates and a turbulent atmosphere has always been a hazardous place for humanity. Some have chosen to take greater risks by living in exposed locations, and have suffered disasters as a result, not least from the presence of snow and ice. It is appropriate, therefore, that a volume on snow and ice should be a component of this Elsevier series on Hazards, Risks, and Disasters which comprises the following volumes:

Hydro-Meteorological Hazards, Risks, and Disasters
Volcanic Hazards, Risks, and Disasters
Landslide Hazards, Risks, and Disasters
Earthquake Hazard, Risk, and Disasters
Coastal and Marine Hazards, Risks, and Disasters
Snow and Ice-Related Hazards, Risks, and Disasters
Wildfire Hazards, Risks, and Disasters
Biological and Environmental Hazards, Risks, and Disasters
Hazards, Risks, and Disasters in Society

Although some short-term, locally based ice and snow hazards, such as avalanches, icebergs, and blizzards, from time to time reach the media due to the scale of the disaster, many existing and potential, often long-term cryogenic hazards, such as rising sea-levels from melting ice sheets, have received relatively little exposure outside specialist academic departments. This situation is changing: ice and snow are particularly vulnerable in the context of the current period of rapid global warming. Fundamental changes are taking place throughout the cryogenic system, and are substantially increasing the risk of cryogenic hazards at all spatial and temporal scales worldwide.

This volume is arranged in three sections. In the introductory chapter the editors define the cryosphere, highlight its various constituent parts with their specific functions in environmental systems, and discuss the temporal and spatial scales involved and their significance for human living conditions. Several chapters are then devoted to explaining key elements of the cryospheric system and its interaction with other global systems, such as the biosphere and hydrosphere, as well as socioeconomic aspects involved. Having established the systems and processes which underpin the cryospheric system, further chapters are devoted to exploring the challenges presented to humanity in the form of hazards, risks, and disasters associated with specific cryogenic environments; ice sheets, cold lowlands, high mountains, ice on lakes, rivers, and seas.

The editors of this volume are particularly grateful to the large international network of leading experts who have collaborated under tight time constraints to produce a collection of papers, which effectively highlight many of the issues that face humanity in the rapidly changing, cryogenic part of our global environmental system. Their contributions have been appraised by external reviewers, and the volume editors, who take this opportunity to thank all those who gave freely of their time and expertise to enhance the quality of the contributions to this volume. The editors also wish to extend their gratitude to the managing and editorial staff at Elsevier, in particular to Louisa Hutchins and Mohanambal Natarajan for her prompt and efficient response to all questions.

Finally we hope that this volume of work on ice- and snow-related hazards, risks, and disasters, will be instrumental in raising the profile of this rapidly changing subject and provide its due share of publicity. Not only the academic community, but also hazard managers, administrators, and the public at large should gain much from this timely volume.

Wilfried Haeberli
Colin Whiteman
March 2014

Reviewers of individual chapters of Snow- and Ice-Related Hazards, Risks, and Disasters:

Etienne Berthier, France
Helgi Björnsson, Iceland
Robin Brown, Canada
Ross Brown, Canada
Christopher Burn, Canada
Alton Byers, USA
Luke Copland, Canada
Carolyn Driedger, USA
Mats Eriksson, Sweden
Bernd Etzelmüller, Norway
Dave Evans, UK
Mark Fahnestock, USA
Sven Fuchs, Austria
Hilmar Gudmundsson, UK
Tristram Hales, UK
Bernard Hallet, USA
Charles Harris, UK
Kenneth Hewitt, Canada
Regine Hock, USA

Martin Hoelzle, Switzerland
Per Holmlund, Sweden
Jo Jacka, Australia
Bruce Jamieson, Canada
Hester Jiskoot, Canada
Margreth Keiler, Switzerland
Anders Leverman, Germany
Norikazu Matsuoka, Japan
Dave McClung, Canada
Humphrey Melling, Canada
Martin Mergili, Austria
Johannes Oerlemans, Netherlands
Christian Rixen, Switzerland
Olav Slaymaker, Canada
Martin Schneebeli, Switzerland
Jean Schneider, Switzerland
Jean-Claude Thouret, France
One anonymous reviewer

Chapter 1

Snow and Ice-Related Hazards, Risks, and Disasters: A General Framework

Wilfried Haeberli [1] and Colin Whiteman [2]
[1] Department of Geography, University of Zurich, Switzerland,
[2] School of Environment and Technology, University of Brighton, UK

ABSTRACT

Snow and ice constitute the cryosphere on Earth and influence human activities at various scales of time and space. Through their proximity to phase-change thresholds, they are strongly linked to climatic conditions and presently subject to rapid changes induced by ongoing trends of global warming. Hazards, risks, and disasters related to snow and ice not only result from direct impacts on humans and their infrastructure by, for instance, snow avalanches, floods from glacial lakes, or accelerated erosion of permafrost coasts, they are also a consequence of the expansion of human activities into previously avoided dangerous regions, such as new shipping routes in the polar ocean, and tourist installations in cold mountains that are becoming ice free. The loss of goods and benefits from reducing or even vanishing cryosphere components constitutes serious threats to human well-being through, for example, diminishing meltwater supply in high-mountain rivers during dry seasons or rising global sea level. Further protection, mitigation, and adaptation procedures, combined with modern observational technologies will be required to anticipate, monitor, and deal with the challenges created by complex and highly interconnected geo- and ecosystems under conditions of growing disequilibrium.

1.1 INTRODUCTION

Snow and ice are common components of natural and human environments on Earth (UNEP, 2007; Singh et al., 2011; Williams and Ferrigno, 2012). They constitute the cryosphere (Figure 1.1). Via climatic conditions and the water cycle, their existence and variability affects humans around the globe, even in places where this may not be directly recognized or obvious. Dramatic changes in ice volumes during most recent Earth history, the Pleistocene ice ages, created an important cryogenic heritage in many landscapes, including

Snow and Ice-Related Hazards, Risks, and Disasters. http://dx.doi.org/10.1016/B978-0-12-394849-6.00001-9

FIGURE 1.1 **The cryosphere.** *From UNEP (2007). Free download.*

Snow

Sea ice

Ice shelves

Ice sheets

Glaciers and ice caps

Permafrost, continuous

Permafrost, discontinuous

Permafrost, isolated

lakes and rivers, vegetation and fauna (Swift et al., 2014). As with all hazardous components of the global environment, living with snow and ice involves costs and benefits.

Before introducing the range of hazards associated with snow and ice that are covered in this volume, it is important to differentiate clearly between the concepts of hazard, risk and disaster because these terms do not refer to the same idea and yet are sometimes used indiscriminately. Essentially, a *hazard* is a feature or situation that can be expected to impact negatively on the life, health, property, or environment of humans if an event (e.g., an avalanche) or significant change (e.g., loss of Arctic sea ice) occurs. *Risk* is a more complex concept involving not only the probability that a hazardous event or significant change will occur, but also the expected loss or cost and the degree to which this can be mitigated. Risk related to cryospheric hazards must therefore be considered from the points of view of both the physical hazard and the human response (Whiteman, 2011). Thus risk increases as human contact with the cryosphere becomes more frequent and more extensive. Population growth, especially in naturally hazardous regions, such as alpine mountains, automatically raises the level of risk, and the development of infrastructure may both enhance the accessibility of more hazardous locations and increase the value of potential losses. In any particular event or situation, the scale of loss of both life and property may be of such great magnitude that the outcome is described as a disaster.

Physical impacts from snow and ice can be both primary/direct (avalanches, for example), and secondary/indirect (for example, sea-level rise). Corresponding hazards are largely related to the inherent physical characteristics of snow and ice (Arenson et al., 2014). These undergo melting and freezing, they move (creep, slide, fall), float, are heavy en masse, are sometimes hard and sometimes brittle or ductile. Each of these characteristics may contribute, in isolation or in combination, to the type and intensity of ice- and snow-related hazards, and they often interact with other components of the environment. Threats and damage can arise over short as well as long time periods, at local to regional, continental, and even global scale.

Snow and ice are especially sensitive to climate change. Impacts of ongoing and potentially accelerating human-induced global warming are a serious concern (e.g., IPCC, 2013) and increasingly predominate in discussions about hazard assessment and risk reduction by adaptation, mitigation, and prevention measures. A critical aspect is the relatively high melt/freeze threshold of ice, only about 15 °C below the mean temperature of the Earth's atmosphere. Considering the variability of atmospheric temperatures around this mean, the high frequency with which the melt/freeze threshold is crossed is not surprising, and obviously contributes to frequent and extensive hazard events.

Snow and ice are also often of major importance well beyond their geographical extent, producing meltwater for irrigation, industry, and households during dry seasons (Seibert et al., 2014). In colder regions, river and lake ice may facilitate traffic in remote areas. Thus changes in the temporal or spatial distribution of snow and ice may have a wide range of unwelcome consequences. Although the exact details of future climatic conditions are still difficult to predict, climate change will strongly if not dramatically change snow and ice conditions on Earth, increasing the impacts of many hazards in the short, medium, and long term, even though some hazards and some hazardous locations may cease to exist as ice disappears completely from some locations. The environmental changes caused by the ongoing loss of snow and ice may constitute an equally strong if not even stronger long-term challenge. The different responses to climate change of the various environmental components induce growing disequilibria in complex environmental geo- and ecosystems. The rate of change as influenced by human interference with the climate system will be critical in terms of the scale of hazards and our ability to adapt in time to new conditions. The slower the rate of change, the more degrees of freedom that will remain for difficult decisions to be taken and policies to be implemented to accommodate environments in which the prevalence of snow and ice is greatly reduced.

1.2 COSTS AND BENEFITS: LIVING WITH SNOW AND ICE

The components of the cryosphere—continental ice sheets, glaciers, river and lake ice, sea ice, seasonal snow, and frozen ground—strongly differ with respect to their volume, spatial extent, and occurrence in time (Table 1.1). These three aspects govern their primary environmental functions, the economic and social benefits they provide and the degree of hazard they may pose to humans, with their associated risks and potential for disasters (UNEP, 2007).

The largest ice bodies on Earth are the continental *ice sheets*. With their enormous mass, white/cold surfaces (Figure 1.2), and the direct ice contact of their margins with the sea, ice sheets are drivers of the global environment, actively influencing physical and living conditions worldwide (Bentley et al., 2007). Via atmospheric and ocean circulation, the presence of a large ice sheet covering the continent of Antarctica at the South Pole for the past millions of years, and for millions of years to come, has a strong cooling influence on the global climate (Ohmura, 2014). This cooling effect, together with long-term fluctuations of incoming solar radiation due to variations in the Earth's orbit around the sun, enables the development of ice ages, with their dramatic effects. One of these effects concerns the formation and disappearance of other ice sheets, and associated large changes in sea level (Allison et al., 2014). The growth of the Laurentide Ice Sheet over large parts of North America alone lowered sea level by at least 100 m and profoundly changed hydrography and coastlines at continental and global scales.

Intercontinental migration of flora and fauna, including humans, especially between Eurasia and the Americas, became possible due to exposure of land (e.g., Bering Strait). Conversely, rapid volume losses of these great ice sheets

TABLE 1.1 Cryosphere Components

Components of the Cryosphere	Area Covered (Million Square Kilometers)	Ice Volume (Million Cubic Kilometers)	Potential Sea-Level Rise (Centimeters)
Snow on land (Northern Hemisphere) (annual minimum–maximum)	1.9–45.2	0.0005–0.005	0.1–1
Sea ice, Arctic and Antarctic (annual minimum–maximum)	19–27	0.019–0.025	0
Ice shelves	1.5	0.7	0
Ice sheets (total)	14.0	27.6	6390
Greenland	1.7	2.9	730
Antarctica	12.3	24.7	5660
Glaciers and ice caps (lowest and [highest] estimates)	0.51 [0.54]	0.05 [0.13]	15 [37]
Permafrost (Northern Hemisphere)	22.8	4.5	~7
River and lake ice	(n/a)	(n/a)	(n/a)

Source: From UNEP, 2007.

FIGURE 1.2 **View of part of the eastern Greenland Ice Sheet, where it covers a rugged mountain topography.** *Photograph: W. Haeberli (1998).*

submerged these temporary land bridges and displaced people. Together with thermal ocean expansion, glacier melting, and other factors, mass losses of the current ice sheets in Antarctica and Greenland could cause sea level to rise over timescales of decades to a few centuries by 1−2 m, an amount that would be disastrous for populations and infrastructure in coastal regions (Church et al., 2007). Especially sensitive parts of ice sheets are their floating ice shelves and their outlet glaciers, the latter flowing over beds that lie below sea level and commonly with a pronounced adverse slope (Vieli, 2014).

The ice layers in these ice sheets contain deep-frozen information about past climatic and environmental conditions on Earth. The results of ice-core analyses from deep drilling are now the best quantitative documentation relating to the evolution of the greenhouse effect under natural conditions during the past hundreds of thousands of years and clearly exhibit increasing human impacts on atmospheric chemistry (Figure 1.3; Lipenkov, 2006; cf. Jouzel et al., 2013).

The much smaller *glaciers* in cold high-latitude and high-altitude regions (Figure 1.4) react passively to fluctuations in global climate and have, therefore, been known as unique indicators of global climatic and environmental change (WGMS, 2008). The largest areas covered with glaciers and other small ice masses exist in the North American mountains, on islands of the Canadian Arctic, peripheral to the Greenland ice sheet, and in Central Asia. Smaller glacier volumes in high-altitude mountain chains of densely populated

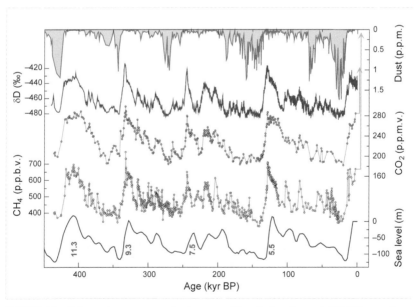

FIGURE 1.3 **Information on atmospheric composition during the past 450,000 years from the Vostok ice core, Antarctica.** The δD values (red curve) reflect temperatures, which can be compared with the concentrations of the greenhouse gases CH_4 and CO_2 (green and blue curves). *From Lipenkov (2006) with permission.*

FIGURE 1.4   Glaciers at high latitude on Ellesmere Island *((a) Photograph: J. Noetzli (2008))* and at high altitude/low latitude in the Cordillera Blanca, Peru. *(Huascarán; (b) Photograph: W. Haeberli (2012)).*

mid- and low-latitude regions, such as the South American Andes, high mountains in central Asia or the European Alps store amounts of water, which can be especially important for human livelihood; they often constitute an essential source of (clean) domestic water, water for hydropower production, or irrigation water for food production in large surrounding lowlands under dry seasonal or interannual conditions. Where they exist, glaciers strongly affect landforms and scenery, attracting tourists and associated infrastructure. Under wet, maritime-type conditions, highly active temperate glaciers like those in Southern Chile, on the west slope of the New Zealand Alps or in southern

Alaska, reach down to permafrost-free and often forested land, in places at sea level. Glaciers in regions with dry, continental-type climatic conditions are cold to polythermal, exchange much less mass, flow less actively, and often terminate far above timberline in terrain with widespread permafrost; examples are the glaciers of the Brooks Range in Alaska, the ones on Disko Island in Greenland (Citterio et al., 2009), or those in the mountains of the Tibetan Plateau (Zemp et al., 2007).

Significant disasters have primarily been related to floods and debris flows from ice-dammed lakes or the breaching of moraine dams (Clague and O'Connor, 2014). Internationally coordinated monitoring of glaciers as part of climate observation goes back to 1894 (Haeberli, 2007) and documents a striking and continued, if not accelerating, global trend of rapid loss in glacier area and volume (Figure 1.5). Continuation of this trend and even complete deglaciation of entire mountain chains is probable under conditions of anthropogenically enhanced greenhouse effects (e.g., Zemp et al., 2006; Radić et al., 2014). Such developments involve drastic changes in landscape diversity and the scenic attraction of cold mountains and a decreasing to vanishing meltwater supply during dry/warm seasons for many regions at their foot (Haeberli et al., 2013). The formation of numerous new lakes to some degree compensates for the loss of attractive mountain scenery and is of interest for hydropower production. However, it also involves an increasing risk of far-reaching flood waves from large rock/ice avalanches impacting them (Haeberli, 2013).

Lake and river ice at high latitudes and high altitudes is a key component of cold regions river and lake systems. It mainly occurs in the Northern Hemisphere and affects an extensive portion of the global hydrological system, including 7 out of 15 of the world's largest rivers and 11 of the 15 largest lakes (Prowse et al., 2007). Floating ice forming during the cold season on lakes and rivers can also serve as an indicator of climate change. In contrast to the fluctuations of mountain glaciers, which primarily provide evidence about summer conditions and high altitudes, lake and river ice primarily reflect winter conditions, especially at low altitudes (Kubat et al., 2014). Long-term observations (Figure 1.6) document a clear trend for fall freezeup to occur later, spring breakup to occur earlier and the duration of ice cover to become correspondingly shorter. The processes involved are, however, highly complex and difficult to model or predict. Aquatic ecosystem dynamics closely follow the variability of freshwater ice condition, and human traffic in remote areas is often facilitated by the formation of a solid ice cover at the surface of such water bodies. The interaction of floating ice plates during breakup—sometimes even during winter—may lead to disastrous river-ice jams (Figure 1.7; Beltaos, 1995) and large-scale flooding. Such floods are often significantly more extreme and cause greater damage than open-water floods, and constitute the predominant threat to many subarctic and arctic settlements, which are typically situated at river confluences or on lakeshores. Continued

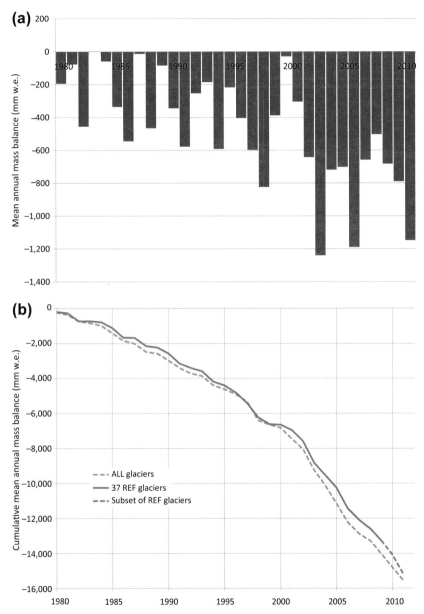

FIGURE 1.5   Annual (a) and cumulative (b) mass balance of glaciers with long observational time series. *Data from the World Glacier Monitoring Service. Free download.*

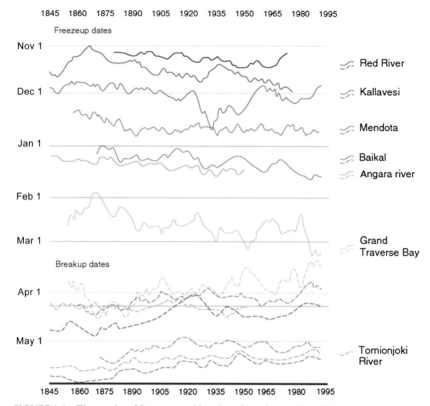

FIGURE 1.6   **Time series of freezeup and breakup dates from selected Northern Hemisphere lakes and rivers (1846–1995).** *From UNEP (2007). Free download.*

FIGURE 1.7   **Ice jam at Galena, Yukon River, Alaska, 27 May 2013.** *Photograph: Keystone.*

impacts of atmospheric warming on lake and river ice are expected to have strong effects on living conditions, especially at high latitudes.

Sea ice is frozen ocean water, which grows to characteristic thicknesses of up to a few meters at the surface of cold polar seas (Figure 1.8). It plays a key role in the climate system (Gerland et al., 2007). At the North Pole, atmospheric circulation drives a dense cover of sea ice around the surface of the deep Arctic Ocean, while at the South Pole, sea ice drifts around the margins of the Antarctic continent. With its large area and snow cover of long duration, sea ice influences the Earth's surface albedo and the global radiation balance. Large masses of sea ice leave the Arctic Ocean, carried by the East Greenland current, and mix with warm highly saline water from the Gulf Stream, affecting the formation of deep ocean water in the North Atlantic, which in turn drives the global ocean circulation (conveyor belt). At the margins of the continental ice sheets, icebergs from calving glaciers are often intermixed with

FIGURE 1.8   **Ice floes and leads in the Arctic Ocean.** *Photograph: W. Haeberli (July 2008).*

frozen ocean water. Their irregular geometry contrasts with that of large tabular icebergs, sometimes of huge dimensions, which break from ice shelves. Unlike wind-driven sea ice, large icebergs mostly respond to ocean currents. The existence of floating ice at the ocean surface also strongly influences high-latitude ecosystems and livelihood. The extent and concentration of sea ice at both poles undergoes large seasonal variations, but superimposed on these are two contrasting long-term trends documented since the beginning of satellite observations. While the extent of sea ice around the Antarctic continent has been stable or slowly increasing, sea ice in the Arctic Ocean shows a strong if not accelerating trend of reduced extent, thickness and volume (Figure 1.9). Extraordinary record lows of spring/summer sea ice extent have been observed especially in 2007 and 2012. The Northern Sea Route north of the Russian Coast and the Northwest Passage through the Canadian Archipelago are

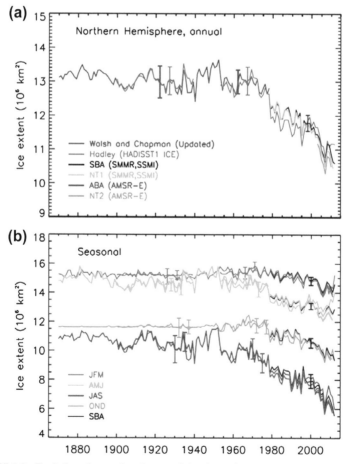

FIGURE 1.9   **Evolution of annual and seasonal Arctic sea ice cover.** *From IPCC (2013).*

becoming increasingly open connections between the Atlantic and Pacific Oceans, thereby significantly shortening traffic distances for shipping. Continued shrinking or even complete disappearance of an arctic summer sea ice cover would not only have very strong impacts on the global and regional climate systems, especially with respect to albedo, atmospheric humidity and ocean circulation, but also on polar ecosystems, food chains, and livelihood (Kubat et al., 2014). The dependence of the polar bear on the extent and evolution of arctic sea ice has become a widely used icon for climate change impacts on living conditions. The disappearance of winter ice could be a key tipping point in the climate system for long time periods to come.

Seasonal snow could be called "a nervous interface" between the atmosphere and the Earth surface (cf. Barry et al., 2007). With its temporarily large winter surface area—especially in the northern hemisphere (Figure 1.10)—and high albedo, it is closely linked to the global radiation balance. Higher global temperature reduces the extent of snow in space and time and decreases global average albedo. Such lowering of global albedo strengthens warming trends in the climate system, which further reduces the extent of snow cover and so on. This self-reinforcing effect (positive feedback cycle) is one of the primary reasons for the much more pronounced climate changes observed in snow-affected high latitudes and high altitudes than elsewhere.

Seasonal snow covers floating ice and protects it from rapid melting, but is itself protected by floating ice from rapid melting in open water. Important interactions also exist between snow, other cryosphere components, the hydrosphere and biosphere in cold environments. The mass balance of continental ice sheets and glaciers depends on snow accumulation. Equally, the energy fluxes at the surface of frozen ground are essentially governed by the

FIGURE 1.10   **Spring snow over western Norway, Svartisen ice cap.** *Photograph: W. Haeberli (March 1996).*

high albedo and efficient thermal insulating capacity of the snow cover. Seasonal snow in cold regions is a fundamentally important temporary storage component in river runoff and water supply regimes and is a decisive source of soil water for vegetation in springtime, a primary factor in the food chain of animals (Seibert et al., 2014; Callaghan and Johansson, 2014).

Winter snow is a great attraction for tourism and sports but can be a vexation for traffic and repeatedly causes hazardous avalanches to run down steep slopes into inhabited areas. Snow indeed interacts with almost everything at higher latitudes and altitudes, but has a high variability in space and time since it depends on short-term weather conditions—both precipitation and temperature—and can easily be eroded, transported and redeposited by winds. Such high variability and the difficulty of mapping it from satellites in forested regions or mountain shadows limit the possibilities of defining clear trends in the long-term climate-related evolution of snow cover. Indications nevertheless become more and more clear, that northern hemisphere snow cover extent has started to decrease (Figure 1.11). Continuation of this trend of snow cover reduction in space and time will dramatically impact large and complex cold regions geo- and ecosystems, likely reducing simultaneously both hazards and risks, and important economic and social benefits provided by snow.

Frozen ground can occur at short to long timescales and penetrate correspondingly to shallow or very great depths. Daily and seasonal freeze/thaw cycles govern not only near-surface frost weathering of natural rocks but also of roads and other human infrastructure in cold regions. Permafrost is subsurface material that remains at temperatures below 0 °C throughout the year and is a widespread key element of cold environments (Romanovsky et al., 2007). Temperature, thickness, and ice content of permafrost vary strongly in space as a function of climate (air temperature, snow cover, incoming radiation), subsurface conditions (rock/sediment type, groundwater), and freezing processes. In high-latitude lowlands, permafrost depth can reach up to several

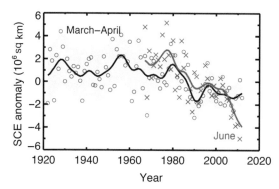

FIGURE 1.11 **Evolution of snow cover extent (SCE) (deviation from mean) in the Northern Hemisphere.** *From IPCC (2013).*

hundreds of meters and many high-mountain summits like the famous Matterhorn in the European Alps are frozen throughout. The volume of ice contained in permafrost can in places far exceed the pore volume of the host rock or soil (Figure 1.12), exerting a fundamental influence on the hydraulic and geotechnical characteristics of the frozen material. Frost heave, thaw settlement, creep, slope failure, and enhanced coastal erosion lead to spectacular geomorphic landforms, such as patterned ground, pingos, rock glaciers, or thermokarst lakes and constitute special challenges for the construction and maintenance of human infrastructure (mines, pipelines; Figure 1.13), houses, roads, bridges, airports, etc (Streletskiy et al., 2014). Under the cold and

FIGURE 1.12   Ice wedge in gravel pit near Fairbanks, Alaska. *Photograph: W. Haeberli (1978).*

FIGURE 1.13   **Trans Alaska Pipeline north of the Brooks Range, Alaska.** *Photograph: W. Haeberli (2008).*

dry-continental conditions of Alaska, Canada or Russia, the hardness and greatly reduced permeability of ice-rich permafrost affects forest and tundra vegetation, soil humidity, surface hydrography and river flow over very large areas.

Borehole observations provide clear evidence that permafrost at high latitudes has been rather rapidly warming during the past decades (Figure 1.14; Romanowsky et al., 2010), while the information from high-mountain sites (Haeberli et al., 2011), where climate-related monitoring commenced only recently, is somewhat less clear. Snow, vegetation, and water interact with frozen ground in a sensitive but complex way. This makes accurate prediction of future ecosystem evolution under the influence of global warming difficult. As heat diffusion at greater depth is a very slow process strongly further retarded by latent heat set free by the melting of ground ice, the response of thick permafrost to climate change is likely to be extremely slow. The effects of a continued warming trend will therefore lead farther and farther away from the equilibrium-like conditions of the thermally relatively stable Holocene, and deeply affect conditions in cold environments over many decades, centuries, or even millennia to come. Besides significant negative effects on the biosphere, the hydrosphere, living conditions and human infrastructure, the decomposition of organic matter previously frozen in permafrost may set free important amounts of greenhouse gases (Ohmura, 2014; Streletskiy et al., 2014). This self-reinforcing effect (positive feedback cycle) of global warming, permafrost thawing and enhanced greenhouse gases needs to be included in climate-scenario calculations and global climate policy but so far remains difficult to assess precisely. In rugged topography, permafrost warming and

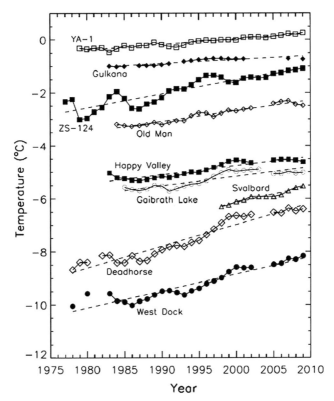

FIGURE 1.14   **Evolution of borehole temperatures at high-latitude sites of the Northern Hemisphere.** *From IPCC (2013).*

degradation could initiate a long-term destabilization of steep and icy mountain peaks (Deline et al., 2014), a particular challenge with respect to local/regional hazards, risks, and disasters.

1.3 SMALL AND LARGE, FAST AND SLOW, LOCAL TO GLOBAL: DEALING WITH CONSTRAINTS

The cryosphere is present at all latitudes, altitudes, and gradients, on land and sea and in the air (Bolch and Christiansen, 2014). Consequently, the wide variety and global extent of hazards associated with ice and snow should not be a surprise. All continents are at some risk from cryospheric hazards, although the scale of impact varies. Least affected are the continents of Africa and Australia, which are broadly located at lower latitudes, lack extensive areas of high altitude, and possess relatively low population densities. In contrast, Asia has a large latitudinal range, one of the largest areas of high altitude on Earth that includes the world's highest mountain, and is relatively densely populated. Short-term variability, medium-term fluctuations and long-term changes challenge humans to adapt and to develop flexible concepts of hazard prevention and risk reduction. Risks must be assessed and managed in an integrative way (Bründl and Margreth, 2014), as well as in the context of other risks and constraints of society (Carey et al., 2014). Concerning future conditions in a warmer world, the question of tipping points in the climate system has become an increasingly important aspect of policy-related discussions (Levermann et al., 2011).

Small and rapid events of limited dimensions are usually important at local to regional scales. They can be extreme events caused by short-term, quasi-stochastic variability in environmental conditions or occur as a consequence of a medium to long-term cumulative evolution. This difference in the governing process has a profound influence on concepts of hazard assessment and risk reduction. Snow avalanches (Figure 1.15; Schweizer et al., 2014) typically involve time intervals of seconds to minutes and horizontal reaches of hundreds of meters. Potential for their formation is built up through heavy snowfall as part of short-term variability in meteorological conditions. Even under future warmer conditions, extrapolation of statistical information on frequency of occurrence together with experience and the results from model simulations can be used with reasonable safety, at least as long as climate models remain uncertain in predicting extremes. Glacier surges (Harrison et al., 2014), with ice margins advancing over kilometers during months to a few years, follow quiescent time intervals during which ice builds up over decades. Even though not related to short-term variability in environmental conditions, extrapolation of documented information on

FIGURE 1.15   **Artificially released powder snow avalanche at Walenstadt (Schattenbach) Switzerland.** *Photograph: A. Aschwanden (March 2003).*

past surge events and their return periods can still provide useful indications of potential hazards.

Ice/rock avalanches (Evans and Delaney, 2014) can travel within minutes over trajectories of kilometers. They usually occur on steep slopes and are the result of slope stability systematically decreasing over time periods of centuries and millennia through processes of weathering, erosion, increasing englacial temperatures, loss of surface ice, and permafrost degradation. Anticipation of possible future events in this case cannot simply extrapolate information on past events alone but must be based on scenario modeling with respect to long-term changes in complex conditions of surface/subsurface ice and topography. The same is true for outbursts of glacial and periglacial lakes, which have comparable time and space scales and result from systematic changes in glacier geometry. Early recognition of these hazardous events requires well organized and focused monitoring. Glacier surges, lake outbursts and ice/rock avalanches can affect infrastructure at the regional scale where process chains are induced (cf. below). Preventive measures include temporary to permanent evacuation of people and artificial defense structures. The latter are especially economical and widespread in the case of repetitive processes like snow avalanches.

Small and slow processes are primary concerns of long-term planning, disaster prevention, and adaption to hazards and risks at local to regional scales. Cumulative changes in hazard conditions and in possible risks related to the destabilization of icy mountain peaks, to the formation of new lakes or to continued erosion of permafrost coasts (Figure 1.16), for instance, occur over large areas but must be judged at local to regional levels in connection with existing human infrastructure and expansion into remote areas. In contrast to the triggering of sudden events, such as those discussed in the

FIGURE 1.16   **Coastal erosion in permafrost at Prudhoe Bay.** *Photograph: W. Haeberli (June 2008).*

previous section, long- and short-term changes in the occurrence of hazard events must also be considered. In the short term, the occurrence of hazardous events can be strongly influenced by the impact of continued climate change on long-term ice-related factors. Primary local, long-term factors affecting the stability of steep icy rock walls are: (1) the geological setting (mainly lithology and structure); (2) topography (mainly slope inclination and vertical extent); and (3) ice conditions (glaciers, permafrost). As changes in ice conditions are now most rapid and significant, they have a predominant effect on the short-term occurrence of present-day rock/ice avalanches (Deline et al., 2014). Changes in ice conditions produce different response characteristics, because glaciers are likely to vanish within decades, whereas thick permafrost in deeply frozen mountain peaks will degrade over future centuries. Understanding and interpreting the corresponding disequilibrium conditions with respect to hazardous events constitutes a major scientific challenge and requires integrated modeling of surface and subsurface ice.

The formation of attractive, potentially useful but also dangerous lakes with continued deglaciation in cold mountains is another process, which takes place over years to decades, can easily be monitored using remote sensing techniques and—on the basis of high-resolution digital terrain information—can even be predicted for large regions still covered by glaciers today (Linsbauer et al., 2012). The shrinking and vanishing of their glacier cover also affects the hazard potential of erupting ice-clad volcanoes, for instance in South America (Delgado Granados et al., 2014). Complex interactions affect the development of coastal erosion in arctic permafrost. With the climate-induced retreat and loss of arctic summer sea ice, the length of time of coastal exposure to wave action in open water increases. Growing areas of ice-free water also induce higher waves, which can reach up to higher ground with rising seas. Moreover, increasingly ice-free water may produce more precipitation and thicker snow cover, which can accelerate permafrost warming and degradation. This last case clearly illustrates the fact that changes in complex and highly interconnected systems come into play.

Large and rapid processes in high mountains are often linked to chains of causality (Clague and O'Connors, 2014; Deline et al., 2014; Evans and Delaney, 2014) leading to hazards, risks, and disasters at regional scales. Large rock/ice avalanches from icy mountain peaks such as Huascarán (Cordillera Blanca, 1970 event), Dzhimarai-Khokh (Caucasus, 2002 event) or Hualcán/ Laguna 513 (Cordillera Blanca, 2010 event) can impact a lake, produce a significant impact wave, which may transform into a debris flow damming a river further down and producing another lake, which, in turn, may overflow and erode its landslide dam, flooding valleys below containing settlements and important infrastructure over tens to even a few hundreds of kilometers (Evans and Delaney, 2014). The development of corresponding hazard potentials is especially important in connection with climate change and its profound impacts on high-mountain geosystems and processes. The formation of new lakes at the foot of slowly destabilizing icy mountain peaks and their flanks is thereby especially important as such lakes constitute multipliers with respect to the frequency, reach, and damage potential of hazardous processes. They, hence, systematically increase the risks in extended down-valley sections of river courses. Anticipating such hazard potentials and growing risks requires holistic consideration of entire systems and their potential interactions, from the highest icy mountain peaks down to distant populated areas. Careful anticipation is especially important, because the changes are progressing rapidly and resulting effects are likely to persist for many generations. Newly available high-resolution digital terrain information has made it possible to develop corresponding computer models and model chains (Schneider et al., 2014). These not only help with analyzing the sensitivities of model results with respect to assumptions and simplifications to be made about individual processes and phenomena in the chain reaction, but also facilitate visualizing possible effects, estimating hazard and risk levels, defining hazard zones and planning preventive measures in a *trans*-disciplinary, participative way with and for the responsible authorities and affected people (Carey et al., 2014). Such preventive measures include early warning systems or artificial—especially flood retention—structures. Flood retention by lake level lowering and overflow protection has been successfully applied, for instance in Peru or in the European Alps (cf. Clague and O'Connor, 2014). Flood, avalanche or debris-flow protection by retention dams farther down in flat sections of valleys involves not only more important investments but also induces marked and sometimes controversial landscape changes (Figures 1.17

FIGURE 1.17   **Laguna Parón, Cordillera Blanca, Peru. Lake level was artificially regulated for combined flood protection and hydropower production but serious conflicts now exist at this site.** *Photograph: W. Haeberli (June 2009).*

FIGURE 1.18 **Altitudinal belts at Mont Blanc and Chamonix in the French Alps. Note Taconnaz antiavalanche defense measure at right margin.** *Photograph: W. Haeberli (June 2011).*

and 1.18, cf. Carey et al., 2012a; Bründl and Margreth, 2014) However, retention dams become more appropriate for long-term protection as multi-purpose projects in combination with provision of water supply and hydro-power production develop. Careful balancing of the interests and concerns of various parts of society is necessary, as the famous case of Laguna Parón in the Peruvian Cordillera Blanca (Figure 1.17) shows. New technological infrastructures can change power distribution and management philosophy among stakeholders, which in turn may lead to serious unintended conflicts (Carey et al., 2012a). As another example of large, rapid events, extreme snow storms lasting hours to days can cause heavy impacts on traffic, avalanche danger and living conditions in general over entire regions, even up to (sub-) continental scales.

Large and slow processes at continental to global scales may have the strongest impacts on living conditions on Earth including the potential of disastrous changes over extended time periods. Disappearance of arctic sea ice increasingly opens the possibility of considerably shortening traffic routes for shipping between the Atlantic and Pacific oceans. This, together with the prospecting for, and future exploitation of, natural resources is likely to introduce the dangers of pollution and heavy ecological disturbances to currently pristine arctic and subarctic environments. The reduction in snowfall/snow-cover duration, together with the disappearance of glaciers, will drastically impact the seasonality of runoff and freshwater supply for densely populated regions surrounding mountain chains at low and midlatitudes. The lack of meltwater from snow and glaciers may indeed strongly constrain the range of possibilities for agriculture, energy production, etc. during dry/warm seasons. Vanishing surface and subsurface ice is likely to affect the stability of high-mountain slopes at a worldwide scale, causing at many places growing risks related to landslides and flood waves from new lakes impacted by them. Sea-level rise related to vanishing glaciers and the destabilizing (Figure 1.19) and melting of ice sheets could far exceed a few decimeters over the coming centuries and constitute a fundamental threat to coastal settlements and infrastructure with drastic impacts on near sea-level ecosystems and river morphology (Allison et al., 2014; Church et al., 2007). Continued thawing of ice-rich permafrost in arctic regions will strongly influence ground stability for infrastructure, surface/subsurface water and growth conditions for vegetation (Streletskiy et al., 2014). Thawing is also likely to set free greenhouse gases so far sequestered within the permafrost and thereby enhance global warming. Over a rather longer timescale, and in relation to future glacial cycles, re-positories for radioactive waste in some midlatitude countries (Fischer et al., 2014) must be designed to be safe for 10^5-10^6 years under the effects of glacial erosion, permafrost formation and strongly changed groundwater flow conditions, often to great depths. Concerning such long-term climatic cycles, a fundamental question is how fast the excess atmospheric greenhouse gases presently being accumulated by mankind will be removed by the Earth system

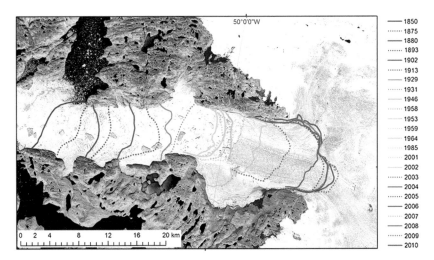

FIGURE 1.19   **Retreat of Jakobshavn Isbre, Greenland west coast.** *Based on Csatho et al. (2008) updated using Landsat images. Courtesy of B. Csatho and J. Briner, University of Buffalo.*

to return the atmosphere to preindustrial conditions, and whether these changes in atmospheric composition could be tipping elements in the climate system, leading to switching from one quasi-equilibrium stage to another.

The discussion about *tipping elements* in the climate system, as noted earlier, has become a focal point of policy-related discussions (Levermann et al., 2011). The concept involves various aspects but primarily that of irreversibility with respect to human/political time dimensions. Once larger glaciers are gone, there will be no political decision for many future generations, which could, immediately or over the short term bring their meltwater back for water supply in the dry season. At present, lost summer ice in the Arctic reforms in the following winter and is no tipping element, but the loss of the winter pack-ice cover with its effect on polar snow and global albedo would probably be irreversible over a very long time span. Thermal disturbance of permafrost from twentieth century warming already affects depths ranging from 50 to 100 m below the surface and will—together with effects from future surface warming and following the physical laws of heat diffusion—slowly but continuously penetrate to greater depth, affecting groundwater and slope stability for coming decades to centuries if not millennia. Calving instabilities of large tidal outlet glaciers with beds far below sea level and the positive feedback between mass balance and altitude could make major parts of the two continental ice sheets, Greenland and Antarctica, disappear over the coming millennia with corresponding dramatic effects on the atmosphere and the global ocean.

1.4 BEYOND HISTORICAL EXPERIENCE: MONITORING AND MANAGING RAPID CHANGES

Hazards related to snow and ice strongly depend on climatic conditions. There can be little doubt that the quasi-periodic climatic fluctuations of a relatively stable Holocene are being rapidly replaced systematically by a less stable climatic regime beyond preindustrial variability ranges of recent millennia. The corresponding impacts on environmental conditions necessitate new concepts to be applied for hazard anticipation and risk assessment. Technological innovation is opening important new possibilities. At the same time, vulnerabilities, damage potentials and, hence, risks due to snow and ice are directly linked to societal developments such as the continuing expansion of human activity and infrastructure into hazardous areas. Integrated views, scenario-based thinking and participative planning have become key-terms concerning mid- to long-term hazard assessments, risk reduction and disaster prevention. Environmental conditions change under the influence of the rising energy content of the climate system. The involved geo- and ecosystems are highly interconnected and complex. Strongly differing response characteristics of their components with respect to the imposed forcing will unavoidably lead them into pronounced disequilibrium conditions. Altitudinal belts (forest, meadows, frozen ground, glaciers; Figure 1.18) in high-mountain regions, for instance, will not only shift but also deform and change. Snow and ice thereby play different roles. Seasonal snow, with its important influence on the energy and water balance of cold environments, reacts in a stochastic way on timescales of hours to days, weeks and months, directly depends on short-term meteorological conditions, and thus creates basic uncertainty in calculations of future scenarios. Glaciers, with their pronounced memory function, are key indicators of changing energy contents in the climate system and react in a cumulative way within years to decades to corresponding developments. Permafrost also shows cumulative changes but rather on timescales of decades, centuries to millennia. In many regions of the world, glacier and permafrost environments are already different from what they have been within the historical lifetime of settlements and infrastructures. Under such conditions, the historical-empirical knowledge base, increasingly loses its significance and applicability for anticipating future glacier and permafrost hazards. Moreover, grave accidents very often imply complex system reactions and process chains (Figure 1.20). This is the reason why comprehensive "holistic" views are so important.

Technological innovations have revolutionized geoscience and hazard research, especially during the past two or three decades. Digital terrain information in combination with high-resolution satellite imagery, numerical models, and geographical information system applications have opened a rich

FIGURE 1.20 Broken and collapsed debris cone of ice-avalanche deposits from the advancing Giétro Glacier after the sudden outburst of an ice-dammed lake in 1818 (16 June, 20 million cubic meters of water escaped) and devastating flood wave at Mauvoisin/Bagnes/ Rhone valley, Swiss Alps; painted probably in 1818 by Hans Conrad Escher von der Linth. This historical glacier catastrophe at regional scale cannot happen again due to glacier retreat and construction of a hydropower reservoir in the valley. *From Rasmo et al. (1981); cf. Haeberli, (2007) with permission.*

field of spatial analyses concerning phenomena, processes, and conditions related to hazardous events. As an example, the probability of permafrost occurrence in the complex topography of rugged mountain regions can at least roughly be assessed at various scales using modern modeling approaches (Boeckli et al., 2012; Gruber, 2012). Retreat scenarios of glaciers and the occurrence of overdeepenings in their beds with their potential for future lake formation can be realistically modeled though still with large uncertainties concerning their exact size (Figure 1.21; Linsbauer et al., 2012). Easy access to such data and methodologies greatly facilitates communication (Figure 1.22) but also has its problematic side: everybody can do hazard assessments for almost every spot on Earth and make it available to the whole world via the Internet, but without quality control. As a consequence, responsible authorities risk losing control over information for the public about highly delicate questions. The International Working Group for Glacier and Permafrost Hazards (GAPHAZ) was established to help governments by offering international expertise in critical cases. In a rapidly changing world, monitoring of critical situations is becoming a key approach for timely hazard recognition and assessment. Radar and laser technologies have now reached resolutions and precisions, which allow for detection of very small movements in remote and inaccessible areas over even short-time intervals (Strozzi et al., 2010). Together with signal transmission from sensor networks and dataloggers, they will strongly enhance the possibility of early detection and warning related to destabilizing mountain flanks but may be costly. Like digital elevation model differencing, space-born information from such techniques provides fundamentally important quantitative information concerning changes of thousands of glaciers for time intervals of decades down to years (Figure 1.23). Daily information about Arctic and Antarctic sea ice extent and concentration are publicly available on the Internet. Similarly, information and warnings about snow avalanches are made available for the public in mountain tourist regions via all modern media, and some new lakes forming at retreating glaciers can be publicly observed using Webcams.

Changes in societal conditions can locally outweigh other effects, including impacts from climate change. A primary factor is the expansion of infrastructure (housing, roads, tourist installations, etc.) and human activity into previously avoided potentially dangerous areas. Examples of avalanches of ice, rock, and snow are Mattmark (1965) and Brenva (1997), both in the European Alps. At Mattmark, an ice avalanche from Allalin glacier killed 88 workers on a temporary construction site for the Mattmark hydropower dam (Figure 1.24). Two scientific experts had been involved in the preceding planning stage—one for snow avalanches and one for glacier hydrology—but neither one was responsible for glacier avalanches. In the still highly controversial lawsuit following the tragedy, both experts were treated as witnesses and the accused engineers received an acquittal. The Swiss Federal Government, however, established a working group on glacier

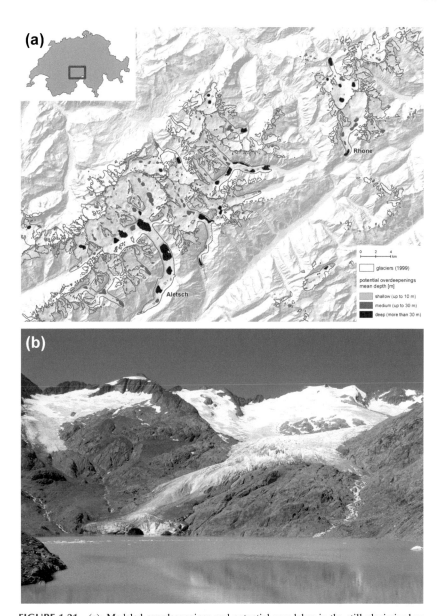

FIGURE 1.21   (a): Modeled overdeepenings and potential new lakes in the still glacierized region of the central Swiss Alps. Aletsch Glacier is in the lower left, and Rhone Glacier is in the upper right corner. Gauli Glacier with its new lake (Figure 1.21(b)) and another probably soon-forming lake is indicated with a green circle; the model run was done using a digital elevation model in which the lower lake did not exist yet but was covered by glacier ice. *(From Haeberli and Linsbauer (2013))*. (b): New lake which recently formed in a pronounced bed overdeepening of Gauli Glacier, Bernese Alps, Switzerland, as a consequence of continued glacier retreat. Another lake is likely to form in the coming years to a few decades in the probably overdeepened bed part indicated by the less inclined glacier surface above the bedrock sill with the present steep/thin glacier tongue (cf. Figure 1.21a for model simulation/position and Frey et al. (2010) for morphological indications of bed overdeepenings). *Photograph: M. Bütler (August 2012).*

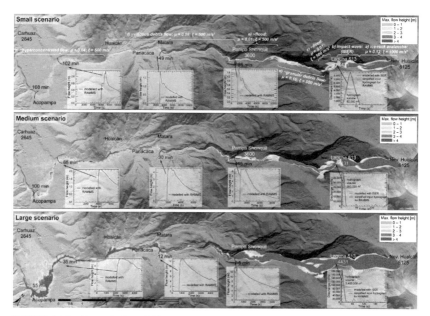

FIGURE 1.22   **Numerical modeling results of three scenarios (small, medium, and large) of potential mass movement process chains at Nevado Hualcán, Cordillera Blanca, Peru.** Flow durations indicated in minutes along the flow path are essential for warning and evacuation aspects. *From Schneider et al. (2014) with permission.*

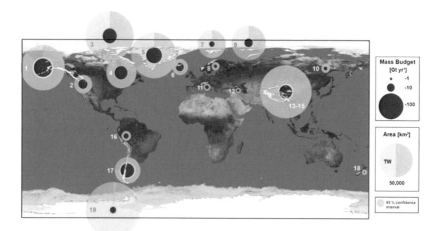

FIGURE 1.23   **Regional glacier mass budgets and areas from GRACE and ICESat data.** Red circles show 2003–2009 regional glacier mass budgets, and pale blue/green circles show regional glacier areas with tidewater basin fractions (the extent of ice flowing to termini in the ocean) in blue shading. Peach-colored halos surrounding red circles show the 95 percent confidence intervals in mass change estimates, but can only be seen in regions that have large uncertainties. *From Gardner et al. (2013). Reprinted with permission from AAAS.*

FIGURE 1.24   Deposits of the ice avalanche from Allalin glacier, which killed 88 workers at the construction site for the Mattmark dam (hydropower reservoir), Saas Valley, Southern Swiss Alps on August 30, 1965. *Comet-Photograph: Zürich (August 31, 1965).*

hazards, which started systematic efforts for improving the knowledge base with respect to anticipating and assessing dangerous situations (Haeberli et al., 2006). The enormous powder snow avalanche triggered by a rock avalanche from the permafrost-affected Brenva rock slope on the Italian side of Mont Blanc killed two skiers on an open ski run. Here, several people including the geology expert and the employee responsible for the safety of the ski run, were judged guilty and received a conditional penalty. These examples illustrate that careful and integrative hazard consideration is necessary where human activity follows retreating glaciers and also that "worst case" scenarios, especially concerning process chains, must be taken into consideration. After the disaster, caused by the outburst in 1941 of Laguna Palcacocha in the Cordillera Blanca, which destroyed the center of town and thousands of lives in Huaraz, Peru, the center was rebuilt. Moving away from the hazard zone was obviously not a politically realistic option (Carey, 2010; cf. Carey et al., 2012b concerning a comparable situation at nearby Carhuaz). As a consequence, expensive measures have to be taken in the source region of this disaster with its rapid ongoing changes. This necessitates constant observation and periodically repeated assessments by experts who must apply not only the existing but also the rapidly evolving understanding and technological possibilities in the field of snow and ice-related hazards. Analysis must be comprehensive, systemic, and cover the entire chain of possible processes from the very source of possible events down to possible socioeconomic impacts generally quite far away from it (cf. IPCC, 2012). In a remarkable effort to help in guaranteeing the necessary quality of hazard and risk assessments, GAPHAZ—the Standing

Scientific Working Group of the International Association of Cryospheric Sciences and the International Permafrost Association—issued a set of recommendations as follows:

1. Global change
   Climate change can induce disturbance in glacier and permafrost equilibrium and can shift hazard zones beyond historical limits. In many regions, human settlements and activities increasingly extend toward endangered zones increasing local vulnerability. As a result, historical data alone are not sufficient any more for hazard assessments and have to be combined with new observation and modeling approaches.
2. Chain reactions and interactions
   Glacier- and permafrost-related disasters often include a combination of processes and chain reactions. Hazard assessments therefore have to be integrative and consider such variety and interaction of processes.
3. Monitoring
   Due to the accelerated change of high-mountain environments, hazard assessments must be undertaken routinely and regularly, combined with appropriate monitoring.
4. Integrative risk assessments
   Integrative hazard assessments should be achieved by interdisciplinary cooperation of experts, and the application of modern observation and modeling techniques designed for such integrative approaches. Managing glacier and permafrost hazards requires risk assessments. For that purpose hazard assessments have to be combined with vulnerability assessments.
5. Remote sensing
   Modern space technologies enable initial estimation of hazard potentials to be performed by virtually everyone and everywhere, independent of political and geographical restrictions. This fundamental "democratization" process related to high-mountain (and other) hazards involves a number of new opportunities, dangers and responsibilities, for the public, the authorities in charge, and the experts involved.
6. Socioeconomic context
   The transfer and dissemination of expert hazard assessments to the authorities and to the public, and thus the efficiency of assessments, is to a large degree dependent on the socioeconomic context and the hazard perception of the endangered population. Communication of results from glacier and permafrost hazard assessments should consider these circumstances.

In this volume "Snow and Ice-Related Hazards, Risks and Disasters," leading experts in the field provide a modern knowledge basis concerning the involved challenges, possibilities and needs for further improvement to save lives and goods in snow and ice-affected environments on Earth.

REFERENCES

Allison, I., Colgan, W., King, M., Paul, F., 2014. Ice sheets, glaciers and sea level rise. In: Haeberli, W., Whiteman, C. (Eds.), Snow and Ice-related Hazards, Risks and Disasters. Elsevier, pp. 713−747.

Arenson, L., Colgan, W., Marshall, H.P., 2014. Physical, thermal and mechanical properties of snow, ice and permafrost. In: Haeberli, W., Whiteman, C. (Eds.), Snow and Ice-related Hazards, Risks and Disasters. Elsevier, pp. 35−75.

Barry, R.G., Armstrong, R., Callaghan, T., Cherry, J., Gearheard, Sh, Nolin, A., Don Russell, D., Zöckler, C., 2007. Snow. In: UNEP (Ed.), Global Outlook for Ice & Snow. UNEP/GRID, Arendal, pp. 39−62.

Beltaos, S., 1995. River Ice Jams. Water Resources Publications, LLC, Colorado.

Bentley, ChR., Thomas, R.H., Velicogna, I., 2007. Ice sheets. In: UNEP (Ed.), Global Outlook for Ice & Snow. UNEP/GRID, Arendal, pp. 99−113.

Boeckli, L., Brenning, A., Gruber, S., Noetzli, J., 2012. Permafrost distribution in the European Alps: calculation and evaluation of an index map and summary statistics. Cryosphere 6, 807−820. http://dx.doi.org/10.5194/tc-6-807-2012.

Bolch, T., Christiansen, H.H., 2014. Large-scale settings: the physiography of the cryosphere. In: Haeberli, W., Whiteman, C. (Eds.), Snow and Ice-related Hazards, Risks and Disasters. Elsevier, pp. 201−217.

Bründl, M., Margreth, S., 2014. Integrative risk management: the example of snow avalanches. In: Haeberli, W., Whiteman, C. (Eds.), Snow and Ice-related Hazards, Risks and Disasters. Elsevier, pp. 263−301.

Callaghan, T.V., Johansson, M., 2014. Snow, ice and the biosphere. In: Haeberli, W., Whiteman, C. (Eds.), Snow and Ice-related Hazards, Risks and Disasters. Elsevier, pp. 139−165.

Carey, M., 2010. In the Shadow of Melting Glaciers: Climate Change and Andean Society. Oxford University Press, New York.

Carey, M., French, A., O'Brien, E., 2012a. Unintended effects of technology on climate change adaptation: an historical analysis of water conflicts below Andean Glaciers. J. Hist. Geogr. 38, 181−191.

Carey, M., Huggel, C., Bury, J., Portocarrero, C., Haeberli, W., 2012b. An integrated socio-environmental framework for glacier hazard management and climate change adaptation: lessons from Lake 513, Cordillera Blanca, Peru. Climatic Change 112 (3), 733−767.

Carey, M., McDowell, G., Huggel, C., Jackson, M., Portocarrero, C., Reynolds, J.M., Vicuña, L., 2014. Integrated approaches to adaptation and disaster risk reduction in dynamic socio-cryospheric systems. In: Haeberli, W., Whiteman, C. (Eds.), Snow and Ice-related Hazards, Risks and Disasters. Elsevier, pp. 219−261.

Church, J.A., Nicholls, R., Hay, J.E., Gornitz, V., 2007. Ice and sea level change. In: UNEP (Ed.), Global Outlook for Ice & Snow. UNEP/GRID, Arendal, pp. 153−180.

Citterio, M., Paul, F., Ahlstrøm, A.P., Jepsen, H.F., Weidick, A., 2009. Remote sensing of glacier change in West Greenland: accounting for the occurrence of surge-type glaciers. Ann. Glaciol. 50 (53), 70−80.

Clague, J.J., O'Connor, J.E., 2014. Glacier-related outburst floods. In: Haeberli, W., Whiteman, C. (Eds.), Snow and Ice-related Hazards, Risks and Disasters. Elsevier, pp. 487−519.

Csatho, B., Schenk, T., van der Veen, C.J., Krabill, W.B., 2008. Intermittent thinning of Jakobshavn Isbræ, West Greenland, since the little ice age. J. Glaciol. 54 (184), 131−144.

Delgado Granados, H., Julio Miranda, P., Ontiveros Gonzáles, G., Cortés Ramos, J., Carrasco Núñez, G., Pulgarín Alzate, B., Mothes, P., Moreno Roa, H., Cáceres Correa, B.E., 2014.

Hazards at ice-clad volcanoes. In: Haeberli, W., Whiteman, C. (Eds.), Snow and Ice-related Hazards, Risks and Disasters. Elsevier, pp. 607−646.

Deline, P., Gruber, S., Delaloye, R., Fischer, L., Geertsema, M., Giardino, M., Hasler, A., Kirkbride, M., Krautblatter, M., Magnin, F., McColl, S., Ravanel, L., Schoeneich, P., 2014. Ice loss and slope stability in high-mountain regions. In: Haeberli, W., Whiteman, C. (Eds.), Snow and Ice-related Hazards, Risks and Disasters. Elsevier, pp. 303−344.

Evans, S.G., Delaney, K.B., 2014. Catastrophic mass flows in the mountain glacial environment. In: Haeberli, W., Whiteman, C. (Eds.), Snow and Ice-related Hazards, Risks and Disasters. Elsevier, pp. 563−606.

Fischer, U.H., Bebiolka, A., Brandefelt, J., Follin, S., Hirschorn, S., Jensen, M., Keller, S., Kennell, L., Näslund, J.-O., Normani, S., Selroos, J.-O., Vidstrand, P., 2014. Radioactive waste under conditions of future ice ages. In: Haeberli, W., Whiteman, C. (Eds.), Snow and Ice-related Hazards, Risks and Disasters. Elsevier xx-xx (in this volume).

Frey, H., Haeberli, W., Linsbauer, A., Huggel, C., Paul, F., 2010. A multi level strategy for anticipating future glacier lake formation and associated hazard potentials. Nat. Hazard Earth Sys. Sci. 10, 339−352.

Gardner, A.S., Moholdt, G., Cogley, J.G., Wouters, B., Arendt, A.A., Wahr, J., Berthier, E., Hock, R., Pfeffer, T., Kaser, G., Ligtenberg, S.R.M., Bolch, T., Sharp, M.J., Hagen, J.O., van den Broeke, M.R., Frank Paul, F., 2013. A reconciled estimate of glacier contributions to sea level rise: 2003 to 2009. Science 340 (6134), 852−857. http://dx.doi.org/10.1126/science.1234532.

Gerland, S., Aars, J., Bracegirdle, T., Eddy Carmack, E., Hop, H., Hovelsrud, G.K., Kovacs, K., Lydersen, C., Perovich, D.K., Richter-Menge, J., Rybråten, S., Strøm, H., Turner, J., 2007. Ice in the sea. In: UNEP (Ed.), Global Outlook for Ice & Snow. UNEP/GRID, Arendal, pp. 63−96.

Gruber, S., 2012. Derivation and analysis of a high-resolution estimate of global permafrost zonation. Cryosphere 6, 221−233. http://dx.doi.org/10.5194/tc-6-221-2012.

Haeberli, W., 2007. Changing views of changing glaciers. In: Orlove, B., Wiegandt, E., Luckman, B.H. (Eds.), Darkening Peaks − Glacial Retreat. Science and Society. University of California Press, pp. 23−32.

Haeberli, W., 2013. Mountain permafrost − research frontiers and a special long-term challenge. Cold Regions Sci. Technol. 96, 71−76. http://dx.doi.org/10.1016/j.coldregions.2013.02.004.

Haeberli, W., Linsbauer, A., 2013. Global glacier volumes and sea level − small but systematic effects of ice below the surface of the ocean and of new local lakes on land. Brief communication. Cryosphere 7, 817−821. http://dx.doi.org/10.5194/tc-7-817-2013.

Haeberli, W., Huggel, C., Kaeaeb, A., Gruber, S., Noetzli, J., Zgraggen-Oswald, S., 2006. Development and perspectives of applied research on glacier and permafrost hazards in high-mountain regions − the example of Switzerland. In: Proceedings of the International Conference on High Mountain Hazard Prevention, Vladikavkaz/Moscow, June 23-26, 2004, pp. 219−228.

Haeberli, W., Huggel, C., Paul, F. Zemp, M., 2013. Glacial response to climate change. In: Shroder, J. (Editor in Chief), James, L.A., Harden, C.P., Clague, J.J. (Eds.), Treatise on Geomorphology. Academic Press, San Diego, CA. In: Geomorphology of Human Disturbances, Climate Change, and Natural Hazards, vol. 13, pp. 152−175.

Haeberli, W., Noetzli, J., Arenson, L., Delaloye, R., Gärtner-Roer, I., Gruber, S., Isaksen, K., Kneisel, C., Krautblatter, M., Phillips, M., 2011. Mountain permafrost: development and challenges of a young research field. J. Glaciol. 56 (200) (special issue), 1043−1058.

Harrison, W.D., Osipova, G.B., Nosenko, G.A., Espizua, L., Kääb, A., Fischer, L., Huggel, C., Burns, P.A.C., Truffer, M., Lai, A.W., 2014. Glacier surges. In: Haeberli, W., Whiteman, C. (Eds.), Snow and Ice-related Hazards, Risks and Disasters. Elsevier, pp. 437−485.

IPCC, 2012. Managing the Risks of Extreme Events and Disasters to Advance Climate Change Adaptation. A Special Report of Working Groups I and II of the Intergovernmental Panel on Climate Change. In: Field, C.B., Barros, V., Stocker, T.F., Qin, D., Dokken, D.J., Ebi, K.L., Mastrandrea, M.D., Mach, K.J., Plattner, G.-K., Allen, S.K., Tignor, M., Midgley, P.M. (Eds.). Cambridge University Press, Cambridge, UK, and New York, NY, USA, p. 582.

IPCC, 2013. Climate Change 2013: The Physical Science Basis. Working Group I Contribution to the Fifth Assessment Report of the Intergovernmental Panel on Climate Change. Cambridge University Press, New York, etc.

Jouzel, J., Lorius, C., Reynaud, D., 2013. The White Planet: The Evolution and Future of Our Frozen World. English translation, adapted and revised. Princeton University Press, Princeton.

Kubat, I., Sayed, M., Fowler, C.D., 2014. Floating ice and the ice pressure challenge to ships. In: Haeberli, W., Whiteman, C. (Eds.), Snow and Ice-related Hazards, Risks and Disasters. Elsevier, pp. 437−485.

Levermann, A., Bamber, J.L., Drijfhout, S., Ganopolski, A., Haeberli, W., Harris, N.R.P., Huss, M., Krüger, K., Lenton, T.M., Lindsay, R.W., Notz, D., Wadhams, P., Weber, S., 2011. Potential climatic transitions with profound impact on Europe − review of the current state of six 'tipping elements of the climate system'. Climatic Change 110 (3−4), 845−878. http://dx.doi.org/10.1007/s10584-011-0126-5.

Linsbauer, A., Paul, F., Haeberli, W., 2012. Modeling glacier thickness distribution and bed topography over entire mountain ranges with GlabTop: application of a fast and robust approach. J. Geophys. Res. 117, F03007. http://dx.doi.org/10.1029/2011JF002313.

Lipenkov, V., on behalf of the Vostok project members, 2006. Vostok ice core project. Pages News 14 (1), 29−31.

Ohmura, A., 2014. Snow and ice in the climate system. In: Haeberli, W., Whiteman, C. (Eds.), Snow and Ice-related Hazards, Risks and Disasters. Elsevier, pp. 77−98.

Prowse, T.D., Bonsal, B.R., Duguay, C.R., Hessen, D.O., Vuglinsky, V.S., 2007. River and lake ice. In: UNEP (Ed.), Global Outlook for Ice & Snow. UNEP/GRID, Arendal, pp. 201−213.

Radić, V., Bliss, A., Beedlow, A.C., Hock, R., Miles, E., Cogley, J.G., 2014. Regional and global projection of twenty first century glacier mass changes in response to climatic scenarios from global climate models. Climate Dynamics 42, 37−58.

Rasmo, N., Röthlisberger, M., Ruhmer, E., Weber, B., Wied, A., 1981. Die Alpen in der Malerei. Rosenheimer, Rosenheim.

Romanovsky, V.E., Gruber, S., Instanes, A., Jin, H., Marchenko, S.S., Smith, S.L., Trombotto, D., Walter, K.M., 2007. Frozen ground. In: UNEP (Ed.), Global Outlook for Ice & Snow. UNEP/GRID, Arendal, pp. 115−152.

Romanowsky, V.E., Smith, S.L., Christiansen, H.H., 2010. Permafrost thermal state in the polar northern hemisphere during the international polar year 2007−2009: a Synthesis. Permafrost Periglacial Process. 21, 181−200.

Schneider, D., Huggel, C., Cochachin, A., Guillén, S., García, J., 2014. Mapping hazards from glacier lake outburst floods based on modelling of process cascades at Lake 513, Carhuaz, Peru. Adv. Geosci. 35, 145−155. http://dx.doi.org/10.5194/adgeo-35-145-2014.

Schweizer, J., Bartelt, P., van Herwijnen, A., 2014. Snow avalanches. In: Haeberli, W., Whiteman, C. (Eds.), Snow and Ice-related Hazards, Risks and Disasters. Elsevier, pp. 395−436.

Seibert, J., Jenicek, M., Huss, M., Ewen, T., 2014. Snow and ice in the hydrosphere. In: Haeberli, W., Whiteman, C. (Eds.), Snow and Ice-related Hazards, Risks and Disasters. Elsevier, pp. 99−137.

Singh, V.P., Singh, P., Haritashya, U.K., 2011. Encyclopedia of Snow, Ice and Glaciers. Springer, 1253 pp.

Streletskiy, D., Anisimov, O., Vasiliev, A., 2014. Permafrost degradation. In: Haeberli, W., Whiteman, C. (Eds.), Snow and Ice-related Hazards, Risks and Disasters. Elsevier, pp. 303−344.

Strozzi, T., Delaloye, R., Kääb, A., Ambrosi, C., Perruchoud, E., Wegmüller, U., 2010. Combined observations of rock mass movements using satellite SAR interferometry, differential GPS, airborne digital photogrammetry, and airborne photography interpretation. J. Geophys. Res.: Earth Surface (2003−2012). 115, F01014.

Swift, D.A., Cook, S., Heckmann, T., Moore, J., Gärther-Roer, I., Korup, O., 2014. Ice and snow as landforming agents. In: Haeberli, W., Whiteman, C. (Eds.), Snow and Ice-related Hazards, Risks and Disasters. Elsevier, pp. 167−199.

UNEP, 2007. Global Outlook for Ice & Snow. UNEP/GRID, Arendal, 235 pp.

Vieli, A., 2014. Retreat instability of tidewater glaciers and marine ice sheets. In: Haeberli, W., Whiteman, C. (Eds.), Snow and Ice-related Hazards, Risks and Disasters. Elsevier, pp. 677−712.

WGMS, 2008. Global glacier changes: facts and figures. In: Zemp, M., Roer, I., Kääb, A., Hoelzle, M., Paul, F., Haeberli, W. (Eds.), UNEP, World Glacier Monitoring Service. University of Zurich, Switzerland.

Whiteman, C.A., 2011. Cold Region Hazards and Risks. Wiley-Blackwell, 366 pp.

Williams, R.S., Ferrigno, J., 2012. State of the Earth's Cryosphere at the Beginning of the 21st Century: Glaciers, Global Snow Cover, Floating Ice, and Permafrost and Periglacial Environments. Satellite Image Atlas of Glaciers of the World. U.S. Geological Survey Professional Paper 138−A, 496 pp.

Zemp, M., Haeberli, W., Hoelzle, M., Paul, F., 2006. Alpine glaciers to disappear within decades? Geophys. Res. Lett. 33, L13504. http://dx.doi.org/10.1029/2006GL026319.

Zemp, M., Haeberli, W., 21 contributing authors, 2007. Glaciers and ice caps. In: UNEP (Ed.), Global Outlook for Ice & Snow. UNEP/GRID, Arendal, pp. 115−152.

Chapter 2

Physical, Thermal, and Mechanical Properties of Snow, Ice, and Permafrost

Lukas U. Arenson [1], William Colgan [2] and Hans Peter Marshall [3]

[1] *BGC Engineering Inc., Vancouver, BC, Canada,* [2] *Geological Survey of Denmark and Greenland, Copenhagen, Denmark,* [3] *Department of Geosciences and Center for Geophysical Investigation of the Shallow Subsurface, Boise State University, ID, USA*

ABSTRACT

The physical, thermal, and mechanical properties of snow, ice, and permafrost, specifically frozen ground, are responsible for the structure and behavior of these cryospheric materials. A fundamental understanding of these properties is therefore key to evaluating associated cryospheric hazards. The material properties of snow, ice, and permafrost vary over many orders of magnitude. Generally, ice has the smallest range in physical and thermal properties (e.g., density or thermal conductivity), while properties of snow and frozen ground have much larger variations. Frozen ground has the largest absolute range in thermal properties, and snow has the largest relative range in physical properties. The mechanical response of each material, including its strength and creep behavior, also varies widely, depending on stress, temperature, and loading rate. Rapid loading under cold conditions results in brittle behavior with high peak strength and small critical strain, whereas slow loading under warm conditions results in ductile behavior with low peak strength and large strain. A full appreciation of the hazards associated with snow, ice, and permafrost requires an understanding of how the basic properties of these cryospheric materials are influenced by environmental conditions.

2.1 INTRODUCTION

The cryosphere, which encompasses all frozen water on Earth, consists of snow, ice (including lake and river ice, sea ice, glaciers, and ice sheets), and permafrost (including seasonally frozen ground). The physical, mechanical, and thermal properties of snow, ice, and permafrost control the dynamics of associated cryospheric hazards. These properties are vital to the numerical modeling and forecasting of natural hazards involving snow, ice, and permafrost.

Snow and Ice-Related Hazards, Risks, and Disasters. http://dx.doi.org/10.1016/B978-0-12-394849-6.00002-0

Snow cover area varies dramatically throughout the year. In North America, for example, snow covers nearly 50 percent (45 million km^2) of land in the winter, and less than 10 percent (3 million km^2) of land in the summer (e.g., Armstrong and Brodzick, 2001). Over one billion people, between one-sixth and one-third of the global population, rely on water at least partially supplied by seasonal snow or glacial melt (Beniston et al., 2003; Barnett et al., 2005). Rain-on-snow and snowmelt can cause rapid increases in streamflow, and are often major contributors to spring floods and debris flows in mountainous terrains. Extreme winter weather conditions are a major cause of death, accounting for 18 percent of natural hazard deaths in the US between 1970 and 2004 (approximately 100 out of 570 hazards-related deaths every year). This winter mortality is similar to the mortality stemming from hazards such as extreme heat (20 percent), severe summer weather (28 percent), floods (14 percent), and lightning (12 percent) (Borden and Cutter, 2008). Snow avalanches caused an average of 27 deaths per year in the Western US alone between 1998 and 2008 (Spencer and Ashley, 2011). In some states, avalanches are the leading source of death due to natural hazards. Although the property damage resulting from avalanches is typically low in the US, due to low population density in mountainous areas, elsewhere avalanches can have significant economic consequences. In Switzerland, avalanches are a critical natural hazard, and therefore significant investment has been made in better understanding snow mechanics as well as infrastructure for mitigation, adaptation, and forecasting (e.g., Fuchs et al., 2007).

Nearly 70 percent of Earth's freshwater, which is only about 3 percent of the total water on Earth, is stored in glaciers and ice sheets that are fed by snowfall (IPCC, 2013). Glaciers and ice sheets, which together comprise the most abundant form of terrestrial ice by mass, are controlled by a balance of snow accumulation and ice ablation. By definition, glaciers and ice sheets develop thermomechanical flow, driven by gravity, whereby ice ablating at lower elevations is replenished by snowfall accumulating at higher elevations (Cogley et al., 2011). The response of glaciers and ice sheets to contemporary climate change is a subject of substantial interest for longer-term hazards, such as changing water resources and sea level rise (Vaughan and Comiso, 2013, Chapter 20). Differences in individual glacier mass, elevation, geometry, thermal regime, and tidewater forcing, however, result in complex and sometimes asynchronous responses to common climatic forcing (Joughin et al., 2010). In addition to climate-driven glacier retreat, increasing global resource demand and improved mining and prospecting techniques are making cold regions mineral development more feasible. Mining projects seeking to exploit periglacial, proglacial, and even subglacial mineral deposits are being confronted with unique glacier management challenges, such as open ice pit excavation and maintenance (Colgan and Arenson, 2013; Colgan, 2014). Calving tidewater glaciers and ice shelves creates icebergs that pose a significant hazard to boats and ships as well as pipelines on the seabed. Also, extraction of oil in the Arctic has created new challenges due to rapidly changing sea ice conditions.

Frozen ground, which includes both permanently frozen ground (or permafrost) as well as seasonally frozen ground, occupies approximately 54 million km^2 of the land area of the Northern Hemisphere (Zhang et al., 2003). Permafrost alone covers 20–25 percent of Earth's land surface, making it the largest component of the cryosphere by area. During boreal summer, the extent of permafrost is more than twice that of snow cover. The genesis and dynamics of ground ice can be very different from the ice in glaciers or ice sheets. Glaciogenic ground ice ultimately originates from glacier ice, whereas cryogenic ground ice forms within the ground through a combination of thermophysical, physicochemical, and physicomechanical processes (Shur and Jorgenson, 1998; Stephani et al., 2010). As the structure of glaciogenic and cryogenic ground ice differs, the mechanical response of each ice type depends on its morphology. Under contemporary climate change and projections of the substantial degradation of near-surface permafrost over the next century, thawing of frozen ground presents a real subsidence hazard for existing northern infrastructure and population centers (Nelson et al., 2001; ACIA, 2005; Lawrence and Slater, 2005; Callaghan et al., 2011; IPCC, 2013).

A fundamental understanding of the properties of cryospheric materials is critical to evaluating associated cryospheric hazards. In this chapter, the physical, thermal, and mechanical properties of snow, ice, and frozen ground are described. We also briefly mention dynamic and electrical properties of each material, as knowledge of these properties is useful in geophysical surveying, as well as for understanding material response to solar radiation. Although an exhaustive review is beyond the scope of this chapter, we have cited a comprehensive range of references to assist the reader in seeking more detailed material-specific information.

2.2 DENSITY AND STRUCTURE

2.2.1 Snow

The structure of snow is very complex because of the sensitivity of new snow structure to the atmospheric conditions through which it falls, as well as the environmental conditions it experiences on the ground. Postdeposition, snow structure changes rapidly and continuously depending on air temperature, wind, and pressure conditions. Hence, nascent snow structure is almost irrelevant to the characteristics of a mature snowpack. Wind tends to degrade complex snow grain structure, producing small rounded grains, which can sinter to form strong bonds. Under low temperature gradient conditions, curvature effects dominate metamorphism, and snow grains rapidly round and form strong bonds due to microscale vapor pressure gradients. When temperature gradients exceed approximately 0.1 K/cm, metamorphism is dominated by vapor pressure gradients aligned with the temperature gradient, causing anisotropy and faceting (e.g., McClung and Schaerer, 1993). Snow microstructure is a major control on both thermal (e.g., Schneebeli and Sokratov, 2004; Satyawali and Singh, 2008) and mechanical (e.g., Schweizer, 1999; Petrovic, 2003) properties, and influences the resulting crystal structure of ice formed from multiyear snow.

Snow microstructure, although it is the major control on thermal and mechanical behavior, remains extremely difficult and time consuming to measure. The most accurate microstructure observations to date have been performed using X-ray tomography or microcomputed tomography (micro-CT) including observations under various temperature gradients (e.g., Schneebeli and Sokratov, 2004) and mechanical tests (Schleef and Löwe, 2013; Wang and Baker, 2013). This new scale of observation has increased our understanding of the dynamics of snow microstructure, and will likely facilitate more sophisticated model parameterizations to represent microscale processes. High-resolution snow penetrometers are capable of measuring microstructure in the field (Marshall and Johnson, 2009; Löwe and Van Herwijnen, 2012) providing input for in situ mechanical properties. Several other techniques have shown promise, including near-infrared (NIR) photography of snow pit walls (e.g., Matzl and Schneebeli, 2006) and millimeter-scale cone penetrometry (e.g., Schneebeli and Johnson, 1998; Johnson and Schneebeli, 1999). NIR photography can provide estimates of specific surface area, and penetrometer measurements have been used to classify snow type (e.g., Satyawali et al., 2009; Havens et al., 2013).

Although it is much less variable than microstructure, natural snow density varies more than the density of most Earth materials, ranging over an order of magnitude in seasonal snow ($10-550$ kg/m^3), and increasing by almost another factor of two in perennial snowfields due to refrozen meltwater and/or compressive forces generating multiyear firn and eventually glacial ice. Firn is partially compacted snow from the past season, in which snow crystals have metamorphosed, creating a material denser than fresh snow. New snow density depends on air temperature, relative humidity, and wind conditions between its formation in clouds and its deposition on the Earth's surface. In maritime environments, new snow densities typically range from approximately 100 to 300 kg/m^3. In continental environments, however, new snow densities typically range between approximately 10 and 100 kg/m^3. Modeling fresh snow density remains highly uncertain, due to the large number of factors and range of environmental conditions through which snow falls before reaching the ground. Once deposited on the Earth's surface, snow and firn density increases through metamorphism, eventually approaching the density of ice. Metamorphism is a combination of both physical and thermal processes (e.g., Shapiro et al., 1997) (Figures 2.1 and 2.2).

Uncertainty in snow density can be a major complication when estimating both the water stored in snow or firn and the snow load or potential impact

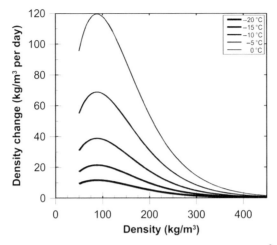

FIGURE 2.1   Daily snow density change under weight of 30 cm of 100 kg/m^3 snow.

FIGURE 2.2   Snowpack layering. *Left: photo: J. Bay, right: photo: K. Klassen.*

forces in avalanche zones. In mature snowpacks, for example, the formation of a saturated layer at the base of the snowpack may increase snow load by more than 100 kg/m^2. Although dry snow densification at wind-protected sites can be modeled reasonably well, wind loading, which can enhance densification significantly, is difficult to simulate. Similarly, liquid water percolation is extremely heterogeneous, both horizontally and vertically, making the collection of accurate density observations costly and time consuming over any appreciable area. In comparison to vertical variability in density, bulk snowpack density or depth-averaged density is easier to measure and likely less spatially variable. Although time consuming to measure, snow density and structure are correlated with other snow properties (e.g., thermomechanical, electrical, and acoustic properties) and are key controls in both avalanche and snow-load hazards and much easier to measure than snow microstructure.

2.2.2 Ice

In perennial snowfields, the snowfall that undergoes multiannual densification into ice is generally not regarded as true ice until the pore spaces between crystals have sufficiently deformed to prevent airflow. In the absence of substantial meltwater, this typically happens around a density of $830 \, kg/m^3$ (sometimes called the "close-off" density), which may be regarded as the theoretical minimum density of meteoric ice (Cogley et al., 2011). The theoretical maximum density of all freshwater ice types is $917 \, kg/m^3$ (Table 2.1). Although centimeter-scale samples recovered from several meters depth in a glacier may approach this ideal value (Butkovich, 1954, 1959), effective bulk density is often modified by nonglacier ice impurities, such as macroporosity (i.e., ice fractures or bubbles), refrozen water and/or sediment (Figure 2.3). In glaciers, bulk ice porosity also arises from surface and basal crevasses, as well as components of the glacier hydrology system (Flowers and Clarke, 2002). Observed values of bulk ice porosity range between fractional volumes of 0.004 and 0.013 in mountain glaciers (Fountain and Walder, 1998). The bulk ice porosity in heavily crevassed glaciers can exceed a fractional volume of 0.2, resulting in local bulk densities of $\sim 700 \, kg/m^3$ (Meier et al., 1994). Alternatively, the entrained sediment content of ice within some portions of a glacier can exceed a fractional volume of 0.5, blurring the distinction between frozen ground and ice, resulting in local bulk glacier densities of $\sim 2000 \, kg/m^3$ (Swinzow, 1962).

For most practical purposes, such as the thermodynamic modeling of the mass and energy balances of ice, and converting geodetic glacier volume changes into mass changes, bulk ice density is taken as between 900 and

TABLE 2.1 Physical Properties of Snow, Ice and Frozen Ground: Density (kg/m^3). Note the Rapid Change in Snow Density Under Load (Figure 2.1)

Snow, fresh	100–300 (maritime) 15–100 (arctic, continental)	Snow, old	200–550 (dry) 400–650 (wet)
Firn	550–830	Air	1.15–1.42
Ice	830–917	Water (0 °C)	999.87

Frozen Ground (Volumetric Ice Content)

Ice-poor (<20 percent)	1900–2300	Intermediate (20–50 percent)	1500–2300
Ice-rich (50–80 percent)	1000–1800	Dirty ice (>80 percent)	700–1300

Note: See text for references.

FIGURE 2.3   A nearly vertical ice outcrop from beneath a moraine of the Greenland ice sheet, near Kangerlussuaq, Greenland. Note the multiple fractures, some of which have been filled by refrozen water. Given the physical and thermal properties of ice, such vertical ice faces are highly transient. *Photo: W. Colgan.*

910 kg/m^3, with local density excursions ignored in time and space (Price et al., 2011; Zwally et al., 2011; Colgan et al., 2012). Ice density decreases very slightly with increasing temperature, with a linear thermal expansion coefficient of -5.1×10^{-5}/K, and also increases very slightly with increasing pressure, with a linear elastic compressibility coefficient of 37.1×10^{-6}/MPa (Shumskiy, 1959; Bader, 1964; Gow and Williamson, 1972; Gammon et al., 1983). These effects typically only become significant in ice-sheet settings, where ice thicknesses are kilometers scale. Even in these settings, however, the tendency for both temperature and pressure to increase with depth produces counteracting effects on ice density (e.g., Glow, 1970). Due to the extremely

subtle nature of this ice density dependence on temperature and pressure, ice is generally regarded as an incompressible material, similar to water, in virtually all natural settings.

2.2.3 Frozen Ground

The structure and density of frozen ground can vary significantly (Andersland and Ladanyi, 2004), encompassing a range that spans from dirty ice (e.g., Hooke et al., 1972) to frozen bedrock with ice-filled fissures and joints (Matsuoka and Sakai, 1999; Davies et al., 2001). As substantial air bubbles may be present in some permafrost soils, the minimum density of frozen ground may be 500 kg/m^3 (e.g., Arenson and Springman, 2005a). The maximum density of frozen ground primarily depends on the density of the soil or rock content. For frozen soils, the maximum density may be as high as 2200 kg/m^3, but permafrost density can be much higher for metal-rich frozen rocks (Table 2.1). Frozen ground density is therefore highly site-specific, and can only be accurately determined when the volumetric contents and densities of individual constituents (i.e., ice, air, water, rock, and soil) are known.

Between endmembers of unfrozen soil and ice, the structure of frozen ground can be differentiated into four different classes based on volumetric ice content (Arenson et al., 2007). Ice-poor soils are characterized by having a low ice content, whereby only pore ice exists and soil grains remain in contact. In these soils, the volumetric ice content is smaller than the porosity of the unfrozen material. Ice-rich soils typically have volumetric ice contents between 50 and 80 percent. In these soils, the volumetric ice content is larger than the soil porosity in the unfrozen state (Figure 2.4). The ice in such soils is classified as excess ice. Between ice-poor and ice-rich soils, frozen ground may be classified as containing an intermediate ice content. When volumetric ice content is greater than 80 percent, the frozen ground is typically considered as

FIGURE 2.4 Ice-rich coarse colluvium. *Photo: L. Arenson.*

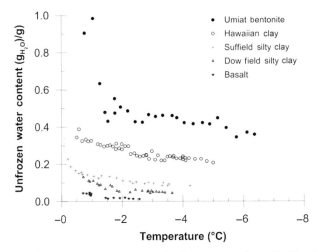

FIGURE 2.5   Unfrozen water content curves for five representative soils. The fines content increases from the coarse basalt to the very fine bentonite, clay consisting mostly of montmorillonite. *After Anderson and Morgenstern (1973); Andersland and Ladanyi (2004).*

dirty ice. In terms of relative soil density, which is determined based on a ratio between the maximum, minimum, and actual void ratio, frozen soils with more than 50 percent volumetric ice content (ice-rich soils) have no relative density as the soil particles are loosely distributed within the ice matrix with interrupted particle-to-particle contact. On the other hand, ice-poor soils, if saturated, have relative densities close to 100 percent of the density of unfrozen soil.

As indicated by Williams (1967a; 1967b) and many others (e.g., Anderson and Tice, 1972; Wettlaufer and Worster, 2006) trace water within a frozen soil remains unfrozen even at temperatures below 0 °C. The water may either be present as adsorbed water, a thin liquid layer around the solid particles, or as free water within the pores. The amount of unfrozen water depends on the characteristics of the soil particles, and is typically higher for fine-grained soils (Figure 2.5). Freezing-point depression due to pore water chemistry and pressure can further increase the amount of unfrozen water in a frozen soil. Because the strength of frozen and partially frozen soils depends on effective stress, pore pressure and the presence of unfrozen water are important controls on the stability of frozen ground (Terzaghi, 1925).

2.3 THERMAL PROPERTIES

Hazards are mainly controlled by the strength of the cryospheric material, which is often a function of temperature. In order to assess a cryospheric hazard and evaluate how it changes with time, the thermal properties of the material need to be known. The thermal conductivity, the heat capacity, and latent heat do control how quickly the temperatures in the material change at

certain distances from the source of a temperature change, such as the atmosphere, with time.

2.3.1 Snow

The thermal properties of snow vary primarily with density, but are also
dependent on microstructure and temperature, and often exhibit significant
anisotropy. The thermal conductivity of snow ranges from 0.04 to 1 W/m/K
over the density range of 100−550 kg/m^3 (e.g., Sturm and Johnson, 1992;
Sturm et al., 1997) (Table 2.2). At a given snow density, variations in microstructure can change thermal conductivity by a factor of two, as can variations
in crystal anisotropic orientation (e.g., Schneebeli and Sokratov, 2004;
Kaempfer et al., 2005; Satyawali and Singh, 2008; Riche and Schneebeli, 2010;
Shertzer and Adams, 2011). More recent work (Riche and Schneebeli, 2010;
Calonne et al., 2011) has indicated that needle-probe observations can be
biased, due to structural and thermal effects from the probe insertion and length
of time required for temperature to become stable. Accurate laboratory observations including high-resolution microstructure with X-ray tomography are
allowing anisotropic models of thermal conductivity to be developed and tested
(e.g., Calonne et al., 2011; Löwe et al., 2013; Riche and Schneebeli, 2013).
Figure 2.6 shows the variation of thermal conductivity with density, as
expressed by a summary of observed data and a conventional model (Sturm
et al., 1997), as well as newer models that incorporate high-resolution

TABLE 2.2 Thermal Properties of Snow, Ice, Frozen Ground, and Water:
Thermal Conductivity and Specific Heat Capacity. Snow Parameters are
Highly Density Dependent (e.g., Andersland and Ladanyi, 2004; Sturm
et al., 1997; Oldroyd et al., 2013)

	Snow	Ice	Frozen Ground	Water
Thermal conductivity (W/m/K)	0.04−1.0 (~proportional to the square of density)	2.10−2.76	Sand and gravel: 0.5−4.0 Silt and clay: 0.1−2.2 Peat: 0.03−1.25	0.562
Specific heat capacity (kJ/kg/K)	2.09	1.741−2.097	Silt, clay, sand, and gravel: 0.70−1.3 Peat: 0.7−2.2	4.217
Thermal diffusivity (10^{-7} m^2/s)	1−8 (~proportional to density)	1	0.1−80	1.33

Note: See text for references.

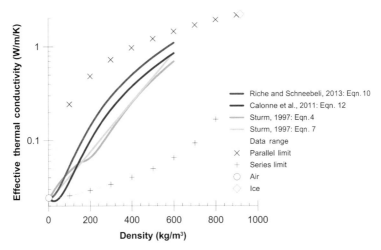

FIGURE 2.6 Variation of thermal conductivity with density, the range of data in the literature summarized by Sturm et al. (1997) and several proposed models.

microstructure information (Calonne et al., 2011; Riche and Schneebeli, 2013). In theory, the limiting cases of thermal conductivity in snow and ice correspond to heat transfer through ice structures aligned in parallel and series with respect to the temperature gradient (Sturm et al., 1997). Recent modeling studies have shown that density observations at 10 cm resolution do not resolve thin low conductivity layers that can have a significant effect on estimated temperature profiles near the surface (Dadic et al., 2008).

Latent heat from rain-on-snow events, surface hoar formation, and sublimation from the snow surface can have a significant effect on the thermal state of the snow cover (e.g., Marks et al., 2001; Marks and Winstral, 2001; Lehning et al., 2002). Large temperature gradients, caused by thermal conductivity variations near ice layers, cause faceting that leads to weak layer formation and is a major cause of avalanches (e.g., McClung and Schaerer, 1993; Schweizer, 1999). Thermal heat released from the snow cover can cool the snow surface to temperatures far below the boundary layer air temperature, causing deposition of vapor and surface hoar formation. Surface hoar, the solid equivalent of dew, is not only a potentially important weak layer in snow slope stability, but it also documents a major change in snow structure. On glaciers and perennial snowfields, hoar formation in the early autumn results in visual changes in ice cores that are used for counting annual layers.

2.3.2 Ice

The specific heat capacity of ice is approximately half that of water, ranging from 2.097 J/g/K at 0 °C down to 1.741 J/g/K at −50 °C (Table 2.2). The thermal conductivity of ice, which is approximately four times greater than water, ranges from 2.10 W/m/K at 0 °C up to 2.76 W/m/K at −50 °C (Yen, 1981). For the purposes of modeling ice temperatures, however, these relatively minor temperature dependences are typically ignored, even in the most complex thermodynamic glacier community models (Price et al., 2011). Instead, a constant specific heat capacity of 2.009 J/g/K and a thermal conductivity characteristic of the ice mass under consideration are typically employed. The latent heat of fusion associated with the change of phase between water and ice is universally taken as 333.5 kJ/kg whereas the latent heat of vaporization is 2260 kJ/kg (Clarke et al., 1977; Funk et al., 1994; Payne, 1995). At one atmosphere of pressure, the phase change between ice and water occurs at 273.15 K (0 °C). This pressure-melting point is depressed by pressure, decreasing by 0.0742 K/MPa, or 0.87 K/km of overburden ice (Lliboutry, 1976). The distinction of whether or not ice is at the pressure-melting point is particularly important in glacial studies, as glaciers in which ice is at the pressure-melting point, develop a hydrologic network capable of routing liquid water, as well as enhancing sliding, and consequently erosion, at the glacier bed (Payne, 1995; Fountain and Walder, 1998; Anderson et al., 2004; Colgan et al., 2011). Glacial research has been a key driver for understanding ice properties, as glaciers and ice sheets are the most common form of terrestrial ice.

In a glacier, the temperature of a given parcel of ice is dependent on a conventional energy balance comprised of conduction, advection, and production terms. Heat production stems from the energy released by strain deformation. The relative influence of these three terms varies by glacier setting. Conduction is influenced by both the mean atmospheric temperature at the ice surface, as well as the geothermal heat flux at the ice bed (Hooke, 1977; Zhang et al., 2012). Advection and production are both influenced by stress and velocity fields within the ice, which ultimately depend on glacier geometry (Clarke et al., 1977; Funk et al., 1994). The thermal response timescales of glaciers and ice sheets historically have been assumed to be governed by advection, vertical advection in particular, as gradients in ice velocity are typically larger in the vertical than horizontal (Hooke, 1977). Thus, the volume timescale of an ice mass, estimated as the thickness scale of the ice mass divided by the surface balance scale, has been taken as characteristic of its thermal response timescale: decadal to centurial for glaciers, centurial to millennial for ice sheets (Jóhannesson et al., 1989; Greuell, 1992; Haeberli and Hoelzle, 1995; Payne, 1995). Emerging research, however, suggests that the latent heat contained in small quantities of water entering an ice mass may be sufficient to warm glaciers and ice sheets to the pressure-melting point on a decadal timescale (Phillips et al., 2010). Some enthalpy-based modeling approaches now explicitly acknowledge this process, termed cryo-hydrologic warming, as a fourth term introduced to the conventional energy balance. There is presently limited direct observational (Bader and Small, 1955; Jarvis and Clark, 1974) and indirect inferential (Van der Veen et al., 2011; Phillips et al., 2013) support for cryo-hydrologic warming.

2.3.3 Frozen Ground

The thermal properties of a frozen soil depend on the thermal properties of both the ice and soil content. The thermal properties of ice are described above. Sand and gravel aggregates have specific heat capacities of about 0.89 J/g/K, whereas quartz and granite have specific heat capacities of about 0.73 and 0.80 J/g/K, respectively (Andersland and Ladanyi, 2004). Close to the melting point, unfrozen water is also present in otherwise frozen soils (Williams, 1967a, 1967b). Various methods to calculate the thermal conductivity of frozen soils have been reviewed and summarized by Farouki (1981). He concluded that the method by Johansen (1977) generally provides a reasonable result for frozen and unfrozen, coarse- or fine-grained soil at a degree of saturation greater than 0.1. Kersten (1949) provided equations to calculate the frozen and unfrozen thermal conductivity for various soils, which Farouki (1981) assessed as sufficient for most practical applications. Harlan and Nixon (1978) summarized the equations of Kersten (1949) in a graphical format on the basis of soil type, dry density, water content, and degree of saturation (Figure 2.7). Ranges of thermal conductivities for different ice saturations are provided in Table 2.2.

Côté and Konrad (2005) evaluated empirical thermal conductivity models introduced by Kersten (1949) and Johansen (1977). As unsatisfactory results

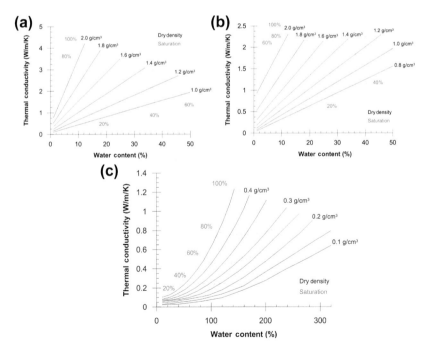

FIGURE 2.7   Average frozen thermal conductivity for sands and gravels (a), silt and clays (b), and peat (c). *After Harlan and Nixon (1978).*

were obtained with these conventional models for most of the unfrozen and frozen samples studied, a revised model was proposed for predicting the thermal conductivity of dense and broadly graded coarse materials used in pavements. This revised Côté and Konrad (2005) model utilizes the geometric mean method to compute the thermal conductivity for the solid soil particles and the saturated materials, a modified form of the geometric mean method to predict the thermal conductivity of the dry materials and empirical correlations.

2.4 MECHANICAL PROPERTIES

2.4.1 Brittle Behavior

2.4.1.1 Strength of Snow

Snow is composed of an ice lattice, and therefore the mechanical properties of ice ultimately determine the mechanical properties of snow. The fractional volume of ice and microstructural arrangement of grains and bonds control the macroscale response of snow under an applied stress. Due to the wide range of snow, firn, and ice densities, from 10 to 917 kg/m^3, the mechanical properties of these materials vary by orders of magnitude (Figure 2.8). Fresh snow will undergo rapid initial densification due to metamorphism, compaction due to the weight of subsequent snow, and possibly become saturated during melt. As snow density and microstructure change significantly on short timescales, so does the strength of snow.

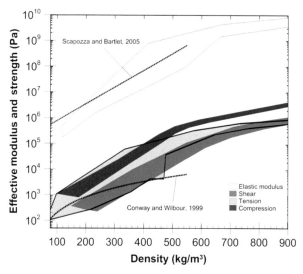

FIGURE 2.8   Elastic modulus, shear strength, tensile strength, and compressive strength for snow as a function of density. *Updated after Mellor (1975) and Shapiro et al. (1997).*

Like ice, the mechanical response of snow can be characterized as visco-elastic, with deformation depending on stress and rate of stress (e.g., Glen, 1958). Brittle failure, however, does occur frequently. Although snow is a porous lattice of ice, it is often modeled as a cellular solid (e.g., Gibson and Ashby, 1997). At strain rates of greater than $10^{-4}-10^{-3}s^{-1}$, depending on temperature, density, and microstructure (Narita, 1983; Schweizer, 1998), snow behaves nearly linear-elastically, until the strength is exceeded and brittle failure occurs (e.g., Marshall and Johnson, 2009). Lower strain rates ($\sim 10^{-6}s^{-1}$) cause viscous deformation (Theile et al., 2011; Schleef and Löwe, 2013), which is described in the following section.

Snow strength is correlated with density, but a wide range of values is observed at any given density, as strength is ultimately controlled by snow microstructure (Figure 2.8). The characteristics of the bonds between snow grains are primarily responsible for tensile, compressive, and shear strength. Microstructure remains very difficult to measure in the field, with detailed accurate measurements only possible with micro-CT laboratory measurements on small samples (e.g., Schneebeli and Sokratov, 2004; Wang and Baker, 2013; Chandel et al., 2014). Macroscale tests are typically used to assess snow stability in the field. Recent research using millimeter-scale cone penetration tests has demonstrated strong correlations between millimeter-scale cone penetrometer measurements and stability, as well as snow strength, but this remains an area of research and is not yet an applied technique (e.g., Schneebeli and Johnson, 1998; Bellaire et al., 2009; Marshall and Johnson, 2009; Pielmeier and Marshall, 2009). Triaxial tests, both confined and un-confined, have been performed in laboratory settings primarily on sieved or artificial snow (Shapiro et al., 1997; Bartelt and von Moos, 2000; Scapozza and Bartelt, 2003; Theile et al., 2011; Schleef and Löwe, 2013), and three-point bending tests were recently performed with success on samples of nat-ural snow slabs to measure fracture energy (Sigrist and Schweizer, 2007).

The order of magnitude of new snow strength can be approximated from density, and models of snow densification are used operationally to estimate the evolution of snow strength and stability during storms based on estimates of rates of strengthening and loading (e.g., Conway and Wilbour, 1999). Buried weak faceted snow layers, caused by temperature gradient meta-morphism, are commonly weak in shear but strong in compression, causing them to persist (e.g., Shapiro et al., 1997). The strength of these layers remains challenging to model (Hagenmuller et al., 2014) or measure, and hence forms a dangerous component of snow slope stability prediction (e.g., Schweizer, 1999; Schweizer et al., 2003). Figure 2.8 shows the elastic modulus, shear strength, tensile strength, and compressive strength for snow as a function of density. The shaded regions indicate the range of measurements in the liter-ature (as reviewed by Shapiro et al., 1997). The large range of values at a given density is believed to be primarily related to differences in microstructure, and only secondarily related to differences in snow temperature.

Fracture propagation in weak snow layers is an important process capable of propagating over hundreds of meters or more in seconds. Avalanches are often remotely triggered, even from flat terrain, eventually releasing in a distant location where tensile stresses exceed tensile strength of the slab. Current research has provided evidence for propagation in shear (e.g., McClung, 2011), as well as compressive failure causing a collapse of the overlying slab and the bending moment of the slab as the source of energy propagation (Heierli et al., 2008).

2.4.1.2 Strength of Ice

Over different stress environments and timescales, terrestrial ice exhibits both brittle and ductile behaviors. Brittle behavior, perhaps most evident in the presence of crevasses on glaciers, is directly modulated by ice strength (Figure 2.9). The uniaxial compressive strength of ice is similar to that of ultra-low density rocks. Indeed, for the purposes of excavation, glacier ice may be regarded as analogous to coal (Colgan and Arenson, 2013). Unlike brittle rocks, however, in which strength is primarily dependent on grain size and orientation (Lopez Jimeno et al., 1995), ice strength is primarily dependent on temperature. The compressive strength of meteoric ice demonstrates a significant decrease with temperature, from a practical maximum of ~ 40 MPa at $-50\,^{\circ}$C to a practical minimum of ~ 3 MPa at $0\,^{\circ}$C. The tensile strength of meteoric ice is an order of magnitude less than compressive strength (between ~ 1 and 3 MPa), and exhibits a negligible temperature dependence in comparison. The dependence of compressive strength on grain size and orientation

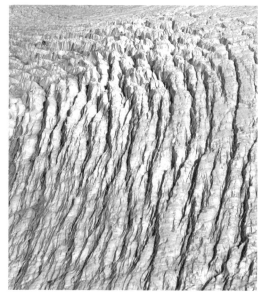

FIGURE 2.9 An ice rumple on Sermeq Avannarleq, near Ilulissat, Greenland. Surface crevassing occurs when the tensile strength of ice is exceeded, and can substantially reduce bulk glacier density. *Photo: W. Colgan.*

is also negligible. Empirical evidence, however, suggests a slight increase in tensile strength with decreasing crystal size ($\sim$ 1 MPa over the grain size range 10 to 1 mm), as well as a slight dependence on the angle between grain c-axes and the tensile load (Schulson and Duval, 2009). Empirical observations suggest that ice is less stiff than many common materials such as glass. A shear modulus of 3.8 GPa, a bulk modulus of 8.7 GPa, and a Young's modulus of 8.3 GPa are characteristic of meteoric ice at $-5\,°C$ (Gold, 1958; Hobbs, 1974; Cuffey and Paterson, 2010).

While ice is not strictly a linear elastic material, as it experiences nonlinear viscous deformation, linear elastic fracture mechanics are often employed to describe crack propagation in glacier ice (Van der Veen, 1998). The application of linear elastic fracture mechanics to a viscous elastic fluid is underpinned by the assumption that any region of plastic deformation at the fracture tip is small in comparison to the fracture length, which allows the elastic stresses near the fracture tip to be described by a stress-intensity factor. The critical stress-intensity factor required to propagate a fracture in glacier ice, or the mode I ("opening") fracture toughness, has been measured as between 0.05 and 0.40 MPa$^{1/2}$ (Fischer et al., 1995; Rist et al., 1999; Schulson and Duval, 2009). Although seemingly independent of ice temperature, this fracture toughness likely decreases with increasing bulk porosity or decreasing density. As a paucity of observational data prevents modeling variations in ice fracture toughness, the above range may be used in "high" and "low" bracketing assessments (Van der Veen, 1998, 2007). Depending on local ice fracture toughness, tensile stresses of 30–80 kPa are sufficient for a single/isolated crevasse to open. Crevasse fields, multiple closely spaced crevasses, are believed to require tensile stresses in excess of 300 kPa (Vaughan, 1993; Van der Veen, 1998). The presence of water can greatly enhance fracture propagation in ice masses; a crevasse continually supplied with water can penetrate through the entire thickness of an ice mass in stress settings that are normally not conducive to crevasse formation, based on the relative differences between ice and water density (Van der Veen, 2007).

2.4.1.3 Strength of Frozen Ground

The strength of a frozen soil depends strongly on the applied strain rate and can, similar to ice, be divided into four general responses (Gold, 1970), and summarized in Figures 2.10 and 2.11:

1. ductile behavior with strain hardening;
2. dilatant behavior with strain softening;
3. brittle behavior with brittle failure just after yield point; and
4. brittle failure.

Triaxial compression tests on various ice-rich frozen soils confirm this multimodal behavior (Arenson and Springman, 2005a, 2005b; Yamomoto, 2013; Yamamoto and Springman, in press). These tests further confirm the

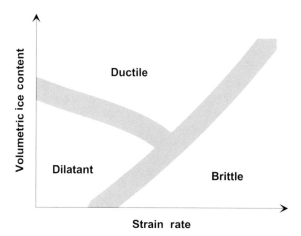

FIGURE 2.10   Strength response of frozen soils as a function of the strain rate and the volumetric ice content. *After Vyalov (1963).*

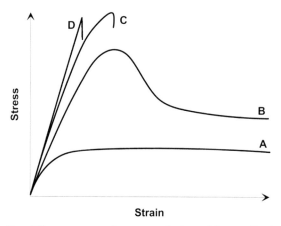

FIGURE 2.11   Four different stress–strain responses for ice and frozen soil *(After Arenson et al., 2007).* (A) Ductile behavior with strain hardening; (B) dilatant behavior with strain softening; (C) brittle behavior with failure just after yield point; (D) brittle failure.

influence of volumetric ice content on the strength of frozen soil. As volumetric ice content decreases, the interaction between the solid particles increases, typically strengthening the frozen soil. Even though brittle behavior may occur in heavily compacted unfrozen soils, the difference between dilatant and ductile behavior is generally interpreted as a function of soil density or void ratio (Taylor, 1948; Bolton, 1986). Following a first resistance peak at small strains, the stress–strain behavior of frozen soil exhibits a plateau (Sayles, 1974; Da Re et al., 2003). Andersen et al. (1995) investigated this first yield region, and found that the ice matrix, the strain rate, and the ground temperature all influence the form and intensity of the mechanical first resistance of frozen ground. Although the Young's modulus of frozen ground is independent of strain rate and temperature, it increases slightly with density and decreases marginally with increasing confining pressure. In contrast, the upper yield stress is independent of density, only slightly dependent on confining pressure, but strongly dependent on strain rate and temperature. In addition to small strain effects, various additional influences are still poorly understood and the focus of current research. This includes volumetric strain response and the self-healing potential or refreezing of broken.

Based on available observations, it appears that the strength of frozen soils at large strain is similar to the residual strength of an unfrozen soil, as all contributing ice bonds are broken and the measured ice cementation is lost. This notion is confirmed by direct shear tests on frozen sands, in which an expansion of the shear zone was noted due to crack propagation along the shear zone, as well as mobilization of ice granules and particles acting and dilating together (Yasufuku et al., 2003). Very large strains are necessary for this behavior, which may not be attained during laboratory investigations using conventional triaxial and direct shear testing. These large strains, however, are likely to form within creeping slopes under natural conditions (Haeberli et al., 2006). The presence of liquid water also affects the strength of a frozen soil, as an increased liquid water content reduces ice bonding and extraparticle suction, which reduces the apparent cohesion, and ultimately the shear resistance, of frozen soils. In some settings, however, as suction develops, the effective stress counteracting the reduction in apparent cohesion actually strengthens the frozen soil. In summary, unfrozen water can either strengthen or weaken frozen soil, and thus its effect needs to be evaluated on a site-specific basis.

Here, we briefly review the influence of soil temperature, strain rate, ice content, air content, confining pressure, salinity, dynamic load, and resintering on the mechanical properties of frozen ground. Generally, the strength of frozen ground increases as soil temperature decreases. In addition, as discussed above, variations in liquid water content can change the stress–strain behavior from ductile to brittle at the same strain rate. Although the strength of frozen ground increases with strain rate, the mechanical response of frozen ground changes toward brittle behavior at higher strain rates (Figure 2.10). Under very low strain rates, creep dominates the deformation of frozen ground, because complete relaxation takes place concurrently with small increases in stress.

The strength of a frozen soil increases as the volumetric ice content decreases, due to the structural hindrance that develops as soil particles interact with each other. Within a medium- to dense-frozen soil, ice is the dominant bonding (cohesive) agent. Relative to unfrozen soils, the presence of ice results in a higher resistance at low confining stresses. Some researchers report a slight decrease in the relative strength for dirty ice (Hooke et al., 1972; Ting et al., 1983). Discrete element modeling of solid-ice mixtures confirm this latter observation, by inferring that failure planes follow ice-solid boundaries (Arenson and Sego, 2005).

Only a few studies have investigated the effect of air content on the strength behavior of frozen soils (e.g., Gagnon and Gammon, 1995; Arenson and Springman, 2005a, 2005b; Yamamoto and Springman, in press). These tests suggest that the air within the soil matrix suppresses dilatancy, by providing significant opportunity for elimination of the air voids under pressure.

Confining stress has a minor effect on frozen ground strength at stresses below the melting pressure (Jones, 1982). When volumetric ice content is below ~60 percent, the strength of frozen soil increases with increasing confining pressure. Increasing interaction between soil particles increases strength, as frictional dissipation of the work component is a function of confinement stress.

The strength of a frozen soil decreases as the salinity increases, due to an increase in liquid water content as a consequence of freezing-point depression (Sego et al., 1982; Nixon and Lem, 1984; Hivon and Sego, 1995). Additionally, the ice crystal structure in frozen soils with saline pore fluid is more fragile than in similar nonsaline frozen soils (Arenson and Sego, 2006). A typical change in the unconfined compressive strength of frozen ground as a function of strain rate and salinity is presented in Figure 2.12. Linear correlations exist between the unconfined compression strength, or the secant modulus, of frozen ground and soil temperature, strain rate, and salinity (Nguyen et al., 2010).

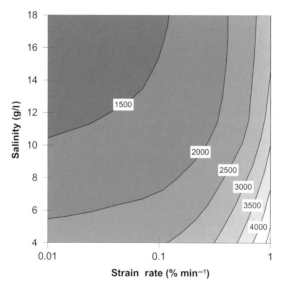

FIGURE 2.12   Unconfined compressive strength (kPa) of frozen soil as a function of strain rate and salinity for a temperature of −10 °C. *After Nguyen et al. (2010).*

Progress in the dynamic characteristics of frozen soils was summarized by Zhao et al. (2003). Various tests have shown that a critical strain rate exists at which the dynamic strength of frozen ground is similar to its static strength. At higher strain rates, the dynamic strength exceeds the static strength, whereas static strength exceeds dynamic strength at lower strain rates. We note, however, that only a few studies have examined the role of dynamic load on frozen ground strength.

The effect of resintering broken ice bonds in frozen soils has been investigated by temporarily pausing triaxial compression tests, to allow broken bonds within the ice matrix to sinter (Arenson and Springman, 2005a). Immediately after reloading resintered frozen ground, resistance exceeded the values observed under analogous continuous loading. Following additional shearing, resistance converged toward values associated with continuous loading. The extra resistance of resintered frozen ground, relative to continuously loaded frozen ground, has been speculated to depend on the time allowed for the self-healing process (Arenson and Springman, 2005a).

2.5 DUCTILE BEHAVIOR

The mechanical behavior of snow, ice, and frozen soils depends on strain rate, or both strain and time (Figure 2.10). For very slow loading rates, the mechanical response is generally ductile, and may be described as a creep response. Creep defines the continued deformation of a material with time under constant stress. Based on various laboratory studies, three distinct creep stages are apparent: primary, secondary (or steady state), and tertiary creep (Figure 2.13). The transitions between these creep stages are dependent on the material, and typically independent of the time required to attain a certain threshold strain. These three creep stages exist for snow, ice, and frozen ground.

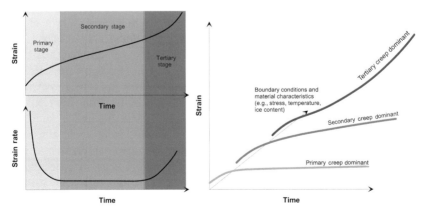

FIGURE 2.13   Basic creep behavior. *After Arenson et al. (2007).*

2.5.1 Creep of Snow

Viscous strain of snow under low stress rates causes densification, and also moves snow downhill under gravity. Densification rates of near-surface snow vary exponentially with snow temperature, density, and liquid water content (e.g., Kojima, 1967; Yamazaki et al., 1993; Shapiro et al., 1997). Dry snow viscous strain rates due to metamorphism alone vary over three orders of magnitude, and liquid water from rain or snowmelt can increase rates by three to five orders of magnitude (e.g., Marshall et al., 1999). Viscous densification in snow is controlled by microstructure, and is caused by a combination of grain boundary sliding and intracrystalline deformation of the bonds, with recent results indicating the latter process most likely dominates in low-density snow (e.g., Theile et al., 2011). Densification rates vary exponentially with density, and with the exception of thick ice layers caused by refrozen liquid water, overburden stress is required to attain snow densities greater than $350 \, \text{kg/m}^3$. The threshold density between snow and firn is approximately $550 \, \text{kg/m}^3$, which corresponds to the mass and volume of packed spheres of ice and agrees with field measurements (Table 2.1).

Compaction due to overburden stress and/or refrozen liquid water is required to increase snow density from $550 \, \text{kg/m}^3$ to the density of ice ($917 \, \text{kg/m}^3$). Seasonal snow is rarely exposed to sufficient stress for sufficient time to achieve ice density via metamorphism, and typically a saturated snowpack does not refreeze due to the latent heat associated with water. Snow that persists through a summer becomes firn, and forms on ice sheets, glaciers, and perennial snowfields. Increased overburden pressure from multiple winters of accumulation and/or exposure to low temperatures after becoming saturated with liquid water, allows firn to slowly increase its density to that of solid ice (e.g., Herron and Langway, 1980).

Snowmelt dynamics prior to saturation are extremely complex, including preferential flow in vertical columns (Colbeck and Davidson, 1972; Schneebeli, 1995; Waldner et al., 2004) and lateral flow along stratigraphic layers (e.g., Eiriksson et al., 2013). Densification due to melt in unsaturated snow is challenging to model because of the large heterogeneity in liquid water content, and required stratigraphic and microstructural properties that are typically not available. In contrast, energy balance and calibrated temperature index models of snowmelt typically perform well for estimating the temporal evolution of melt during spring. But only after a homogenous saturated wetting front has reached the ground and largely removed major stratigraphic boundaries and preferential flow paths, and in the absence of additional new snow to increase the albedo.

The combination of terrain curvature and snow creep can cause concentrations of tensile stress at convexities and compressive stress at concavities, often causing these inflection points in the snowpack profile to have the lowest slope stability (e.g., McClung and Schaerer, 1993; Schweizer et al., 2003). When tensile strength is exceeded during creep, a surface crack forms, which can propagate to the ground, allowing the lower snowpack to glide at the snow−ground interface. Glide crack avalanches typically form in deep maritime snowpacks on rock slabs with low friction, and they are hard to forecast and do not typically respond to explosives. Snow creep can also cause substantial static loads on structures, which requires careful engineering design.

2.5.2 Creep of Ice

Although refrozen ice may accumulate in the lowest layers of perennial snowfields, it is the presence of deformational flow that distinguishes glaciers from snowfields (Cogley et al., 2011). Over long timescales, ice deforms and flows under the force of gravity as a non-Newtonian fluid. This ductile behavior of ice, evident in the structured velocity fields of a glacier, can result from deviatoric stresses that are only a fraction of ice strength. Historically, "1 bar" (100 kPa) was taken as the characteristic basal shear stress required to induce viscous deformation in an ice mass (Glen, 1954; Weertman, 1957). In practice, local basal shear stresses within individual glaciers fluctuate around characteristic mean values of up to 200 kPa, depending on mass flux and stress setting (Haeberli and Hoelzle, 1995). Inverting stress as a function of strain rate may be used to infer bed topographies beneath glaciers (Huss and Farinotti, 2012). This allows digital terrain models to be estimated underneath glaciers, which can be used to project potential lake-formation sites following deglaciation (Linsbauer et al., 2012).

Glen's law (sometimes called "Steinemann−Glen's law") is the most commonly used first-order description of the relation between deviatoric stress and strain rate underlying the deformation of ice. Glen's law assumes that ice is an incompressible non-Newtonian fluid, similar to polycrystalline metals (Glen, 1954, 1958; Steinemann, 1954):

$$\dot{\epsilon}_{XZ} = EA\tau_{XZ}{}^{n}$$

where $\dot{\epsilon}_{XZ}$ is strain rate and τ_{XZ} is shear stress. In a linear viscous fluid, n would equal 1 and $1/A$ would represent fluid viscosity.

As ice is a non-Newtonian fluid, $1/A$ is often referred to as "effective viscosity" as its units differ from a true viscosity (i.e., $Pa^{-n}s^{-1}$ instead of $Pa^{-1}s^{-1}$). Given the dependence of ice strength on temperature, A varies by five orders of magnitude over the ice temperature range 0 to $-30\,°C$, and is often described by an Arrhenius-type relation dependent on ice temperature (e.g., Marshall, 2005). In addition to impurities, effective ice viscosity can be substantially reduced by crystal orientation, liquid water content, and, to a lesser degree, crystal size. These cumulative "softening" effects are typically recognized through the incorporation of a dimensionless enhancement factor (E), where E ranges from 1 for isotropic ice to as high as 10 for ice in tertiary shear creep with a strong single maximum fabric (i.e., c-axes aligned perpendicular to the basal plane; Budd et al., 2013). More sophisticated treatments of the deformation and flow of an ice mass involve solving additional higher order terms of the Navier−Stokes equations describing fluid mechanics (LeMeur et al., 2004; Zwinger and Moore, 2009).

Although observed values of n in Glen's law range between 1.5 and 4.2 (Weertman, 1973), most studies of ice flow adopt a constant empirical value of $n = 3$. This third-order stress−strain relation, interpreted to represent ice flow dominated by dislocation glide on the basal plane, is rate-limited by dislocation climb (Pettit and Waddington, 2003). Empirical evidence suggests that $n \to 1$ in low-stress environments (<30 kPa), such as ice-sheet interiors, shallow ice shelves, and near-surface temperate ice (Mellor and Testa, 1969; MacAyeal and Holdsworth, 1986; Budd and Jacka, 1989; Alley, 1992; Durham et al., 2001; Marshall et al., 2002). In these low gravitational driving stress settings, diffusion creep, Harper−Dorn creep or grain boundary sliding likely become the dominant mechanisms of linear flow. At ice temperatures below $-12\,°C$, the suppression of dynamic recrystallization likely contributes to a decrease in effective n (Pimienta and Duval, 1987). Conversely, limited empirical evidence suggests that n increases above 3 at stresses above ~ 1 MPa (Barnes et al., 1971). Although such high basal shear stresses are never realized in natural settings, the emergence of open ice pit mining, in which glacier geometry is modified into unnatural combinations of ice slope and ice thickness, offers the potential to achieve previously inconceivable basal shear stresses (Colgan, 2014).

2.5.3 Creep of Frozen Ground

The creep behavior of a frozen soil is not only dependent on temperature, but also on the soil type and density (Andersland and Ladanyi, 2004). Medium- to high-density sands and silts under medium to high stress will typically show primary and secondary creep, as observed in ice-rich permafrost slopes or rock glaciers (Haeberli et al., 2006) (Figure 2.13). Tertiary creep of granular frozen soils only occurs at high stress levels and temperatures close to the melting point. In the absence of excess ice, creep deformations may be completely lacking. Ice-rich silts and clays exhibit a short primary creep stage, but an extensive secondary creep, and tertiary creep at very large strains. Under moderate conditions, secondary (steady state) creep dominates and primary creep deformations are generally ignored in stability assessment.

Steinemann−Glen's law may also be used to describe the creep of ice-rich frozen soils, using adjusted A and n parameters. Using ice-rich frozen coarse samples under triaxial creep, Arenson and Springman (2005a) concluded that

1. Minimum axial creep strain rates increase as a power function with the applied deviatoric stress.

2. The creep susceptibility decreases with decreasing volumetric ice content, even though this effect was mainly noted for volumetric ice contents lower than about 65 percent. At higher volumetric ice contents, the creep susceptibility seems to be independent of the ice content.
3. The volumetric air content influences the volumetric response and mechanisms. Primary creep phases increase in terms of length and resulting net strain with increasing air content.
4. Tertiary creep was achieved only for selected samples within the duration of the creep stage (up to 4 weeks) for deviatoric stresses larger than about 400 kPa, which exceeds the values typically expected in naturally occurring frozen slopes.
5. Creep strain rates do not appear to be influenced by the confining pressure of up to 400 kPa.

Bray (2013) presented an empirical relaxation-based approach to approximate the secondary creep of frozen ground. Once secondary creep conditions are initiated, a continuously decreasing stepped creep test can be used to approximate secondary creep. With increasing liquid water content, parameters A and n approach the values observed for polycrystalline ice. At temperatures lower than $-3\,°C$, the creep rates of polycrystalline ice are greater than for the undisturbed frozen soils. At temperatures higher than $-2\,°C$ and stresses between 200 kPa and 1000 kPa, however, ice-rich soils creep at a greater rate than polycrystalline ice. This enhanced deformational flow is attributed to the presence of substantial liquid water within the frozen ground matrix. Andersland and Ladanyi (2004) provided secondary creep parameters for a range of different frozen soils. Although the parameter A depends on numerous factors, such as soil temperature and composition, and may vary by more than one order of magnitude, the value of n is typically more constrained. Observed values of n range from 1.28 for frozen Ottawa sand, with a dry density of $1.67\ kg/cm^3$, to 3.70 for colluvial loam, with a dry density of $1.50\ kg/cm^3$. Observations support a linear correlation between the volumetric ice content and the secondary creep parameter, n for frozen coarse materials (Arenson and Springman, 2005a).

2.6 DYNAMIC AND ELECTROMAGNETIC PROPERTIES

The dynamic and electromagnetic properties of snow, ice, and frozen ground are of relevance in the interpretation of geophysical measurements, response to electromagnetic radiation, as well as the assessment of the mechanical response to dynamic loads, such as earthquakes, explosions, or machine vibrations. Similar to the diverse static mechanical properties of cryospheric materials, a wide range occurs in the dynamic and electromagnetic properties of snow, ice, and permafrost (Table 2.3).

TABLE 2.3 Dynamic and Electromagnetic Properties of Snow, Ice, and Frozen Ground

	P-Wave Velocity	Electrical Resistivity	Electrical Permittivity	Electromagnetic Travel Velocity
Snow	~1 km/s	$10^6 - 10^7 \, \Omega m$	1−3	0.23 m/ns
Ice	~3.8 km/s	$10^4 - 10^8 \, \Omega m$	3−4	0.167 m/ns
Frozen ground	2.4−4.4 km/s	$10^3 - 10^6 \, \Omega m$	4−8	0.11−0.15 m/ns
Water	1.45−1.58 km/s	10−300 Ωm	80	0.033 m/ns
Air	0.33 km/s	∞	1	0.3 m/ns

Note: See text for references.

2.6.1 Snow

The magnetic permeability of snow is equal to that of air, and the conductivity is close to zero unless it is wet and has dust. At frequencies below ~100 GHz, the real part of the relative dielectric permittivity of ice is frequency independent, and in snow varies approximately linearly with density, ranging from a value near that in air at 50 kg/m^3 (1.08) to 2.15 at 550 kg/m^3, with dense firn approaching the value for ice (3.00). In dry snow, the imaginary part of the dielectric permittivity is below 0.005 for frequencies below 10 GHz, and therefore penetration of tens to thousands of meters is possible. In wet snow however, the imaginary part of the dielectric permittivity is a strong function of frequency, and penetration is limited to less than 1 m for frequencies above 5 GHz or wetness greater than 1−2 percent by volume. In the optical range, the real and imaginary parts of dielectric permittivity vary strongly with frequency; scattering and absorption are significant; and penetration depth is less than 50 cm, but varies with wavelength and grain size.

Standard commercial ground penetrating radars (GPRs) are commonly used to map snow depth and snow-water equivalent, as well as to estimate liquid water content (Lundberg and Thunehed, 2000; Harper and Bradford, 2003; Bradford et al., 2009). Repeat GPR measurements from the snowpack base have been used to track snowpack development and the evolution of melt (e.g., Heilig et al., 2010). To maximize resolution in snow for detailed studies of dry snow stratigraphy, custom-built, frequency-modulated, continuous-wave radar systems have been used (Marshall and Koh, 2008). This technique allows ultra-broadband microwave radar measurements, providing vertical resolutions of <1 cm. Observations at C-, X-, and Ku-band frequencies (2−20 GHz) can be made above the highest GPR frequency commercially available, and therefore can be used to measure volume scattering from snow grains and simulate radar remote sensing observations. X- and Ku-band backscatter from snow is currently the proposed mission concept for measuring snow-water equivalent from space. Low-frequency electromagnetic measurements have recently been used to measure flow of liquid water in snow using self-potential (Kulessa et al., 2012).

Snow is a very efficient absorber of acoustic energy (e.g., Shapiro et al., 1997; Maysenhölder et al., 2012), and ultrasonic frequencies are rapidly attenuated. The standard automated snow-depth measurement uses the reflection from an ultrasonic pulse in air to track the snow surface at a fixed location. Large acoustic signals, such as those caused by avalanches, have been measured on seismic arrays and using infrasound (e.g., Ulivieri et al., 2011; Lacroix et al., 2012). Swept-frequency acoustic signals have been used to measure snow depth and snow-water equivalent at low frequencies (Kinar and Pomeroy, 2009). Seismic P-wave velocities vary with density and liquid water content, and cover a wide range that is currently poorly constrained, but measurements indicate velocities between air and ice (e.g., Booth et al., 2013). Shear wave and fracture velocities have been measured in only a few studies.

2.6.2 Ice

GPR, also called ice penetrating radar or radio-echo sounding in the glaciology community, is an important tool for retrieving spatial profiles of not only ice thickness, but also englacial stratigraphy or layers. Relatively low radar frequencies (1−250 MHz) can resolve ice thickness up to several kilometers (e.g., Gogineni et al., 2001), whereas relatively high radar frequencies (250−1000 MHz) can resolve decimeter-scale near-surface internal layers (e.g., Kanagaratnam et al., 2001; Eisen et al., 2003).

The real part of the dimensionless relative permittivity, or dielectric constant, of glacier ice is in the range 3−4 percent, or ∼5 percent of that of freshwater. Similarly, the electrical conductivity of glacier ice is ∼0.01 mS/m, or ∼2 percent of that of freshwater (Hubbard and Glasser, 2005). Variations in both these properties with ice temperature and impurity or water content are often neglected. Given that resolving depth by radar return is dependent on two-way travel time, a critical dynamic property of ice for GPR is the propagation velocity of radar waves in ice, which is generally taken as ∼167 m/μs (Dowdeswell and Evans, 2004). Although radar velocity increases slightly with radar frequency, and decreases slightly with both water content and ice temperature, radar velocity is often assumed to be uniform with ice depth. Site-specific radar velocity may be adjusted from the characteristic 167 m/μs to account for site-specific conditions (e.g., Conway et al., 2009).

Primarily due to the substantially high dielectric constant of water, the presence of even relatively small amounts of englacial water in glacier ice can substantially decrease the velocity of radar waves and increase the attenuation of radar energy. Although strongly frequency dependent, the attenuation of radar energy can substantially increase with ice temperature. For example, at 2 MHz, attenuation increases from 0.005 dB/m at −20 °C to 0.031 dB/m at −5 °C (Conway et al., 2009). A characteristic attenuation value for radar energy in ice is often taken as 0.01 dB/m, or ∼10 percent of that of freshwater (Hubbard and Glasser, 2005). Although there is substantial uncertainty associated with inferring the water content of ice layers based on radar returns (Murray et al., 2007), it is possible to infer the presence or absence of water at the base of kilometer-scale radar profiles (Oswald and Gogineni, 2008).

Seismology remains an essential tool in determining not only ice thickness, but also basal properties such as the presence of deformable till or subglacial water storage, which cannot be identified via GPR alone (Smith, 1997; Peters et al., 2007). Empirical evidence suggests that primary shock wave (P-wave) velocity in ice is independent of detonation velocity or explosive mass (e.g., Ingram, 1960). P-wave velocity (v_p in m/s), which is generally around ∼3800 m/s in glacier ice, is slightly dependent on ice temperature (Kohnen, 1974):

$$v_p = -(2.30 \pm 0.17)T + 3795$$

where T is ice depth-averaged temperature, and the $\pm$ coefficient range is derived from a survey of observed values, and is in agreement with laboratory values. The secondary, or shear, wave (S-wave) velocity exhibits only approximately half the temperature dependence of P-wave velocity:

$$v_s = -(1.20 \pm 0.58)T + 1915$$

where v_s is S-wave velocity (in m/s; Kohnen, 1974). As the S-wave is of lesser importance in determining ice thickness, fewer in situ observations are available to constrain the v_s temperature−velocity relation, especially at low temperatures. Observed seismic attenuation values in glacier ice range from 0.15×10^{-3} to 0.46×10^{-3}/m in meteoric glacier ice, but can be substantially higher at ice-shelf locations, where the accretion of brackish marine basal ice is known to influence the dynamic properties of ice (Smith, 1997; McGrath et al., 2012). Seismic attenuation is highly sensitive to englacial ice temperature, especially close to the pressure-melting point. Recent comparisons between englacial temperature and seismic attenuation suggest that seismic attenuation may actually serve as a proxy for englacial temperature (Peters et al., 2012).

Direct comparisons of the concurrent application of GPR and seismic profiling do exist (Benjumea et al., 2003; Navarro et al., 2005; Shean and Marchant, 2010). Given that GPR can be recorded from moving aircraft or ground vehicles, it offers relatively easy and rapid data acquisition in comparison to seismic profiling. As radar is attenuated by the presence of even relatively small quantities of liquid water, however, seismic may often be the only viable method to retrieve bed information where ice is at the pressure-melting point with appreciable englacial water. Additionally, seismic techniques appear to be preferable to radar for identifying the firn–ice transition depth within a glacier, and less sensitive to vertically oriented internal reflectors within the ice that create parabolic diffractions in radar profiles. Perhaps as a consequence of the logistical difficulties associated with transporting explosives and the requirement of stationary seismic measurements, GPR is the more frequently used of the two ice-sounding methods.

A less common use of explosives in glacier ice is for the express purpose of ice excavation. Ice excavation by blasting is seldom more economical than by mechanical excavation, given the inherent ability of ice to absorb a relatively large proportion of blast energy through deformation without loss of cohesion (Colgan and Arenson, 2013). Explosive avalanche control must overcome a similar response in snow, and often trees or rock outcrops are targeted, or explosives are deployed along a tramway suspended above the surface to maximize the affected area. Consequently, peak shock-pressure values in ice are much lower than comparable values in brittle rock (Ingram, 1960). Given a diminishing return on removed ice-in-place with increasing detonation velocity, excavating glacier ice through blasting is most successful when deflagrating, or low-brisance, explosives with subsonic explosion velocities are employed (Rausch, 1959).

2.6.3 Frozen Ground

In frozen ground, most dynamic parameters are temperature and ice content dependent. In particular, close to the melting point of ice, where unfrozen water is present in the frozen soil, the dynamic and electromagnetic properties can change significantly with only minor changes in ground temperature. The seismic velocity depends on the elastic modulus and density of frozen ground. Typical P-wave velocities for frozen soils vary between 2.2 km/s and 4.5 km/s (Hecht, 2000; Hauck and Kneisel, 2008). As this velocity range overlaps with that of many rocks, as well as glacier ice, frozen soils cannot be clearly distinguished from rocks and ice using seismic measurements alone.

The electrical resistivity of ice is higher than the electrical resistivity of most other geomaterials. Unfrozen clays have an electrical resistivity of $1-100\,\Omega\text{m}$, and unfrozen sands and gravels have electrical resistivity of between 100 and $5000\,\Omega\text{m}$. In comparison, frozen soils have an electrical resistivity of between 10^3 and $10^7\,\Omega\text{m}$ (French et al., 2006; Fortier et al., 2008; Hauck and Kneisel, 2008). This wide range of electrical resistivity is primarily a function of unfrozen water content and ground temperature, and, for buried ice of glacial origin, also the state of the firn-to-ice metamorphosis (Haeberli and Vonder Mühll, 1996). The electrical resistivity of frozen ground can therefore be considered $10-100$ times greater than the electrical resistivity of comparable unfrozen ground (Barnes, 1963).

A last important dynamic and electromagnetic property of frozen ground is electromagnetic travel velocity. A large contrast exists between the electromagnetic properties of air, water, ice, and soil particles (Table 2.3). Frozen soils have a velocity of 0.11−0.15 m/ns, which is slightly higher than the value for loose, unfrozen sediments, similar to the velocity in ice and slightly lower than that in snow, but less than half of the velocity in air, and more than three times higher than the velocity in water (Davis and Annan, 1989).

2.7 SUMMARY

Cryospheric hazards are controlled by the physical, mechanical, and thermal properties of snow, ice, and frozen ground. A hazard is typically caused if the strength of the material is exceeded, or if deformation is too large. Within a cryospheric material, sufficient changes in temperature and stress, often over short timescales, can cause material failure. When assessing a cryospheric risk, a good understanding of the governing mechanical and thermal properties of the snow, ice, or frozen ground is critical, as the ranges of these properties can be extremely wide, spanning many orders of magnitude, and overlapping. Changes in material boundary conditions, such as those due to construction activities or climatic changes, must be assessed carefully and their impacts on the mechanical properties with time evaluated (Figure 2.14). Since snow and ice surfaces may not always be smooth, and ground ice occurrences extremely complex, simplifications and assumptions must often be made (Figure 2.15). This chapter can only provide an overview of the breadth of the physical, mechanical, and thermal properties of cryospheric materials, with the hopes of

FIGURE 2.14 An abandoned access road to the Greenland ice sheet, near Kangerlussuaq, Greenland. A landscape shaped by snow, ice, and permafrost makes for a hostile engineering environment, especially under transient climate forcing as a result of global warming. *Photo: W. Colgan.*

FIGURE 2.15 Glacier with penitentes surface located in the high, arid Chilean Andes. They are formed as tall thin blades of compact snow or ice, with the blades oriented toward the general direction of the Sun. *Photo: L. Arenson.*

providing insight on the challenges associated quantifying cryospheric hazards. The wealth of research publications on the complex properties of cryospheric materials, although introduced here, is always available for further consultation in order to evaluate more detailed properties of snow, ice, and permafrost.

ACKNOWLEDGMENT

The authors acknowledge the critical reading and valuable input from Martin Schneebeli and T.H. (Jo) Jacka on an earlier version of the chapter. W.C. thanks Daniel McGrath and Atsuhiro Muto for a collegial review of the portions of this chapter that discuss ice. We also like to thank the Editors for their significant contribution in improving the text.

REFERENCES

ACIA, 2005. Arctic Climate Impact Assessment. Cambridge University Press. Available from: http://www.acia.uaf.edu.
Alley, R., 1992. Flow-law hypothesis for ice-sheet modeling. J. Glaciol. 38, 245−256.

Andersen, G.R., Swan, C.W., Ladd, C.C., Germaine, J.T., 1995. Small-strain behavior of frozen sand in triaxial compression. Can. Geotech. J. 32, 428−451.

Andersland, O.B., Ladanyi, B., 2004. Frozen Ground Engineering. John Wiley & Sons.

Anderson, D.M., Tice, A.R., 1972. Predicting Unfrozen Water Contents in Frozen Soils from Surface Area Measurements. Highway Research Record No. 393: 12−18.

Anderson, D.M., Morgenstern, N.R., 1973. Physics, chemistry and mechanics of frozen ground: a review. In: Permafrost: The North American Contribution to the Second International Conference. National Academy of Sciences, Washington, DC, pp. 257−288.

Anderson, R., Anderson, S., MacGregor, K., Waddington, E., O'Neel, S., Riihimaki, C., Loso, M.G., 2004. Strong feedbacks between hydrology and sliding of a small alpine glacier. J. Geophys. Res. 109, F03005.

Arenson, L.U., Sego, D.C., 2005. Modelling the strength of composite materials using discrete elements. In: Proceedings of the 58th Canadian Geotechnical Conference. Saskatchewan, Saskatoon.

Arenson, L.U., Sego, D.C., 2006. The effect of salinity on the freezing of coarse grained sands. Can. Geotech. J. 43, 325−337.

Arenson, L.U., Springman, S.M., 2005a. Triaxial constant stress and constant strain rate tests on ice-rich permafrost samples. Can. Geotech. J. 42, 412−430.

Arenson, L.U., Springman, S.M., 2005b. Mathematical descriptions for the behaviour of ice-rich frozen soils at temperatures close to 0 °C. Can. Geotech. J. 42, 431−442.

Arenson, L.U., Springman, S.M., Sego, D.C., 2007. The rheology of frozen soils. Appl. Rheol. 17, 1−14.

Armstrong, R.L., Brodzick, M.J., 2001. Recent northern hemisphere snow extent: a comparison of data derived from visible and microwave satellite sensors. Geophys. Res. Lett. 28, 3673.

Bader, H., Small, F., 1955. Sewage Disposal at Ice Cap Installations. Snow Ice Permafrost Research Establishment Report 21: 1−4.

Bader, H., 1964. Density of Ice as a Function of Temperature and Stress. CRREL Special Report 64. Cold Regions Research and Engineering Laboratory, Hanover, NH, 1−6.

Barnes, D.F., 1963. Geophysical methods for delineating permafrost. In: Proceedings of the International Conference on Permafrost. Publication No. 1287. National Academy of Sciences, Lafayette, Indiana, pp. 349−355.

Barnes, P., Tabor, D., Walker, J., 1971. The friction and creep of polycrystalline ice. Proc. R. Soc. Lond. Ser. A 324, 127−155.

Barnett, T.P., Adam, J.C., Lettenmaier, D.P., 2005. Potential impacts of a warming climate on water availability in snow-dominated regions. Nature 438, 303−309.

Bartelt, P., von Moos, M., 2000. Triaxial tests to determine a microstructure-based snow viscosity law. Ann. Glaciol. 31, 457−462.

Bellaire, S., Pielmeier, C., Schneebeli, M., Schweizer, J., 2009. Stability algorithm for snow micro-penetrometer measurements. J. Glaciol. 55, 805−813.

Beniston, M., Keller, F., Koffi, B., Goyette, S., 2003. Estimates of snow accumulation and volume in the Swiss Alps under changing climatic conditions. Theor. Appl. Climatol. 76, 125−140.

Benjumea, B., Macheret, Y., Navarro, F., Teixido, T., 2003. Estimation of water content in a temperate glacier from radar and seismic sounding data. Ann. Glaciol. 37, 317−324.

Bolton, M.D., 1986. The strength and dilatancy of sands. Géotechnique 36, 65−78.

Booth, A.D., Mercer, A., Clark, R., Murray, T., Jansson, P., Axtell, C., 2013. A comparison of seismic and radar methods to establish the thickness and density of glacier snow cover. Ann. Glaciol. 54 (64), 73−82.

Borden, K.A., Cutter, S.L., 2008. Spatial patterns of natural hazards mortality in the United States. Int. J. Health Geogr. 7 (1), 64.

Bradford, J.B., Harper, J.T., Brown, J., 2009. Complex dielectric permittivity measurements from ground- penetrating radar data to estimate snow liquid water content in the pendular regime. Water Resour. Res. 45, W08403.

Bray, M.T., 2013. Secondary creep approximations of ice-rich soils and ice using transient relaxation tests. Cold Reg. Sci. Technol. 88, 17–36.

Budd, W.F., Jacka, T.H., 1989. A review of ice rheology for ice sheet modelling. Cold Reg. Sci. Technol. 16, 107–144.

Budd, W.F., Warner, R.C., Jacka, T.H., Jun, L., Treverrow, A., 2013. Ice flow relations for stress and strain-rate components from combined shear and compression laboratory experiments. J. Glaciol. 59 (214), 374–392.

Butkovich, T., 1954. Ultimate Strength of Ice. Snow Ice Permafrost Research Establishment. Research Paper 11, 1–12.

Butkovich, T., 1959. Some Physical Properties of Ice from the TUTO Tunnel and Ramp, Thule, Greenland. Technical Report 47. Cold Regions Research and Engineering Laboratory, Hanover, NH, 1–17.

Callaghan, T.V., Johansson, M., Prowse, T.D., Olsen, M.S., Reiersen, L.-O., 2011. Arctic cryosphere: changes and impacts. Ambio 40, 3–5.

Calonne, N., Flin, F., Morin, S., Lesaffre, B., du Roscoat, S.R., Geindreau, C., 2011. Numerical and experimental investigations of the effective thermal conductivity of snow. Geophys. Res. Lett. 38, L23501.

Chandel, C., Srivastava, P.K., Mahajan, P., 2014. Micromechanical analysis of deformation of snow using X-ray tomography. Cold Reg. Sci. Technol. 101, 14–23.

Clarke, G., Nitsan, U., Paterson, W., 1977. Strain heating and creep instability in glaciers and ice sheets. Rev. Geophys. Space Phys. 15, 225–237.

Cogley, J., Hock, R., Rasmussen, L., Arendt, A., Bauder, A., Braithwaite, R.J., Jansson, P., Kaser, G., Möller, M., Nicholson, L., Zemp, M., 2011. Glossary of Glacier Mass Balance and Related Terms. IHP-VII Technical Documents in Hydrology No. 86, IACS Contribution No. 2. UNESCO-IHP, Paris.

Colbeck, S., Davidson, G., 1972. Water percolation through homogeneous snow. In: Proceedings of the Banff Symposium. IAHS, No.107, pp. 242–257.

Colgan, W., 2014. Considering the ice excavation required to establish and maintain an open ice pit. J. Cold Reg. Eng. 28 (3), 04014003. http://dx.doi.org/10.1061/(ASCE)CR.1943-5495. 0000067.

Colgan, W., Arenson, L.U., 2013. Open pit glacier ice excavation: a brief review. J. Cold Reg. Eng. 27, 223–243.

Colgan, W., Pfeffer, W., Rajaram, H., Abdalati, W., Balog, J., 2012. Monte Carlo ice flow modeling projects a new stable configuration for Columbia Glacier, Alaska by c. 2020. Cryosphere 6, 1395–1409.

Colgan, W., Rajaram, H., Anderson, R., Steffen, K., Phillips, T., Joughin, I., Zwally, H.J., Abdalati, W., 2011. The annual glaciohydrology cycle in the ablation zone of the Greenland ice sheet: part 1. hydrology model. J. Glaciol. 57, 697–709.

Conway, H., Wilbour, C., 1999. Evolution of snow slope stability during storms. Cold Reg. Sci. Technol. 30, 67–77.

Conway, H., Smith, B., Vaswani, P., Matsuoka, K., Rignot, E., Claus, P., 2009. A low-frequency ice-penetrating radar system adapted for use from an airplane: test results from Bering and Malaspina Glaciers, Alaska, USA. Ann. Glaciol. 50, 93–97.

Côté, J., Konrad, J., 2005. Thermal conductivity of base-course materials. Can. Geotech. J. 78, 61−78.

Cuffey, K., Paterson, W.S.B., 2010. The Physics of Glaciers, fourth ed. Butterworth-Heinemann, Elsevier, Academic Press. 704 pp.

Da Re, G., Germaine, J.T., Ladd, C.C., 2003. Triaxial testing of frozen sand: equipment and example results. J. Cold Reg. Eng. 17, 90−118.

Dadic, R., Schneebeli, M., Lehning, M., Hutterli, M.A., Ohmura, A., 2008. Impact of the microstructure of snow on its temperature: a model validation with measurements from summit, Greenland. J. Geophys. Res. Atmos. 113, D14303.

Davies, M.C.R., Hamzaand, O., Harris, C., 2001. The effect of rise in mean annual temperature on the stability of rock slopes containing ice filled discontinuities. Permafrost Periglac. Process. 12, 137−144.

Davis, J.L., Annan, A.P., 1989. Ground-penetrating radar for high-resolution mapping of soil and rock stratigraphy. Geophys. Prospect. 37, 531−551.

Dowdeswell, J., Evans, S., 2004. Investigations of the form and flow of ice sheets and glaciers using radio-echo sounding. Rep. Prog. Phys. 67, 1821−1861.

Durham, W., Stern, L., Kirby, S., 2001. Rheology of ice I at low stress and elevated confining pressure. J. Geophys. Res. 106, 11,031−11,042.

Eiriksson, D., Whitson, M., Luce, C.H., Marshall, H.P., Bradford, J., Benner, S.G., Black, T., Hetrick, H., McNamara, J.P., 2013. An evaluation of the hydrologic relevance of lateral flow in snow at hillslope and catchment scales. Hydrol. Process. 27, 640−654.

Eisen, O., Nixdorf, U., Keck, L., Wagenbach, D., 2003. Alpine ice cores and ground-penetrating radar: combined investigations for glaciological and climatic interpretations of a cold Alpine ice body. Tellus 55B, 1007−1017.

Farouki, O.T., 1981. Thermal properties of Soils. CRREL Monograph 81−1. Cold Regions Research and Engineering Laboratory, Hanover, NH.

Fischer, M., Alley, R., Engelder, T., 1995. Fracture toughness of ice and firn determined from the modified ring test. J. Glaciol. 41, 383−394.

Flowers, G., Clarke, G., 2002. A multicomponent coupled model of glacier hydrology 1. Theory and synthetic examples. J. Geophys. Res. 107, 2287.

Fountain, A., Walder, J., 1998. Water flow through temperate glaciers. Rev. Geophys. 36, 299−328.

Fortier, R., LeBlanc, A.-M., Allard, M., Buteau, S., Calmels, F., 2008. Internal structure and conditions of permafrost mounds at Umiujaq in Nunavik, Canada, inferred from field investigations and electrical resistivity tomography. Can. J. Earth Sci. 45, 367−387.

French, H.K., Binley, A., Kharkhordin, I., Kulessa, B., Krylov, S.S., 2006. Cold regions hydrogeophysics: physical characterisation and monitoring. In: Vereecken, H., et al. (Eds.), Applied Hydrogeophysics. Springer, Dordrecht, The Netherlands.

Fuchs, S., Thöni, M., McAlpin, M., Gruber, U., Bründl, M., 2007. Avalanche hazard mitigation strategies assessed by cost effectiveness analyses and cost benefit analyses—evidence from Davos, Switzerland. Nat. Hazards 41, 113−129.

Funk, M., Echelmeyer, K., Iken, A., 1994. Mechanisms of fast flow in Jakobshavn Isbrae, West Greenland: part II. modeling of englacial temperatures. J. Glaciol. 40, 569−585.

Gagnon, R.E., Gammon, P.H., 1995. Triaxial experiments on iceberg and glacier ice. J. Glaciol. 41, 528−540.

Gammon, P.H., Kiefte, H., Clouter, M.J., Denner, W.W., 1983. Elastic constants of artificial and natural ice samples by Brillouin spectroscopy. J. Glaciol. 29, 433−460.

Gibson, L.J., Ashby, M.F., 1997. Cellular Solids: Structure and Properties, Second ed. Cambridge University Press.

Glen, J., 1954. The stability of ice-dammed lakes and other water-filled holes in glaciers. J. Glaciol. 2, 316−318.

Glen, J., 1958. The flow law of ice: a discussion of the assumptions made in glacier theory, their experimental foundations and consequences. IAHS 47, 171−183.

Glow, A., 1970. Preliminary results of studies of ice cores from the 2164 m deep drill hole, Byrd Station, Antarctica. IASH 86, 78−90.

Gogineni, S., Tammana, D., Braaten, D., Leuschen, C., Akins, T., Legarsky, J., Kanagaratnam, P., Stiles, J., Allen, C., Jezek, K., 2001. Coherent radar ice thickness measurements over the Greenland ice sheet. J. Geophys. Res. 106, 33761−33772.

Gold, L.W., 1958. Some observations on the dependence of stain on stress for ice. Can. J. Phys. 3610, 1265−1275.

Gold, L.W., 1970. Process of failure in ice. Can. Geotech. J. 7, 405−413.

Gow, A., Williamson, T., 1972. Linear compressibility of ice. J. Geophys. Res. 77, 6348−6352.

Greuell, W., 1992. Hintereisferner, Austria: mass-balance reconstruction and numerical modelling of the historical length variations. J. Glaciol. 38, 233−244.

Haeberli, W., Hoelzle, M., 1995. Application of inventory data for estimating characteristics of and regional climate-change effects on mountain glaciers: a pilot study with the European Alps. Ann. Glaciol. 21, 206−212.

Haeberli, W., Vonder Mühll, D., 1996. On the characteristics and possible origins of ice in rock glacier permafrost. Zeitschrift für Geomorphologie N.F 104, 43−57.

Haeberli, W., Hallet, B., Arenson, L., Elconin, R., Humlum, O., Kääb, A., Kaufmann, V., Ladanyi, B., Matsuoka, N., Springman, S., Vonder Mühll, D., 2006. Permafrost creep and rock glacier dynamics. Permafrost Periglac. Process. 17, 189−214.

Hagenmuller, P., Theile, T.C., Schneebeli, M., 2014. Numerical simulation of microstructural damage and tensile strength of snow. Geophys. Res. Lett. 41 (1), 86−89.

Harlan, R.L., Nixon, J.F., 1978. Ground thermal regime. In: Andersland, O.B., Anderson, D.M. (Eds.), Geotechnical Engineering for Cold Regions. McGraw-Hill Ryerson, pp. 103−163.

Harper, J., Bradford, J., 2003. Snow stratigraphy over a uniform depositional surface: spatial variability and measurement tools. Cold Reg. Sci. Technol. 37, 289−298.

Hauck, C., Kneisel, C., 2008. Applied Geophysics in Periglacial Environments, vol. 256. Cambridge University Press, Cambridge.

Havens, S., Marshall, H.P., Pielmeier, C., Elder, K., 2013. Automatic grain type classification of snow micro penetrometer signals with random forests. IEEE Trans. Geosci. Remote Sens. 51, 3328−3335.

Hecht, S., 2000. Fallbeispiele zur Anwendung refraktionsseismischer Methoden bei der Erkundung des oberflächennahen Untergrundes. Zeitschrift für Geomorphologie. N.F (Supplementband 123), 111−123.

Heierli, J., Gumbsch, P., Zaiser, M., 2008. Anticrack nucleation as triggering mechanism for snow slab avalanches. Science 321, 240−243.

Heilig, A., Eisen, O., Schneebeli, M., 2010. Temporal observations of a seasonal snowpack using upward- looking GPR. Hydrol. Process. 24, 3133−3145.

Herron, M.M., Langway Jr, C.C., 1980. Firn densification: an empirical model. J. Glaciol. 25, 373−385.

Hivon, E.G., Sego, D.C., 1995. Strength of frozen soils. Can. Geotech. J. 32, 336−354.

Hobbs, P., 1974. Ice Physics. Oxford Clarendon Press, New York.

Hooke, R.L., 1977. Basal temperatures in polar ice sheets: a qualitative review. Quat. Res. 7, 1−13.

Hooke, R.L., Dahlin, B.B., Kauper, M.T., 1972. Creep of ice containing dispersed fine sand. J. Glaciol. 11, 327−336.

Hubbard, B., Glasser, N.F., 2005. Field Techniques in Glaciology and Glacial Geomorphology. John Wiley & Sons, 412 pp.

Huss, M., Farinotti, D., 2012. Distributed ice thickness and volume of all glaciers around the globe. J. Geophys. Res. 117, F04010.

Ingram, L., 1960. Measurements of explosion-induced shock waves in ice and snow, Greenland, 1957 and 1958. Army Eng. Waterw. Exp. Sta. Misc. Pap. 21, 2−399.

IPCC, 2013. Climate change 2013: the physical science basis. In: Stocker, T.F., Qin, D., Plattner, G.-K., Tignor, M., Allen, S.K., Boschung, J., Nauels, A., Xia, Y., Bex, V., Midgley, P.M. (Eds.), Contribution of Working Group I to the Fifth Assessment Report of the Intergovernmental Panel on Climate Change. Cambridge University Press, Cambridge, United Kingdom and New York, NY, USA, p. 1535.

Jarvis, G., Clark, G., 1974. Thermal effects of crevassing on Steele glacier, Yukon Territory, Canada. J. Glaciol. 13, 243−254.

Jóhannesson, T., Raymond, C., Waddington, E., 1989. A simple method for determining the response time of glaciers. In: Oerlemans, J. (Ed.), Glacier Fluctuations and Climatic Change, pp. 343−352.

Johansen, O., 1977. Thermal Conductivity of Soils. CRREL Technical Translation 637. Cold Regions Research and Engineering Laboratory, Hanover, NH.

Johnson, J.B., Schneebeli, M., 1999. Characterizing the microstructural and micromechanical properties of snow. Cold Reg. Sci. Technol. 30 (1), 91−100.

Jones, S.J., 1982. The confined compressive strength of polycrystalline ice. J. Glaciol. 28, 171−177.

Joughin, I., Smith, B., Howat, I., Scambos, T., Moon, T., 2010. Greenland flow variability from ice-sheet-wide velocity mapping. J. Glaciol. 56, 415−430.

Kanagaratnam, P., Gogineni, S., Gundestrup, N., Larsen, L., 2001. High-resolution radar mapping of internal layers at the North Greenland ice core project. J. Geophys. Res. 106, 33799−33811.

Kaempfer, T., Schneebeli, M., Sokratov, S., 2005. A microstructural approach to model heat transfer in snow. Geophys. Res. Lett. 32, L21503.

Kersten, M.S., 1949. Laboratory Research for the Determination of the Thermal Properties of Soils. Minnesota University, Minneapolis Engineering Experiment Station.

Kinar, N., Pomeroy, J.W., 2009. Automated determination of snow water equivalent by acoustic reflectometry. IEEE Trans. Geosci. Remote Sens. 47, 3161−3167.

Kohnen, H., 1974. The temperature dependence of seismic waves in wave. J. Glaciol. 13, 144−147.

Kojima, K., 1967. Densification of seasonal snow cover. In: Oura, H. (Ed.), Physics of Snow and Ice, Proc. Int. Conf. on Low Temp. Sci, vol. 1. Hokkaido Univ., Sapporo, pp. 929−952.

Kulessa, B., Chandler, D., Revil, A., Essery, R., 2012. Theory and numerical modeling of electrical self-potential signatures of unsaturated flow in melting snow. Water Resour. Res. 48, W09511.

Lacroix, P., Grasso, J.R., Roulle, J., Giraud, G., Goetz, D., Morin, S., Helmstetter, A., 2012. Monitoring of snow avalanches using a seismic array: location, speed estimation, and relationships to meteorological variables. J. Geophys. Res. Earth Surf. 117, F01034.

Lawrence, D., Slater, A., 2005. A projection of severe near-surface permafrost degradation during the 21st century. Geophys. Res. Lett. 32, L24401.

Lehning, M., Bartelt, P., Brown, B., Fierz, C., Satyawali, P., 2002. A physical snowpack model for the Swiss avalanche warning part ii: snow microstructure. Cold Reg. Sci. Technol. 35, 147−167.

LeMeur, E., Gagliardini, O., Zwinger, T., Ruokolainen, J., 2004. Glacier flow modelling: a comparison of the shallow ice approximation and the full-Stokes solution. C. R. Phys. 5, 709−722.

Linsbauer, A., Paul, F., Haeberli, W., 2012. Modeling glacier thickness distribution and bed topography over entire mountain ranges with Glab-Top: application of a fast and robust approach. J. Geophys. Res. 117, F03007.

Lliboutry, L., 1976. Physical processes in temperate glaciers. J. Glaciol. 16, 151−158.

Lopez Jimeno, C., Lopez Jimeno, E., Ayala Carcedo, F., 1995. Drilling and Blasting of Rocks. Geomining Technological Institute of Spain, 391.

Löwe, H., Van Herwijnen, A., 2012. A Poisson shot noise model for micro-penetration of snow. Cold Reg. Sci. Technol. 70, 62−70.

Löwe, H., Riche, F., Schneebeli, M., 2013. A general treatment of snow microstructure exemplified by an improved relation for thermal conductivity. Cryosphere 7 (5), 1473−1480.

Lundberg, A., Thunehed, H., 2000. Snow wetness influence on impulse radar snow surveys theoretical and laboratory study. Nord. Hydrol. 31, 89−106.

MacAyeal, D., Holdsworth, G., 1986. An investigation of low-stress ice rheology on the Ward-Hunt ice Shelf. J. Geophys. Res. 91, 6347−6358.

Marks, D., Link, T., Winstral, A., Garen, D., 2001. Simulating snowmelt processes during rain-on-snow over a semi-arid mountain basin. Ann. Glaciol. 32, 195−202.

Marks, D., Winstral, A., 2001. Comparison of snow deposition, the snow cover energy balance, and snowmelt at two sites in a semiarid mountain basin. J. Hydrometeorol. 2, 213−227.

Marshall, S., 2005. Recent advances in understanding ice sheet dynamics. Earth Planet. Sci. Lett. 240, 191−204.

Marshall, H.P., Johnson, J.B., 2009. Accurate inversion of high-resolution snow penetrometer signals for microstructural and micromechanical properties. J. Geophys. Res. Earth Surf. 114, F04016.

Marshall, H.P., Koh, G., 2008. FMCW radars for snow research. Cold Reg. Sci. Technol. 52, 118−131.

Marshall, H., Conway, P., Rasmussen, L.A., 1999. Snow densification during rain. Cold Reg. Sci. Technol. 30, 35−41.

Marshall, H.P., Harper, J.T., Pfeffer, W.T., Humphrey, N.F., 2002. Depth-varying constitutive properties observed in an isothermal glacier. Geophys. Res. Lett. 29, 61-1−61-4.

Matsuoka, N., Sakai, H., 1999. Rockfall activity from alpine cliff during thawing periods. Geomorphology 28, 309−328.

Matzl, M., Schneebeli, M., 2006. Measuring specific surface area of snow by near-infrared photography. J. Glaciol. 52 (179), 558−564.

Maysenhölder, W., Heggli, M., Zhou, X., Zhang, T., Frei, E., Schneebeli, M., 2012. Microstructure and sound absorption of snow. Cold Reg. Sci. Technol. 83-84, 3−12.

McClung, D.M., 2011. The critical size of macroscopic imperfections in dry snow slab avalanche initiation. J. Geophys. Res. Earth Surf. 116, F03003.

McClung, D.M., Schaerer, P., 1993. The Avalanche Handbook. The Mountaineers, Seattle, WA, 271 pp.

McGrath, D., Steffen, K., Scambos, T., Rajaram, H., Casassa, G., Rodriguez Lagos, J., 2012. Basal crevasses and associated surface crevassing on the Larsen C ice shelf, Antarctica, and their role in ice-shelf instability. Ann. Glaciol. 53, 10−18.

Meier, M., Lundstrom, S., Stone, D., Kamb, B., Engelhardt, H., Humphrey, N., Dunlap, W.W., Fahnestock, M., Krimmel, R.M., Walters, R., 1994. Mechanical and hydrologic basis for the rapid motion of a large tidewater glacier 1. Observations. J. Geophys. Res. 99, 15,219−15,229.

Mellor, M., Testa, R., 1969. Creep of ice under low stress. J. Glaciol. 8, 147–152.

Mellor, M., 1975. A review of basic snow mechanics. In: The International Symposium on Snow Mechanics. IAHS-AISH Publication 114, Grindelwald, Switzerland, pp. 251–291.

Murray, T., Booth, A., Rippin, D., 2007. Water-content of glacier-ice: limitations on estimates from velocity analysis of surface ground-penetrating radar surveys. J. Environ. Eng. Geophys. 106, 10.2113/JEEG12.1.87.

Narita, H., 1983. An experimental study on tensile fracture of snow. Contrib. Inst. Low Temp. Sci. 32 (A), 1–37.

Navarro, F., Macheret, Y., Benjumea, B., 2005. Application of radar and seismic methods for the investigation of temperate glaciers. J. Appl. Geophys. 57, 193–211.

Nelson, F., Anisimo, O., Shiklomanov, N., 2001. Subsidence risk from thawing permafrost. Nature 410, 889–890.

Nguyen, A.D., Sego, D.C., Arenson, L.U., Biggar, K.W., 2010. The dependence of strength and modulus of frozen saline sand on temperature, strain rate and salinity. In: Proceedings of the 63th Canadian Geotechnical Conference. Alberta, Calgary, pp. 467–475.

Nixon, J.F., Lem, G., 1984. Creep and strength of frozen saline fine-grained soils. Can. Geotech. J. 21, 518–529.

Oldroyd, H.J., Higgins, C.W., Huwald, H., Selker, J.S., Parlange, M.B., 2013. Thermal diffusivity of seasonal snow determined from temperature profiles. Adv. Water Resour. 55, 121–130.

Oswald, G., Gogineni, S., 2008. Recovery of subglacial water extent from Greenland radar survey data. J. Glaciol. 54, 94–102.

Payne, A., 1995. Limit cycles in the basal thermal regime of ice sheets. J. Geophys. Res. 100, 4249–4263.

Peters, L., Anandakrishnan, S., Alley, R., Smith, A., 2007. Extensive storage of basal meltwater in the onset region of a major West Antarctic ice stream. Geology 35, 251–254.

Peters, L., Anandakrishnan, S., Alley, R., Voigt, D., 2012. Seismic attenuation in glacial ice: a proxy for englacial temperature. J. Geophys. Res. 117, F02008.

Petrovic, J.J., 2003. Mechanical properties of ice and snow. J. Mater. Sci. 3, 1–6.

Pettit, E., Waddington, E., 2003. Ice flow at low deviatoric stress. J. Glaciol. 49, 359–369.

Phillips, T., Rajaram, H., Colgan, W., Steffen, K., Abdalati, W., 2013. Evaluation of cryo-hydrologic warming as an explanation for increased ice velocities in the wet snow zone, Sermeq Avannarleq, West Greenland. J. Geophys. Res. 118, 1–16.

Phillips, T., Rajaram, H., Steffen, K., 2010. Cryo-hydrologic warming: a potential mechanism for rapid thermal response of ice sheets. Geophys. Res. Lett. 37, L20503.

Pielmeier, C., Marshall, H.P., 2009. Rutschblock-scale snowpack stability derived from multiple quality- controlled snowmicropen measurements. Cold Reg. Sci. Technol. 59, 178–184.

Pimienta, P., Duval, P., 1987. Rate controlling processes in the creep of polar glacier ice. J. Phys. 48, 243–248.

Price, S., Payne, A., Howat, I., Smith, B., 2011. Committed sea-level rise for the next century from Greenland ice sheet dynamics during the past decade. PNAS 108, 8978–8983.

Rausch, D., 1959. Studies of ice excavation. Q. Colo. Sch. Mines 54, 1–90.

Riche, F., Schneebeli, M., 2010. Microstructural change around a needle probe to measure thermal conductivity of snow. J. Glaciol. 56 (199), 871–876.

Riche, F., Schneebeli, M., 2013. Thermal conductivity of snow measured by three independent methods and anisotropy considerations. Cryosphere 7, 217–227.

Rist, M., Sammonds, P.R., Murrell, S.A.F., Meredith, P.G., Doake, C.S.M., Oerter, H., Matsuki, K., 1999. Experimental and theoretical fracture mechanics applied to Antarctic ice and surface crevassing. J. Geophys. Res. 104 (B2), 2973–2987.

Sayles, F.H., 1974. Triaxial Constant Strain Rate Tests and Triaxial Creep Tests on Frozen Ottawa Sand. CRREL Technical Report 190. United States Army Corps of Engineers.

Satyawali, P.K., Singh, A., 2008. Dependence of thermal conductivity of snow on microstructure. J. Earth Syst. Sci. 117, 465−475.

Satyawali, P.K., Schneebeli, M., Pielmeier, C., Stucki, T., Singh, A.K., 2009. Preliminary characterization of alpine snow using snowmicropen. Cold Reg. Sci. Technol. 55, 311−320.

Scapozza, C., Bartelt, P., 2003. Triaxial tests on snow at low strain rate. Part II: constitutive behaviour. J. Glaciol. 49, 91−101.

Schleef, S., Löwe, H., 2013. X-ray microtomography analysis of isothermal densification of new snow under external mechanical stress. J. Glaciol. 59 (214), 233−243.

Schneebeli, M., 1995. Development and Stability of Preferential Flow Paths in a Layered Snowpack. IAHS Publication 228, 89−95, 2 plates.

Schneebeli, M., Johnson, J.B., 1998. A constant-speed penetrometer for high-resolution snow stratigraphy. Ann. Glaciol. 26, 107−111.

Schneebeli, M., Sokratov, S., 2004. Tomography of temperature gradient metamorphism of snow and associated changes in heat conductivity. Hydrol. Process. 18, 3655−3665.

Schulson, E., Duval, P., 2009. Creep and Fracture of Ice. Cambridge University Press.

Schweizer, J., 1998. Laboratory experiments on shear failure of snow. Ann. Glaciol. 26, 97−102.

Schweizer, J., 1999. Review of dry snow slab avalanche release. Cold Reg. Sci. Technol. 30, 43−57.

Schweizer, J., Jamieson, J.B., Schneebeli, M., 2003. Snow avalanche formation. Rev. Geophys. 41, 1016. http://dx.doi.org/10.1029/2002RG000123, 4.

Sego, D.C., Schultz, T., Banasch, R., 1982. Strength and deformation behaviour of frozen saline sand. In: Third International Symposium on Ground Freezing. U.S. Army Corps of Engineers, Hanover, New Hampshire, USA, pp. 11−17.

Shapiro, L., Johnson, J.B., Sturm, M., Blaisdell, G.L., 1997. Snow Mechanics: Review of the State of Knowledge and Applications. CRREL Technical Report 97−3, 36 pp.

Shean, D., Marchant, D., 2010. Seismic and GPR surveys of Mullins glacier, McMurdo dry valleys, Antarctica: ice thickness, internal structure and implications for surface ridge formation. J. Glaciol. 56, 48−64.

Shertzer, R.H., Adams, E.E., 2011. Anisotropic thermal conductivity model for dry snow. Cold Reg. Sci. Technol. 69, 122−128.

Shumskiy, P., 1959. Density of glacier ice. J. Glaciol. 3, 568−573.

Shur, Y., Jorgenson, M.T., 1998. Cryostructure development on the floodplain of Colville River Delta, Northern Alaska. In: Proceedings of the 7th International Conference on Permafrost. Yellowknife, N.W.T., Canada, pp. 993−999.

Sigrist, C., Schweizer, J., 2007. Critical energy release rates of weak snowpack layers determined in field experiments. Geophys. Res. Lett. 34, L03502.

Smith, A., 1997. Basal conditions on Rutford ice stream, West Antarctica, from seismic observations. J. Geophys. Res. 102, 543−552.

Spencer, J., Ashley, W., 2011. Avalanche fatalities in the western United States: a comparison of three databases. Nat. Hazards 58, 31−44.

Steinemann, S., 1954. Results of preliminary experiments on the plasticity of ice crystals. J. Glaciol. 2, 404−412.

Stephani, E., Fortier, D., Shur, Y., 2010. Applications of cryofacies approach to frozen ground engineering − case study of a road test site along the Alaska highway (Beaver Creek, Yukon, Canada). In: Proceedings of the 63rd Canadian Geotechnical Conference. Alberta, Calgary, pp. 476−483.

Sturm, M., Johnson, J., 1992. Thermal-conductivity measurements of depth hoar. J. Geophys. Res. Solid Earth 97, 2129−2139.

Sturm, M., Holmgren, J., Konig, M., Morris, K., 1997. The thermal conductivity of seasonal snow. J. Glaciol. 43, 26−41.

Swinzow, G., 1962. Investigation of shear zones in the ice sheet margin, Thule Area, Greenland. J. Glaciol. 4, 215−229.

Taylor, D.W., 1948. Fundamentals of Soil Mechanics. Wiley, New York.

Terzaghi, C., 1925. Principles of soil mechanics. Eng. News-Rec. 95 (19−27), 740−1068.

Theile, T., Löwe, H., Theile, T.C., Schneebeli, M., 2011. Simulating creep of snow based on microstructure and the anisotropic deformation of ice. Acta Mater. 59 (18), 7104−7113.

Ting, J.M., Martin, R.T., Ladd, C.C., 1983. Mechanisms of strength for frozen sand. J. Geotech. Eng. Div. ASCE 109, 1286−1302.

Ulivieri, G., Marchetti, E., Ripepe, M., Chiambretti, I., De Rosa, G., Segor, V., 2011. Monitoring snow avalanches in Northwestern Italian Alps using an infrasound array. Cold Reg. Sci. Technol. 69 (2−3), 177−183.

Van der Veen, C., 1998. Fracture mechanics approach to penetration of surface crevasses on glaciers. Cold Reg. Sci. Technol. 27, 31−47.

Van der Veen, C., 2007. Fracture propagation as means of rapidly transferring surface meltwater to the base of glaciers. Geophys. Res. Lett. 34, L01501.

Van der Veen, C., Plummer, J., Stearns, L., 2011. Controls on the recent speed-up of Jakobshavn Isbræ, West Greenland. J. Glaciol. 57, 770−782.

Vaughan, D., 1993. Relating the occurrence of crevasses to surface strain rates. J. Glaciol. 39, 255−266.

Vaughan, D., Comiso, J., 2013. Observations: cryosphere. In: Bamber, J., Huybrechts, P., Lemke, P. (Eds.), Working Group I Contribution to the IPCC Fifth Assessment Report (AR5), Climate Change 2013: The Physical Science Basis.

Vyalov, S.S., 1963. Rheology of frozen soil. In: Proceedings of the International Conference on Permafrost, National Academy of Sciences: National Research Council. Publication 1287, Lafayette, Indiana, USA, pp. 332−339.

Waldner, P.A., Schneebeli, M., Schultze-Zimmermann, U., Flühler, H., 2004. Effect of snow structure on water flow and solute transport. Hydrol. Process. 18 (7), 1271−1290.

Wang, X., Baker, I., 2013. Observation of the microstructural evolution of snow under uniaxial compression using X-ray computed microtomography. J. Geophys. Res. Atmos. 118 (22), 12371−12382.

Weertman, J., 1957. On the sliding of glaciers. J. Glaciol. 3, 33−38.

Weertman, J., 1973. Creep of ice. In: Whalley, E., Jones, S., Gold, L. (Eds.), Physics and Chemistry of Ice. Royal Society of Canada, Ottawa, pp. 320−337.

Wettlaufer, J.S., Worster, M.G., 2006. Premelting dynamics. Annu. Rev. Fluid Mech. 38, 427−452.

Williams, P.J., 1967a. Unfrozen Water Content of Frozen Soils and Soil Moisture Suction. Norwegian Geotechnical Institute, Publication No. 72, 11−26.

Williams, P.J., 1967b. Unfrozen Water in Frozen Soils: Pore Size − Freezing Temperature − Pressure Relationship. Norwegian Geotechnical Institute, Publication No. 72, 37−48.

Yamazaki, T., Kondo, J., Sakuraoka, T., Nakamura, T., 1993. A one-dimensional model of the evolution of snow-cover characteristics. Ann. Glaciol. 18, 22−26.

Yamomoto, Y., 2013. Instabilities of alpine permafrost: strength and stiffness in a warming regime (ETH dissertation No. 21596). Institute for Geotechnical Engineering ETH Zurich, Switzerland. 500 pp.

Yamamoto Y., Springman S.M. 2014. Axial compression stress path tests on artificial frozen soil samples in a triaxial device at temperatures just below 0 °C. Canadian Geotechnical Journal, 51. http://dx.doi.org./10.1139/cgj-2013-0257.

Yasufuku, N., Springman, S.M., Arenson, L.U., Ramholt, T., 2003. Stress-dilatancy behaviour of frozen sand in direct shear. In: Proceedings of the 8th International Conference on Permafrost. Zurich, Switzerland, pp. 1253−1258.

Yen, Y., 1981. Review of Thermal Properties of Snow, Ice and Sea Ice. CRREL Technical Report 81−10. Cold Regions Research and Engineering Laboratory, Hanover, NH.

Zhang, T., Barry, R.G., Knowles, K., Ling, F., Armstrong, R.L., 2003. Distribution of Seasonally and Perennially Frozen Ground in the Northern Hemisphere. In: Proceedings of the 8th International Conference on Permafrost, vol. 2, 1289−1294.

Zhang, T., Xiao, C., Colgan, W., Qin, X., Du, W., Sun, W., Liu, Y.S., Ding, M.H., 2012. Observed and modelled ice temperature and velocity along the main flowline of East Rongbuk glacier, Qomolangma (Mount Everest), Himalaya. J. Glaciol. 59, 438−448.

Zhao, S.P., Zhu, Y.L., He, P., 2003. Recent progress in research on the dynamic response of frozen soil. In: Proceedings of the 8th International Conference on Permafrost. Zurich, Switzerland, pp. 1301−1306.

Zwally, H., Li, J., Brenner, A., Beckley, M., Cornejo, H., DiMarzio, J., Giovinetto, M.B., Neumann, T.A., Robbins, J., Saba, J.L., Yi, D.H., Wang, W.L., 2011. Greenland ice sheet mass balance: distribution of increased mass loss with climate warming; 2003−07 versus 1992−2002. J. Glaciol. 57, 88−102.

Zwinger, T., Moore, J., 2009. Diagnostic and prognostic simulations with a full stokes model accounting for superimposed ice of Midtre Lovénbreen, Svalbard. Cryosphere 3, 217−229.

Chapter 3

Snow and Ice in the Climate System

Atsumu Ohmura

Institute for Atmospheric and Climate Science, Swiss Federal Institute of Technology (E.T.H.), Zurich, Switzerland

ABSTRACT

The status of the cryosphere at the end of the 20th century is presented with respect to their relationships to climate. Firstly, the physical condition of the earth's surface is analysed to understand how the earth alone among all planets has a favourable condition for the existence of water in all phases. Then the distributions of the four main components of the cryosphere, the snow cover, the sea ice, the permafrost and the glaciers are interpreted in relation to the climates. The geographic distributions of the main cryospheric components are interpreted in light of the energy balance. The principle of the energy balance allows one to estimate the future changes in the cryosphere under the on-going climate warming.

3.1 INTRODUCTION

The cryosphere owes its existence mainly to the absolute energy emission of the sun, the orbit of the earth, and the chemical composition of the atmosphere. Because of the favorable combination of these three components, the Earth alone among all planets in the solar system enjoys the thermal condition that comprises a temperature range crossing the freezing/melting point of water as illustrated in Figure 3.1. The figure indicates the temperature ranges on the earth's surface expressed as a function of the surface atmospheric pressure. Further, the figure shows the lines of the phase diagram of water separating the Pressure—Temperature field into three fields of Solid, Liquid, and Gas from the left to the right. The convergence of the S/L and L/G lines at 273.16 K and 6.11 hPa is the triple point of water. At the earth's surface, the monthly mean temperatures plotted against the surface pressure are distributed in the shape of a slightly deformed rectangle. At the sea surface, the monthly mean temperatures range between a minimum of $-37\,°C$, the August temperature at the two coldest regions at the sea level in the Antarctic, Belgrano, and Little America, and $35\,°C$,

Snow and Ice-Related Hazards, Risks, and Disasters. http://dx.doi.org/10.1016/B978-0-12-394849-6.00003-2

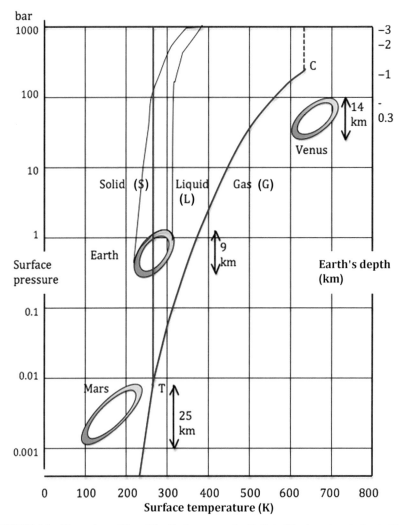

FIGURE 3.1 Thermal condition of the Earth, compared with that of the two closest planets. The figure shows how the Earth alone has water in three phases. The areas surrounded by closed lines indicate the monthly mean temperature and pressure at the surfaces of the planets. Three temperature/pressure regions of the three phases of water are indicated by S (solid), L (liquid), and G (gas) defined by thick lines of the phase transitions. "T" indicates the triple point, and "k" the critical point. The two thin curves extending upward from the "Earth" indicate the estimated distribution of the mean temperature of the lithosphere.

the July temperature at the hottest spot in the world, Djibouti. In the vertical dimension, the surface pressure ranges from 1041 hPa at Jericho to the estimated 300 hPa at the top of Mount Everest. The line of freezing/melting temperature cuts through the range of the Earth's surface temperature, proving the necessary condition not only for the existence of the cryosphere on the Earth's surface but also the water in all three phases. This condition is a unique feature of the Earth, not found on other planets. This basic thermal condition of the Earth's surface extends into the lithosphere. Under the colder areas in Polar regions, the rising ground temperature crosses the pressure melting point at about $1-2$ km below the surface, indicating the possible maximum depth of permafrost.

Let us compare this terrestrial situation with those on the two closest neighboring planets, Mars and Venus. Mars is located at a distance of 1.52 Astronomical Units (AU) from the sun, and it receives only 43 percent of the solar radiation in comparison with the earth. Further, the extremely thin atmosphere gives little greenhouse effect. At the lowest altitudes of Mars' surface, the temperature ranges between $-125\,°C$ and $-40\,°C$, far below the triple point of water. At the highest mountain peak on Mars, the pressure and temperature are estimated at 10^{-3} atm (1 hPa) and $-150\,°C$. This condition also provides the necessary temperature for ice, but excludes the possibility of the existence of liquid water. Venus, on the other hand orbits at only 0.72 AU around the sun, and the solar constant is almost twice that of the earth. Further, the thick atmosphere with CO_2 produces a strong greenhouse effect, causing a high surface temperature. The surface temperature of Venus ranges from $320\,°C$ to $450\,°C$ and crosses the critical temperature of water at $374\,°C$ (647 K). The Temperature—Pressure field on Venus lies entirely on the right-hand side of the L (Liquid)/G (Gas) line, indicating no possibility for the existence of liquid or solid water.

Thus, the Earth alone has the unique situation of having water in all three phases, providing not only the cryosphere but also setting the cryosphere in a constant exchange relationship with the hydrosphere and the atmosphere, which favors the existence of life.

3.2 PHYSICAL EXTENT OF THE CRYOSPHERE

The mean total area of the cryosphere is estimated at 68×10^6 km^2, corresponding to 13 percent of the surface area of the earth (Ohmura, 2004; Hock et al., 2009). The area of the cryosphere fluctuates in the course of a year between 78×10^6 km^2 in the northern winter and 58×10^6 km^2 in the northern summer. The surface area and ice volume of the main cryospheric components, snow cover, sea ice, permafrost, and glaciers are presented in Table 3.1. The shortest-lived component of the cryosphere is the seasonal snow cover, which occupies the largest area, and is mostly in the Northern Hemisphere. The area of the snow cover fluctuates between 46×10^6 km^2 in the late northern winter and 4×10^6 km^2 in the northern summer. Significant year-to-year variations are observed around the southern flank of the snow cover, while the persistent snow cover lasting until late spring tends to appear in the northeastern Eurasian Continent (Rikiishi et al., 2004). These areas are on land, and the snow cover on glaciers is accounted for as a part of glaciers. Toward the end of the northern winter, about 30 percent of the global land area is covered by snow cover. The minimum area of 4×10^6 km^2 in the northern summer is

TABLE 3.1 Surface Area and Volume of the Four Main Cryospheric Components, Seasonal Snow Cover, Sea Ice, Permafrost, and Ice Sheets/Glaciers

	Cryosphere	
	Area in (10^6 km^2)	Ice Volume
Seasonal snow cover	4 (19)−46(58)	0.5(1)−5(6) × 10^3 km^3
Sea ice	19−27	23−32 × 10^3 km^3
Permafrost	23	0.4 × 10^6 km^3 (11−35 × 10^3 km^3)
Ice Sheets and Glaciers	17	28.43 × 10^6 km^3
Total	58−78 × 10^6 km^2	28.8 × 10^6 km^3
Ice Sheets and Glaciers	**(10^6 km^2)**	**(10^6 km^3)**
Antarctic ice sheet	13.86 (82.0%)	25.24 (89.3%)
Greenland ice sheet	1.70 (10.1%)	2.91 (10.3%)
Glaciers and Ice Caps	1.34 (7.9%)	0.27 (0.3%)
Total	16.89 × 10^6 km^2	28.43 × 10^6 km^3

The values in the brackets include the snow cover on sea ice (January 2014).
Sources are Ohmura (2004) and Hock et al. (2009).

mostly the snow cover in southern South America near the end of the southern winter. The snow cover areas in brackets in Table 3.1 are those including the snow cover on sea ice. Volumewise, the seasonal snow cover is the smallest among all components of the cryosphere.

Sea ice has the second shortest mean life of about 1 year. In the course of a year, the global sea ice extent fluctuates between 19 and 27 × 10^6 km^2, around a mean of 23 × 10^6 km^2 (Peng et al., 2013). The relatively small fluctuation in global total sea ice extent is due to the phase difference of about half a year between the Arctic and the Antarctic sea ice fields. The courses of the seasonal change in the sea ice extent are however different in the Arctic and Antarctic. The main difference between the two Polar regions is seen during the receding processes. In the Arctic, the sea ice extent recedes steadily over six months from March to September, while in the Antarctic, the sea ice extent diminishes rather abruptly over three months from November to January. The abrupt nature of the Antarctic sea ice disappearance in the austral summer is due to the mechanical dispersion of the sea ice field by the summer westerlies, rather than the gradual melt, which is the main cause for the disintegration of the Arctic sea ice (Enomoto and Ohmura, 1990). The mean sea ice volume is one order of magnitude larger than the snow cover, 28 (±4) × 10^3 km^3, making its mean thickness to be 1.2 m. The sea ice thickness shows a sign of quick decline (Wadhams and Davis, 2000).

Permafrost is the most widely distributed cryospheric component. Most permafrost is observed in the polar regions and high mountains. There is also submarine permafrost on the continental shelf off the Eurasian and North American continents. The total surface area of permafrost is estimated at 27×10^6 km^2, and is mostly distributed in the Northern Hemisphere. In the Southern Hemisphere, minor areas of permafrost occur on the ice-free zones on Antarctica, on Sub-Antarctic islands, and the southern Andes. Permafrost is conventionally classified into continuous, discontinuous, sporadic, and isolated permafrost. The first three types occupy the majority of the permafrost area roughly in similar proportions. A detailed description of the global permafrost distribution is presented in Gruber (2012), and in Barry and Gan (2011).

Glaciers and ice sheets occupy an estimated 17×10^6 km^2, with an ice volume of 28×10^6 km^3. This indicates that they held as much as 80 percent of the fresh water on the earth. With respect to ice mass, 89 percent is in Antarctica, 10 percent in Greenland, and the remaining 0.3 percent is distributed among all mountain glaciers and ice caps outside the continental ice sheets of Antarctica and Greenland. The distribution of individual glaciers is not restricted to polar and midlatitudes. They are found virtually at any latitude in the high mountains of arid regions, the tropics, and equatorial regions. The rapid decrease in the glacier mass since the beginning of the twentieth century is one of the major changes on the Earth's surface and in the hydrological cycle, causing sea level rise and a variety of catastrophes, which are among the main subjects of this book. The areas and volumes of glaciers and ice sheets are summarized for the 30-year period from 1980 to 2010 in the lower half of Table 3.1.

The total area of the cryosphere in Table 3.1 is not the arithmetic sum of the surface areas of the four cryospheric components. In winter, most permafrost areas lie under the snow cover. This overlapping relationship is taken into account to calculate the total cryosphere surface area.

3.3 CLIMATIC CONDITIONS OF THE CRYOSPHERE

The four major components of the cryosphere, snow cover, sea ice, permafrost, and glaciers have different forming and decaying processes. Therefore, the climatic conditions under which they form and disappear are also different. For example, the snow cover is formed by the deposition of the ice crystals condensed in the atmosphere, and deposited on the Earth. The decay of the snow cover is the melt on the surface, and the result of the energy balance. Consequently, it is strongly influenced by temperature and solar radiation. In both processes, however, snow formation and disappearance happen at a temperature not far from the melting point. In the following section, the climatic conditions associated with the four cryospheric components are summarized.

3.3.1 Snow Cover

Snow cover plays a very important role in the climate system. It has a high albedo, which works to cool the surface. It has a large latent heat of fusion, which shifts the effect of solar heating. Snow cover is a natural reserve of fresh water, delaying melt−water discharge by up to half a year. Snow cover is an excellent thermal insulator, keeping the ground temperature warmer than the air temperature during cold seasons. This effect influences the growth of sea ice and the depth of permafrost and its active layer (Goodrich, 1982; Holtsmark, 1952; Shesterikov, 1963).

The surface temperature during snowfall is usually at the subfreezing point, but can slightly exceed the freezing point. It is, however, extremely rare to observe snowfall at a surface air temperature >2.5 °C. For climatology it is, however, the statistical values that are mostly needed. Figure 3.2 shows the fraction of solid precipitation out of all monthly precipitation expressed as a function of the monthly mean temperature. It indicates that the precipitation is 100 percent snow when the monthly mean temperature is below −6 °C. Likewise, the probability of snow is zero if the monthly mean temperature is >6 °C. Since the monthly precipitation is rarely published after separating into frozen and liquid precipitations, such statistics are necessary for estimating the amount of monthly snowfall. The snow cover melts whenever an excessive energy is supplied to its surface after the temperature attains the melting point. The mean energy balance during the snowmelt is summarized in Table 3.2.

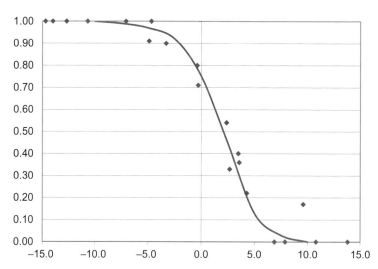

FIGURE 3.2 A fraction of monthly total solid precipitation in water equivalent (ordinate), expressed as a function of monthly mean air temperature (abscissa). The statistics are based on the 10-year monthly precipitation observed at Aklavik, Frobisher Bay, and Resolute in the Canadian Arctic.

TABLE 3.2 Energy Balance of Snow Cover Under Various Conditions

Melt Period E. balance	Global	Reflected	Albedo	Net Solar	Long In	Long Out	Net radiation	Sensible Flux	Latent Flux	Subsurface Flux	Melt	Source
Central Polar Ocean	210	−105	0.50	105	291	−322	73	−6	−6	−6	−55	Vowinckel and Orvig (1966, 1970)
Beaufort sea[3]	246	−168	0.68	78	292	−311	59	2	−2	−7	−51	Persson et al. (2002); H. Huwald (EPFL)[3]
Dry snow zone[1], Greenland	305	−253	0.83	53	214	−255	12	1	−7	−7	0	Ohmura and Raschke (2005)
Accumulation area, Greenland	346	−294	0.85	52	241	−285	8	15	−13	−6	0	Ambach (1977); Ambach and Markl (1983)
Equilibrium line, Greenland	288	−223	0.77	65	261	−305	21	16	−6	−8	−30	Ohmura et al. (1994)
Ablation area, Greenland	272	−152	0.55	120	285	−313	91	29	−20	−11	−90	Ambach (1963)
Interior Arctic tundra	257	−136	0.53	121	283	−320	85	−22	−30	−9	−24	Ohmura (1982)
Coastal Arctic tundra[2]	245	−136	0.56	106	270	−309	67	−13	−5	−15	−35	Weller and Holmgren (1974)
Boreal forest	230	−32	0.14	198	358	−392	163	−67	−78	−10	−9	Pugsley (1970)
Midlatitude grassland	253	−132	0.52	121	270	−313	79	−21	−19	−7	−32	Takeuchi et al. (1992)
Midlatitude city (Ottawa)	180	−54	0.30	126	298	−344	80	27	−79	0	−28	Gold and Williams (1961)

Unit is watts per square meter, except for albedo.
[1]Indicates the measurement at the Summit where the surface does not melt. This station is added for the sake of comparison.
[2]Indicates the measurements with net radiation whose components are not given. They are substituted with the measurements of each component carried out at nearby stations.
[3]Compiled by H. Huwald based on the original data from SHEBA.

The table enables one to compare the energy sources during snowmelt on different surfaces and in different climates. Eleven different surfaces were chosen for comparison: snow on pack ice in the Central Polar Ocean, and also pack ice but in the Marginal Arctic Sea, snow covers in various ice sheet zones on Greenland including the dry snow zone, the wet accumulation area, the equilibrium line, and the ablation area, two regions on the tundra, one in the interior of a Canadian Arctic Island, and the other on the coast of northern Alaska facing the Beaufort Sea, one site in the boreal forest zone of Canada, and snow covers on the midlatitude grassland in Hokkaido and on a parking lot in the city of Ottawa. The general climatic conditions and the radiative and thermal characteristics of the underlying surfaces of these sites are very different. Despite these differences, a common mechanism is working during the melt periods. There are virtually two main energy sources for the melt. The most important is the terrestrial radiation from the atmosphere. The terrestrial atmospheric radiation is three to four times greater than the second largest source, the absorbed solar radiation. The availability of these two irradiances essentially determines the speed of the melt. Since the terrestrial atmospheric radiation received at the earth's surface originates from the atmospheric layer very close to the surface, the conventionally measured surface temperature at approximately 2 m above the surface is a good indicator of the terrestrial atmospheric radiation (Ohmura, 2001). This is the reason why the temperature index model for the melt proves to be realistic. If one wishes to attain a melt simulation of a higher accuracy, the absorbed solar radiation must be taken into account, as this component is responsible for one-quarter to one-third of the total energy source for the melt.

The seasonal snow cover reacts most quickly to climate change in comparison with sea ice, glaciers, and permafrost. Figure 3.3 presents the change of the annual number of days with a measurable snow cover at 50 stations in the Swiss Alps. On an average, the number of days with a measurable snow cover decreased by 3 days in a decade. This trend is mostly created in the warmer seasons of spring and summer (Marty and Meister, 2012). The decreasing tendency is the fastest at higher elevations, as the rate of temperature increase is the most prominent at the highest reaches in the Alps. The physical reason for the warming amplification at high altitudes was presented by Ohmura (2012a). One of the main diabatic processes the release of latent heat of condensation at elevated atmospheric layers is responsible for the stronger warming at high altitudes. The temperature to the fourth exponent of the Stefan−Boltzmann equation also contributes to the larger temperature variation by a given change in a flux in the energy budget equation.

3.3.2 Sea Ice

Unlike glaciers, which represent the accumulated history of the past, the existence of sea ice is more strongly influenced by transient local conditions.

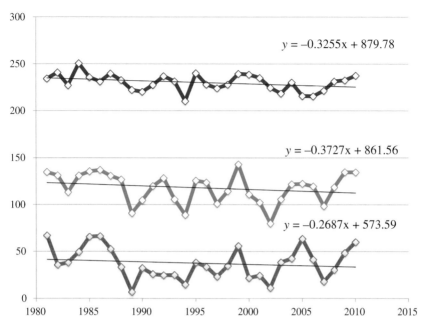

$$y = -0.3255x + 879.78$$

$$y = -0.3727x + 861.56$$

$$y = -0.2687x + 573.59$$

FIGURE 3.3   Change in the annual number of days with a measurable snow cover in three Alpine zones in Switzerland for the period from 1980 to 2010. The data are provided by the Swiss Federal Office for Meteorology and Climatology (MeteoSwiss) through "idaweb". The ordinate indicates the number of days in a year. Dark blue: altitude >1500 m asl; Light blue: altitude between 800 and 1500 m asl; red line: altitude <800 m asl.

Essentially, the existence of sea ice is the result of the local energy balance in the uppermost layer of the sea, and its movement if the ice is in motion. The sea ice energy balance is the budget of heat of an impure and heterogeneous ice that is usually covered with snow or paddled with melt water, and is hence extremely complex. The movement of the sea ice is the fastest speed among cryosphere components on the Earth's surface, and this adds to more complications. Nevertheless, a convenient rule of thumb concerns the sea ice edge and the climate in the Northern Hemisphere. The air temperature at the growing ice edge, that is from October to March, tends to fall between −8 and −10 °C, while on the melting and receding ice edge from May to August, the temperature at the ice edge coincides with an air temperature between 2 and 4 °C. In September, the month of the minimum ice extent, the ice edge coincides very well with −2 °C isothermal lines (Gorshkov, 1983).

This acute asymmetry between the prevailing temperatures during the freezing and melting periods arises from the fact that the seawater is difficult to freeze in comparison with fresh water. The main reason for the slow freezing process of the seawater is the current salt content of the seawater. In the Arctic Ocean where the salt content tends to be lower, the salinity nevertheless does not drop below 30 per mil. The water with salinity >25 per mil has the freezing point at the temperature of the highest density. Therefore, the cooled surface water becomes heavier than the water below causing convection. As the cooled and heavier surface water sinks, upwelling warmer water emerges from beneath, until the rate of the convective heat supply is surpassed by the surface cooling rate. This causes an enormous delay in ice formation in autumn, in comparison with the spring melt.

An assessment of the annual energy balance was made for various parts of the Arctic Ocean by Vowinckel and Orvig (1970). The series of works by E. Vowinckel and S. Orvig was made based on their synoptic energy balance method, and included sea ice. Year-round measurements of the full energy balance on sea ice are rare. On the ice floe, the author is aware of the observations made by Surface Heat Budget of the Arctic Ocean Experiment (SHEBA) in the Marginal Sea of the Beaufort Sea and Chukchi Sea. Table 3.3 presents the monthly and annual energy balance on the sea ice according to Vowinckel and Orvig (1970) for the Central Polar Ocean and Persson et al. (2002) for the Beaufort and Chukchi seas. The melt season for the Central Polar Ocean ends in early September while it continues to late September in the marginal seas. The energy balance, hence the ice formation in fall and winter, is entirely dominated by long-wave radiation. In the main, the energy loss requirement for the freezing of sea ice is long-wave net radiation, both in the Central Polar Ocean and in the marginal seas. The melt season begins in April in the marginal seas, whereas in the Central Polar Ocean, it begins a month later. During the melt, solar radiation and long-wave atmospheric radiation virtually dominate the surface energy balance, providing the necessary heat for the melt. Solar radiation in summer in the Arctic, especially in the Central Polar Ocean, is the strongest in the entire Northern Hemisphere. The high albedo of the sea ice, however, cuts 75 percent of the incoming solar radiation. The ratio of the long-wave atmospheric radiation to the absorbed solar radiation is remarkably constant for both oceans. The long-wave atmospheric radiation is 5–6 times stronger than the absorbed solar radiation. The importance of the long-wave atmospheric radiation as the heat source for melting the sea ice makes the sea ice susceptible to the changes in the greenhouse effect of the atmosphere.

Sea ice is affected by snow cover in various ways. Snow cover is an effective thermal insulator that reduces the freezing speed. This process is treated numerically by Frolov et al. (2005). During the melt season, the snow cover often forms superimposed ice providing an additional ice layer of higher albedo. The superimposed ice formation from the snow cover can also occur before the onset of the melt. If the snow cover becomes sufficiently massive to lower the surface of the sea ice of marine origin, the flooding seawater, together with the snow cover, can add to the sea ice. If the snowfall happens in late summer, the deposited snow in the ocean can become a part of the sea ice (Shesterikov, 1963).

TABLE 3.3 Monthly and Annual Energy Balance of the Sea Ice in the Central Polar Ocean and the Marginal Arctic Ocean in the Beaufort and Chukchi Seas

Central Polar Ocean

	Jan	Feb	Mar	Apr	May	Jun	Jul	Aug	Sep	Oct	Nov	Dec	Annual	Sources
Global	0	0	12	131	269	310	219	126	33	2	0	0	92	GEBA
Reflected	0	0	-10	-107	-221	-236	-142	-86	-27	-2	0	0	-69	GEBA
Albedo	0.84	0.83	0.82	0.82	0.76	0.65	0.68	0.82	0.83				0.75	GEBA
Short-wave net	0	0	2	24	48	74	77	40	6	0	0	0	23	GEBA
Long-wave in	161	144	155	175	240	274	305	300	262	231	177	166	216	Vowinckel
Long-wave out	-198	-180	-203	-246	-280	-316	-334	-327	-289	-261	-219	-207	-255	
Long-wave net	-38	-36	-48	-71	-40	-41	-28	-27	-27	-30	-42	-41	-39	
Net radiation	-38	-36	-45	-48	8	33	48	13	-21	-30	-42	-41	-17	GEBA
Sensible HF	2	2	1	0	-3	-4	0	-3	-2	0	1	1	0	Adjusted to GEBA
Latent HF	1	1	0	-2	-1	-3	-1	-3	-2	-1	1	1	-1	
Subsurface HF	34	34	44	50	0	0	0	0	25	31	40	40	25	
Heat of melt	0	0	0	0	-4	-26	-47	-8	0	0	0	0	-7	
Subsurface HF and melt	34	34	44	50	-4	-26	-47	-8	25	31	40	40	18	

TABLE 3.3 Monthly and Annual Energy Balance of the Sea Ice in the Central Polar Ocean and the Marginal Arctic Ocean in the Beaufort and Chukchi Seas—cont'd

SHEBA Site, Beaufort and Chukchi Seas, 74°–81° N, 143°–168 °W, 0 m asl, Sea Ice 1997–1998

	Jan	Feb	Mar	Apr	May	Jun	Jul	Aug	Sep	Oct	Nov	Dec	Annual
Global	0	5	46	142	249	280	211	111	75		0	0	102
Reflected	0	−4	−39	−120	−204	−200	−136	−78	−61		0	0	−76
Albedo		0.85	0.85	0.85	0.82	0.71	0.64	0.7	0.81				0.75
Short-wave net	0	1	7	22	44	80	75	33	14		0	0	25
Long-wave in	170	164	201	220	246	283	300	299	296		210	152	229
Long-wave out	−198	−190	−222	−242	−274	−308	−314	−311	−304		−227	−185	−251
Long-wave net	−27	−26	−21	−22	−28	−26	−15	−11	−7		−18	−33	−22
Net radiation	−27	−26	−14	−1	16	54	61	22	7		−18	−33	3
Sensible HF	6	9	3	0	−2	0	3	−2	−1		2	6	2
Latent HF	0	0	−1	−2	−2	−2	−1	−2	0		0	0	−1
Subsurface HF	21	18	11	0	−11	−18	4	−9	−6	0	16	27	4
Heat of melt	0	0	0	0	−1	−34	−67	−9	0	0	0	0	−9
Subsurface HF and melt	21	18	11	0	−12	−52	−63	−18	−6	0	16	27	−5

The unit is watts per square meter, except for albedo

Source: Persson et al. (2002); H. Huwald (ETH) and composed from Vowinekel and Oring (1970), and GEBA data

3.3.3 Permafrost

Permafrost is defined in terms of the ground temperature. The relationships between the frozen ground and the relevant climatic elements, however, are complex. First, the ground is not necessarily in a frozen state by simply being at or below the freezing point. This is due to the freezing point depression caused by the density of chemical solutions and hydrostatic pressure. Further, the state of the ground is influenced by a number of local surface conditions, such as topography, surface cover, and the depth and duration of the snow cover. The relationship between the air temperature and the local topographic conditions was systematically studied by Volken (2008) in the Alps. The thermal characteristics of the subsurface especially close to the surface play an important role. The characteristics and the surface conditions of the ground in permafrost and near-permafrost regions are summarized in Table 3.4. Macroscopically viewed, continuous permafrost appears north of the annual mean air temperature of $-8\,°C$. The southern boundary of discontinuous permafrost in North America tends to coincide with the isothermal line of the annual mean air temperature of -0.8 to $-0.2\,°C$, while the lower limit of the discontinuous Alpine permafrost occurs in the zone of an annual mean air temperature between -2.5 and $-3\,°C$. Fujii and Higuchi (1978) pointed out that the air temperature associated with high-altitude permafrost tends to be slightly lower in comparison with high-latitude and low-altitude discontinuous permafrost. They attributed the difference to a possibly larger solar radiation at higher altitudes. This speculation can be supported by recent observations of global radiation in Northern Canada and the European Alps. The annual mean global solar radiation on the southern boundary of discontinuous permafrost is close to 135 W/m^2 in Canada, whereas the lower boundary of the permafrost in the Alps coincides with the annual mean global radiation of 170 W/m^2. The vertical gradient of solar radiation is presently known only in the Alps, and its annual mean value is 1.26 $W/m^2/100$ m (Marty et al., 2002).

Since the existence of permafrost is not as obvious as that of snow cover or glaciers, various methods for detecting permafrost have been developed. Harris (1983) used, in addition to annual mean temperature, freezing and thawing indices to distinguish the boundary of continuous to discontinuous permafrost. Fujii and Higuchi (1978) added the coldest and warmest monthly mean temperatures as a good indicator to separate the southern boundaries of continuous and discontinuous permafrost and the lower boundary of Alpine permafrost. To detect the local occurrence of permafrost in the Alps, Haeberli (1973) and Haeberli et al. (1998) suggested using the winter temperature at the bottom of the snow cover, which has proven to be highly effective in detecting permafrost.

The depth of an active layer varies greatly even within a small area. Rouse (1983) showed that permafrost with different characteristics and topography on the west coast of the Hudson Bay were remarkably similar in net radiation

TABLE 3.4 Characteristics of Permafrost Near the Southern (Lower) Boundaries of Continuous and Discontinuous Permafrost

Site	Region	Latitude	Altitude	Characteristics	Thickness	AMAT °C	AMGT °C	Active Layer	Snow Cover	Remarks	Sources
Outwash plane	Axel Heiberg island	79° 25′N	35 m	Outwash plane	>500 m	−18	−13.3	70–80 cm	30 cm		(1)
Resolute	NWT	74° 42′N	15 m	Uplifted beach	330 m	−16	−12.6				(11)
Churchill	Manitoba	58° 45′N	35 m	Continuous (surface: Bare rock)	30–70 m	−7.3	−2.9	760 cm	45 cm	Limit of continuous permafrost	(2)
Churchill	Manitoba	58° 45′N	23 m	Continuous (38 cm organic layer)		−7.3	−0.9	50 cm	50 cm		(2)
Churchill	Manitoba	58° 45′N	20 m	No permafrost (154 cm organic layer)		−7.3	0.4	Irrelevant	75 cm	No permafrost	(2)
Tarfala	Sweden	67° 56′N	1,500 m	Continuous	>100 m	−6	−4	1.3 m			(3)
Tarfala	Sweden	67° 56′N	1,200–1,500 m	Discontinuous		−5	−4 to −2	2–4 m			(3)
Tarfala, T25	Sweden	67° 56′N	1,130 m	No permafrost	Irrelevant	−2	1.5	Irrelevant		Relict permafrost	(3)
Galdhöppiggen	Norway	61° 38′N	2,430 m	Continuous	100–200 m	−6.1	−6	1.1 m			(3)
Juvasshytta	Norway	61° 41′N	1,830 m	Discontinuous		−3.5	−3.5	1.5 m			(3)
Hochebenekar	Austria	46° 50′N	2,600–2,700 m	Discontinuous		−3.5 to −2.9	−6.7	2–3 m			(4)
Schönwieshütte	Austria	46° 50′N	2,270 m	No permafrost	Irrelevant	−1.1	−0.1	Irrelevant	120 cm		(4)

National Park	Switzerland	46° 50'N	2,100-2,200	Discontinuous		-4 to -3				(5)
Central Swiss Alps	Switzerland	46° 30'N	3,300 m	Continuous		-7				(6)
Murtèl-corvatsch	Switzerland	46° 22'N	2,670 m	Discontinuous	53 m	-2.3		3.4 m		(6)
Stockhorn	Southern Switzerland	46° 02'N	3,410 m	Discontinuous	100 m	-2.6		3.0 m		(6)
Fujisan	Japan	35° 21'N	2,800-2,900 m	Discontinuous	50-60 m	-1.4 to -1.8		0.5-1 m	30-300 cm	(7)
Muli	Qilianshan, China	38° 10'N	4,089 m	Discontinuous	60 m	-3	-2.3 to -0.6	3 m		(8)
Rushui	Qilianshan, China	37° 20'N	3,300 m	Discontinuous	32.5 m	-2.9	-0.8	2.5 m		(9)
Kuulungshankou	Tibet, China	35° 45'N	4 m350-4,560 m	Discontinuous	40-100 m	-3.6	-1.0 to -4.0	1.0-4.0 m		(9)
Luanhaizi	Tibet, China	34° 43'N	4,677-4,715 m	Discontinuous	70-155 m	-2.7	-1.5 to -4.4	1.1-4.5 m		(9)
Temengela	Tibet, China	32° 51'N	4,100 m	Discontinuous	67 m	-2.6	-1.7			(9)
Tangula	Tibet, China	32° 21'N	4,780 m	Continuous	20-130 m	-5.6	-1.7 to -4.5	1.5-3.2 m		(9)
North slope of the Himalayas	Tibet, China	28° 13'N	5,200-5,400 m	Discontinuous		-2.8 to -3.4				(9)
Khumbu Himal	Himalayas	27° 55'N	4,900-5,000 m	Discontinuous		-2.8 to -3.4	1.5 to 2.8			(10)

Two cases were added from the area of very thick permafrost (Axel Heiberg Island and Resolute) far away from the southern border. Three cases of non-permafrost regions are added for comparison as these regions are very close to the southern (lower) boundary of the discontinuous permafrost.

Sources: (1) Ohmura (1982); (2) Brown (1978); (3) King (1984); (4) Haeberli and Patzelt (1982); (5) Haeberli (1978); (6) Haeberli et al. (1979); (7) Fujii and Higuchi (1972, 1978); (8) Tong and Li (1983), Wang and Li (1983), Wu (1983); (9) Guo and Cheng (1983); (10) Fujii and Higuchi (1976); (11) Geological Survey of Canada (1967).

and ground heat flux. Features of the subsurface, such as the soil thermal characteristics and water content, are responsible for the wide variety of the summer maximum depth of the active layer. In the case of much longer time scales, such as decades, Smith and Rieseborough (1983) investigated the effect of the climate change on degradation of permafrost. They found water content and the depth of the snow cover to be the most influential factors in determining the sensitivity of the permafrost to the climate change. Since the permafrost contains a large amount of methane, the decay of permafrost is expected to enhance the greenhouse effect (Streletskiy et al., 2014). The melt of permafrost may further contribute to additional methane emission by creating an anaerobic environment in the melt depressions. The importance of the long-term permafrost monitoring and the technical proposal for such a monitoring were proposed by Haeberli et al. (1993). The Global Terrestrial Network for Permafrost is today responsible for the monitoring of permafrost within the Global Climate Observing System.

3.3.4 Glaciers

Glaciers exist when the Earth's surface appears above the equilibrium-line altitude. The position of the equilibrium line on glaciers is determined by mass balance, which is commonly parameterized by the annual precipitation and the temperature of the melting season. This type of parameterization succeeds especially for the extraequatorial glaciers, where annual accumulation resembles the precipitation (P), and air temperature (T) is a good indicator for summer melt. We call this type of expression a P/T diagram; it was initially proposed by Ahlman (1924) and revised by scientists of later generations (Andrews, 1975; Ohmura et al., 1992). The work by Ohmura et al. (1992) has been revised with more glaciological and updated meteorological data and is presented in Figure 3.4. The physical basis for adopting air temperature is its capacity to predict long-wave incoming radiation, which is the most important single energy source for the melt (Ohmura, 2001). Further, the adoption of the second-most important parameter, that is, solar radiation, contributes substantially to systematizing the diverse equilibrium line altitude (ELA) conditions with respect to easily available meteorological data. The air temperature is that of the free atmosphere, as the external air temperature is more influential than the local temperature on glaciers (Lang, 1968). The choice of the free atmospheric temperature also frees one from the cumbersome work of downscaling. Figure 3.4 indicates that the equivalent effects of temperature, precipitation, and solar radiation are $1\,°C$, 350 mm w.e., and $7\,W/m^2$, respectively. The present observational uncertainties are estimated at $0.2\,°C$, 30 mm, and $4\,W/m^2$, for temperature, precipitation, and radiation, respectively.

The change in mass balance can be monitored by the ELA. Figure 3.5 presents the change of ELAs from 17 representative glacier regions of the world. In the second half of the twentieth century, the ELA climbed with a

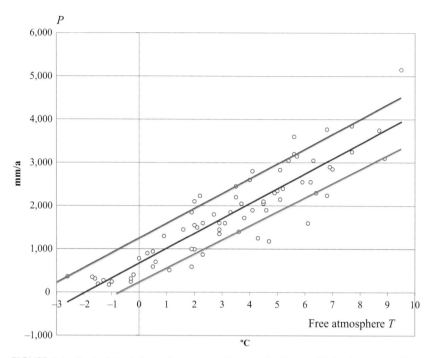

FIGURE 3.4 Revised Precipitation/Temperature diagram showing the likely climatic condition for the location of equilibrium line altitude. P = winter balance (Bw) + measured summer precipitation; T = free atmosphere T for JJA (DJF); The blue line indicates the observed, mean-absorbed, global solar radiation of the melt season JJA (DJF) of 100 W/m^2, whereas the red and brown lines indicate higher and lower bounds of the radiation at 125 and 75 W/m^2, respectively. The scatter of the (P, T) for various glaciers is systematically captured by introducing the third variable of global solar radiation. The data for P and T are based on ECHAM Interim.

global mean rate of 5 m/a. Regionally, the steepest ELA ascent was observed in the Caucasus with a rate of 13 m/a. The ELA climbed with the slowest speed of 0.8 m/a in Svalbard. The glaciers in the Alps, on the North American Cordillera, and in the Canadian Arctic all showed ascending rates of 6 m/a. The glaciers in Scandinavia show during 1944−2014 almost a zero trend of the ELA change. This series, however, is made of two periods of contrasting trends before and after the late 1980s. Before this time, the ELA slowly decreased at 0.4 m/a, but after the late 1980s, it has been ascending at one of the fastest rates of 12 m/a. Globally averaged, the acceleration of the net mass loss with the 5 m/a ELA ascent is equivalent to 8 mm/a^2, superposed on the mean net mass loss of 250 mm/a. The particulars of the mass balance acceleration are due to an increase in the accumulation of 2 mm/a and ablation of −10 mm/a. During the same period, the global land temperature rose at a rate of 0.25 K/Decade. The increasing rates of global solar radiation, and long-wave incoming radiation, were 0.1 W/m^2/Decade and 1.6 W/m^2/Decade, respectively (Ohmura, 2012b). These independently obtained values converge

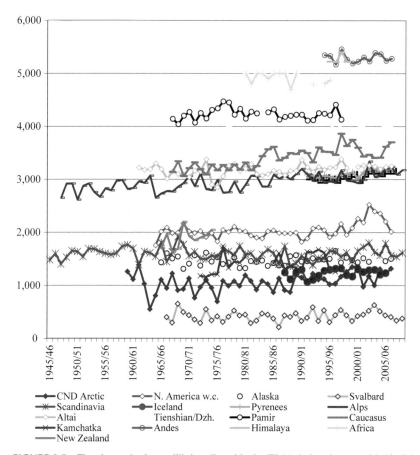

FIGURE 3.5 The change in the equilibrium line altitude (ELA) during the second half of the twentieth century for 17 glacierized regions of the world. Data are due to Kasser (1973), Müller (1977), Haeberli (1985), and subsequent Fluctuations of Glaciers.

in explaining the source of the energy accelerating the melt. The radiative increase of 1.7 W/m^2/Decade must be first expended for increasing the temperature by 0.25 K/Decade that is equivalent to 1.1 W/m^2/Decade. The difference of 0.6 W/m^2/Decade over the mean ablation period of about 60 days is 3.1 MJm^2, which is sufficient to cause a -9 mm/a^2 acceleration of the ablation. This calculation based on radiation observation matches well with the ablation acceleration obtained from mass balance observations. Therefore, 95 percent of the current acceleration of the glacier ablation is caused by the increasing greenhouse effect, and 5 percent by slightly brightening solar radiation.

This is a brief overview of the relationship between the climate and the cryosphere. The detailed relationship and its role in cryosphere-related catastrophes are the subjects of the following chapters.

REFERENCES

Ahlman, H. W. son, 1924. Le niveau de glaciation comme function de láccumulation d'humidité sous forme solide. Geogr. Ann. 6, 223−272.

Ambach, W., 1963. Untersuchung zum Energieumsatz in der Ablationszone des Grönländischen Inlandeises (Camp IV-EGIG, 69° 40′05″N, 49° 37′58″W). Expédition Glaciologique Internationale au Groenland 1957−1960, 4, (4). Medd. Groenl. 174 (4), 311 pp.

Ambach, W., 1977. Untersuchung zum Energieumsatz in der Akkumulationszone des Grönländischen Inlandeises. Expédition Glaciologique Internationale au Groenland 1967−1968, 4, (7). Medd. Groenl. 187 (7), 44 pp.

Ambach, W., Markl, G., 1983. Untersuchungen zum Strahlungshaushalt in der Akkumulationszone des Grönländischen Inlandeises (Station Carrefour 69 49′25″N, 47 25′57″W, 1850 m). Expédition Glaciologique Internationale au Groenland 1967−1968, 4 (6). Medd. Groenl. 187 (6), 61 pp.

Andrews, J.T., 1975. Glacier Systems—an Approach to Glaciers and Their Environments. Environment Systems Series, Duxbury Press, North Scituate, 191 pp.

Barry, R.G., Gan, T.Y., 2011. The Global Cryosphere, Past, Present, and Future. Cambridge University Press, Cambridge, 474 pp.

Brown, R.J.E., 1978. Influence of climate and terrain on ground temperature in the continuous permafrost zone of Northern Manitoba and Keewatin District, Canada. Proc. Third Intern. Conf. Permafrost, July 10−13, 1978, Edmonton, Alberta, Canada Nat. Res. Counc. Can. 1, 15−21.

Enomoto, H., Ohmura, A., 1990. The influence of atmospheric half-year cycle on the sea ice extent in the Antarctic. J. Geophys. Res. 95, 9497−9511.

Frolov, I.E., Gudkovich, Z.M., Radionov, V.F., Shirochkov, A.V., Timokhov, L.A., 2005. The Arctic Basin, Results from the Russian Drifting Stations. Springer, Berlin, Heidelberg and New York, 272 pp.

Fujii, Y., Higuchi, K., 1972. On the permafrost at summit of Mt. Fuji. Seppyo 13, 175−186.

Fujii, Y., Higuchi, K., 1976. Ground Temperature and its Relation to Permafrost Occurrences in the Kumbu Himal and Hidden Valley, Nepal Himalayas. Seppyo 38 (Special Issue): Glaciers and Climates of Nepal Himalayas—Report of the Glaciological Expedition to Nepal, 125−128.

Fujii, Y., Higuchi, K., 1978. Distribution of Alpine permafrost in the Northern Hemisphere and its relation to air temperature. Proc. Third Intern. Conf. Permafrost, July 10−13, 1978, Edmonton, Alberta, Canada Nat. Res. Counc. Can. 1, 366−371.

Geological Survey of Canada, 1967. Permafrost in Canada. A Map, 1246A, first ed. Geol. Surv. Canada, Ottawa 1, 7,603,200.

Gold, L.W., Williams, G.P., 1961. Energy balance during the snow melt period at an Ottawa site. IASH, ICSI, Publ. 54, 288−294.

Goodrich, L.E., 1982. The influence of snow cover on the ground thermal regime. Can. Geotech. J. 19, 421−432.

Gorshkov, S.G., 1983. Arctic ocean. In: Gorshkov, S.G. (Ed.), World Ocean Atlas, vol. 3. Pergamon, Oxford, p. 184.

Gruber, S., 2012. Derivation and analysis of a high-resolution estimate of global permafrost zonation. Cryosphere 6, 221−233.

Guo, P., Cheng, G., 1983. Zonation and Formation History of Permafrost in Qilian Mountains of China. In: Permafrost Fourth International Conference Proceedings. National Academy Press, Washington, D.C., pp. 395−400.

Haeberli, W., 1973. Die Basis-Temperatur der winterlichen Schneedecke als mölicher Indikator füdie Verbreitung von Permafrost in den Alpen. Z Gletscherkd. Glazialgeol. 9, 221−227.

Haeberli, W., 1978. Special aspects of high mountain permafrost methodology and zonation in the Alps. Proc. Third Intern. Conf. Permafrost, July 10−13, 1978, Edmonton, Alberta, Canada Nat. Res. Counc. Can. 1, 378−384.

Haeberli, W., 1985. Fluctuations of Glaciers 1975−1980, vol. V. International Association of Hydrological Sciences/UNESCO, Paris.

Haeberli, W., Patzelt, G., 1982. Permafrostkartierung im Gebiet der Hochebenkar-Blockgletscher, Obergurgl, Ötztaler Alpen. Z. Gletscherkd. Glazialgeol. 18, 127−150.

Haeberli, W., Barsch, D., Brown, J., Guodong, C., Corte, A.E., Dramis, F., Evin, M., Gorbunov, A.P., Guglielmin, M., Harris, C., Harris, S.A., Hoelzle, M., Kääb, A., King, L., Lieb, G.K., Matsuoka, N., Ødegaard, R.S., Shroder, J.F., Sollid, J.L., Trombotto, D., Von der Mühll, D., Shaoling, W., Xiufeng, Z., 1998. Mapping, modelling and monitoring of mountain permafrost: a review of ongoing programs. Mountain permafrost working group report. Proc. Seventh Int. Conf. Permafrost, Yellowknife—Canada, 233−246.

Haeberli, W., Hoelzle, M., Keller, F., Schmid, W., Vonder Mühl, D.S., Wagner, S., 1993. Monitoring the long-term evolution of mountain permafrost in the Swiss Alps. Monitoring the long-term evolution of mountain permafrost in the Swiss Alps. Proc. Sixth Intern. Conf. Permafrost, Beijing 1, 214−219.

Haeberli, W., King, L., Flotron, A., 1979. Surface movement and lichen-cover studies at the active rock glacier near the Grubengletscher, Wallis, Swiss Alps. Arct. Alp. Res. 11, 421−441.

Harris, S.A., 1983. Comparison of the climatic and geomorphic methods of predicting permafrost distribution in western Yukon Territory. In: Proc. Fourth Intern. Conf. Permafrost. National Academy Press, Washington, D.C., pp. 450−455.

Hock, R., de Woul, M., Radic, V., Dyurgerov, M., 2009. Mountain glaciers and ice caps around Antarctica make a large sea-level contribution. Geophys. Res. Lett. 36, L07501. http://dx.doi.org/10.1029/2008GL037202.

Holtsmark, B.E., September 22−27, 1952. Insulating effect of a snow cover of young sea ice. In: Proceedings of the Third Alaskan Science Conference, pp. 60−65.

Kasser, P., Comp, 1973. Fluctuations of Glaciers 1965−1970. International Association of Hydrological Sciences/UNESCO, Paris.

King, L., 1984. Permafrost in Scandinavia. Geogr. Inst. Univ. Heiderberg, 174 pp.

Lang, H., 1968. Relations between glacier runoff and meteorological factors observed on and outside the glacier. IAHS Publ. 79, 429−439.

Marty, C., Meister, R., 2012. Long-term snow and weather observations at Weissfluhjoch and its relation to other high-altitude observatories in the Alps. Theor. App. Climatol. 110, 573−583.

Marty, C., Philipona, R., Fröhlich, C., Ohmura, A., 2002. Altitude dependence of surface radiation fluxes and cloud forcing in the Alps: results from Alpine Surface Radiation Budget network. Theor. Appl. Climatol. 72, 137−155.

Müller, F., Comp, 1977. Fluctuations of Glaciers 1970−1975, vol. III. International Association of Hydrological Sciences/UNESCO, Paris.

Ohmura, A., 1982. Climate and energy balance on the arctic tundra. J. Climatol. 2, 65−84.

Ohmura, A., Kasser, P., Funk, M., 1992. Climate at the equilibrium line of glaciers. J. Glaciol. 38 (130), 397−411.

Ohmura, A., 2001. Physical basis for the temperature/melt-index method. J. Appl. Meteor. 40, 753−761.

Ohmura, A., 2004. Cryosphere during the Twentieth Century. Geophys. Monogr. 150, 239−257. Am. Geophys. Union.

Ohmura, A., 2012a. Enhanced temperature variability in high-altitude climate change. Theor. Appl. Climatol. http://dx.doi.org/10.1007/s00704-012-0687-x.

Ohmura, A., 2012b. Present status and variations in the Arctic energy balance. Polar Sci. 6. http://dx.doi.org/10.1016/j.polar.2012.03.003.

Ohmura, A., Konzelmann, T., Rotach, M., Forrer, J., Wild, M., Abe-Ouchi, A., Toritani, H., 1994. Energy balance for the Greenland ice sheet by observation and model computation. In: Jones, H.G., Davies, T.D., Ohmura, A., Morris, E.M. (Eds.), Snow and Ice Covers; Interactions with the Atmosphere and Ecosystems. IAHS Publ. No. 223, pp. 85−94.

Ohmura, A., Raschke, E., 2005. Energy budget at the Earth's surface. Chap. 10. In: Hantel, M. (Ed.), Observed Global Climate, Group V: Geophysics, Landolt−Börnstein Numerical and Functional Relationships in Science and Technology, New Series, vol. 6. Springer Verlag, Berlin, Heidelberg, New York, pp. 10.1−10.28.

Peng, G., Meier, W., Scott, D., Savoie, M., 2013. A long-term and reproducible passive microwave sea ice concentration data record for climate studies and monitoring. Earth Syst. Sci. Data 5, 311−318.

Persson, P.O.G., Fairall, C.W., Andreas, E.L., Guest, P.S., Perovich, D.K., 2002. Measurements near the atmospheric surface flux group tower at SHEBA: near surface conditions and surface energy budget. J.Geophys. Res. 107 (C10), 8045. http://dx.doi.org/10.1029/2000JC000705.

Pugsley, W.I., 1970. The surface energy budget of Central Canada. Publ. Meteorol. 96, 58. Dept. Meteorol. McGill Univ. Montreal.

Rikiishi, K., Hashiya, E., Imai, M., 2004. Linear trends of the length of snow-cover season in the Northern Hemisphere as observed by the satellites in the period 1972−2000. Ann. Glaciol. 38, 229−237.

Rouse, W.R., 1983. Active layer energy exchange in wet and dry tundra of the Hudson Bay lowlands. In: Permafrost Fourth International Conference Proceedings. National Academy Press, Washington, D.C., pp. 1089−1094.

Shesterikov, N.P., 1963. Some peculiarities of ice growth in the area of Mirny. Probl. Arkt. Antarkt. 13, 37−65.

Smith, M.W., Rieseborough, D.W., 1983. Permafrost sensitivity to climate change. In: Permafrost Fourth International Conference Proceedings. National Academic Press, Washington, D.C., pp. 1178−1187.

Streletskiy, D., Anisimov, O., Vasiliev, A., 2014. Permafrost degradation. In: Haeberli, W., Whiteman, C. (Eds.), Snow and Ice-related Hazards, Risks and Disasters. Elsevier, pp. 303−344.

Takeuchi, Y., Kodama, Y., Ishikawa, N., Kobayashi, D., 1992. The effect of snow cover on the thermal characteristics of the surface layer. Low Temp. Sci., Ser. A 51, 63−76.

Tong, B., Li, S., 1983. Some characteristics of permafrost on Qinghai-Xizang plateau and a few factors affecting them. In: Professional Papers on Permafrost Studies of Qinghai-Xizang Plateau. Lanzhou Inst. Glaciol. Cryopedol, pp. 1−11.

Volken, D., 2008. Mesoklimatische Temperaturverteilung im Rhone- und Vispertal (diss. Nr. 17705). E.T.H. Zurich, 128 pp.

Vowinckel, E., Orvig, S., 1966. The heat budget over the Arctic Ocean. Arch. Meteorol. Geophys. Bioklimatol. 14, 303−325.

Vowinckel, E., Orvig, S., 1970. The climate of the North Polar Basin. In: Orvig (Ed.), 1970: Climates of Polar Regions, World Survey of Climatology, vol. 14. Elsevier, Amsterdam, London, New York, pp. 129−252.

Wang, J., Li, S., 1983. Analysis of geothermal conditions near permafrost base along Qinghai-Xizang Plateau. Lanzhou Inst. Glaciol. Cryopedol. In: Professional Papers on Permafrost Studies of Qinghai-Xizang Plateau, pp. 38−43.

Wadhams, P., Davis, N.R., 2000. Further evidence of ice thinning in the Arctic Ocean. Geophys. Res. Lett. 27, 3973—3975.

Weller, G., Holmgren, B., 1974. The microclimates of the arctic tundra. J. Appl. Meteorol. 13, 854—862.

Wu, Z., 1983. The basic types of ground temperature curves in plateau type permafrost and their application. Lanzhou Inst. Glaciol. Cryopedol. In: Professional Papers on Permafrost Studies of Qinghai-xizang Plateau, pp. 185—194.

Chapter 4

Snow and Ice in the Hydrosphere

Jan Seibert [1,2], Michal Jenicek [1,3], Matthias Huss [4,5] and Tracy Ewen [1]

[1] *Department of Geography, University of Zurich, Switzerland,* [2] *Department of Earth Sciences, Uppsala University, Uppsala, Sweden,* [3] *Department of Physical Geography and Geoecology, Faculty of Science, Charles University in Prague, Czech Republic,* [4] *Laboratory of Hydraulics, Hydrology and Glaciology (VAW), ETH, Zurich, Switzerland,* [5] *Department of Geosciences, University of Fribourg, Switzerland*

ABSTRACT

In large areas of the world, runoff and other hydrological variables are controlled by the spatial and temporal variation of the $0\,°C$ isotherm, which is central for the temporal storage of precipitation as snow or ice. This storage is of crucial importance for the seasonal distribution of snow and ice melt, a major component of the movement of water in the global water cycle. This chapter provides an introduction to the role of snow and ice in the hydrosphere by discussing topics including snowpack characteristics, snow observation approaches, the energy balance of snow-covered areas, and modeling of snowmelt. Furthermore, the role of glaciers and glacial mass balances, including modeling glacier discharge, is discussed. Finally, an overview of the hydrology of snow- and ice-covered catchments is given, and the influence of snow, glaciers, river ice, and frozen soils on discharge is discussed.

4.1 INTRODUCTION

Snow and ice are important components of the hydrosphere, and many processes related to snow and ice are dominated by the important threshold behavior caused by the large amount of energy, which is released or consumed during the phase change between liquid and frozen water. In the Northern Hemisphere, about one-quarter of the area experiences mean annual temperatures that are below $0\,°C$, and for more than half of the area, temperatures are below $0\,°C$ during at least one month of the year (Brown and Goodison, 2005; Davison and Pietroniro, 2005). Hence, a permanent or seasonal snow cover exists in large parts of the world, even if the snow cover extent in the Southern Hemisphere is much less ($\sim 2\%$ of that in the Northern Hemisphere). The annual snowfall

Snow and Ice-Related Hazards, Risks, and Disasters. http://dx.doi.org/10.1016/B978-0-12-394849-6.00004-4

fraction increases with elevation and latitude, where the amount of solar energy varies significantly over the year. Overall, the climate determines where snowfall typically occurs, and drivers at the global, regional, and local scales all play a role. The characteristically small difference in temperature that determines whether precipitation falls as rain or snow can have large effects on ecosystems. In regions where there is either a permanent or seasonal snow cover, snow significantly influences all aspects of the cryosphere and many environmental variables, as well as social and economic patterns.

The existence of snow cover on the ground directly changes the partitioning of the incoming energy. Snow has a high albedo, which means that a large fraction of the incoming solar energy is reflected back into the atmosphere. The albedo of new snow can be up to about 0.9 and decreases to about 0.5 as snow ages, whereas typical values for areas without snow cover are 0.1−0.3 (Table 4.1). This difference explains the observation that mean air temperatures are significantly lower ($\sim 5\,°C$) if the ground is covered by snow, which implies that there is a positive feedback that can enhance the accumulation of snow in early winter and maintain the snowpack longer into late spring and summer.

Snow also has a high insulation capacity, which means that energy fluxes between the atmosphere and the ground are significantly reduced by snow. It has, for instance, been observed that seasonal soil freezes to deeper depths if there is no snow present compared to the situation with an insulating snow cover (Hayashi, 2013). Wintertime snowpack also acts as insulation for the local environment, providing an insulating cover for vegetation and hibernating animals.

TABLE 4.1 Albedo of Various Surfaces for Short-Wave Radiation

Surface	Typical Range in Albedo (−)
New snow	0.80−0.90
Old snow	0.60−0.80
Melting snow (porous → fine grained)	0.40−0.60
Snow ice	0.30−0.55
Coniferous forest with snow	0.25−0.35
Green forest	0.10−0.20
Bare soils	0.10−0.30
Water	0.05−0.15

Based on Hendriks, 2010; Maidment, 1992.

The spatial and temporal variation of the $0\,°C$ isotherm is central for the temporal storage of precipitation as snow or ice. This storage is of crucial importance for the seasonal distribution of snow and ice melt, a major component of the movement of water in the global water cycle. In many areas, snow and ice melt is the main contribution to annual river runoff and can affect water supply, hydropower production, agricultural irrigation, and transportation. In extreme cases, large snowmelt rates, sometimes combined with heavy springtime rainfall, may lead to disastrous flooding. Rapid snowmelt can also trigger landslides and mass wasting of hill slopes. Too little snowmelt, on the other hand, can lead to low river flows in summer, or even drought, which can have significant economic, environmental, and ecological consequences. Especially in many arid regions of the world, the temporal storage of snow in nearby mountainous regions is of critical importance for water supply.

In this chapter, we provide an overview of the role of snow and ice in the water cycle. In the following section on snow accumulation and snowpack properties, we address topics including snowpack characteristics, snow observation approaches, energy balance of snow-covered areas, and modeling of snowmelt. In the section that follows, we discuss glaciers and glacial mass balance, including modeling glacier discharge. In the final section, an overview of the hydrology of snow- and ice-covered catchments is given, and the influence of snow, glaciers, river ice, and frozen soils on discharge is discussed. The present chapter thus mainly provides a background to the other chapters in this volume, which discuss the different aspects of snow- and ice-related natural hazards.

4.2 SNOW ACCUMULATION AND MELT

4.2.1 Snowpack Description

A snowpack can be described in many different ways. Although snow depths can be easily measured, the snow water equivalent (S_{WE} in mm) is the more relevant property of a snowpack for most snow hydrological questions since the S_{WE} is the water content in snow that directly contributes to runoff. The S_{WE} is defined as the amount of liquid water that would be obtained upon complete melting of the snowpack per unit ground surface area. S_{WE} (mm), snow depth (d (m)), snow density (ρ_s (kg m^{-3})), and water density ($\rho_w = \sim 1{,}000$ kg m^{-3}) are directly related (Eqn (4.1)). The density of snow can vary considerably (Table 4.2) (DeWalle and Rango, 2008; Fierz et al., 2009; Maidment, 1992; Singh and Singh, 2001). New snow generally has the lowest densities with about 100 kg m^{-3}, and densities increase with aging snow due to metamorphism to about 350–400 kg m^{-3} for dry old snow and up to 500 kg m^{-3} for wet old snow. Firn, snow that has not melted during the past summer, has densities typically ranging from 550 to 800 kg m^{-3}.

TABLE 4.2 Typical Snow Density for Different Snow Types

Snow Type	Density (kg m^{-3})
New snow (immediately after falling in calm conditions and at very low temperature)	10–30
New snow (immediately after falling in calm conditions)	50–70
Wet new snow	100–200
Settled snow	200–300
Depth hoar	200–300
Wind packed snow	350–400
Wet snow	350–500
Firn	500–830
Glacier ice	850–917

Based on DeWalle and Rango, 2008; Fierz et al., 2009; Maidment, 1992; Singh and Singh, 2001.

$$S_{WE} = 1000d\,\frac{\rho_s}{\rho_w}. \tag{4.1}$$

Different types of precipitation also impact the density of the snowpack. These include the type of snow (wet or dry), graupel, sleet (or ice pellets), freezing rain, freezing drizzle, rain, and drizzle. The type of snow crystal, determined during formation by the temperature and moisture content of the atmosphere, also affects the density of the snowpack. The most complex crystal structures tend to form light, low-density snowpacks due to the large spacing in between their branches. Smaller crystals such as plate crystals, needles, and columns, tend to form heavier, high-density snowpacks because they pack together very efficiently, leaving little space for air pockets in between. Changes in snowpack density due to the increase in the pressure of the top layers are caused by the mass of newly fallen snow on the older snow. The increase in the snow density causes the increase in the thermal conductivity of the snowpack. For a further discussion of snow density related to snow microstructure, see (Arenson et al., 2014) (this volume).

At the catchment scale, the amount of snow stored in a catchment is determined by the spatial distribution of S_{WE}, which is largely controlled by the spatial distribution of snow depths. Another important variable is the fractional snow cover. Snow depletion curves describe the S_{WE} as a function of fractional snow cover and are important for forecasting seasonal runoff during periods with a gradual decrease in snow cover. In alpine areas, the curves are commonly s-shaped, reflecting the area–elevation curve, which is typically steep in the low and higher elevations and flatter in the middle elevations (Seidel and Martinec, 2010).

4.2.2 Snow Accumulation

4.2.2.1 Spatial Distribution of Snow

Large-scale variations of snow accumulation are controlled by temperature, which in turn is related to elevation and latitude (Essery, 2003; Jost et al., 2007) and exposition, whereas at smaller scales, the effects of topography on snow redistribution and of forests on snow accumulation and snowmelt can be pronounced. The forest affects snow interception, wind speed, and the density of the snowpack, depending mainly on the type of forest (i.e., coniferous, deciduous, or mixed species), forest age, and canopy cover. The role of the forest is also important during the melting period because it reduces the amount of short-wave solar radiation reaching the surface and thus snowmelt rates (Holko et al., 2009; Jeníček et al., 2012; Jost et al., 2007, 2009; López-Moreno et al., 2008).

Stähli and Gustafsson (2006) evaluated the effect of forests on snow accumulation in a small prealpine catchment in Switzerland. They found that as a long-term average, annual maximum S_{WE} was about 50 percent less in the forest than in open areas. The importance of this effect decreased with increasing snowfall. They also found that the variability of S_{WE} increased throughout the winter season and could be explained by varying snow accumulation and melt at different altitudes and exposures. Similar findings were obtained in several paired catchment studies after clear-cutting. For a catchment in Northern Sweden, for instance, it was observed that the S_{WE} increased after clear-cutting by about 30 percent at the end of the accumulation period, snowmelt occurred earlier and spring flood runoff increased significantly in some years (Schelker et al., 2013). Overall, clear-cutting increases the snowpack and the S_{WE} directly since snow is no longer intercepted by the vegetation canopy. Clear-cutting will also have a local effect on the wintertime energy budget through an increase in albedo.

4.2.2.2 Snow Observations

Snowfall can be measured by precipitation gauges in the same way as rainfall, but the errors due to wind effects are commonly significant and systematic underestimates of ≥ 50 percent have been reported (Rasmussen et al., 2012; Sevruk et al., 2009). Adaptations of precipitation gauges to improve snowfall measurements include heated gauges, which melt the snow and allow the S_{WE} of the snowfall to be measured directly, or gauges with large inverted cone shields that can improve measurements in open, windy areas. Without heating, snowfall commonly accumulates on top of the gauge, resulting in erroneous measurements (Figure 4.1). Unlike rainfall, snow remains on the ground and can also be measured in situ.

FIGURE 4.1   Nonfunctioning precipitation gauge after heavy snowfall (Mumsarby/Dannemora, Sweden).

Snow depth, either the entire snowpack or the amount of newly fallen snow, is generally measured using a snow stake, which is simply a height measurement taken at the snow surface with the zero point either at the ground surface or last snow layer. Snow depth can also be measured using automatic sensors that usually apply an ultrasonic beam reflecting from the ground level. For low-cost measurements carried out in many different localities, time-lapse photography can be used (e.g., Garvelmann et al., 2013). This solution enables not only snow depth to be measured but it also allows information about site conditions during the observing period to be recorded, such as rain-on-snow events, cloudiness, and snow interception on the canopy.

The S_{WE} can be measured directly by taking a snow core with a snow tube, and then weighing the snow and dividing by the area of the opening of the snow tube. As an alternative, the S_{WE} can be computed from measurements of snowpack depth (d (m)) and estimates of snow density (ρ_s) using Eqn (4.1) and the density of liquid water ($\sim 1{,}000$ kg m^{-3}). Because snow depth generally varies spatially more than snow density does, and is also easier to measure, it is usually a good strategy to measure snow depth at a frequency at least 10 times greater than snow density. Statistical estimation methods of snow density and its evolution during winter can be used in cases where only snow depth data are available (Jonas et al., 2009). Such approaches generally give similar results to more complex physically based modeling of S_{WE} (Bavera et al., 2014).

For continuous S_{WE} measurements, a snow pillow or snow scales are commonly used (Figure 4.2). Such observations are based on measuring the mass of the snow lying on the installed pillow or sensor plate with a specific area (usually ~ 9 m^2). Another way of measuring S_{WE} in situ is a snow permittivity measurement of the snowpack (Egli et al., 2009). This measurement is based on measuring the dielectric constant of three main components of the snowpack: ice, water, and air. The measurement is carried out along a ribbon placed in the snowpack (usually placed parallel or diagonal to the

FIGURE 4.2   Climate station with snow scales in Modrava, Czech Republic (left), and snow pit close to the Furka Pass, Switzerland (right). *Photograph: M. Jenicek.*

ground). With this system, it is possible to measure the S_{WE}, snow density, and liquid water content.

For mapping S_{WE} over larger areas, promising results have been achieved using ground-based or airborne ground-penetrating radar (Lundberg et al., 2006; Sold et al., 2013). An alternative is the measurement of gamma radiation from the earth (Peck et al., 1971). The method is based on the attenuation of natural gamma radiation by the water accumulated in the snowpack with a detector fixed either a few meters from the ground surface or read from an aircraft. Gamma radiation is then measured over time with and without snow cover. A more recent technique is the highly accurate remote sensing of the snowpack on the ground and on glaciers using laser scanners (Egli et al., 2012; Grünewald et al., 2010; Prokop, 2008; Sold et al., 2013). Sensors can be placed in the aircraft (light detection and ranging, LIDAR) or can be used as a terrestrial device placed on the tripod (terrestrial laser scanning). Due to their construction and method of use, scanners have enabled the remote sensing of areas that are impossible to measure directly, for example, distant locations or steep slopes prone to ice or snow avalanches or rock falls (Egli et al., 2012; Grünewald et al., 2010). Present sensors are highly accurate (<1 cm both horizontally and vertically). However, not all sensors can be used for snow sensing due to the reflection properties of the snow cover (Prokop, 2008). Because of the high acquisition cost of laser scanners, a laser range finder is in some cases used to determine the snow depth (Hood and Hayashi, 2010). The disadvantages of these devices are mainly (1) poorer resolution of the snow depth grid; (2) generally only a small area can be covered from one location; and (3) measurements take more time than with a terrestrial laser scanner. Despite these disadvantages, the device is highly portable and a potentially improved alternative to expensive laser scans.

In recent years, much progress has been made using remote airborne and space-borne sensing methods to determine snow characteristics (Molotch and Margulis, 2008; Tait, 1998; Vander Jagt et al., 2013). These methods provide information about snow cover, generally for large areas, and basically on-line. Methods for remotely determining the S_{WE} based on the repeated monitoring of snow cover area (SCA) have been developed (Farinotti et al., 2010; Molotch and Margulis, 2008). The SCA can be measured by both optical and radar instruments. Radar is also suitable for observations of other parameters such as the amount of liquid water content (Storvold et al., 2006), and passive microwave sensors can also be used to derive S_{WE} (Molotch and Margulis, 2008; Tait, 1998).

4.2.3 Snow Redistribution, Metamorphism, and Ripening Process

4.2.3.1 Snow Redistribution by Wind

Snow transport, or the deposition of falling snow, is controlled by the wind field of the surface boundary layer, which is modified (in speed and direction) by local surface topography (Lehning et al., 2008). Variability in the wind field ultimately causes snow to be nonuniformly distributed or redistributed over the landscape. The spatial variation of S_{WE} is commonly strongly influenced by the effects of topography on snow redistribution and the timing of the snow disappearance is in turn controlled by this redistribution, in addition to differences in melt rates (Anderton et al., 2004) (Figure 4.3). The spatial redistribution of snow affects the basin-averaged snowmelt to different degrees; the

FIGURE 4.3   Spatial distribution of snow as a result of varying snow accumulation and melt and snow redistribution (Furkapass, Switzerland, June 2013). Snow on the south-facing side of the valley had already melted, whereas on the north-facing side, snow was still present in areas that had gained snow due to snow redistribution.

effects of snow redistribution were found to be very important for Reynolds Creek (Idaho, US; Luce et al., 1998), whereas only local effects were reported for an area in the German Alps (Bernhardt et al., 2012). For Reynolds Creek, the spatial redistribution of snow in drifts sustained streamflow later into the spring and summer compared to a more uniform snowpack.

4.2.3.2 Snow Metamorphism

The process of snow crystal metamorphism begins immediately after snowpack accumulation on the ground. In nature, most crystal changes are related to pressure and temperature changes in the snowpack (DeWalle and Rango, 2008). The three main types of snowpack metamorphism are briefly discussed below (see the Chapter by Arenson et al. (2014) for a discussion of metamorphism in the context of snow microstructure).

Equitemperature metamorphism is based on the migration of vapor from convex to concave ice surfaces because of higher vapor pressure on convex surfaces of the crystals than on concave surfaces. Snow crystals became more rounded due to this process. More rounded crystals increase snowpack stability due to their higher compaction ability and thus reduce the risk of avalanches.

Temperature-gradient metamorphism occurs as a result of changes in snowpack temperature and is the most important premelt densification process for the snowpack. The physical principle is based on the higher vapor pressure in a warmer snowpack than in a cooler snowpack. This gradient causes an upward migration of water vapor within the snowpack, from the warmer ground surface to the cooler snow surface. This process causes the formation of a new layer with large faceted crystals (depth hoar) that are poorly connected to each other, and thus, snowpack stability on the slope is lower and more prone to avalanches (e.g., Fierz et al., 2009). This depth hoar layer can be recognized only by snow pit analysis, which is infrequently carried out due to the time and effort required, and it thus represents a potential danger.

Melt-freeze metamorphism is typical during the spring period when air and snowpack temperatures increase due to the higher solar radiation. Snow on the surface of the snowpack tends to melt first, typically the small snow grains because they have lower melt temperatures compared to large snow grains. The melted snow can trickle into the colder middle portion of the snowpack where it refreezes and leads to the formation of a well-connected, large-grained snowpack. A specific amount of liquid water is generally present in the snowpack before snowmelt; typically from 2 to 5 percent with a maximum of 10 percent (e.g., Fierz et al., 2009; Techel and Pielmeier, 2011). A higher portion of liquid water can lead to the specific type of snow and water movement called slush flow that commonly also activates a top soil layer (Eckerstorfer and Christiansen, 2012). Melt-freeze metamorphism generally increases snowpack stability. On the other hand, the avalanche danger increases with an increase in the volume of liquid water stored in the snowpack.

For an in-depth discussion on snowpack stability and avalanches, the reader is referred to Schweizer et al. (2014) (this volume).

4.2.4 Snowpack Development

In regions where there is a seasonal snowpack, the snowpack develops over the course of the season starting with an accumulation phase, when precipitation falling as snow accumulates, increasing the S_{WE} until the melt period begins with an increased input of solar radiation. At first, the melt is characterized by a warming phase, when the average snowpack temperature increases more or less steadily until the snowpack becomes isothermal, that is, when the temperature over the snow profile reaches 0 °C. At this point, melting within the snowpack is possible and marks the beginning of the ripening phase. The melt within the snowpack is retained until the liquid, water-holding capacity is exceeded. Once exceeded, any further input of energy (from warmer air and/or soil and/or liquid precipitation) gained by the snowpack will then be used for phase changes from solid to liquid, producing more melt. The output phase begins when water output occurs at the base of the snowpack. It is important to note that a snowpack, as a porous medium, can store considerable amounts of liquid water. As a rule of thumb, snow can hold up to about 10 percent of the snowpack S_{WE} by capillary forces against gravitation.

4.2.4.1 Cold Content

The cold content of a snowpack is the total heat that is necessary to warm the snowpack to 0 °C over the entire vertical profile and can be expressed as the amount of liquid water (in (m)), which must be frozen in the snowpack to warm the snowpack to isothermal conditions (DeWalle and Rango, 2008). This variable of the energy state of a snowpack is important for snowmelt processes and their modeling and can be calculated as

$$C_C = \frac{\rho_s \cdot c_i \cdot d \cdot (273.16 - T_s)}{\rho_w \cdot L_f}, \tag{4.2}$$

where C_C (m) is the snowpack cold content, ρ_s is the snowpack density (kg m^{-3}), c_i is the specific heat of ice (J kg^{-1} K^{-1}), d is the snowpack depth (m), T_s is the snowpack temperature (K), ρ_w is the water density (kg m^{-3}), and L_f is the latent heat of fusion (J kg^{-1}).

Cold content is commonly used by hydrologists when simulating snowmelt runoff. Generally, it can be calculated for the whole profile or for specific snowpack layers, which is useful for modeling avalanches. The evolution of the snowpack and snowpack temperatures in specific layers can be used to calculate the cold content over the accumulation and melt periods (Figure 4.4).

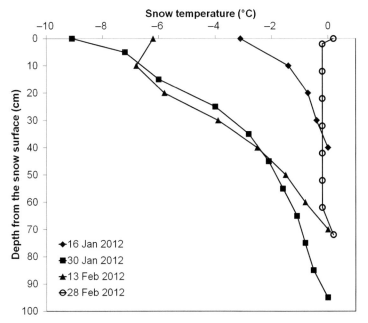

FIGURE 4.4   Snow-temperature evolution in a snow pit during the winter of 2011/2012; Krušné Mountains, Czech Republic. January 16 and January 30, 2012, represent typical winter conditions with low air temperatures. The third date, February 13, 2012, represents the beginning of spring warming (note the snow-temperature inversion in the top 10 cm of the snowpack). February 28, 2012, represents isothermal conditions in the snowpack during snowmelt. *Data from the Charles University in Prague, Faculty of Science.*

4.2.5   Snowmelt

4.2.5.1   Energy Balance

The development of the snowpack depends on the amount of energy available to the snowpack for melting snow that can be expressed in terms of the snowmelt energy balance. The energy balance of the snowpack represents the basic approach to snowmelt modeling and calculates all heat fluxes between the atmosphere, snow, and soil (Figure 4.5). The snowmelt energy balance can be defined by

$$Q_m = Q_{ns} + Q_{nl} + Q_h + Q_e + Q_p + Q_g + Q_i, \qquad (4.3)$$

where all Q refer to heat fluxes (W m^{-2}). Q_m is the total heat flux (positive or negative) available for snow melting, Q_{ns} is the heat flux due to short-wave radiation, Q_{nl} is the heat flux due to long-wave radiation, Q_h is the sensible heat flux, Q_e is the latent heat flux caused by water phase changes, Q_p is the heat supplied by precipitation, Q_g is the heat supplied by the ground, and Q_i is the change in the internal energy of the snowpack.

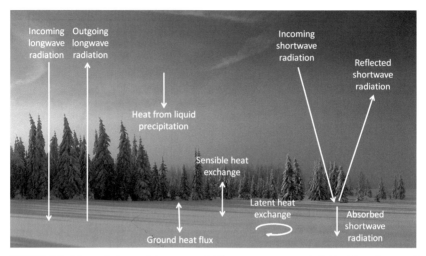

FIGURE 4.5   Energy balance of the snowpack. The arrows represent individual heat fluxes and interactions between the atmosphere, snow, and soil environments. *Photograph: M. Jenicek.*

The short-wave and long-wave radiation balance of the snowpack Q_{nr} can be computed according to

$$Q_{nr} = (1 - \alpha) \cdot S_i + (L_i - L_o), \qquad (4.4)$$

where Q_{nr} (W m^{-2}) is the total radiation (sum of Q_{ns} and Q_{nr}), α (−) is the albedo, L_i is the incoming long-wave radiation, L_o is the outgoing long-wave radiation, and S_i is the incoming short-wave radiation (all in W m^{-2}) (Singh and Singh, 2001). The amount of short-wave radiation absorbed by snow depends mainly on latitude, season (sun inclination), atmospheric diffusion, slope, aspect, and obstacles with shadow effects like forest cover (see also Figure 4.6).

Long-wave radiation (in the range of 6.8−100 μm) is both incoming (from the atmosphere, surrounding terrain, and vegetation) and outgoing from the snow surface (Singh and Singh, 2001). The radiation from the surface is largely absorbed by the atmosphere, largely by absorption from carbon dioxide and water vapor.

A convective sensible heat transfer between the air and the snow is driven by differences between air and snowpack temperatures. The amount and the direction of the heat flux is given by

$$Q_h = \rho_a \cdot C_{pa} \cdot (T_a - T_s / r_h), \qquad (4.5)$$

where ρ_a is the density of air (kg m^{-3}), C_{pa}, the specific heat of air at constant pressure (1,010 J kg^{-1} °C^{-1}), T_a and T_s (°C) are the air and snow surface temperatures, respectively, and r_h is a term to describe the resistance to a heat flux between the snow surface and the overlying air (s m^{-1}), which is a function of surface roughness and wind speed. For the sensible heat flux, large

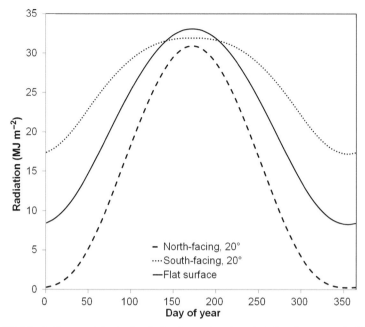

FIGURE 4.6   Daily potential solar irradiation for north and south-facing slopes at an inclination of 20° and a flat surface as a function of day of the year (i.e., Julian date) at a latitude of 45°N. The calculation considers clear sky conditions. *Based on the method presented in Dingman (2008).*

differences between the snow surface and air in spring and early summer cause an energy gain of the snowpack. However, the snowpack temperature is commonly higher than the air temperature during spring nights, and this situation represents a snowpack energy loss.

The latent heat transfer represents the transfer of water vapor between the atmosphere and the snowpack due to water phase changes, namely, evaporation and sublimation. The water vapor flux and its direction are determined by the vapor-pressure gradient and the intensity of the turbulence. Evaporation and the sublimation from the snow represent a snowpack energy loss; condensation and the deposition of the water vapor on the snow surface represent a snowpack energy gain. The latent heat flux is given by

$$Q_e = L_v \cdot (\rho_{va} - \rho_{vs})/r_e, \qquad (4.6)$$

where L_v is the latent heat of vaporization (2.47×10^6 J kg^{-1}), ρ_{va} and ρ_{vs} are the vapor densities of air and snow surfaces, respectively (kg m^{-3}), and r_e (s m^{-1}) is a term to describe the resistance to water vapor transport between the air and snow surface, which is a function of surface roughness and wind speed.

Liquid precipitation on snow, known as rain-on-snow events, represents an additional heat flux into the snowpack that is caused by the higher temperature of the precipitation compared to the temperature of the snow. The heat supplied to the snowpack equals the difference of snow energy before and after precipitation, once temperature equilibrium is reached. Equation (4.7) expresses the daily amount of transferred energy Q_p (kJ m^{-2} d^{-1}) depending on the precipitation P_r (mm d^{-1}) and the temperature difference (Singh and Singh, 2001),

$$Q_p = \frac{\rho_w \cdot C_p \cdot (T_r - T_s) \cdot P_r}{1000} + \rho_w \cdot L_f \cdot P_r, \qquad (4.7)$$

where ρ_w refers to the water density ($\sim$1000 kg m^{-3}), C_p refers to the specific heat capacity of water (4.20 kJ kg^{-1} °C^{-1}), T_r is the temperature of liquid precipitation, T_s is the temperature of the snowpack (both in (°C)), and L_f is the latent heat of fusion (334 kJ kg^{-1}). Equation (4.7) is also valid when the snowpack is below freezing, and the rain freezes within the snowpack ($L_f > 0$).

The total amount of heat from the precipitation is generally quite low in comparison with other heat sources. However, precipitation heat input becomes more important when shorter time steps (less than one day) are considered. For example, a daily precipitation of $P_r = 30$ mm, rain temperature of $T_r = 10$ °C, and snow temperature of $T_s = 0$ °C results in a total heat input of 1,260 kJ m^{-2} d^{-1}, or 14.6 W m^{-2}. If we consider the same amount of precipitation falling in 3 h, we obtain 116.7 W m^{-2}, which is a significant part of the energy balance for that 3 h period, and is comparable to the amount of global short-wave radiation during cloudy and rainy days.

Ground heat flux generally plays a minor role because of the small heat conductivity of the ground and, in the case of a higher snow depth, the lack of solar radiation reaching the soil. The ground heat causes the slow ripening of the snowpack during winter and may cause slow snowmelt. The ground heat flux can be expressed as

$$Q_g = K_g \cdot \partial T / \partial z, \qquad (4.8)$$

where K_g is the thermal conductivity of the soil (W m^{-1} °C^{-1}) and $\partial T / \partial z$ (°C m^{-1}) is the temperature gradient in the soil.

The internal energy of the snowpack depends on snow temperature and can be expressed as the sum of the internal energy of the three snowpack components: ice, water, and water vapor. The change in the internal energy can play a significant role for glaciers and can delay the onset of melt for snow.

Calculation of the energy balance of the snowpack is a physically based approach for modeling the snowmelt. Based on the availability of meteorological data, a wide range of approximations of the energy balance can be used. The advantage of methods based on the energy budget is their broad range of use under different climatic conditions. The energy budget approach enables snow accumulation, transformation, and melting processes to be physically represented. The main disadvantage of the energy budget method is the difficulty in obtaining the input data necessary for the parameterization, calibration, and validation of the model. It is thus difficult to use this approach

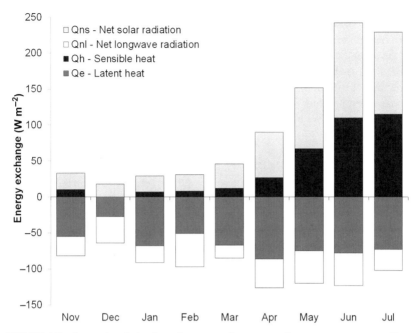

FIGURE 4.7   Seasonal variation in major snowpack energy exchange components at a Sierra Nevada alpine ridge site in 1986. *Data from Marks and Dozier, 1992.*

for calculating snow accumulation and snowmelt for ungauged catchments. Figure 4.7 shows an example of the seasonal variation in the major snowpack energy exchange components.

Energy fluxes can be converted into an amount of melted snow, as described in DeWalle and Rango (2008), where

$$M = Q_{\mathrm{m}}/\left(\rho_{\mathrm{w}} \cdot L_{\mathrm{f}} \cdot B\right), \tag{4.9}$$

where M (m d^{-1}) is the amount of melt water, Q_{m} (kJ m^{-2} d^{-1}) is the positive daily output from the energy budget, ρ_{w} is the water density ($\sim 1{,}000$ kg m^{-3}), L_{f} is the latent heat of fusion (334 kJ kg^{-1}), and B ($-$) is the thermal quality of snow, which is defined as the energy necessary to melt a certain mass of snow relative to the energy necessary to melt the same mass of ice at 0 °C.

4.2.5.2 Degree-Day Method

The calculation of the individual items of the snowpack energy balance is complex and needs a lot of data that are commonly not available. To overcome the lack of data, the so-called index methods are characteristically used. These methods calculate the snowmelt using some accessible and easily measurable variable that is generally related to the energy balance. The air temperature is commonly used because of its high correlation with snow and glacier melt, and availability (Braithwaite and Olesen, 1989). Other studies have also shown the importance of long-wave radiation and sensible heat flux on the energy budget. Both heat fluxes generally provide 75 percent of the entire energy balance, and the air temperature plays a key role in their calculation (Hock, 2003; Ohmura, 2001; Pohl et al., 2006). Approaches incorporating air temperature are referred to as temperature-index methods.

In the temperature-index method, the S_{WE} decreases according to the melt, M (mm d^{-1}) and is calculated as

$$M = m_f(T - T_T), \tag{4.10}$$

where m_f (mm °C^{-1} d^{-1}) is the melt or degree-day factor, T_T is the threshold temperature for snowmelt to start, and T is the mean air temperature (both in C). The values for T_T typically vary from 0 to 2 °C. The degree-day factor generally has values between 1 and 7 mm °C^{-1} d^{-1}, with lower values for forested areas (Seibert, 1999). The temperature index method has been developed for a daily time step, but has also been used for shorter or longer time intervals.

The initial storage of melt water within the snowpack can be conceptualized by a liquid water storage, which has to exceed a certain fraction of the S_{WE} before drainage from the snowpack occurs (Lindström et al., 1997). If temperatures decrease, this liquid water might refreeze, which can be simulated by an equation similar to Eqn (4.10), but with a refreezing coefficient C_{FR} (Eqn (4.11)). This coefficient typically has a value of 0.05. This means that the refreezing is 20 times less efficient than the melt, which reflects the difference that the melt occurs at the snow surface, whereas the refreezing occurs when the liquid water is distributed over the snowpack and mainly insulated against the cold.

$$R = C_{FR} \cdot m_f \cdot (T_T - T). \tag{4.11}$$

The temperature index method dates back to work by Finsterwalder and Schunk (1887) in a glaciological study in the Alps and is probably still the most widely used snowmelt method among hydrologists. Hock (2003) summarizes the advantages of the temperature index method as (1) the availability of air temperature data; (2) the relatively simple spatial distribution of the air temperature and its predictability; (3) the simplicity of the computational procedure; and (4) satisfactory results of the model despite its simplicity. Although there are clear advantages of this method, Beven (2001) formulated shortcomings of this method as (1) the accuracy of the method decreases with the increasing temporal resolution; (2) the intensity of the snowmelt has a large spatial variability depending on topographic conditions, mainly slope, aspect and land cover. This variability is very difficult to express using temperature index methods.

The shortcomings of the models based on the temperature-index method with a daily time step are apparent, mainly in cases where air temperature fluctuations are around the melting point. The mean daily air temperature indicates no snowmelt; however, the positive air temperature that occurs during the day could cause snowmelt (Hock, 2003). In mountain areas, it is necessary to consider the change in air temperature depending on the elevation; therefore, the basin is generally divided into several elevation zones (Essery, 2003).

Several enhanced and spatially distributed temperature-index models have been developed (e.g., Hock, 1999; Pellicciotti et al., 2005) that address some of the shortcomings of the classical degree−day model. In these approaches, the degree-day factor varies as a function of potential solar radiation, and the formulation is closer to the principles of the energy balance.

The melt factor m_f is the key parameter in Eqn (4.10) and is influenced by several factors (Table 4.3). Martinec (1975) derived an empirical relation between the melt factor and the snow density as

$$m_f = 11 \frac{\rho_s}{\rho_w},\qquad(4.12)$$

where m_f is the melt factor (mm $°C^{-1}$ d^{-1}), ρ_s is the snow density, and ρ_w is the water density (both (kg m^{-3})). This equation reflects the tendency for the melt factor to increase in the spring, together with an increase in snow density due to the ripening process. Kuusisto (1980) derived an empirical relation between the melt factor and the snow density separately for forest and open areas:

$$\text{forest: } m_f = 0.0104\rho_s - 0.70,\qquad(4.13)$$

$$\text{open areas: } m_f = 0.0196\rho_s - 2.39.\qquad(4.14)$$

Forests cause a decrease in the amount of direct solar radiation that reaches the surface and therefore the snowmelt in periods without precipitation. Different melt factors were derived by Federer et al. (1972) for the northwest of the USA. They experimentally derived the melt factor 4.5−7.5 mm $°C^{-1}$ d^{-1} for open areas, 2.7−4.5 mm $°C^{-1}$ d^{-1} for deciduous forests, and 1.4−2.7 mm $°C^{-1}$ d^{-1} for coniferous forests (approximate ratio 3:2:1). Kuusisto (1980) expressed variations in the melt factor dependent on the relative canopy cover of coniferous forests, C_{canopy} (−) with typical values of 0.1−0.7, as

$$m_f = 2.92 - 1.64C_{canopy}.\qquad(4.15)$$

4.3 GLACIERS AND GLACIAL MASS BALANCE

4.3.1 Glacier Mass Balance

The glacier mass balance provides information about the amount of water stored or released by a glacier within a given time period (Cogley et al., 2011). The glacier mass budget is characteristically evaluated over one hydrological year (October 1−September 30) and is reported as water equivalent per unit area per

TABLE 4.3 Factors Influencing Melt Factor

Factor	Cause	Influence on Melt Factor
Seasonal influence	Decrease of cold content and albedo, increase in short-wave radiation and snow density	Melt factor increases
Open area versus forest	Shading and wind protection	Lower melt factor and its spatial variability in the forest
Topography (slope, aspect)	Variability of short-wave radiation and wind exposure	Higher melt factor on south-facing slopes
Snow-cover area	Spatial snowmelt variability	Melt factor decreases in the basin with larger SCA
Snowpack pollution	Dust and other pollution cause lower albedo	Higher melt factor
Precipitation	Rainfall supplies sensible heat, clouds decrease solar radiation	Generally lower melt factor on rainy days due to lower radiation. But precipitation itself causes higher melt factor.
Snow versus ice	Glacial ice has lower albedo than snow	Higher melt factor in glaciated basins
Meteorological conditions for certain air temperature	Higher snowmelt with higher wind speed, higher radiation, or higher moisture for the same temperature	Higher melt factor

Based on DeWalle and Rango, 2008.

year $(m\, a^{-1})$. Mass balance b $(m\, a^{-1})$ at one point on the glacier surface is defined as the sum of accumulation c $(m\, a^{-1})$ and ablation a $(m\, a^{-1})$ as

$$b = c + a. \tag{4.16}$$

By integrating b over the glacier surface S (Cogley et al., 2011), the glacierwide, surface mass balance B_{sfc} corresponding to the mean thickness of the water equivalent added or removed can be calculated as

$$B_{\text{sfc}} = \frac{1}{S} \int_{S} b \, dS. \tag{4.17}$$

Basal ice melt and mass loss due to frontal ablation (calving, ice break-off) add to the total mass balance ΔM, but these components are comparatively small or almost negligible for most glaciers in mid- and low-latitude mountain ranges.

Mass balances can be determined using a variety of methods ranging from direct measurements on the glacier surface to techniques using remote sensing. The measurement of glacial mass changes at annual to seasonal time scales is mostly based on the glaciological method applying spatial interpolation and extrapolation of melt and accumulation measurements at a number of ablation stakes and snow pits (Kaser et al., 2003; Østrem and Stanley, 1969; Zemp et al., 2013). At time scales of a few years to several decades, the comparison of repeated information on glacier surface elevation provides accurate data on long-term volume changes of large and inaccessible glaciers (e.g., Bauder et al., 2007; Bolch et al., 2013; Paul and Haeberli, 2008). Recently, the application of satellite-based changes in the Earth's gravity field has become increasingly popular for evaluating glacier mass changes for entire mountain ranges (e.g., Gardner et al., 2013).

Most mid- and high-latitude mountain glaciers exhibit clearly separated accumulation and ablation periods (Figure 4.8(a)). Glacier storage changes are positive during the winter season with precipitation mainly falling as snow, and limited or absent melt. The ablation season is commonly concentrated over a few months with the highest air temperatures and strong solar radiation (Figure 4.8(c)). Glaciers thus store most of the winter precipitation, which is often strongly enhanced due to orographic uplift, and release it during a few summer months (Jansson et al., 2003; Stenborg, 1970). This behavior has a significant effect on the hydrological regime of glacierized drainage basins that are dominated by the components of snow and ice melt (Figure 4.8(e)).

4.3.2 The Glacial Drainage System

Glaciers are an important storage element in the hydrological cycle and are characterized by their water retention potential at different temporal and

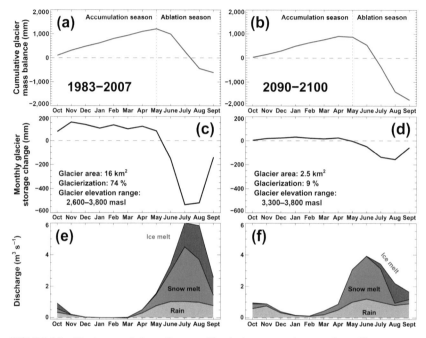

FIGURE 4.8 Glacier mass balance B_{sfc}, monthly glacier storage change and runoff components in the catchment of Findelengletscher, Switzerland, for the periods 1983–2007 and 2090–2100 based on glaciohydrological modeling *(Huss et al., 2014)*. (a,b) Cumulative glacier surface mass balance indicating limited mass losses in 1983–2007 and a strongly negative mass balance in 2090–2100. (c,d) Monthly glacier storage change relative to the water balance of the entire catchment. (e,f) Simulated components of catchment discharge (bare-ice melt, snowmelt, and rain minus evapotranspiration).

spatial scales (Jansson et al., 2003). Whereas water retention in the ice volume refers to periods of decades to centuries, englacial and subglacial water storage is highly important at shorter time scales of a few hours to days. Melt water derived from bare-ice surfaces is generally collected in supraglacial streams and is transported relatively rapidly to a moulin connecting surface water flow with the glacier bed (Figure 4.9). Melt generated in the accumulation area, however, infiltrates into porous firn and snow where it might be stored for days to months until it runs off, or where it refreezes again.

The characteristics and the development of the englacial and subglacial drainage systems determine water flow through the ice (see Fountain and Walder, 1998; for a review), as well as the hydraulic head within the glacier. Englacial water pressure exhibits a strong connection with surface ice flow over its close link to basal sliding (e.g., Iken and Bindschadler, 1986). Subglacial water flow occurs either in a rather inefficient, distributed, and interlinked system of small cavities or in individual channels at the ice—bedrock interface that promote an efficient drainage (e.g., Flowers, 2002; Röthlisberger

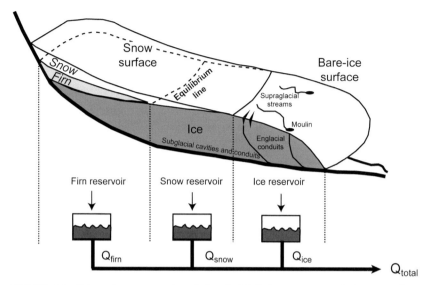

FIGURE 4.9 Schematic representation of the glacial drainage system and the approach to modeling it using linear reservoirs. *Redrawn after Hock and Jansson, 2005.*

and Lang, 1987, Figure 4.9). The subglacial drainage system develops throughout the melting season with sustained high melt water input and water pressures. Consequently, the water retention capacity of a glacier decreases with time (Hewitt et al., 2012; Hock and Hooke, 1993) and the diurnal amplitudes of runoff increase over the summer (Figure 4.10). Quantitative knowledge about the properties of the glacial drainage system and its state of evolution can be gained, for example, by dye tracing experiments (see, e.g., Finger et al., 2013; Werder and Funk, 2009).

4.3.3 Modeling Glacier Discharge

Over the last few decades, many different approaches to simulate glacier discharge have been developed. They focus on the spatiotemporal evolution of the englacial and subglacial drainage systems, the water retention capacity of glaciers, as well as on the long-term hydrological impacts of glaciers and their changes (Hock and Jansson, 2005; Hock, 2005). In the following, a brief overview of some modeling approaches is provided.

The physically based modeling of the subglacial drainage system has been attempted by a number of studies (e.g., Flowers, 2002; Hewitt et al., 2012; Schoof, 2010). By describing the processes of water cavity growth and drainage channel enlargement as functions of time-variable water pressure, water supply, the thermodynamic conditions and the ice flow, the development of the drainage system in time and space can be simulated, and the properties of subglacial water flow be inferred. Many hydrological studies however apply

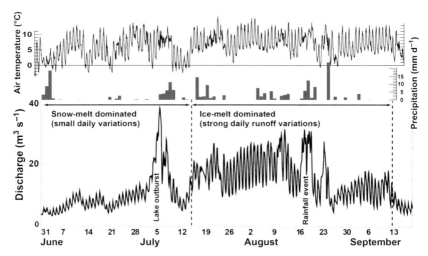

FIGURE 4.10   Runoff from the catchment of Gornergletscher (66% glacierized), Switzerland, between June and September 2004. Temperature and precipitation series of a station close to the glacier are shown for comparison while the runoff peak in July is driven by the outburst of the glacier-dammed lake, two runoff peaks in August can be attributed to rainfall events.

simpler, empirical approaches that estimate water retention and glacial runoff using linear reservoir models (Hock and Jansson, 2005). In this widely applied approach, discharge Q for each time step t is obtained from a linear relation with the volume V of hypothetical reservoirs for melt water originating from snow, firn, and bare-ice surfaces (Hock and Noetzli, 1997, Figure 4.9) as

$$Q(t) = kV(t). \tag{4.18}$$

The retention constant k is characterized by the water retention capacity of the respective reservoir and is determined by calibrating the model with field data (see, e.g., Hock, 1999).

The modeling of glacier hydrology necessarily involves the calculation of snow and ice melt as a function of the surface energy balance. Numerical methods for simulating glacier melt are similar to those for snowmelt but need to further take into account the significant albedo difference between snow and ice. Simulating runoff from glaciers over time scales of years to decades requires the description of all components of the high-alpine water balance, the snow and ice melting processes, the routing of runoff through the glacial and periglacial system. Semiempirical to process-based models have been proposed and applied with different spatial discretizations (see, e.g., Farinotti et al., 2012; Horton et al., 2006; Schaefli et al., 2007). Glaciohydrological studies over periods with strong glacier changes furthermore need to take into account the evolution of glacier area and ice thickness by physical ice-flow modeling or simplified approaches (e.g., Huss et al., 2010; Immerzeel et al., 2012).

4.4 HYDROLOGY OF SNOW- AND ICE-COVERED CATCHMENTS

4.4.1 Influence of Snow on Discharge

The most obvious feature of runoff in snow-dominated catchments is the large seasonality. Commonly, there is a more-or-less pronounced, spring flood, during which a large fraction of the annual runoff occurs during a few weeks. In highly glacierized catchments, the seasonality is even more pronounced with the runoff basically following the available energy for melt. The runoff regime of catchments without snow influence characteristically shows the opposite pattern, that is, low runoff during summer due to energy being available for evaporation. In mountainous areas, the importance of snow, and therefore also the seasonal differences, increases with elevation and the annual maximum streamflow is observed later for higher elevations (Figure 4.11(a)). In lowland catchments, a similar pattern can be observed when going from lower to higher latitudes (Figure 4.11(b)). For larger river catchments, this means that the runoff from snow and glacial melt in upstream mountainous headwater catchments can play a more important role in relative terms during summer months than on the annual average (Viviroli and Weingartner, 2004). This can be illustrated by discharge along the River Rhine (Figure 4.12): the runoff regime of the upper part is clearly dominated by snow and glacial melt, and the high summer runoff from this part of the catchment balances the low summer runoff contributions downstream.

During late summer, glacial melt contributes far more than one might expect from the proportion of the area covered by glaciers. For the month of August, for instance, glacier storage change contributes on average about 7 percent of the streamflow in River Rhine at the Andernach station, despite glaciers covering only 0.23 percent of the catchment area (Huss, 2011). The relative contribution can vary significantly and can reach >20 percent for warm summers, when both evaporation in the lower parts of the catchment is high and glacial melt is above the long-term average.

Another feature of runoff from snow-covered catchments, which is even more evident for glacierized catchments, as discussed below, is the observation of clear diurnal variations in runoff with maximum values typically in the late afternoon (Figures 4.10 and 4.13).

Despite this seasonal variation with the highest runoff in spring (on average), in smaller catchments, peak flows are generally caused by rainfall, rather than by snowmelt, because the latter is a comparatively slow process (Figures 4.13 and 4.14). Only in larger catchments do the highest flow peaks originate from snowmelt, because melting simultaneously occurs over large areas.

Rain-on-snow events are of special importance for flooding (Eiriksson et al., 2013; Floyd and Weiler, 2008; Juras et al., 2013). A common

(a)

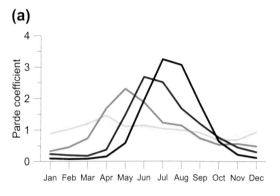

(b)

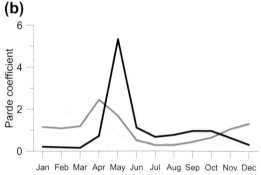

FIGURE 4.11   Seasonal runoff variation for catchments with varying snow dominance. The Pardé coefficient is the long-term monthly mean runoff divided by the annual mean. The runoff regime varies with elevation (see (a)), five catchments in Switzerland from low (light gray) to high (black) elevations: River Töss (mean elevation 650 m asl), River Goldach (830 m), River Sense (1068 m), River Minster (1351 m), River Dischmabach (2372 m), and River Rhone at Gletsch (2719 m). *(Data from Weingartner and Aschwanden, 1992)* and latitude (see (b)), three lowland catchments in Sweden from the South to the North: River Sege (light gray), River Fyris (dark gray), and River Vuoddasbäcken (black). *Data from SMHI, 1968–2012.*

misunderstanding is that the rain provides large amounts of energy to melt the snow. Energy balance calculations (see above) show that the energy added to the snowpack through rain water is small in comparison to the other energy fluxes on a daily or longer time scale. However, if rain falls on an already melting snowpack, the rainfall cannot be stored within the snowpack and it quickly percolates downward. The actual snowmelt during rain-on-snow events might be small (Mazurkiewicz et al., 2008), but an on-going snow-melt can provide wet antecedent conditions, which can lead to high flow peaks.

The above-mentioned facts cause the quick response of the catchment to the liquid precipitation (Figure 4.14). The rain-on-snow floods are character-ized by a steep rising limb of the hydrograph and high flood peaks in com-parison to spring floods caused by snowmelt without additional liquid

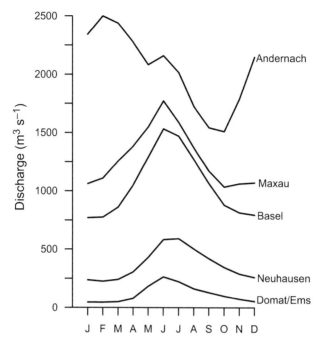

FIGURE 4.12   Seasonal runoff variation for different locations along River Rhine. *Data from Global Runoff Data Center, GRDC.*

precipitation. The total flood volume depends on the volume of liquid precipitation and melting snow. However, catastrophic flood events are most commonly caused by a combination of several unfavorable circumstances; a relatively high S_{WE} at both higher and lower elevations, air temperatures significantly above the melting point at all elevations, high liquid precipitation, windy conditions causing turbulent heat exchange, ripe snowpack (isothermal conditions), and, before the rain-on-snow event, antecedent soil moisture, frozen soils, and existing river ice.

One of the biggest rain-on-snow events and resulting floods in Central Europe on a large scale occurred in the Danube River basin during the spring of 2006. The flood mainly affected the area of the upper Danube River basin and some northern tributaries such as the Morava River, Váh River, and Tisza River. The main reasons for the flood were the unusually high quantities of snow at medium elevations, together with high air temperature and liquid precipitation. Although the liquid precipitation was important but still not extreme on a large scale (it ranged from 50 mm to around 100 mm depending on the subbasin location), the high snow storage caused extreme flood peaks with return periods up to 100 years for some subbasins, such as the Morava River and Danube River at Budapest (Wachter, 2007).

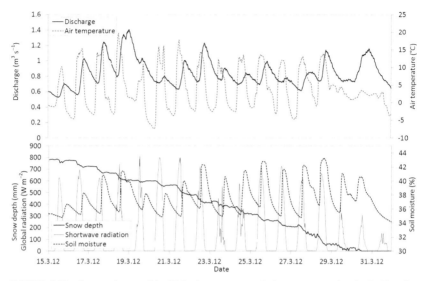

FIGURE 4.13 Snowmelt caused by high air temperature during sunny weather in the Byst̆rice River basin, Krušné Mts, Czech Republic. Diurnal variation of the air temperature, snow depth, soil moisture, and discharge is typical for these kinds of situations. The soil was not frozen during this period. Typical diurnal variations of the snow depth are caused by freezing conditions during night time, which temporarily stops the snowmelt. *Data from the Charles University in Prague, Faculty of Science.*

The occurrence of rain-on-snow events will probably change in the future due to the increase in air temperature and precipitation during winter. Surfleet and Tullos (2013) assessed historical and predicted runoff data for the Santiam River basin in Oregon and concluded that the peak flows associated with rain-on-snow events will decrease at lower elevations, whereas there will be a significant rise in the frequency of peak daily flows <5−10 year return period at medium to high altitudes. However, the overall frequency of rain-on-snow events was expected to decrease in the studied basin due to generally smaller amounts of snow.

4.4.2 Influence of Glaciers on Discharge

Glaciers are important elements of the hydrological cycle in alpine environments and are known to strongly influence the runoff regime of glacierized drainage basins at local to regional and larger scales, even with only a minor degree of glacier coverage (Hock, 2005; Kaser et al., 2010). The significant influence of glaciers on the hydrological regime of mountainous catchments has been well documented by a number of studies (e.g., Barnett et al., 2005; Fountain and Tangborn, 1985; Kuhn and Batlogg, 1998; Meier and Tangborn, 1961). Hock (2005) identified five specific characteristics of glacierized basins: (1) During years with glacier mass gain, specific runoff from the

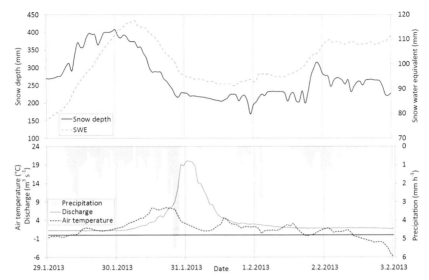

FIGURE 4.14   Rain-on-snow event from January 29, 2013, to February 3, 2013, at the Modrava climate station, Šumava Mountains, Czech Republic. The different timing of peaks for the displayed variables (snow depth, S_{WE}, air temperature, discharge) can be shown to indicate the "sponge" effect of the snowpack with metamorphic changes of ice crystals and possible accumulation of liquid water in the snowpack (a decrease of snow depth with increasing S_{WE}). *Data from the Charles University in Prague, Faculty of Science.*

catchment is reduced while in years with mass loss, additional water from long-term glacial storage contributes to discharge; (2) The runoff regime in basins with a glacierization of more than a few percent shows strong seasonal variations with small discharge over 6–8 months and peak runoff during the melting season; (3) Runoff is characterized by distinct diurnal variations with large amplitudes (Figure 4.10); (4) Basins with an intermediate glacierization exhibit reduced interannual runoff fluctuations as melt and precipitation compensate for each other (Farinotti et al., 2012; Lang, 1986); (5) Glacier runoff is positively correlated with air temperature, whereas it is negatively correlated with precipitation.

4.4.2.1 *Quantifying Glacier Contribution to Runoff*

Many recent studies have addressed the contribution of past, present, and future glacier melt to runoff (e.g., Huss, 2011; Immerzeel et al., 2010; Kaser et al., 2010; Koboltschnig and Schöner, 2011; Weber et al., 2010). The percentage of glacial water in streamflow is considered a good proxy for future runoff changes related to the potential disintegration of glaciers with ongoing atmospheric warming. As summarized by Radić and Hock (2014), considerable disagreement occurs in the literature about the definition of glacier contribution to runoff leading to major difficulties in comparing published estimates.

Among the six different concepts for quantifying glacier contribution detected by Radić and Hock (2014), two major groups can be separated: Some studies consider only melt water originating from the ablation area (i.e., bare-ice melt) as glacier runoff and explicitly exclude snowmelt over the glacier (e.g., Jost et al., 2012; Koboltschnig and Schöner, 2011; Racoviteanu et al., 2013; Stahl et al., 2008; Weber et al., 2010). Other papers define glacier contribution based on a water balance approach, and thus take into account all water derived from glacierized surfaces (Huss, 2011; Kaser et al., 2010; Lambrecht and Mayer, 2009; Neal et al., 2010). Evaluations of glacier contribution to runoff thus need to detail the applied approach in order to provide useful values for water resource management.

A straightforward and independent approach to determine the importance of snow and ice melt water in stream-flow runoff is the measurement of stable chemical isotopes that allow the direct quantification of the fractional contribution of the different runoff components in the analyzed sample (e.g., Collins, 1977; Finger et al., 2013; Racoviteanu et al., 2013; Taylor et al., 2001). This method is purely observational and does not require hydrological modeling or water-balance calculations for evaluating the melt water contribution. An immediate comparison to the above approaches is, however, difficult as chemical analysis only refers to particle concentrations, whereas daily to seasonal runoff variations also propagate by differences in the hydraulic head, for example, through lakes. A reliable determination of glacier contribution by chemical analysis is thus only possible in the proglacial stream, but not in large-scale catchments.

4.4.2.2 Glacier Runoff and Climate Change

Strong changes in the storage and runoff characteristic of glacierized drainage basins are expected in response to climate change over the twenty-first century (Barnett et al., 2005; Braun et al., 2000; Hock et al., 2005; Huss et al., 2008; Stahl et al., 2008). Due to glacier recession, a significant decrease in glacier storage capacity is probable—glaciers store smaller amounts of water in winter and thus release less during the melting season (Figure 4.8(d)). This reduction in the natural and beneficial capability of glaciers to smooth out runoff variability by their anticyclic behavior will be crucial in terms of water resource management (e.g., Chen and Ohmura, 1990; Schaefli et al., 2007), as well as regarding the effects of extreme years such as the European heat wave of 2003 (Zappa and Kan, 2007). Far-reaching impacts due to glacier retreat in many of the Earth's mountain ranges are thus expected to be most severe in regions with summer-dry conditions (Hagg and Braun, 2005; Immerzeel et al., 2010; Juen et al., 2007; Sorg et al., 2012). The example of Findelengletscher, Switzerland, shows that drainage basin runoff during the summer months might decrease by ≥ 50 percent due to a lack of snow and ice melt compared to the present situation (Figure 4.8).

The recent decline of glacier ice volume observed in all regions around the globe (Gardner et al., 2013) is a major concern regarding sea-level rise (see also the Chapter by Allison et al. (2014)). Until the end of the twenty-first century, melting glaciers are expected to raise global sea level by 0.1–0.2 m (Marzeion et al., 2012; Radić et al., 2014). From a hydrological perspective, it is, however, important to mention that for several reasons not all glacial melt water will immediately contribute to sea-level rise (Haeberli and Linsbauer, 2013): (1) In polar regions, glacier ice is partly grounded below sea level and is thus not a net contribution when removed. This factor accounts for a 5–10 percent reduction in sea-level rise estimates from glaciers at a global scale. (2) In a deglaciating landscape the formation of new proglacial lakes can be observed (Haeberli et al., 2012). These lake basins store some of the glacial melt water, but might also represent a serious hazard potential (Clague and O'Connor, 2014; Haeberli and Whiteman, 2014). (3) Some glacierized regions (mostly in High Mountain Asia) drain into endorheic basins that have no direct connection to the ocean. Furthermore, glacial melt water might also contribute to groundwater recharge. Overall, these factors lead to a small but systematic decrease in the sea-level rise contribution from melting glacier ice and thus need to be accounted for.

4.4.3 River Ice

River ice (Figure 4.15) is an important feature of many rivers, which can contribute to extreme events. The cooling of water during fall and winter differs for flowing river water compared to that for lakes. While in lakes after an initial cooling to about 4 °C, a thermal stratification develops with only the upper water layer cooling down further, generally the entire water column cools down at the same rate in rivers due to the mixing of the flowing water. The cooling is mainly caused by the negative, long-wave radiation budget, the convective heat transfer to the atmosphere, and the cooling effect of evaporation. Precipitation can have some cooling effect, especially in the case of

FIGURE 4.15   Water, snow, and ice in the Roseg Valley, Pontresina, Switzerland.

snow when energy is consumed for melting. Groundwater is characteristically the main contribution to river flow and can be a significant source of energy due to the higher temperatures. When river water cools down to 0 °C, various types of ice can form (Prowse, 2005). These include border ice along the river banks and frazil ice. The latter is made up of loose ice crystals that start forming when the water is supercooled (<-0.05 °C), and both grow and aggregate into larger ice pieces. With continued cooling eventually a complete ice cover can develop. This ice cover can breakup depending on factors such as ice thickness and strength, river geometry, flow velocity, and water levels. In cold climates, ice breakup occurs usually as one major spring event. The importance of this type of event in cold climates is also indicated by many historical records, which have been used as climatological indicators, especially for times before measurements started (Kajander, 1993; Magnuson et al., 2000; Rannie, 1983). In more temperate climates, there might be several freeze—breakup cycles during one winter season. During ice breakup, high water levels can be caused by ice dams, and severe flooding might be due to failure of such ice dams (Prowse, 2005). Although river ice can cause high flows, the more surprising observation is also that runoff in winter and spring can be affected by temporary water storage within rivers driven by hydraulic friction in situations with river ice (Prowse and Carter, 2002). Winter runoff thus decreases, and spring runoff is enhanced.

4.4.4 Permafrost and Seasonally Frozen Soils

The freezing of soil, both permanently and seasonally, can have significant impacts on runoff generation processes (Hayashi, 2013). Soil freezing starts when the temperatures at the soil surface decrease below zero degrees, which causes a temperature gradient toward the surface from the so-called freezing front. The soil energy balance is highly dependent on the presence of a snow cover. Due to the low thermal conductivity of snow, snow has an insulating effect, and a snow cover of 20—30 cm is enough to inhibit further freezing from above (Hirota et al., 2006), whereas freezing from below in permafrost regions occurs independently of snow cover. In ice-rich frozen soils, conductive heat transfer is dominant because the ice prevents advective transfer by liquid water or vapor, which would otherwise be more important. The largest effect of soil frost on runoff generation is the changed infiltration capacity. Detailed modeling of soil frost and its effects on hydraulic conductivity is challenging, and approaches such as using two flow domains (slow and quick flow) have to be used to better predict water fluxes in frozen soils (Stähli et al., 1996, 1999). A general observation, however, is that the soil water content at the time of freezing largely influences the resulting infiltration capacity. In the case of wet soils at the beginning of freezing, the frozen soil can form an almost impermeable layer. This can generate a complex interaction between climate, snow, and runoff. For instance, in boreal catchments, higher runoff peaks might be observed after snow-poor winters, compared to after snow-rich winters, which might seem counterintuitive at first. The explanation is that during winters with less snow, there is less insulation and deeper frozen soil will develop, which in turn would lead to restricted infiltration and thus higher runoff peaks. Because snow depth and soil frost are generally inversely correlated, the presence of soil frost might seldom lead to extreme floods (Lindström et al., 2002). However, the interplay between snow accumulation and soil frost is complex, and the conditions during early winter are most important (Bayard et al., 2005; Stähli et al., 2001). In many areas, especially within forests, soil frost does not generally fully stop infiltration, but rates can be significantly reduced. In wetlands, on the other hand, soil frost can generate an impermeable layer and cause overland flow (Laudon et al., 2007). On a larger scale, the effect of other factors commonly masks a clear relationship between snow accumulation, soil frost, and runoff (Shanley and Chalmers, 1999).

Permafrost can have even larger effects on hydrological processes. Permafrost is subsurface ground material (soil, sediment, bedrock, etc.) that has a temperature below $0\,^{\circ}C$ for consecutive years and exists in about $20-25$ percent of the Earth's land surface (Dankers, 2008). In areas with permafrost, the upper parts of the ground generally thaw and refreeze seasonally. This part of the ground is termed the "active layer" and begins to form during spring snowmelt and reaches its greatest depth in late summer/early fall. This means that the ground portion that is available for infiltration and liquid water storage increases during the summer and then again decreases during winter. Ice-rich permafrost can decrease the hydraulic conductivity of the ground dramatically, so that an almost impermeable layer can be formed. In this case, groundwater (or saturated conditions) can be found both seasonally in the active layer above the perennially frozen part of the ground and below this zone. With the impermeable layer, permafrost can result in increased soil moisture in the active layer, and might even result in perched lakes. Where water infiltrates into the deeper groundwater in areas with higher elevations, confined aquifers with artesian wells can be formed in neighboring areas of lower elevation (Hinzman et al., 2005). When permafrost thaws due to climate change, the active layer becomes deeper, and the streamflow dynamics might change. For a catchment in Northern Sweden, observed changes in streamflow dynamics quantified by recession coefficients could be linked to permafrost thawing rates (Lyon et al., 2009). Increasing temperatures also cause permafrost degradation and surface subsidence (Streletskyi, 2014). One important consequence of melting permafrost in mountainous areas is an increased risk for slope failures due to a decrease in slope stability (Deline and Gruber, 2014).

4.5 CONCLUDING REMARKS

In this chapter, we discussed various aspects of snow and ice in the water cycle. In cold climates, the freezing—thawing threshold is crucial for many processes influencing the hydrology at different scales. Due to this temperature threshold, snow- and glacier-dominated catchments might react very sensitively to climate variations, especially for those parts of the year where temperatures today are around 0 °C. In general, it is expected that higher temperatures will shorten the snow season, and reduce glacier storage capacity, leading to changes in seasonal runoff. Although the exact impacts on the length of the snow cover period will vary, one can estimate that with a mean temperature increase of 3 °C, the snow-cover period can be easily shortened by about a month. Besides direct impacts on the hydrology, this also has indirect impacts through possible changes in the vegetation, because the extension of the part of the year without snow cover implies a prolongation of the growing season. As discussed above, with a decreased snow cover in winter, higher temperatures might paradoxically lead to an increased cooling of soils due to the missing insulation of the snow. For permafrost areas, the deeper active layer might result in increased infiltration rates and storage and thus reduced overland flow. A good understanding of the role of snow and ice in the hydrology of cold region catchments is crucial for addressing natural hazards and water management in these sensitive systems, both today and in the future. Due to the characteristically difficult conditions, field observations are still limiting our knowledge of the hydrological processes in these regions, and further studies combining field observations and modeling are needed for an improved understanding.

REFERENCES

Allison, I., Colgan, W., King, M., Paul, F., 2014. Ice sheets, glaciers and sea-level rise. In: Haeberli, W., Whiteman, C. (Eds.), Snow and Ice-Related Hazard, Risks and Disasters. Elsevier, pp. 713−747.

Anderton, S.P., White, S.M., Alvera, B., 2004. Evaluation of spatial variability in snow water equivalent for a high mountain catchment. Hydrol. Process. 18, 435−453.

Arenson, L.U., Colgan, W., Marshall, H.P., 2014. Physical, mechanical and thermal properties of snow, ice and permafrost. In: Haeberli, W., Whiteman, C. (Eds.), Snow and Ice-Related Hazard, Risks and Disasters. Elsevier, pp. 35−75.

Barnett, T.P., Adam, J.C., Lettenmaier, D.P., 2005. Potential impacts of a warming climate on water availability in snow-dominated regions. Nature 438, 303−309.

Bauder, A., Funk, M., Huss, M., 2007. Ice-volume changes of selected glaciers in the Swiss Alps since the end of the 19th century. Ann. Glaciol. 46, 145−149.

Bavera, D., Bavay, M., Jonas, T., Lehning, M., De Michele, C., 2014. A comparison between two statistical and a physically-based model in snow water equivalent mapping. Adv. Water Resour. 63, 167−178.

Bayard, D., Stähli, M., Parriaux, A., Flühler, H., 2005. The influence of seasonally frozen soil on the snowmelt runoff at two Alpine sites in southern Switzerland. J. Hydrol. 309, 66−84.

Bernhardt, M., Schulz, K., Liston, G.E., Zängl, G., 2012. The influence of lateral snow redistribution processes on snow melt and sublimation in alpine regions. J. Hydrol. 424-425, 196−206.

Beven, K.J., 2001. Rainfall-runoff Modelling: The Primer. Wiley, Chichester.

Bolch, T., Sandberg Sørensen, L., Simonsen, S.B., Mölg, N., Machguth, H., Rastner, P., Paul, F., 2013. Mass loss of Greenland's glaciers and ice caps 2003−2008 revealed from ICESat laser altimetry data. Geophys. Res. Lett. 40, 875−881.

Braithwaite, R.J., Olesen, O.B., 1989. Calculation of glacier ablation from air-temperature, west Greenland. In: Oerlemans, J. (Ed.), Glacier Fluctuations and Climatic Change, Glaciology and Quaternary Geology. Kluwer Academic Publishers, Dordrecht, pp. 219−233.

Braun, L.N., Weber, M., Schulz, M., 2000. Consequences of climate change for runoff from Alpine regions. Ann. Glaciol. 31, 19−25.

Brown, R.D., Goodison, B.E., 2005. Snow cover. In: Anderson, M.G. (Ed.), Encyclopedia of Hydrological Sciences. John Wiley and Sons, Chichester, UK, pp. 1−12.

Chen, J., Ohmura, A., 1990. Estimation of Alpine glacier water resources and their change since the 1870s. In: Hydrology in Mountainous Regions. I—Hydrological Measurements; the Water Cycle (Proceedings of Two Lausanne Symposia, August 1990). IAHS Publ, pp. 127−136. No. 193.

Clague, J.J., O'Connor, J.E., 2014. Lake outbursts. In: Haeberli, W., Whiteman, C. (Eds.), Snow and Ice Related Hazard, Risks and Disasters. Elsevier, pp. 487−519.

Cogley, J.G., Hock, R., Rasmussen, L.A., Arendt, A.A., Bauder, A., Braithwaite, R.J., Jansson, P., Kaser, G., Nicholson, L., Zemp, M., 2011. Glossary of Glacier Mass Balance and Related Terms, IHP-VII Technical Documents in Hydrology No. 86. IACS Contribution No. 2, Paris.

Collins, D.N., 1977. Hydrology of an Alpine glacier as indicated by the chemical composition of meltwater. Z. für Gletscherkd. Glazialgeol. 13, 219−238.

Dankers, R., 2008. Arctic and snow hydrology. In: Bierkens, M.F.P., Dolman, A.J., Troch, P.A. (Eds.), Climate and the Hydrological Cycle. International Association of Hydrological Sciences, pp. 137−156.

Davison, B., Pietroniro, A., 2005. Hydrology of snow covered basins. In: Anderson, M.G. (Ed.), Encyclopedia of Hydrological Sciences. John Wiley and Sons, Chichester, UK, pp. 1−19.

Deline, P., Gruber, S., et al., 2014. Ice loss and slope stability. In: Haeberli, W., Whiteman, C. (Eds.), Snow and Ice Related Hazard, Risks and Disasters. Elsevier, pp. 521−561.

DeWalle, D.R., Rango, A., 2008. Principles of Snow Hydrology. Cambridge University Press, Cambridge.

Dingman, S.L., 2008. Physical Hydrology, Second ed. Waveland Pr Inc.

Eckerstorfer, M., Christiansen, H.H., 2012. Meteorology, topography and snowpack conditions causing two extreme mid-winter slush and wet slab avalanche periods in high Arctic maritime Svalbard. Permafr. Periglac. Process. 23, 15−25.

Egli, L., Jonas, T., Grünewald, T., Schirmer, M., Burlando, P., 2012. Dynamics of snow ablation in a small Alpine catchment observed by repeated terrestrial laser scans. Hydrol. Process. 26, 1574−1585.

Egli, L., Jonas, T., Meister, R., 2009. Comparison of different automatic methods for estimating snow water equivalent. Cold Reg. Sci. Technol. 57, 107−115.

Eiriksson, D., Whitson, M., Luce, C.H., Marshall, H.P., Bradford, J., Benner, S.G., Black, T., Hetrick, H., McNamara, J.P., 2013. An evaluation of the hydrologic relevance of lateral flow in snow at hillslope and catchment scales. Hydrol. Process. 27, 640−654.

Essery, R., 2003. Aggregated and distributed modelling of snow cover for a high-latitude basin. Glob. Planet. Change 38, 115−120.

Farinotti, D., Magnusson, J., Huss, M., Bauder, A., 2010. Snow accumulation distribution inferred from time-lapse photography and simple modelling. Hydrol. Process. 24, 2087−2097.

Farinotti, D., Usselmann, S., Huss, M., Bauder, A., Funk, M., 2012. Runoff evolution in the Swiss Alps: projections for selected high-alpine catchments based on ENSEMBLES scenarios. Hydrol. Process. 26, 1909−1924.

Federer, C.A., Pierce, R.S., Hornbeck, J.W., 1972. Snow management seems unlikely in the Northeast. In: Proceedings Symposium on Watersheds in Transition. American Water Resources Association, pp. 212−219.

Fierz, C., Armstrong, R.L., Durand, Y., Etchevers, P., Greene, E., McClung, D.M., Nishimura, K., Satyawali, P.K., Sokratov, S.A., 2009. The International Classification for Seasonal Snow on the Ground. IHP-VII Technical Documents in Hydrology N°83. IACS Contribution N°1, UNESCO-IHP, Paris.

Finger, D., Hugentobler, A., Huss, M., Voinesco, A., Wernli, H., Fischer, D., Weber, E., Jeannin, P.-Y., Kauzlaric, M., Wirz, A., Vennemann, T., Hüsler, F., Schädler, B., Weingartner, R., 2013. Identification of glacial meltwater runoff in a karstic environment and its implication for present and future water availability. Hydrol. Earth Syst. Sci. 17, 3261−3277.

Finsterwalder, S., Schunk, H., 1887. Der Suldenferner. Z. des Dtsch. Österreichischen Alpenvereins 18, 72−89.

Flowers, G.E., 2002. A multicomponent coupled model of glacier hydrology 1. Theory and synthetic examples. J. Geophys. Res. 107, 2287.

Floyd, W., Weiler, M., 2008. Measuring snow accumulation and ablation dynamics during rain-on-snow events: innovative measurement techniques. Hydrol. Process. 22, 4805−4812.

Fountain, A.G., Tangborn, W.V., 1985. The effect of glaciers on streamflow variations. Water Resour. Res. 21, 579−586.

Fountain, A.G., Walder, S., 1998. Water flow through temperate glaciers. Rev. Geophys, 299−328.

Gardner, A.S., Moholdt, G., Cogley, J.G., Wouters, B., Arendt, A.A., Wahr, J., Berthier, E., Hock, R., Pfeffer, W.T., Kaser, G., Ligtenberg, S.R.M., Bolch, T., Sharp, M.J., Hagen, J.O., van den Broeke, M.R., Paul, F., 2013. A reconciled estimate of glacier contributions to sea level rise: 2003 to 2009. Science (80) 340, 852−857.

Garvelmann, J., Pohl, S., Weiler, M., 2013. From observation to the quantification of snow processes with a time-lapse camera network. Hydrol. Earth Syst. Sci. 17, 1415−1429.

Grünewald, T., Schirmer, M., Mott, R., Lehning, M., 2010. Spatial and temporal variability of snow depth and ablation rates in a small mountain catchment. Cryosphere 4, 215−225.

Haeberli, W., Linsbauer, A., 2013. Global glacier volumes and sea level—small but systematic effects of ice below the surface of the ocean and of new local lakes on land. Cryosphere 7, 817−821.

Haeberli, W., Whiteman, C., 2014. In: Haeberli, W., Whiteman, C. (Eds.), Snow and Ice-Related Hazards, Risks and Disasters - A general Framework. Elsevier, pp. 1−34.

Haeberli, W., Schleiss, A., Linsbauer, A., Künzler, M., Bütler, M., 2012. Gletscherschwund und neue Seen in den Schweizer Alpen. Wasser Energ. Luft 104, 93−102.

Hagg, W., Braun, L., 2005. The influence of glacier retreat on water yield from high mountain areas: comparison of Alps and Central Asia. In: de Jong, C., Collins, D., Ranzi, R. (Eds.), Climate and Hydrology of Mountain Areas. John Wiley & Sons, Ltd, Chichester, UK, pp. 263−275.

Hayashi, M., 2013. The cold vadose zone: hydrological and ecological significance of frozen-soil processes. Vadose Zo. J., 1−8.

Hendriks, M.R., 2010. Introduction to Physical Hydrology. Oxford University Press.

Hewitt, I.J., Schoof, C., Werder, M.A., 2012. Flotation and free surface flow in a model for subglacial drainage. Part 2. Channel flow. J. Fluid Mech. 702, 157−187.

Hinzman, L.D., Kane, D.L., Woo, M., 2005. Permafrost hydrology. In: Anderson, M.G. (Ed.), Encyclopedia of Hydrological Sciences. John Wiley and Sons, Chichester, UK, pp. 1−15.

Hirota, T., Iwata, Y., Hayashi, M., Suzuki, S., Hamasaki, T., Sameshima, R., Takayabu, I., 2006. Decreasing soil−frost depth and its relation to climate change in Tokachi, Hokkaido, Japan. J. Meteorol. Soc. Jpn. 84, 821−833.

Hock, R., 1999. A distributed temperature-index ice- and snowmelt model including potential direct solar radiation. J. Glaciol. 45, 101−111.

Hock, R., 2003. Temperature index melt modelling in mountain areas. J. Hydrol. 282, 104−115.

Hock, R., 2005. Glacier melt: a review of processes and their modelling. Prog. Phys. Geogr. 29, 362−391.

Hock, R., Hooke, R.L., 1993. Evolution of the internal drainage system in the lower part of the ablation area of Storglaciären. Sweden. Geol. Soc. Am. Bull. 105, 537−546.

Hock, R., Jansson, P., 2005. Modeling glacier hydrology. In: Anderson, M.G. (Ed.), Encyclopedia of Hydrological Sciences. John Wiley and Sons, Chichester, UK, pp. 1−9.

Hock, R., Noetzli, C., 1997. Areal melt and discharge modelling of Storglaciaren. Sweden. Ann. Glaciol. 24, 211−216.

Hock, R., Jansson, P., Braun, L., 2005. Modelling the response of mountain glacier discharge to climate warming. Glob. Chang. Mt. Reg, 243−252.

Holko, L., Škvarenina, J., Kostka, Z., Frič, M., Staroň, J., 2009. Impact of spruce forest on rainfall interception and seasonal snow cover evolution in the western Tatra mountains, Slovakia. Biologia (Bratisl) 64, 594−599.

Hood, J.L., Hayashi, M., 2010. Assessing the application of a laser rangefinder for determining snow depth in inaccessible alpine terrain. Hydrol. Earth Syst. Sci. 14, 901−910.

Horton, P., Schaefli, B., Mezghani, A., Hingray, B., Musy, A., 2006. Assessment of climate-change impacts on alpine discharge regimes with climate model uncertainty. Hydrol. Process. 20, 2091−2109.

Huss, M., 2011. Present and future contribution of glacier storage change to runoff from macro-scale drainage basins in Europe. Water Resour. Res. 47, 1−14.

Huss, M., Farinotti, D., Bauder, A., Funk, M., 2008. Modelling runoff from highly glacierized alpine drainage basins in a changing climate. Hydrol. Process. 22, 3888−3902.

Huss, M., Jouvet, G., Farinotti, D., Bauder, A., 2010. Future high-mountain hydrology: a new parameterization of glacier retreat. Hydrol. Earth Syst. Sci. 14, 815−829.

Huss, M., Zemp, M., Joerg, P.C., Salzmann, N., 2014. High uncertainty in 21st century runoff projections from glacierized basins. J. Hydrol. 510, 35−48.

Iken, A., Bindschadler, R.A., 1986. Combined measurements of subglacial water pressure and surface velocity of Findelengletscher, Switzerland: conclusions about drainage system and sliding mechanism. J. Glaciol. 32, 101−119.

Immerzeel, W.W., Shrestha, A.B., Bierkens, M.F.P., Beek, L.P.H., Konz, M., 2012. Hydrological response to climate change in a glacierized catchment in the Himalayas. Clim. Change 110, 721−736.

Immerzeel, W.W., van Beek, L.P.H., Bierkens, M.F.P., 2010. Climate change will affect the Asian water towers. Science (80) 328, 1382−1385.

Jansson, P., Hock, R., Schneider, T., 2003. The concept of glacier storage: a review. J. Hydrol. 282, 116−129.

Jeníček, M., Beitlerová, H., Hasa, M., Kučerová, D., Pevná, H., Podzimek, S., 2012. Modeling snow accumulation and snowmelt runoff—present approaches and results. Acta Univ. Carolinae, Geogr. 47, 15−24.

Jonas, T., Marty, C., Magnusson, J., 2009. Estimating the snow water equivalent from snow depth measurements in the Swiss Alps. J. Hydrol. 378, 161−167.

Jost, G., Dan Moore, R., Weiler, M., Gluns, D.R., Alila, Y., 2009. Use of distributed snow measurements to test and improve a snowmelt model for predicting the effect of forest clearcutting. J. Hydrol. 376, 94−106.

Jost, G., Moore, R.D., Menounos, B., Wheate, R., 2012. Quantifying the contribution of glacier runoff to streamflow in the upper Columbia River Basin, Canada. Hydrol. Earth Syst. Sci. 16, 849−860.

Jost, G., Weiler, M., Gluns, D.R., Alila, Y., 2007. The influence of forest and topography on snow accumulation and melt at the watershed-scale. J. Hydrol. 347, 101−115.

Juen, I., Kaser, G., Georges, C., 2007. Modelling observed and future runoff from a glacierized tropical catchment (Cordillera Blanca, Perú). Glob. Planet. Change 59, 37−48.

Juras, R., Pavlásek, J., Děd, P., Tomášek, V., Máca, P., 2013. A portable simulator for investigating rain-on-snow events. Z. für Geomorphol. 57 (Suppl), 73−89.

Kajander, J., 1993. Methodological aspects on river cryophenology exemplified by a tricentennial break-up time series from Tornio. Geophysical 29, 73−95.

Kaser, G., Fountain, A., Jansson, P., 2003. A Manual for Monitoring the Mass Balance of Mountain Glaciers. IHP-VI Technical Documents in Hydrology, Paris.

Kaser, G., Grosshauser, M., Marzeion, B., 2010. Contribution potential of glaciers to water availability in different climate regimes. Proc. Natl. Acad. Sci. U.S.A 107, 20223−20227.

Koboltschnig, G.R., Schöner, W., 2011. The relevance of glacier melt in the water cycle of the Alps: the example of Austria. Hydrol. Earth Syst. Sci. 15, 2039−2048.

Kuhn, M., Batlogg, N., 1998. Glacier runoff in Alpine headwaters in a changing climate. In: Hydrology, Water Resources and Ecology in Headwaters (Proceedings of the HeadWater'98 Conference Held at Meran/Merano, Italy, April 1998). IAHS Publ., pp. 79−88. No. 248.

Kuusisto, E., 1980. On the values and variability of degree-day melting factor in Finland. Nord. Hydrol. 11, 235−242.

Lambrecht, A., Mayer, C., 2009. Temporal variability of the non-steady contribution from glaciers to water discharge in western Austria. J. Hydrol. 376, 353−361.

Lang, H., 1986. Forecasting meltwater runoff from snow-covered areas and from glacier basins. In: Kraijenhoff, D.A., Moll, J.R. (Eds.), River Flow Modelling and Forecasting Water Science and Technology Library, Volume 3. Springer, Netherlands, pp. 99−127.

Laudon, H., Sjöblom, V., Buffam, I., Seibert, J., Mörth, M., 2007. The role of catchment scale and landscape characteristics for runoff generation of boreal streams. J. Hydrol. 344, 198−209.

Lehning, M., Löwe, H., Ryser, M., Raderschall, N., 2008. Inhomogeneous precipitation distribution and snow transport in steep terrain. Water Resour. Res. 44, 1−19.

Lindström, G., Bishop, K., Löfvenius, M.O., 2002. Soil frost and runoff at Svartberget, northern Sweden—measurements and model analysis. Hydrol. Process. 16, 3379−3392.

Lindström, G., Johansson, B., Persson, M., Gardelin, M., Bergström, S., 1997. Development and test of the distributed HBV-96 hydrological model. J. Hydrol. 201, 272−288.

López-Moreno, J.I., Stähli, M., Lopezmoreno, J., Stahli, M., 2008. Statistical analysis of the snow cover variability in a subalpine watershed: assessing the role of topography and forest interactions. J. Hydrol. 348, 379−394.

Luce, C.H., Tarboton, D.G., Cooley, K.R., 1998. The influence of the spatial distribution of snow on basin-averaged snowmelt. Hydrol. Process. 12, 1671−1683.

Lundberg, A., Richardson-Näslund, C., Andersson, C., 2006. Snow density variations: consequences for ground-penetrating radar. Hydrol. Process. 20, 1483−1495.

Lyon, S.W., Destouni, G., Giesler, R., Humborg, C., Mörth, M., Seibert, J., Karlsson, J., Troch, P.A., 2009. Estimation of permafrost thawing rates in a sub-arctic catchment using recession flow analysis. Hydrol. Earth Syst. Sci. Discuss. 6, 63−83.

Magnuson, J.J., Robertson, D.M., Benson, B.J., Wynne, R.H., Livingstone, D.M., Arai, T., Assel, R.A., Barry, R.G., Card, V., Kuusisto, E., Granin, N.G., Prowse, T.D., Stewart, K.M., Vuglinski, V.S., 2000. Historical trends in lake and river ice cover in the northern hemisphere. Science (80) 289, 1743−1746.

Maidment, D.R., 1992. Handbook of Hydrology. McGraw-Hill Professional, New York.

Marks, D., Dozier, J., 1992. Climate and energy exchange at the snow surface in the Alpine region of the Sierra Nevada: 2. Snow cover energy balance. Water Resour. Res. 28, 3043−3054.

Martinec, J., 1975. Snowmelt−runoff model for stream flow forecasts. Nord. Hydrol. 6, 145−154.

Marzeion, B., Jarosch, A.H., Hofer, M., 2012. Past and future sea-level change from the surface mass balance of glaciers. Cryosphere 6, 1295−1322.

Mazurkiewicz, A.B., Callery, D.G., McDonnell, J.J., 2008. Assessing the controls of the snow energy balance and water available for runoff in a rain-on-snow environment. J. Hydrol. 354, 1−14.

Meier, M.F., Tangborn, W.V., 1961. Distinctive characteristics of glacier runoff. US Geol. Surv. Prof. Pap. 424-B, 14−16.

Molotch, N.P., Margulis, S.A., 2008. Estimating the distribution of snow water equivalent using remotely sensed snow cover data and a spatially distributed snowmelt model: a multi-resolution, multi-sensor comparison. Adv. Water Resour. 31, 1503−1514.

Neal, E.G., Hood, E., Smikrud, K., 2010. Contribution of glacier runoff to freshwater discharge into the Gulf of Alaska. Geophys. Res. Lett. 37, 1−5.

Ohmura, A., 2001. Physical basis for the temperature-based melt-index method. J. Appl. Meteorol. 40, 753−761.

Østrem, G., Stanley, A., 1969. Glacier Mass-balance Measurements: A Manual for Field and Office Work. Canadian Department of Energy, Mines and Resources.

Paul, F., Haeberli, W., 2008. Spatial variability of glacier elevation changes in the Swiss Alps obtained from two digital elevation models. Geophys. Res. Lett. 35, 1−5.

Peck, E.L., Bissell, V.C., Jones, E.B., Burge, D.L., 1971. Evaluation of snow water equivalent by airborne measurement of passive terrestrial gamma radiation. Water Resour. Res. 7, 1151−1159.

Pellicciotti, F., Brock, B., Strasser, U., Burlando, P., Funk, M., Corripio, J., 2005. An enhanced temperature-index glacier melt model including the shortwave radiation balance: development and testing for Haut Glacier d'Arolla, Switzerland. J. Glaciol. 51, 573−587.

Pohl, S., Marsh, P., Liston, G.E., 2006. Spatial-temporal variability in turbulent fluxes during spring snowmelt. Arct. Antarct. Alp. Res. 38, 136−146.

Prokop, A., 2008. Assessing the applicability of terrestrial laser scanning for spatial snow depth measurements. Cold Reg. Sci. Technol. 54, 155−163.

Prowse, T., Carter, T., 2002. Significance of ice-induced storage to spring runoff: a case study of the Mackenzie River. Hydrol. Process. 16, 779−788.

Prowse, T.D., 2005. River-ice hydrology. In: Anderson, M.G. (Ed.), Encyclopedia of Hydrological Sciences. John Wiley and Sons, Chichester, UK, pp. 1−21.

Racoviteanu, A.E., Armstrong, R., Williams, M.W., 2013. Evaluation of an ice ablation model to estimate the contribution of melting glacier ice to annual discharge in the Nepal Himalaya. Water Resour. Res. 49, 5117−5133.

Radić, V., Bliss, A., Beedlow, A.C., Hock, R., Miles, E., Cogley, J.G., 2014. Regional and global projections of twenty-first century glacier mass changes in response to climate scenarios from global climate models. Clim. Dyn. 42, 37−58.

Radić, V., Hock, R., 2014. Glaciers in the Earth's hydrological cycle: Assessments of glacier mass and runoff changes on global and regional scales. In The Earth's Hydrological Cycle (pp. 813−837). Springer Netherlands.

Rannie, W.F., 1983. Breakup and freezeup of the Red River at Winnipeg, Manitoba Canada in the 19th century and some climatic implications. Clim. Change 5, 283−296.

Rasmussen, R., Baker, B., Kochendorfer, J., Meyers, T., Landolt, S., Fischer, A.P., Black, J., Thériault, J.M., Kucera, P., Gochis, D., Smith, C., Nitu, R., Hall, M., Ikeda, K., Gutmann, E., 2012. How well are we measuring snow: the NOAA/FAA/NCAR Winter precipitation test bed. Bull. Am. Meteorol. Soc. 93, 811−829.

Röthlisberger, H., Lang, H., 1987. Glacial hydrology. In: Gurnell, M., Clark, M.J. (Eds.), Glacio-fluvial Sediment Transfer: An Alpine Perspective. John Wiley & Sons, Ltd, Chichester, UK, pp. 207−284.

Schaefli, B., Hingray, B., Musy, A., 2007. Climate change and hydropower production in the Swiss Alps: quantification of potential impacts and related modelling uncertainties. Hydrol. Earth Syst. Sci. 11, 1191−1205.

Schelker, J., Kuglerová, L., Eklöf, K., Bishop, K., Laudon, H., 2013. Hydrological effects of clear-cutting in a boreal forest—snowpack dynamics, snowmelt and streamflow responses. J. Hydrol. 484, 105−114.

Schoof, C., 2010. Ice-sheet acceleration driven by melt supply variability. Nature 468, 803−806.

Schweizer, J., Bartelt, P., van Herwijnen, A., 2014. Snow avalanches. In: Haeberli, W., Whiteman, C. (Eds.), Snow and Ice-Related Hazards, Risks and Disasters. Elsevier, pp. 395−436.

Seibert, J., 1999. Regionalisation of parameters for a conceptual rainfall-runoff model. Agric. For. Meteorol. 98-99, 279−293.

Seidel, K., Martinec, J., 2010. Remote Sensing in Snow Hydrology—Runoff Modelling, Effect of Climate Change. Springer Verlag.

Sevruk, B., Ondrás, M., Chvíla, B., 2009. The WMO precipitation measurement intercomparisons. Atmos. Res. 92, 376−380.

Shanley, J.B., Chalmers, A., 1999. The effect of frozen soil on snowmelt runoff at Sleepers River, Vermont. Hydrol. Process. 13, 1843−1857.

Singh, P., Singh, V.P., 2001. Snow and Glacier Hydrology. Kluwer Academic Publishers, London.

Sold, L., Huss, M., Hoelzle, M., Andereggen, H., Joerg, P.C., Zemp, M., 2013. Methodological approaches to infer end-of-winter snow distribution on alpine glaciers. J. Glaciol. 59, 1047−1059.

Sorg, A., Bolch, T., Stoffel, M., Solomina, O., Beniston, M., 2012. Climate change impacts on glaciers and runoff in Tien Shan (Central Asia). Nat. Clim. Change 2, 725−731.

Stahl, K., Moore, R.D., Shea, J.M., Hutchinson, D., Cannon, a. J., 2008. Coupled modelling of glacier and streamflow response to future climate scenarios. Water Resour. Res. 44, 1−13.

Stähli, M., Gustafsson, D., 2006. Long-term investigations of the snow cover in a subalpine semi-forested catchment. Hydrol. Process. 20, 411−428.

Stähli, M., Jansson, P.-E., Lundin, L.-C., 1996. Preferential water flow in a frozen soil—a two-domain model approach. Hydrol. Process. 10, 1305−1316.

Stähli, M., Jansson, P.-E., Lundin, L.-C., 1999. Soil moisture redistribution and infiltration in frozen sandy soils. Water Resour. Res. 35, 95−103.

Stähli, M., Nyberg, L., Mellander, P.-E., Jansson, P.-E., Bishop, K.H., 2001. Soil frost effects on soil water and runoff dynamics along a boreal transect: 2. Simulations. Hydrol. Process. 15, 927−941.

Stenborg, T., 1970. Delay of run-off from a glacier basin. Geogr. Ann. Ser. A, Phys. Geogr. 52, 1−30.

Storvold, R., Malnes, E., Larsen, Y., Høgda, K.A., Hamran, S.E., Müller, K., Langley, K.A., 2006. Sar remote sensing of snow parameters in Norwegian areas—current status and future perspective. J. Electromagn. Waves Appl. 20, 1751−1759.

Streletskiy, D., 2014. Permafrost degradation. In: Haeberli, W., Whiteman, C. (Eds.), Snow and Ice-Related Hazards, Risks and Disasters. Elsevier, pp. 303−344.

Surfleet, C.G., Tullos, D., 2013. Variability in effect of climate change on rain-on-snow peak flow events in a temperate climate. J. Hydrol. 479, 24−34.

Tait, A.B., 1998. Estimation of snow water equivalent using passive microwave radiation data. Remote Sens. Environ. 64, 286−291.

Taylor, S., Feng, X., Kirchner, J.W., Osterhuber, R., Klaue, B., Renshaw, C.E., 2001. Isotopic evolution of a seasonal snowpack and its melt. Water Resour. Res. 37, 759−769.

Techel, F., Pielmeier, C., 2011. Point observations of liquid water content in wet snow—investigating methodical, spatial and temporal aspects. Cryosphere 5, 405−418.

Vander Jagt, B.J., Durand, M.T., Margulis, S.A., Kim, E.J., Molotch, N.P., 2013. The effect of spatial variability on the sensitivity of passive microwave measurements to snow water equivalent. Remote Sens. Environ. 136, 163−179.

Viviroli, D., Weingartner, R., 2004. The hydrological significance of mountains: from regional to global scale. Hydrol. Earth Syst. Sci. 8, 1017−1030.

Wachter, K., 2007. The Analysis of the Danube Floods 2006 (Vienna).

Weber, M., Braun, L., Mauser, W., Prasch, M., 2010. Contribution of rain, snow-and ice melt in the upper Danube discharge today and in the future. Geogr. Fis. Dinam. Quat. 33, 221−230.

Weingartner, R., Aschwanden, H., 1992. Abflussregimes als Grundlage zur Abschätzung von Mittelwerten des Abflusses. In: Hydrologischer Atlas Der Schweiz. Federal Office for the Environment FOEN, Bern p. Tafel 5.2.

Werder, M.A., Funk, M., 2009. Dye tracing a jökulhlaup: II. Testing a jökulhlaup model against flow speeds inferred from measurements. J. Glaciol. 55, 899−908.

Zappa, M., Kan, C., 2007. Extreme heat and runoff extremes in the Swiss Alps. Nat. Hazards Earth Syst. Sci. 7, 375−389.

Zemp, M., Thibert, E., Huss, M., Stumm, D., Rolstad Denby, C., Nuth, C., Nussbaumer, S.U., Moholdt, G., Mercer, A., Mayer, C., Joerg, P.C., Jansson, P., Hynek, B., Fischer, A., Escher-Vetter, H., Elvehøy, H., Andreassen, L.M., 2013. Reanalysing glacier mass balance measurement series. Cryosphere 7, 1227−1245.

Chapter 5

Snow, Ice, and the Biosphere

Terry V. Callaghan [1,2,3] and Margareta Johansson [4]

[1] *Royal Swedish Academy of Sciences, Lilla Frescativägen, Stockholm, Sweden,* [2] *Department of Animal and Plant Sciences, University of Sheffield, Sheffield, UK,* [3] *Department of Botany, National Research Tomsk State University, Tomsk, Siberia, Russia,* [4] *Department of Physical Geography and Ecosystem Science, Lund University, Lund, Sweden*

ABSTRACT

Snow and ice are defining environmental characteristics of the polar and alpine regions that are rapidly changing. They provide habitats for some biota and mediate biogeochemical processes. They also modify winter habitats. Snow insulates soil, vegetation, and animals that live under the snow, and it isolates plants from herbivores, and small herbivores from predators. Snow and ice also have physical and mechanical properties that present challenges and opportunities to some plants and animals. In addition, snow and ice affect spring and summer habitats of plants because their duration affects the timing and length of the growing/active period and because snow stores winter precipitation that is released in spring as a water supply for soil organisms and plants. Nutrients and pollutants, accumulated in winter snow, are released during the spring thaw when they have their effects on the biosphere, including on people. However, snow and ice conditions are changing during current global warming resulting in damaging events to plants and animals in many Arctic areas. On the other hand, changing vegetation is affecting snow conditions.

Ice cover also provides a habitat and a platform for species such as marine mammals. It isolates marine and freshwater organisms in the water column from light, both beneficial photosynthetically active radiation and harmful ultraviolet-B. On land, permafrost exerts a major influence on biogeochemical cycling, particularly storage and release of carbon, while freeze—thaw cycles generate soil movement that selects for specifically adapted plant roots and limits agriculture.

5.1 INTRODUCTION

Snow and ice are defining environmental characteristics of the polar and alpine regions, although they occur in other regions and even sporadically in climatically hot areas. As a cover on the landscape, both snow and ice are extremely variable in time and space and during the current era of climate change, they follow complex trends. Snow cover duration in the Northern Hemisphere is declining, particularly in spring, but snow depth is increasing in some areas (Derksen and Brown, 2012). The frequency and thickness of ice layers in the snow pack on land in many areas of the Arctic are increasing as warmer winters lead to unseasonal snow melt events. Sea, lake, and river ice cover duration is also declining (Meier et al., 2011; Prowse et al., 2011). Ice, and snow, in particular, have numerous effects on plants, animals, and microbes—as well as on biogeochemical cycling—in managed and natural ecosystems. They act as filters for organisms that can survive freezing and snow cover for a long period of the year. Chernov (1985) recognized that snow cover, at least in the Arctic, is the "main and decisive factor in the lives of the majority of animals and plants". Consequently, organisms that live in areas where snow and ice are important environmental features must avoid or adapt to these conditions. While snow and ice affect organisms, some organisms and particularly plants significantly affect snow conditions, although there is little evidence that they affect ice conditions.

Knowledge is essential to understand changes in ecosystem services resulting from the current and predicted impacts of changing snow and ice conditions on the biosphere of the cold regions. Knowledge integrated over the entire north is required to estimate the impact of changes in regulatory ecosystem services, such as fluxes of greenhouse gases, on the global community, whereas a knowledge of provisioning ecosystem services, such as availability of food, is essential for local people who must adapt to changing conditions.

This contribution introduces the reader to the effects of snow and ice on the biosphere and the effects of the biosphere on snow. It mainly focuses on northern high-latitude land areas where dramatic changes in snow and ice are being recorded, land use changes are less extensive than elsewhere, and it is possible to draw on recent assessments of changing snow and ice conditions (Callaghan et al., 2011; Prowse et al., 2011; Meier et al., 2011). Although this contribution gives priority to snow, some of the more important relationships between changing ice conditions and the biosphere are summarized. These include the effects of freezing on cells and organisms and adaptations to freezing, the effects of changes in permafrost on biogeochemical cycling, the modification of light regimes by floating sea and freshwater ice and the limitations of plant growth by soil movement during freeze—thaw cycles. However, temporarily frozen ground in temperate or other latitudes is not considered here.

5.2 SNOW AND ICE AS HABITATS

While snow and ice are winter characteristics of many regions of the Earth and periodically affect the habitat for plants, animals that do not migrate in winter, and microbes, it is also a habitat per se. An example is glacial ice that is a habitat for glacial ecosystems. These ecosystems have been described from the polar regions (Hodson et al., 2008) and also probably occur in high-alpine areas. Algae, invertebrates, and microbes have been found to thrive in glaciers and in the snow pack. Three glacial ecosystems have been classified: one inhabiting the glacier surface (the supraglacial ecosystem), one at the ice-bed interface (the subglacial ecosystem), and a third englacial ecosystem. The supraglacial ecosystem has a diversity of bacteria, algae, phytoflagellates, fungi, viruses and occasional rotifers, tardigrades, and diatoms within the snow pack, supraglacial streams, and melt pools. The subglacial system is dominated by aerobic/anaerobic bacteria and most probably viruses in basal ice/till mixtures and subglacial lakes. The englacial ecosystem within glacier ice contains relatively inactive microbes. In contrast, the supraglacial and subglacial ecosystems have a significant effect upon the dynamics, composition, and abundance of nutrients in glacial melt water (Hodson et al., 2008).

Another example of this type of habitat is the sea ice that contains organisms and food webs of four layers (Meier et al., 2011). Microbial ecosystems develop when organisms are incorporated into the forming ice together with organic compounds and inorganic nutrients. On melt, these organisms and nutrients contribute to the formation of algal blooms at the retreating ice edge where members of other trophic layers aggregate to use the basic energy source.

A third example of this habitat type is land-based permafrost. Some land-based permafrost is 3 million years old and contains *living* microbes (Gilichinsky et al., 2001). Millions of viable cells per gram of soils in permafrost exist and bacteria, green algae, cyanobacteria, and fungi can be found. In addition, enzymes and pigments exist. Some of the organisms in ancient permafrost are relicts of the Cenozoic era. Some could be pathogens not yet experienced by "modern" species and mankind. Unfortunately, no one knows the potential impacts if/when ancient organisms trapped in permafrost are released.

5.3 SNOW AND ICE AS MODERATORS OF HABITAT

When snow and ice are seasonally present, they directly affect the winter habitat of those organisms that do not migrate, in many ways. Snow and ice insulate soil, vegetation, and animals that live under them, and snow isolates plants from herbivores, and small herbivores from predators. Snow and ice also have physical and mechanical properties that present challenges and opportunities to some plants and animals. In addition, snow and ice affect spring and summer habitats of plants and animals because their duration affects the growing/active period and because snow stores winter precipitation that is released in spring as a water supply to soil organisms and plants. Nutrients and pollutants, accumulated in winter snow, are released during the spring thaw when they have their effects on the biosphere. However, snow and ice conditions are changing during current global warming, and they lead to damaging events to some plants and animals. Nonwinter habitat is also moderated by snow and ice, which control the start date and length of the snow- and ice-free periods. Snow also controls soil moisture in the beginning of the growing season directly through its winter accumulation and spring release.

5.3.1 Modification of Winter Habitat

Snow and ice moderate winter habitat for plants and animals in many ways (Figures 5.1 and 5.2).

5.3.1.1 Insulation and Isolation

Snow has important "insulating" properties that enhance the winter survival—and even growth—of plants, the survival and reproduction of animals, such as voles and lemmings, and the development of the active layer in permafrost areas. Lemmings and other microtine small rodents spend winter in subnival cavities, that is, tunnels and holes under the snow (Emanuelsson, 1984) (Figure 5.2). Here, they have access to the shoots of plants for food, and they take advantage of the insulating properties of snow to nest and breed. Occupied nests of Taiga Voles can be 25 °C higher than ground temperatures (Wolff, 1980).

Some plant species, particularly some species of moss and lichen, also use cavities under snow to grow when light conditions permit before the snow-free season starts. They experience higher temperatures with smaller extremes below the snow and relatively high levels of CO_2 and humidity that enhance

FIGURE 5.1 Snow isolates animals and plants from their essential resources and provides a structural burden for plants and insecure ground for animals. When adaptations occur, snow can provide protection from harsh winter weather; plants from herbivores; and herbivores from predators. *Photograph: T.V. Callaghan.*

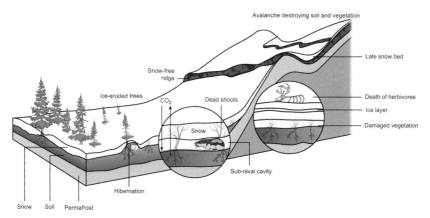

FIGURE 5.2   Schematic representation of snow and ice in the northern winter habitat. The circles are magnifications of successful lemming activity under snow (left) and effects of midwinter thaws and ice layers in the snow. *Source: T.V. Callaghan.*

growth. Photosynthesis under snow can contribute as much as 19 percent of the total annual photosynthesis (Larsen et al., 2007a). However, the same habitat features enhance the activity of soil microbes, and winter respiration can make a major contribution to annual carbon emissions (Fahnestock et al., 1998). An experimental increase in snow depth in birch forest tundra increased annual carbon efflux by between 60 and 160 percent (Larsen et al., 2007b; Nobrega and Grogan, 2007). Thus, a moderate increase in snow depth can enhance winter respiration sufficiently to alter the ecosystem's annual net carbon exchange from a sink to a source.

However, the ecological interactions mediated by snow are complex: although plant survival is increased by snow cover, the synchronous increase in survival and reproduction of voles and lemmings leads to increased herbivory and damage to vegetation that can outweigh any benefits to plant growth. While the animals eat relatively small amounts of vegetation (mainly grasses and mosses) accessible in subnival cavities, the removal of plant shoots leads to the formation of masses of dead material trapped in the snow (Emanuelsson, 1984). This material can be 50 percent of above-ground biomass in lemming peaks (Turchin and Batzli, 2001) (Figure 5.2). At thaw, this material forms piles of dead material on the ground surface.

Snow also "isolates" animals and plants from winter conditions above the snow pack. Animals that live below the snow pack are protected from their predators, and both plants and animals are protected from harsh winter conditions. In extreme, low-temperature environments, the height of plant canopies is relatively uniform and corresponds to the average thickness of the snow pack (Chernov, 1985; Sonesson and Callaghan, 1991). Plants that extend above the snow pack are prone to grazing (Stöcklin and Körner, 1999) and winter desiccation and tissue erosion and loss through exposure to high winds

FIGURE 5.3   Snow cover protects vegetation in the Arctic from low temperatures, high winds, and ice-crystal abrasion. Beyond the tree line, only stunted trees survive such as mountain birch in Swedish Lapland. Exposure can lead to ice-sculptured trees such as white spruce in the Churchill area of Manitoba, Canada (inset). *Photograph: T.V. Callaghan; inset, Sveinbjörnsson et al. (2002).*

and ice-crystal abrasion. In this process, ice crystals are blasted at plant stems by high winds (Sveinbjörnsson et al., 2002). At tree line, this process results in die-back, breakage of tree tops and stems, and tissue loss, which produces a range of plant canopy forms from low-lying dense mats to "flag"-shaped trees (Figure 5.3).

5.3.1.2 Implications of Changing Snow and Ice Conditions for Subnival Cavities and Ground Vegetation

Adaptation to long-term snow conditions has resulted in the coupling of the growth of plants and behavior of animals to predictable snow conditions. However, during current global warming, snow conditions are becoming unpredictable. The length of the snow period is discussed below, but another major characteristic that is changing is the consistency of snow in the snow profile (Johansson et al., 2011). More frequent warm events in winter (which is in general warming faster than summer in the Arctic) lead to partial or total melting of the snow pack and then refreezing. This produces ice layers in the snow (Figure 5.4), some of which encapsulate the ground vegetation when the melt has been total. The frequency of these events is increasing (see the

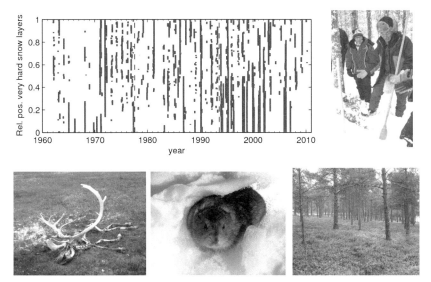

FIGURE 5.4   The position of hard snow and ice layers (red) relative to the depth of snow in a 50-year long data set from sub-Arctic Sweden (Johansson et al., 2011). These layers denote midwinter thaw events that are becoming more frequent and deeper over time. Resulting ice-encrustation of vegetation is a concern to Saami reindeer herders (inset, top right) who probe the snow pack to test the quality of the grazing for their animals. Snow thaw in winter followed by low temperatures reduces wild reindeer and lemming populations (inset bottom left and middle) and damages vegetation (inset, bottom right). *Photographs: top right, bottom left, and bottom right, T.V. Callaghan.*

references in Callaghan et al., 2011), and they lead to species-specific effects in plants including delayed bud development, reduced flower production, increased shoot mortality, and reduced abundance of soil invertebrates (Bokhorst et al., 2012). In contrast, lichens—but not mosses—are tolerant of the events (Bjerke et al., 2011). In one event in December 2007 in coastal northern Norway, plant productivity in summer was reduced by about 30 percent over an area of at least 1,400 km^2 (Bokhorst et al., 2009). These events are also an important factor in the decline of the valuable yellow cedar in areas of Canada and Alaska due to loss of winter dormancy (Hennon et al., 2006) and root death in cold soils unprotected by snow (Beier et al., 2008).

The warming in winter events together with rain-on-snow events kills musk oxen, reindeer, and lemmings and voles (with impacts on their predators and alternative prey). Lemming and vole cycles have been dampened in many areas (e.g., Kausrud et al., 2008). Also, when spring melt is rapid, the small rodents can drown in melt water in the subnival cavities (Chernov, 1985). When their population densities are low, their predators seek alternative prey such as ground-nesting birds whose populations then also decrease (McKinnon et al., 2013). Eventually, the predators also decline. Winter warming and rain-on-snow events have probably (but see Tyler, 2010 for another view) reduced wild reindeer herds in many areas by about 30 percent in 1999−2014 (McRae et al., 2010) with major mortality events on Svalbard (Aanes et al., 2000; Kohler and Aanes, 2004), the Queen Elizabeth Islands (Miller and Barry, 2009), and on the Yamal Peninsula (Forbes and Stammler, 2009; Bartsch et al., 2010). On St Matthew Island, overexploitation of resources by an introduced reindeer herd, extreme winter snowfall, and high winds in 1963 resulted in the extirpation of a herd that had reached 6,000 animals (Klein, 1968). The small Peary's reindeer has been nearly extirpated in High Arctic Canada (Barry et al., 2007), and thousands of musk oxen died on Banks Island in 2003 (Rennert et al., 2009; Grenfell and Putkonen, 2008).

Olofsson et al. (2009) have shown that the experimental exclusion of herbivores from an area of vegetation amplifies the stimulation of plant growth by warming. It is possible therefore that the 37 percent greening of Arctic vegetation recorded by satellites since 1982 (Xu et al., 2013) is a result of the impact of winter snow conditions on herbivores, rather than a direct effect of summer temperature and extension of the snow-free period on the vegetation as often assumed; or the greening could be driven by a combination of both winter and summer processes.

While thaw events in winter have increased in frequency in many areas, Bulygina et al. (2010) have recorded a decrease in the frequency of icing events that damage agricultural crops in Russia. These events have been defined as the presence of an ice crust that is >20 mm thick during a five-day period.

5.3.1.3 Physical and Mechanical Properties of Snow and Ice

The mechanical properties of snow and ice are important for plants, animals, and soils. The weight of snow imposes mechanical constraints on plants. Tree branches must be either structurally strong or flexible to withstand the weight of snow in the canopy (Callaghan and Johansson, 2009), or the canopy architecture must be adapted to shed snow rather than retain it. Tall, narrow canopies, adapted to maximize the receipt of low-angle solar radiation (Miller and Tiezen, 1972), might also be an adaptation to reduce the amount of snow retained. An accumulated amount of snow exceeding $20 \, kg/m^2$ can lead to forest damage (Kilpelainen et al., 2010), and snow loads of 3,000 kg/tree have been recorded in Finland (http://www.metla.fi/index-en.html) (Figure 5.5). The deciduous trees that form the tree line in snow-rich maritime areas may be a response to reducing snow load compared with narrow, evergreen canopies in relatively snow-poor continental tree lines.

The mass of snow is also important on slopes when it becomes unstable and moves. Snow avalanches present a periodic disturbance to vegetation and soils in many cold regions with slopes. The vegetation and soil layers may be completely removed as a result of some avalanches. Less frequently recorded events are slush torrents that form when snow partially thaws on a mountain

FIGURE 5.5 Biomechanical properties of plants related to loads by ice and snow are under-researched and seldom acknowledged. Over 3,000 kg of snow per tree can accumulate (top left), and some plants are capable of withstanding masses of ice surrounding their tissues (bottom left), which also results in oxygen starvation. Some trees such as birch are capable of bending under the weight of snow (right) and reproducing clonally when their apices touch the ground. *Photo credits: left upper and lower, Callaghan and Johansson (2009); right, T.V. Callaghan.*

side but is temporarily retained by an ice or snow dam (Gude and Scherer, 1995). When this dam breaks, the water content—sometimes equivalent to a small lake—flows down the mountain side in minutes and removes vegetation, soil, and rock from the slope and results in accumulations of debris in alluvial fans, which change the landscape.

Snow (Figure 5.6) and floating ice (Figure 5.7) mechanically support animals and people. A deep snow cover in winter allows browsing animals, such as moose, to reach tree bark and stems at greater heights than they can reach in summer (Figure 5.6). They therefore have a store of winter fodder. Thick, floating ice allows people and animals to use frozen lakes, rivers, and coastal areas as transport routes and as platforms for various activities such as hunting. Animals move across snow and ice surfaces and must have behavioral and morphological adaptations to be able to do this. The area of foot relative to body weight is important to move across snow while the hairs and feathers on feet are required for traction, and an adequate supply of blood is necessary to prevent feet from freezing, for example, in penguins. However, adaptation is not always perfect, and ice crust formation on the snow prevents reindeer from

FIGURE 5.6   While a snow cover protects much vegetation from grazing during winter, snow also acts as a platform that elevates browsers such as moose (left). Adult moose can reach 2 m high when standing on snow (msd). The scars on the bark of aspen trees often lead to tree death (right). *Photograph: right, R. Van Bogaert.*

migrating from the tundra in autumn to forest regions with deeper snow and greater lichen production. Such episodes of ice crust formation in the Yamal Region are becoming more frequent (Bartsch et al., 2010).

While the structural properties of snow and ice support the weight of an-imals, they also provide resistance to herbivores that must dig through the snow or break the ice to access vegetation. This process requires the herbivores to know where to dig, probably through smell, as digging at random would waste valuable energy.

One major control on plant population growth in the Arctic and other cold regions is the physical disturbance of the active layer during freeze and thaw cycles (Jonasson and Callaghan, 1992). The active layer is the zone of most biological activity in the ground, and its thickness limits the depth of soil that roots can exploit for nutrients and water or access for anchorage. Conse-quently, those trees that exist in permafrost areas have shallow, disc-shaped root systems, and the trees are vulnerable to wind throw. During freezing of the active layer, the ground expands, and during thawing, the ground contracts. This results in upward and downward displacement of the ground surface (Figure 5.8). Well-adapted plants tolerate such movements by having specially adapted root systems including coiled roots (Bell and Bliss, 1978), or struc-turally elastic roots (Jonasson and Callaghan, 1992). However, young

FIGURE 5.7   Ice cover on lakes provides a platform for animal movement and limits UV-B damage to aquatic organisms. However, the ice layer also limits photosynthetically active radiation and plankton production that fish directly or ultimately depend on. *Photograph: Lake Torneträsk in sub-Arctic Sweden in March. J. Förster.*

seedlings of Arctic plants and the relatively simple tap-root systems of many crop plants are "jacked up" out of the ground and killed (Perfect et al., 1988) (Figure 5.8). This provides a major limitation on the development of agriculture in northern latitudes.

Where permafrost is present, it provides a physical, impermeable barrier to the downward movement of water. In flat areas of the Arctic tundra, this leads to the abundance of lakes and ponds. Initial permafrost degradation leads to slumping and collapse of trees, and surface drainage (in Alaska, Hinzman et al., 2005, Canada, Smol and Douhglas, 2007, and Siberia, Smith et al., 2005) with the possibility for greater plant production. However, further degradation in areas of low precipitation can lead to soil moisture deficits and the formation of steppe-like communities (Callaghan et al., 2005).

The surfaces of snow and ice are very reflective, and the reflection of ul-ultraviolet (UV)-B radiation is important in the biosphere. Arctic peoples, such as the Inuit, used snow goggles made from drift wood or bone to prevent snow blindness. However, Arctic and Antarctic animals that emerge from the polar night to high UV-B conditions must have well-adapted eyes. Antarctic seals have tears with UV-B absorbing compounds that prevent damage, while reindeer actually use UV-B light reflected from snow and vegetation to "see" (Hogg et al., 2011). Plants of polar and alpine regions have various adaptations to high UV-B light, most of which include the induction of pigments to protect DNA. Pigments can be found in leaf hairs, in leaf surface cell layers, or throughout leaf tissues (Semerdjieva et al., 2003). In general, Arctic plants are preadapted to high UV-B regimes as historically many species moved into the

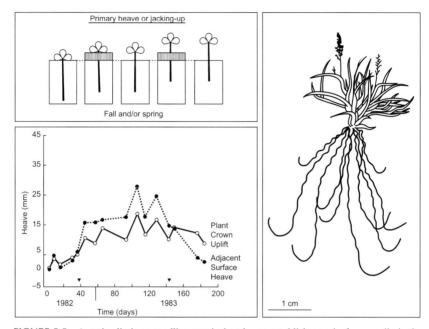

FIGURE 5.8 A major limit on seedling survival and crop establishment in frozen soils is the physical jacking-up of plants with ill-adapted root systems (left). Adaptations to withstand freeze−thaw cycles are elastic and even coiled roots as in the grass *Phippsia algida* (right). *Sources: left, Perfect et al. (1988); Right, Bell and Bliss (1978).*

Arctic from alpine latitudes during the early Holocene (Callaghan et al., 2005). Although Arctic and alpine plants are generally well adapted to high UV-B regimes with little damage experienced, the chemical changes in plant tissues associated with protection affect the herbivores that eat the plants and the decomposer organisms that break down the dead plant material (references in Callaghan et al., 2005).

Animals and plants that live in cold conditions need cells that either avoid freezing or are tolerant of it because ice-crystal formation physically damages cells. In animals, vital organs and tissues must be protected from physical disruption by ice crystals. One means is to produce glycerol, a cryoprotectant. Adult Arctic beetles can survive temperatures that are below $-35\,°C$ (Crawford, 2013) in this way. Another mechanism to cope with freezing is supercooling in which cellular water is purified to an extent that avoids freezing around nucleation centers, so that the water can exist in a liquid state below $0\,°C$ (Crawford, 2013). This mechanism allows some beetles to exist at temperatures below $-40\,°C$.

In the current period of atmospheric temperature rise, and particularly during winter warming, the inability of eggs of some insect pest species to survive particularly low temperatures has profound implications. The autumn moth, *Epirrita autumnata*, is a defoliator of mountain birch forests in northern

Fennoscandia. Its eggs are killed by temperatures lower than $-36\,°C$ (Tenow, 1996), and this is a major limitation on population density. However, populations periodically reach mass outbreak proportions. Warmer winters are therefore expected to lead to higher insect populations in summer and more extensive and intense tree defoliation events (Jepsen et al., 2011).

Plant cells differ from animal cells in that they possess cell walls. Their response to freezing is therefore different. Freeze-tolerant plant cells allow water to pass through cell membranes into the cavities between cell walls where freezing is harmless to the vital internal cell structures (Crawford, 2013). Some trees and shrubs in the north avoid freezing damage by supercooling. Some Arctic plants can survive complete ice encapsulation, which deprives tissues of oxygen, and can continue to flower or grow following thaw (Crawford, 1993; Preece et al., 2012).

5.3.1.4 Snow, Disease, and Health

Melting snow and ice could have many effects on human health but as health is a response to interactions among very many factors, it is difficult to isolate the specific effects of snow and ice. However, snow melt that increases flooding may lead to a breakdown in infrastructure such as sanitation and a spread of infectious disease agents (Parkinson and Butler, 2005). Snow cover also affects the availability of contaminants. Snow can accumulate harmful pollutants (see the references in Callaghan et al., 2011) and because snow is permeable, many chemical reactions take place within the snow pack. These affect nitrogen oxides, halogens, ozone, organic compounds, and mercury. Mercury in particular is toxic and bioaccumulates in the traditional food sources of Arctic peoples. Changing snow and ice conditions lead to the unpredictability of transport conditions and consequently to an increased number of accidents.

Changing snow conditions affect plant and animal diseases. Observations by Sami reindeer herders have indicated changes in winter snow conditions in the Swedish and Norwegian sub-Arctic areas (Riseth et al., 2010). Winters have become generally milder since the 1970s, and snowfall patterns have changed with implications for the biosphere: normally, the snow used to fall on frozen ground (-5 to $-10\,°C$), but currently (the last decades), the snow often falls during warmer conditions ($>0\,°C$). This change in snow results in a wet surface, mold formation, and a reduced ability for the reindeer to smell food beneath the snow. According to the reindeer herders, the snowfall intensity also is greater now—several tens of centimeter fall within single days (Erik Anders Niia, personal communication). The snow mold that the reindeer herders observe can cause mortality of reindeer calves, but its effects on other animals is unknown. Snow molds can also significantly reduce plant growth including trees, agricultural crops, and tundra shrubs (Bokhorst et al., 2012). In an experimental prolongation of the snow season, Olofsson et al. (2011) showed an increase in the plant pathogen *Arwidssonia empetri*, which reduced the growth of evergreen dwarf shrubs. In addition, the increased snow duration made plants more palatable to herbivores (Torp et al., 2010). Snow mold abundance can be increased by moose browsing damage to young pine trees (Stöcklin and Körner, 1999).

However, not all aspects of reducing snow and ice duration have negative effects on people: lower albedo for UV-B radiation will probably decrease the incidence of snow blindness and cancers such as malignant melanoma. Similarly, snow accumulation can be beneficial when nutrients such as nitrogen are deposited. Plants, particularly mosses, can use these nutrients trapped in snow (Tye et al., 2005).

5.3.2 Modification of Nonwinter Habitat: Duration of the Snow-Free and Ice-Free Periods

5.3.2.1 Snow

The duration of snow is a critical determinant of the length of the growing season for plants, and thus, the length of the season when these primary producers can be accessed by herbivores and other dependents. The length of the snow-free period is variable from year to year and has increased in the Arctic by about 13 days between 1972 and 2008 due to global warming (Callaghan et al., 2011; Derksen and Brown, 2012). The length of the snow-free period generally increases from high polar latitudes to temperate latitudes with the exception that high Arctic and Antarctic polar deserts can be very dry with little snow cover. Snow cover also varies with topography, accumulating in late-lying snow bed habitats and being blown away from exposed ridges or fell fields (Figure 5.2).

The patterns of snow accumulation on the landscape are probably the major determinant of plant species and community distributions (Evans et al., 1989). At one extreme, fell field plants must be adapted to withstand low winter temperatures, high winds, and soil water deficits when they are exposed to winter climate. At another extreme, where snow persists into late summer in depressions with deep snow accumulation, plants must be able to complete their seasonal cycles of leaf production and reproduction within a very short time. In contrast, the impact of snow on animals depends on their abilities to migrate. Arctic residents such as musk ox and lemmings experience the complete snow climate, although some animals such as brown bear and ground squirrels hibernate. Short-distance migrants such as reindeer, arctic fox, and ptarmigan are affected by snow conditions within the Arctic and in the winter grazing lands in the forests to the south. Animal species that migrate over long distances to the Arctic in spring (e.g., waders and geese) are influenced by spring snow conditions when they arrive in the Arctic.

Overall, the duration of the growing season affects vegetation production, herbivore population growth, and the success of predators. On Ellesmere Island, short growing seasons reduced plant productivity, decreased populations of the herbivores musk ox and Arctic hare, and reduced the population size of their predator, the wolf (Mech, 2004). The length of the growing season also affects biogeochemistry. For each day earlier, modeled productivity of vegetation over the panArctic Region increases with a carbon drawdown of 9.5 gC/m^2 (Euskirchen et al., 2006) when moisture is nonlimiting.

5.3.2.2 Lake and River Ice

The duration of lake and river ice in the North has decreased between 1846 and 1995, and this trend is predicted to continue in the future (Prowse et al., 2011). This might lead to higher primary production in lakes and rivers as more light is available for photosynthesis and more inputs of nutrient from the catchments is expected under warmer conditions (Vincent et al., 2011). It is also possible that the increasing open water areas can attract more aquatic bird populations (Vincent et al., 2009) and could possibly also favor less-cold-tolerant plants, fish, and other organisms that are likely to invade from the south. On the contrary, these warmer conditions may pose a threat to cold-adapted specialists such as Arctic char (Power et al., 2008) and could even drive some northern species to extinction (Sharma et al., 2007). Also, while increases in photosynthetically active radiation, resulting from earlier lake ice melt, enhance productivity, the removal of ice will give greater exposure of sensitive organisms to harmful UV-B radiation, particularly in temperate alpine lakes with high solar angles and thinner ozone layers.

5.3.2.3 Sea Ice

During 2004−2014, the minimum sea-ice extent in the Arctic Ocean has broken new minimum records several times, especially in 2007 and 2012, and it is now likely that the Arctic will be free of ice during the summer within 30 to 40 years (Meier et al., 2011; Jeffries et al., 2013). The sea ice is a unique ecosystem and home to a range of marine mammals such as the polar bear, walrus, and seals, and for different communities of other biota that live within or in association with the ice (Figure 5.9). Ice margins and gaps in the sea ice (polynias) are sites of high productivity where birds and mammals congregate to feed (Meier et al., 2011). Algal blooms follow the retreating ice edge using new nutrient pools when the polar night is over.

Marine ecosystems have been affected by the decreasing sea ice (Vincent et al., 2011). For example, satellite observations indicate an increase of phytoplankton biomass as a result of the shorter ice duration (Arrigo et al., 2008). Different whale species will most likely be affected in different ways depending on their specialization and flexibility of diet. Migrating whales are, for example, likely to expand their ranges and periods in northern waters (Vincent et al., 2011). Another example of a species that has been affected is the polar bear. A decline in condition and reproductive success of polar bears was reported from western Hudson Bay in Canada (Regehr et al., 2007). Polar bears are very sensitive to changes in sea-ice conditions and drastic declines of

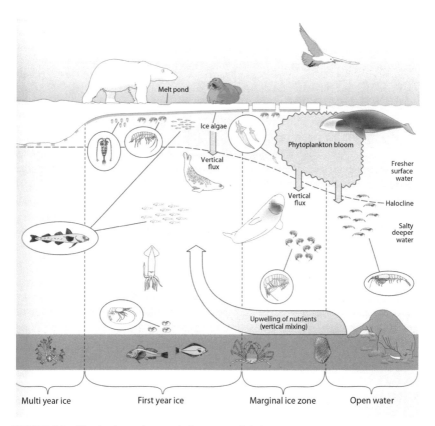

FIGURE 5.9 The Arctic sea ice margin is an area of high productivity on which complex food webs depend. The schematic shows the sea-ice community during the spring ice retreat and melt period. *Source: drawn by Audun Igesund, Norwegian Polar Institute and published by Meier et al. (2011).*

up to two-thirds of their population are predicted for the future (e.g., Durner et al., 2009).

5.3.3 Modification of the Nonwinter Habitat: The Timing of the Spring Melt

In addition to the duration of the snow-free and ice-free periods, the timing of the snow melt and ice break-up events is also of great biological importance (e.g., Wipf and Rixen, 2008).

In the Arctic, there is an asynchrony between the timing of snow melt and solar elevation that determines the amount of energy available for photosynthesis (Lewis and Callaghan, 1976). Thus, plants develop a photosynthetic canopy when the solar angle is declining and in-coming radiation is not optimal. In contrast, when plants senesce in autumn, the solar angle is very low. During current global warming, the snow melt is occurring earlier, and plants are experiencing a greater synchrony and optimization of incoming radiation. This has an impact on biogeochemical cycles in that carbon sequestration increases. In contrast, a delay of autumn when solar angles are low is not expected to lead to increased plant productivity (Callaghan et al., 2005) but, with an increase in soil microorganism activity due to warmer soils, increasing emissions of CO_2 are expected. Surprisingly, snow manipulation experiments have shown that the generalizations of effects of changing timing of the snow-free period are not always validated: considerable increases in snow depth might have little effect on the timing of snow melt where melt is fast (Walker et al., 1999), while an earlier melt might result in earlier senescence and no net increase in plant growth (Starr et al., 2008).

Spring snow cover also affects animals, both residents and migrants. Long-term observations show that a 10 percent decrease in spring snow cover advances egg laying by five to six days in one goose species and increases breeding success by 20 percent (Madsen et al., 2007). Similarly, musk ox population size in northeast Greenland increases as the length of the snow-free period increases and early spring snow cover decreases (Forchhammer et al., 2008).

The timing of spring thaw has major effects on the phenology (timing of events) in both plants (Wipf and Rixen, 2008) and animals. As spring thaw is occurring earlier, many phenological events are also taking place earlier. The significant warming and greening of 37 percent of the Arctic estimated from satellite images since the early 1980s (Xu et al., 2013) have been associated with a considerable seasonality or phenological advancement of the plant-growing season. For example, flowering occurring up to 3 weeks earlier has been reported in Northeast Greenland (Høye et al., 2007). Across six plant species monitored at Zackenberg, the timing of flowering was closely associated with spring snow cover in the year of flowering, with an average 2.5-day advance following a 10 percent reduction in snow cover on June 10 (Forchhammer pers. Comm.). However, no such responses were found in the growth and reproductive output of the same plants. Furthermore, when climate effects, such as timing of snow thaw, are considered across trophic levels, such as consumer–resource interactions, a phenological response of a plant species may have significant effects on its herbivore consumer, as reported in a reindeer population in West Greenland (Post and Forchhammer, 2008), but not necessarily on its own performance (Forchhammer, pers. comm.). Reindeer calf production in Finland increased by about one calf per 100 females for each day of earlier snow melt (Turunen et al., 2009).

5.3.4 Modification of Nonwinter Habitat: Melt–Water Supply

Snow stores winter precipitation in the solid form and releases liquid water in spring. This liquid water is a particularly important source of soil moisture for plants in areas where summer precipitation is low. In the coastal tundra of Alaska, a longer growing season has not been associated with a greening of the vegetation as expected (Gamon et al., 2013). This is because of soil moisture deficits that develop late in the growing season. Further south, the growth of at least some boreal evergreen species depends on moisture from melting snow (Yarie, 2008). As the moisture regime in much of the boreal region is only marginally suitable for forest growth, reduced snow cover has probably led to the forest decline and loss of forest productivity recorded by Lloyd and Bunn (2007) and Goetz et al. (2005). This forest response leads to a spiral of events that lead to further forest destruction: trees weakened by lack of soil moisture are more prone to fire and damage from insect pests such as the European spruce engraver beetle (Bakke, 1989), the North American engraver beetle (McCullough et al., 1998), and the Siberian silk moth (Kharuk et al., 2003). Changes in soil moisture supply from melting snow and its multiple effects are expected to lead in some areas to a change from needle-leaf evergreen trees to deciduous trees that are commercially less valuable (Juday, 2009).

Melt water from snow has major effects on ground water-logging during spring, and this will affect emissions of carbon (both amount and species—CO_2 or CH_4) and N_2O (Callaghan et al., 2011). There is also evidence that the magnitude of tundra fires is increasing (Mack et al., 2011). This has been associated with increases in thunder storms and lightning strikes, but it is not known if changing snow conditions are leading to drier and more fire-vulnerable tundra.

5.3.5 Snow—Permafrost—Vegetation Interactions

Permafrost is a critically important feature of cold regions, but its characteristics and impacts on the biosphere vary greatly between tundra—taiga areas of the north and alpine regions in general. On steep mountain slopes, soils are generally mineral soils, thin and mobile, and permafrost mainly occurs in the bedrock. In contrast, permafrost in the tundra is often present in deep organic soils, while the active layer above supports most of the biological activity. In the alpine regions, thawing permafrost leads to slope instabilities (Callaghan et al., 2011; Deline et al., 2014) that detach soils and vegetation. In high northern latitudes, snow affects permafrost and the biosphere mainly through its insulating properties (Streletskiy et al., 2014). A deeper snow cover increases the thickness of the active layer, provides increased shelter to vegetation and soil, and provides increased soil moisture as a result of snow melt water and water from thawing permafrost (Johansson et al., 2013). In an experimental manipulation of snow amount and duration in a permafrost area, increased snow depth and duration were found to increase the growth and flowering of cotton grass (*Eriophorum vaginatum*), a dominant species of many tundra areas (Johansson et al., 2013) (Figure 5.10).

FIGURE 5.10   Snow affects permafrost and soil moisture in spring and summer. Experimentally increased snow depth in sub-Arctic Sweden (top left) leads to permafrost thaw, increased moisture (top right), and increased growth and flowering of *Eriophorum vaginatum* (bottom). *Photographs: M. Johansson.*

5.3.6 Thawing Permafrost and Biogeochemical Cycles

The thawing of permafrost in Arctic soils affects biogeochemical cycles. Vast stores of carbon sequestered by plants from the atmosphere over millennia have accumulated in Arctic soils because permafrost limits microbial activity. Permafrost and the active layer above it contains about twice as much carbon as is found in the global atmosphere (Schuur et al., 2008). Experimental warming of permafrost soils shows that old carbon is released (Dorrepaal et al., 2009) thereby indicating that naturally warming permafrost soils could release vast quantities of carbon in the future, at least from the surface layers. Where soils are aerated, carbon is released as CO_2, but where the soils have little oxygen, for example, in waterlogged areas, methane is released and this is about 30 times more potent than CO_2 is as a greenhouse gas.

5.4 EFFECTS OF VEGETATION ON SNOW

Vegetation has complex influences on the accumulation and ablation of snow (Pomeroy et al., 2006). It affects both the vertical and horizontal distributions of snow (Callaghan et al., 2011). Tree canopies can intercept and retain large quantities of snow. Although sublimation takes place from these canopies, overall, they reduce the accumulation of snow on the ground (Hedstrom and Pomeroy, 1998), which leads to ground cooling and probably to the protection of permafrost in marginal areas. Vegetation also reduces the redistribution of snow over the landscape resulting in greater accumulation in vegetation patches than in open areas, and it reduces snow density, which increases ground insulation (Sturm et al., 2001).

Vegetation can accelerate or retard snow melt depending on vegetation characteristics (see references in Callaghan et al., 2011). Vegetation shades snow, but because vegetation and its debris on the snow surface absorb and transmit incoming solar radiation more efficiently than snow does, the net effect of the presence of vegetation is increased rates of snow melt, particularly in less dense vegetation and in areas at the edge of patches of vegetation (Essery et al., 2008). Tree temperatures that are more normal to low-angle incoming solar radiation in high latitudes can be considerably higher than those of the snow and above 0 °C. The expansion of shrubs in the tundra could lead to regional warming through a positive feedback loop (Chapin et al., 2005; Myers-Smith et al., 2011). Although increases in shrub advance in the Alaskan Arctic have not yet resulted in warming, it is predicted that an increase of shrubs could increase summer heating of the atmosphere by 3.7 W m^2, which is equivalent to a doubling of CO_2 (Chapin et al., 2005).

5.5 CONCLUSIONS AND PERSPECTIVES

Snow and ice conditions have selected a range of adaptations in the organisms and processes of the cold-region biosphere. As there are numerous and complex interactions among species within the biosphere, the responses of one species to its habitat often affects the behavior and population growth of many more species. Furthermore, some species, such as tree and shrub species, also modify their own snow environment. These interactions among species and between species and their environment, have evolved over millennia and are particularly tuned to predictable dynamics within both the biosphere and environment. However, the environment is rapidly changing due to global warming and its impacts on the biosphere. Although the cold regions have experienced warmer conditions throughout historical time, the current rapid "rate" of change in combination with other anthropogenic impacts on the environment such as more intensive land use, habitat fragmentation and increasing pollution—particularly long-range black carbon particles that alter snow albedo—are generating a transitional period in historically cold regions that is dynamically unstable and quite unpredictable. As the cold regions are historically an area of minimal land use but major conservation value, there is a huge challenge for humanity to plan and implement an appropriate balance between opportunistic use of warming cold regions and the protection of their environment and biosphere. Such decisions need to be knowledge based. This knowledge is essential to understand changes in ecosystem services resulting from the impacts of changing snow and ice conditions on the biosphere of the cold regions. Knowledge integrated over the entire north is required to estimate the impact of changes in regulatory ecosystem services, such as fluxes of greenhouse gases, on the global community, whereas a knowledge of provisioning ecosystem services, such as availability of food, is essential for local people who must adapt to changing conditions.

ACKNOWLEDGMENTS

We wish to thank FORMAS for funding the project "Climate change, impacts and adaptation in the sub Arctic: a case study from the northern Swedish mountains" (214-2008-188). The study also contributes to the EU Framework 7 Infrastructure Project "INTERACT" (www. eu-interact.com). We also wish to thank the reviewers for constructive suggestions and improvements.

REFERENCES

Aanes, R., Saether, B.E., Oritsland, N.A., 2000. Fluctuations of an introduced population of Svalbard reindeer: the effects of density dependence and climatic variation. Ecography 23, 437−443.

Arrigo, K.R., van Dijken, G., Pabi, S., 2008. Impact of a shrinking Arctic ice cover on marine primary production. Geophys. Res. Lett. 35, L19603 http://dx.doi.org/10.1029/2008GL035028.

Bakke, A., 1989. The recent Ips typographus outbreak in Norway—Experiences from a control program. Ecography 12, 515−519.

Barry, R.G., Armstrong, R., Callaghan, T., Cherry, J., Gearheard, S., Nolin, A., Russell, D., Zöckler, C., 2007. Snow. UNEP: Global Outlook for Snow and Ice, pp. 39−62.

Bartsch, A., Kumpula, T., Forbes, B.C., Stammler, F., 2010. Detection of snow surface thawing and refreezing in the Eurasian Arctic using QuikSCAT: implications for reindeer herding. Ecol. Appl. 20, 2346−2358.

Beier, C.M., Sink, S.E., Hennon, P.E., D'Amore, D.V., Juday, G.P., 2008. Twentieth-century warming and the dendroclimatology of declining yellow-cedar forests in southeastern Alaska. Can. J. For. Res.-Rev. Can. Rech. For. 38, 1319−1334.

Bell, K.L., Bliss, L.C., 1978. Root growth in a polar semidesert environment. Can. J. Bot. 56, 2470−2490.

Bjerke, J.W., Bokhorst, S., Zielke, M., Callaghan, T.V., Bowles, F.W., Phoenix, G.K., 2011. Contrasting sensitivity to extreme winter warming events of dominant sub-Arctic heathland bryophyte and lichen species. J. Ecol. 99, 1481−1488.

Bokhorst, S., Bjerke, J.W., Tömmervik, H., Callaghan, T.V., Phoenix, G.K., 2009. Winter warming events damage sub-Arctic vegetation: consistent evidence from an experimental manipulation and a natural event. J. Ecol. 97, 1408−1415.

Bokhorst, S., Bjerke, J.W., Tömmervik, H., Preece, C., Phoenix, G., 2012. Ecosystem response to climatic change: the importance of the cold season. Ambio 41 (Suppl. 3), 246−255.

Bulygina, O.N., Groisman, P.Y., Razuvaev, V.N., Radionov, V.F., 2010. Snow cover basal ice layer changes over Northern Eurasia since 1966. Environ. Res. Lett. 5, 10.

Callaghan, T.V., Johansson, M., 2009. The changing, living tundra: a tribute to Yuri Chernov. In: Golovatch, S.I., Makarova, O.L., Babenko, A.B., Penev, L.D. (Eds.), Species and Communities in Extreme Environments. Festschrift towards the 75th Anniversary and a Laudatio in Honour of Academician Yuri Ivanovich Chernov. Pensoft Publishers, Sofia-Moscow, pp. 13−52.

Callaghan, T.V., Björn, L.O., Chernov, Y., Chapin, F.S., Christensen, T.R., Huntley, B., Ims, R., Jonasson, S., Jolly, D., Matveyeva, N., Panikov, N., Oechel, W.C., Shaver, G.R., et al., 2005.

Tundra and polar desert ecosystems, pp. 243–352. In: ACIA. Arctic Climate Impacts Assessment. Cambridge University Press, 1042 pp.

Callaghan, T.V., Johansson, M., Brown, R.D., Groisman, P.Ya, Labba, N., Radionov, V., 2011. Changing snow cover and its impacts, pp. 4-1 to 4-58. In: Arctic Monitoring and Assessment Programme 2011. Snow, Water, Ice and Permafrost in the Arctic (SWIPA): Climate Change and the Cryosphere. Arctic Monitoring and Assessment Programme, Oslo, Norway. xii + 538 pp.

Chapin III, F.S., Sturm, M., Serreze, M.C., McFadden, J.P., Key, J.R., Lloyd, A.H., McGuire, A.D., Rupp, T.S., Lynch, A.H., Schimel, J.P., Beringer, J., Chapman, W.L., Epstein, H.E., Euskirchen, E.S., Hinzman, L.D., Jia, G., King, C.-L.P., Tape, D., Thompson, C.D.C., Walker, D.A., Welker, J.M., 2005. Role of land-surface changes in arctic summer warming. Science 310, 657–660.

Chernov, Y.I., 1985. The Living Tundra (D. Löve, Trans.). Cambridge University Press, Cambridge, 213 pp.

Crawford, R.M.M., 1993. Plant survival without oxygen. Biologist 40, 110–114.

Crawford, R.M.M., 2013. Tundra–Taiga Biology. Human, Plant and Animal Survival in the Arctic. Oxford University Press, Oxford, 270 pp.

Deline, P., Gruber, S., Delaloye, R., Fischer, L., Geertsema, M., Giardino, M., Hasler, A., Kirkbride, M., Krautblatter, M., Magnin, F., McColl, S., Ravanel, L., Schoeneich, P., 2014. Ice loss and slope stability in high-mountain regions. In: Haeberli, W., Whiteman, C.A. (Eds.), Snow- and Ice-Related Hazards, Risks and Disasters. Elsevier, pp. 607–646.

Derksen, C., Brown, R., 2012. Spring snow cover extent reductions in the 2008–2012 period exceeding climate model projections. Geophys. Res. Lett. 39, L19504. http://dx.doi.org/10.1029/2012GL053387.

Dorrepaal, E., Toet, S., van Logtestijn, R.S.P., Swart, E., van de Weg, M.J., Callaghan, T., Aerts, R., 2009. Climate warming accelerates carbon respiration from deep sub-Arctic peat layers. Nature 460, 616–619.

Durner, G.M., Douglas, D.C., Nielson, R.M., Amstrup, S.C., McDonald, T.L., Stirling, I., Mauritzen, M., Born, E.W., Wiig, Ø., DeWeaver, E., Serreze, M.C., Belikov, S.E., Holland, M.M., Maslanik, J., Aars, J., Bailey, D.A., Derocher, A.E., 2009. Predicting 21st-century polar bear habitat distribution from global climate models. Ecol. Monogr. 179, 25–58.

Emanuelsson, U., 1984. Ecological Effects of Grazing and Trampling on Mountain Vegetation in Northern Sweden (Ph.D. thesis). University of Lund.

Essery, R., Pomeroy, J., Ellis, C., Link, T., 2008. Modelling longwave radiation to snow beneath forest canopies using hemispherical photography or linear regression. Hydrol. Process. 22, 2788–2800.

Euskirchen, E.S., McGuire, A.D., Kicklighter, D.W., Zhuang, Q., Clein, J.S., Dargaville, R.J., Dye, D.G., Kimball, J.S., McDonald, K.C., Melillo, J.M., Romanovsky, V.E., Smith, N.V., 2006. Importance of recent shifts in soil thermal dynamics on growing season length, productivity, and carbon sequestration in terrestrial high-latitude ecosystems. Glob. Change Biol. 12, 731–750.

Evans, B.M., Walker, D.A., Benson, C.S., Nordstrand, E.A., Petersen, G.W., 1989. Spatial interrelationships between terrain, snow distribution and vegetation patterns at an arctic foothills site in Alaska. Holarctic Ecol. 12, 270–278.

Fahnestock, J.T., Jones, M.H., Brooks, P.D., Walker, D.A., Welker, J.M., 1998. Winter and early spring CO_2 efflux from tundra communities of northern Alaska. J. Geophys. Res. Atmos. 103, 29023–29027.

Forbes, B.C., Stammler, F., 2009. Arctic climate change discourse: the contrasting politics of research agendas in the West and Russia. Polar Res. 28, 28—42.

Forchhammer, M.C., Schmidt, N.M., Høye, T.T., Berg, T.B., Hendrichsen, D.K., Post, E., 2008. Population Dynamical Responses to Climate Change. In: Advances in Ecological Research, vol. 40. Elsevier Academic Press Inc., San Diego, 391—419.

Gamon, J.A., Huemmrich, K.F., Stone, R.S., Tweedie, C.E., 2013. Spatial and temporal variation in primary productivity (NDVI) of coastal Alaskan tundra: decreased vegetation growth following earlier snowmelt. Remote Sens. Environ. 129, 144—153.

Gilichinsky, D., Shatilovich, A., Spirina, E., Fazutdinova, R., Gubin, S., Rivkina, E., Vorobyova, E., Soina, V., 2001. How long the life might be preserved? The terrestrial model for astrobiology. In: Paper Presented at the Bridge between the Big Bang and Biology. Consiglio Nazionale delle Ricerche of Italy, Rome, pp. 25—37.

Goetz, S.J., Bunn, A.G., Fiske, G.J., Houghton, R.A., 2005. Satellite-observed photosynthetic trends across boreal North America associated with climate and fire disturbance. Proc. Natl. Acad. Sci. U.S.A. 102, 13521—13525.

Grenfell, T.C., Putkonen, J., 2008. A method for the detection of the severe rain-on-snow event on Banks Island, October 2003, using passive microwave remote sensing. Water Resour. Res. 44. http://dx.doi.org/10.1029/2007WR005929.

Gude, M., Scherer, D., 1995. Snowmelt and slush torrents—preliminary report from a field campaign in Kärkevagge, Swedish Lappland. Geogr. Ann. Ser. A-Phys. Geogr. 77A, 199—206.

Hedstrom, N.R., Pomeroy, J.W., 1998. Measurements and modelling of snow interception in the boreal forest. Hydrol. Process. 12, 1611—1625.

Hennon, P., Amore, D., Wittwer, D., Johnson, A., Schaberg, P., Hawley, G., Beier, C., Sink, S., Juday, G., 2006. Climate Warming. Reduced Snow, and Freezing Injury Could Explain the Demise of Yellow-Cedar in Southeast Alaska. World Resource Review, USA, 427—445.

Hinzman, L.D., Bettez, N.D., Bolton, W.R., Chapin, F.S., Dyurgerov, N.B., Fastie, C.L., Griffith, B., Hollister, R.D., Hope, A., Huntington, H.P., Jensen, A.M., Jia, G.J., Jorgenson, T., Kane, D.L., Klein, D.R., Kofinas, G., Lynch, A.H., Lloyd, A.H., McGuire, A.D., Nelson, F.E., Oechel, W.C., Osterkamp, T.E., Racine, C.H., Romanovsky, V.E., Stone, R.S., Stow, D.A., Sturm, M., Tweedie, C.E., Vourlitis, G.L., Walker, M.D., Walker, D.A., Webber, P.J., Welker, J.M., Winker, K., Yoshikawa, K., 2005. Evidence and implications of recent climate change in northern Alaska and other arctic regions. Clim. Change 72, 251—298.

Hodson, A., Anesio, A.M., Tranter, M., Fountain, A., Osborn, M., Priscu, J., Laybourn-Parry, J., Sattler, B., 2008. Glacial ecosystems. Ecol. Monogr. 78, 41—67.

Hogg, C., Neveu, M., Stokkan, K.A., Folkow, L., Cottrill, P., Douglas, R., Hunt, D.M., Jeffery, G., 2011. Arctic reindeer extend their visual range into the ultraviolet. J. Exp. Biol. 214, 2014—2019.

Høye, T.T., Post, E., Meltofte, H., Schmidt, N.M., Forchhammer, M.C., 2007. Rapid advancement of spring in the High Arctic. Curr. Biol. 17, R449—R451.

Jeffries, M.O., Richter-Menge, J., Overland, J.E. (Eds.), 2013. Arctic Report Card: Update for 2013. http://www.arctic.noaa.gov/reportcard/.

Jepsen, J.U., Kapari, L., Hagen, S.B., Schott, T., Vindstad, O.P.L., Nilssen, A.C., Ims, R.A., 2011. Rapid northwards expansion of a forest insect pest attributed to spring phenology matching with sub-Arctic birch. Glob. Change Biol. 17 (6), 2071—2083.

Johansson, M., Callaghan, T.V., Bosiö, J., Åkerman, H.J., Jackowicz-Korczynski, M., Christensen, T.R., 2013. Rapid responses of permafrost and vegetation to experimentally increased snow cover in sub-Arctic Sweden. Environ. Res. Lett. 8, 035025. http://dx.doi.org/10.1088/1748-9326/8/3/035025.

Johansson, C., Pohjola, V.A., Callaghan, T.V., Jonasson, C., 2011. Changes in snow characteristics in sub-Arctic Sweden. In: Callaghan, T.V., Tweedie, C.E. (Eds.), Multi-decadal Changes in Tundra Environments and Ecosystems: The International Polar Year Back to the Future Project Ambio, 40 (6), pp. 566−574.

Jonasson, S., Callaghan, T.V., 1992. Root mechanical properties related to disturbed and stressed habitats. New Phytol. 122, 179−186.

Juday, G.P., 2009. Boreal Forests and Climate Change. Oxford Companion to Global Change. Oxford University Press, pp. 75−84.

Kausrud, K.L., Mysterud, A., Steen, H., Vik, J.O., Østbye, E., Cazelles, B., Framstad, E., Eikeset, A.M., Mysterud, I., Solhøy, T., Stenseth, N.C., 2008. Linking climate change to lemming cycles. Nature 456, 93−98.

Kharuk, V.I., Ranson, K.J., Kuz'michev, V.V., Im, S., 2003. Landsat-based analysis of insect outbreaks in southern Siberia. Can. J. Remote Sens. 29, 286−297.

Kilpelainen, A., Gregow, H., Strandman, H., Kellomaki, S., Venalainen, A., Peltola, H., 2010. Impacts of climate change on the risk of snow-induced forest damage in Finland. Clim. Change 99, 193−209.

Klein, D.R., 1968. The introduction, increase, and crash of reindeer on St. Matthew Island. J. Wildl. Manag. 32, 350−367.

Kohler, J., Aanes, R., 2004. Effect of winter snow and ground-icing on a Svalbard reindeer population: results of a simple snowpack model. Arct. Antarct. Alp. Res. 36, 333−341.

Larsen, K.S., Grogan, P., Jonasson, S., Michelsen, A., 2007b. Dynamics and microbial dynamics in two subarctic ecosystems during winter and spring thaw: effects of increased snow depth. Arct. Antarct. Alp. Res. 39, 268−276.

Larsen, K.S., Ibrom, A., Jonasson, S., Michelsen, A., Beier, C., 2007a. Significance of cold-season respiration and photosynthesis in a subarctic heath ecosystem in Northern Sweden. Glob. Change Biol. 13, 1498−1508.

Lewis, M.C., Callaghan, T.V., 1976. Tundra. In: Monteith, J.L. (Ed.), Vegetation and the atmosphere, vol. 2. Academic press, London, pp. 399−433.

Lloyd, A.H., Bunn, A.G., 2007. Responses of the circumpolar boreal forest to 20th century climate variability. Environ. Res. Lett. 2, 13.

Mack, M.C., Bret-Harte, M.S., Hollingsworth, T.N., Jandt, R.R., Schuur, E.A.G., Shaver, G.R., Verbyla, D.L., 2011. Carbon loss from an unprecedented Arctic tundra wildfire. Nature 475, 489−492.

Madsen, J., Tamstorf, M., Klaassen, M., Eide, N., Glahder, C., Rigét, F., Nyegaard, H., Cottaar, F., 2007. Effects of snow cover on the timing and success of reproduction in high-Arctic pink-footed geese *Anser brachyrhynchus*. Polar Biol. 30, 1363−1372.

McCullough, D.G., Werner, R.A., Neumann, D., 1998. Fire and insects in northern and boreal forest ecosystems of North America. Annu. Rev. Entomol. 43, 107−127.

McKinnon, L., Berteaux, D., Gauthier, G., Bety, J., 2013. Predator-mediated interactions between preferred, alternative and incidental prey in the arctic tundra. OIKOS 122, 1042−1048.

McRae, L., Zöckler, C., Gill, M., Loh, J., Latham, J., Harrison, N., Martin, J., Collen, B., 2010. Arctic Species Trend Index 2010: Tracking Trends in Arctic Wildlife. CAFFF CBMP Report No. 20. CAFF International Secretariat, Akureyri, Iceland, 39 pp.

Mech, L.D., 2004. Is climate change affecting wolf populations in the High Arctic? Clim. Change 67, 87−93.

Meier, W.N., Gerland, S., Granskog, M.A., Key, J.R., Haas, C., Hovelsrud, G.K., Kovacs, K., Makshtas, A., Michel, C., Perovich, D., Reist, J.D., van Oort, B.E.H., 2011. Chapter 9. Sea ice. In: Arctic Monitoring and Assessment Programme 2011. Snow, Water, Ice and Permafrost in the Arctic (SWIPA): Climate Change and the Cryosphere. Arctic Monitoring and Assessment Programme, Oslo, Norway xii + 538 pp.

Miller, F.L., Barry, S.J., 2009. Long-term control of Peary Caribou numbers by unpredictable, exceptionally severe snow or ice conditions in a non-equilibrium grazing system. Arctic 62, 175—189.

Miller, P.C., Tieszen, L., 1972. A preliminary model of processes affecting primary production in the Arctic tundra. Arct. Alp. Res. 4, 1—18.

Myers-Smith, I.H., Forbes, B.C., Wilmking, M., Hallinger, M., Lantz, T., Blok, D., Tape, K.D., Macias-Fauria, M., Sass-Klaassen, U., Levesque, E., et al., 2011. Shrub expansion in tundra ecosystems: dynamics, impacts and research priorities. Environ. Res. Lett. 6 (4), 045509.

Nobrega, S., Grogan, P., 2007. Deeper snow enhances winter respiration from both plant-associated and bulk soil carbon pools in birch hummock tundra. Ecosystems 10, 419—431.

Olofsson, J., Ericson, L., Torp, M., Stark, S., Baxter, R., 2011. Carbon balance of Arctic tundra under increased snow cover mediated by a plant pathogen. Nat. Clim. Change 1, 220—223.

Olofsson, J., Oksanen, L., Callaghan, T., Hulme, P.E., Oksanen, T., Suominen, O., 2009. Herbivores inhibit climate-driven shrub expansion on the tundra. Glob. Change Biol. 15 (11), 2681—2693.

Parkinson, A.J., Butler, J.C., 2005. Potential impacts of climate change on infectious diseases in the Arctic. Int. J. Circumpolar Health 64, 478—486.

Perfect, E., Miller, R.D., Burton, B., 1988. Frost upheaval of overwintering plants: a qualitative study of displacement processes. Arct. Alp. Res. 20, 70—75.

Pomeroy, J.W., Bewley, D.S., Essery, R.L.H., Hedstrom, N.R., Link, T., Granger, R.J., Sicart, J.E., Ellis, C.R., Janowicz, J.R., 2006. Shrub tundra snowmelt. Hydrol. Process. 20, 923—941.

Post, E., Forchhammer, M.C., 2008. Climate change reduces reproductive success of an Arctic herbivore through trophic mismatch. Philos. Trans. R. Soc. B 363, 2369—2375.

Power, M., Reist, J.D., Dempson, J.B., 2008. Fish in high-latitude lakes. In: Vincent, W.F., Laybourn-Parry, J. (Eds.), Polar Lakes and Rivers—Limnology of Arctic and Antarctic Aquatic Ecosystems. Oxford University Press, UK., pp. 249—269.

Preece, C., Callaghan, T.V., Phoenix, G.K., 2012. Impacts of winter icing events on the growth, phenology and physiology of sub-Arctic dwarf shrubs. Physiol. Plant. 146, 460—472.

Prowse, T.K., Alfredsen, S., Beltaos, B., Bonsal, C., Duguay, A., Korhola, J., McNamara, W.F., Vincent, V., Vuglinsky, Weyhenmeyer, G., 2011. Chapter 6. Changing lake and river ice regimes: trends, effects and implications. In: Arctic Monitoring and Assessment Programme 2011. Snow, Water, Ice and Permafrost in the Arctic (SWIPA): Climate Change and the Cryosphere. Arctic Monitoring and Assessment Programme, Oslo, Norway xii + 538 pp.

Regehr, E.V., Lunn, N.J., Amstrup, S.C., Stirling, I., 2007. Effects of earlier sea ice breakup on survival and population size of polar bears in western Hudson Bay Source. J.Wildl. Manag. 71, 2673—2683.

Rennert, K.J., Roe, G., Putkonen, J., Bitz, C.M., 2009. Soil thermal and ecological impacts of rain on snow events in the circumpolar arctic. J. Climate 22, 2302—2315.

Riseth, J.Å., Tømmervik, H., Helander-Renvall, E., Pohjola, V., Labba, N.T., Labba, N., Niia, E.A., Kuhmunen, H., Schanche, A., Jonasson, C., Johansson, C., Sarri, L.-E., Bjerke, J.W., Malnes, E., Callaghan, T.V., 2010. "Snow and Ice" Sami TEK and science in concert for understanding climate change effects on reindeer pasturing. Polar Rec. 47, 202—217.

Schuur, E.A.G., Bockheim, J., Canadell, J.G., Euskirchen, E., Field, C.B., Goryachkin, S.V., Hagemann, S., Kuhry, P., Lafleur, P.M., Lee, H., Mazhitova, G., Nelson, F.E., Rinke, A., Romanovsky, V.E., Shiklomanov, N., Tarnocai, C., Venevsky, S., Vogel, J.G., Zimov, S.A., 2008. Vulnerability of permafrost carbon to climate change: implications for the global carbon cycle. Bioscience 58, 701—714.

Semerdjieva, S.I., Sheffield, E., Phoenix, G.K., Gwynn-Jones, D., Callaghan, T.V., Johnson, G., 2003. Contrasting strategies for UV-B screening in sub-Arctic dwarf shrubs. Plant Cell Environ. 26, 957−964.

Sharma, S., Jackson, D.A., Minns, C.K., Shuter, B.J., 2007. Will northern fish populations be in hot water because of climate change? Glob. Change Biol. 13, 2052−2064.

Smith, L.C., Sheng, Y., MacDonald, G.M., Hinzman, L.D., 2005. Disappearing Arctic lakes. Science 308, 1429.

Smol, J.P., Douglas, M.S.V., 2007. Crossing the final ecological threshold in High Arctic ponds. Proc. Natl. Acad. Sci. U.S.A. 104, 12395−12397.

Sonesson, M., Callaghan, T.V., 1991. Plants of the Fennoscandian Tundra. Arctic 44, 95−105.

Starr, G., Oberbauer, S.F., Ahlquist, L.E., 2008. The photosynthetic response of Alaskan tundra plants to increased season length and soil warming. Arct. Antarct. Alp. Res. 40, 181−191.

Stöcklin, J., Körner, Ch, 1999. Recruitment and mortality of *Pinus sylvestris* near the Nordic treeline: the role of climatic change and Herbivory. Ecol. Bull. 47, 168−199.

Streletskiy, D., Anisimov, O., Vasiliev, A., 2014. Permafrost degradation. In: Haeberli, W., Whiteman, C.A. (Eds.), Snow- and Ice-Related Hazards, Risks and Disasters. Elsevier, pp. 303−344.

Sturm, M., McFadden, J.P., Liston, G.E., Chapin III, F.S., Racine, C.H., Holmgren, J., 2001. Snow-shrub interactions in Arctic tundra: a hypothesis with climatic implications. J. Climate 14, 336−344.

Sveinbjörnsson, B., Hofgaard, A., Lloyd, A., 2002. Natural causes of the tundra-taiga boundary. Ambio Spec. Rep. 12, 23−29.

Tenow, O., 1996. Hazards to a mountain birch forest—Abisko in perspective. Ecol. Bull. 45, 104−114.

Torp, M., Olofsson, J., Witzell, J., Baxter, R., 2010. Snow-induced changes in dwarf birch chemistry influence level of herbivory and autumnal moth performance. Polar Biol. 33, 692−702.

Turchin, P., Batzli, G.O., 2001. Availability of food and the population dynamics of arvicoline rodents. Ecology 82, 1521−1534.

Turunen, M., Soppela, P., Kinnunen, H., Sutinen, M.L., Martz, F., 2009. Does climate change influence the availability and quality of reindeer forage plants? Polar Biol. 32, 813−832.

Tye, A.M., Young, S.D., Crout, N.M.J., West, H.M., Stapleton, L.M., Poulton, P.R., Laybourn-Parry, J., 2005. The fate of N-15 added to High Arctic tundra to mimic increased inputs of atmospheric nitrogen released from a melting snowpack. Glob. Change Biol. 11, 1640−1654.

Tyler, N.J.C., 2010. Climate, snow, ice, crashes, and declines in populations of reindeer and caribou (*Rangifer tarandus* L.). Ecol. Monogr. 80 (2), 197−219.

Vincent, W., Callaghan, T.V., Dahl-Jensen, D., Johansson, M., Kovacs, K.M., Michel, C., Prowse, T., Reist, J.D., Sharp, M., 2011. Section 11.4 − Impacts of changing snow, water, ice and permafrost on Arctic ecosystems. In: Arctic Monitoring and Assessment Programme 2011. Snow, Water, Ice and Permafrost in the Arctic (SWIPA): Climate Change and the Cryosphere. Arctic Monitoring and Assessment Programme, Oslo, Norway xii + 538 pp.

Vincent, W.F., Whyte, L.G., Lovejoy, C., Greer, C.W., Laurion, I., Suttle, C.A., Corbeil, J., Mueller, D.R., 2009. Arctic microbial ecosystems and impacts of extreme warming during the International Polar Year. Polar Sci. 3, 171−180.

Walker, M.D., Walker, D.A., Welker, J.M., Arft, A.M., Bardsley, T., Brooks, P.D., Fahnestock, J.T., Jones, M.H., Losleben, M., Parsons, A.N., Seastedt, T.R., Turner, P.L., 1999. Long-term experimental manipulation of winter snow regime and summer temperature in arctic and alpine tundra. Hydrol. Process. 13, 2315−2330.

Wipf, S., Rixen, C., 2008. A review of snow manipulation experiments in Arctic and alpine tundra ecosystems. Polar Res. 29, 95–109.

Wolff, J.O., 1980. Social organization of the taiga vole *Microtus xanthognathus*. Biologist (Charleston) 62, 34–45.

Xu, L., Myneni, R.B., Chapin III, F.S., Callaghan, T.V., Pinzon, J.E., Tucker, C.J., Zhu, Z., Bi, J., Ciais, P., Tommervik, H., Euskirchen, E.S., Forbes, B.C., Piao, S.L., Anderson, B.T., Ganguly, S., Nemani, R.R., Goetz, S.J., Beck, P.S.A., Bunn, A.G., Cao, C., Stroeve, J.C., 2013. Temperature and vegetation seasonality diminishment over northern lands. Nat. Clim. Change. http://dx.doi.org/10.1038/NCLIMATE1836.

Yarie, J., 2008. Effects of moisture limitation on tree growth in upland and floodplain forest ecosystems in interior Alaska. Forest Ecol. and Manag. 256, 1055–1063.

Chapter 6

Ice and Snow as Land-Forming Agents

Darrel A. Swift [1], Simon Cook [2], Tobias Heckmann [3], Jeffrey Moore [4], Isabelle Gärtner-Roer [5] and Oliver Korup [6]

[1] *Department of Geography, University of Sheffield, United Kingdom,* [2] *School of Science and the Environment, Manchester Metropolitan University, Manchester, United Kingdom,* [3] *Department of Physical Geography, Catholic University of Eichstätt-Ingolstadt, Germany,* [4] *Department of Geology and Geophysics, University of Utah, Salt Lake City, UT, United States,* [5] *Department of Geography, University of Zürich, Switzerland,* [6] *Institute of Earth and Environmental Science, University of Potsdam, Germany*

ABSTRACT

Many high-latitude and high-altitude regions are covered by ice and snow for substantial parts of the year, and typically, the landscape of such regions bears the strong imprint of ice- and snow-related processes operating over Quaternary to modern timescales. Despite strong research interest in the nature, rate, and efficacy of cold-region geomorphic processes, most research has been devoted to glacier and permafrost phenomena, whereas comparably few studies have quantitatively addressed the role of snow as a land-forming agent. In this chapter, we review the current research on land-forming processes related to glacial erosion and deposition; permafrost and periglacial processes; and snow-related processes such as nivation, snow creep, and snow avalanching. Our objective is to highlight those questions that drive current research and those that seem sufficiently promising to further our understanding of geomorphic form and process in the cryosphere. We do so bearing in mind that such process knowledge is essential for successfully predicting form and process, and hence avoiding snow- and ice-related hazards. We discuss whether certain aspects concerning the role of ice and snow as land-forming agents may have been overrated or underrated, if not overlooked, in the context of comprehensive studies on landscape evolution in polar and high mountainous terrain. We conclude by outlining a number of recommendations for future research in the field.

Snow and Ice-Related Hazards, Risks, and Disasters. http://dx.doi.org/10.1016/B978-0-12-394849-6.00006-8

6.1 GLACIAL PROCESSES AND LANDSCAPES

Glaciation plays a significant role in landscape evolution at regional to continental scales. Glaciers and ice sheets carve spectacular landscapes (e.g., Valla et al., 2011; Ward et al., 2012) and transport prodigious quantities of sediment (e.g., Hallet et al., 1996; Cowton et al., 2012). Glacial processes therefore exert a strong influence on relief and landscape dynamics, and this poses considerable challenges for landscape and geohazard management. Notably, glacial processes result in deep incision, causing the oversteepening of valley slopes, and the erosion of deep basins, which influence ice mass stability and provide locations for the formation of potentially hazardous lakes. A sound understanding of glacial erosional processes, patterns, and rates is therefore essential, and considerable challenges to understanding remain to be addressed.

6.1.1 Mechanisms and Controls

Glaciers erode bedrock through three primary mechanisms: quarrying (or "plucking"), abrasion, and erosion by subglacial water. Quarrying is the removal of bedrock in blocks as ice moves over bedrock steps (Figure 6.1) and is often considered the most important glacial erosional process (e.g., Riihimaki et al., 2005; Cohen et al., 2006; Iverson, 2012; Zoet et al., 2013). Sliding in the presence of steps opens cavities that enhance bedrock fracture by increasing the stresses imposed on areas of bedrock that remain in contact with the overlying ice. Quarrying is favored by fast sliding, which promotes cavitation (Hallet, 1996), and fluctuating high basal water pressures (Iverson, 1991; Cohen et al., 2006), which facilitate the opening and propagation of fractures within the rock mass. Hence, quarrying should be especially effective where the bedrock has preexisting fractures (Dühnforth et al., 2010; Iverson, 2012; Hooyer et al., 2012).

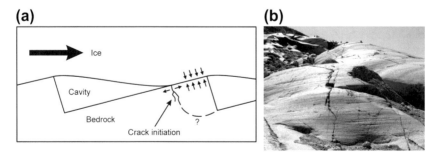

FIGURE 6.1 (a) Stresses and areas of probable fracture initiation when ice flows over bedrock steps *(after Hallet (1996))*. (b) Quarried face on the lee side of a bedrock bump near the terminus of Rhonegletscher, Switzerland (former ice flow direction from left to right). *Photograph: W. Haeberli.*

(a)   **(b)**

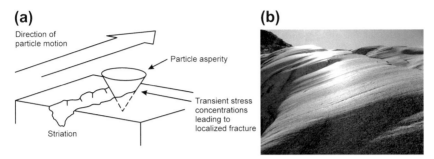

FIGURE 6.2   (a) Simple representation of abrasion, showing the formation of a single striation at the asperity of a moving clast *(modified from Drewry (1986))*. Abrasion can also lead to polishing through the removal of asperities from the bedrock surface. (b) A polished and striated bedrock surface in the forefield of Rhonegletscher, Switzerland. *Photograph: W. Haeberli.*

Abrasion is wear on bedrock by clasts and particles entrained by sliding ice (Figure 6.2). In widely used models (Boulton, 1974, 1979; Hallet, 1979, 1981), abrasion increases with sliding but is also dependent on the force exerted by clasts on the bed. Boulton's model assumes that clasts support the weight of the overlying ice, meaning the force exerted by an individual clast is a function of (a) effective pressure (ice pressure minus subglacial water pressure) and (b) the area of clast–bed contact. Hallet's model recognizes that ice deforms around clasts, meaning the force exerted by an individual clast is a function of (a) the buoyant weight of the clast within the ice and (b) drag exerted on the clast as ice deforms around it to replace ice lost by basal melting. For a given sliding rate, these assumptions have differing implications for the abrasion rate (Table 6.1).

Deep rock-cut glaciofluvial channels and tunnel valleys provide evidence that water in subglacial and proglacial environments may be a very effective erosional agent. The seasonal nature of meltwater production means flow takes place in subglacial channels at high velocities and retains a high unsatisfied sediment transporting capacity within and beyond the glacier system (Alley et al., 1997). Glacial meltwater therefore possesses a prodigious ability to strip bedrock of protective sediment layers and acquire energy and tools for fluvial bedrock erosion. In ice sheet contexts, tunnel valleys may result from rapid fluvial bedrock erosion when stored water is released catastrophically (e.g., Shaw, 2002, 2010; Kehew et al., 2012; Fischer et al., 2014).

Process rates in ice sheet and glacier systems depend on a wide variety of spatially variable factors that exert a strong control on patterns of erosion at landscape scales. Climate is a preeminent controlling factor that dictates glacier thermal regime and therefore the presence of water in the glacier system and of sliding at the glacier bed. Cold climates that are characteristic of continental and polar glaciation restrict sliding to locations where ice is sufficiently thick to promote basal melting, for example, at the base of troughs beneath large ice sheets. Further, for these ice masses, seasonal melting at the

TABLE 6.1 Emphasis on Particular Controlling Variables Shown by Different Abrasion Models and Their Implications for Subglacial Abrasion Rates Under Constant Basal Sliding Velocity.

Effective Pressure (Boulton Model)		Clast Concentration (Hallet Model)	
Low effective pressure	Abrasion rate increases as effective pressure increases because individual clasts impose greater force on the bed	Low clast concentration	Abrasion rate increases as clast concentration increases because this increases the number of clasts in contact with the bed
High effective pressure	Abrasion rate decreases as effective pressure increases because high pressure promotes melting of basal ice layers and deposition of entrained clasts	High clast concentration	Abrasion rate decreases as clast concentration increases because ice deformation around clasts is impeded, reducing the drag exerted on individual clasts

ice surface and the seasonal development of hydraulically efficient channels may be absent or spatially limited.

Where sliding is present, basal sliding velocity is a key control on quarrying and abrasion rates, meaning erosion potential is the highest along the principal axis of glacier flow and along this axis reaches a peak beneath the equilibrium line altitude (ELA) where the total ice flux is the greatest (e.g., Anderson et al., 2006). These patterns have been confirmed in simplified models of flow in valley cross-sections (e.g., Harbor, 1992) and long profiles (e.g., MacGregor et al., 2000; Anderson et al., 2006). Velocity is also influenced by valley topography (e.g., Egholm et al., 2012) and in ice sheet settings is guided strongly by topographic steering of ice flow (e.g., Kessler et al., 2008). An important additional control on velocity is subglacial water pressure, which rises seasonally as surface melt enters the glacial drainage system and reaches the bed. This results in rapid sliding in the ablation zones of glaciers and ice sheets (e.g., Herman et al., 2011), and high variation in basal water pressure produced by diurnal melt is suspected to enhance rates of glacial quarrying (e.g., Hooke, 1991).

Erosion potential is further dependent on the sediment transport capacity of glacial and fluvial transport pathways. This relationship can be complex. For example, erosion products may accumulate as a subglacial till layer that protects the bed from further erosion (e.g., Zemp et al., 2005), but deformation of the till layer, which can promote fast ice flow, can sustain sediment evacuation and thus

FIGURE 6.3   Sediment-charged proglacial outwash at Haut Glacier d'Arolla, Switzerland, emanating from the subglacial drainage system. Measured evacuation rates for suspended sediment alone exceed 500 tons per day during peak melt season and the total annual load is equivalent to around 1.5 mm of bedrock erosion per year (e.g., Swift et al., 2002). *Photograph: M. Hambrey.*

maintain erosion of the bed. Further, in the ablation areas of glaciers and ice sheets, evacuation of erosion products is dominated by the work of subglacial channels (e.g., Alley et al., 1997; Swift et al., 2002, 2005a,b) (Figure 6.3), although evacuation can be limited where water and sediment are required to ascend adverse slopes (Hooke, 1991; Creyts et al., 2013). The wide variety of sediment entrainment and transfer processes in glacial systems has been reviewed extensively elsewhere (e.g., Alley et al., 1997; Knight, 1997).

6.1.2 Landforms and Landscape Evolution

Patterns of ice flux and hydrology and their feedbacks with topography produce distinctive erosional landforms that attest to the efficacy of glacial processes. In addition, glacial and fluvial transport pathways can focus sediment deposition to produce distinctive sedimentary landforms.

Cirques are semicircular or amphitheater-shaped erosional features that are typified by a steep headwall and overdeepened floor located behind a prominent lip (e.g., Barr and Spagnolo, 2013, Figure 6.4). Cirque erosion is initiated during cirque-style glacial conditions that occur where mountainous topography rises above the snowline. The pattern of cirque erosion is controlled by rotational movement about a point at the ice surface that approximates the ELA (e.g., Grove, 1958), meaning the deepest erosion occurs behind the lip. Lowering of the ELA causes ice from multiple cirques to converge (see below) and to thus form a prominent valley step (Anderson et al., 2006, Figure 6.4). Oscillation of the climate between glacial and interglacial states means that the intervening step may experience relatively little glacial erosion (Cook and Swift, 2012, Figure 6.5).

FIGURE 6.4 A mountain cirque (Cwm Cau, Cadair Idris, west Wales) demonstrating a classic amphitheater shape comprising a steep headwall and overdeepened floor. The cirque is separated from the valley floor below by a prominent lip and steep valley step. *Photograph: S. Cook.*

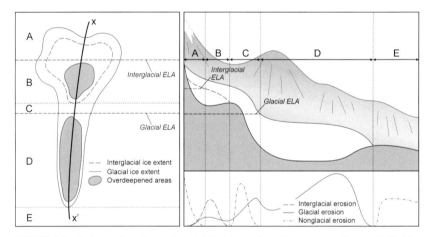

FIGURE 6.5 Patterns of erosion and glacial incision under a time-varying climate. (Left) Plan view of interglacial and glacial ice extent. Horizontal lines indicate ELA positions and divisions of the along-flow transect x−x′. (Right) Profile view of the ice surface and bed topography along transect x−x′ (vertical scale is exaggerated). The total contributions of glacial and interglacial erosional processes (shown below) result in a small high-elevation overdeepened cirque basin and deeply incised low-elevation overdeepened trough. *Adapted from Cook and Swift (2012).*

Below the cirque level, convergence of ice into preglacial valleys enables deep erosion that produces long, characteristically parabolic (or "U-shaped") troughs that frequently exhibit steps associated with mass inputs from smaller tributaries that may appear as "hanging" valleys (e.g., MacGregor et al., 2000,

(a)   **(b)**

FIGURE 6.6   Glacial troughs. (a) A deep valley in the French Pyrénées demonstrating a classic parabolic shape and, in the foreground, an overdeepened valley floor (Gaube Lake). *Photograph: Simon Cook.* (b) Part of the Sognefjord fjord system in Norway. Sognefjord reaches a maximum depth of around 1,300 m below the sea level and, as is characteristic of fjord systems, the bed rises to form a sill at the fjord mouth. *Photograph: D. Swift.*

2009) (Figures 6.5 and 6.6(a)). In ice sheet contexts, topographic steering of ice sheet flow enables the cutting of troughs through mountain ranges to 10^3 m below sea level (e.g., Kessler et al., 2008) (Figure 6.6(b)). Preexisting drainage patterns and geological structures exert a strong control on trough location (e.g., Glasser and Ghiglione, 2009), whereas bedrock strength appears to influence the shape of trough cross-sections (e.g., Swift et al., 2008). Further, troughs in both glacier and ice sheet contexts frequently demonstrate overdeepened morphologies (see below).

Overdeepenings are closed basins in the glacier bed that are common features in glaciated landscapes (Cook and Swift, 2012, Figure 6.7(a)). In large glacier systems, their depth can exceed 1 km and their presence leaves ocean-terminating outlet glaciers vulnerable to rapid retreat (Cook and Swift, 2012; Allison et al., 2014; Vieli, 2014). Ice mass geometry permits ice and water to ascend adverse subglacial slopes and consideration of ice and water fluxes indicates that troughs should tend toward a uniformly overdeepened morphology (Cook and Swift, 2012). However, trough profiles frequently exhibit multiple overdeepenings (Figure 6.7(a)) that indicate the presence of feedback processes that serve to focus deep erosion. Flow convergence at valley confluences appears to explain the location of many overdeepenings, and steering of flow by basins may be an important feedback (e.g., Herman et al., 2011). Nevertheless, the importance of water means that hydrological feedbacks are likely to be very important. Alley et al. (1997, 2003) have emphasized the importance of sediment flushing by water, meaning overdeepening should be focused toward the glacier terminus where subglacial water fluxes reach their maximum (e.g., Figure 6.7(b)). In contrast, Hooke (1991) emphasized the importance of surface melt in promoting effective quarrying, meaning overdeepenings should occur where irregularities in the bed cause crevassing at the ice surface.

(a)   **(b)**

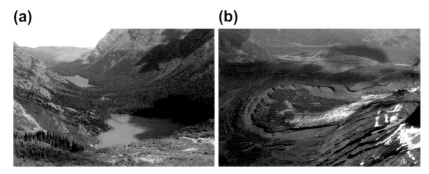

FIGURE 6.7   (a) A series of distinct overdeepenings in the floor of a glacial trough (Grinnell Lake, Josephine Lake, and Lake Sherburne, Glacier National Park, Montana, US). *(Photograph: Darrel Swift)*. (b) A terminal overdeepening that is partially confined by an arcuate moraine ridge (Bagley Icefield, Alaska). *Photograph: D. McCully*.

Moraines and outwash fan heads are prominent depositional landforms that exert an important control on glacier hazards and terminus stability. For ocean- and lake-terminating glaciers, high sediment fluxes and deposition at the grounding line can generate sediment shoals that exert backstress on the glacier front, promoting stability and even glacier advance (e.g., Alley et al., 2007; Anandakrishnan et al., 2007). In mountain glacier contexts, recession and thinning of debris-charged glaciers has led to the development of pro-glacial lakes impounded by terminal moraines or outwash fan heads (e.g., Richardson and Reynolds, 2000; Korup and Montgomery, 2008; Evans et al., 2013; Clague and O'Connor, 2014) (Figure 6.8). Moraine dams are commonly ice cored, meaning their integrity reduces as the ice core degrades and the volume of water in the impounded lake increases. Dam stability can be further reduced by displacement waves produced by mass movement events origi-nating from steep, unstable surrounding slopes (e.g., Hubbard et al., 2005; Korup and Tweed, 2007; Schaub et al., 2013; Haeberli, 2013; Clague and O'Connor, 2014).

At the landscape scale, patterns of glacial erosion and landscape response strongly reflect the scale and topographic context of glaciation as well as the climatic and tectonic regime. In ice sheet contexts that are typical of large-scale continental glaciation, deep troughs that incise through high-elevation landscapes are produced by selective linear erosion (e.g., Sugden, 1968, 1978) (Figure 6.9). Selectivity of erosion is enhanced by topographic steering of ice flow, which focuses flow along preglacial valleys and through cols in mountain ranges (e.g., Harbor, 1992; Kessler et al., 2008), but the climatic context of polar ice sheets means that basal thermal regime is also recognized to be a primary control (e.g., Oerlemans, 1984; Jamieson et al., 2008). In this context, erosion is confined to topographic lows where ice is sufficiently thick to promote basal melting, thereby deepening preglacial valleys and reinforcing the selectivity of erosion through thinning of adjacent ice and isostatic uplift of

FIGURE 6.8   Lakes developing behind moraine dams in Bhutan as glaciers recede and thin. *Photograph: USGS NASA JPL/AGU/Jeffrey Kargel (http://visibleearth.nasa.gov/view.php? id=59561).*

FIGURE 6.9   A deep steep-sided trough produced by selective linear erosion, Penny Ice Cap outlet glacier, Baffin Island, Nunavut, Canada. *Photograph: NASA/Michael Studinger.*

adjacent topography. This enables deep incision, whereas adjacent landscapes demonstrate remarkable degrees of preservation (e.g., Fabel et al., 2002).

In mountain glacier contexts that are characteristic of the glaciation of active orogens, correlation between mountain range elevation and snowline or cirque-floor altitude (Figure 6.10) has been used to support the "glacial buzzsaw" hypothesis (Brozovic et al., 1997) that describes how glacial cirques forming close to the long-term average ELA limit relief by setting the base level for hill-slope processes acting on adjacent mountain peaks (e.g., Oskin and Burbank, 2005; Anders et al., 2010; Egholm et al., 2009). In the absence of active uplift, erosional unloading leads to isostatic uplift of the mountain

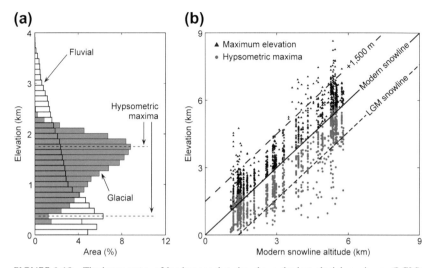

FIGURE 6.10 The hypsometry of landscapes that rise above the last glacial maximum (LGM) snowline indicates the presence of a "glacial buzzsaw". (a) Hypsometry shows distinct maxima at a high elevation, above which the topography is sharply curtailed. (b) At the global scale, summit elevations are confined to altitudes 1,500 m above the local snowline, while hypsometric maxima lie between the modern and LGM snowlines. This pattern is largely independent of tectonic uplift rate, lithology, and general tectonic setting. *Figures redrawn from Egholm et al. (2009).*

range, thereby maintaining mountain height just below the snowline. Nevertheless, glacial deepening and widening of preglacial valleys into glacial troughs (e.g., Harbor, 1992; Amerson et al., 2008), which occurs below cirque level, also contributes substantially to relief production. For example, in the Swiss Alps, Champagnac et al. (2007) have demonstrated that the greatest relief is focused around wide trunk valleys, whereas Norton et al. (2010) demonstrated that trough incision is more significant than cirque erosion in producing "over-steepened" valley profiles.

The deepening and widening of valleys by glacial erosion may be strongly localized in space and time and is likely to influence mountain-scale patterns of isostatic uplift, topographic evolution, and crustal deformation (cf. Herman et al., 2011). Further, during interglacial periods, the concept of paraglacial response (Ballantyne, 2002) describes how the destabilization of sedimentary landforms by retreating ice and the effects of glacial lowering of base levels and steepening of valley sides will result in rapid postglacial landscape change (e.g., Meigs and Sauber, 2000; Hinderer, 2001; Dadson and Church, 2005). Notably, glacial troughs and overdeepenings become important sinks for sediments produced by erosion of adjacent valley sides and mountain peaks (Hinderer, 2001; Straumann and Korup, 2009). The importance of this sediment redistribution and storage is illustrated by Champagnac et al. (2009), who show that anticipated exhumation of the Swiss Alpine Foreland associated with postorogenic decay of the Alpine mountain belt has been negated by the loading effect of postglacial sedimentation in large Quaternary overdeepenings.

6.1.3 Challenges and Uncertainties

Glacial erosion rates are a major concern for geohazard management because erosion of cirques and troughs sets the base level for processes acting on adjacent slopes. Notably, the oversteepening of valley sides by glacial erosion leaves a legacy of slope instability that is itself a powerful agent of landscape modification and a concern for geohazard management (e.g., McColl, 2012). Recent debate has focused on the role of cirque incision, but it is very likely that trough incision (and overdeepening) is an equally or even more important mechanism of stimulating glacial and postglacial erosion of adjacent slopes and peaks (e.g., Amerson et al., 2008; Champagnac et al., 2007; van der Beek and Bourbon, 2008; Norton et al., 2010; Herman et al., 2011; Cook and Swift, 2012). Further, the erosion of overdeepenings presents a very specific and significant problem for the long-term burial of radioactive waste in glaciated regions (Talbot, 1999; Fischer and Haeberli, 2010, 2012; Fischer et al., 2014; Iverson and Person, 2012) because the controls on the location and depth of such focused erosion are poorly understood (Cook and Swift, 2012).

Quantifying rates of glacial erosion is challenging. The most comprehensive data sets of glacial erosion rates in mountainous regions have been derived from measurements of sediment evacuated by subglacial drainage (e.g., Hallet et al., 1996; Koppes and Montgomery, 2009). These data sets reveal that temperate glaciers can erode bedrock at rates in excess of 10 mm/a (Figure 6.11), although rates between 0.1 and 10 mm/a are typical, with lowest rates of erosion ($\sim$0.01 mm/a) exhibited by polar glaciers. In ice sheet contexts, Cowton et al. (2012) have used sediment evacuation data to obtain a mean erosion rate of 2 mm/a for a land-terminating sector of the Greenland ice sheet. A handful of longer-term estimates have also been derived, for example, using thermochronometry. In the European Alps, Valla et al. (2011) have shown a twofold increase in relief since the mid-Pleistocene transition, equating to a minimum mean erosion rate of around 1.0 mm/a. In contrast, in the Lambert glacier system in Antarctica, Thomson et al. (2013) have shown 2 km of incision since 34 Ma, equating to a mean erosion rate of 0.05 mm/a.

Exceptional erosion rates have been demonstrated at rapidly retreating tidewater glaciers (Hallet et al., 1996; Koppes et al., 2010) and glaciers that overlie or advance across unconsolidated sediments (e.g., Motyka et al., 2006; Smith et al., 2007, 2012) where evacuation is influenced strongly by sediment recycling (e.g., Riihimaki et al., 2005; Motyka et al., 2006), fast ice flow (Koppes and Hallet, 2002, 2006), and rapid ablation (e.g., Alley et al., 1997). Consequently, Koppes and Montgomery (2009) warn that high erosion rates documented by sediment evacuation studies may be a transient response to modern climate change. Further, many erosion rate measurements do not

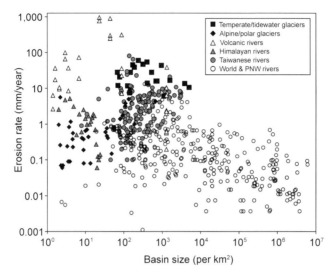

FIGURE 6.11 A compilation of erosion rates for basins worldwide showing higher erosion rates for glaciated basins than for other basins of an equivalent size. Exceptions are mainly basins in very active tectonic and volcanic domains. *Redrawn from Koppes and Montgomery (2009).*

differentiate between the influence of glacial and nonglacial processes, including extraglacial mass movement and periglacial processes. Notably, Delmas et al. (2009) argue that, relative to periglacial processes acting on cirque headwalls, glaciers perform relatively little erosion and serve mainly as agents of sediment transfer. The significance of subglacial fluvial bedrock erosion also remains poorly known, in part because it is extremely difficult to differentiate the relative contributions of subglacial and interglacial fluvial processes (e.g., Montgomery and Korup, 2011).

Timing constraints on the erosion of overdeepenings are even less well known and, though it may be possible to infer erosion rates by analogy with incision rates estimated for glacial troughs, the evolution of overdeepenings and their fundamental limits are poorly known. Ultimately, overdeepening depth must be limited by buoyancy effects and mass balance feedbacks (Cook and Swift, 2012). However, the evacuation of water and sediment from the basin is also crucial, which has focused attention on thresholds that restrict the flow of water along the adverse slope (Hooke, 1991; Alley et al., 1997). Notably, ponding of water should occur when the adverse slope exceeds 11 times the gradient of the ice surface slope above, resulting in conditions suitable for the formation of a subglacial lake (Clarke, 2005). However, the importance of fluvial sediment transport indicates that erosion will effectively cease when the adverse slope gradient exceeds 1.2−1.7 times the ice surface slope because this threshold is associated with a rapid reduction in the hydraulic efficiency of the subglacial drainage system (e.g., Hooke, 1991; Alley et al., 1997, 1998, 2003).

6.2 PERIGLACIAL AND PERMAFROST PROCESSES AND LANDFORMS

Although glaciers produce impressive landscapes, substantial amounts of the Earth's surface are being shaped by ice that is mostly hidden from view in the periglacial zone. The term "periglacial" was first introduced by Lozinski (1909) and defined as the zone peripheral to both former and recent glaciers. Today, the term is more generally used for conditions, processes and landforms associated with cold nonglacial environments (Washburn, 1979). Permafrost is closely related to the periglacial zone, although there is no obligatory overlap between periglacial and permafrost areas: the occurrence of permafrost is an indicator of periglacial conditions, but not an essential periglacial attribute (Thorn, 1992). Periglacial environments in mountains and subpolar regions are characterized by freeze-thaw processes and mechanical frost weathering (French, 1996). These are mainly driven by temperature and time, influencing the depth of frost penetration and, depending on topography, resulting in manifold gravitational movements (Figure 6.12). Although some regions are dominated by diurnal freezing processes, others are shaped by seasonal freeze−thaw cycles and permafrost.

Diurnal, seasonal, and permafrost processes may affect depths of up to 10^{-1}, 10^{0}, and 10^{3} m, respectively. Besides these thermal conditions, moisture and rock-joint distributions additionally affect the susceptibility to weathering. Freeze-thaw-related weathering is assumed to be most intensive near the base

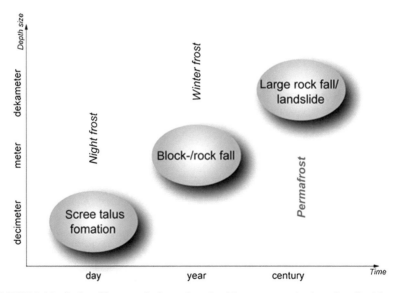

FIGURE 6.12   Scales of frost weathering and gravity-driven mass wasting in rock walls. *Adapted from Haeberli and Gruber (2008).*

of the permafrost active layer (Hallet et al., 1991; Haeberli et al., 2006), whereas the occurrence of gravitational failures in rock walls, such as rock falls or large rock avalanches, results from a complex pattern of thermo-hydro-mechanical processes (Krautblatter et al., 2012).

Thermal conditions as well as the occurrence of ground ice are the most effective drivers for geomorphic processes and landform evolution in periglacial environments. One important process in this context is ice segregation that occurs at the freezing front. Ice crystals attract water from adjacent voids, which freezes on contact and forms larger crystals. Fed mostly by capillary water, the crystals continue to grow and finally form lenses. The vertical pressure resulting from the growing ice lens heaves the surface. This process is very important in terms of weathering and the production of large rock clasts, depending on the thaw-depth regime (Hallet et al., 1991). Frost action is based on the effect of volumetric expansion of freezing water. Volume expansion is mainly related to diurnal frost affecting depths of $10^{-1}-10^{0}$ m. In unconsolidated sediments, the amount of unfrozen water and ground ice in pores, cavities or voids influences the mechanical behavior of freezing or frozen material and leads to shattering, cracking, heaving, settling, or cryoturbation. Mass wasting may occur slowly, involving (geli-)solifluction or permafrost creep processes, and catastrophic detachment slides, retrogressive thaw slumps, debris flows, rock falls, or rock avalanches. Fluvial, aeolian, and coastal processes also transport material in periglacial and permafrost regions.

Changes in this sensitive periglacial/permafrost environment may lead to decaying slope stability (e.g., Haeberli et al., 2010; Deline et al., 2014). Permafrost degradation essentially involves a deepening of the summer thaw, permafrost thawing, or the development of taliks, which induces geomorphic changes in the form of thermokarst formation, active-layer detachment slides, rock falls, or rock glacier motion (Vaughan et al., 2013).

6.2.1 Landforms and Processes Related to Seasonal Frost and Permafrost

Diurnal and seasonal frost processes may lead to solifluction on slopes as low as approximately 2° (Figure 6.13) and sorting processes in flatter areas. Patterned ground (Figure 6.14) includes forms such as circles, polygons, or stripes resulting from the sorting of fine and coarse clasts in regolith cover. Classifications of patterned ground are based on geometric forms and the degree of clast sorting. While the fine material expands on freezing and arches upward, the larger stones are moved radially away from the fine material core (French, 1996). Circles, nets, and polygons characteristically develop on flat surfaces, whereas elongated structures such as stripes develop with increasing slope angles. Recent studies focus on the monitoring of heave and sorting processes in the laboratory, as well as in the field in order to provide a better process understanding (Matsuoka et al., 2003). Ice-wedge polygons are

FIGURE 6.13   Large solifluction lobes at a mountain slope above a seasonal settlement in Kyrgyzstan. *Photograph: T. Bolch.*

FIGURE 6.14   Patterned ground in Svalbard. *Photograph: W. Haeberli.*

another distinct feature of periglacial environments, occurring in cold permafrost, and are well developed in unconsolidated sediments. The initial thermal contraction crack occurs in winter. With the thawing of the active layer, the crack fills with water, which freezes again and expands the crack over an annual cycle. The activity of the ice wedge is based on the frequency of cracking (Christiansen, 2005).

Pingos are prominent landforms that develop and persist only in permafrost (Figure 6.15). They are perennial, ice-cored hills with a common conical shape, up to a 60-m height and <400 m diameter (Mackay, 1998), and they can

FIGURE 6.15   Pingos in Tuktoyuktuk, Canada. *Photograph: W. Haeberli.*

be up to 1000 years old. The two types of pingos are closed-system (hydrostatic) pingos, and open-system (hydraulic) pingos. The former occur in lowland settings within the continuous permafrost zone, and the latter are more common in valley bottoms and foot slopes in both discontinuous and continuous permafrost (Harris and Ross, 2007). Similar but smaller ground ice-related landforms associated with permafrost are lithalsas, mineral palsas, and seasonal ground ice mounds. Palsas are peaty permafrost mounds containing a core of alternating layers of segregated ice and peat or mineral soil material (French, 1996).

Talus and talus cones develop from accumulation of frost-shattered debris, and also commonly occur in periglacial climates. Talus cones have inclinations between 25° and 35°, that is, near the angle of internal friction of the constituting material, and their formation is influenced by lithology and active freeze−thaw cycles driving the mechanical weathering of the source rock face. Where perennially frozen and rich in subsurface ice, talus and talus cones may start to creep and form the rooting zone of rock glaciers. Active rock glaciers result from the long-term cumulative deformation of ice-debris mixtures and are among the most prominent permafrost landforms and indicators in mountain areas (Barsch, 1996; Humlum, 2000; Haeberli et al., 2006, Figure 6.16). Following Barsch (1992, p.176), "active rock glaciers are lobate or tongue-shaped bodies of perennially frozen unconsolidated material supersaturated with interstitial ice and ice lenses that move downslope or downvalley by creep as a consequence of the deformation of ice contained in them and which are, thus, features of cohesive flow." Methods such as geomorphic mapping, terrestrial geodetic surveys, and photogrammetric analyses have been applied to assess the activity and movement of rock glaciers (Roer and Nyenhuis, 2007). In general, the activity of rock glaciers is classified by their ice content and flow behavior (Haeberli, 1985; Barsch, 1996).

FIGURE 6.16   Rock glacier in the area of McCarthy, Alaska, USA. *Photograph: I. Gärtner-Roer.*

Active rock glaciers contain ice and move at rates >0.1 m/a, whereas inactive landforms still contain ice but remain immobile. Relict rock glaciers indicate former permafrost conditions; they lack ice and stopped moving often several hundreds to thousands of years ago (Frauenfelder et al., 2001). The rock glacier definition encapsulates form, material, and process, although the role of active processes remains elusive. Large-scale climatological boundary conditions for rock glacier occurrence include a mean annual air temperature below −1 °C to −2 °C, and an annual precipitation of <2,500 mm. Locally, however, the topographic and meteorological controls on rock glacier formation and growth remain largely unknown (Humlum, 2000).

Rock glaciers commonly occur at the foot of free rock faces (talus rock glaciers) or below moraines (debris rock glaciers), and form tongue- or lobe-shaped bodies with characteristic lengths of 200−800 m (Barsch, 1996); they can be 20−100 m thick. Diagnostic characteristics include steep lateral and frontal slopes at the angle of repose with an apron of coarse blocks at the slope toe built by rock glacier creep (Haeberli, 1985). Commonly, several lobes are superimposed on each other forming a complex topography from different rock glacier generations. The stratigraphy of rock glaciers is often described as a three-layer sequence: The uppermost 1−5 m consists of 0.2−5 m boulders covering an ice-rich permafrost layer with 50−70% of ice, and about 30% of fine-material content (Barsch, 1996). The third and basal layer features larger rocks that were deposited at the rock glacier front and overridden by subsequent advances. This characteristic sorting of the rock glacier material is visible in frontal sections.

The top layer composed of coarse rock fragments forms a microtopography of ridges and furrows that reflect the flow processes and the recent history of surface deformation (Haeberli, 1985; Wahrhaftig and Cox, 1959). Borehole surveys indicate that a large part of the surface deformation occurs in a shear horizon at a depth within the rock glacier (Arenson et al., 2002). During recent decades, a growing number of rock glaciers have shown increases in movement rates, possibly related to rising permafrost temperatures and liquid water content (Kääb et al., 2007; Ikeda et al., 2008). Furthermore, seasonal and interannual changes in horizontal velocities appear to be regionally synchronized, possibly reflecting a short-term response to varying ground surface temperatures (Kääb et al., 2007; Delaloye et al., 2010). In the context of possible permafrost degradation, some rock glaciers show clear signs of acceleration with horizontal velocities of $10^0 - 10^2$ m/a, displaying crevasse-like features at their surface indicative of shear mechanisms similar to those known for landslides (Roer et al., 2008; Springman et al., 2012).

Contemporary climatic change has led to warming permafrost, changes in the amount of frozen and unfrozen water, and concomitant landform changes. In general, permafrost is sensitive to changes in temperature and other controls such as snow cover, ice content, and vegetation (Voigt et al., 2010). Changes in spatial extent, thickness, and temperature of permafrost are used as prime diagnostics of climate change. Projected increases in temperature will very likely lead to continued warming and thawing of permafrost, thus contributing to significant changes in slope stability, groundwater fluxes, and nutrient cycles. Thermokarst topography is diagnostic of permafrost decay in soils and sediments, and features mounds, sink holes, slumps, and lake basins caused by the disturbance of the thermal equilibrium brought about by geomorphic, vegetation, or climate changes. In high mountain areas, rock glacier acceleration and degradation indicate the reaction to changes in the thermohydrological regime, as also active-layer detachment slides and rock falls (Vaughan et al., 2013).

6.3 THE ROLE OF SNOW IN FORMING LANDSCAPES

6.3.1 Rationale: Why Snow Matters

Snow cover in high-altitude and high-latitude regions provides both a resource and hazard for local inhabitants. While many mountain communities rely on stored water in winter snowpacks for drinking and irrigation, others are faced with regularly repeated damage, response and recovery from snow-related hazards, which frequently requires costly engineering intervention. Yet, the role of snow as an agent capable of modifying landscape form receives much less attention, despite a strong need to better understand processes driving snow-related resources and hazards.

Snowpacks are complex multiphase systems that vary rapidly in time and space. Predicting snowpack thickness, morphology, density, and water content is crucial in modern resource planning and disaster prevention (McClung and Schaerer, 1993; Immerzeel et al., 2010). However, the dynamic and changing properties of snow cover have received less attention from those seeking to understand geomorphic work and landscape evolution. For example, near-surface ground temperatures have received much recent attention as a potential driver of bedrock fracturing, rock-fall detachment, and hill-slope and glacial erosion (Sanders et al., 2012; Krautblatter et al., 2013). Yet the variability of snow cover that acts as insulator, moisture reservoir, and habitat, remains challenging to implement in process models (Zenklusen Mutter et al., 2010). Even while ground temperatures are increasingly implicated as a driver of natural hazards in alpine landscapes (Gischig et al., 2011a,b), the influence of snow cover remains difficult to accurately parameterize (Mott and Lehning, 2010), and can lead to large uncertainties in models aiming to simulate ground temperature distributions (Goodrich, 1982). Owing to its low thermal conductivity, high albedo, and air-filled void space, snow is an effective insulator, and even approximately 50-cm-thick snow cover may sufficiently sustain approximately 20 °C temperature differences between the ground and air (Callaghan et al., 2011). This can generate significant warming at depths, inhibiting frost penetration, and increase mean annual temperatures (Goodrich, 1982; Zhang, 2005). The result can affect the distribution and activity of periglacial landforms and sediment transport (Haeberli et al., 2006; Kääb et al., 2007), rates of frost-cracking and bedrock damage (Anderson, 1998; Hales and Roering, 2007), and paraglacial rock-slope displacement (Moore et al., 2011; Hasler et al., 2012).

6.3.2 Investigating Snow-Related Geomorphic Processes

The role of snow processes as a driver of geomorphic change has received some research attention with regard to avalanches as agents of sediment transport. Snow cover on hillslopes is dynamic and subject to movement. While the vertical settlement of snow is a consequence of its own weight and metamorphosis, downslope shear by gravity induces an internal deformation of the snow cover. Together, these movements result in snow creep characterized by very slow velocities at the soil–snow interface and increasing velocities toward the snow surface in the order of $10^0–10^1$ mm/day. In contrast, snow gliding affects the entire snow cover with constant velocities up to 10^2 mm/day with extreme rates of 10^2 mm/h, where free water exists at the base of the snow cover. Snow gliding is facilitated by smooth surfaces on slopes $>15–25°$.

Snow avalanches involve the sudden, rapid (>1 m/s; Bozhinskiy and Losev, 1998) movement of the snow cover. Avalanches as land-forming agents are not new to geomorphology (Allix, 1924). A number of quantitative studies since the late 1950s have revealed their potential contribution to the sediment budget and landscape evolution of subpolar and alpine regions (Rapp, 1960; Luckman, 1977, 1978; Heckmann et al., 2005). Particularly, springtime wet-snow avalanches have a recognized ability to plough, pluck, entrain, and transport coarse sediment and organic debris as they traverse snow-free hillslopes (Gardner, 1970, 1983; Luckman, 1977, 1978; Ackroyd, 1986; Bell et al., 1990; Heckmann et al., 2005; Sass et al., 2010; Freppaz et al., 2010). However, despite field evidence that these effects can be locally significant, snow-avalanche erosion has not received the necessary treatment and parameterization for use in quantitative landscape evolution models. Similarly, the ability of snow avalanches to sculpt bedrock landforms on steep rock cliffs and gullies remains disputed. Few authors provide field evidence of these effects (Allix, 1924; Matthes, 1938; Moore et al., 2013), and we are still far from being able to predict landscape form from first principles of avalanche processes. Much work remains in order to develop useful relationships for the reliable prediction of avalanche erosion in alpine and arctic landscapes.

All types of snow avalanches (except dry powder avalanches) are capable of doing geomorphic work wherever they are in contact with a snow-free surface; this can occur in any zone of the avalanche track (initiation, transit, and deposition). However, full-depth snow avalanches are the most effective land-forming agent (Gardner, 1983) due to the high specific weight of wet snow ($>400 \, \text{kg/m}^3$) and the resulting high hydrostatic pressure at the avalanche base. Slush avalanches are also known for their geomorphic impact (Rapp, 1960). The entrainment of rocks and woody debris further increases the destructiveness of avalanches, which can be highly relevant in natural hazard assessment.

Direct scouring and ploughing, leading to the erosion and entrainment of soil and loose debris is an important geomorphic process. The scouring action of snow glides and avalanches on soil- and scree-covered slopes has been figuratively termed "bulldozing" (Jomelli and Bertran, 2001). According to studies referring to the erosion and entrainment of snow cover by avalanches, shear stress as a linear function of flow height and critical velocity should be among the main factors governing soil erosion by avalanches. Enhanced soil erosion by snow gliding, creep, and avalanches has been reported to occur especially when tussock-forming grasses or bushes become frozen into the snow cover, and snow gliding can pull or lever out the plants and clods of topsoil.

Furthermore, debris entrained in a snow avalanche serves as an erosional tool. The impact of rock clasts generates scratch marks and weakens bedrock structure, thus enhancing weathering and preparing for the detachment of rocks. Abrasion is thought to scale with avalanche velocity, impact pressure, and incidence angle, in addition to entrained sediment content. The detachment of loose rock fragments is driven by the tractive stresses of the moving avalanche and has been termed plucking. Avalanches formed by the collapse of snow cornices have been shown to be an important process in the erosion of plateau edges and cirque headwalls (Eckerstorfer et al., 2013), and the regular formation, persistence, and failure of overhanging cornices enhances both weathering processes and loose sediment evacuation from steep rock walls (Sanders et al., 2013).

Numerical models that are routinely used in predicting avalanche runout for hazard zonation have also found application in predicting areas of high velocity and impact pressure that may control sediment entrainment along avalanche tracks. Moore et al. (2013) used one such model—RAMMS (http://ramms.slf.ch/)—to study avalanche sediment transport in a small catchment in the southern Swiss Alps, and found a good correlation between predicted areas of maximum velocity and pressure with field observations of avalanche damage and scour. Many such models also now allow for erosion of snow at the advancing avalanche front; future experiments may allow these models also to predict hillslope erosion.

6.3.3 Snow-Related Landforms

A number of landforms are conditioned by snow cover as opposed to those generated by the direct action of moving snow (Figure 6.17). Snow cover modulates geomorphic processes through its spatiotemporal distribution, duration, and physical properties (Thorn, 1992), while the weight of snow cover enhances the gliding velocity of shallow landslides (Matsuura et al., 2003). Snowmelt provides large amounts of water that may be converted to or may enhance the formation of overland flow and hence erosion and sediment

FIGURE 6.17   Relevance of snow cover and avalanche erosion in mountainous terrain: (A) large avalanches chutes in the eastern Swiss Alps. (B) Sediment-rich avalanche debris below steep bedrock sluice, Matukituki Valley, Southern Alps, New Zealand; note person for scale. (C) Eroded snow—avalanche bridge with thick cover of organic debris, Flüelabach, eastern Swiss Alps. (D) Remnants of snow—avalanche bridge, Zügenschlucht, eastern Swiss Alps. *Adapted from Korup and Rixen, 2014.*

transport. Particularly late-lying snow patches may induce chemical weathering, and intensify, through the provision of meltwater, both sheet wash and solifluction. A whole suite of geomorphic processes has been assembled within the concept of nivation (Christiansen, 1998), for example, backwall failure, sliding and flow, niveofluvial erosion, development of pronival stone pavements, accumulation of alluvial fans, and pronival solifluction. Consequently, there is a variety of landforms generated by nivation. A frequent assemblage of such landforms is a nivation hollow, meltwater channels that emanate from the hollow, and an alluvial fan built from the outwash.

The spatial pattern of nivation is governed by topography (providing suitable snow accumulation sites), and to a large extent by the wind-driven spatial distribution of snow cover (Christiansen, 1998). Nivation is a self-reinforcing phenomenon, where an initial small disturbance cumulates over time: a seasonal or perennial snow cover will facilitate erosion that deepens existing hollows, in which more snow is likely to accumulate in the future. Another prominent landform type frequently attributed to the influence of snow, including avalanches, is the pronival rampart. Pronival ramparts are 10^1-10^2 m long, ridge-like features typically built up of coarse blocks, with a thickness of typically $<5-10$ m (Matthews et al., 2011; Hedding and Sumner, 2013; and references therein). Debate endures whether protalus ramparts are related landforms or genetically different (Hedding and Sumner, 2013). Clearly, the variety of different processes and forms associated with nivation has bred considerable terminological difficulties, and more rigorous definitions would aid this important aspect of periglacial geomorphology.

Avalanche-transported debris also forms characteristic depositional landforms, and contributes to forming cones or fans that are primarily shaped by other processes. Avalanche debris may consist of all kinds of material mostly derived from upper avalanche tracks and initiation zones (Jomelli and Bertran, 2001). However, avalanches can also rework scree-covered surfaces where the latter are snow free (Gardner, 1983). Avalanches can also have indirect geomorphic effects: The uprooting of trees and removal of vegetation cover, for example, can lead to enhanced and sustained erosion by slope wash, fluvial processes, and landslides (Ackroyd, 1986). The transfer of large masses of snow to lower areas, where it often persists for a longer time, also implies avalanche effects on hydrology and microclimate (Bozhinskiy and Losev, 1998; and references therein).

Irrespective of their sediment content, snow avalanches descending steep avalanche tracks can generate impact landforms (Luckman et al., 1994) termed avalanche plunge pools or pits or avalanche tarns. Such land-forming may occur at impact pressures >600 kN/m^2 generated by avalanches involving $>300,000$ m^3. Plunge pools are accompanied by depositional landforms such as mounds or ramparts (Matthews et al., 2011) that are formed by the excavation of debris, depending on the interaction of the impacting avalanche with rivers or lakes.

Bedrock sections of avalanche tracks are scoured by the plucking and abrasive action of avalanches to form gullies with a characteristic U-shaped crossprofile (unlike V-shaped channels formed by fluvial incision) that have been termed avalanche furrows (Sekiguchi and Sugiyama, 2003); their crossprofile has been described as comparatively shallow (1–3 m deep and 2–4 m wide). Abrasion as an important driver of avalanche gully formation is indicated by boulder impact marks and striae on gully sidewalls (Moore et al., 2013). Similarly, boulders dragged over scree-covered surfaces by avalanches have been reported to form scour marks in the direction of flow (Jomelli and Bertran, 2001).

Soil erosion by snow avalanches may strongly modify soil patterns on affected slopes (Freppaz et al., 2010; and references therein) by removing soil cover and retarding soil development. Despite some evidence for primary erosion of bedrock, and despite the observation of high sediment loads in areas where avalanches erode soils, the major contribution of snow avalanches to the sediment budget of arctic and alpine catchments appears to involve entrainment and transport of sediment from upper avalanche tracks to the depositional zones (Luckman, 1977, 1978; Ackroyd, 1986; Bell et al., 1990; Moore et al., 2013; Korup and Rixen, 2014). Investigations of avalanche-transported sediment suggest that the transport mechanism is unselective, that no sorting occurs during transport, and that the corresponding deposits lack any characteristic fabric (Jomelli and Bertran, 2001). Blikra and Sæmundson (1998) provided an overview of the sedimentology of avalanche-borne sediments. The gradual melt-out of sediment-laden avalanche snow deposits often leaves entrained sediment particles balancing on top of larger rocks and bedrock ledges; this phenomenon of perched boulders is a diagnostic feature of avalanche deposits. Colluvial deposits themselves, however, can be reworked by avalanches, as indicated by characteristic small-scale landforms such as debris tails at the leeward side of large boulders (Luckman, 1977).

Finally, episodic delivery of both snow and rock debris by snow avalanches also potentially drives rock glacier formation (Humlum, 2000). Regular deposition of avalanche-borne sediment may create characteristic landforms such as avalanche boulder tongues (Rapp, 1960; Luckman, 1978) that are characterized by varying thickness (up to 15–25 m), a convex crossprofile, and a concave longitudinal profile (Gardner, 1970), and form part of a continuum of landforms, including rock fall-derived talus cones, talus cones modified by avalanches, and avalanche boulder tongues.

6.3.4 Quantifying Rates of Geomorphic Processes Affected by Snow

Several studies have focused on quantifying rates of sediment transport by snow avalanches. Many of these concluded that the contribution of snow avalanches to the sediment budget of avalanche-prone catchments is important

(Rapp, 1960). However, reported rates of sediment transport by snow avalanches are notoriously variable. Rates have been documented to range from virtually zero up to thousands of tons per event, but time series covering more than a few years of detailed coverage are rare (Heckmann et al., 2005).

While rates of sediment accumulation can be measured with reasonable accuracy from mapping, sampling, and statistical simulation of avalanche deposits (Korup and Rixen, 2014), direct measurement of erosion is more complicated. Direct measurement of avalanche erosion requires visible erosional scars (Heckmann et al., 2002). However, a morphological budgeting approach using multitemporal surveying has, to our knowledge, not yet been implemented in avalanche geomorphology research. Ceaglio et al. (2012) used ^{137}Cs-based estimations of (total) soil erosion on areas prone to snow cover-related erosion in order to compare with field measurements of avalanche sediment yield. More often, erosion rates are estimated by relating measured sediment yield to a contributing area. The comparability of results gained from this methodology is difficult to ascertain because different authors use different areas, for example, the initiation zone of the avalanche, different fractions of the hydrological catchment area, or the entire catchment area. Assuming that the deposited sediment must be related to the contributing source area rather than to the total catchment, Heckmann (2006) spatially modeled the total process area of single avalanches to achieve a more appropriate reference area for the estimation of denudation rates. Another approach has been to estimate the volume of depositional landforms of a known age. The avalanche furrows attributed to avalanche erosion by Sekiguchi and Sugiyama (2003) are 1—3 m deep and 2—4 m wide; assuming a Holocene age, they inferred erosion rates of 0.1—0.3 mm/a.

6.4 CONCLUSIONS AND OUTLOOK

The presence and motion of ice and snow create a myriad of forms that range from continental to microscopic scale. Many of these forms are diagnostic and represent the fundaments of glacial and periglacial geomorphologies, while touching upon neighboring fields such as permafrost and snow science. The rates at which ice and snow sculpt landscapes encompass a significant range, and some debate remains about the overall efficacy of glacial, periglacial, and snow erosion, transport, and deposition. This uncertainty severely affects our ability to predict these processes and any concomitant hazard potential to people and infrastructure in close vicinity to pervasive ice and snow cover.

From our review, we have identified numerous research themes regarding ice and snow as land-forming agents. Many of these themes overlap, and highlight the many process linkages in the cryosphere that deserve more detailed research attention. Future research avenues on the geomorphic role of ice and snow may include:

- Better distinction between rates of glacial, fluvial, and periglacial components in erosion rate estimates, particularly those from glacial sediment budgets.
- Elaboration of ways to extrapolate and predict glacial erosion rate estimates beyond the measurement intervals.
- Empirical analyses of landscape-scale patterns of erosion and deposition and development of numerical glacier and ice sheet models to explain the focused nature of trough incision and overdeepening.
- Derivation of reliable remote sensing-based proxies of the stability of glacial lakes at the regional scale, particularly those situated in high mountainous terrain (the same applies to detailed on-site exploration).
- Robust modeling of glacier and landscape responses to climate change with a view to detecting newly arising threats (natural hazards) and opportunities (natural resources).
- Elucidation of, in more quantitative form, the formation conditions, flow, and evolution of rock glaciers as classic permafrost indicators.
- Development of further numerical and physical models that link permafrost dynamics to slope stability.
- Identification of key environmental and mechanical controls on the process of soil and bedrock erosion by snow avalanches, and inclusion of these processes in dynamic runout modeling and hazard analyses.

ACKNOWLEDGMENTS

This work is partly funded by the Potsdam Research Cluster for Georisk Analysis, Environmental Change and Sustainability (PROGRESS). We thank D. Evans, W. Haeberli, and C. Whiteman for constructive reviews of this contribution.

REFERENCES

Ackroyd, P., 1986. Debris transport by avalanche, Torlesse range, New Zealand. Z. Geomorph. 30, 1−14.

Alley, R.B., Anandakrishnan, S., Dupont, T.K., Parizek, B.R., Pollard, D., 2007. Effect of sedimentation on ice-sheet grounding-line stability. Science 315, 1838−1841.

Alley, R.B., Cuffey, K.M., Evenson, E.B., Strasser, J.C., Lawson, D.E., Larson, G.J., 1997. How glaciers entrain and transport basal sediment: physical constraints. Quat. Sci. Rev. 16, 1017−1038.

Alley, R.B., Lawson, D.E., Evenson, E.B., Strasser, J.C., Larson, G.J., 1998. Glaciohydraulic supercooling: a freeze-on mechanism to create stratified, debris-rich basal ice: II. Theory. J. Glaciol. 44, 563−569.

Alley, R.B., Lawson, D.E., Larson, G.J., Evenson, E.B., Baker, G.S., 2003. Stabilizing feedbacks in glacier-bed erosion. Nature 424, 758−760.

Allison, I., Colgan, W., King, M., Paul, F., 2014. Ice sheets, glaciers and sea level rise. In: Haeberli, W., Whiteman, C. (Eds.), Snow and Ice-related Hazards, Risks and Disasters. Elsevier, pp. 713−747.

Allix, A., 1924. Avalanches. Geographical Rev. 14, 519−560.

Amerson, B.E., Montgomery, D.R., Meyer, G., 2008. Relative size of fluvial and glaciated valleys in central Idaho. Geomorphology 93, 537–547.

Anandakrishnan, S., Catania, G.A., Alley, R.B., Horgan, H.J., 2007. Discovery of till deposition at the grounding line of Whillans ice stream. Science 315, 1835–1838.

Anders, A.M., Mitchell, S.G., Tomkin, J.H., 2010. Cirques, peaks, and precipitation patterns in the Swiss Alps: connections among climate, glacial erosion, and topography. Geology 38, 239–242.

Anderson, R.S., 1998. Near-surface thermal profiles in alpine bedrock: implications for the frost weathering of rock. Arct. Alp. Res. 30, 362–372.

Anderson, R.S., Molnar, P., Kessler, M.A., 2006. Features of glacial valley profiles simply explained. J. Geophys. Res. Earth Surf. 111, F01004.

Arenson, L., Hoelzle, M., Springman, S., 2002. Borehole deformation measurements and internal structure of some rock glaciers in Switzerland. Permafrost Periglacial Process. 13, 117–135.

Ballantyne, C.K., 2002. Paraglacial geomorphology. Quat. Sci. Rev. 21, 1935–2017.

Barr, I.D., Spagnolo, M., 2013. Palaeoglacial and palaeoclimatic conditions in the NW Pacific, as revealed by a morphometric analysis of cirques upon the Kamchatka Peninsula. Geomorphology 192, 15–29.

Barsch, D., 1992. Permafrost creep and rockglaciers. Permafrost Periglacial Process. 3, 175–188.

Barsch, D., 1996. Rockglaciers: Indicators for the Present and Former Geoecology in High Mountain Environments. Springer, Berlin, 331pp.

Bell, I., Gardner, J.S., de Scally, F., 1990. An estimate of snow avalanche debris transport, Kaghan Valley, Himalaya, Pakistan. Arct. Alp. Res. 22, 317–321.

Blikra, L.H., Sæmundson, T., 1998. The Potential of Sedimentology and Stratigraphy in Avalanche-Hazard Research. I, vol. 203. NGI Publication, pp. 60–64.

Boulton, G.S., 1974. Processes and patterns of subglacial erosion. In: Coates, D.R. (ed.) Glacial Geomorphol. (State University of New York, Binghamton), 41–87.

Boulton, G.S., 1979. Processes of glacier erosion on different substrata. J. Glaciol. 23, 15–37.

Bozhinskiy, A.N., Losev, K.S., 1998. The fundamentals of avalanche science. SLF Davos Mitt. 55, 280.

Brozovic, N., Burbank, D.W., Meigs, A.J., 1997. Climatic limits on landscape development in the northwestern Himalaya. Science 276, 571–574.

Callaghan, T.V., Johansson, M., Brown, R.D., et al., 2011. Multiple effects of changes in Arctic snow cover. Ambio 40, 32–45.

Ceaglio, E., Meusburger, K., Freppaz, M., Zanini, E., Alewell, C., 2012. Estimation of soil redistribution rates due to snow cover related processes in a mountainous area (Valle d'Aosta, NW Italy). Hydrol. Earth Syst. Sci. 16, 517–528.

Champagnac, J.D., Molnar, P., Anderson, R.S., Sue, C., Delacou, B., 2007. Quaternary erosion-induced isostatic rebound in the western Alps. Geology 35, 195–198.

Champagnac, J.-D., Schunegger, F., Norton, K., von Blanckenburg, F., Abbühl, L.M., Schwab, M., 2009. Erosion-driven uplift of the modern Central Alps. Tectonophysics 474, 236–249.

Christiansen, H.H., 1998. Nivation forms and processes in unconsolidated sediments, NE Greenland. Earth Surf. Process. Landforms 23, 751–760.

Christiansen, H.H., 2005. Thermal regime of ice-wedge cracking in Adventdalen, Svalbard. Permafrost Periglacial Process. 16, 87–98.

Clague, J.J., O'Connor, J.E., 2014. Glacier-related outburst floods. In: Haeberli, W., Whiteman, C. (Eds.), Snow and Ice-related Hazards, Risks and Disasters. Elsevier, pp. 487–519.

Clarke, G.K.C., 2005. Subglacial processes. Annu. Rev. Earth Planet. Sci. 33, 247–276.

Cohen, D., Hooyer, T.S., Iverson, N.R., Thomason, J.F., Jackson, M., 2006. Role of transient water pressure in quarrying: a subglacial experiment using acoustic emissions. J. Geophys. Res. Earth Surf. 111 (F3).

Cook, S.J., Swift, D.A., 2012. Subglacial basins: their origin and importance in glacial systems and landscapes. Earth-Sci. Rev. 115, 332−372.

Cowton, T., Nienow, P., Bartholomew, I., Sole, A., Mair, D., 2012. Rapid erosion beneath the Greenland ice sheet. Geology 40, 343−346.

Creyts, T.T., Clarke, G.K.C., Church, M., 2013. Evolution of subglacial overdeepenings in response to sediment redistribution and glaciohydraulic supercooling. J. Geophys. Res. Earth Surf. 118, 423−446.

Dadson, S.J., Church, M., 2005. Postglacial topographic evolution of glaciated valleys: a stochastic landscape evolution model. Earth Surf. Process. Landforms 30, 1387−1403.

Delaloye, R., Lambiel, C., Gärtner-Roer, I., 2010. Overview of rock glacier kinematics research in the Swiss Alps: seasonal rhythm, interannual variations and trends over several decades. Geographica Helv. 65, 135−145.

Deline, P., Gruber, S., Delaloye, R., Fischer, L., Geertsema, M., Giardino, M., Hasler, A., Kirkbride, M., Krautblatter, M., Magnin, F., McColl, S., Ravanel, L., Schoeneich, P., 2014. Ice loss and slope stability in high-mountain regions. In: Haeberli, W., Whiteman, C. (Eds.), Snow and Ice-related Hazards, Risks and Disasters. Elsevier, pp. 303−344.

Delmas, M., Calvet, M., Gunnell, Y., 2009. Variability of Quaternary glacial erosion rates − a global perspective with special reference to the Eastern Pyrenees. Quat. Sci. Rev. 28, 484−498.

Dühnforth, M., Anderson, R.S., Ward, D., Stock, G.M., 2010. Bedrock fracture control of glacial erosion processes and rates. Geology 38, 423−426.

Drewry, D.J., 1986. Glacial Geologic Processes. Edward Arnold, London.

Eckerstorfer, M., Christiansen, H.H., Rubensdotter, L., Vogel, S., 2013. The geomorphological effect of cornice fall avalanches in the Longyeardalen valley, Svalbard. Cryosphere 7, 1361−1374.

Egholm, D.L., Nielson, S.B., Pedersen, V.K., Lesemann, J.E., 2009. Glacial effects limiting mountain height. Nature 460, 884−888.

Egholm, D.L., Pedersen, V.K., Knudsen, M.F., Larsen, N.K., 2012. Coupling the flow of ice, water, and sediment in a glacial landscape evolution model. Geomorphology 141, 47−66.

Evans, D.J.A., Rother, H., Hyatt, O.M., Shulmeister, J., 2013. The glacial sedimentology and geomorphological evolution of an outwash head/moraine-dammed lake, South Island, New Zealand. Sediment. Geol. 284-285, 45−75.

Fabel, D., Stroeven, A.P., Harbor, J., Kleman, J., Elmore, D., Fink, D., 2002. Landscape preservation under Fennoscandian ice sheets determined from in situ produced [10]Be and [26]Al. Earth Planet. Sci. Lett. 201, 397−406.

Fischer, U.H., Haeberli, W., 2010. In: Glacial Erosion Modelling: Results of a Workshop Held in Unterägeri, Switzerland, 29 April−1 May 2010. Nagra Arbeitsbericht NAB 10−34.

Fischer, U.H., Haeberli, W., 2012. In: Glacial Overdeepening: Results of a Workshop Held in Zürich, Switzerland, 20−21 April 2012. Nagra Arbeitsbericht NAB 12−48.

Fischer, U.H., Bebiolka, A., Brandefelt, J., Follin, S., Hirschorn, S., Jensen, M., Keller, S., Kennell, L., Näslund, J.-O., Normani, S., Selroos, J.-O., Vidstrand, P., 2014. Radioactive Waste Under Conditions of Future Ice Ages. In: Haeberli, W., Whiteman, C. (Eds.), Snow and Ice-related Hazards, Risks and Disasters. Elsevier, pp. 345−393.

Frauenfelder, R., Haeberli, W., Hoelzle, M., Maisch, M., 2001. Using relict rockglaciers in GIS-based modelling to reconstruct Younger Dryas permafrost distribution patterns in the Err-Julier area, Swiss Alps. Norw. J. Geogr. 55, 195−202.

French, H.M., 1996. The Periglacial Environment. Longman, Essex, 341 pp.

Freppaz, M., Godone, D., Filippa, G., Maggioni, M., Lunardi, S., Williams, M.W., Zanini, E., 2010. Soil erosion caused by snow avalanches a case study in the Aosta valley (NW Italy). Arct. Antarct. Alp. Res. 42, 412—421.

Gardner, J.S., 1970. Geomorphic significance of avalanches in the Lake Louise area, Alberta, Canada. Arct. Alp. Res. 2, 135—144.

Gardner, J.S., 1983. Observations on erosion by wet snow avalanches, Mount Rae area, Alberta, Canada. Arct. Alp. Res. 15, 271—274.

Gischig, V.S., Moore, J.R., Evans, K.F., Amann, F., Loew, S., 2011a. Thermomechanical forcing of deep rock slope deformation — part 1: conceptual study of a simplified slope. J. Geophys. Res. Earth Surf. 116, F04010.

Gischig, V.S., Moore, J.R., Evans, K.F., Amann, F., Loew, S., 2011b. Thermomechanical forcing of deep rock slope deformation — part 2: the Randa rock slope instability. J. Geophys. Res. Earth Surf. 116, F04011.

Glasser, N.F., Ghiglione, M.C., 2009. Structural, tectonic and glaciological controls on the evolution of fjord landscapes. Geomorphology 105, 291—302.

Goodrich, L.E., 1982. The influence of snow cover on the ground thermal regime. Can. Geotech. J. 19, 421—432.

Grove, J.M., 1958. Some structures associated with rotational flow in compound and composite cirque glaciers. IAHS Publ. 47, 306—312.

Haeberli, W., 1985. Creep of mountain permafrost: internal structure and flow of alpine rockglaciers. Mitt. VAW/ETH Zürich 77, 119.

Haeberli, W., 2013. Mountain permafrost—research frontiers and a special long-term challenge. Cold Reg. Sci. Technol. 96, 71—76.

Haeberli, W., Gruber, S., 2008. Research challenges for permafrost in steep and cold terrain: an Alpine perspective. In: Kane, D.L., Hinkel, K.M. (Eds.), Ninth International Conference on Permafrost (Fairbanks, Alaska), vol. 1, pp. 1—9.

Haeberli, W., Hallet, B., Arenson, L., Elconin, R., Humlum, O., Kääb, A., Kaufmann, V., Ladanyi, B., Matsuoka, N., Springman, S., Von der Mühll, D., 2006. Permafrost creep and rock glacier dynamics. Permafrost Periglacial Process. 17, 189—214.

Haeberli, W., Noetzli, J., Arenson, L., Delaloye, R., Gärtner-Roer, I., Gruber, S., Isaksen, K., Kneisel, C., Krautblatter, M., Phillips, M., 2010. Mountain permafrost: development and challenges of a young research field. J. Glaciol. 56, 1043—1058.

Hallet, B., 1979. A theoretical model of glacial abrasion. J. Glaciol. 23, 321—334.

Hallet, B., 1981. Glacial abrasion and sliding: their dependence on the debris concentration in basal ice. Ann. Glaciol. 2, 23—28.

Hallet, B., 1996. Glacial quarrying: a simple theoretical model. Ann. Glaciol. 22, 1—8.

Hallet, B., Hunter, L., Bogen, J., 1996. Rates of erosion and sediment evacuation by glaciers: a review of field data and their implications. Global Planet. Change 12, 213—235.

Hallet, B., Walder, J.S., Stubbs, C.W., 1991. Weathering by segregation ice growth in microcracks at sustained subzero temperatures: verification from an experimental study using acoustic emissions. Permafrost Periglacial Process. 2, 283—300.

Hales, T.C., Roering, J.J., 2007. Climatic controls on frost cracking and implications for the evolution of bedrock landscapes. J. Geophys. Res. Earth Surf. 112, F02033.

Harbor, J.M., 1992. Numerical modeling of the development of u-shaped valleys by glacial erosion. Geol. Soc. Am. Bull. 104, 1364—1375.

Harris, C., Ross, N., 2007. Pingos and pingo scars. In: Encyclopedia of Quaternary Science, vol. 3, pp. 2200—2206.

Hasler, A., Gruber, S., Beutel, J., 2012. Kinematics of steep bedrock permafrost. J. Geophys. Res. 117, F01016.

Heckmann, T., 2006. Untersuchungen zum Sedimenttransport durch Grundlawinen in zwei Einzugsgebieten der Nördlichen Kalkalpen: Quantifizierung, Analyse und Ansätze zur Modellierung der geomorphologischen Aktivität. Profil-Verlag, Diss., München.

Heckmann, T., Wichmann, V., Becht, M., 2002. Quantifying sediment transport by avalanches in the Bavarian Alps — first results. Z. Geomorphol. 127, 137−152.

Heckmann, T., Wichmann, V., Becht, M., 2005. Sediment transport by avalanches in the Bavarian Alps revisited — a perspective on modeling. Z. Geomorphol. 138, 11−25.

Hedding, D.W., Sumner, P.D., 2013. Diagnostic criteria for pronival ramparts: site, morphological and sedimentological characteristics. Geogr. Ann. 95A, 315−322.

Herman, F., Beaud, F., Champagnac, J.-D., Lemieux, J.-M., Sternai, P., 2011. Glacial hydrology and erosion patterns: a mechanism for carving glacial valleys. Earth Planet. Sci. Lett. 310, 498−508.

Hinderer, M., 2001. Late Quaternary denudation of the Alps, valley and lake fillings and modern river loads. Geodin. Acta 14, 231−263.

Hooke, R.L., 1991. Positive feedbacks associated with erosion of glacial cirques and over-deepenings. Geol. Soc. Am. Bull. 103, 1104−1108.

Hooyer, T.S., Cohen, D., Iverson, N.R., 2012. Control of glacial quarrying by bedrock joints. Geomorphology 153, 91−101.

Hubbard, B., Heald, A., Reynolds, J.M., Quincey, D., Richardson, S.D., Luyo, M.Z., Portilla, N.S., Hambrey, M.J., 2005. Impact of a rock avalanche on a moraine-dammed proglacial lake: Laguna Safuna Alta, Cordillera Blanca, Peru. Earth Surf. Process. Landforms 30, 1251−1264.

Humlum, O., 2000. The geomorphic significance of rock glaciers: estimates of rock glacier debris volumes and headwall recession rates in West Greenland. Geomorphology 35, 41−67.

Ikeda, A., Matsuoka, N., Kääb, A., 2008. Fast deformation of perennially frozen debris in a warm rock glacier in the Swiss Alps: an effect of liquid water. J. Geophys. Res. 113 (F1), F01021.

Immerzeel, W.W., van Beek, L.P., Bierkens, M.F., 2010. Climate change will affect the Asian water towers. Science 328, 1382−1385.

Iverson, N.R., 1991. Potential effects of subglacial water-pressure fluctuations on quarrying. J. Glaciol. 37, 27−36.

Iverson, N.R., 2012. A theory of glacial quarrying for landscape evolution models. Geology 40, 679−682.

Iverson, N., Person, M., 2012. Glacier-bed geomorphic processes and hydrologic conditions relevant to nuclear waste disposal. Geofluids 12.

Jamieson, S.S.R., Hulton, N.R.J., Hagdorn, M., 2008. Modelling landscape evolution under ice sheets. Geomorphology 97, 91−108.

Jomelli, V., Bertran, P., 2001. Wet snow avalanche deposits in the French Alps: structure and sedimentology. Geogr. Ann. 83A, 15−28.

Kääb, A., Frauenfelder, R., Roer, I., 2007. On the response of rockglacier creep to surface temperature increase. Global Planet. Change 60, 172−187.

Kehew, A.E., Piotrowski, J.A., Jorgensen, F., 2012. Tunnel valleys: concepts and controversies - a review. Earth-Sci. Rev. 113, 33−58.

Kessler, M.A., Anderson, R.S., Briner, J.P., 2008. Fjord insertion into continental margins driven by topographic steering of ice. Nat. Geosci. 1, 365−369.

Knight, P.G., 1997. The basal ice layer of glaciers and ice sheets. Quat. Sci. Rev. 16, 975−993.

Koppes, M.N., Hallet, B., 2002. Influence of rapid glacial retreat on the rate of erosion by tidewater glaciers. Geology 30, 47−50.

Koppes, M., Hallet, B., 2006. Erosion rates during rapid deglaciation in Icy Bay, Alaska. J. Geophys. Res. Earth Surf. 111 (F2).

Koppes, M.N., Montgomery, D.R., 2009. The relative efficacy of fluvial and glacial erosion over modern to orogenic timescales. Nat. Geosci. 2, 644–647.

Koppes, M., Sylwester, R., Rivera, A., Hallet, B., 2010. Variations in sediment yield over the advance and retreat of a calving glacier, Laguna San Rafael, North Patagonian Icefield. Quat. Res. 73, 84–95.

Korup, O., Montgomery, D.R., 2008. Tibetan plateau river incision inhibited by glacial stabilization of the Tsangpo gorge. Nature 455, 786–789.

Korup, O., Rixen, C., 2014. Soil erosion and organic carbon export by wet snow avalanches. Cryosphere Discuss. 8, 1–19.

Korup, O., Tweed, F., 2007. Ice, moraine, and landslide dams in mountainous terrain. Quat. Sci. Rev. 26, 3406–3422.

Krautblatter, M., Funk, D., Günzel, F.K., 2013. Why permafrost rocks become unstable: a rock–ice-mechanical model in time and space. Earth Surf. Process. Landforms 38, 876–887.

Krautblatter, M., Huggel, C., Deline, P., Hasler, A., 2012. Short communication: research perspectives on unstable high-alpine bedrock permafrost: measurement, modelling and process understanding. Permafrost Periglacial Process. 23, 80–88.

Lozinski, W., 1909. Über die mechanische Verwitterung der Sandsteine im gemäßigten Klima. In: Bulletin international de l'Academie des Sciences de Cracovie, Classe des Sciences Mathémathiques et Naturelles, vol. 1, pp. 1–25.

Luckman, B.H., 1977. The geomorphic activity of snow avalanches. Geogr. Ann. 59A, 31–48.

Luckman, B.H., 1978. Geomorphic work of snow avalanches in the Canadian Rocky mountains. Arct. Alp. Res. 10, 261–276.

Luckman, B., Matthews, J., Smith, D., McCarroll, D., McCarthy, D., 1994. Snow-avalanche impact landforms: a brief discussion of terminology. Arct. Alp. Res. 26, 128–129.

MacGregor, K.R., Anderson, R.S., Anderson, S.P., Waddington, E.D., 2000. Numerical simulations of glacial-valley longitudinal profile evolution. Geology 28, 1031–1034.

MacGregor, K.R., Anderson, R.S., Waddington, E.D., 2009. Numerical modeling of glacial erosion and headwall processes in alpine valleys. Geomorphology 103, 189–204.

Mackay, J.R., 1998. Pingo growth and collapse, Tuktoyaktuk Peninsula Area, Western Arctic Coast, Canada: a long-term field study. Geogr. Phys. Quatern. 52, 271–323.

Matsuoka, N., Abe, M., Ijiri, M., 2003. Differential frost heave and sorted patterned ground—field measurements and a laboratory experiment. Geomorphology 52, 73–85.

Matthes, F.E., 1938. Avalanche sculpture in the Sierra Nevada of California. Int. Assoc. Sci. Hydrol. Bull. 23, 631–637.

Matthews, J.A., Shakesby, R.A., Owen, G., Vater, A.E., 2011. Pronival rampart formation in relation to snow-avalanche activity and Schmidt-hammer exposure-age dating (SHD): three case studies from southern Norway. Geomorphology 130, 280–288.

Matsuura, S., Asano, S., Okamoto, T., Takeuchi, Y., 2003. Characteristics of the displacement of a landslide with shallow sliding surface in a heavy snow district of Japan. Eng. Geol. 69, 15–35.

McColl, S.T., 2012. Paraglacial rock-slope stability. Geomorphology 153, 1–16.

McClung, D., Schaerer, P., 1993. The Avalanche Handbook. The Mountaineers Books, Seattle, 271 pp.

Meigs, A., Sauber, J., 2000. Southern Alaska as an example of the long-term consequences of mountain building under the influence of glaciers. Quat. Sci. Rev. 19, 1543–1562.

Montgomery, D.R., Korup, O., 2011. Preservation of inner gorges through repeated Alpine glaciations. Nat. Geosci. 4, 62–67.

Moore, J.R., Egloff, J., Nagelisen, J., Hunziker, M., Aerne, U., Christen, M., 2013. Sediment transport and bedrock erosion by wet snow avalanches in the Guggigraben, Matter Valley, Switzerland. Arct. Antarct. Alp. Res. 45, 350−362.

Moore, J.R., Gischig, V., Katterbach, M., Loew, S., 2011. Air circulation in deep fractures and the temperature field of an Alpine rock slope. Earth Surf. Process. Landforms 36, 1985−1996.

Mott, R., Lehning, M., 2010. Meteorological modeling of very high-resolution wind fields and snow deposition for mountains. J. Hydrometeorol. 11, 934−949.

Motyka, R.J., Truffer, M., Kuriger, E.M., Bucki, A.K., 2006. Rapid erosion of soft sediments by tidewater glacier advance: Taku Glacier, Alaska, USA. Geophys. Res. Lett. 33.

Norton, K., Abbühl, L., Schlunegger, F., 2010. Glacial conditioning as an erosional driving force in the Central Alps. Geology 38, 655−658.

Oerlemans, J., 1984. Numerical experiments on large-scale glacial erosion. Z. Gletscherkd. Glazialgeol. 20, 107−126.

Oskin, M., Burbank, D.W., 2005. Alpine landscape evolution dominated by cirque retreat. Geology 33, 933−936.

Rapp, A., 1960. Recent development of mountain slopes in Kärkevagge and surroundings, northern Sweden. Geogr. Ann. 42A, 71−200.

Richardson, S., Reynolds, J., 2000. An overview of glacial hazards in the Himalayas. Quat. Int. 65, 31−45.

Riihimaki, C.A., MacGregor, K.R., Anderson, R.S., Anderson, S.P., Loso, M.G., 2005. Sediment evacuation and glacial erosion rates at a small alpine glacier. J. Geophys. Res. Earth Surf. 110 (F3).

Roer, I., Nyenhuis, M., 2007. Rockglacier activity studies on a regional scale: comparison of geomorphological mapping and photogrammetric monitoring. Earth Surf. Process. Landforms 32, 1747−1758.

Roer, I., Haeberli, W., Avian, M., Kaufmann, V., Delaloye, R., Lambiel, C., Kääb, A., 2008. Observations and considerations on destabilizing active rockglaciers in the European Alps. In: Kane, D.L., Hinkel, K.M. (Eds.), Ninth International Conference on Permafrost (Fairbanks, Alaska), vol. 2, pp. 1505−1510.

Sanders, J.W., Cuffey, K.M., MacGregor, K.R., Collins, B.D., 2013. The sediment budget of an alpine cirque. Geol. Soc. Am. Bull. 125, 229−248.

Sanders, J.W., Cuffey, K.M., Moore, J.R., MacGregor, K.R., Kavanaugh, J.L., 2012. Periglacial weathering and headwall erosion in cirque glacier bergschrunds. Geology 40, 779−782.

Sass, O., Heel, M., Hoinkis, R., Wetzel, K.F., 2010. A six year record of debris transport by avalanches on a wildfire slope (Arnspitze, Tyrol). Z. Geomorphol. 54, 181−193.

Schaub, Y., Haeberli, W., Huggel, C., Künzler, M., Bründl, M., 2013. Landslides and new lakes in deglaciating areas: a risk management framework. The second World Landslide Forum. In: Margottini, C., et al. (Eds.), Landslide Science and Practice, vol. 7, pp. 31−38.

Sekiguchi, T., Sugiyama, M., 2003. Geomorphological features and distribution of avalanche furrows in heavy snowfall regions of Japan. Z. Geomorphol. NF 130, 117−128.

Shaw, J., 2002. The meltwater hypothesis for subglacial bedforms. Quat. Int. 90, 5−22.

Shaw, J., 2010. Defending and testing hypotheses: a response to John Shaw's paper 'in defence of the meltwater (megaflood) hypothesis for the formation of subglacial bedform fields' reply. J. Quat. Sci. 25, 824−825.

Smith, A.M., Bentley, C.R., Bingham, R.G., Jordan, T.A., 2012. Rapid subglacial erosion beneath Pine island glacier, west Antarctica. Geophys. Res. Lett. 39.

Smith, A.M., Murray, T., Nicholls, K.W., Makinson, K., Aoalgeirsdottir, G., Behar, A.E., Vaughan, D.G., 2007. Rapid erosion, drumlin formation, and changing hydrology beneath an Antarctic ice stream. Geology 35, 127−130.

Springman, S., Arenson, L., Yamamoto, Y., Maurer, H., Kos, A., Buchli, T., Derungs, G., 2012. Multidisciplinary investigations on three rock glaciers in the Swiss Alps: legacies and future perspectives. Geogr. Ann. A94, 215−243.

Straumann, R., Korup, O., 2009. Quantifying sediment storage at the mountain-belt scale. Geology 37, 1079−1082.

Sugden, D.E., 1968. The selectivity of glacial erosion in the Cairngorm Mountains, Scotland. Trans. Inst. Br. Geogr. 45, 79−92.

Sugden, D.E., 1978. Glacial erosion by the Laurentide ice sheet. J. Glaciol. 20, 367−391.

Swift, D.A., Nienow, P.W., Hoey, T.B., 2005a. Basal sediment evacuation by subglacial meltwater: suspended sediment transport from Haut Glacier d'Arolla, Switzerland. Earth Surf. Process. Landforms 30, 867−883.

Swift, D.A., Nienow, P.W., Hoey, T.B., Mair, D.W.F., 2005b. Seasonal evolution of runoff from Haut Glacier d'Arolla, Switzerland and implications for glacial geomorphic processes. J. Hydrol. 309, 133−148.

Swift, D.A., Nienow, P.W., Spedding, N., Hoey, T.B., 2002. Geomorphic implications of subglacial drainage configuration: rates of basal sediment evacuation controlled by seasonal drainage system evolution. Sediment. Geol. 149, 5−19.

Swift, D.A., Persano, C., Stuart, F.M., Gallagher, K., Whitham, A., 2008. A reassessment of the role of ice sheet glaciation in the long-term evolution of the East Greenland fjord region. Geomorphology 97, 109−125.

Talbot, C.J., 1999. Ice ages and nuclear waste isolation. Eng. Geol. 52, 177−192.

Thomson, S.N., Reiners, P.W., Hemming, S.R., Gehrels, G.E., 2013. The contribution of glacial erosion to shaping the hidden East Antarctic landscape. Nat. Geosci. 6, 203−207.

Thorn, C.E., 1992. Periglacial geomorphology: what, where, when? In: Dixon, J.C., Abrahams, A.D. (Eds.), Periglacial Geomorphology, Proceedings of the 22nd Ann. Binghampton Symp. In Geomorphology. Wiley and Sons, Chichester, pp. 1−30.

Valla, P.G., Shuster, D.L., van der Beek, P.A., 2011. Major increase in relief of the European Alps during Mid-Pleistocene glaciations. Nat. Geosci. 4, 688−692.

Van der Beek, P., Bourbon, P., 2008. A quantification of the glacial imprint on relief development in the French western Alps. Geomorphology 97, 52−72.

Vaughan, D.G., Comiso, J.C., Allison, I., Carrasco, J., Kaser, G., Kwok, R., Mote, P., Murray, T., Paul, F., Ren, J., Rignot, E., Solomina, O., Steffen, K., Zhang, T., 2013. Observations: cryosphere. In: Climate Change 2013: The Physical Science Basis. Contribution of Working Group I to the Fifth Assessment Report of the Intergovernmental Panel on Climate Change. Cambridge University Press, Cambridge, United Kingdom and New York, NY, USA.

Vieli, A., 2014. Retreat instability of tidewater glaciers and marine ice sheets. In: Haeberli, W., Whiteman, C. (Eds.), Snow and Ice-related Hazards, Risks and Disasters. Elsevier, pp. 677−712.

Voigt, T., Füssel, H.M., Gärtner-Roer, I., Huggel, C., Marty, C., Zemp, M. (Eds.), 2010. Impacts of Climate Change on Snow, Ice, and Permafrost in Europe: Observed Trends, Future Projections, and Socioeconomic Relevance. European Topic Centre on Air and Climate Change. Technical Paper 2010/13: 117 pp.

Wahrhaftig, C., Cox, A., 1959. Rock glaciers in the Alaska Range. Geol. Soc. Am. Bull. 70, 383−436.

Washburn, A.L., 1979. Geocryology. A Survey of Periglacial Processes and Environments. Arnold, London, 406 pp.

Ward, D.J., Anderson, R.S., Haeussler, P.J., 2012. Scaling the Teflon peaks: rock type and the generation of extreme relief in the glaciated western Alaska Range. J. Geophys. Res. Earth Surf. 117, F01031.

Zhang, T., 2005. Influence of the seasonal snow cover on the ground thermal regime: an overview. Rev. Geophys. 43, RG4002.

Zemp, M., Kääb, A., Hoelzle, M., Haeberli, W., 2005. GIS-based modelling of the glacial sediment balance. Z. Geomorphol. NF 138, 113−129.

Zenklusen Mutter, E., Blanchet, J., Phillips, M., 2010. Analysis of ground temperature trends in Alpine permafrost using generalized least squares. J. Geophys. Res. Earth Surf. 115 (F4). http://dx.doi.org/10.1029/2009JF001648.

Zoet, L.K., Alley, R.B., Anandakrishnan, S., Christianson, K., 2013. Accelerated subglacial erosion in response to stick-slip motion. Geology 41, 159−162.

Chapter 7

Mountains, Lowlands, and Coasts: the Physiography of Cold Landscapes

Tobias Bolch [1,2] and Hanne H. Christiansen [3,4]

[1] *Department of Geography, University of Zurich, Switzerland,* [2] *Institute for Cartography; Technische Universität Dresden, Dresden, Germany,* [3] *Arctic Geology Department, The University Centre in Svalbard, UNIS, Longyearbyen, Norway,* [4] *Center for Permafrost, CENPERM; Department of Geoscience and Natural Resource Management, University of Copenhagen, Denmark*

ABSTRACT

Large parts of the terrestrial area of planet Earth belong to the cryosphere. The distribution is mainly governed by temperature, precipitation, and wind. Hence, snow and ice are predominant in high latitudes, but are restricted to high altitudes in mid-latitudes and low latitudes. Here, we first give an overview of the physiography of high mountains, cold lowlands, and cold coasts, and then focus on glaciers and permafrost, and their interaction, as the most important and widespread components of the terrestrial cryosphere.

7.1 INTRODUCTION

The cryosphere (including seasonal snow, lake and river ice, sea ice, glaciers, continental ice sheets, permafrost, and seasonally frozen ground) influences a total area of about 68 million km^2 (~ 13 percent of the surface area; Barry and Gan, 2011; Ohmura, 2014). The occurrence and distribution of the cryosphere depend predominantly on temperature, precipitation, and wind. In general, the temperature decreases with latitude and altitude, while precipitation decreases with increasing continentality. These general patterns are, especially for precipitation and also for wind, strongly altered by the occurrence and distribution of high mountains. In cold climatic areas, the wind redistributes the solid precipitation during large parts of the year, enabling glaciers to grow, and allowing significant cooling of snow-free areas of the landscape, which allows permafrost to exist. High mountains are usually characterized by relatively high precipitation and glacier cover while the cold lowlands are predominantly influenced by the occurrence of permafrost (Barry and Gan, 2011; French, 2007). The general characteristics of the physiography of the cold regions on Earth are important background information to understand the distribution of processes associated with the cryosphere, such as glacier or permafrost-related hazards. Glaciers and permafrost comprise an important part of the cryosphere. Their interaction is especially important to consider in cold-dry regions (Haeberli, 2005). The aim of this contribution is to present background information about the occurrence and characteristics of high mountains, lowlands, and coasts with respect to the cryosphere, and then to provide more detailed information about glaciers and permafrost and their interactions.

7.2 PHYSIOGRAPHY OF THE TERRESTRIAL CRYOSPHERE

7.2.1 High Altitudes/Mountains

About 10 percent of the land surface ($\sim$3 percent of the Earth) are >2,000 m asl., and can therefore be counted as high altitudes (Richter, 2002). "Mountains" can, however, not be defined in a simple way as several plains exist at altitudes >2,000 m (e.g., the Tibetan Plateau in Asia and the Altiplano in South America) while a mountainous landscape characterized by valleys and peaks has a certain vertical distance but need not necessarily extend into high altitudes, for example, when they are located in coastal regions. Thompson (1964) suggested a value of 600 m to distinguish mountains from hills. A widely used definition for (high) mountains introduced by Carl Troll (1956, 1973) includes landscape features: high mountains extend above the upper treeline, are characterized by cryogenic processes, and are currently glacierized or have formerly been glaciated. The lower limit of the high mountain belt is located close to the sea in the arctic region and reaches its highest values in the mountains of Tibet and South America. The Cordilleras of North and South America are the most extensive mountain ranges stretching from north to south, while the Himalaya and its adjacent ranges are the most wide-ranging east−west mountain chains. An overview of the topography and the location of the major mountain ranges of Earth is given in Figure 7.1.

The mountainous landscape is special in several aspects: it is characterized by steep slopes and is, hence, subject to extreme gravitational mass movements, such as rock falls, debris flows, and snow or ice avalanches. Topographic conditions can vary significantly within a short distance and so do climate parameters like solar radiation or precipitation and surface characteristics such as vegetation and snow cover. Glaciers, permafrost, and glacial and periglacial landforms can exist even in low latitudes due to high elevation and respective cold temperatures. The population density is quite high in several mountain ranges, and even large cities are located in mountain valleys

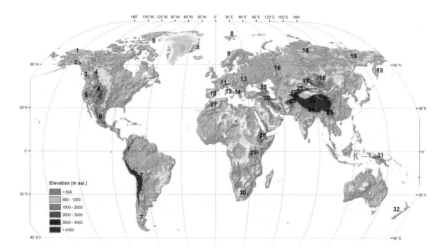

FIGURE 7.1   Topography of the Earth based on the SRTM30 DEM with the major mountain ranges (1 Brooks Range, 2 Alaska Range, 3 Coast Mountains, 4 Rocky Mountains, 5 Sierra Nevada, 6 Sierra Madre, 7 Andes, 8 Mountains in the Arctic, 9 Scandinavian mountains, 10 Pyrenees and Spanish Mountains, 11 Alps, 12 Apennines, 13 Carpathians, 14 SE-European Mountains, 15 Caucasus, 16 Urals, 17 Altai, 18 Siberian Mountains, 19 Kamchatka, 20 Mountains of the Middle East, 21 Tien Shan, 22 Pamir, 23 Hindu Kush−Karakoram, 24 Himalaya, 25 Hengduan Shan, 26 Mountains of Tibet, 27 Atlas, 28 Ethiopian Highlands, 29 Equatorial African Mountains, 30 South African Mountains, 31 Mountains of New Guinea, and 32 Southern Alps) and the extensive lowland (green); cartography: T. Bolch.

or at the margins of the mountains. While the population benefits from the resources of the mountains, such as fresh water and its natural beauty for recreation, the negative aspects of this zone are the natural hazards that endanger life and infrastructure.

The distribution of temperature and precipitation is a result of complex interaction between global circulation patterns and the oceans while these general patterns are strongly altered by high mountains (Figures 7.2 and 7.3). The largest amount of precipitation occurs especially at the south-eastern edge of High Mountain Asia during the Monsoon season where the Himalayas form a natural barrier and force the clouds to release moisture as rain or snow within a short distance (Figure 7.3, Barry and Chorley, 2003). The same region is dry during the non-monsoon period. In contrast, the Tarim Basin with the Takla-makan Desert located north of the Himalaya is almost entirely surrounded by the mountains of high Asia and is therefore one of the driest regions on Earth. Mountains located at the coast, such as the Cordilleras of South and North America or the Scandinavian mountains, usually receive a high amount of precipitation with sharply decreasing values inland and leeward of the mountain ranges (Figure 7.3). However, cold ocean currents close to conti-nents, such as the Humboldt Current or the Californian Current, modulate these general rules and lead to low moisture content of the air. Hence, adjacent

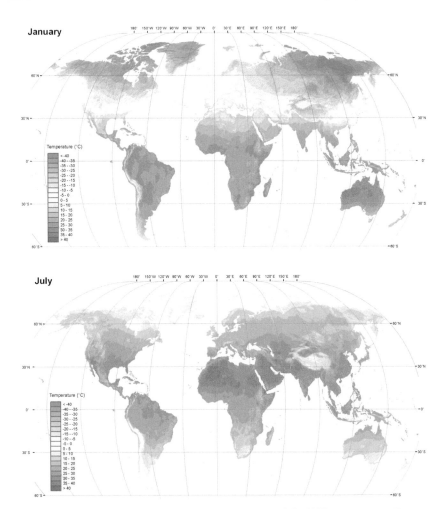

FIGURE 7.2   Mean temperature in January and July (period 1950−2000). *Data source: Hijmans et al. 2005; cartography: T. Bolch.*

coasts are dry and even high mountains close to these coasts (e.g., the Andes of the Atacama or the Cordilleras in California) are relatively arid (Barry and Chorley, 2003).

7.2.2 Cold Lowlands

Large lowlands are characteristic elements of each continent and are typically drained by large rivers. Overall, lowlands with elevations below 500 m asl. cover about 50 percent of the terrestrial parts of the Earth. The lowlands generally represent accumulation of sediments eroded, often by periglacial processes, and transported fluvially from the highlands. The largest cold

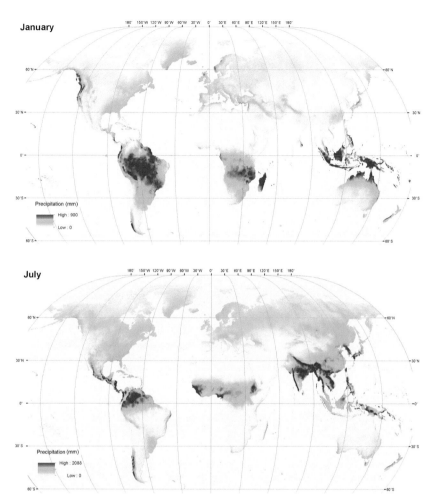

FIGURE 7.3   Mean precipitation in January and July (period 1950–2000). *Data source: Hijmans et al. 2005; cartography: T. Bolch.*

lowland areas are the Canadian Shield and the Siberian lowlands, where permafrost is widespread and the lowest temperatures on Earth are measured (Figures 7.2 and 7.4).

Overall, seasonal snow cover occupies large areas in the Northern Hemisphere and varies between about 46.5 million square kilometers in the late northern winter and about four million square kilometers in the northern summer (Barry and Gan, 2011; Ohmura, 2014). The Canadian Shield and the Siberian lowlands receive relatively low amounts of precipitation especially during winter (Figure 7.3). Wind redistribution of the solid snow precipitation means that the snow cover is usually thin, thus allowing the occurrence of thick permafrost (cf. Section 7.4). In contrast, continental summers are

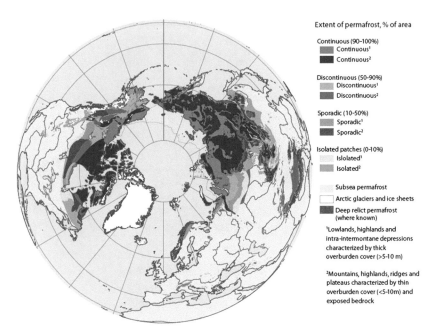

FIGURE 7.4   The distribution of the different permafrost types on the Northern Hemisphere as compiled by the International Permafrost Association, IPA. *Source: AMAP (2011) based on Brown et al. (1997).*

relatively warm causing thick active layers, and enabling even forests to grow on continuous permafrost.

7.2.3 Cold Coasts

A substantial portion of coasts is affected by the cryosphere. In polar areas, sea ice and ground ice both limit and enhance erosion processes, especially as >65 percent of the Arctic coastline is composed of unconsolidated material (Lantuit et al., 2011), much of which is relatively ice-rich sedimentary lowlands. The polar coastlines are characterized by being ice bound for most of the year, thus not geomorphologically active, and therefore relatively protected from erosion and deposition. Despite shallow shelf water, a high ground-ice content, and generally fine-grained unconsolidated material found in lowland sedimentary coasts, such coasts are rather sensitive to waves and storm surges during the short summer period (Lantuit et al., 2011). These coasts can therefore experience both relatively high annual erosion and deposition rates.

As in other coastal environments, cold coasts undergo normal coastal geomorphological changes controlled by a combination of climate and local coastal currents, although, mainly in the summer period. Likewise, sea-level changes can cause the development of raised marine beaches that can be found to cover the lower parts of the landscape (Figure 7.5(a)). Sediment

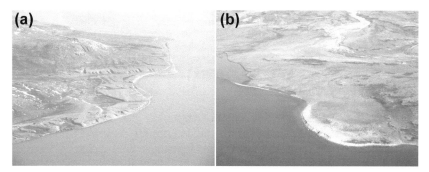

FIGURE 7.5   (a): Raised marine beaches with sub-recent coastal bluffs, and present coastal spits at Daneborg, NE Greenland, September 2013. (b): Coastal bluff eroded into Quaternary sediments (center foreground). Long-shore sediment transport has then led to the accumulation of a coastal spit (center left). *From the Zackenberg area, NE Greenland, August 2013 (photographs: H. Christiansen).*

transport takes place in both long-shore and cross-shore directions primarily in sedimentary lowland coasts, and the net result will vary along the coast, with some areas undergoing erosion, while others receive sediment. This leads to the development of distinct coastal bluffs indicating coastal erosion, and associated coastal spits (Figure 7.5(b)) or deltas typical for depositional coastal areas. In wintertime, the cold coasts are, however, not very geomorphologically active due to the sea ice cover. Typically, an ice foot starts forming as the coast cools during autumn, and by mid to late winter, this has reached its maximum size, clearly preventing wave action on the coasts even if open water occurs.

Bedrock coastlines can be compared to sedimentary coasts that are less subject to significant short-term erosion or deposition. The coasts of Antarctica, Greenland, and Arctic and Antarctic Islands are affected by glaciers. Many glaciers, ice caps, and outlet glaciers from the ice sheets in these regions terminate directly in the ocean. This is especially true for Antarctica, where most of the coast is formed of ice from the ice sheet, the ice shelves or some local glaciers (88 percent of the entire coastline) (ADD, version 4.1: www.add.scar.org). The non-glacierized coast (12 percent of the entire coastline) can mainly be found where high mountains exist: at McMurdo Sound and around the Antarctic Peninsula. Arctic marine-terminating glaciers are also common in Ellesmere Island, Greenland, Svalbard, Novaya Zemlya, Franz-Joseph Land, and Severnaya Zemlya (Carr et al., 2013), but do not dominate lengthwise. High precipitation also causes glaciers at lower latitudes to terminate in the sea, such as in southern Alaska, the Southern Andes and South Georgia. The entire area of ocean-terminating glaciers (excluding the ice sheets) is approximately 280,000 km^2 ($\sim$38 percent of the entire ice cover, Pfeffer et al., 2014). However, specific numbers for the length of the coast formed by ice outside Antarctica are not available.

Climate change and related changes in ocean currents have led to a significant retreat of most of the tidewater glaciers and marine-terminating outlet glaciers from the ice sheets and thus to significant changes along the coast (e.g., Rasmussen et al., 2011; Bjørk et al., 2012; Carr et al., 2013).

7.3 GLACIERS AND ICE SHEETS: EXTENT AND DISTRIBUTION

Glaciers form an important and visually very clear part of the cryosphere. Overall, an area of about 14,680,000 km^2 is currently covered by land ice ($\sim$10 percent of the terrestrial land). The largest area is occupied by the two continental ice sheets of Greenland and Antarctica, which have an extent of about 1,680,000 km^2 (Rastner et al., 2012) and 12,273,000 km^2 (ADD, version 4.1: www.add.scar.org), respectively. The glaciers and ice caps take up about 727,000 km^2 ($\sim$5 percent of the total ice cover) with a volume of approximately 170,000 km^3 globally (Table 7.1) and vary from small valley or cirque glaciers to large ice caps (Pfeffer et al., 2014).

The ice cover is subject to large variations reflecting climate variability. Large parts of North America and North−central Europe were covered by ice sheets during the Last Glacial Maximum around 20−18 ka BP. The Siberian Lowlands were, in contrast, relatively free of perennial surface ice (cf. Figure 10.1 in Streletskiy et al., 2014).

The global distribution of glaciers and ice sheets is constrained by climate and topography and, hence, reflects altitudinal and latitudinal trends and also the precipitation amount (Figures 7.1−7.3). The highest glacier coverage (and glacier volume) can be found in the Arctic (Arctic Canada, Greenland, Svalbard, and the Russian Arctic) and Antarctic regions where approximately 44 percent ($\sim$53 percent) and approximately 18 percent ($\sim$22 percent), respectively, of the Earth's glacier cover (glacier volume) is located (Table 7.1, Huss and Farinotti, 2012; Pfeffer et al., 2014). High Mountain Asia contains approximately 16 percent of the glacier cover ($\sim$120,000 km^2) but <6 percent of the glacier volume. Very little ice can be found in the dry Middle East and the tropical Low Latitudes with together <0.5 percent of the global land ice cover-located at the highest elevations in these regions. Tidewater glaciers account for 39 percent of the global glaciers' extent (Pfeffer et al., 2014).

The equilibrium line altitude (ELA), the line on a glacier where net accumulation equals net ablation equals zero, is located at a relatively low elevation in humid-maritime regions with both high accumulation in the upper reaches of the glacier and high ablation in the lower parts. These regions are characterized by temperate glaciers with firn and ice at the pressure melting point. The glaciers have a high mass turnover, are flowing relatively fast, and react strongly to a temperature increase. Maritime glaciers usually have a larger vertical extent and reach lower altitudes than do continental ones. In contrast, in dry–continental regions with low precipitation and short melting periods, the mean glacier elevation is located relatively

TABLE 7.1 Glacier Coverage of the Entire Earth and the 19 Sub-regions According to the Randolph Glacier Inventory

Region	Number	All Glaciers					Tidewater Glaciers
		Area (km²)	Area (Percent of Total)	Volume (km³)	Volume (Percent of Total)	Mean Thickness (m)	Area (km²)
Alaska	26,944	86,715 ± 4,596	12.0	20,402 ± 1,501	12.0	234	11,781
Western Canada and United States	15,215	14,559 ± 1,383	2.0	1,025 ± 84	0.6	70	0
Arctic Canada north	4,538	104,873 ± 3,356	14.4	9,814 ± 1,115	5.8	94	49,911
Arctic Canada south	7,347	40,894 ± 2,004	5.6	34,399 ± 4,699	20.2	841	3,030
Greenland periphery	19,323	89,721 ± 4,486	12.3	19,042 ± 2,655	11.2	212	31,106
Iceland	568	11,060 ± 288	1.5	4,441 ± 370	2.6	402	0
Svalbard	1,615	33,922 ± 1,187	4.7	9,685 ± 922	5.7	286	14,884
Scandinavia	2,668	2,851 ± 265	0.4	256 ± 19	0.2	90	0
Russian Arctic	1,069	51,592 ± 1,445	7.1	16,839 ± 2,205	9.9	326	33,435
North Asia	4,403	3,430 ± 353	0.5	140 ± 15	0.1	41	0
Central Europe	3,920	2,063 ± 215	0.3	117 ± 10	0.1	57	0
Caucasus and Middle East	1,386	1,139 ± 114	0.2	61 ± 6	0.0	54	0

(Continued)

TABLE 7.1 Glacier Coverage of the Entire Earth and the 19 Sub-regions According to the Randolph Glacier Inventory—cont'd

Region	Number	All Glaciers					Tidewater Glaciers
		Area (km²)	Area (Percent of Total)	Volume (km³)	Volume (Percent of Total)	Mean Thickness (m)	Area (km²)
Central Asia	4,6543	62,606 ± 5,259	8.6	5,026 ± 503	3.0	80	0
South Asia West	22,822	33,859 ± 2,607	4.7	3,241 ± 287	1.9	96	0
South Asia East	14,095	21,799 ± 1,809	3.0	1,312 ± 119	0.8	60	0
Low latitudes	2,863	2,346 ± 246	0.3	144 ± 16	0.1	61	0
Southern Andes	16,046	29,333 ± 1,731	4.0	6,674 ± 507	3.9	228	7,004
New Zealand	3,537	1,162 ± 142	0.2	70 ± 5	0.0	60	0
Antarctic and sub-Antarctic	2,752	132,867 ± 2,524	18.3	37,517 ± 8,402	22.0	282	131,192
Total	197,654	726,792 ± 34,159	100.0	170,214 ± 20,688	100.0	234	281,543

Source: Pfeffer et al. 2014, Huss and Farinotti, 2012.

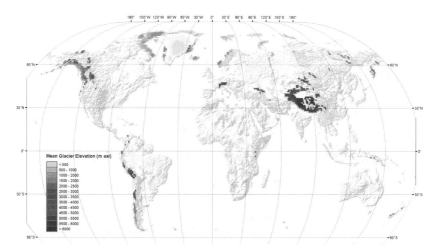

FIGURE 7.6   Distribution of glaciers (excluding Antarctica) and their mean elevation. *Calculated by F. Paul, data source: SRTM30 and DCW, cartography: T. Bolch.*

high (Haeberli and Burn, 2002; Shumskiy, 1964). These continental glaciers are polythermal, or even cold, with firn and ice below pressure melting temperature, and have a smaller vertical extent and a low mass turnover, and are often surrounded by permafrost (Etzelmüller and Hagen, 2005). Overall, mean glacier altitude, that can be calculated based on glacier outlines and a digital elevation model, provides a reasonable estimate of the ELA (cf. Braithwaite and Raper, 2009), and rises from almost sea level at high latitudes to >5,000 m in the subtropics (Figure 7.6). Glaciers terminating in water bodies, especially tidewater glaciers, show different behavior from land-terminating glaciers as they exhibit a stronger mass loss due to enhanced melting and calving (Vieli, 2014). Further special glaciers are surge-type and pulsating glaciers that usually experience periodically rapid advances (surges) due to mechanical instabilities (Harrison et al., 2014). Also, debris-covered glaciers can react differently to climate: thin debris enhances melt while thick debris can significantly reduce melt and delay the reaction of the glaciers to climate (e.g., Nakawo and Young, 1981). Supraglacial and proglacial lakes are typical of debris-covered glaciers with negative mass balance, which in turn increases melt. Hence, debris-covered glaciers with lakes can have an even higher area and mass loss than debris-free glaciers do (Bolch et al., 2011; Basnett et al., 2013).

7.4 PERMAFROST TYPES, EXTENT, AND DISTRIBUTION

Permafrost is soil, rock, sediment, or other earth material that remains at or below 0 °C for two or more consecutive years (van Everdingen, 1998). Thus, it is solely defined on the basis of temperature and duration. Permafrost exists in large areas of the terrestrial polar regions (French, 2007, Figure 7.4), while coastal and subsea permafrost exists around Arctic coastlines. Typically, permafrost does not occur beneath glaciers, as they isolate the ground from the necessary atmospheric cooling, but permafrost can exist under thin cold-based glaciers or along the margins of polythermal glaciers.

Permafrost is classified according to its areal extent. If it occupies not <90 percent of the area, it is considered to be continuous. It is discontinuous if it occupies from 50 to 90 percent, sporadic if it occupies from 10 to 50 percent, and is considered to occur in isolated patches if it occupies <10 percent (Figure 7.4).

Above the permafrost, the active layer freezes and thaws seasonally, and below the active layer is the transient layer, which thaws only at the decadal to centennial scale (Shur et al., 2005). Most subsea permafrost is not subject to the seasonal variation seen on land, since its upper temperature is controlled by bottom seawater temperatures. Active-layer thickness is influenced by climate and local factors and can vary from <0.5 m in vegetated, organic terrain to >10 m in areas of exposed bedrock. Most biological and hydrological activities in Arctic soils are confined to this layer of seasonal thaw (Hinzman et al., 2005). Terrestrial permafrost thickness ranges from a few decimeters at the southern limit of the permafrost zone to about 1,500 m in the north of the Arctic region (Figure 7.7). The thickest permafrost is found in areas that have not recently been covered by glaciers, such as Siberia, where ground cooling for a longer time has allowed for >1,000-m-thick permafrost to develop. Areas that have been glacier covered during the last glaciation typically do not have >200- to 500-m-thick permafrost (French, 2007).

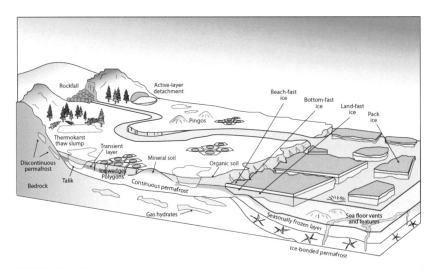

FIGURE 7.7 Distribution of permafrost types and associated landforms along a conceptual transect from sub-Arctic to high arctic conditions on the continental shelf. *Source: AMAP (2011).*

Permafrost is controlled by climate and also by a combination of several local factors (Streletskiy et al., 2014). Thus, the permafrost thermal regime is controlled by the exchanges of heat and moisture between the atmosphere and the Earth's surface, and by the thermal properties of the underlying ground (Williams and Smith, 1989). The existence of permafrost depends on past and present states of energy fluxes through the active layer. This layer experiences seasonal variations in ice/water content, thermal conductivity, density, mechanical properties, and solute redistribution. Other important factors include snow cover, vegetation, soil organic layer thickness, soil moisture, ice content, and drainage conditions controlled by the local geomorphology.

Overall, permafrost exists in about 25 percent of the terrestrial parts of the Earth (French, 2007). Permafrost regions currently occupy about 16.7 million square kilometers in Eurasia, and 10.2 million square kilometers in North America. Canada and Russia contain the most extensive areas of permafrost, approximately 50 percent and 66 percent of their territories, respectively. Large permafrost areas exist in China and the Tibetan Plateau covering 1.6 million square kilometers or 17 percent of the country (Ran et al., 2012), and 0.53 million square kilometers in Mongolia. Besides Greenland, permafrost in Europe is found in all the Scandinavian countries (0.18 million square kilometers), and even in the Alps (0.013 million square kilometers). High Mountain Asia (especially in India, Tajikistan, Pakistan, Kazakhstan, and on the Tibetan Plateau), South America (primarily in Argentina and Chile), New Zealand, Antarctica, and sub-Antarctic islands all have some permafrost present (Gruber, 2012).

This global distribution clearly shows that permafrost is primarily found in the very extensive Arctic lowlands, while mountain permafrost is, globally, significantly less extensive (Figure 7.4). However, in the mountains, all the different types of permafrost can exist as an altitudinal gradient if the mountains are high enough. This means that continuous permafrost can be found also in relatively low latitude high mountains.

The upper parts of permafrost can experience freezing and thawing at centennial to millennial scales. French and Shur (2010) conclude that permafrost can be stable under fluctuating climatic conditions if the ground is protected by a high ground-ice content during warm periods. Such stability of the permafrost toward climatic fluctuations is a consequence of a layer of the ground that, although a part of the active layer during warm summers, under normal climatic conditions is the upper part of the permafrost. If this layer has a high ice content, it provides thermal inertia. The net result is that permafrost can have a relatively low sensitivity to atmospheric temperature rise, or anthropogenic disturbance, when the top permafrost is ice-rich (Shur et al., 2005). This is called the transient layer (Shur et al., 2005). The transient layer experiences high and quasi-uniform ice content and undergoes freezing/thawing at decadal to century scales. Obviously, the most important condition in the permafrost is its temperature. This is typically monitored in boreholes to varying depths in the ground in different landforms. Permafrost temperatures vary from being very close to 0 °C at the southern extent of permafrost, to being down to −15 °C in the high Arctic (Romanovsky et al., 2010).

Permafrost also occurs in the seabed (Figure 7.4) as demonstrated by several geophysical studies including drilling into the Siberian, Chukchi, and Beaufort shelves, but data are mostly unpublished. Subsea permafrost occurs in the continental shelves of the Arctic Ocean adjacent to Russia, Alaska, Canada, Greenland, and Svalbard (Osterkamp, 2001) where it is likely to exist beneath depths of water of up to 100 m (Hinz et al., 1998). Due to changes in the sea level and to shoreline erosion, the present configuration of subsea permafrost in the Beaufort Sea resembles a wedge or tabular sheet extending from the Arctic Ocean coast down to a 100-m water depth, a distance of up to 100 km (Osterkamp and Fei, 1993). On the wide Siberian shelf, the 100-m water depth may be over 700-km offshore (Romanovskii et al., 1998).

7.5 GLACIER–PERMAFROST INTERACTIONS

Glaciers are perennial surface ice consisting mainly of snow, firn, and ice, while permafrost with its perennial subsurface ice can be both very ice-rich to almost completely ice-free (Arenson et al., 2014). Permafrost dominates where it is cold and dry, while glaciers predominantly occur in wet and cold conditions. Direct glacier–permafrost interactions are important to consider as they can coexist in cold regions and are crucial for understanding glacial and periglacial processes and geohazards (Etzelmüller and Hagen, 2005). For example, permafrost below and beneath polythermal glaciers has a significant influence on glacier flow. Steep cold or polythermal hanging glaciers can become unstable with temperature increase and permafrost thaw (Haeberli, 2005). In addition, changes in glaciers directly impact permafrost in that receding glaciers expose new land to permafrost formation, if the deglaciated areas are located in an environment cold enough to allow significant ground cooling (French, 2007). Land uplift due to isostatic adjustment as glaciers recede likewise creates possibilities for permafrost to form. Also, stagnant dead glacier ice such as found in ice-cored moraines, with a significant debris cover, can be preserved if located in a permafrost environment, and the debris cover is thicker than the active layer (Houmark-Nielsen et al., 1994). Thermokarst landforms can also develop from thawing of ice-cored moraines. Such a landscape is traditionally considered to be still undergoing areal deglaciation and is thus characterized as glacial, despite, in some respects, being considered as part of the permafrost environment, or transitioning to periglacial conditions. In addition, as glaciers become thinner, their thermal characteristics might change and they could become cold-based, thus increasing the potential for permafrost to form beneath the glacier. Creep of perennially frozen morainic material, sometimes containing remains of former glaciers or ice patches, is a classical example of direct interaction between glacial deposits and permafrost (Humlum, 1997).

REFERENCES

AMAP, 2011. Snow, Water, Ice and Permafrost in the Arctic (SWIPA). Climate Change and the Cryosphere. Arctic Monitoring and Assessment Programme (AMAP), Oslo, Norway xii + 538 pp.

Arenson, L., Colgan, W., Marshall, H.P., 2014. Physical, thermal and mechanical properties of snow, ice and permafrost. In: Haeberli, W., Whiteman, C. (Eds.), Snow and Ice-related Hazards, Risks and Disasters. Elsevier, pp. 35−75.

Barry, R.G., Chorley, R., 2003. Atmosphere, Weather and Climate, eight ed. Routledge.

Barry, R.G., Gan, T.Y., 2011. The Global Cryosphere. Past, Present, and Future. Cambridge University Press.

Basnett, S., Kulkarni, A.V., Bolch, T., 2013. Influence of debris-cover and glacial lakes on the recession of glaciers in Sikkim Himalaya, India. J. Glaciol. 59, 1035−1046.

Bjørk, A.A., Kjær, K.H., Korsgaard, N.J., Khan, S.A., Kjeldsen, K.K., Andresen, C.S., Box, J.E., Larsen, N.K., Funder, S., 2012. An aerial view of 80 years of climate-related glacier fluctuations in southeast Greenland. Nat. Geosci. 5, 427−432.

Bolch, T., Pieczonka, T., Benn, D.I., 2011. Multi-decadal mass loss of glaciers in the Everest area (Nepal, Himalaya) derived from stereo imagery. The Cryosphere 5, 349−358.

Braithwaite, R.J., Raper, S., 2009. Estimating equilibrium-line altitude (ELA) from glacier inventory data. Ann. Glaciol. 50, 127−132.

Brown, J., Ferrians Jr., O.J., Heginbottom, J.A., Melnikov, E.S., 1997. Circum-Arctic map of permafrost and ground-ice conditions. U.S. Geological Survey in Cooperation with the Circum-Pacific Council for Energy and Mineral Resources, Circum-Pacific Map Series CP-45, scale, Washington, DC, 1:10,000,000, 1 sheet.

Carr, J.R., Stokes, C.R., Vieli, A., 2013. Recent progress in understanding marine-terminating Arctic outlet glacier response to climatic and oceanic forcing: twenty years of rapid change. Prog. Phys. Geogr. 37, 436−467.

Etzelmüller, B., Hagen, J.O., 2005. Glacier−permafrost interaction in Arctic and alpine mountain environments with examples from southern Norway and Svalbard. In: Harris, C., Murton, J.B. (Eds.), Cryospheric Systems: Glaciers and Permafrost. Geological Society, London, pp. 11−27. Special Publication 242.

van Everdingen, R. (Ed.), 1998 revised May 2005. Multi-language Glossary of Permafrost and Related Ground-ice Terms. National Snow and Ice Data Center/World Data Center for Glaciology, Boulder, CO.

French, H.M., 2007. The Periglacial Environment, Third ed. Wiley. 458 pp.

French, H.M., Shur, Y., 2010. The principles of cryostratigraphy. Earth Sci. Rev. 101, 190−206.

Gruber, S., 2012. Derivation and analysis of a high-resolution estimate of global permafrost zonation. The Cryosphere 6, 221−233.

Haeberli, W., 2005. Investigating glacier−permafrost relationships in high mountain area: historical background, selected examples and research needs. In: Harris, C., Murton, J.B. (Eds.), Cryospheric Systems: Glaciers and Permafrost. Geological Society, London, pp. 29−37. Special Publication 242.

Haeberli, W., Burn, C., 2002. Natural hazards in forests—glacier and permafrost effects as related to climate changes. In: Sidle, R.C. (Ed.), Environmental Change and Geomorphic Hazards in Forests, 9. IUFRO Research Series, pp. 167−202.

Harrison, W.D., Osipova, G.B., Nosenko, G.A., Espizua, L., Kääb, A., Fischer, L., Huggel, C., Burns, P.A.C., Truffer, M., Lai, A.W., 2014. Glacier surges. In: Haeberli, W., Whiteman, C. (Eds.), Snow and Ice-related Hazards, Risks and Disasters. Elsevier, pp. 437−485.

Hijmans, R., Cameron, S., Parra, J., Jones, P.J.A., 2005. Very high resolution interpolated climate surfaces for global land areas. Int. J. Climatol. 25, 1965−1978.

Hinz, K., Delisle, G., Block, M., 1998. Seismic evidence for the depth extent of permafrost in shelf sediments of the Laptev Sea, Russian Arctic. In: Paper presented at the Seventh International Conference on Permafrost. Collection Nordicana, Yellowknife, Canada, pp. 453−457.

Hinzman, L.D., Bettez, N.D., Bolton, W.R., Chapin III, F.S., Dyurgerov, M.B., Fastie, C.L., Griffith, B., Hollister, R.D., Hope, A., Huntington, H.P., Jensen, A.M., Jia, G.J., Jorgenson, T., Kane, D.L., Klein, D.R., Kofinas, G., Lynch, A.H., Lloyd, A.H., McGuire, A.D., Nelson, F.E., Oechel, W.C., Osterkamp, T.E., Racine, C.H., Romanovsky, V.E., Stone, R.S., Stow, D.A., Sturm, M., Tweedie, C.E., Vourlitis, G.L., Walker, M.D., Walker, D.A., Webber, P.J., Welker, J.M., Winker, K., Yoshikawa, K., 2005. Evidence and implications of recent climate change in northern Alaska and other arctic regions. Clim. Change 72, 251−298.

Houmark-Nielsen, M., Hansen, L., Jørgensen, M., Kronborg, C., 1994. Stratigraphy of a Late Pleistocene ice-cored moraine at Kap Herschell, central East Greenland. Boreas 23, 505−512.

Humlum, O., 1997. Origin of rock glaciers: observations from Mellemfjord, Disko island, central west Greenland. Permafrost Periglac. Processes 7, 1−20.

Huss, M., Farinotti, D., 2012. Distributed ice thickness and volume of all glaciers around the globe. J. Geophys. Res. 117, F04010. http://dx.doi.org/10.1029/2012JF002523.

Lantuit, H., Overduin, P.P., Couture, N., Are, F., Atkinson, D., Brown, J., Cherkashov, G., Drozdov, D., Forbes, D., Graves-Gaylord, A., Grigoriev, M., Hubberten, H.-W., Jordan, J., Jorgenson, T., Odegard, R., Ogorodov, S., Pollard, W.H., Rachold, V., Sedenko, S., Solomon, S., Steenhuisen, F., Streletskaya, I., Vasiliev, A., 2011. The arctic coastal dynamics Database: a new classification scheme and statistics on arctic permafrost coastlines. Estuar. Coasts 16, 1−18.

Nakawo, M., Young, G.J., 1981. Field experiments to determine the effect of a debris layer on ablation of glacier ice. Ann. Glaciol. 2, 85−91.

Ohmura, A., 2014. Snow and ice in the climate system. In: Haeberli, W., Whiteman, C. (Eds.), Snow and Ice-related Hazards, Risks and Disasters. Elsevier, pp. 77−98.

Osterkamp, T.E., 2001. Subsea permafrost. In: Steele, J.H., Thorpe, S.A., Turekian, K.K. (Eds.), Encyclopedia of Ocean Sciences. Elsevier, pp. 2902−2912.

Osterkamp, T.E., Fei, T., 1993. Potential Occurrence of Gas Hydrates in the Continental Shelf Near Lonely, Alaska. In: Paper Presented at Sixth International Conference on Permafrost, pp. 500−505. Beijing, China.

Pfeffer, W., Arendt, A.A., Bliss, A., Bolch, T., Cogley, J.G., Gardner, A.S., Hagen, J.-O., Hock, R., Kaser, G., Kienholz, C., Miles, E.S., Moholdt, G., Mölg, N., Paul, F., Radić, V., Rastner, P., Raup, B.H., Rich, J., Sharp, M.J., et al., 2014. The Randolph Glacier Inventory: a globally complete inventory of glaciers. J. Glaciol. 60 (211), 537−552.

Ran, Y., Li, X., Cheng, G., Zhang, T., Wu, Q., Jin, H., Jin, R., 2012. Distribution of permafrost in China: an overview of existing permafrost maps. Permafrost Periglac. Processes 23, 322−333.

Rasmussen, L.A., Conway, H.B., Krimmel, R.M., Hock, R., 2011. Surface mass balance, thinning, and iceberg production, Columbia Glacier, Alaska, 1948−2007. J. Glaciol. 57 (203), 431−440.

Rastner, P., Bolch, T., Mölg, N., Machguth, H., Le Bris, R., Paul, F., 2012. The first complete inventory of the local glaciers and ice caps on Greenland. The Cryosphere 6, 1483−1495.

Richter, M., 2002. Hypsometrie der Kontinente. Petermanns Geogr. Mitt. 146, 58−59.

Romanovskii, N.N., Gavrilov, A.V., Kholodov, A.L., Pustovoit, G.P., Huberten, H.-W., Nissen, F., Kassens, H., 1998. Map of Predicted Off-shore Permafrost Distribution on the Laptev Sea Shelf. In: Paper Presented at Seventh International Conference on Permafrost. Collection Nordicana, Yellowknife, Canada, pp. 967−973.

Romanovsky, V.E., Smith, S.L., Christiansen, H.H., 2010. Permafrost Thermal state in the polar northern hemisphere during the international polar year 2007—2009: a synthesis. Permafrost Periglac. Processes 21, 106—116.

Shumskiy, P.A., 1964. Principles of Structural Glaciology. Transl. D. Kraus. Dorer, New York.

Shur, Y., Hinkel, K.M., Nelson, F.E., 2005. The transient layer: implications for geocryology and climate-change science. Permafrost Periglac. Processes 16, 5—18.

Streletskyi, D., Anisimov, O., Vasiliev, A., 2014. Permafrost degradation. In: Haeberli, W., Whiteman, C. (Eds.), Snow and Ice-related hazards, risks and disasters. Elsevier, pp. 303—344.

Thompson, W.F., 1964. How and why to distinguish between mountains and hills. Prof. Geogr. 16, 6—8.

Troll, C., 1956. Uber das Wesen der Hochgebirgsnatur. Jahrbuch des Deutschen Alpenvereins 80, 142—157.

Troll, C., 1973. High mountain belts between the polar ice caps and the equator: their definition and lower limit. Arctic Alpine Res. 5, A19—A27.

Vieli, A., 2014. Retreat instability of tidewater glaciers and marine ice sheets. In: Haeberli, W., Whiteman, C. (Eds.), Snow and Ice-related Hazards, Risks and Disasters. Elsevier, pp. 677—712.

Williams, P.J., Smith, M.W., 1989. The Frozen Earth. Fundamentals of Geocryology. Studies in Polar Research. Series. xvi + 306 pp. Cambridge University Press, Cambridge, New York, Port Chester, Melbourne, Sydney.

Chapter 8

Integrated Approaches to Adaptation and Disaster Risk Reduction in Dynamic Socio-cryospheric Systems

Mark Carey [1], Graham McDowell [2], Christian Huggel [3], Jerrilynn Jackson [4], César Portocarrero [5], John M. Reynolds [6] and Luis Vicuña [3]

[1] *Robert D. Clark Honors College, University of Oregon, USA,* [2] *Department of Geography, McGill University, Montral, Canada,* [3] *Department of Geography, University of Zurich, Switzerland,* [4] *Department of Geography, University of Oregon, USA,* [5] *Independent Consultant, Huaraz, Peru,* [6] *Reynolds International Ltd, Mold, UK*

ABSTRACT

Cryospheric hazards in mountain ranges, at high latitudes, and around ice-covered volcanoes can adversely affect people by generating disasters such as glacial lake outburst floods, rock-ice landslides, lahars, and iceberg instability, as well as risks related to glacier runoff variability. These dangers are not simply biophysical; rather they are environmental events embedded within dynamic socioecological systems. To recognize the specific social and biophysical elements of cryospheric risks and hazards, in particular, this chapter introduces the concept of the socio-cryospheric system. To improve adaptive capacity and effectively grapple with diverse risks and hazards in socio-cryospheric systems, integrated approaches that span the natural sciences, engineering and planning, and the social sciences are needed. The approach outlined here involves three elements: (1) understanding cryospheric risks and hazards through scientific investigation and the accumulation of environmental knowledge regarding the biophysical basis of the hazardous stimuli; (2) preventing the natural events from occurring through risk management and engineering strategies; and (3) reducing susceptibility to harm by addressing the socioeconomic, political, and cultural factors that influence vulnerability to risks, hazards, and disasters. This chapter analyzes several case studies of particular hazards (in particular places), including glacier and glacial lake hazards in Peru (Cordillera Blanca and Santa Teresa) and Nepal; volcano-ice hazards in Colombia and Iceland; glacier runoff and melt water-related hazards in Nepal and Peru; and coastal hazards in Greenland. These case studies help illustrate achievements and limitations of the three-pronged approach to adaptation, while

Snow and Ice-Related Hazards, Risks, and Disasters. http://dx.doi.org/10.1016/B978-0-12-394849-6.00008-1
219

revealing opportunities for greater symbiosis among scientific/knowledge-based, risk management/engineering-based, and vulnerability-based approaches to adaptation and disaster risk reduction in socio-cryospheric systems.

8.1 INTRODUCTION

Cryospheric hazards affect populations worldwide, in high mountains such as the Andes, Alps, and Himalaya as well as in high-latitude areas such as the Arctic and Patagonia. Ice loss, glacial lake outburst floods (GLOFs), avalanches, rock-ice landslides, lahars on glacierized volcanoes, and both seasonal and long-term glacier runoff variability can lead to dramatic or even catastrophic impacts on regional populations (IPCC, 2012). Even people who do not live adjacent to ice can be affected by cryospheric hazards through impacts on downstream water use, tourism economies, energy production, infrastructure, agriculture and food supplies dependent on glacier melt water, and disaster recovery costs. Moreover, glacier and ice sheet loss due to climate change will affect coastal populations worldwide through sea level rise. Cryospheric hazards thus affect populations within and beyond high-mountain and high-latitude regions through low-frequency, high-magnitude events such as GLOFs and avalanches, as well as through slow-onset processes with major impacts, such as sea level rise and hydrologic variability in glacier-fed watersheds. These events, as Haeberli and Whiteman, 2014 note in this book's introduction, can be slow or fast, and they can be small or large, thereby acting on various spatial and temporal scales. Because these hazards occur in dynamic and distinct environmental and social contexts, reducing vulnerability and promoting resilience requires a comprehensive approach attentive to the unique factors that affect socioecological systems in cold regions—what we refer to as "socio-cryospheric systems."

In this chapter, we articulate an integrative three-pronged approach for improving adaptation in dynamic socio-cryospheric systems. We show that successful adaptation to cryospheric hazards hinges on (1) understanding cryospheric hazards through scientific investigation and the accumulation of environmental knowledge regarding the biophysical basis of the hazard; (2) preventing the natural events from occurring through risk management and engineering strategies; and (3) reducing susceptibility to harm by addressing the socioeconomic, political, and cultural factors that condition vulnerability to risks and hazards related to the cryosphere. We present several case studies from around the world highlighting the challenges, successes, and limitations of past and ongoing adaptation and risk reduction efforts in cold regions as well as opportunities for greater symbiosis among scientific/knowledge-based, risk management/engineering-based, and vulnerability-based approaches to adaptation in socio-cryospheric systems. The chapter is organized around several distinct types of cryospheric hazards, including glacier and glacial lake hazards, volcano—ice hazards, glacier runoff and water-use hazards, and coastal hazards.

8.2 INTEGRATED ADAPTATION IN DYNAMIC SOCIO-CRYOSPHERIC SYSTEMS

Research on disaster risk reduction tends to define risk as the combination of the likelihood of a natural event occurring, human exposure to those events, and vulnerability to hazardous events. According to IPCC's "Managing the Risks of Extreme Events and Disasters to Advance Climate Change Adaptation (SREX)" report, "disaster risk management and adaptation to climate change focus on reducing exposure and vulnerability and increasing resilience to the potential adverse impacts of climate extremes, even though risks cannot fully be eliminated" (IPCC, 2012). Clearly, risk reduction and adaptation involve linked agendas focusing on both society and environment. In turn, most of the research on risk and hazards now recognizes that, while "natural hazards" such as earthquakes or floods are environmental events, disasters are social and political rather than "natural" (Maskrey, 1993; Wisner et al., 2004). Consequently, to understand and manage cryospheric risks and hazards—or any kind of risk or hazard—it is essential to comprehend the interplay between various societal forces and environmental processes. Building on conceptual work by Folke (2006), Young et al. (2006), and Turner et al. (2003), we use the phrase "socio-cryospheric systems" to conceptualize these dynamic socioecological interactions in the specific contexts of cryospheric risks and hazards.

Resilience scholarship is useful for examining snow- and ice-related risks, especially as resilience research broadens to include more nuanced treatment of the societal conditions affecting system change and as it sheds its previous emphasis on system stability. At the most basic level, resilience is the ability to absorb disturbances without experiencing a fundamental loss of system structure or function (Folke et al., 2010). Most resilience studies have until recently emphasized the ecological dimensions and scientific approaches to systems, with a lack of attention to social contingency, including power relations, resource conflicts, and cultural factors. Resilience research now increasingly considers both external and internal drivers of change, with growing attention to the significance of socioeconomic and political dynamics in socioecological systems (Brown, 2014). This work no longer conceptualizes resilience as the persistence of system stability but rather allows for directed system change through adaptation and transformation (Adger, 2000; Folke et al., 2010; Gallopín, 2006). This conceptualization of resilience is useful for analyzing cryospheric hazards because it recognizes that system adaptability (not stability) is crucial. Further, it suggests for this chapter that shocks are as likely to stem from cryospheric or climatic changes as they are from other forces of global change such as neoliberal political-economic reforms, power struggles, socioeconomic inequality, or divergent cultural values.

To understand the vulnerability-based component of adaptation and to include insights from risk management-/engineering-based approaches, we also draw from vulnerability research, which emphasizes the ability of individuals, households, communities, and/or institutions to address, plan for, or adjust to cryospheric hazards. Adaptive capacity includes a focus not only on risk management and engineering, such as draining glacial lakes to prevent GLOFs or building reservoirs to regulate water flows, but also on societal aspects related to exposure, sensitivity, and vulnerability. These societal variables are diverse and can include aspects of economics, technology, social capital, knowledge and information, institutions, governance, and culture (Engle, 2011). When adaptive capacity across some or all of these variables is insufficient, vulnerability increases (consistent with Smit and Wandel, 2006; Ford and Smit, 2004). Similarities exist between efforts to reduce risk to natural hazards and efforts to adapt to climate change (Mercer, 2010). In fact, in recent years, substantial efforts have been undertaken to integrate the scientific communities of climate change adaptation and disaster risk reduction, notably in the aforementioned SREX report (IPCC, 2012). Adaptive measures are now considered critical for cryosphere-related hazards, such as GLOF threats and diminishing water supplies (Adger et al., 2007), with increased adaptive capacity and resilience expected to reduce vulnerability. Integrated adaptation, as the following cases about specific types of cryospheric risks and hazards will demonstrate, emphasizes the resilience of socio-cryospheric systems by increasing the effectiveness of adaptation to potentially harmful snow- and ice-related events.

8.3 GLACIER AND GLACIAL LAKE HAZARDS

Abrupt, low-frequency, high-magnitude, cryospheric hazards can include GLOFs, avalanches, and "jökulhlaups," an Icelandic word literally meaning "glacier running" (or "glacier leap") that refers to the hydrostatic jacking of a glacier by subglacial volcanically derived melt water, which commonly occurs instantaneously and results in an outburst flood of melt water and debris from beneath the glacier. GLOFs are an increasingly recognized hazard in mountain ranges worldwide. They occur when moraine or ice dams rupture, thereby releasing significant amounts of lake water in a matter of minutes or hours. Sometimes, these outbursts can create catastrophic GLOFs, destroying communities and infrastructure and killing people, such as the 1941 Lake Palcacocha GLOF in Peru's Cordillera Blanca that killed an estimated 5,000 people (Carey, 2005). Glacier landslides or avalanches can also occur abruptly and cause widespread destruction, with the 1970 Mount Huascarán event that killed an estimated 8,000−15,000 people among the most deadly in world history (Evans et al., 2009; Oliver-Smith, 1986). The dynamic conditions created by climate change and glacier shrinkage can clearly pose a variety of risks to human populations—and it is challenging for researchers to monitor and assess these changing situations worldwide (Haeberli et al., 2013). In each case of glacier or glacial lake hazard, the socioeconomic contexts in particular places and periods affect exposure, adaptive capacity, and resilience. Thus, it is critical to examine these hazards from an integrated approach that examines the larger socio-cryospheric system from scientific, engineering, and social perspectives, as demonstrated by the following case studies in Peru and Nepal.

8.3.1 Cordillera Blanca, Peru

Peru has experienced some of the world's most deadly glacier-related catas-trophes, and these disasters date back many decades (Ames, 1998; Ames Marquez and Francou, 1995; Carey, 2005, 2010; Reynolds, 1992; Zapata Luyo, 2002). Since the catastrophic Lake Palcacocha GLOF in 1941, Peruvian engineers and government officials have been working to reduce glacier hazards in the country's most glaciated mountain range, the Cordillera Blanca (Figure 8.1). Local residents below the glaciated mountains have also insisted on government action to protect them from GLOFs and glacial landslides that have killed thousands in the region during the last half century. In 1951, in response to another destructive GLOF, the government created the Control Commission of Cordillera Blanca Lakes to study, monitor, drain, and dam dangerous glacial lakes in the region. The agency remains in existence today as the Glaciology and Hydrological Resources Unit, though its scope and personnel are severely reduced from past decades, despite the continuation of glacier hazards in the Cordillera Blanca and elsewhere.

To date, Peru's responses to glacier and glacial lake hazards have concentrated on risk-management/engineering approaches (as opposed to knowledge- or vulnerability-based approaches). These approaches include, among other things, the partial drainage and damming of 35 unstable Cordillera Blanca glacial lakes that significantly reduced the likelihood of GLOF disasters (Carey, 2010). There has also been notable work, but with much less depth, in scientific/knowledge-based approaches to adaptation in Peru. Achievements in the vulnerability-based realm have been the least successful in the region, though this aspect is increasingly the focus in recent discussions about adaptive capacity and resilience (e.g., Hegglin and Huggel, 2008; Hill, 2013). But adaptive capacity in Peru has historically been restricted in part because the country has had limited economic, technical, and scientific resources. Although the national economy has expanded markedly during 2004−2014, the income has not been equally distributed to all parts of the country or among all constituencies, characteristically leaving highland areas with few resources or funds despite national or private company economic growth. Further, efforts at vulnerability reduction before 2014 largely failed due to local resistance among the vulnerable populations themselves (Oliver-Smith, 1986; Carey, 2010). In the 1980s and early 1990s, violent civil conflict related to the revolutionary group Shining Path also made engineering projects at glacial lakes too dangerous. Nevertheless, Peruvians have extensive and successful experiences with glacier hazard reduction in a densely populated

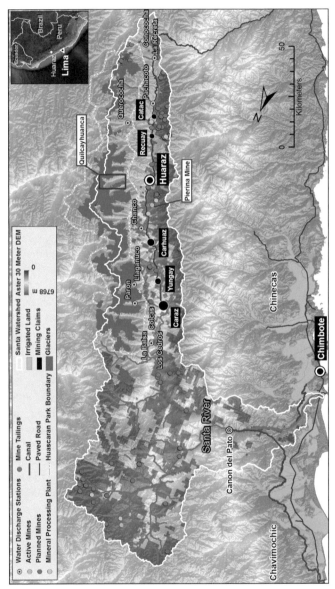

FIGURE 8.1   Map of Cordillera Blanca and Santa River watershed, Peru. *Map by J. Bury.*

and dynamic socio-cryospheric system. Peruvian experts have traveled to Nepal and elsewhere to share the tremendous knowledge and experience gained in the Andes with other countries and specialists around the world facing cryospheric hazards, particularly GLOFs.

Engineering solutions have been the most effective and most widely practiced in the Cordillera Blanca. These strategies involved reducing the volume of water in glacial lakes by digging open trenches through the natural moraine dams to reduce the water level. To avoid a subsequent rise in the lake level, engineers installed drainage pipes at the lower lake level. They then built up earthen or cement dams above those pipes to fill in the cuts in moraines and to protect the dam against waves when ice from avalanches or calving glaciers crashed into lakes. These strategies were initiated in the 1940s and have continued into the twenty-first century (Lliboutry et al., 1977; Carey, 2010; Reynolds, 1992, 1993; Zapata Luyo, 2002). Of course lake sizes, volumes, and other characteristics have changed over time, and the major earthquake in 1970 that damaged many artificial glacial lake dams and triggered the deadly Huascarán rock and ice sturzstrom have all inspired technicians and government officials to reassess their disaster risk reduction plans. In some cases, previous projects were removed and reconstructed with improved technologies; in other cases, entirely new strategies were developed to stabilize dangerous glacial lakes (Carey et al., 2012b; Reynolds et al., 1998; Vilímek et al., 2005).

Yet glacier hazards persist, even from previously stabilized lakes such as Lake 513 that caused a GLOF in 2010, or Lake Palcacocha (Figure 8.2) that has an increasing volume of water and has expanded due to glacier shrinkage

FIGURE 8.2   Lake Palcacocha, Peru, with two security dams in the foreground (dam on the left clearly damaged by 2003 lake overflow). *Photograph by M. Carey.*

to a point where ice and rock avalanches can fall directly into the lake (Carey et al., 2012b; Emmer and Vilímek, 2013; Emmer et al., 2014). In both these cases, the engineering projects significantly reduced the GLOF risk before catastrophe struck, showing both the importance of advance planning and action as well as the difficulty or impossibility of completely eliminating risk. The 2010 Lake 513 event occurred at a glacial lake that had previously been drained enough to leave a 20-m freeboard and a bedrock dam (Reynolds et al., 1998). When an avalanche of approximately 400,000 m^3 of rock and ice landed in the lake, however, the displacement waves overtopped the dam and caused a GLOF that released a large amount of debris that damaged downstream areas but fortunately caused no loss of human life. Had the earlier mitigation works not have been carried out, the consequences of the 2010 event would have been significantly worse with possibly several thousand fatalities. However, the design concept for the original engineering works was based on likely instability of the rapidly retreating and very steep valley hanging above the lake, rather than on deterioration of the stability of the uppermost mountain flanks that is occurring increasingly and with increasing frequency in high-mountain regions worldwide as a result of climate change (Kääb et al., 2005). This event, among others, shows that cryospheric hazard-reduction techniques must always be assessed and reevaluated due to ever-changing environmental conditions such as those arising from climate change. Scientific studies to detect potential threats and reduce downstream vulnerability must be maintained and improved in such a dynamic socio-cryospheric system as the Cordillera Blanca.

Lake Palcacocha is another glacial lake creating renewed risks despite a history of engineering projects since it produced the 1941 GLOF that destroyed a third of the city of Huaraz. At that point, the lake had an estimated 10−12 million cubic meters of water, most of which was removed when the lake drained catastrophically in 1941. Engineers later drained more water and artificially dammed the lake in the 1950s and 1970s, leaving the lake with only 0.5 million cubic meters of water in 1974. A wave triggered by the 2003 landslide from a lateral moraine overtopped and partially destroyed a security dam at the lake, demonstrating the dynamic and unstable nature of the glacial lake environment (Carey, 2005; Vilímek et al., 2005). A 2009 bathymetry study then revealed that the lake held approximately 17 million meters of water, more water than when the lake produced the deadly 1941 GLOF and 34 times more water than it had in 1974. As the water volume expanded drastically due to the shrinking glacier that created more space between the moraine dam and the glacier snout, the population inhabiting the city of Huaraz grew from 12,000 people in 1941 to >100,000 people today—and thousands or even tens of thousands inhabit the floodplain hazard zone. Currently, short-term technical measures are underway to lower the lake level with siphons and thus expand the freeboard. But this is not a long-term solution to remove water from the lake or to reduce vulnerability and enhance resilience in local communities such as Huaraz. Some groups are studying the basin to better understand trigger mechanisms for potential GLOFs (Emmer and Vilímek, 2013; Emmer et al., 2014). But to date, most of the emphasis has been on lowering Palcacocha's lake level, which could be successful, though not a guarantee (recall the Lake 513 example above) because conditions change so quickly in these glaciated landscapes. Lacking, then, is a vulnerability approach that would involve working with the Huaraz population on plans such as early warning systems, evacuation procedures, education programs, building codes and zoning, or other measures to reduce human vulnerability and build adaptive capacity.

8.3.2 Santa Teresa, Peru

The case of Santa Teresa shows the complexities of a vulnerability-based approach to cryospheric hazards, as well as underscoring the importance of considering local perspectives and the need to analyze local societal conditions in multiple ways that involve a detailed understanding of social, cultural, political, and economic contexts in particular places and times (Adger et al., 2013). The town of Santa Teresa—not to be confused with the district or catchment also named Santa Teresa—was founded in its current location after a large debris flow in 1998 forced the relocation of the town from its original site (Municipalidad Distrital de Santa Teresa, n.d.). This event originated at approximately 4500 m in the high-mountain area of the Sacsara catchment; it was triggered by the massive mobilization of poorly consolidated sediment most likely of glacial origin. The 1998 Sacsara debris flow with a volume of several million cubic meters destroyed the town of Santa Teresa. A month later, the Ahobamba debris flow again inundated Santa Teresa. It killed five inhabitants and destroyed the Machupicchu hydroelectric station, 224 employee homes, 10 bridges, and 47 km of railroad track between Machu-picchu and Quillabamba (PREDES, 2007). Afterward, Santa Teresa was relocated to a higher site, thereby representing at first glance a successful adaptation to natural hazards.

On one level, a successful technical strategy was employed to protect the population: new plots and homes allowed people to inhabit supposedly safer places, not exposed to the kinds of debris flows that destroyed the original town in 1998. Moreover, the debris flow later brought new economic opportunities to the community by opening up an alternative access road via Santa Teresa to Peru's top tourist attraction, Machupicchu, which tourists previously reached through the town of Aguascalientes. Today, both the Santa Teresa and Aguascalientes roads function. Prior to the debris flow, Santa Teresa inhabitants survived primarily on agriculture, but after the disaster, the newly located community emerged as an important tourist site that today offers a wide range of services such as hotels, restaurants, guides, and transportation. Further, major hydropower projects as well as financial contributions through the "Canon Gasífero" (Sociedad Nacional de Minería Petróleo y Energía, 2013) supported the town's development and helped transform it into an attractive place not only for tourists but also for local and non-local investors.

A critical need remains, however, for more knowledge about the likelihood of similar debris flow events occurring in the region, especially considering the rate of glacier retreat that produces new hazards and the influx of people and money that generate societal variability in the rapidly changing socio-cryospheric system. The question remains: can the nearby glaciated mountains trigger another debris flow of equal or larger magnitude? Considering this last question, it is necessary to also ask who is at risk and to what extent. This involves understanding vulnerability and implementing successful adaptation practices, which entails deciphering who is building in the "unsafe" locations and who defines those sites as "safe," "unsafe," or "safe enough" for living—that is, in the sense of these classic questions by Starr (1969). These steps are critical because the local government has not implemented or enforced building codes to prevent building in the zones scientists believe are exposed to potential hazards.

Other aspects of the socioeconomic landscape must be examined in Santa Teresa to effectively implement vulnerability-based approaches to glacier hazards. The economic benefits associated with investments and the tourism economy are not equally shared among all residents, social groups, or stakeholders in the region. Additionally, economic opportunities and successes may also lead to increased exposure to natural high-mountain and cryospheric hazards for certain actors. For others, expanded economic capital may occur and thus enhanced security or reduced vulnerability. Still others may experience improved social capital (network of the villagers, nearby relatives, etc.) due to living in Santa Teresa, which can reduce people's vulnerability in certain ways (e.g., emotional support, family and friend networks, financial assistance in the case of illness or other unexpected events, including natural hazards).

Overall, the Santa Teresa case may appear as a success story at first glance. But given the complexity and variation among these various issues, a clear need exists to understand vulnerability on a much deeper level than solely analyzing direct physical exposure to the natural hazards (relocation). The improved tourism economy may also bring new hazards and vulnerabilities not present prior to the 1998 debris flow and subsequent relocation. Perceptions of the glacier hazards also influence vulnerability. On the one hand, relocation offered a suitable risk management approach to the hazard because it reduced the likelihood of a debris flow inflicting damage on the community. But, on the other hand, more knowledge of the hazard origins and trigger mechanisms are necessary to better understand the likelihood of future floods. Moreover, it is crucial to understand peoples' perceptions of their own vulnerabilities and the vulnerabilities within the larger population at risk (e.g., in the "unsafe" places).

Their perceptions are crucial for prioritizing risks—all risks including social, cultural, economic, and political risks, not just natural hazards (Jurt, 2009).

Despite what initially appears as successful community relocation and helpful new economic opportunities that triggered financial growth in Santa Teresa, deeper analysis raises important considerations about the success of these measures and outcomes, which thus challenges a simplistic portrayal of community resilience in the face of cryospheric hazards. This Santa Teresa case shows that perceptions differ considerably among different social groups. Without accounting for these diverse perceptions and vulnerabilities—which likely change over time—and without taking measures beyond the physical relocation of exposed populations, the community's adaptive capacity may be reduced. These categories are important but have to be put into the wider context of the site's history as well as the different actors' perspectives and roles.

8.3.3 Nepal

GLOFs were brought poignantly to the attention of the Nepali and international authorities when a GLOF destroyed the Austrian Government-funded Dig Tsho Hydroelectric Power scheme in 1985. Thereafter, the Nepali Government initiated a series of studies of glacial lakes with Japanese colleagues (Yamada, 1993), and through subsequent years, many other studies have been conducted in Nepal and throughout the broader Himalayan region, including many by the International Center for Integrated Mountain Development (ICIMOD) in Kathmandu. In 1991, Chubung, a small glacial lake in the upper reaches of Rolwaling, 110 km northeast of Kathmandu, burst and the ensuing flood damaged downstream areas and killed an elderly deaf woman unable to hear warnings as she worked in a flour mill. The villagers were alarmed by the devastation from this one small lake. They also recognized that the potential destruction from the much larger neighboring glacial lake, Tsho Rolpa, would be significantly greater should its dam fail. These villagers held many meetings (Figure 8.3) about the hazards, discussed possible evacuation, and appealed to embassies in Kathmandu for assistance. In 1993, the Nepali Government's Water and Energy Commission Secretariat (WECS) undertook preliminary observations in the Rolwaling valley and around Tsho Rolpa, but little was deduced about the lake's potential hazards.

It was not until the following year that the first formal glacial hazard assessment of Tsho Rolpa was undertaken, which subsequently led to the first engineering efforts to reduce GLOF threats in the region (Reynolds, 1998). The Nepal Netherlands Friendship Association and a Dutch pipe manufacturer (Wavin Overseas BV) agreed to fund and install specially designed and manufactured pipes to act as trial siphons, the design of which was based on that used successfully in Peru (Reynolds, 1992; Richardson and Reynolds, 2000). The first set of siphons was installed in 1995 and operated as required, thereby demonstrating for the first time in the Himalayas that such

FIGURE 8.3   Rolwaling Sherpas engage in traditional dancing following one of the regular liaison meetings with them in relation to the mitigation works at Tsho Rolpa, Rolwaling, Nepal. *Photograph by J. Reynolds.*

technology—which had proved to be enormously successful in Peru—could be used at an altitude of 4500 m, with the pipes being transported by porters and the siphons maintained by local villagers. These siphons were augmented in 1997 with additional pipes that functioned for several months (Reynolds, 1999). A key factor in the successful use of these siphons was the involvement of the local Rolwaling Sherpas in the discussions over what could be done and how, and how they might be involved, before any major new projects were initiated. This included discussions on how to involve local labor in the subsequent Tsho Rolpa GLOF Risk Reduction Project and how to address local religious concerns about the way the work was to be conducted (BPC Hydroconsult, 1997). Clearly, successful GLOF prevention required an array of local, national, and international actors—from local residents and porters to government officials to engineers with experience in Peru and Nepal, which also shows the necessity of holistic, integrated approaches to risk reduction in socio-cryospheric systems.

In 1996, the responsibility for managing glacial hazards in Nepal was transferred by the National Planning Commission of Nepal from WECS to the Department of Hydrology and Meteorology (DHM). Also, in the same year, with additional support from the UK government, negotiations between the Nepali and the Dutch governments commenced with a view to the Dutch government funding emergency remediation works at Tsho Rolpa. It also became clear that the very large number of siphons required to draw down the lake level sufficiently would have been impractical. Subsequently, an open channel with sluice gates design was developed in 1997 for which $2.9 million funding was approved in 1998 (Tsho Rolpa GLOF Risk Reduction Project; Rana et al., 2000). An early warning system funded by the

World Bank in 1998 was installed by 1999 to provide villages, and a hy-droelectric power installation downstream, notice of any impending GLOF. Construction of the open canal started at Tsho Rolpa in May 1999 and was completed in June 2000. The drawdown of the lake by 3.5 m was achieved successfully by July 2000. However, the total required drawdown to achieve a long-term solution was determined to be at least 20 m. The intention of the initial 3.5-m drawdown was to provide a temporary reprieve during which time a Tsho Rolpa GLOF Permanent Remediation Project could be formu-lated, and, hopefully, the remaining 17.5-m lowering of the lake could be funded, designed, and built (Rana et al., 2000). To date, no progress has been made to start this project.

Yet cryospheric conditions remain unstable and the GLOF hazard persists. A large mass of stagnant glacier ice, the width of the lake and thought to be up to 80 m thick and several hundred meters long, is currently grounded at the western end of the lake and is partly integrated into the northwest corner of the terminal moraine. As the submerged ice continues to melt, there might come a point when the ice may become instantaneously buoyant, creating a massive displacement wave and destroying part of the terminal moraine. This process, albeit on a much smaller scale thus far, has been observed to occur at the lake (Richardson and Reynolds, 2000). This could lead to the failure of the sub-stantive part of the 150-m-high terminal moraine and a breach >50 m deep, resulting in a very high peak flow rate and drainage in excess of 130 million cubic meters of water plus debris. A GLOF of this magnitude, especially when compared with previously modeled hypothetical GLOFs from Tsho Rolpa (BPC Hydroconsult, 1997), could have devastating consequences downstream for communities, the existing Khimti hydroelectric station, and the recently constructed infrastructure along the Tama Kosi River that supports the Lamabaga hydroelectric project on the Nepalese–Tibetan border. Although GLOF risk remains and the knowledge of the hazards has increased through awareness and scientific studies, there has still been little action to reduce threats at the lake or decrease vulnerability in downstream communities. This situation is endemic throughout Nepal and the Himalayan region, which stands in sharp contrast to Peru's historical record of more successful engineering projects to reduce disaster risk in socio-cryospheric systems.

Political conflicts, media manipulation, poor accessibility necessitating high costs of logistical support, difficulty in involving enough locals, and rushed or careless research have all thwarted more effective GLOF prevention approaches in Nepal to enhance community resilience and reduce cryospheric hazards. While there was significant local support for these projects, the high-profile Nepali government's Tsho Rolpa project attracted increasing attention internationally and from Maoist guerrillas in the late 1990s and into the early 2000s. Indeed, individual monitoring installations and key components of the early warning system were vandalized or stolen by people from outside the Rolwaling Valley, ultimately rendering it inoperable. However, DHM has been

able to maintain low-key operations at Tsho Rolpa since 2000, and the political conflicts have subsided.

Media manipulation played a key role in diverting attention to another
hazard, as well as influencing the type of science conducted and the com-
mitments donors made (or rescinded) in the region. During the early stages of
the Tsho Rolpa project and especially between 1997 and 2000, DHM and its
international partners received a great deal of criticism in the media from a
number of local institutions that rejected the original scientific basis for the
project. These same opponents have also acted to divert attention from the
developing hazards associated with Tsho Rolpa to the more easily accessible
Imja Tsho in the Solukhumbu, which may be less of a threat (e.g., Hambrey
et al., 2008; S. Takenaka, personal communication 2009). At the same time,
representatives from some vulnerable communities have protested that they
have not been consulted. Indeed, in July 2012, following a TV news report
about the risk of a GLOF, many surprised Solukhumbu residents with little
information fled their homes overnight (BBC News, 2012).

The backlash from local people following the media alarms in 2008 and
2012, as well as these various other issues, raises several important questions.
The first is how to engage realistically and practically between government
agencies, local communities, and international organizations to ensure that
projects are formulated responsibly and executed properly (Reynolds, 2013).
The second is that with the ease of access to Imja Tsho coupled with international publicity about glacial hazards, in general, many more people have
been attracted to visit the place and express opinions, whether or not qualified
to do so. The third is the way in which donor organizations become involved in
both disaster risk reduction programs, such as GLOF prevention, and in media
coverage, often then diverting funds away from one project to other, perhaps to
more strategic projects to appease the media.

The science associated with glacial hazard assessment has made significant advances in understanding the biophysical processes (e.g., Ames, 1998;
Fernández Concha, 1957; Haeberli et al., 1989, 2001; Lliboutry et al., 1977;
Portocarrero, 1995; Trask, 1953; Zapata Luyo, 2002), to being able to use
remote-sensing techniques to aid those assessments (Quincey et al., 2005,
2007; Kargel et al., 2011) and in modeling GLOF behavior (Westoby et al.,
2014). In parallel, there have been significant advances in methods of mitigation through technology and engineering projects, with Peru leading the
way and the Himalayan region starting on this road in the late 1990s. Many
challenges remain, however, including the need to integrate diverse knowledge sources and the science focusing on different world regions; the need to
share scientific conclusions and insights with people potentially affected by
the biophysical processes; and the need for policy makers, scientists, and
engineers to consider the changing high-mountain environment and climate
while bringing multiple benefits to the local communities through more
holistic risk reduction strategies (Reynolds, 2010). It is likely that direct

liaison with the local communities on a regular basis could heighten awareness of nearby glacial lake hazards, thereby also potentially improving the level of cooperation in mitigation works. The media, at times, can misconstrue scientific conclusions and frustrate local residents. Furthermore, a major disadvantage for engineering-based mitigation of hazardous glacial lakes in Nepal, when compared with those in Peru, is the generally greater remoteness and larger distances separating those in the Himalayas from the vulnerable communities downstream. Unless significant local labor, which is often not available, can be employed and managed safely and efficiently, the use of expensive helicopter support becomes essential. Thus, the financial costs of mitigating hazardous glacial lakes in the Himalayas can be significantly greater than for a comparably sized project in the Cordillera Blanca in Peru, for instance.

While these cases from Peru and Nepal demonstrate the need for integrated or holistic approaches to cryospheric hazards, they are not the only world regions where societies have grappled—often successfully—with cryospheric hazards. In the European Alps, for example, there have been examples of scientific developments, risk reduction engineering, and vulnerability-based approaches to cryospheric hazards during recent decades (Haeberli et al., 1989). Research by Huggel et al. (2004) showed various innovations regarding the assessment of GLOFs and ice avalanches in the Swiss Alps, employing a variety of scientific measurements and methods, including historical data on past events, to project approximations of potential events, probability of occurrence, and likely maximum magnitude (Huggel et al., 2004). But the science of these projections requires continued refinement and data collection, while the population and infrastructure vulnerable to ice avalanches and glacial floods are also constantly changing though only rarely analyzed in the context of cryospheric hazards. Successful combinations of scientific and engineering approaches have also been employed to reduce the risk of GLOFs in Switzerland, such as the 25 years of photogrammetric (including aerial photography) analysis and other onsite field studies of periglacial lakes and permafrost in Valais, which ultimately concluded with effective artificial lowering of water levels in dangerous lakes. The initial lowering of a lake level in this area was conducted after a 1970 outburst, but it was necessary to continue the engineering projects at several periglacial lakes during the 1980s and 1990s due to continued lake instability, biophysical changes, and the vulnerable population in the Saas Valley below. Engineering works over the decades included using siphons, water pumping, tunneling, drilling, and digging v-shaped ditches on moraine surfaces to lower water levels and reduce the GLOF risk (Haeberli et al., 2001). Such scientific approaches and engineering strategies have been applied to cryospheric hazards elsewhere and considerably more often than vulnerability-based approaches or research with a societal emphasis, though exceptions exist (e.g., He et al., 2012; Carey, 2010; Takenaka et al., 2012).

8.4 VOLCANO−ICE HAZARDS

Glacier-covered volcanoes can cause catastrophic disasters, such as lahars and jökulhlaups that form when volcanic activity melts ice abruptly and rapidly, thereby unleashing debris flows and floods down slope. Volcano−ice interactions threaten populations worldwide, such as in the US Pacific Northwest where volcanoes like Mount Rainier pose lahar risks to Seattle and Tacoma, and Mount St Helens that erupted in 1980 with large lahars that flowed into the Columbia River, stopped ocean-going ships, and caused significant infrastructural damage in southern Washington and northern Oregon. Volcano−ice hazards can cause unforeseen impacts, but they can also be predicted, as the following Colombia case shows, through scientific research and monitoring of volcanic and glacier activity, which is also occurring in Iceland. Scientific monitoring and studying of volcanoes help understand volcanic processes and detect environmental conditions before disasters occur. Moreover, both governmental and non-governmental institutions can reduce vulnerability by strengthening adaptive capacity and resilience. But protecting populations adjacent to ice-covered volcanoes (and other hazards) requires effective interactions among scientists, institutions, and the vulnerable communities. It is thus imperative to link scientific approaches to cryospheric risks with engineering-based approaches in potential lahar impact zones and vulnerability-based approaches such as building socio-institutional capacity and building local trust to facilitate resilience. Without effective and consistent cooperation in this socio-institutional context, adaptive capacity and resilience may be insufficient to reduce susceptibility to harm, leading to major disasters as in the 1985 Colombian tragedy of Nevado del Ruiz. It is therefore important to critically analyze and document past experiences, both negative and positive, to learn how to successfully build socio-institutional capacity, as the following examples from Colombia and Iceland indicate.

8.4.1 Colombia

The Colombian case of Nevado del Ruiz provides a remarkable example of the role of socio-institutional factors in cryospheric hazards and disasters. Nevado del Ruiz, an active composite stratovolcano (5320 masl) with an ice cap on top, erupted on November 13, 1985. The eruption was moderate (Volcanic Explosivity Index VEI = 3, with a scale from 0 to 8), but it triggered a series of enormous lahars, predominantly by the interaction of pyroclastic density currents with snow and ice (Huggel et al., 2007; Thouret, 1990). The lahars traveled with velocities of 5−15 m/s, with an enormous peak discharge up to 48,000 m^3/s and total volumes of about 90×10^6 m^3 toward the East into the Magdalena Valley in the Department of Tolima (Pierson et al., 1990). The city of Armero (Figure 8.4) was hit at night by the lahars about 2 h after the eruption, killing approximately 25,000 people (Naranjo et al., 1986).

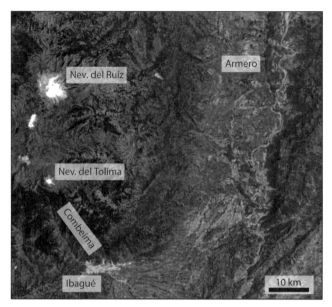

FIGURE 8.4   Satellite map showing the location of Nevados del Ruíz and Tolima, the city of Ibagué, and Armero before it was destroyed in the 1985 disaster. *Satellite image derived from Google Earth.*

It is important to recognize that this relatively small eruption resulted in one of the world's most deadly volcanic and cryospheric disasters, signaling that exposure and vulnerability were dominant aspects of the magnitude of risk and loss. Poor institutional communication and coordination at various levels account for other key factors explaining the Ruiz disaster (Hall, 1992). Many institutions and government agencies were involved in crisis management before the disaster, including several city administrations such as Armero and Manizalez; regional governments (Tolima, Caldas); the national government of Colombia; technical federal agencies such as the geological survey (INGEOMINAS, since 2011 the Colombian Geological Survey (SGC)); and, to some extent, international institutions including UN organizations. Volcanic unrest on Nevado del Ruiz started about one year before the 1985 catastrophe, and scientists from INGEOMINAS rapidly assessed the hazards based on indications of similar historical lahar events in 1595 and 1845. Only a few weeks before the disaster, a corresponding hazard map was released where Armero was clearly marked as being located in lahar runout zones (Parra and Cepeda, 1990). The hazard map triggered a political debate due to potential devaluation of property and land. The local and regional governments of the Tolima Department remained poorly prepared, emergency management plans were not implemented in a timely fashion, and no alarm system was installed for Armero. General political reluctance also occurred to bear the economic or political costs of early evacuation or a false alarm (Voight, 1990). Armero was thus hit unprepared by the lahars.

The relationship between local people and institutions, especially government institutions, can play a major role in vulnerability issues. Lack of trust in government institutions, for instance, can exacerbate vulnerability, and thus risk (e.g., Carey, 2005). In the aftermath of the 1985 Armero catastrophe, surviving people were permanently relocated to other cities, and the area of the former Armero was banned for permanent settlements. Many of the relocated people from Armero were sent to Ibagué, the capital of the Departamento de Tolima. They were brought to areas in the outskirts of Ibagué that were poor and highly exposed to floods from the Combeima River and lahars from the ice-capped Nevado del Tolima. In addition, these city areas have also become populated by internally displaced people from various regions in Colombia due to the decade-long armed conflict. Nowadays, it is a district with a particularly high crime rate, and disaster-prevention programs are difficult to implement in this impoverished, socially disrupted district of Ibagué (Figure 8.5). People's limited trust in government institutions has also been found in the upstream regions of the Combeima valley, which are also endangered by landslide hazards (Künzler et al., 2012). Local people have to

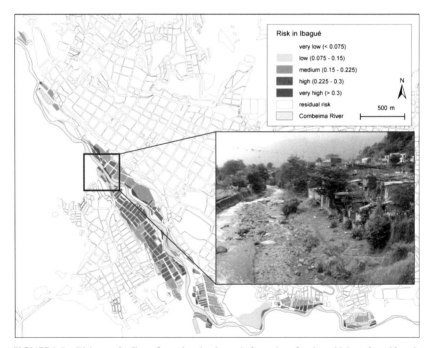

FIGURE 8.5   Risk map for Ibagué as related to hazards from river floods and lahars from Nevado del Tolima. The inset picture shows a city quarter, highly exposed and vulnerable to floods and lahars of the Combeima river. *Adapted from Künzler et al. (2012).*

weigh various risks, from poverty and violence to their physical location in hazard zones and ability to comply with early warning or evacuation systems. If they lack faith in institutions charged with helping them, then resilience and adaptive capacity may be decreased.

According to surveys of the local populations (Thomas and Peñuela, 2008), residents in the Combeima valley express the highest confidence in local institutions such as schools, the church, police, and emergency bodies. Technical-scientific agencies including INGEOMINAS (now SGC) and the Meteorological and Hydrological Service (IDEAM) are less well known, and confidence in them is lower. The lowest confidence was found for institutions of the municipality and the regional government of Tolima. These levels of trust complicate disaster preparedness and reduce long-term adaptive capacity (Beck, 1986, 2000; Giddens, 1999; Bogardi, 2004; O'Brien et al., 2006). Limited participation in community collaboration activities or in political decision-making processes (e.g., elections) increases distrust and erodes confidence in official institutions. In the Combeima valley, Thomas and Peñuela (2008) found that participation is low, with negative implications for disaster prevention. Low confidence in government institutions and low participation in the political process are rooted in a history of personal and collective experiences in Colombia, just as in many other countries.

Beyond the regional dynamics around Nevado del Ruiz, this Colombian case also reveals how extreme events and disasters drive adaptation agendas at national-scale institutions and policies. As a response to the tragedy, the Colombian Government developed and implemented the National System for Disaster Response and Prevention. The backbone of this system includes disaster response and prevention committees at the national, regional, and local levels, and also includes technical agencies, such as the SGC and IDEAM, and non-governmental organizations such as the Red Cross. Following an international and regional trend, the system has been transformed from a disaster response focus into a more integrative risk management perspective to enhance adaptive capacity and community resilience in the face of continued volcano−ice hazards. The system proved particularly effective in the recent case of the Nevado del Huila volcano, where eruptions and volcano−ice interactions in 2007 and 2008 resulted in lahars of similar and even larger proportions than those at Ruiz in 1985, but casualties were kept at a minimum (Worni et al., 2012).

8.4.2 Iceland

Jökulhlaups are often triggered with little warning by a series of variables including subglacial volcanic eruption, the bursting of a dam or moraine, or geothermal heating. Initially coined by Icelandic farmers but now commonly used synonymously (but mistakenly) with any glacially derived outburst flood, jökulhlaups are perhaps more common in Iceland than in any other places on

FIGURE 8.6   Southern Iceland: the A1 highway, which rings the island, is regularly damaged from jökulhlaups. In the background of the photograph is the new replacement road and bridge. *Photograph by M. Jackson.*

Earth due in part to the interconnected relationship between glaciers and volcanoes (Björnsson and Pálsson, 2008). However, although Iceland has suffered significantly from glacier-related hazards, it has also made important achievements in adapting to them. Many of Iceland's volcanoes, including Grímsvötn, Eyjafjallajökull, and Katla, are covered with large glaciers and icecaps. Recent and historical eruptions (Grímsvötn in 2011, Eyjafjallajökull in 2010, and Katla in 1918) have caused those ice bodies to discharge some of the largest and most devastating floods in Icelandic history (Björnsson and Pálsson, 2008). However, these are not the first and only Icelandic ice-related hazards (Figure 8.6). Icelanders have been living with jökulhlaups since the island was first settled during the ninth century. Efforts to create more integrated approaches to jökulhlaup hazards—programs, for example, that consider the socio-cryospheric system as a whole—have accelerated since the twentieth century. The Icelandic system, although far from complete, represents a case of attempting to bridge the physical and social components needed for successful adaptation, which encompasses the scientific/knowledge-based, risk management/engineering-based, and vulnerability-based approaches to adaptation.

Currently, scientific investigation and monitoring of jökulhlaups contribute to an islandwide early warning system designed and maintained by the Icelandic Meteorological Office (MET, 2013). The network includes >115 automatic weather stations, 100 manned stations, 170 hydrological gauges, 55 seismic stations, and 70 GPS stations (MET, 2013). Stations are operated by a variety of entities including the Ministry of Industry and the National Energy Authority; however, reports are generated daily and transferred to the MET for evaluation by scientists (Sigurðsson et al., 2011). If the MET detects any unusual activity or possible hazards, it can issue a warning to the Civil Protection Authorities, which could then trigger various civilian hazard management procedures. The MET receives additional information to detect and evaluate cryospheric hazards, including regular reports of glacier depth, elevation, and condition of known glacial lakes and flood zones (MET, 2013). As extensive as this physical monitoring system appears, Icelanders recognize that it alone is ineffective in reducing the impacts of jökulhlaup hazards.

Multiple studies reviewing both the physical and social components of Iceland's hazards warning system call for increased attention on human vulnerability to GLOFs and jökulhlaups. To make Iceland's cryospheric, hazard-reduction system more successful, it must incorporate vital social factors including risk and vulnerability studies, public outreach and education, and emergency management response planning and implementation. Natural hazards on the Icelandic volcano Katla (Figure 8.7) reveal the need for greater attention to vulnerability-based approaches to adaptation. The 1918 eruption of Katla produced such a large jökulhlaup that the southern coast of Iceland was extended by 5 km due to the sheer volume of flood deposits (Maria et al., 2000). Katla typically erupts twice per century (Gudmundsson et al., 2007). Ominously, the volcano has remained quiet since the 1918 eruption and scientists, public officials, and communities near Katla such as Vik and Alftaver have continued careful monitoring of the volcano. Potential jökulhlaups from a Katla eruption could have far-reaching impacts on the area, including loss of life, destruction of infrastructure, and damage to property (Jónsdóttir, 2012). To enhance vulnerability-based approaches to these hazards, many studies have examined local risk perceptions in Vik and Alftaver (e.g., Bird, 2010; Bird et al., 2010a,b, 2009; Jóhannesdóttir and Gísladóttir, 2010; Jóhannesdóttir, 2005). Jónsdóttir (2012) noted that although general knowledge about Katla and its potential to produce jökulhlaups was high, a lack existed of both preparedness and mitigation efforts. Some residents expressed frustration with current preparation and mitigation plans, though some did indicate their faith in scientists' monitoring efforts (Bird et al., 2010a). Some residents also conveyed that animals and property were not sufficiently taken into consideration in hazard-reduction planning (Bird et al., 2010a). Jóhannesdóttir and Gísladóttir (2010) and Bird et al. (2010a) concluded in separate studies that to maximize the reduction of jökulhlaup hazards from Katla, it is crucial to expand community participation and utilize local environmental knowledge.

Despite limitations, the hazard-reduction programs developed for Katla have offered a successful example for cryospheric hazards elsewhere in Iceland. As Bird and Gísladóttir (2012) noted, during the 2010 eruption of the volcano Eyjafjallajökull, Icelandic officials closely monitored the eruption for potential jökulhlaup hazards. Evacuation exercises and procedures created and practiced for Katla were transferred to the Eyjafjallajökull area during the eruption and were credited with improving safety and response effectiveness.

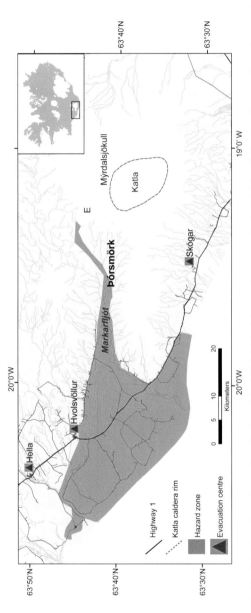

FIGURE 8.7 The jökulhlaup hazard zone from Katla and the Mýrdalsjökull ice cap in southern Iceland is identified with subsequent hazard evacuation centers located Hella, Hvolsvöllur, and Skógar. *From Bird et al. (2010b).*

Iceland maintains an extensive monitoring network and early warning system that could be beneficially replicated in other regions with similar conditions and potential hazards. It demonstrates in particular the achievements of the scientific and technical approaches to natural hazard reduction in socio-cryospheric systems. Coupling the country's sophisticated and extensive biophysical monitoring system with a more socially focused vulnerability approach would provide a more integrative adaptation strategy. Improved public participation could increase the effectiveness of the Icelandic Civil Protection Office and other agencies in minimizing potential jökulhlaup hazards (Bird and Gísladóttir, 2012). Creating and implementing local hazard response plans that are guided by both community members and national-level strategies could result in enhanced safety for populations and improved resilience in the face of ongoing but generally unpredictable cryospheric hazards.

8.5 GLACIER RUNOFF, HYDROLOGIC VARIABILITY, AND WATER-USE HAZARDS

Significant strides have been made in the ability to describe large-scale biophysical changes in the alpine cryosphere (IPCC, 2013). Yet understanding of how these changes affect complex socio-cryospheric systems is still in its infancy even though sensitivity to hydrologic variability stemming from cryospheric change is thought to be significant worldwide. It is extremely difficult to project how discharge will change in the future, even if snow and ice volumes are known. Moreover, how much water is consumed throughout each watershed, and the extent to which factors such as ecological conditions and riparian plant communities influence water flow in glacierized basins, is highly uncertain in many regions. Knowing too little about these biophysical dimensions of hydrological change, as well as the ways in which changes propagate through complex socio-cryospheric systems, has profound implications for our capacity to develop meaningful adaptation interventions as well as our ability to identify and support effective autonomous adaptations (i.e., adaptation plans or actions initiated by affected populations). As a case study of climate-related hydrological change and human water use in the Khumbu region of Nepal illustrates, a priori assumptions about what biophysical changes in the cryosphere are relevant to highland communities can be misguided, a discovery that once again illustrates the need to consider integrated approaches to adaptation research and planning in glacierized mountain regions.

8.5.1 Nepal

Khumbu is located in northeast Nepal, approximately 140 km from the capital city of Kathmandu (Figure 8.8). Its hydrology is heavily influenced by the

FIGURE 8.8   Scene from the Khumbu region. *Photograph by G. McDowell.*

Indian monsoon and the temporal redistribution effects of glacier and snow melt (Thapa and Shakya, 2008). Livelihoods in Khumbu are based on agriculture, pastoralism, and tourism-related services (the Nepalese side of Mount Everest is located in Khumbu) (Sherpa and Bajracharya, 2009). Agriculturalists grow a limited number of crops capable of surviving the region's harsh environment (Sherpa and Bajracharya, 2009). Staple crops are minimally or non-irrigated and therefore highly precipitation-dependent (Shrestha and Aryal, 2011). Selling excess yield provides an important source of cash income for goods not produced by families. Pastoralists engage in transhumance practices, where livestock are grazed in highland pastures in the summer before being brought to lower elevations once winter snowfall covers high-elevation rangelands (Sherpa and Kayastha, 2009; Brower, 1991). As a result of tourism, communities along popular trekking routes have seen substantial economic opportunity with attendant social changes (Nepal, 2005; HKKH, 2009; Sherpa and Bajracharya, 2009). Much of the region has access to electricity via a network of micro-hydro stations (Ives, 2004). Due to the region's remoteness and current limitations of climate and hydrological models, a dearth occurs of detailed regional- to local-scale information about climate-related hydrological changes (Sharma et al., 2009). This section (based on McDowell et al., 2013) helps explain the pathways through which hydrological changes affect Khumbu's population.

Given the scientific discourse at the time of this study in 2010—which focused heavily on the potential effects of glacial change on Himalayan rivers and adjacent populations—it was thought that local residents would report glacial change as a salient hydrological exposure. However, although 78 percent of respondents ($n = 80$) indicated that hydrological conditions were changing, glacial change was not cited as a concerning issue from a water-resource perspective. Instead, decreasing winter snowfall was cited (73 percent) as the primary driver of socially relevant hydrological change (consistent with Ives, 2013). Residents indicated that they do not access water from the region's large glacially influenced rivers due to their high discharge, unstable banks, and excessive turbidity. Instead, they collect water from smaller streams originating in lower-elevation, snow-dominated catchments. Residents thus expressed their highest concern for changes in winter snowfall regimes, as contemporary snowfall dynamics were linked by respondents to decreasing winter and spring stream discharge, declining crop yields, and changing rangeland conditions. Although sensitivity to decreased snowfall was observed across the region, adaptability to this change was found to be highly variable due primarily to socioeconomic differentiation within Khumbu (Figure 8.9).

Most households citing declining winter snowfall also indicated sensitivity to reduced stream discharge (86 percent). Three adaptation strategies were employed in attempts to ameliorate this stress: water rationing; collecting water from streams with higher discharge volumes, which were always further away than traditional access points; and installing roof-water collection systems. Households that lacked physical health, family support (e.g., single-headed

FIGURE 8.9   Spatial distribution of dominant livelihood activities within Khumbu. *Map by Adam Bonnycastel. From McDowell et al., 2013.*

households), and/or expendable income were constrained in their ability to effectively augment seasonal water shortage. Agriculturalists cited negative effects of snowfall change on crop yield, with implications for food provisioning and their ability to generate cash income. Effective adaptations were lacking among this group, as changing staple crop varieties—an approach occurring in lowland locales—was considered impractical due to Khumbu's harsh environment (consistent with observations of Howden et al., 2007). Pastoralists were adapting by grazing livestock at high-elevation rangelands into early winter due to reduced snow cover, though purchasing excess hay to compensate for reduced rangeland grass productivity was also mentioned. Some communities were receiving too little electricity generation for winter heating and cooking needs as a result of reduced stream flow. Affected residents respond by collecting fuel wood to satisfy household requirements and, for those catering to tourists, guests' heating preferences; this response was acknowledged an unsustainable coping strategy for the long term.

Direct and indirect water demand driven by tourism was found to play a significant role in enhancing adaptation challenges, although increased income also helps address some issues. For example, locals running tourist lodges must cater to guests' expectations for high water availability (e.g., shower facilities, flushing toilets), thereby broadening the gap lodge owners must overcome between water supply and water demand. Further, economic incentives to grow poorly adapted, irrigation-dependent crop varieties for tourists motivate agricultural households to increase their water use. The influence of "western" ideas is producing other more water-intensive cultural norms among Khumbu residents, such as using additional water to meet enhanced expectations of personal cleanliness. These observations illustrate that people's experiences of and responses to cryospheric change are not a "natural" or predictable outcome, but rather involve place-specific conditions emerging from the interplay of environmental changes and constantly evolving socioeconomic conditions.

The various adaptation strategies evident across Khumbu reveal the agency and inventiveness of humans vis-à-vis environmental change. Notwithstanding, 78 percent of those who had observed climate-related hydrological changes indicated that current changes were having a negative impact on their ability to support themselves and their families. Indeed, vulnerability to reduced water access for household uses, declining crop yields, reduced ability to meet the high water demands of tourists, and reduced hydroelectricity generation each affected at least 50 percent of the region's population (see McDowell et al., 2013 for a comprehensive discussion).

Analysis of these vulnerabilities demonstrates that science-driven assessments of cryospheric change must be augmented by place-based studies of socio-cryospheric systems to identify socially relevant cryospheric changes as well as the factors that condition how societies experience and respond to relevant stimuli. Regional institutions such as ICIMOD, and other place-based studies of change in the Himalayas, have adopted and demonstrated the utility of this approach (Byers et al., 2013). Long-term monitoring and modeling of physical environmental changes can play vital roles in the development of informed adaptation initiatives. But linking these top-down perspectives with bottom-up approaches to understand cryospheric changes and adaptation provides a powerful foundation for reducing vulnerability. As well, it is evident that engineering efforts could play a larger role in adaptation in Khumbu. For example, more efficient micro-hydroelectricity generation may help counteract electricity generation losses from reduced stream flow, and basic infrastructure developments could help deliver water from distant sources to communities. There is no silver bullet for successful adaptation to environmental change in complex socio-cryospheric systems. Instead, a "silver buckshot" approach combining insights from vulnerability studies, scientific assessments of cryospheric change, and strategic engineering projects is needed. Finally, because adaptation efforts may lead to unintended ecological degradation (Turner et al., 2010; Carey et al., 2012a), more holistic thinking is needed in adaptation planning, particularly in sensitive alpine environments such as those of the high Himalayas.

8.5.2 Peru

Glacier-hydrology hazards in Peru's Cordillera Blanca and Santa River basin offer another compelling case for the need to consider multiple, intersecting human and environmental variables—the complex socio-cryospheric system—in any effort to enhance adaptive capacity in glacier-fed watersheds. Approximately 70 percent of Cordillera Blanca's glacier melt water drains into the Santa River that drains into the Pacific Ocean (Figure 8.1). Events at Lake Parón in the Cordillera Blanca demonstrate the challenges and complexities of successful, long-term adaptive capacity and the ways in which GLOFs and water-use concerns intersect. Local residents, engineers, and scientists had considered Parón a dangerous lake since as early as 1941. A lake drainage project began in the late 1960s and was completed in 1985, when a 1.1-km drainage tunnel was bored into the lake bed and the water level lowered by 45 m, removing 58 million cubic meters of the lake's water (Carey et al., 2012a). In 1992, floodgates were installed where the drainage tunnel left the lake. The floodgates could be opened and closed to regulate water flow downstream daily or seasonally, thereby transforming Parón into a reservoir to enhance hydroelectricity generation at Cañón del Pato (Reynolds, 1993). Although initially managed by the state-run power company Electroperú, Cañón del Pato was consolidated by the private US company Duke Energy by 1998 through the country's neoliberal reforms that privatized the energy sector.

During the following decade, Duke Energy managed the Parón reservoir to maximize energy production, altering stream flow rates in ways that aggravated many local residents in the rural community of Cruz de Mayo and the city of Caraz, among other areas. They complained that water in the Lullán River flowing out of Parón came at the wrong time and often at flow levels so high they inundated local irrigation canals, caused siltation and other problems in the Caraz water treatment plant, and eroded or otherwise damaged the riparian environment. Moreover, many local residents complained that Parón, when drained in the dry season, was ugly and not a suitable tourist site, which is increasingly important for the regional economy. Some also believed that Duke's annual drainage of Parón would restrict local capacity to deal with future glacier shrinkage and the ensuing dry season water reduction (Carey et al., 2012a; Lynch, 2012). Perceptions of change thus mattered as much as actual hazards and hydrologic variability. In 2008, local residents had grown so frustrated that they seized control of the Parón reservoir. Protests against both Duke and the Peruvian government for allowing Duke's supposed mismanagement continued for years, but the case still remains unresolved today, despite several attempts to reach an agreement suitable to locals and Duke Energy, while also maintaining the lake level at a safe elevation to avoid a catastrophic outburst flood.

The Lake Parón case shows the complexities of adaptation in socio-cryospheric systems in the face of two intertwined hazards: a GLOF and hydrologic variability. The initial lake drainage tunnel was highly successful for preventing a GLOF from Parón, and the well-intentioned engineers in the 1980s had no way of knowing that neoliberal reforms under President Fujimori would radically alter reservoir management more than a decade later. Privatization of the hydroelectric station was the most important force fueling the post-2008 Parón conflict, but those management changes also occurred in an era of rapid glacier shrinkage and growing concerns about water supplies in glacier-fed watersheds, which influenced both perceptions and actions among the various stakeholders. Moreover, the stability of Lake Parón could come into question in the future, especially if a new lake above Parón, which some refer to as Artesoncocha Alta, grows large enough and becomes unstable—and depending on how local communities and the hydroelectric company manage Parón's lake level. The issue also exists of the distinctive and massive moraine threshold of the debris-covered Hatunraju glacier, which actually dams Lake Parón and could be affected by future earthquake activity. At the same time, efforts to reduce local vulnerability to glacier hazards in the region—particularly through hazard zoning to keep people outside potential flood and avalanche paths—have largely been rejected by local residents, being perceived as government-imposed, top-down policies meant to usurp local autonomy rather than protecting them (Bode, 1990; Oliver-Smith, 1986). Any future plans to reduce vulnerability to GLOFs and hydrologic variability below Lake Parón will require an integration of engineering projects, scientific understanding of climate-glacier-water dynamics, and local involvement to manage water, retain and improve their livelihoods, protect their culture, expand their decision-making capacity (power and autonomy), and empower them as stakeholders in the upper Santa River watershed.

Farther downstream in the Santa River watershed, other concerns have occurred about the impact of glacier shrinkage and melt water variability on large-scale, agro-industrial irrigation, particularly for the massive Chavimochic Irrigation Project in the La Libertad Department (Figure 8.10).

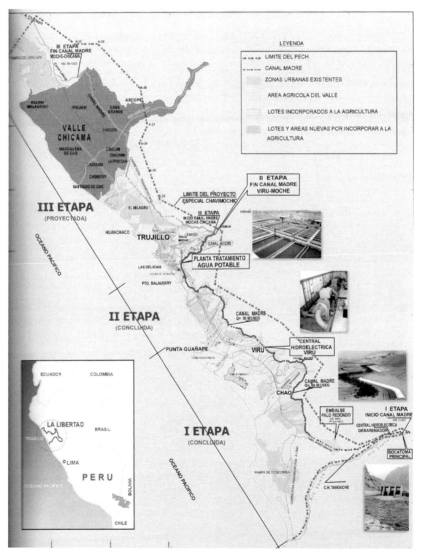

FIGURE 8.10   Chavimochic Irrigation Project showing the three stages (*etapas*), the water intake (*bocatoma principal*) from the Santa River at the bottom, and the water treatment plant for Trujillo (*planta tratamiento agua potable*). *Map by Chavimochic Project (http://www.chavimochic.gob.pe/).*

The project diverts Santa River water north into La Libertad for a 154-km mother canal to deliver the following: glacier-fed Santa River water into La Libertad, generation of 8 MW of hydroelectricity, potable water to nearly 800,000 people around Trujillo, irrigation of 46,000 ha of new land in the desert that was previously just sand, and improvement of irrigation on 28,000 ha that was previously agricultural land but now has more and more stable water supplies for agro-industrial development. Stage III, which is supposed to begin in 2014, will involve constructing the massive 360 million cubic meter Palo Redondo reservoir, extending the mother canal another 113 km from the Moche Valley to the Chicama Valley north of Trujillo, and installing a 60-MW hydroelectric station, among other infrastructure and irrigation components in La Libertad (Chavimochic, 2012). Chavimochic has become one of Peru's largest and most lucrative agro-industrial projects (Shimizu, 2006; Valcárcel, 2002). The Ministry of Economy and Finance estimates that Chavimochic will yield $1.5 US billion annually for Peru once Stage III is completed (El Comercio, 2013). But Chavimochic depends on Cordillera Blanca glacier runoff and sustained water in the Santa River.

The adaptive capacity for Chavimochic—as well as other large- and small-scale irrigators and farmers—hinges on three variables: understanding the socio-cryospheric system, engineering projects to maintain water flow, and reducing vulnerability to hydrologic variability and diminished water supplies. First, scientists need an adequate understanding of the glaciological and hydrological conditions influencing water supply in the Santa River basin. Cordillera Blanca glacier runoff accounts for up to two-thirds of the river's water flow during the dry season; in the wet season, that percentage is significantly lower, indicating that glacier shrinkage will have a particularly important impact on dry season water supplies (Juen et al., 2007; Mark et al., 2005). Some disagreement also exists about the effect and timing of glacier shrinkage on water supplies, with the newest research suggesting that peak water has already been passed, whereas studies had projected the steep decline in glacier runoff to come perhaps decades in the future (Baraer et al., 2012). It is also critical to recognize not only the role of glaciers in the system but also the ways in which human land- and water-use influences downstream water availability. The farther one travels downstream from glaciers, the more human variables affect the amount of water (Bury et al., 2013; Carey et al., 2013). Given that Chavimochic is in the lower Santa watershed, it is thus necessary to consider human and biophysical forces—the comprehensive socio-cryospheric system—when trying to understand the relationship between glacier runoff and the quantity of water available at the Chavimochic intake.

Second, adaptive capacity depends upon engineering efforts both to divert Santa River water north into La Libertad and to maintain the integrity and functionality of the intake station ("bocatoma") on the Santa River. The

Chavimochic mother canal extends 154 km and will expand another 113 km with Stage III. It passes from the Santa watershed through the Chao, Virú, and Moche valleys today and will later extend into the Chicama watershed with Stage III. Moving water through various watersheds in the desert is a major engineering feat that requires constant vigilance to maintain its effectiveness. The intake station on the Santa River is obviously the most critical piece of infrastructure because, without it, no water enters the mother canal to flow north. This intake, however, was severely damaged when water levels rose markedly during the 1997–1998 El Niño Southern Oscillation (ENSO) event, thus indicating the potential weakness of this infrastructure, especially because GLOFs have occurred repeatedly through recent history in the Cordillera Blanca, and these GLOFs generally drain into the Santa River, maintaining significant force even as they flow past the current site of the Chavimochic intake (Carey, 2005).

Third, integrated adaptation depends upon adaptive capacity among the human population within Chavimochic and throughout the Santa River watershed. In this area, land use and agricultural productivity, water laws, international and national markets, transportation infrastructure, adequate living and working conditions, cooperation among various stakeholders throughout the Santa River watershed, governance structures, non-governmental organizations and governmental institutions, cultural values, and available technologies all influence vulnerability to changes in glacier runoff. A shift in international food demands, for example, could trigger a shift from growing asparagus to another crop, perhaps a crop with a different water requirement. Shifts in national water laws or watershed commission management priorities could also affect water-use patterns in the region, thereby increasing or reducing vulnerability to declining Santa River water supplies. Such interactions among various societal forces alongside changing dynamics in the cryospheric and hydrologic systems obviously make adaptive capacity dependent upon a host of socioecological variables within and well beyond the watershed, making a case for a broadly conceptualized socio-cryospheric system that extends across many scales, even internationally.

8.6 COASTAL RESOURCES AND HAZARDS

Cryospheric conditions in the Arctic present hazards distinct from high-mountain regions but nonetheless offer important insights into the need for integrated efforts to enhance adaptive capacity in socio-cryospheric systems. Temperatures in the Arctic are rising at approximately twice the global average, with cascading effects on the northern cryosphere widely documented (IPCC, 2013; AMAP, 2011). Due to changing sea ice regimes—in the context of evolving global energy markets, rising global demand for oil and gas, and improving exploration and extraction technologies—interest in the Arctic's offshore oil and gas resources has intensified (Nuttall, 2010; IEA, 2012). This section on hydrocarbon development in the Disko Bay region of Greenland

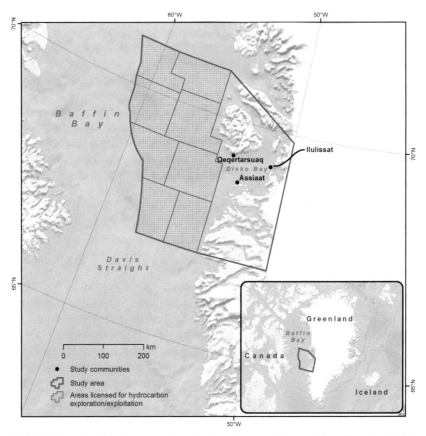

FIGURE 8.11   The Disko Bay region. *Map by Adam Bonnycastel. From McDowell and Ford, 2014.*

highlights several salient considerations for coastal resources development in the context of cryospheric change (see McDowell and Ford, 2014 for more).

The Disko Bay region is located on the central west coast of Greenland, with marine areas located in southeastern Baffin Bay and Disko Bay proper (Figure 8.11). The area is a recognized hub of Arctic biodiversity (Gensbol, 2004) and supports productive fisheries as well as subsistence hunting and a burgeoning tourism industry (Nuttall and Nutall, 2009; Ford and Goldhar, 2012). Greenland's government is keen to initiate hydrocarbon-based development in the region, with 2007, 2008, and 2011 licensing rounds leading to exploration licenses for an offshore area of 92,901 km^2 (Government of Greenland, 2013). Although government- and industry-funded impact assessments have been conducted, McDowell and Ford (2014) argue that environmental change is insufficiently analyzed in these assessments, leading decision makers and stakeholders to endorse hydrocarbon development

FIGURE 8.12   Icebergs in Disko Bay. *Photograph by G. McDowell.*

activities based on information that may underreport uncertainty and the extent of potential risks.

Icebergs discharged from marine-terminating outlet glaciers of the Greenland Ice Sheet are a conspicuous feature of the regional seascape (Figure 8.12). A growing body of literature on the effects of climate change on the dynamics of marine-terminating glaciers dynamics, however, suggests that iceberg discharge into regions such as Disko Bay could increase as a result of subsurface ice degradation and attendant feedbacks (e.g., Rignot et al., 2010; Holland et al., 2008; Nick et al., 2009; Andresen et al., 2011). Indeed, if climate-driven, ocean—glacier interactions prove to be a significant feature of cryospheric change in the region, it is conceivable that iceberg concentrations may increase over the course of development activities, with implications for drilling and shipping safety. Such eventualities could compromise not only industry interests and personnel but also the ecological foundations of the region's marine-dependent economy. Accordingly, anticipating new risks, and how existing risk may change under novel climate conditions, should be fundamental to the assessment of industrial activities proposed in quickly changing Arctic locales.

Consideration of offshore oil and gas development in the Disko Bay region vis-à-vis issues like changing iceberg fluxes again makes it clear that integrated approaches to understanding cryospheric change and adaptation are necessary to reduce vulnerability to cryosphere-related risks. For example, although engineering-based risk reduction adaptations (e.g., iceberg towing) may be sufficient under current conditions, scientific monitoring and modeling projections are needed to understand the nature of the future iceberg environment. Moreover, vulnerability assessments are required to determine how risks associated with increasing iceberg hazards might propagate through local communities as well as how issues of power, marginalization, and difference condition susceptibility to harm. Research in 2013 showed that differentiation in the distribution of opportunities (e.g., new employment prospects) and risks (e.g., erosion of environment-dependent livelihoods) related to hydrocarbon development, is closely tied to socioeconomic and political conditions such as occupation, community of residence, age, and education. For example, the perceived need for and means of adaptation to climate change and industrial development is different for subsistence hunters, those working in seafood-processing plants, residents interested in resource extraction careers, and public servants involved in education and administration. Adaptations drawing on insights from vulnerability studies, scientific assessments, and strategic engineering capabilities must be supported.

8.7 DISCUSSION AND CONCLUSIONS

The ongoing influence of cryospheric risks and hazards in mountain and coastal regions are of growing interest to scientists, policy makers, and populations exposed to the associated risks. Research to date has typically focused on the science of hazards such as GLOFs, avalanches, and landslides—particularly from biophysical, not human, perspectives. The rapid changes observed in and projected for the future of socio-cryospheric systems, however, require both expansion and changes to ways in which these and other cryospheric hazards are analyzed and prepared for. Changing water resources may be perceived locally or even nationally as more important and immediate risks than GLOFs with relatively low frequencies. Further, sea level rise introduces a new type of cryospheric hazard that is projected to gain rapid importance through the coming century.

The various changes in the cryosphere are thus occurring within specific socio-cryospheric contexts, and also on various, commonly intersecting spatial and temporal scales that make resilience difficult for societies worldwide. On temporal scales, there are short-term seasonal variations in water flow, decadal changes to moraine-dammed glacial lakes, century-scale loss of ice that influences sea level, and the millennium time scale of tectonic shifts that affects volcanic activity and the deglaciation of ice sheets. At the same time, these temporal scales can interact as a glacial lake may form over decades but unleash a GLOF in a matter of minutes, or when a lahar descends in minutes after centuries or millennia of volcanic and tectonic activity. Spatial scales can also interact as an avalanche may affect local populations while also destroying a hydroelectric facility that generates power for the national grid, or an ice sheet melting in Greenland that contributes to global sea level rise, or glacier runoff variability that influences agriculture for international export and crop prices globally. All these hazards thus exist within unique environmental, social, economic, political, cultural, and technological contexts that vary through space and time. Thus, it is important to consider socio-cryospheric systems not only as local or regional areas—such as glacier-fed watersheds, alpine slopes and valleys, fjords and bays with tidewater glaciers—but also at national and international levels.

Given the complexities of these multiple, intersecting spatial and temporal scales, as well as the dynamic and ever-changing nature of socioecological systems, it is essential to approach cryospheric hazards from a three-pronged framework that includes scientific/knowledge-based approaches, risk management/engineering-based approaches, and vulnerability-based approaches to increase resilience and adaptive capacity. The cases discussed here illuminate achievements and limitations with these approaches, particularly showing why the often-overlooked vulnerability approach is so essential for integrated adaptation to cryospheric hazards.

For glacier and glacial lake hazards, engineering efforts in Peru's Cordillera Blanca have been successful since the 1940s and up to the present. Yet with continued socio-cryospheric changes in the region, the hazards persist—even at glacial lakes that have previously been partially drained and artificially dammed—and the population in nearby valleys has increasingly moved into potential avalanche and GLOF paths. Both these issues reveal that the engineering approach alone is insufficient because there is a continued need for studies to understand GLOF and avalanche triggers and for continued monitoring of the hazards. Additionally, with an ever-larger population exposed to these hazards, it is necessary to reduce their exposure and vulnerability through zoning, building codes, education, early warning systems, and broader efforts to ensure socioeconomic well-being—efforts that have been attempted but often resisted by local populations in the past. In Santa Teresa, Peru, relocation of the population after the debris flow initially reduced vulnerability and even bolstered the local economy through new tourism opportunities. However, the economic benefits were unevenly distributed and some segments of the population have again become vulnerable to future debris flows and cryospheric hazards.

The Colombian case of volcano—ice hazards demonstrates the importance of effective institutions—and public trust in those institutions—that act as intermediaries between the scientists monitoring hazards and the communities vulnerable to lahars and other risks. Effective communication among governmental and non-governmental agencies, scientists, and communities can help increase socio-institutional adaptive capacity. In Iceland, a successful glacier hazard monitoring system emerged, as well as an early warning system and evacuation plans for local people, but people remain vulnerable because their livestock, crops, infrastructure, and property are exposed to hazards. Icelandic approaches would be strengthened through more community participation and recognition of local perspectives about hazards.

Glacier runoff and water-use hazards in Nepal and Peru show how a failure to analyze downstream societies in glacier-fed watersheds can misportray their vulnerability. The impact of glacier shrinkage on the communities in Nepal analyzed in this study have been overestimated because they depend on snow-fed waterways (rather than glaciated basins) and because their water-use practices are shaped by many forces well beyond water availability, such as tourism and western cultural influences. In Peru, on the other hand, the energy industry and large-scale irrigation projects are vulnerable to hydrologic variability, as well as to other global forces, especially neoliberal political-economic reforms that privatized hydroelectricity and altered watershed management significantly. It is thus critical to understand climate—glacier—water flow rates, to adequately engineer waterways and intakes, and to appreciate how human vulnerability is affected not only by the glacier melt water but also by energy and food demands and prices, laws, and many other regional, national, and international forces. In coastal areas subject to hazards such as icebergs, oil and gas extraction may be impacted by ocean-ice dynamics in ways that could adversely affect shipping and drilling activities, local communities, and Greenland's economy. In all these cases, it is clear that enhancing adaptive capacity in the face of cryospheric hazards requires, first, the conceptualization of dynamic socio-cryospheric systems and, second, the implementation of an integrated approach to disaster risk reduction that focuses on knowledge/scientific monitoring, engineering/risk management strategies, and vulnerability reduction plans that consider both place-specific and far-reaching variables acting on multiple temporal and spatial scales.

ACKNOWLEDGMENTS

This article is based on work supported by the US National Science Foundation under grant #1253779, the Swiss Agency for Development and Cooperation, the Canadian Social Sciences and Humanities Research Council, and McGill University Department of Geography and Faculty of Science. We also acknowledge the collaboration with various local communities, national water and disaster risk institutions, as well as non-governmental organizations such as CARE and research groups such as TARN (Transdisciplinary Andean Research Network).

REFERENCES

Adger, W.N., 2000. Social and ecological resilience: are they related? Progr. Hum. Geogr. 24, 347—364.

Adger, W.N., Agrawala, S., Mirza, M.M.Q., Conde, C., O'Brien, K., Pulhin, J., Pulwarty, R., Smit, B., Takahashi, K., 2007. Assessment of adaptation practices, options, constraints and capacity. In: Parry, M., Canziani, O., Palutikof, J., Van Der Linden, P., Hanson, C. (Eds.), Climate Change 2007: Impacts, Adaptation and Vulnerability. Contribution of Working Group II to the Fourth Assessment Report of the Intergovernmental Panel on Climate Change. Cambridge University Press, New York.

Adger, W.N., Barnett, J., Brown, K., Marshall, N., O'Brien, K., 2013. Cultural dimensions of climate change impacts and adaptation. Nat. Clim. Change 3, 112—117.

AMAP, 2011. Snow, Water, Ice, and Permafrost in the Arctic (SWIPA): Climate Change and the Cryosphere. Arctic Council, Oslo. Arctic Monitoring and Assessment Programme (AMAP).

Ames, A., 1998. A documentation of glacier tongue variations and lake development in the Cordillera Blanca, Peru. Z. Gletscherkunde Glazialgeologie 34, 1—36.

Ames Marquez, A., Francou, B., 1995. Cordillera Blanca glaciares en la historia. Bull. L'Institut Francais d'Études Andines 24, 37−64.

Andresen, C.S., McCarthy, D.J., Dylmer, C.V., Seidenkrantz, M.S., Kuijpers, A., Lloyd, J.M., 2011. Interaction between subsurface ocean waters and calving of the Jakobshavn Isbræ during the late Holocene. Holocene 21, 211−224.

Baraer, M., Mark, B.G., McKenzie, J., Condom, T., Bury, J., Huh, K.I., Portocarrero, C., Gómez, J., Rathay, S., 2012. Glacier recession and water resources in Peru's Cordillera Blanca. J. Glaciol. 58, 134−150.

BBC News, September 12, 2012. Everest Sherpas in Glacial Lake Study Warning. http://www.bbc. co.uk/news/science-environment-19569256.

Beck, U., 1986. Risikogesellschaft: Auf dem Weg in eine andere Moderne. Suhrkamp, Frankfurt.

Beck, U., 2000. Risk society revisited: theory, politics and research programmes. Risk Soc. Crit. Issues Soc. Theory, 211−229.

Bird, D., 2010. Social Dimensions of Volcanic azards, Risk and Emergency Response Procedures in Southern Iceland (Ph.D. dissertation). Faculty of Life and Environmental Sciences, University of Iceland, Reykjavík.

Bird, D.K., Gísladóttir, G., 2012. Residents' attitudes and behaviour before and after the 2010 Eyjafjallajökull eruptions—a case study from southern Iceland. Bull. Volcanol. 74 (6), 1263−1279.

Bird, D., Gísladóttir, G., Dominey-Howes, D., 2009. Resident perception of volcanic hazards and evacuation procedures. Nat. Hazards Earth Syst. Sci. 9, 251−266.

Bird, D., Gísladóttir, G., Dominey-Howes, D., 2010a. Residents' perception of and response to volcanic risk mitigation strategies in a small rural community, southern Iceland. An unpublished paper included in a PhD dissertation by: Deanne K. Bird. Social dimensions of volcanic hazards, risk and emergency response procedures in southern Iceland. Faculty of Life and Environmental Sciences, University of Iceland, Reykjavík.

Bird, D., Gísladóttir, G., Dominey-Howes, D., 2010b. Volcanic risk and tourism in southern Iceland: implications for hazard, risk and emergency response education and training. J. Volcanol. Geotherm. Res. 189, 33−48.

Björnsson, H., Pálsson, F., 2008. Icelandic glaciers. Jökull 58, 365−386.

Bode, B., 1990. No Bells to Toll: Destruction and Creation in the Andes. Paragon House, New York.

Bogardi, J.J., 2004. Hazards, risks and vulnerabilities in a changing environment: the unexpected onslaught on human security? Global Environ. Change 14, 361−365.

BPC Hydroconsult, 1997. Tsho Rolpa GLOF Risk Reduction Project Formulation Mission Final Report. BPC Hydroconsult, Kathmandu, Nepal.

Brower, B., 1991. Sherpa of Khumbu: People, Livestock, and Landscape. Oxford University Press, Delhi.

Brown, K., 2014. Global environmental change I: a social turn for resilience? Prog. Hum. Geogr. 38, 107−117.

Bury, J., Mark, B.G., Carey, M., Young, K., McKenzie, J., Baraer, M., French, A., Polk, M., 2013. New geographies of water and climate change in Peru: coupled natural and social transformations in the Santa river watershed. Ann. Assoc. Am. Geogr. 103, 363−374.

Byers, A.C., McKinney, D.C., Somos-Valenzuela, M., Watanabe, T., Lamsal, D., 2013. Glacial lakes of the Hinku and Hongu valleys, Makalu Barun National Park and Buffer zone, Nepal. Nat. Hazards 69, 115−139.

Carey, M., 2005. Living and dying with glaciers: people's historical vulnerability to avalanches and outburst floods in Peru. Global Planet. Change 47, 122−134.

Carey, M., 2010. In the Shadow of Melting Glaciers: Climate Change and Andean Society. Oxford University Press, New York.

Carey, M., Baraer, M., Mark, B.G., French, A., Bury, J., Young, K.R., McKenzie, J., 2013. Toward hydro-social modeling: merging human variables and the social sciences with climate-glacier runoff models (Santa river, Peru). J. Hydrol. http://dx.doi.org/10.1016/j.jhydrol.2013.11.006.

Carey, M., French, A., O'Brien, E., 2012a. Unintended effects of technology on climate change adaptation: an historical analysis of water conflicts below Andean glaciers. J. Hist. Geogr. 38, 181—191.

Carey, M., Huggel, C., Bury, J., Portocarrero, C., Haeberli, W., 2012b. An integrated socio-environmental framework for glacier hazard management and climate change adaptation: lessons from Lake 513, Cordillera Blanca, Peru. Climatic Change 112, 733—767.

Chavimochic, 2012. Chavimochic en cifras 2000—2010. Gobierno Regional La Libertad/Proyecto Especial Chavimochic, Trujillo.

El Comercio, 2013. Proyecto Chavimochic aportará US$1.500 millones anuales al PBI nacional, El Comercio (15 Julio 2013). http://elcomercio.pe/economia/1604227/noticia-proyecto-chavimochic-aportara-us1500-millones-anuales-al-pbi-nacional (accessed 05.09.13.).

Emmer, A., Vilímek, V., 2013. Review article: lake and breach hazard assessment for moraine-dammed lakes: an example from the Cordillera Blanca (Peru). Nat. Hazards Earth System Sci. 13, 1551—1565.

Emmer, A., Vilímek, V., Klimes, J., Cochachin, A., 2014. Glacier retreat, lakes development and associated natural hazards in Cordillera Blanca, Peru. In: Shan, W., Guo, Y., Wang, F., Marui, H., Strom, A. (Eds.), Landslides in Cold Regions in the Context of Climate Change. Springer, New York.

Engle, N.L., 2011. Adaptive capacity and its assessment. Global Environ. Change 21, 647—656.

Evans, S.G., Bishop, N.F., Smoll, L.F., Murillo, P.V., Delaney, K.B., Oliver-Smith, A., 2009. A re-examination of the mechanism and human impact of catastrophic mass Flos originating on Nevado Huascarán, Cordillera Blanca, Peru in 1962 and 1970. Eng. Geol. 108, 96—118.

Fernández Concha, J., 1957. El problema de las lagunas de la Cordillera Blanca. Boletín de la Sociedad Geológica del Perú 32, 87—95.

Folke, C., 2006. Resilience: the emergence of a perspective for social-ecological systems analyses. Global Environ. Change 16, 253—267.

Folke, C., Carpenter, S.R., Walker, B., Scheffer, M., Chapin, T., Rockström, J., 2010. Resilience thinking: integrating resilience, adaptability and transformability. Ecol. Society 15. Article 20. http://www.ecologyandsociety.org/vol15/iss4/art20/.

Ford, J.D., Smit, B., 2004. A framework for assessing the vulnerability of communities in the Canadian arctic to risks associated with climate change. Arctic 57, 389—400.

Ford, J.D., Goldhar, C., 2012. Climate change vulnerability and adaptation in resource dependent communities: a case study from West Greenland. Climate Res. 54.

Gallopín, G.C., 2006. Linkages between vulnerability, resilience, and adaptive capacity. Global Environ. Change 16, 293—303.

Gensbol, B., 2004. A Nature and Wildlife Guide to Greenland. Gyldendal Publishers, Copenhagen.

Giddens, A., 1999. Risk and responsibility. Mod. Law Rev. 62, 1—10.

Government of Greenland, 2013. List of Mineral and Petroleum Licences in Greenland. Bureau of Minerals and Petroleum, Nuuk.

Gudmundsson, M.T., Hoganadottir, P., Kristinsson, A.B., Gudbjornsson, S., 2007. Geothermal activity in the subglacial Katla caldera, Iceland, 1999—2005, studied with radar altimetry. Ann. Glaciol. 45 (1), 66—72.

Haeberli, W., Whiteman, C., 2014. Snow and Ice-related Hazards, Risks and Disasters — A General Framework, pp. 1–34.

Haeberli, W., Alean, J.-C., Müller, P., Funk, M., 1989. Assessing risks from glacier hazards in high mountain regions: some experiences in the Swiss Alps. Ann. Glaciol. 13, 96–102.

Haeberli, W., Huggel, C., Paul, F., Zemp, M., 2013. Glacial responses to climate change. In: Shroder, J.F. (Ed.), Treatise on Geomorphology, vol. 13. Academic Press, San Diego, California.

Haeberli, W., Kääb, A., Vonder Mühll, D., Teysseire, P., 2001. Prevention of outburst floods from periglacial lakes at Grubengletscher, Valais, Swiss Alps. J. Glaciol. 47, 111–122.

Hall, M.L., 1992. The 1985 Nevado del Ruiz eruption—scientific, social, and governmental response and interaction before the event. In: Mc Call, G.J.H., Laming, D.J.C., Scott, S.C. (Eds.), Geohazards, Natural and Man-made. Chapman & Hall, London, pp. 43–52.

Hambrey, M.J., Quincey, D.J., Glasser, N., Reynolds, J.M., Richardson, S.D., Clemmens, S., 2008. Sedimentological, geomorphological and dynamic context of debris-mantled glaciers, Mount Everest (Sagarmatha) region, Nepal. Quat. Sci. Rev. 27, 2361–2389. http://dx.doi.org/10.1016/j.quascirev.2008.08.010.

He, Y., Wu, Y.F., Liu, Q.F., 2012. Vulnerability assessment of areas affected by Chinese cryospheric changes in future climate change scenarios. Chinese Sci. Bull. 57, 4784–4790.

Hegglin, E., Huggel, C., 2008. An integrated assessment of vulnerability to glacial hazards: a case study in the Cordillera Blanca, Peru. Mountain Res. Dev. 28, 299–309.

Hill, M., 2013. Adaptive capacity of water governance: cases from the Alps and the Andes. Mountain Res. Dev. 33, 248–259.

HKKH, 2009. Situation Analysis of Sagarmatha National Park and Buffer Zone: A Review of Park, Tourism Management and Stakeholder Participation. Hindu Kush-Karakoram-Himalaya Partnership, Kathmandu.

Holland, D.M., Thomas, R.H., De Young, B., Ribergaard, M.H., Lyberth, B., 2008. Acceleration of Jakobshavn Isbrae triggered by warm subsurface ocean waters. Nat. Geosci. 1, 659–664.

Howden, S.M., Soussana, J.F., Tubiello, F.N., Chhetri, N., Dunlop, M., Meinke, H., 2007. Adapting agriculture to climate change. Proc. Natl. Acad. Sci. 104, 19691.

Huggel, C., Ceballos, J.L., Pulgarin, B., Ramirez, J., Thouret, J., 2007. Review and reassessment of hazards owing to volcano–glacier interactions in Colombia. Ann. Glaciol. 45, 128–136.

Huggel, C., Haeberli, W., Kääb, A., Bieri, D., Richardson, S., 2004. An assessment procedure for glacial hazards in the Swiss Alps. Can. Geotech. J. 41, 1068–1083.

IEA, 2012. World Energy Outlook. International Energy Agency, Paris.

IPCC, Intergovernmental Panel on Climate Change, 2012. Managing the Risks of Extreme Events and Disasters to Advance Climate Change Adaptation. A Special Report of Working Groups I and II of the Intergovernmental Panel on Climate Change. Cambridge University Press, New York.

IPCC, 2013. Climate Change 2013: The Physical Science Basis. Working Group I Contribution to the Fifth Assessment Report of the Intergovernmental Panel on Climate Change. Cambridge University Press.

Ives, J.D., 2004. Himalayan Perceptions: Environmental Change and the Well-being of Mountain Peoples. Psychology Press.

Ives, J.D., 2013. Sustainable Mountain Development: Getting the Facts Right. Himalayan Association for the Advancement of Science, Lalitpur, Nepal.

Jóhannesdóttir, G., 2005. Við tölum aldrei um Kötlu hér. Mat íbúa á hættu vegna Kötlugoss (M.Sc. thesis in Environmental Sciences). University of Iceland, Reykjavík.

Jóhannesdóttir, G., Gísladóttir, G., 2010. People living under threat of volcanic hazard in southern Iceland: vulnerability and risk perception. Nat. Hazards Earth Syst. Sci. 10, 407–420.

Jónsdóttir, Á., 2012. Experiencing Environmental Changes in Vík, Southern Iceland. CoastAdapt Report. Institute for Sustainability Studies, University of Iceland. Retrieved from: http://s3.amazonaws.com/academia.edu.documents/30885864/Experiencing_change_in_Vik.pdf? AWSAccessKeyId=AKIAIR6FSIMDFXPEERSA%26Expires=1382306415%26Signature=q TIiikxNF6Q7WdtN4q53W1n8wO0%3D%26response-content-disposition=inline.

Juen, I., Kaser, G., Georges, C., 2007. Modelling observed and future runoff from a glacierized tropical catchment (Cordillera Blanca, Perú). Global Planet. Change 59, 37−48.

Jurt, C., 2009. Perceptions of Natural Hazards in the Context of Social, Cultural, Economic and Political Risks. A Case Study in South Tyrol. PhD Dissertation.

Kääb, A., Reynolds, J.M., Haeberli, W., 2005. Glacier and permafrost hazards in high mountains. In: Huber, U.M., Bugmann, H.K.M., Reasoner, M.A. (Eds.), Global Change and Mountain Regions: An Overview of Current Knowledge. Springer, Dordrecht, Netherlands, pp. 225−234.

Kargel, J.S., Furfaro, R., Kaser, G., Leonard, G.J., Fink, W., Huggel, C., Kääb, A., Raup, B.H., Reynolds, J.M., Wolfe, D., Zapata, M.L., 2011. ASTER imaging and analysis of glacier hazards (Chapter 15). In: Ramachandran, B., Justice, C.O., Abrams, M. (Eds.), Land Remote Sensing and Global Environmental Change: NASA'S Earth Observing System, and the Science of Terra and Aqua. Springer, New York, USA, pp. 325−373.

Künzler, M., Huggel, C., Ramírez, J.M., 2012. A risk analysis for floods and lahars: case study in the Cordillera Central of Colombia. Nat. Hazards 64, 767−796.

Lliboutry, L., Morales, A., Benjamín, Pautre, A., Schneider, B., 1977. Glaciological problems set by the control of dangerous lakes in Cordillera Blanca, Peru. I. Historical failures of Morainic dams, their causes and prevention. J. Glaciol. 18, 239−254.

Lynch, B.D., 2012. Vulnerabilities, competition and rights in a context of climate change toward equitable water governance in Peru's Rio Santa Valley. Global Environ. Change 22, 364−373.

Maria, A., Carey, S., Sigurdsson, H., Kincaid, C., Helgadóttir, G., 2000. Source and dispersal of jökulhlaup sediments discharged to the sea following the 1996 Vatnajökull eruption. Geol. Soc. Am. Bull. 112 (10), 1507−1521.

Mark, B.G., McKenzie, J.M., Gómez, J., 2005. Hydrochemical evaluation of changing glacier melt-water contribution to stream discharge: Callejon de Huaylas, Peru. Hydrol. Sci. J. 50, 975−987.

Maskrey, A. (Ed.), 1993. Los Desastres No Son Naturales. La Red de Estudios Sociales en Prevención de Desastres en América Latina, Bogotá, Colombia.

McDowell, G., Ford, J.D., Lehner, B., Berrang-Ford, L., Sherpa, A., 2013. Climate-related hydrological change and human vulnerability in remote mountain regions: a case study from Khumbu, Nepal. Reg. Environ. Change, 1−12.

McDowell, G., Ford, J.D., 2014. The socio-ecological dimensions of hydrocarbon development in the Disko Bay region of Greenland: opportunities, risks, and tradeoffs. Appl. Geogr. 46, 98−110.

Mercer, J., 2010. Disaster risk reduction or climate change adaptation: are we reinventing the wheel? J. Int. Dev. 22, 247−264.

MET, 2013. Icelandic Meteorological Office. Retrieved from: http://en.vedur.is/about-imo/mission/.

Municipalidad Distrital de Santa Teresa, n.d. Morfología y Población. Población de Santa Teresa. http://www.peru.gob.pe/Nuevo_Portal_Municipal/portales/Municipalidades/767/pm_inicio.asp (accessed 15.12.13.).

Naranjo, J.L., Sigurdsson, H., Carey, S.N., Fritz, W., 1986. Eruption of the Nevado del Ruiz volcano, Colombia, on 13 November 1985: tephra fall and lahars. Science 233, 961−963.

Nepal, S., 2005. Tourism and remote mountain settlements: spatial and temporal development of tourist infrastructure in the Mt Everest region, Nepal. Tourism Geogr. 7, 205−227.

Nick, F.M., Vieli, A., Howat, I.M., Joughin, I., 2009. Large-scale changes in Greenland outlet glacier dynamics triggered at the terminus. Nat. Geosci. 2, 110−114.

Nuttall, M., 2010. Pipeline Dreams: People, Environment, and the Arctic Energy Frontier. International Work Group for Indigenous Affairs Copenhagen.

Nuttall, A.D., Nutall, M., 2009. Europe's northern dimension: policies, co-operation, frameworks. In: Dey Nuttall, A., Nutall, A. (Eds.), Canada's and Europe's Northern Dimensions. University of Oulu Press, Oulu.

O'Brien, G., O'Keefe, P., Rose, J., Wisner, B., 2006. Climate change and disaster management. Disasters 30, 64−80.

Oliver-Smith, A., 1986. The Martyred City: Death and Rebirth in the Andes. University of New Mexico Press, Albuquerque.

Parra, E., Cepeda, H., 1990. Volcanic hazard maps of the Nevado del Ruiz volcano, Colombia. J. Volcanol. Geotherm. Res. 42, 117−127.

Pierson, T.C., Janda, R.J., Thouret, J.C., Borrero, C.A., 1990. Perturbation and melting of snow and ice by the 13 November 1985 eruption of Nevado del Ruiz, Colombia, and consequent mobilization, flow and deposition of lahars. J. Volcanol. Geotherm. Res. 41, 17−66.

Portocarrero, C., 1995. Retroceso de glaciares en el Perú: consecuencias sobre los recursos hídricos y los riesgos geodinámicos. Bull. L'Institut Francais d'Études Andines 24, 697−706.

PREDES, 2007. Plan Regional de Prenvención y Atención a los Desastres de la Región Cuzco. http://www.indeci.gob.pe/planes_proy_prg/p_estrategicos/nivel_reg/prpad_cusco.pdf.

Quincey, D.J., Lucas, R.M., Richardson, S.D., Glasser, N.F., Hambrey, M.J., Reynolds, J.M., 2005. Optical remote sensing techniques in high-mountain environments: application to glacial hazards. Progr. Phys. Geogr. 29 (4), 475−505.

Quincey, D.J., Richardson, S.D., Luckman, A., Lucas, R.M., Reynolds, J.M., Hambrey, M.J., Glasser, N.F., 2007. Early recognition of glacial lake hazards in the Himalaya using remote sensing datasets. Global Planet. Change 56, 137−152.

Rana, B., Shrestha, A.B., Reynolds, J.M., Aryal, R., Pokhrel, A.P., Budhathoki, K.P., 2000. Hazard assessment of the Tsho Rolpa glacier lake and ongoing remediation measures. J. Nepal Geol. Soc. 22, 563−570.

Reynolds, J.M., 1992. The identification and mitigation of glacier-related hazards: examples from the Cordillera Blanca, Peru. In: McCall, G.J.H., Laming, D.C.J., Scott, S. (Eds.), Geohazards. Chapman & Hall, London, pp. 143−157.

Reynolds, J.M., 1993. The development of a combined regional strategy for power generation and natural hazard risk assessment in a high-altitude glacial environment: an example from the Cordillera Blanca, Peru. In: Merriman, P.A., Browitt, C.W.A. (Eds.), Natural Disasters: Protecting Vulnerable Communities. Thomas Telford Ltd, London, pp. 38−50.

Reynolds, J.M., 1998. High altitude glacial lake hazard assessment and mitigation: a Himalayan perspective. In: Maund, J., Eddleston, M. (Eds.), Geohazards in Engineering Geology. Geological Society Engineering Group Special Publication No. 15, pp. 25−34.

Reynolds, J.M., 1999. Photographic feature: glacial hazard assessment at Tsho Rolpa, Rolwaling, Central Nepal. Quart. J. Eng. Geol. 32 (3), 209−214.

Reynolds, J.M., 2010. Managing glacial hazards in a changing climate. In: International Glaciological Conference VICC 2010, Ice and climate change: a view from the South, Valdivia, Chile, 1−3 February 2010. Abstract Book, 80, CECS, Valdivia, Chile.

Reynolds, J.M., 2013. On the Geopolitics of Managing Glacial Hazards. Mountain Hazards 2013, Kyrgyzstan, 16−18 September 2013, p. 129.

Reynolds, J.M., Dolecki, A., Portocarrero, C., 1998. The construction of a drainage tunnel as part of glacial lake hazard mitigation at Hualcán, Cordillera Blanca, Peru. In: Maund, J.G.,

Eddleston, M. (Eds.), Geohazards in Engineering Geology. The Geological Society, London, pp. 41−48.

Richardson, S.D., Reynolds, J.M., 2000. An overview of glacial hazards in the Himalayas. Quat. Int. 65/66 (1), 31−47.

Rignot, E., Koppes, M., Velicogna, I., 2010. Rapid submarine melting of the calving faces of West Greenland glaciers. Nat. Geosci. 3, 187−191.

Sharma, E., Chetri, N., Tse-ring, K., Shrestha, A., Jing, F., Mool, P., Eriksson, M., 2009. Climate Change Impacts and Vulnerability in the Eastern Himalayas. ICIMOD.

Sherpa, L.N., Bajracharya, B., 2009. View of a High Place: Natural and Cultural Landscape of Sagarmatha National Park. HKKH partnership Project/ICIMOD, Kathmandu.

Sherpa, Y.D., Kayastha, R.B., 2009. A study of livestock management patterns in Sagarmatha National Park, Khumbu region: trends as affected by socio-economic and climate change. Kathmandu University J. Sci. Eng. Technol. 5, 110−120.

Shimizu, T., 2006. Expansion of Asparagus Production and Exports in Peru. Institute of Developing Economies (IDE). Discussion Paper No. 73.

Shrestha, A.B., Aryal, R., 2011. Climate change in Nepal and its impact on Himalayan glaciers. Reg. Environ. Change 11, 65−77.

Sigurðsson, O., Sigurðsson, G., Björnsson, B.B., Pagneux, E.P., Zóphóníasson, S., Einarsson, B., Þórarinsson, Ó., Jóhannesson, T., 2011. Flood Warning System and Jökulhlaups − Eyjafjallajökull. Icelandic Meteorological Society. http://en.vedur.is/hydrology/articles/nr/2097.

Smit, B., Wandel, J., 2006. Adaptation, adaptive capacity and vulnerability. Global Environ. Change 16, 282−292.

Sociedad Nacional de Minería Petróleo y Energía, 2013. Reporte Canon Gasífero. Transferencias Primer semestre. http://www.snmpe.org.pe/informes-y-publicaciones-snmpe/canon/cuadros-estadisticos/reporte-canon-gasifero-2013.html.

Starr, C., 1969. Social benefit versus technological risk. Science 165, 1232−1238.

Takenaka, S., Satoh, T., Lhamo, S., 2012. A social survey for GLOF disaster mitigation in Bhutan. Global Environ. Res. 16, 77−82.

Thapa, K.B., Shakya, B., 2008. Integrated Study on Hydrology and Meteorology of Khumbu Region with Climate Change Perspectives. WWF Nepal, Kathmandu.

Thomas, M., Peñuela, F. Integración comunitaria e institucional, en la reducción del riesgo de resastres naturales, en la cuenca del Río Combeima, Municipio de Ibagué, Tolima. Ibagué, Colombia, unpublished report.

Thouret, J.C., 1990. Effects of the November 13, 1985 eruption on the snow pack and ice cap of Nevado del Ruiz volcano, Colombia. J. Volcanol. Geotherm. Res. 41, 177−201.

Trask, Parker D., 1953. El problema de los aluviones de la Cordillera Blanca. Boletín de la Sociedad Geográfica de Lima LXX, 5−75.

Turner, B.L., et al., 2003. Illustrating the coupled human−environment system for vulnerability analysis: three case studies. Proc. Natl. Acad. Sci. 100, 8080−8085.

Turner, W.R., Bradley, B.A., Estes, L.D., Hole, D.G., Oppenheimer, M., Wilcove, D.S., 2010. Climate change: helping nature survive the human response. Conservation Lett. 3, 304−312.

Valcárcel, M., 2002. Agroexportación no tradicional, sistema esparraguero, agricultura de contrata y ONG. Debate Agrario 34, 29−44.

Vilímek, Vít, Zapata Luyo, M., Klimeš, J., Patzelt, Z., Santillán, N., 2005. Influence of glacial retreat on natural hazards of the Palcacocha Lake area, Peru. Landslides 2, 107−115.

Voight, B., 1990. The 1985 Nevado del Ruiz volcano catastrophe: anatomy and retrospection. J. Volcanol. Geotherm. Res. 42, 151−188.

Westoby, M.J., Glasser, N.F., Brasington, J., Hambrey, M.J., Quincey, D.J., Reynolds, J.M., 2014. Modelling outburst floods from moraine-dammed glacial lakes. Earth Sci. Rev. 134, 137—159.

Wisner, B., Piers, B., Cannon, T., Davis, I., 2004. At Risk: Natural Hazards, People's Vulnerability and Disasters. Routledge, New York.

Worni, R., Huggel, C., Stoffel, M., Pulgarín, B., 2012. Challenges of modeling current very large lahars at Nevado del Huila Volcano, Colombia. Bull. Volcanol. 74, 309—324.

Yamada, T., 1993. Glacier Lakes and Their Outburst Floods in the Nepal Himalayas. WECS Report, No. 4/1/291191/1/1, Seq. No. 387.

Young, O.R., Berkhout, F., Gallopin, G.C., Janssen, M.A., Ostrom, E., van der Leeuw, S., 2006. The globalization of socio-ecological systems: an agenda for scientific research. Global Environ. Change 16, 304—316.

Zapata Luyo, M., 2002. La dinámica glaciar en lagunas de la Cordillera Blanca. Acta Montana (Czech Republic) 19, 37—60.

Chapter 9

Integrative Risk Management: The Example of Snow Avalanches

Michael Bründl and Stefan Margreth

WSL Institute for Snow and Avalanche Research SLF, Davos, Switzerland

ABSTRACT

Snow avalanches can pose a risk to persons, buildings, and infrastructure in mountainous regions. As long as people have settled in the Alps, they had to deal with avalanche hazard. Structural mitigation measures, protection forests, and land-use planning reduced the hazard and the risk in the past decades and contributed thereby to the development of mountain regions. In many countries, decisions on dealing with the consequences of avalanches are increasingly based on the assessment of risks and not only on the reduction of hazard. In the first part of this chapter, we present different elements of an integrative risk management of avalanches with a focus on Switzerland. In the second part, we present an example of how this concept can be applied to handle the avalanche risk in the City of Juneau, Alaska.

9.1 INTRODUCTION

Snow avalanches are fascinating phenomena in mountainous regions. In contrast to the beauty of nature, avalanches can also be a hazard. Inhabitants of mountainous regions and tourists of winter resorts face the consequences of avalanche danger almost every winter. Due to the large impact, snow avalanches pose a potential risk for recreational activities in mountainous winter terrain and they also threaten buildings, infrastructure, and traffic routes, including persons using them. Dealing with avalanche danger and risk is as old as the life of humans in mountain regions.

Strategies for dealing with hazard and risk have changed over the centuries; however, the general principles remained the same: preventing the release of avalanches, influencing their direction of movement or avoiding endangered areas either in the long term or in the short term.

Snow and Ice-Related Hazards, Risks, and Disasters. http://dx.doi.org/10.1016/B978-0-12-394849-6.00009-3
263

Much research has been done on hazard and risk management of snow avalanches all over the world in recent decades (e.g., Mears, 1992; Jamieson et al., 2002; McClung and Schaerer, 2006; Whiteman, 2011). For Europe, a part of the progress was achieved in several European research projects in recent years (CADZIE, 2003; SATSIE, 2006; IRASMOS, 2008).

This chapter is organized into two parts: In the first part, we start with a brief overview of terms related to snow avalanche hazard and risk before we present methods for calculating, and criteria for evaluating, risk. We continue with presenting measures for mitigating avalanche risk in settlements and infrastructures and put the focus on the methods and instruments used in Switzerland because they serve as a basis for measures in many countries and mountainous regions of the world (e.g., Chile, Alaska, Iceland, Russia, or India). In the second part, we present a case study to suggest how avalanche risk could be managed in Juneau, Alaska.

The term "risk" has been used since the 1960s to describe and to quantify negative outcomes as a consequence of complex technical installations in the energy and infrastructure sector (e.g., Farmer, 1967; Starr, 1969). Kaplan and Garrick (1981) posed three basic questions, which form the cornerstones of the risk concept:

- "What might happen?"
- "What is allowed to happen?"
- "What needs to be done?" and "Which options are available?"

The question "what might happen?" is being answered in a risk analysis, consisting of a hazard analysis, an exposure analysis, a consequence analysis, and the calculation of risk. The question "what is allowed to happen?" is the guiding question of risk evaluation for comparing risk to risk levels, which reflect what a society is willing to accept. Risks exceeding a societally accepted risk level have to be reduced. How and by which measures this could be achieved is the subject of the third part "what needs to be done?" and "which options are available?" Communication among the involved parties and to the target audience (e.g., stakeholders and/or public) is crucial for the success of the whole process; thus, risk communication belongs to the risk concept and improves social capacity building against natural hazards in communities and regions (Höppner et al., 2012).

From 1980 onwards and especially toward the turn of the millennium, the risk concept has been increasingly adapted and introduced as a systematic approach for dealing with natural hazards in several countries (e.g., Gardner, 1982; Fell and Hartford, 1997; Hollenstein, 1997; Wilhelm, 1997; Fuchs et al., 2004; Bründl et al., 2009). The risk concept is only one part of an integrated risk management approach. Integrated risk management denotes a systematic approach for managing uncertainties related to minimizing potential harm and loss (UNISDR, 2009) and includes measures before an event (prevention), during an event (intervention), and after an event (recovery). In avalanche risk management, most measures realized in recent decades are preventive measures including avalanche warning. Identifying appropriate measures for specific risk situations with given financial resources is the pivotal argument for the application of the risk concept, which is explained in the following sections. However, for a holistic management of avalanche risk and (natural) risks in general, all available prevention, intervention, and recovery measures have to be taken into account to achieve economically, ecologically, and socially viable risk reduction strategies. Several authors have depicted this interplay of prevention, intervention, and recovery with a circle (Figure 9.1), highlighting also that a damage event could have a governing or learning effect on methods, instruments, or measures used for the management of technically, societally, or naturally driven hazards (e.g., Carter, 1991; Ammann, 2006; Jaecklin, 2007; Bründl, 2015).

According to the wide range of fields for which the terms hazard and risk are used (e.g., Hatfield and Hipel, 2002), numerous definitions are given. Based on UNISDR terminology, we use hazard as a term for a dangerous phenomenon—here snow avalanches—that may cause loss of life, injury, loss of livelihoods and services, and social and economic disruption, or environmental damage (UNISDR, 2009). Risk is "the combination of the probability of an event and its negative consequences" (UNISDR, 2009) or "a measure of

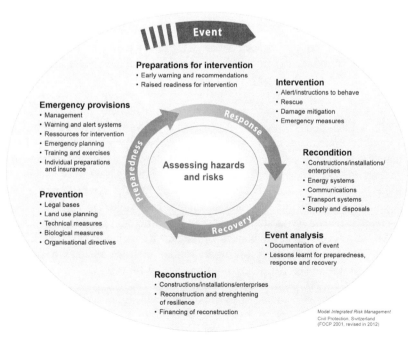

FIGURE 9.1   Circle of integrated risk management. *Federal Office for Civil Protection.*

the probability and severity of loss to the elements at risk, usually expressed for a unit area, object, or activity, over a specified period of time" (Crozier and Glade, 2005).

9.2 RISK ANALYSIS

9.2.1 Hazard Analysis

A hazard analysis is the fundamental basis of a risk analysis, which defines the probability of the hazard occurrence and intensities. The potential course of a hazard (here avalanches) is defined as scenarios with different probabilities based on existing data. The results of a hazard analysis are intensity maps, which reflect the possible courses and intensities of avalanches in the chosen scenarios. In a further step, intensity maps can be used for creating a hazard map, which is the basis for land-use planning in many countries (e.g., Egli and Stucki, 2013).

All possible courses are described by scenarios, which include the hypothetical courses of a damage event, which simplify the real-world situation (Merz, 2006). It is essential that all information used in hazard mapping is provided to make the process transparent.

Before the steps in a hazard analysis are described in detail, some approaches for deriving the probability of occurrence are presented.

9.2.1.1 Probability of Occurrence

Probability of occurrence can be defined as the return period of events within a certain time period, as the frequency of an event, or as encounter probability. The probability of events expressed as the exceedance probability can be calculated from a probability density function of hazard occurrence. A probability density function can also be used to calculate probabilities for magnitude intervals. An example of a probability density function is shown in Figure 9.2. Prerequisites for deriving the probability of occurrence from avalanche records are stable meteorological conditions. Investigations by Laternser and Schneebeli (2002) reveal that no significant trend of avalanche activity could be noticed between 1950 and 2000, whereas newer investigations from Canada assume that changed precipitation patterns might influence avalanche formation and activity (Bellaire et al., 2013).

The frequency of an avalanche is defined as the period of time in which a number of occurrences with a certain runout distance reaches a specific threshold location. The frequency is calculated by Eqn (9.1):

$$p(e) = \frac{\sum_{i=1}^{n} \tau_i}{n} \tag{9.1}$$

with $p(e)$ as the frequency of events, n as the considered observation period, and τ as the number of observed events in time period n. If, as an example,

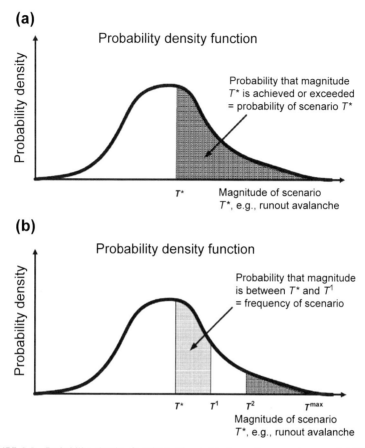

FIGURE 9.2 Probability density functions. Figure (a) shows the exceedance probability and figure (b) the frequency of scenarios that is the probability of magnitude intervals. *Modified according to BUWAL, BABS, BWG, 2005.*

four avalanches crossing a certain road section were observed within 50 years, the annual probability of occurrence (or frequency) at this road section can be estimated by $p(e) = 4/50 = 0.08$, which means that every 12.5 years, one event is expected to cross this specific road section (Figure 9.3). The return period T and the frequency of events $p(e)$ are thus linked by:

$$T = \frac{1}{p(e)} \tag{9.2}$$

The encounter probability $p(E)$ is the probability that an event at a certain location occurs at least once during a given period of observation n. Under the assumption that the events in each year are statistically independent of those in previous years, the encounter probability $p(E)$ can be calculated

FIGURE 9.3 Time period with events reaching a threshold location. The ratio of observed events to the duration of the observation period yields the estimated probability of occurrence. In this example, the process with the runout distance 200 m has an annual probability of 0.08 (4 events in 50 years), which means that every 12.5 years one event with this run-out distance is expected.

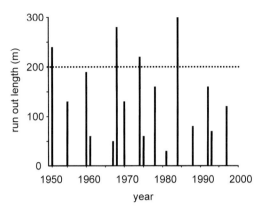

from the return period T and the length of the observation period n according to (Eqn (9.3)):

$$p(E) = 1 - \left[1 - \left(\frac{1}{T} \right) \right]^n \qquad (9.3)$$

If the observation period n is equal to the return period T, the encounter probability of an event at a given location is about $p(E) = 0.65$, whereas if $n = 2T$, the encounter probability is ≈ 0.9. For example, if the expected return period is 100 years, then 200 years of careful observations would be required to be 90 percent confident that the 100-year avalanche has been recorded. This example shows why long-term observations are required to predict long-return periods with reliable confidence.

These calculations hold only for stable natural systems with comparable precipitation patterns, vegetation, and terrain characteristics during the observation period. For changing systems, for example, due to permafrost degradation or glacier retreat (Künzler et al., 2010; Schaub et al., 2013) and for non-recurring processes, for example, rock avalanches, they are not valid.

9.2.1.2 Event Analysis

An event analysis for snow avalanches consists of the following steps:

- Analysis of existing avalanche inventories and historical information;
- Analysis of the terrain, and development of the map of the phenomena;
- Analysis of the meteorological conditions;
- Estimation of avalanche frequency and/or derivation of scenarios.

The first step is routine in event analysis for every natural hazard process (e.g., Zimmermann, 2013). For the definition of scenarios, all available information of past events has to be analyzed. Direct, long-term observation of avalanche runouts enables the definition of frequency at certain locations in the area under consideration. If likely, additional scenarios that were not observed in the past can also be taken into account.

The second step is terrain analysis, including slope angle, aspect, terrain characteristics, roughness, vegetation, existing protection measures, and indications of past avalanche events such as tree damage or ground erosion along the avalanche path. The main goal is to define release areas, which fit to terrain characteristics and existing knowledge. Dendrochronological data sometimes add information to determine the return period of avalanches, especially for areas with no avalanche inventory (Stoffel and Bollschweiler, 2008; Casteller et al., 2011; Schläppy et al, 2014; Decaulne et al., 2014). The results of the terrain analysis are documented in the map of phenomena (Heinimann et al., 1998).

The third step addresses the meteorological conditions and the return period of certain release depths of avalanches. The release depth is related to the sum of snow depth precipitation in the three days before the event, or the three-day snow fall depth, P72 (Burkard and Salm, 1992). The three-day snow fall depth P72 is calculated by evaluating the "increase in snow pack depth during a period of three consecutive days of snow fall" (Sovilla, 2002; Barbolini et al., 2004). The value of P72 for every year is taken as the highest observed value of the positive difference in snow depth calculated using a three-days-wide window, moving by one-day steps (Barbolini et al., 2004). This is evaluated with respect to a flat area and then properly modified for local slope conditions and snow drift overloads (Salm et al., 1990; Barbolini et al., 2002, 2003). If no data from nearby stations are available, the release depth can be interpolated based on data from remote stations (e.g., Bocchiola et al., 2006).

Finally, this information is used to define the avalanche frequency and the relevant scenarios, which are considered in a risk analysis. For these scenarios, the extent and the impact pressure (and/or velocity) are determined by avalanche modeling. The results of avalanche dynamic models depend strongly on the choice of the input parameters such as release volume or friction coefficients. Therefore, a careful judgment of the modeling results is crucial to get useful runout estimates for a specific scenario (Figure 9.4).

9.2.1.3 Avalanche Modeling and Intensity Analysis

The dynamics of avalanches are complex, involving aspects of fluid, particle and soil mechanics. Commonly, the limited amount of data available from real events makes it difficult to evaluate or calibrate models. Various models have been proposed and applied in practice, such as RAMMS (Christen et al., 2010), ELBA+ (Jörg et al., 2012), or SAMOS-AT (Sampl and Zwinger 2004; Sampl and Granig, 2009). They are based on empirical procedures using topographical/statistical information and comparative models for runout distance computations, or are dynamic models simulating avalanche motion. The latter describe either the internal dynamics of the material at certain stages of the motion that is the dynamics of the moving mass as a whole from initiation to stop, or they are combinations of both. Advanced dynamics models also

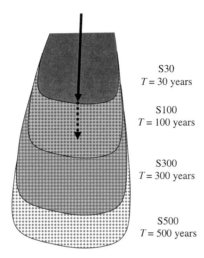

FIGURE 9.4   Example for modeling scenarios as complementary scenarios. The frequency of a scenario can be calculated as the difference between the exceedance probability of two adjacent scenarios.

provide flow height and velocity information during the event. This information is critical for calculating impact pressures and load distribution.

The physical impact of an avalanche is commonly expressed as impact force per area (i.e., pressure). Impact forces (I) due to moving snow are usually assumed to be proportional to the avalanche speed squared (V) and the flow density of the moving material (ρ):

$$I \propto \rho \cdot V^2 \tag{9.4}$$

Impact forces of avalanches can range from relatively harmless blasts from powder clouds to the highly destructive forces of a full-scale, dry flowing avalanche capable of destroying reinforced concrete structures. Generally, dry flowing avalanches have the highest product of velocity squared and flow density so their destructive potential is generally the highest.

Table 9.1 provides a categorization of intensity classes, which is, besides the return period, one criterion for hazard mapping (Section 9.4.3).

Intensity maps are simplifications of the real-world situation and therefore include some uncertainties. This means that they include all possible flow directions and runout lengths of a scenario (Figure 9.5). Including the whole area of an intensity map into a risk calculation, however, would lead to an overestimation of risk. Therefore, commonly, only a part of an intensity area is included in risk analysis expressed with a factor, spatial probability (between 0 and 1), which better reflects the reality (see Eqn (9.5); Wilhelm, 1997). This factor is also used to consider the specific location of an object (e.g., behind a row of buildings), which is—up to now—not reflected in most intensity maps.

TABLE 9.1 Typical Intensity Classes According to Swiss Regulations (BFF and SLF, 1984) in Avalanche Hazard Mapping

Intensity Class	Impact Pressure
Weak	$I < 3$ kPa
Moderate	3 kPa $< I < 30$ kPa
Strong	$I > 30$ kPa

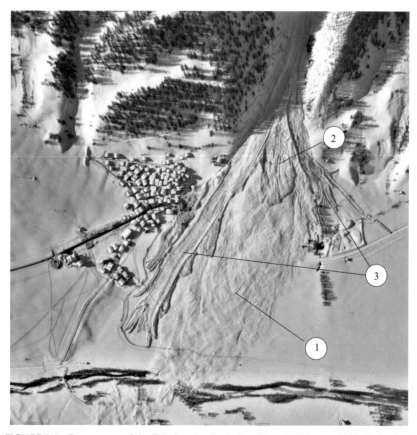

FIGURE 9.5 Runout area of the Trützi avalanche in Geschinen in February 1999, Switzerland (note the houses on the left side of the image as scale). The numbers indicate various avalanche events at different dates. The runout area is in most cases only partially affected by the deposition of a single event. This effect is considered in risk analyses by integrating a spatial occurrence probability. *Federal Office for Topography, coordination unit for aerial photography.*

9.2.2 Exposure and Vulnerability Analysis

In an exposure analysis, the vulnerable objects at risk (damage potential) have to be identified for each scenario. Depending on the scope of the risk analysis, which objects have to be included must be determined. For large areas, taking all elements into account requires great financial and personnel resources; therefore, which level of detail is necessary to meet the goal of a risk analysis has to be carefully checked (Bell and Glade, 2004). It might be worthwhile to concentrate on those elements at risk that are supposed to generate the highest risks. In the following sections, categories of vulnerable objects, which can be taken into account for risk analyses, are described.

9.2.2.1 Buildings and Infrastructure

The location of objects such as buildings, infrastructure, and agricultural and forest areas is assessed by analyzing topographical or zoning maps, aerial photographs, object registers, and/or by verifying their location and characteristics in field campaigns. Additional information, such as type and year of construction and replacement values, can be obtained from insurance data or by estimating their value using reconstruction values (e.g., Papathoma-Köhle et al., 2012; FOEN, 2014). Damage potential and therefore risk is not constant over space and time. Results from studies on the development of damage potential in alpine settlements confirm that the spatial distribution of damage potential is substantially influenced by the historical growth of settlements (e.g., Fuchs et al., 2005; Keiler et al., 2005) and that development of risk depends on variations in the damage potential (e.g., Fuchs et al., 2004). Typical objects included as damage potential in risk analyses are:

- buildings, for example, houses, hotels, industrial buildings, shopping centers, and schools;
- roads and traffic lines, for example, highways, regional roads, bridges, and railways;
- infrastructure, for example, power lines, water reservoirs, and transportation lines;
- communication lines, for example, antenna poles, and telephone lines;
- agricultural or forest land.

Damage to structure and assets is expressed in monetary units per object or in relation to a unit of an object (e.g., 1 m of power line or 1 ha of forest). Since all these data have a spatial dimension, the use of geographical information systems (GISs) is the state-of-the-art technique to handle these data.

An aspect of identification and characterization of exposed buildings and infrastructures is the assessment of vulnerability. Vulnerability is a key concept used in many different contexts (Cutter, 1996; Birkmann, 2006; Thywissen, 2006; Fuchs et al., 2012; IPCC, 2012). In this chapter, we define vulnerability as structural vulnerability of objects; it can be defined as degree of expected loss for an object at risk as the consequence of a hazard impact (Varnes, 1984; Fell, 1994).

In recent years, considerable work has been done on structural vulnerability, however, mainly for water-related processes (e.g., Fuchs et al., 2007; Papathoma-Köhle et al., 2012; Schwendtner et al., 2013; Rheinberger et al., 2013). Fewer contributions deal with structural vulnerability due to snow avalanches (Bründl et al., 2010; Papathoma-Köhle, 2011). Eckert et al. (2012) claimed in a recent state-of-the-art report of existing quantitative risk approaches that data in the snow avalanche field remain scarce and imprecise. Nearly all vulnerability functions for snow avalanches in use today are closely related to each other and go back to one of the original works summarized by Cappabianca et al. (2008), such as the vulnerability values used for benefit—cost analyses in Switzerland (FOEN, 2014). Therefore, risk analyses for snow avalanches based on these vulnerability functions include a certain degree of uncertainty. However, this uncertainty is low in relation to other factors, especially compared to the accuracy of intensity maps (Schaub and Bründl, 2010). Bertrand et al. (2010) investigated by numerical modeling the response of reinforced concrete structures to snow avalanche load. They concluded that damage functions of structures cannot have a general definition and that they depend on technology, construction materials, and geometry. To obtain an idea about the structural response of a given structure, preliminary studies would be required, which is not possible in practice due to limited finance and personnel (Eckert et al., 2012). Table 9.2 gives typical impact pressures expected for avalanches in relation to their destructive effects.

9.2.2.2 Persons and Mobile Objects at Risk

Damage to persons is a key matter in many risk analyses. High impact forces of avalanches can endanger people in buildings if a certain pressure is

TABLE 9.2 Correlation between Avalanche Impact Pressure and Potential Damages (McClung and Schaerer, 2006)

Impact Pressure (kPa)	Potential Damages
1	Break windows
5	Push in doors
30	Destroy wood-frame structures
100	Uproot mature spruce
1,000	Move reinforced concrete structures

exceeded and the structural vulnerability is too high. As an example, avalanches with an impact pressure ≥ 30 kPa can destroy an unreinforced masonry building (BFF and SLF, 1984) and cause injuries or even the death of people inside the building. The vulnerability of persons is also denoted as mortality rate. People in vehicles, trains, or outdoors are more vulnerable to avalanches than in buildings, meaning that the mortality rate of people differs depending on their current location during a hazard event. Empirical data on the mortality of persons in buildings and infrastructure are scarce. Cappabianca et al. (2008) and Papathoma-Köhle (2011) have provided some data in an overview. As an example, mortality for persons inside concrete buildings affected by a dense flow avalanche with 30 kPa is about 0.35 (Cappabianca et al., 2008); Wilhelm suggested 0.46 as average mortality value for persons in buildings (Wilhelm, 1997). Other empirical values and expert judgments are available in the documentation of the software EconoMe (FOEN, 2014).

Another crucial factor for determining damage to people is their number and probability of presence in the endangered area. The number of permanent residents can be derived from residential statistics. For example, the Swiss housing statistics indicate 2.24 persons per residential unit in the year 2000 (BfS, 2008). The number of exposed persons in hotels, guest houses, and hospitals can be roughly quantified by the number of beds and the degree of occupancy. Investigations by Keiler and colleagues in the tourist destination of Galtür, Austria, show that the number of endangered persons depends on the tourist season and varies significantly over time (Keiler et al., 2005).

The number of exposed persons can be accounted for either as an average statistical value, or, in the case of strong variation over time, they can be modeled with situations of exposure. Situations of exposure can be understood as "snapshots," for which we assume that an event happens at the same time a large number of people are present. The exposure factor is calculated as the quotient of the duration of the situation of exposure and the total time a person could potentially be exposed. For example, assume that a winter festival with 1,000 persons takes place in a tourist resort for 6 h once per year. If an avalanche occurs exactly at the same time as the event, a high number of fatalities have to be expected. The resulting exposure factor for this example is 0.00068 (6 h divided by 8,760 h per year if the frequency of avalanches refers to one year; see Bründl et al. (2008) for further explanations). The exposure factor has to be considered in the risk calculation additionally to the probability of occurrence of the scenario. The high number of exposed persons and the high number of expected fatalities in such a situation yield valuable information for managing the risk. Analyzing the damage in situations of exposure allows for effectively controlling a high degree of damage within short time frames and taking appropriate mitigation measures.

9.2.3 Consequence Analysis and Calculation of Risk

In the final step of a risk analysis, endangered areas in the considered scenarios and exposed persons and objects are combined to calculate the expected damage and resulting risk. Meyer et al. (2013) recognize the following categories of damage or costs:

- Direct costs: damage to property due to direct physical contact with the hazard.
- Business interruption: occurs in areas directly affected by the hazard because business connections (e.g., transport of goods) are interrupted.
- Indirect costs: costs induced by direct damage or business interruption often with a time lag.
- Intangible costs: costs that refer to all goods and services, which cannot be measured with monetary units because no market exists, for example, environmental impacts.

Following these categories, we focus on the calculation of direct damage and the resulting risk.

9.2.3.1 Societal Risk of Objects

The damage to an object i in scenario j is calculated following the equation (e.g., Fuchs et al., 2007; Bründl et al., 2008; 2009):

$$D_{ij} = p(o)_{ij} \cdot p(s)_j \cdot A_i \cdot V_{ij} \qquad (9.5)$$

with $D_{i,j}$ as the damage to an object i in scenario j, $p(o)_{i,j}$ as the probability that an object i is present while scenario j is occurring (probability of presence), $p(s)_j$ as the spatial probability that an object i is actually hit in a scenario j, A_i as the value of an object or the number of persons exposed (currency or number), and V_{ij} as the vulnerability of an object i to the impact of a scenario j (alternatively for person factor mortality). The presence probability assumes that persons are not permanently present (e.g., 12 h per day during the year yields $p(o) = 0.5$). Damage as the negative outcome of an event is expressed either as monetary value (e.g., € or $) or as the number of fatalities and/or injured persons.

The risk to an object i in scenario j is the product of the frequency of the scenario p_j (y^{-1}) and the damage according to:

$$R_{ij} = p_j \cdot D_{ij} \qquad (9.6)$$

The total risk R for an investigated area, such as a settlement, is determined as the sum of all object risks in a scenario and of all considered scenarios, such that:

$$R = \sum_j \sum_i R_{ij} \qquad (9.7)$$

The result is the total risk including all objects and persons endangered in an area under investigation, which is the societal risk (or sometimes named as collective risk).

9.2.3.2 Societal Risk on Traffic Routes

One of the first approaches to determine the interaction between snow avalanches and vehicles on roads was developed by Schaerer (1989) in Canada. The avalanche hazard index is a function of the size and type of avalanche, the frequency of avalanche occurrences, the number of avalanche paths and the distance between them, the total length of the highway exposed, the traffic volume, the traffic speed, and the type of vehicles; it considers both moving and waiting traffic. Other approaches are based on variations of Eqns (9.5) and (9.6) and relate the probability that a person is exposed in an endangered section to the mean daily traffic, the length of the section, and vehicle velocity. Risk is calculated (e.g., Wilhelm, 1999; Margreth et al., 2002; Bründl et al., 2008; Rheinberger et al., 2009) according to:

$$R_{ij} = p_j \cdot \frac{\text{MDT} \cdot g_j}{v} \cdot \beta_i \cdot \lambda_{ij} \tag{9.8}$$

with MDT as mean daily traffic per day (1/day), g_j as the length of the endangered road in a scenario j, v as the velocity of cars (km/h), β_i as the number of persons in a vehicle i, and λ_{ij} as the mortality of persons in vehicle i in scenario j. Note that units for g_j are often (m) and for v (km/h) so that v has to be multiplied by 24,000 to get equal units. Hendrikx and Owen (2006) slightly adapted this equation by taking different avalanche types and the resulting probabilities of death into account. However, these approaches (except that of Schaerer, 1989) neglect the effect of waiting traffic, which might result in a likely underestimation of risk (Hendrikx and Owen, 2008). Kristensen et al., (2003) and also Hendrikx and Owen (2008) addressed this problem by taking into account the number of vehicles waiting in adjacent avalanche paths and the duration of vehicle exposure. Zischg et al. (2005) highlighted the importance of temporal variability of vehicles for risk analyses on road networks by showing that short-term risk on roads can be up to 50 times higher than the statistical accident risk. Voumard et al. (2013) recently identified further shortcomings of existing approaches, which do not take into account speed modifications due to winding roads, speed reduction resulting from dense traffic, or vehicle tailbacks forming in front of traffic lights. They suggest a new method with dynamic risk parameters and conclude that the resulting risk is more realistic than that calculated with a static approach.

9.2.3.3 Individual Risk to Persons

In addition to the total risk to persons in an area (societal risk to persons), risk to individual persons has to be calculated to check whether their risk exceeds existing safety goals. Societal risk and individual risk are closely linked to each other as shown, for example, in Jonkman (2007). A simple approach is to divide the total societal risk to persons by its number. This yields an average value and makes no difference that some persons have to bear a much higher risk than others. Comparing individual risks with risk acceptance criteria requires the calculation of the maximum individual risk in an area, for example, of a person permanently present ($p(o)$ close to 1) or of a bus driver passing an endangered road section several times per day. Derived from Eqn (9.8), individual risk is calculated by replacing MDT by the number of movements (Wilhelm, 1999; Bründl et al., 2008; Hendrikx and Owen, 2008).

9.2.3.4 Business Interruption, Indirect and Intangible Consequences

In addition to the direct damage of a natural hazard, damage and costs due to business interruption, indirect consequences, and intangible costs can arise (for definitions, see Meyer et al. (2013)). Business interruptions can be caused by interruption of traffic routes, interruption of lifeline infrastructure (e.g., power lines), or damage to technical infrastructure of a company (e.g., machinery). This induces indirect consequences such as losses due to reduced or even broken production and delivery. If interruptions last for several days or even weeks, this can have long-term effects on the competitiveness of a company. Examples of such calculations are available, for example, from NELAK (2013) and Lehmann and Schaub (2013). Case studies from the avalanche winter 1999 in Switzerland (Wilhelm et al., 2000; SLF, 2000) showed that especially small- and medium-sized enterprises suffer from such interruptions up to the point that business had to be given up (Laternser, 2000). However, other studies from the tourist sector in Switzerland indicate that the long-term effect of avalanche periods causing high levels of damage is low and the number of overnight stays return to a level comparable to that before the avalanche period (Nöthiger, 2003; Nöthiger et al., 2005). Meyer et al. (2013) summarized approaches for quantifying intangible effects used in the assessments of environmental hazards.

9.3 RISK EVALUATION

According to Kaplan and Garrick's question "what is allowed to happen," calculated risk has to be compared with the societally accepted risk level denoted as safety goals. Discussion about risk acceptability is highly controversial, especially regarding life and limb (e.g., Klinke and Renn, 2002; Renn et al., 1992; Siegrist and Cvetkovich, 2000; Sjoberg, 2000; Slovic, 2000). For the evaluation of risks, two notions have to be proved: (1) the individual risk, expressing the probability of a single person dying during a specified time period (mostly one year) by a hazard or, as generally defined by Bottelberghs (2000), "…the probability that an averaged unprotected person permanently present at that point location, would get killed due to an accident at the hazardous activity"; (2) the societal risk, which comprises all risks of objects and material assets in an area under investigation.

9.3.1 Evaluation of Individual Risk

Although individual risk can be checked for both persons and material assets, it has become common to evaluate individual risk only for persons. As the risk concept was originally developed for dealing with risks from technical infrastructures (see Section 9.1 in this chapter), concepts of safety goals for individual risk were also developed in this field (e.g., BUWAL, 1991; Vrijling et al., 1995; Bohnenblust and Slovic, 1998). The general idea behind the formulation of individual risk thresholds is that risk to an individual person should not exceed the lowest natural mortality rate of a specific group of a society (e.g., 10^{-4}/year for the group of 6- to 20-year-old males, Vrijling et al., 1995) by a certain degree, for example, by 1 percent or more. Jonkman et al. (2003) give an overview of safety goals for individual risks based on this general principle. Thresholds for individual risk reported in the literature range between 1×10^{-4} and 1×10^{-5} (Hendrikx and Owens, 2008). In Switzerland, for example, the safety goal for individual risk in the field of natural hazards is set to 10^{-5} per year (Borter, 1999; Bründl, 2009; FOEN, 2014).

9.3.2 Evaluation of Societal Risk

For the evaluation of societal risks, two approaches are prevalent: (1) engineering approach and (2) economic efficiency approach. In the engineering approach, risks are depicted in so-called Frequency—Number (FN) curves (Evans and Verlander, 1997; Hirst, 1998; Stallen et al., 1996), which have the following properties: (1) they are always flat or falling; (2) generally, they are mapped on double logarithmic scales to allow mapping of so-called low probability—high consequences risks; and (3) the smaller the area under one curve within an FN diagram (Figure 9.6) is, compared to the area under another curve, the lower is the risk. For the purpose of risk evaluation and risk management, falling lines in the double logarithmic diagram differentiate nontolerable from risks "as low as reasonably possible" (ALARP-principle) and negligible (tolerable) risks.

The second approach is based on the economic efficiency of safety measures, which has its roots in the paradigm of the expected utility theory. The basic idea is that risk reduction must gain more utility than a financial unit would gain for any other application. In simple terms, the amount of risk reduction must be larger than the amount of money that is needed to achieve this risk reduction. This relationship is used in benefit—cost analyses of mitigation strategies (Fuchs, 2013), which have become an important—but not the only—criterion in risk management of natural hazards (Wilhelm, 1997, 1999; Gamper et al., 2006; Fuchs et al., 2007; Bründl et al., 2009). Such

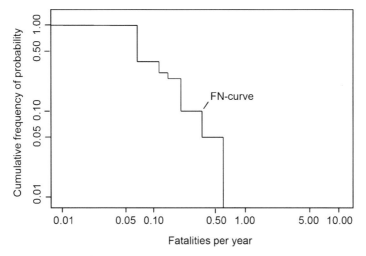

FIGURE 9.6   Example of a frequency−number (FN) diagram.

studies also provide an idea of how much society is willing to pay to avert fatalities (e.g., Leiter and Pruckner, 2009; Rheinberger, 2011; Olschewski et al., 2012). In some countries, cost-benefit analyses have become very common or even a standard for the economic evaluation of mitigation measures (e.g., Larsson-Kråik, 2012; BLFUW, 2006; Bründl et al., 2009; Teich and Bebi, 2009; FOEN, 2014).

9.4 MITIGATION OF RISK

9.4.1 Meaning of Mitigation of Risk

Accidents and damage from interactions of snow avalanches with human activities and infrastructure are best prevented with an integrated approach consisting of a combination of active and passive mitigation measures with a short- or long-term effect (Wilhelm, 1997; SLF, 2000; Rudolf-Miklau and Sauermoser, 2009; Table 9.3). Active methods such as technical measures, protective forests, or artificial triggering directly prevent the formation of avalanches or reduce their damaging effect. These methods typically decrease the probability, the runout length, and/or intensity of an avalanche. Passive methods diminish the effect of avalanches by reducing the number of objects at risk with hazard maps or they decrease the vulnerability by a reinforcement of endangered objects or by organizational measures such as warnings or alert systems. The catastrophic avalanche winter of 1951 when 98 people were killed in Switzerland was the catalyst to develop a systematic avalanche control with comprehensive structural measures. Since then about 1.5 billion Euros have been invested in avalanche control structures (SLF, 2000). Hazard maps have been applied since about 1960 and mitigation measures with a

TABLE 9.3 Classification of Avalanche Mitigation Measures

Mitigation Measures		Long-term Effect	Short-Term Effect
Active methods	Influencing the release of avalanches	• Biological measures such as protective forests or afforestations • Snow-supporting structures • Snow drift measures	• Artificial release of avalanches
	Influencing the effect of avalanches	• Avalanche dams • Braking mounds • Snow sheds and tunnels	• Closure • Evacuation
Passive methods	—	• Hazard map • Land-use planning • Reinforcement of buildings • Emergency management	• Information • Avalanche warning • Safety services • Alert systems

short-term effect were encouraged especially after the catastrophic avalanche winter of 1999. Active long-term avalanche mitigation measures are typically planned where the damage potential is high and cheaper control methods such as explosive control or alert systems are not applicable. The assessment of the effectiveness of mitigation measure is a basic input factor for benefit−cost analysis (Margreth and Romang, 2010). The outset risk without safety measures, the residual risk, and the impact of risk reduction are calculated. The quantification of costs and risk reduction allows a sound benefit−cost comparison of mitigation measures.

9.4.2 Technical Avalanche Mitigation Measures

In the starting zone of avalanches, snow-supporting structures and snow drift structures are applied. Snow-supporting structures have covered the most important release areas in the Alps for >50 years (Figure 9.7). The 3- to 5-m high rigid steel bridges or flexible snow nets are arranged in long lines on slopes steeper than 30°. Permanent supporting structures made of steel with a service life of 100 years and temporary supporting structures made of wood with a service life of about 30 years are distinguished. The design of snow-supporting structures is described in guidelines (Margreth, 2007). Wind has a significant influence on avalanche formation by transporting snow masses into release areas. Snow-drift structures consisting of 2- to 5-m-high vertical fences with a porosity of 50 percent disturb the local wind field in such a way that a controlled deposit of wind transported snow is achieved outside of the release zone. The effect of snow drift measures depends strongly on the local topography and wind conditions.

FIGURE 9.7   Avalanche release beside an area controlled with snow-supporting structures in February 1999. The maximal fracture depth varied between 3 and 4 m (Schweifinen, Zermatt, Switzerland). *Photograph S. Margreth, SLF.*

In the track of avalanches where the velocities and impact pressures are very high, deflecting structures such as walls, dams, or snow sheds change the flow direction of the moving snow masses and protect the objects at risk. Deflecting dams are generally massive earth-filled structures with a height between 5 and 25 m. A steep embankment enhances the deflecting effect. The height of a dam depends mainly on the flow velocity and the deflection angle. New design methods are based on the shock wave theory (Johannesson et al., 2009). Experience has shown that deflection angles should not exceed about 20–30° to prevent overtopping. Snow sheds are the standard method to protect transportation corridors. Their structural design depends mainly on the flow height and velocity of avalanches as well on the height of snow deposits (Margreth and Platzer, 2008). Crucial for their effectiveness is to plan the length of a shed correctly.

If there is sufficient space in the runout zone, catching dams (Figure 9.8) or braking mounds reduce the avalanche velocity and stop dense flow avalanches. When impacting a catching dam, the avalanches move upward along the embankment until the kinetic energy is dissipated (Baillifard et al., 2007). The layout of catching dams is similar to deflecting dams, but they are built at right angles to the avalanche flow and the required height, usually between 15 and 25 m, is much higher than for deflecting dams. Velocity should not exceed about 20 m/s. Avalanche dams often have a double function as protection against snow avalanches and debris flows (Figure 9.8). Individual objects situated in the runout zone can be designed for expected avalanche impact loads. Protecting buildings ranks among the oldest means of defence against avalanches. Widely applied and very effective are avalanche splitters, which divide the avalanche flow, or reinforced impact

FIGURE 9.8 Combined catching dam against avalanches and debris flows in Klosters, Switzerland. The dam height is 17 m, and the retention volume is 260,000 m^3. *Photograph S. Margreth, SLF.*

walls made of concrete without openings (Holub and Hübl, 2008; Suda and Rudolf-Miklau, 2012; Figure 9.9).

9.4.3 Land-Use Planning

Avalanche hazard maps indicate the settlements that are at risk, based typically on the frequency and intensity of events in the relevant area (Gruber and Margreth, 2001). They serve land-use planning purposes and are

FIGURE 9.9 Reinforced residential building in St Antönien, Switzerland, which is situated in a blue avalanche hazard zone. The back wall consists of a massive concrete wall designed on avalanche actions. *Photograph S. Margreth, SLF.*

an important instrument of emergency planning. According to the Swiss Guidelines (BFF and SLF, 1984), a red zone indicates an area that is exposed to considerable danger. In the event of an avalanche being released, buildings are likely to be destroyed. The impact loads generated by a 300-year avalanche can exceed 30 kPa. In the blue zones, rare avalanches are accompanied by only small impact loads of <30 kPa. People inside buildings are fairly safe, but a danger to human life exists outdoors. New building zones can be approved on the basis of a structural reinforcement. In the white zone, the danger is negligible. Avalanche hazard mapping criteria for Italy, Canada, and Austria are provided in Table 9.4. The Icelandic regulation on snow and landslide hazard zoning is based on individual risk (Arnalds et al., 2004).

Land-use planning is considered to be a cost-effective way to manage danger due to avalanches. However, the problems associated with endangered areas settled a long time before a hazard map was established can hardly be solved.

9.4.4 Biological Measures

Forests afford effective and inexpensive protection against avalanches (Weir, 2002). Measured by land surface area, forests are the most extensive form of

TABLE 9.4 Criteria for Hazard Mapping in Italy (Barbolini, 2007), Canada (CAA, 2002) and Austria (BMLFUW, 2011)

	Italy	Canada		Austria
Red zone	$T = 30$ years and $P > 3$ kPa or $T = 100$ years and $P > 15$ kPa	$T < 30$ years and $P > 0$ kPa, $T = 30$ years and $P > 3$ kPa or $T = 300$ years and $P > 30$ kPa	Red zone	$P \geq 10$ kPa
Blue zone	$T = 30$ years and $P < 3$ kPa or $T = 100$ years and $3 < P < 15$ kPa	$T = 30$ years and $1 < P < 3$ kPa or $T = 300$ years and $1 < P < 30$ kPa	Yellow zone	$1 < P < 10$ kPa
Yellow zone	$T = 100$ years and $P < 3$ kPa or $T = 300$ years			

avalanche protection in the Alps. Around one-third of the Swiss Alpine region is forested. Trees offer support to the snowpack, and the snow cover is shallower because, when snow falls, some of it is deposited on the canopy (Bründl et al., 1999). In addition, a more moderate climate prevails in the forest. Tree crown density, gaps in the forest, and the number of trunks per hectare are key criteria that are used to describe the extent of protection afforded (Margreth, 2004). Avalanches that are released above the forest line destroy trees, and the snow masses encounter very little resistance. Biological measures are commonly combined with technical avalanche mitigation measures.

9.4.5 Organizational Measures

Organizational measures are taken at short notice and according to the extent of the current snow and avalanche situation (Stoffel and Schweizer, 2008). The principal organizational measures taken to guard against potential impacts avalanches are warnings, closures, evacuations, alert systems, and the artificial triggering of avalanches (Gubler, 1977; Stoffel, 2001; Haeberli and Whiteman, 2014, Figure 15). Warnings, closures, and evacuations are adopted for endangered areas by local safety authorities. The size and probability of an avalanche release are estimated on the basis of the current weather and avalanche situation. The relevant action is then taken according to the findings. Artificial triggering of avalanches plays a major role in protecting traffic routes, winter sports resorts and, given favorable outline conditions, settlements as well (Stoffel and Margreth, 2009). An avalanche is triggered by detonating an explosive charge in the starting zone, which induces a shock wave and thus imposes an additional load on the snowpack. Numerous methods are applied, including hand charge delivery, the dropping of charges from a helicopter, cannons, and permanently installed triggering devices.

9.5 METHODS AND TOOLS FOR RISK ASSESSMENT AND EVALUATION OF MITIGATION MEASURES

During recent years, a trend from a hazard-oriented approach to a risk-based approach could be observed. Hazard-based in this context means that focus is on dealing with the process, whereas the risk-based approach means that in addition to the probability of occurrence, elements at risk are quantified to yield an estimation of the probability that a certain level of damage is reached (Bründl, 2013). Therefore, new methods and tools for risk assessment (including both risk analysis and risk evaluation) and the evaluation of mitigation measures were developed.

One of the first electronic tools for estimating potential losses from natural disasters using GIS was the software Hazus of the Federal Emergency

Management Agency FEMA (FEMA, 2014). However, this tool is only able to handle earthquakes, floods, and hurricanes and was developed for situations in the United States. RiskScape, a GIS-based tool for risk assessment of earthquakes, flooding (river), tsunami, volcanic ashfall, and windstorm was specifically developed for conditions in New Zealand (RiskScape, 2014). In Switzerland, a general guideline for risk analysis of all Alpine hazards (including snow avalanches) was published in 1999 (Borter, 1999; Borter and Bart, 1999). Wilhelm (1999) elaborated a specific guideline for risk analysis and economic evaluation of mitigation measures against snow avalanches along roads, and Winkler and Burkard (2005) for railways. In 2009, the National Platform for Natural Hazards, PLANAT, issued a guideline "risk concept for natural hazards" (Bründl, 2009; Bründl et al., 2009) and the Federal Office for Roads published a specific method for risk assessment along national highways (ASTRA, 2009).

These methods and guidelines are the basis for several software tools currently in operational use in Switzerland. Since 2008, the online-tool EconoMe (www.econome.admin.ch) has been used for assessing the benefit−cost relationship of mitigation measures against all Alpine hazards (Bründl et al., 2009). Since users cannot change calculation factors in EconoMe because of comparability requirements of the Federal Office for the Environment (FOEN), the tool "EconoMe-Develop" (www.econome-develop.ch) was elaborated on the same platform. EconoMe-Develop allows the user to choose a number of assumptions and calculation parameters according to the specific remit of a risk analysis (Bründl, 2012). For addressing the specific demands for risk analyses along railway lines, the tool EconoMe Railway (www.econome.ch/eco_rail) was introduced to practice in 2010. EconoMe Railway builds on the same framework as EconoMe and covers additional railway-specific damage features (Bründl et al., 2012). In the following years, EconoMe and EconoMe Railway will be merged into one tool. For risk assessment along roads, the Federal Office for Roads initiated the tool RoadRisk (www.roadrisk.ch), which has been applied for risk assessment of the whole highway network in Switzerland (Utelli et al., 2012). In all these software tools, the processes, snow avalanche, debris flow, flood, rock fall, and landslide can be handled simultaneously in one mitigation project, which makes them multirisk instruments.

For presenting and visualizing spatial processes, GISs are state of the art. Cappabianca et al. (2008) depicted calculated specific avalanche risks to objects, and Grêt-Regamey and Straub (2006) linked Bayesian Networks to a GIS to visualize their results on avalanche risk assessment in Davos. The EconoMe tools and RoadRisk also allow for import and export of spatial data for visualization in GIS. We assume that identification of hot spots of risk and for strategic planning, spatial visualization of risk will gain more importance in the future.

9.6 CASE STUDY "EVALUATION OF AVALANCHE MITIGATION MEASURES FOR JUNEAU, ALASKA"

9.6.1 Introduction

Juneau, Alaska's state capital, is situated at sea level in a glacial valley in Southeast Alaska. Steep-sided mountains surround the city. Small avalanches reach the most exposed houses nearly every winter. Large avalanches are rare. Juneau is considered to be one of the largest municipal avalanche hazard areas in the United States (Margreth and Mattice, 2012). However, to date, there are still no structural protection measures in place, which can be attributed to the complexity of the avalanche situation. In 2011, the WSL Institute for Snow and Avalanche Research SLF was mandated to investigate mitigation measures to decrease the avalanche risk (SLF, 2011). Hereafter, we focus on the Behrends Avenue avalanche path, which has the highest avalanche risk.

9.6.2 Avalanche Situation

Site-specific characteristics of an avalanche path are decisive for an optimal planning of mitigation measures. The starting zone of the Behrends Avenue avalanche path is situated on the southwest flank of Mt Juneau (Figure 9.10). The main starting zone is situated between an elevation of 970 and 500 m a.s.l. The maximal width of the starting zone is 500 m. The potential starting zone with a mean inclination of 40° is very large with an area of 25 ha. The flow of large avalanches is rather unconfined. At 300 m a.s.l. the width of frequent avalanches is captured by the tree damage. No big trees occur along a 270-m-wide path. Below the elevation of 150 m the track is <30° and over the last 150 m above the settlement, the slope inclination is 15°. Such slope angles do not retard a dry snow avalanche. The favorable characteristics of the Behrends Avenue avalanche path is that the starting zone is relatively well structured into different smaller pockets and that cold temperatures combined with unusually deep snowfalls are relatively rare. Given these factors, the release of small avalanches is much more likely compared to the release of the whole starting zone in an extreme avalanche.

The avalanche history of the Behrends Avenue avalanche path is relatively well documented. Between 1890 and 2011, an avalanche reached the sea arm three times and the settlement area nine times. The avalanche of 1890, which deposited hundreds of tons of snow on the road along the sea arm, was the largest recorded event. The most destructive and best-documented avalanche in recent years occurred in 1962. Following a period of heavy precipitation arriving from the northeast, a slab avalanche, developing into a powerful powder snow avalanche, broke loose. The main damage to the houses was caused by the powder blast and by impacts of logs or other debris transported by the avalanche. Approximately 35 houses were damaged (Figure 9.11).

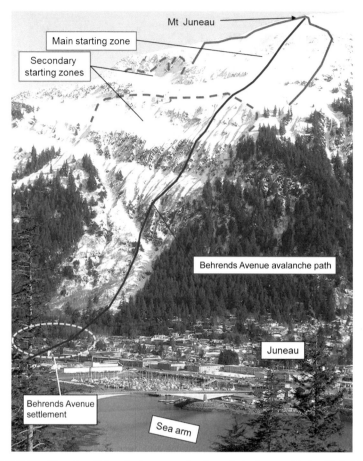

FIGURE 9.10 Overview Behrends Avenue avalanche path in Juneau, Alaska, United States. *Photograph David Kent, April 3, 2007, www.westjuneau.com.*

The climate and snow conditions are important for a hazard analysis as outlined in Section 9.2.1. Juneau lies within an area of maritime influence, which prevails over the coastal areas of southeastern Alaska. The area has little sunshine, generally moderate temperatures and abundant precipitation. The predominant wind direction is from the south. Periods of comparatively severe cold temperatures occur, which are caused by strong northerly winds. On Mt Juneau, these winds can cause important snow drift accumulations in the upper starting zones of the Behrends Avenue avalanche path. The snow line often fluctuates between sea level and the elevation of the starting zones, which causes a wide range of snow conditions in the avalanche paths. Normally, the snowpack consists of thawed and refrozen layers. This favors a higher frequency of wet snow avalanche conditions as opposed to extreme dry snow avalanche conditions, which explains the rare occurrence of extreme dry snow

FIGURE 9.11   View of Behrends Avenue subdivision with the Sea channel in the background after the 1962 avalanche. The powder snow avalanche destroyed several buildings. *Photograph Alaska Mountain Safety Center.*

avalanches. The extreme snow heights in the starting zones on Mt Juneau are estimated to vary between 6 and 8 m, particularly for depressions where even larger snow heights must be expected. We assume that an extreme avalanche (return period up to 300 years) on Mt Juneau might have an average fracture depth of 1.5–2.0 m.

9.6.3 Hazard Analysis

The evaluation of the avalanche hazard is based on terrain analysis, on-site inspections by foot, study of the avalanche history, and avalanche dynamics calculations (Chapter 2.1). We applied the two-dimensional avalanche simulation program RAMMS (Christen et al., 2010) and the one-dimensional avalanche dynamics program AVAL-1D (Christen et al., 2002). AVAL-1D was mainly applied to calculate the effect of powder snow avalanches. The most important input parameters for the avalanche dynamics calculations are the slab thickness, the release area, the friction parameters, and the digital terrain data. Due to the rather specific climatic situation of Juneau (low elevation, coastal climate, abundant precipitation, cold snow conditions at the elevation of the starting zone, wet snow conditions at sea level), we adapted the elevation limits controlling the friction parameters. We performed avalanche dynamics calculations for return periods of 10, 30, and 300 years.

   According to the RAMMS simulations the 10-year dense flow avalanche does not reach the settlement (Figure 9.12). The 30-year dense flow avalanche stops, according to RAMMS simulations, beside the sea. The northwestern part of the settlement is in a zone with avalanche impacts of >30 kPa. At an

FIGURE 9.12 RAMMS-simulation results of the 10-year Behrends Avenue avalanche (red = impact pressure >30 kPa and blue = impact pressure <30 kPa). The simulated avalanche stops directly before the settlement. *Source: RAMMS, SLF; map: City and Borough of Juneau.*

elevation of 100 m, the avalanche velocity varies between 19 and 24 m/s. The maximal pressure of the 30-year powder snow avalanche is 3–4 kPa. The 300-year dense flow avalanche reaches the sea arm with an intensity of >30 kPa (Figure 9.13). Most of the settlement is in the 30-kPa zone; here, the velocities are up to 28 m/s with a flow height of 3.6 m. Such avalanche intensities are capable of completely destroying massive buildings. The 300-year powder snow avalanche develops a pressure of 6–9 kPa.

9.6.4 Consequence Analysis and Risk Evaluation

The damage potential in the Behrends Avenue avalanche path is huge. Forty-five residential buildings, one hotel, one school, a boat harbor, and the East–West highway, which connects the city with the hospital and airport, are within the reach of a 300-year avalanche. The approximate value of all residential properties was estimated to be >$20 million USD. A large avalanche could destroy every building in the avalanche zone, sweep cars off the highways, and damage or destroy boats in the harbor. Such large slides could also block the main highway, hindering emergency response, and possibly blocking road access to the hospital. The City and Borough of Juneau has an avalanche forecaster on staff to deliver daily avalanche bulletins to the community and to

FIGURE 9.13 RAMMS-simulation results of the 300-year Behrends Avenue avalanche (red = impact pressure >30 kPa and blue = impact pressure <30 kPa). According to the simulation the avalanche reaches the sea channel. *RAMMS, SLF; map: City and Borough of Juneau.*

educate the public about living in a community with avalanche problems. The avalanche bulletins notify the public of times when avalanche areas are highly endangered and should be avoided. The City and Borough of Juneau does not issue orders to evacuate hazard zones. According to current standards, the risk is considered to be far beyond an acceptable level. Therefore, additional mitigation measures were studied.

9.6.5 Protection Measures

The finding of optimal mitigation measures is based on the on-site findings, the RAMMS simulations and the experience of the project experts. As a first step, the measures were described only qualitatively with a rough dimensioning, cost estimation, and a review of the varying usefulness of each. The goal was the evaluation of different types of measures in general for a basic discussion and not a detailed benefit–cost analysis.

9.6.5.1 Artificial Release of Avalanches

At first, we looked at the artificial release of avalanches especially because new methods (e.g., the GAZEX exploder or Wyssen tower) have been developed over the past few years. Autonomous devices are relatively cheap and allow remote triggering of avalanches independently of visibility and with a good detonation effect. In general, artificial release above settlements should be applied with extreme caution and should remain an exception (Stoffel and Margreth, 2009). The main risk of artificial release above settlements is in triggering an avalanche that is too large to manage and results in damage. The terrain features on Mt Juneau are not considered to be very favorable for the artificial release of avalanches. This is mainly because the potential starting zone is very large and because the topography is very steep, which even favors small avalanches to reach the settlement. The damage potential is huge, especially because most of the buildings have no protection measures and seem to be very vulnerable to avalanche impacts. We concluded that it cannot be guaranteed to limit the size of an artificially released avalanche in the Behrends Avenue to a volume that is harmless to the settlement. Therefore, we did not recommend applying the artificial release of avalanches under the current conditions.

9.6.5.2 Snow-Supporting Structures

The area that would require snow-supporting structures in the Behrends Avenue slide path is 25 ha and therefore very large. Since there are no long-term snow measurements available, we estimated that the extreme snow depth may vary between 5 and 8 m, which would correspond to a structure height of up to 6.2 m. We estimated that in total 10,800 m of structures would be required with a total cost of at least $32 million USD. As there is no experience with the construction, behavior, and design of snow-supporting structures in Alaska, it would be advisable to install a small test site on Mt Juneau prior to embarking upon such a huge project. Additionally, the slope of the starting zone is considered to be unstable, which would require expensive foundations and high maintenance cost. Therefore, we did not recommend the construction of snow-supporting structures.

9.6.5.3 Avalanche Dams

We studied different variants of dams. Because the design velocity is very high at 30 m/s, a catching dam has to have a height of 25–35 m. The main advantage of a catching dam would be that both the settlement and the highway along the sea channel could be protected. Besides the visual impact and the high cost estimated at $12 million USD for a fill volume of 400,000 m^3, our main concern was that a catching dam could not stop completely a large powder snow avalanche. The height of a deflecting dam was estimated to vary between 18 and 25 m depending on the deflecting angle. The length of the dam would have to be around 300 m and the fill volume between 120,000 and 200,000 m^3. The main disadvantage of deflecting dams is that in the direction of flow, the risk is much increased, which is unwanted in

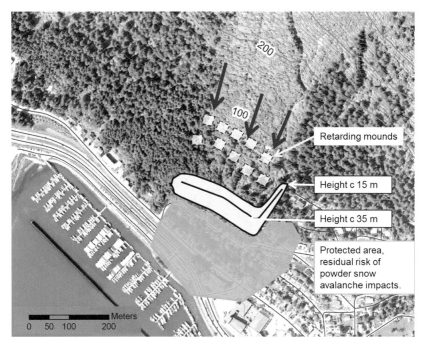

FIGURE 9.14   Sketch of a catching dam in the runout of the Behrends Avenue avalanche path with two possible lines of retarding mounds. *RAMMS, SLF; map: City and Borough of Juneau.*

a densely populated area (Figure 9.14). Further, a deflecting dam cannot stop a large powder snow avalanche. Therefore, we did not recommend planning such an avalanche dam.

9.6.5.4 Direct Protection of Individual Buildings

The goal of a direct protection is to shelter an individual building exposed to avalanches. The most frequently applied forms are the direct reinforcement of a building with a concrete back wall without openings, and avalanche splitters. In the present situation both possibilities are hardly feasible. There is insufficient space to build a wall and avalanche splitters would increase the risk to the neighboring buildings. Because the costs of such direct protection are estimated to be even higher than the value of the building to be protected, we concluded that direct protection of buildings is not recommendable.

9.6.5.5 Buyout of Houses

The most effective ways to reduce the avalanche risk in the settlement were found to be the buyout of the endangered homes by the Government, prohibition of new constructions, and mandatory reinforcement of new buildings in the blue zone. With regard to this complex avalanche situation, where the avalanche risk can hardly effectively be reduced with traditional protection measures, the buyout of the endangered homes was supposed to be a reasonable mitigation strategy.

9.6.6 Conclusions

The avalanche problem is very serious in the runout zone of Behrends Avenue avalanche path. We think that the unacceptable risk to the residents in the hazard zones can only be managed in the short term if the City and Borough of Juneau would order evacuations and close endangered areas during periods of high avalanche danger. As the buildings in the hazard zones have no structural reinforcement, people inside these buildings are not safe. The reduction of the avalanche risk with structural protection measures is prohibitively expensive and therefore not recommended. Further, the artificial release of avalanches is not advisable, mainly because of the danger to people, property, and homes. We are of the opinion that the buyout of endangered homes in the avalanche path by the government is the optimal way to effectively reduce avalanche risk in the long term. The buyout of homes would ensure a permanent solution to the avalanche problem. Following our advice, the City and Borough of Juneau established a buyout plan for endangered homes, which was presented to the homeowners.

9.7 FINAL REMARKS

In this chapter, we have shown that dealing with avalanches in settlements and on traffic routes has changed from a hazard-based to a risk-based approach. The main reason for this shift is that a risk analysis allows not only the identification of areas with the highest risk but also a comparison of potential mitigation measures by benefit–cost criteria. Mitigation measures with a benefit–cost ratio above 1 are economical, and those with the highest benefit–cost ratio are supposed to be realized first. However, a risk analysis and therefore the result of a benefit–cost analysis are subject to uncertainties and should be considered as only one criterion for decision making.

The highest uncertainties stem from hazard analysis. Realistically, modeling the runout of avalanches for certain scenarios depends to a high degree whether a sound validation of model results is possible, for example, by comparing them with historic data. But historic data must also be carefully checked to determine whether earlier conditions are still comparable with the conditions nowadays.

A further, and very important, source of uncertainty is the vulnerability of elements at risk. As our argumentation revealed, only less empirical data are available to validate vulnerability functions; additionally, uncertainties are high when vulnerability functions derived at one location are applied at other locations (see, e.g., Jóhannesson and Arnalds, 2001).

A third argument in support of the careful interpretation of benefit−cost analysis is that they most often do not take all potential damage into consideration for methodological reasons. For example, business interruption and indirect cost are most often neglected but can significantly contribute to the overall risk. Further research is needed to provide methods that can be used in risk analyses in an operational manner.

Last, but not least, the acceptance of mitigation strategies by the local population and politicians is a pivotal argument in decision making as shown by the Juneau case study. If measures are proved to be economically viable but are not widely accepted, they would most probably not be realized in democratic societies. Therefore, well-established risk communication and risk dialog including all involved parties, such as community representatives, engineers, and residents, are of considerable importance for the successful management of avalanche risk.

REFERENCES

Ammann, W.J., 2006. The risk concept. In: Dannenmann, S., Ammann, W.J., Vulliet, L. (Eds.), Risk 21-Coping with Risks Due to Natural Hazards in the 21st Century. Taylor & Francis, London, pp. 3−23.

Arnalds, þ., Jónasson, K., Sigurðsson, S., 2004. Avalanche hazard zoning in Iceland based on individual risk. Ann. Glaciol. 38, 285−290.

ASTRA, 2009. Risikokonzept Naturgefahren Nationalstrassen. Methodik für eine risikobasierte Beurteilung, Prävention und Bewältigung von gravitativen Naturgefahen auf Nationalstrasse. Bundesamt für Strassen, Bern.

Baillifard, M., Kern, M., Margreth, S., 2007. Anleitung zur Dimensionierung von La-wi-nen-auf-fang-dämmen. WSL-Institut für Schnee- und Lawinenforschung SLF, Davos, 36 pp.

Barbolini (Ed.), 2007. Hazard Mapping of Extremely Rapid Mass Movement in Europe. State-of-the-art methods in practice. Deliverable D3.1 FP6-Project "IRASMOS". EU Commission, Davos. http://irasmos.slf.ch.

Barbolini, M., Cappabianca, F., Savi, F., 2003. A new method for the estimation of avalanche distance exceeded probabilities. Surv. Geophys. 24 (5−6), 587−601.

Barbolini, M., Cappabianca, F., Savi, F., 2004. Risk assessment in avalanche-prone areas. Ann. Glaciol. 38, 115−122.

Barbolini, M., Natale, L., Savi, F., 2002. Effects of release conditions uncertainty on avalanche hazard mapping. Nat. Hazards 25, 225−244.

Bell, R., Glade, T., 2004. Quantitative risk analysis for landslides − examples from Bildudalur, NW Iceland. Nat. Hazards Earth Syst. Sci. 4 (1), 117−131.

Bellaire, S., Jamieson, B., Stetham, G., 2013. Relating Avalanche Activity to Climate Change and Coupled Ocean-Atmospheric Phenomena. Davos Atmosphere and Cryosphere Assembly DACA-13, Davos, Switzerland.

Bertrand, D., Naaim, M., Brun, M., 2010. Physical vulnerability of reinforced concrete buildings impacted by snow avalanches. Nat. Hazards Earth Syst. Sci. 10 (7), 1531−1545.

BFF and SLF, 1984. Richtlinien zur Berücksichtigung der Lawinengefahr bei raumwirksamen Tätigkeiten. EDMZ, Bern.

BfS, 2008. Bevölkerung − Die wichtigsten Zahlen. Bern.

Birkmann, J. (Ed.), 2006. Measuring Vulnerability to Natural Hazards: Towards Disaster Resilient Societies. United Nations University Press, New York.

BLFUW, 2006. Richtlinien für die Wirtschaftlichkeitsuntersuchung und Priorisierung von Massnahmen der Wildbach- und Lawinenverbauung gemäss 3 Abs. 2 Z 3 Wasserbautenförderungsgesetz 1985. Bundesministerium für Land- und Forstwirtschaft, Umwelt und Wasserwirtschaft, Sektion Forstwesen, Wien.

BMLFUW, 2011. Richtlinie für die Gefahrenzonenplanung. BMLFUW-LE.3.3.3/0185-IV/5/2007. Fassung vom 4. February 2011, die.wildbach. Lebensministerium, Wien.

Bocchiola, D., Medagliani, M., Rosso, R., 2006. Regional snow depth frequency curves for avalanche hazard mapping in central Italian Alps. Cold Reg. Sci. Technol. 46 (3), 204−221.

Bohnenblust, H., Slovic, P., 1998. Integrating technical analysis and public values in risk-based decision making. Reliab. Eng. Syst. Saf. 59, 151−159.

Borter, P., 1999. Risikoanalysen bei gravitativen Naturgefahren − Methode. 107/I. Bundesamt für Umwelt, Wald und Landschaft, BUWAL, Bern.

Borter, P., Bart, R., 1999. Risikoanalysen bei gravitativen Naturgefahren-107/II-Fallbeispiele und Daten. 107/II. Bundesamt für Umwelt,Wald und Landschaft, BUWAL, Bern.

Bottelberghs, P.H., 2000. Risk analysis and safety policy developments in the Netherlands. J.Hazard. Mat. 71 (1−3), 59−84.

Bründl, M. (Ed.), 2009. Risikokonzept für Naturgefahren. Einzelprojekt A1.1: Leitfaden. Nationale Plattform Naturgefahren PLANAT, Bern, p. 420.

Bründl, M., 2012. EconoMe-Develop − a software tool for assessing natural hazard risk and economic optimisation of mitigation measures. In: Proceedings International Snow Science Workshop "a merging of theory and practice". ISSW 2012, Anchorage, Alaska, pp. 639−643.

Bründl, M., 2013. Dealing with natural hazard risks in Switzerland − the influence of hazard mapping on risk-based decision making. In: Bollschweiler, M., Stoffel, M., Rudolf-Miklau, F. (Eds.), Tracking Torrential Processes on Fans and Cones. Advances in Global Change Research. Springer, Heidelberg, New York, London, pp. 355−365.

Bründl, M., 2015. Analyses of natural disasters and their contribution to changes in natural hazard management in Switzerland. In: Egner, H., Schorch, M., Voss, M. (Eds.), Learning and Calamities. Practices, Interpretations, Patterns. Routledge, New York, London, pp. 199−215.

Bründl, M., Barbolini, M., Bischof, N., Eidsvig, U., Fuchs, S., Korup, O., Kvelsvik, V., Rheinberger, C., Romang, H., Sandersen, F., 2008. Integral Risk Management of Snow Avalanches, Rock Avalanches and Debris Flows in Europe − Technical Report. Deliverable D5.2 FP6-Project "IRASMOS". EU Commission, Davos. http://irasmos.slf.ch.

Bründl, M., Bartelt, P., Schweizer, J., Keiler, M., Glade, T., 2010. Snow avalanche risk analysis − review and future challenges. In: Alcántara-Ayala, I., Goudie, A. (Eds.), Geomorphological Hazards and Disaster Prevention. Cambridge University Press, Oxford, pp. 49−61.

Bründl, M., Romang, H.E., Bischof, N., Rheinberger, C.M., 2009. The risk concept and its application in natural hazard risk management in Switzerland. Nat. Hazards Earth Syst. Sci. 9 (3), 801−813.

Bründl, M., Schneebeli, M., Flühler, H., 1999. Routing of canopy drip in the snowpack below a spruce crown. Hydrol. Processes 13 (1), 49−58.

Bründl, M., Winkler, C., Baumann, R., 2012. "EconoMe-Railway". A new calculation method and tool for comparing the effectiveness and the cost-efficiency of protective measures along railways. In: Koboltschnig, G., Hübl, J., Braun, J. (Eds.), 12th Congress INTERPRAEVENT. International Research Society INTERPRAEVENT, Grenoble, pp. 933−943.

Burkard, A., Salm, B., 1992. Die Bestimmung der mittleren Anrissmächtigkeit d0 zur Berechnung von Fliesslawinen. Bericht Nr. 668. Eidg. Institut für Schnee- und Lawinenforschung SLF, Davos.

BUWAL, 1991. Handbuch I zur Störfallverordnung. Richtlinien für Betriebe mit Stoffen, Erzeugnissen oder Sonderabfällen. Bundesamt für Umwelt, Wald und Landschaft BUWAL, Bern, 74 pp.

CADZIE, 2003. Catastrophic Avalanche: Defence Structures and Zoning in Europe. http://cadzie. grenoble.cemagref.fr (last access 10.03.14).

CAA, 2002. In: McClung, D.M., Stethem, C.J., Schaerer, P.A., Jamieson, J.B. (Eds.), Guidelines for Snow Avalanche Risk Determination and Mapping in Canada. Canadian Avalanche Association (CAA), Revelstoke, BC, p. 24.

Cappabianca, F., Barbolini, M., Natale, L., 2008. Snow avalanche risk assessment and mapping: a new method based on a combination of statistical analysis, avalanche dynamics simulation and empirically-based vulnerability relations integrated in a GIS platform. Cold Reg. Sci. Technol. 54 (3), 193−205.

Carter, W.N., 1991. Disaster Management: A Disaster Manager's Handbook. Asian Development Bank, Manila.

Casteller, A., Villalba, R., Araneo, D., Stöckli, V., 2011. Reconstructing temporal patterns of snow avalanches at Lago del Desierto, southern Patagonian Andes. Cold Reg. Sci. Technol. 67 (1−2), 68−78.

Christen, M., Bartelt, P., Gruber, U., 2002. AVAL-1D: an avalanche dynamics program for the practice. In: International Congress Interpraevent 2002 in the Pacific Rim − Matsumo, Japan, vol. 2. Congress publication, pp. 715−725.

Christen, M., Kowalski, J., Bartelt, P., 2010. RAMMS: numerical simulation of dense snow avalanches in three-dimensional terrain. Cold Reg. Sci. Technol. 63 (1−2), 1−14.

Crozier, M., Glade, T., 2005. Landslide hazard and risk: issues, concepts and approach. In: Glade, T., Anderson, M., Crozier, M.J. (Eds.), Landslide Hazard and Risk. John Wiley & Sons, Chichester, pp. 1−40.

Cutter, S., 1996. Vulnerability to environmental hazards. Prog. Hum. Geogr. 20 (4), 529−539.

Decaulne, A., Eggertsson, Ó., Laute, K., Beylich, A.A., 2014. A 100-year extreme snow-avalanche record based on tree-ring research in upper Bødalen, inner Nordfjord, western Norway. Geomorphology 218, pp. 3−15.

Eckert, N., Keylock, C.J., Bertrand, D., Parent, E., Faug, T., Favier, P., Naaim, M., 2012. Quantitative risk and optimal design approaches in the snow avalanche field: review and extensions. Cold Reg. Sci. Technol. 79−80, 1−19.

Egli, T., Stucki, M., 2013. Hazard mapping and land-use planning − a Swiss perspective. In: Bollschweiler, M., Stoffel, M., Rudolf-Miklau, F. (Eds.), Tracking Torrential Processes on Fans and Cones. Advances in Global Change Research. Springer, Heidelberg, New York, London, pp. 367−374.

Evans, A.W., Verlander, N.Q., 1997. What is wrong with criterion FN-lines for judging the tolerability of risk? Risk Anal. 17 (2), 157−168.

Farmer, F., 1967. Reactor safety and siting: a proposed risk criterion. Nucl. Saf. 8 (6), 539−548.

Fell, R., Hartford, D., 1997. Landslide risk management. In: Cruden, D., Fell, R. (Eds.), Landslide Risk Assessment. Proceedings of the International Workshop on Landslide Risk Assessment − Honolulu, Hawaii, USA, February 19−21, 1997. Balkema, Rotterdam, pp. 51−109.

Fell, R., 1994. Landslide risk assessment and acceptable risk. Can. Geotech. J. 31, 261−272.

FEMA, 2014. Hazus − The Federal Emergency Management Agency's (FEMA's) Methodology for Estimating Potential Losses from Disasters. Federal Emergency Management Agency, Washington, USA.

FOEN, 2014. EconoMe-Wirtschaftlichkeit von Schutzmassnahmen gegen Naturgefahren. Federal Office for the Environment, Bern.

Fuchs, S., 2013. Cost-benefit analysis of natural hazard mitigation. In: Bobrowsky, P. (Ed.), Encyclopedia of natural hazards. Springer, Dordrecht, pp. 121—125.

Fuchs, S., Birkmann, J., Glade, T., 2012. Vulnerability assessment in natural hazard and risk analysis: current approaches and future challenges. Nat. Hazards 64 (3), 1969—1975.

Fuchs, S., Bründl, M., Stötter, J., 2004. Development of avalanche risk between 1950 and 2000 in the municipality of Davos, Switzerland. Nat. Hazards Earth Syst. Sci. 4 (2), 263—275.

Fuchs, S., Heiss, K., Hübl, J., 2007. Towards an empirical vulnerability function for use in debris flow risk assessment. Nat. Hazards Earth Syst. Sci. 7 (5), 495—506.

Fuchs, S., Keiler, M., Zischg, A., Bründl, M., 2005. The long-term development of avalanche risk in settlements considering the temporal variability of damage potential. Nat. Hazards Earth Syst. Sci. 5, 893—901.

Gamper, C.D., Thöni, M., Weck-Hannemann, H., 2006. A conceptual approach to the use of cost benefit and multi criteria analysis in natural hazard management. Nat. Hazards Earth Syst. Sci. 6 (2), 293—302.

Gardner, J., 1982. The role of new technologies in risks from natural hazards. In: Lind, N. (Ed.), Technological Risk. Proceedings of a Symposium on Risk in New Technologies. University of Waterloo Press, Waterloo, Ontario, pp. 153—172.

Grêt-Regamey, A., Straub, D., 2006. Spatially explicit avalanche risk assessment linking Bayesian networks to a GIS. Nat. Hazards Earth Syst. Sci. 6 (6), 911—926.

Gruber, U., Margreth, S., 2001. Winter 1999: a valuable test of the avalanche-hazard mapping procedure in Switzerland. Ann. Glaciol 32, 328—332.

Gubler, H., 1977. Artificial release of avalanches by explosives. J. Glaciol. 19 (81), 419—429.

Haeberli, W., Whiteman, C., 2014. Snow and Ice-Related Hazards, Risks and Disasters — A General Framework. Elsevier, pp. 1—34.

Hatfield, A., Hipel, K., 2002. Risk and systems theory. Risk Anal. 22 (6), 1043—1057.

Heinimann, H.R., Hollenstein, K., Kienholz, H., Krummenacher, B., Mani, P., 1998. Methoden zur Analyse und Bewertung von Naturgefahren. Eine risikobasierte Betrachtungsweise. BUWAL, Bern, 247 pp.

Hendrikx, J., Owens, I., 2006. Avalanche risk evaluation with practical suggestions for risk minimisation: a case study of the Milford Road, New Zealand. In: International Snow Science Workshop ISSW. International Snow Science Workshop, Telluride, Colorado, USA, pp. 757—767.

Hendrikx, J., Owens, I., 2008. Modified avalanche risk equations to account for waiting traffic on avalanche prone roads. Cold Reg. Sci. Technol. 51 (2—3), 214—218.

Hirst, I.L., 1998. Risk assessment: a note on F—N curves, expected numbers of fatalities, and weighted indicators of risk. J. Hazard. Mat. 57 (1—3), 169—175.

Höppner, C., Whittle, R., Bründl, M., Buchecker, M., 2012. Linking social capacities and risk communication in Europe: a gap between theory and practice? Nat. Hazards 64, 1753—1778.

Hollenstein, K., 1997. Analyse, Bewertung und Management von Naturrisiken. vdf Hochschulverlag an der ETH, Zürich, 220 pp.

Holub, M., Hübl, J., 2008. Local protection against mountain hazards—state of the art and future needs. Nat. Hazards Earth Syst. Sci 8, 81—99.

IPCC, 2012. Managing the risks of extreme events and disasters to advance climate change adaptation. In: A Special Report of Working Groups I and II of the Intergovernmental Panel on Climate Change. Cambridge University Press, Cambridge, UK and New York, NY, USA.

IRASMOS, 2008. Integral Risk Management of Rapid Mass Movements. http://irasmos.slf.ch (last access 10.03.14).

Jaecklin, A., 2007. Voll integriertes Risikomanagement. Manage. Qualität 11, 21−23.

Jamieson, J.B., Stetham, C.J., Schaerer, P.A., McClung, D.M. (Eds.), 2002. Land Managers Guide to Snow Avalanche Hazards in Canada. Canadian Avalanche Association, Revelstoke, BC.

Jóhannesson, T., Arnalds, þ., 2001. Accidents and economic damage due to snow avalanches and landslides in Iceland. Jökull 50, 81−94.

Jóhannesson, T., Gauer, P., Issler, D., Lied, K., 2009. The Design of Avalanche Protection Dams. Recent Practical and Theoretical Developments. Project Reports EUR23339. European Communities, Brussels, 205 pp.

Jonkman, S.N., 2007. Loss of Life Estimation in Flood Risk Assessment. Theory and Applications. Delft, 354 pp.

Jonkman, S.N., van Gelder, P., Vrijling, J.K., 2003. An overview of quantitative risk measures for loss of life and economic damage. J. Hazard. Mat. 99 (1), 1−30.

Jörg, P., Granig, M., Bühler, Y., Schreiber, H., 2012. Comparison of measured and simulated snow avalanche velocities. In: Koboltschnig, G., Hübl, J., Braun, J. (Eds.), 12th Congress INTERPRAEVENT. International Research Society INTERPRAEVENT, Grenoble, pp. 169−178.

Kaplan, S., Garrick, B., 1981. On the quantitative definition of risk. Risk Anal. 1 (1), 11−27.

Keiler, M., Zischg, A., Fuchs, S., Hama, M., Stötter, J., 2005. Avalanche related damage potential: changes of persons and mobile values since the mid-twentieth century, case study Galtür. Nat. Hazards Earth Syst. Sci. 5, 49−58.

Klinke, A., Renn, O., 2002. A new approach to risk evaluation and management: risk-based, precaution-based, and discourse-based strategies. Risk Anal. 22 (6), 1071−1094.

Kristensen, K., Harbitz, C.B., Harbitz, A., 2003. Road traffic and avalanches − methods for risk evaluation and risk management. Surveys Geophys. 24 (5−6), 603−616.

Künzler, M., Huggel, C., Linsbauer, A., Haeberli, W., 2010. Emerging risks related to new lakes in deglaciating areas of the Alps. In: Malet, J.-P., Glade, T., Casagli, N. (Eds.), Mountain Risks: Bringing Science to Society. Proceedings of the 'Mountain Risk' International Conference, November 24−26, 2010, Firenze, Italy. CERG Editions, Strasbourg, France, pp. 453−458.

Larsson-Kråik, P.-O., 2012. Managing avalanches using cost−benefit-risk analysis. Inst. Mech. Eng. Proc. Part F: J. Rail Rapid Transit 226 (6), 641−649.

Laternser, M., 2000. Der Lawinenwinter 1999. Fallstudie Goms (Kanton Wallis). Versorgungslage, Bewältigung der Krisensituation und wirtschaftliche Auswirkungen. Eidg. Institut für Schnee- und Lawinenforschung SLF, Davos.

Laternser, M., Schneebeli, M., 2002. Temporal trend and spatial distribution of avalanche activity during the last 50 years in Switzerland. Nat. Hazards 27, 201−230.

Lehmann, Th., Schaub, Y., 2013. Neue Gletscherseen im Alpenraum − Schaden- und Nutzen-potential für den Schweizer Tourismus. In: Thomas, Bieger, Pietro, Beritelli, Christian, Laesser (Eds.), Schweizer Jahrbuch für Tourismus. Erich Schmidt Verlag, Berlin, pp. 111−126.

Leiter, A.M., Pruckner, G.J., 2009. Proportionality of willingness-to-pay to small changes in risk: the impact of attitudinal factors in scope tests. Environ. Resour. Econ. 42, 169−186.

Margreth, S., 2004. Die Wirkung des Waldes bei Lawinen. Schutzwald und Naturgefahren. Forum Wissen 2004, 21−26.

Margreth, S., 2007. Defense Structures in Avalanche Starting Zones. Technical guideline as an aid to enforcement. Environment in Practice no. 0704. Federal Office for the Environment, Bern; WSL Institute for Snow and Avalanche Research SLF, Davos. 134 pp.

Margreth, S., Mattice, T., 2012. Re-evaluation of avalanche mitigation measures for Juneau. In: Proceedings International Snow Science Workshop "a merging of theory and practice". ISSW 2012, Anchorage, Alaska, pp. 150−156.

Margreth, S., Platzer, K., 2008. New findings on the design of snow sheds. In: Jóhannesson, T., Eiriksson, G., Hestnes, E., Gunnarsson, J. (Eds.), International Symposium on Mitigative measures against Snow Avalanches. Egilsstadir, Iceland, March 11−14, 2008, pp. 32−37.

Margreth, S., Romang, H., 2010. Effectiveness of mitigation measures against natural hazards. Cold Reg. Sci. Technol 64, 199−207.

Margreth, S., Stoffel, L., Wilhelm, C., 2002. Winter opening of high alpine pass roads − analysis and case studies from the Swiss Alps. In: Stevens, J. (Ed.), International Snow Science Workshop 2002, Penticton BC, Canada, pp. 59−66.

McClung, D., Schaerer, P., 2006. The Avalanche Handbook, 3rd ed. The Mountaineers. Seattle, USA.

Mears, A.I., 1992. Snow-Avalanche Hazard Analysis for Land-Use Planning and Engineering. Colorado Geological Survey Bulletin 49. Colorado Geological Survey, Dept. of Natural Resources, 55 pp.

Merz, B., 2006. Hochwasserrisiken − Möglichkeiten und Grenzen der Risikoabschätzung. Schweizerbart'sche Verlagsbuchhandlung.

Meyer, V., Becker, N., Markantonis, V., Schwarze, R., van den Bergh, J.C.J.M., Bouwer, L.M., Bubeck, P., Ciavola, P., Genovese, E., Green, C., Hallegatte, S., Kreibich, H., Lequeux, Q., Logar, I., Papyrakis, E., Pfurtscheller, C., Poussin, J., Przyluski, V., Thieken, A.H., Viavattene, C., 2013. Review article: assessing the costs of natural hazards − state of the art and knowledge gaps. Nat. Hazards Earth Syst. Sci. 13 (5), 1351−1373.

NELAK, 2013. In: Haeberli, W., Bütler, M., Huggel, C., Müller, H., Schleiss, A. (Eds.), Neue Seen als Folge des Gletscherschwundes im Hochgebirge − Chancen und Risiken. Formation des nouveux lacs suite au recul des glaciers en haute montagne − chances et risques. Forschungsbericht NFP 61. vdf Hochschulverlag AG an der ETH Zürich, Zürich, p. 300.

Nöthiger, C., Bürki, R., Elsasser, H., 2005. Naturgefahren und Schäden für den Tourismus in den Alpen. Geogr. Helv. 60 (1), 26−34.

Nöthiger, C.J., 2003. Naturgefahren und Tourismus in den Alpen untersucht am Lawinenwinter 1999 in der Schweiz. Eidgenössisches Institut für Schnee- und Lawinenforschung, Davos, 245 pp.

Olschewski, R., Bebi, P., Teich, M., Wissen Hayek, U., Grêt-Regamey, A., 2012. Avalanche protection by forests − a choice experiment in the Swiss Alps. For. Policy Econ. 15 (0), 108−113.

Papathoma-Köhle, M., Kappes, M., Keiler, M., Glade, T., 2011. Physical vulnerability assessment for alpine hazards: state of the art and future needs. Nat. Hazards 58 (2), 645−680.

Papathoma-Köhle, M., Keiler, M., Totschnig, R., Glade, T., 2012. Improvement of vulnerability curves using data from extreme events: debris flow event in South Tyrol. Nat. Hazards 64 (3), 2083−2105.

Renn, O., 1992. Concepts of risk: a classification. In: Krimsky, S., Golding, D. (Eds.), Social Theories of Risk. Praeger, London, pp. 53−79.

Rheinberger, C.M., 2011. A mixed logit approach to study preferences for safety on Alpine roads. Environ. Resour. Econ. 49 (1), 121−146.

Rheinberger, C.M., Bründl, M., Rhyner, J., 2009. Dealing with the white death: avalanche risk management for traffic routes. Risk Anal. 29 (1), 76−94.

Rheinberger, C.M., Romang, H., Bründl, M., 2013. Proportional loss functions for debris flow events. Nat. Hazards Earth Syst. Sci. 13, 2147−2156.

RiskScape, 2014. Easy-to-use Multi-Hazard Impact and Risk Assessment Tool.

Rudolf-Miklau, F., Sauermoser, S. (Eds.), 2009. Handbuch Technischer Lawinenschutz. Ernst & Sohn, Berlin, p. 464.

Salm, B., Burkard, A., Gubler, H.U., 1990. Berechnung von Fliesslawinen − Eine Anleitung für den Praktiker. Mitteilung Nr. 47. Eidg. Institut für Schnee- und Lawinenforschung SLF, Davos.

Sampl, P., Granig, M., 2009. Avalanche simulation with SAMOS-AT. In: International Snow Science Workshop 2009, Davos, Switzerland, pp. 519−523.

Sampl, P., Zwinger, T., 2004. Avalanche simulation with SAMOS. Ann. Glaciol. 38 (1), 393−398.

SATSIE, 2006. Avalanche Studies and Model Validation in Europe. http://www.leeds.ac.uk/satsie (last access 10.03.14).

Schaerer, P., 1989. The avalanche hazard index. Ann. Glaciol. 13, 241−247.

Schaub, Y., Bründl, M., 2010. Zur Sensitivität der Risikoberechnung und Massnahmenbewertung von Naturgefahren. Schweiz. Z. das Forstwes. 161 (2), 27−35.

Schaub, Y., Haeberli, W., Huggel, C., Künzler, M., Bründl, M., 2013. Landslides and new lakes in deglaciating areas: a risk management framework. The Second World Landslide Forum. In: Margottini, C., et al. (Eds.), Landslide Science and Practice, vol. 7, pp. 31−38. http://dx.doi. org/10.1007/978-3-642-31313-4_5.

Schläppy, R., Eckert, N., Jomelli, V., Stoffel, M., Grancher, D., Brunstein, D., Naaim, M., Deschatres, M., 2014. Validation of extreme snow avalanches and related return periods derived from a statistical-dynamical model using tree-ring techniques. Cold Reg. Sci. Technol. 99 (0), 12−26.

Schwendtner, B., Papathoma-Köhle, M., Glade, T., 2013. Risk evolution: how can changes in the built environment influence the potential loss of natural hazards? Nat. Hazards Earth Syst. Sci. 13 (9), 2195−2207.

Siegrist, M., Cvetkovich, G., 2000. Perception of hazards: the role of social trust and knowledge. Risk Anal. 20 (5), 713−719.

Sjoberg, L., 2000. Factors in risk perception. Risk Anal. 20 (1), 1−11.

SLF, 2000. Der Lawinenwinter 1999. Ereignisanalyse. Davos. Eidg. Institut für Schnee- und Lawinenforschung SLF, p. 588.

SLF, 2011. Avalanche Mitigation Study: Behrends Avenue Avalanche Path and White Subdivision Avalanche Path, Juneau, Alaska.

Slovic, P., 2000. The Perception of Risk. Earthscan, London.

Sovilla, B., 2002. Input Parameters for Dense Snow Avalanches Simulation. Course of avalanche dynamics, December 4−5, 2002, Bormio. Eidg. Institut für Schnee- und Lawinenforschung SLF, Davos.

Stallen, P.J.M., Geerts, R., Vrijling, H.K., 1996. Three conceptions of quantified societal risk. Risk Anal. 16 (5), 635−644.

Starr, C., 1969. Social benefit versus technological risk: what is our society willing to pay for safety? Science 165, 1232−1238.

Stoffel, L., 2001. Künstliche Lawinenauslösung. Praxishilfe, 2. überarbeitete Auflage. 53. Mitt. Eidg. Institut für Schnee- und Lawinenforschung. 66 pp.

Stoffel, M., Bollschweiler, M., 2008. Tree-ring analysis in natural hazards research − an overview. Nat. Hazards Earth Syst. Sci. 8 (2), 187−202.

Stoffel, L., Margreth, S., 2009. Artificial avalanche release above settlements. In: Schweizer, J., Van Herwijnen, A. (Eds.), International Snow Science Workshop. September 27 to October 2, 2009, Davos, Switzerland. Proceedings. Birmensdorf, Swiss Federal Institute for Forest, Snow and Landscape Research, pp. 572−576.

Stoffel, L., Schweizer, J., 2008. Guidelines for avalanche control services: organization, hazard assessment and documentation — an example from Switzerland. In: International Snow Science Workshop 2008, Proceedings. September 21–27, 2008. Whistler, BC, CAN, pp. 483–489.

Suda, J., Rudolf-Miklau, F. (Eds.), 2012. Bauen und Naturgefahren. Handbuch für konstruktiven Gebäudeschutz. Springer Verlag, Wien, New York, 510 pp.

Teich, M., Bebi, P., 2009. Evaluating the benefit of avalanche protection forest with GIS-based risk analyses — a case study in Switzerland. For. Ecol. Manage. 257 (9), 1910–1919.

Thywissen, K., 2006. Core terminology of disaster reduction: a comparative glossary. In: Birkmann, J. (Ed.), Measuring Vulnerability to Natural Hazards. United Nations University Press, pp. 448–496.

UNISDR, 2009. 2009 UNISDR Terminology on Disaster Risk Reduction. United Nations International Strategy for Disaster Reduction, Geneva.

Utelli, H.-H., Arnold, P., Hunziger, L., Gruner, U., Kipfer, A., Perren, B., Cajos, J., 2012. Management von gravitativen Naturgefahren auf Nationalstrassen in der Schweiz. Methodik Risikokonzept und dessen Anwendung. In: Koboltschnig, G., Hübl, J., Braun, J. (Eds.), 12th Congress INTERPRAEVENT. International Research Society INTERPRAEVENT, Grenoble, pp. 1115–1126.

Varnes, D., 1984. Landslide hazard zonation: a review of principles and practice. Nat. Hazards 3. UNESCO, Paris.

Voumard, J., Caspar, O., Derron, M.H., Jaboyedoff, M., 2013. Dynamic risk simulation to assess natural hazards risk along roads. Nat. Hazards Earth Syst. Sci. 13 (11), 2763–2777.

Vrijling, J.K., van Hengel, W., Houben, R.J., 1995. A framework for risk evaluation. J. Hazard. Mat. 43, 245–261.

Weir, P., 2002. Snow Avalanche Management in Forested Terrain. Res. Br., B.C. Min. For., Victoria., B.C. Land Manage. Handb. No. 55.

Whiteman, C.A., 2011. Cold Region Hazards and Risks. Wiley-Blackwell, Chicester, 366 pp.

Wilhelm, C., 1997. Wirtschaftlichkeit im Lawinenschutz Methodik und Erhebungen zur Beurteilung von Schutzmassnahmen mittels quantitativer Risikoanalyse und ökonomischer Bewertung. Eidg. Institut für Schnee- und Lawinenforschung, Davos, 309 pp.

Wilhelm, C., 1999. Kosten-Wirksamkeit von Lawinenschutz-Massnahmen an Verkehrsachsen 1999 Vorgehen, Beispiele und Grundlagen der Projektevaluation. Vollzug Umwelt. Praxishilfe. Buwal, Bern, 110 pp.

Wilhelm, C., Wiesinger, T., Bründl, M., Ammann, W., 2000. The avalanche winter 1999 in Switzerland — an overview. In: Int. Conf. Snow Science Workshop. ISSW, Big Sky, Montana, pp. 8–15.

Winkler, C., Burkard, A., 2005. Pflichtenheft Risikoanalysen Naturgefahren SBB. Interne Arbeitsgrundlage. Version 20.5.2005. Schweizerische Bundesbahnen SBB, Brig.

Zimmermann, M., 2013. Hazard assessment. In: Bollschweiler, M., Stoffel, M., Rudolf-Miklau, F. (Eds.), Tracking Torrential Processes on Fans and Cones. Advances in Global Change Research. Springer, Heidelberg, New York, London, pp. 343–353.

Zischg, A., Fuchs, S., Keiler, M., Stötter, J., 2005. Temporal variability of damage potential on roads as a conceptual contribution towards a short-term avalanche risk simulation. Nat. Hazards Earth Syst. Sci 5 (2), 235–242.

Chapter 10

Permafrost Degradation

Dmitry Streletskiy [1], Oleg Anisimov [2] and Alexander Vasiliev [3]

[1] *Department of Geography, George Washington University, Washington DC, USA,* [2] *State Hydrological Institute, St. Petersburg, Russia,* [3] *Earth Cryosphere Institute RAS, Moscow, Russia*

ABSTRACT

Climatic changes over the last 50 years resulted in a decrease of permafrost extent, an increase of permafrost temperature, and deepening of the active layer in numerous locations across the Arctic and High Mountainous environments. Permafrost degradation poses serious impacts ranging from local changes in topographic and hydrologic conditions, impacts on infrastructure and sustainability of northern communities, changes to vegetation and wildlife dynamics, and to global impacts on climate system. Hazards associated with permafrost degradation are exacerbated in areas of human activities, especially in large settlements with developed infrastructure in the Arctic. Unlike smaller communities, which have higher mobility, large population centers have to build in situ adaptive capacity to face environmental changes. Permafrost degradation can have severe socioeconomic consequences as most of the existing infrastructure will require expensive engineering solutions to maintain economic activities on permafrost.

10.1 INTRODUCTION

Permafrost plays an important role in global climate change, in the balance of greenhouse gases, arctic environment ecosystems, and human activities in the Polar Regions (ACIA, 2005; AMAP, 2011; Hinzman et al., 2005; Romanovsky et al., 2010a; Shakhova et al., 2010). Changes in climatic parameters, particularly the air temperature, snow depth, and duration of the warm period over the last 50 years have resulted in an increase of permafrost temperature and deepening of the active layer in numerous locations across the Arctic. Several locations along the southern permafrost boundary have lost permafrost completely, whereas in others, upper portions of the permafrost have thawed to depths below that of seasonal freezing.

Permafrost degradation poses serious impacts ranging from local changes in topographic and hydrologic conditions (Hinzman et al., 2005; Shiklomanov and Nelson, 2013; Shur et al., 2005; Woo, 2012), impacts on infrastructure and sustainability of northern communities (Anisimov and Reneva, 2006; Anisimov et al., 2010; Grebenets et al., 2012; Nelson et al., 2001), and changes to vegetation and wildlife dynamics (Jorgenson et al., 2013), and to global impacts on changes in greenhouse gas emissions (Tarnocai et al., 2009; Wisser et al., 2011; Zimov et al., 2006). Permafrost degradation is a spatially heterogeneous process, meaning that permafrost characteristics, such as temperature, thickness, or extent, may react differently in different climatic zones.

Multiple indications exist that modern climatic changes in the 20th to the early twenty-first century have caused permafrost degradation over the large areas in North America and Eurasia (Romanovsky et al., 2010a). From the geological prospective, this is not unique. Climatic transitions between glacial and interglacial periods have always been associated with changes in permafrost. The last glacial maximum of Late Pleistocene 20,000−18,000 years before present is very illustrative for understanding the relation between climate and alternative glaciation forms, that is, glaciers and permafrost. At this time, the Arctic was almost 20 °C colder, and the sea level was 120−130 m lower, leading to permafrost aggradation on exposed continental shelves of the Arctic seas (Figure 10.1). Much of the territory that received sufficient amounts of precipitation was covered by glaciers, including the zone from the Appalachians to the Rocky Mountains in North America, Greenland, Arctic islands, Ireland, the British Isles, Scandinavia, Polar Urals, and Taimyr. In contrast to this, territories with cold and dry climate, including parts of France, Germany, Poland, the Czech Republic, Hungary, Ukraine, the majority of the European part of Russia, southwestern Siberia, and northern Kazakhstan, were occupied by permafrost until around 12,000−18,000 years before present (French et al., 2009; Romanovskii and Hubberten, 2001; Romanovskii et al., 2005; Velichko and Faustova, 2009). During the early Holocene climatic optimum (8,000−6,000 years before present) permafrost retreated northward in North America (Zoltai, 1995), and disappeared in Europe and substantial areas of Northern Eurasia, including in large portions of West Siberia south of the Arctic Circle. Permafrost did not thaw completely in areas with high ice content and is still present at great depths (so-called "relict" permafrost). Subsequent cooling (5,000−3,000 years before present) resulted in permafrost aggradation that allowed the new Holocene permafrost to reach the top of the old (Pleistocene) permafrost located north of the Arctic Circle. In locations where Pleistocene permafrost thawed to a depth of 150−200 m, the newly formed Holocene permafrost was not able to reach the old permafrost. This created three-layered permafrost profiles, in which relatively young Holocene permafrost is separated from the Late Pleistocene permafrost by a zone of unfrozen ground. Smaller advances and retreats of the permafrost boundary occurred during the last 3,000 years. The last notable permafrost advance corresponded to the Little Ice Age (1550−1850 AD), when temperatures were about 1 °C colder than in the present, resulting in permafrost aggradation up to 25 m thick and the advance of permafrost to locations further south from the present terminus. Permafrost from the Little Ice Age occurs in

(a)   **(b)**

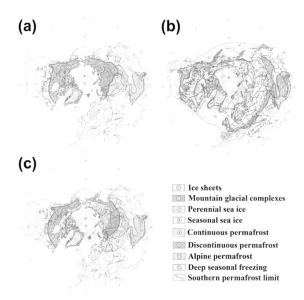

(c)

☐ Ice sheets
☐ Mountain glacial complexes
☐ Perennial sea ice
☐ Seasonal sea ice
☐ Continuous permafrost
☐ Discontinuous permafrost
☐ Alpine permafrost
☐ Deep seasonal freezing
☐ Southern permafrost limit

FIGURE 10.1   Reconstructions of permafrost extent (Velichko and Faustova, 2009; Velichko and Nechaev, 2009): (a) Riss-Würm interglacial 125 ka BP, (b) Glacial maximum 20−18 ka BP, (c) Holocene optimum 6−8 ka BP.

locations where a thick peat layer has been able to preserve it despite warmer air temperatures.

The difference in present permafrost changes and the past events is threefold. First, the rate of contemporary climate change is unprecedented. Air temperatures in selected locations in the Arctic have increased up to 5 °C since the early twentieth century (Anisimov and Vaughan, 2007). Second, the impact of climate change on permafrost is exacerbated by technogenic factors and land use. Lastly, the current situation differs dramatically from the past due to the presence of substantial populations in regions currently occupied by permafrost. Economic activities and infrastructure development in cold regions are largely affected by the presence of permafrost. Whether constructing a cableway in the Swiss Alps (Haeberli, 1992), a railroad on the Tibetan Plateau (Ma et al., 2011), or a metallurgy plant in Siberia, changing permafrost conditions should be taken into account. Failure to do so may result in the deterioration of the natural environment resulting in hazardous conditions for human life and infrastructure.

To understand the hazards associated with permafrost degradation and their impacts on infrastructure, society, and environment, we first need to understand what makes the permafrost system unique, and what factors are responsible for its changes. Physical and mathematical representations of these factors allow for the construction of permafrost models capable of scaling local observations collected over limited observational areas to larger regions.

Combined with climate models, they are used to predict future changes to permafrost and to evaluate the impacts of these changes on other natural, human, and economic systems.

10.2 PERMAFROST AND RECENT CLIMATE CHANGE

Recent climatic changes are pronounced in permafrost regions. Since 1980 the Arctic has been warming at approximately twice the global rate, demonstrating the strongest temperature changes ($\sim 1^\circ$C/decade) in winter and spring, and the smallest changes in autumn (AMAP, 2011; IPCC, 2013). In the period 1976–2012, spring temperatures over large regions in Siberia and Chukotka have been rising by 0.8–1.2 °C per decade (Roshydromet, 2014). Sea ice declined at an unprecedented rate throughout all seasons reaching the absolute minimum of 3.41 million square kilometers in September 2012, which is 18 percent lower than in 2007, when the previous record of 4.17 million square kilometers was recorded (Jeffries et al., 2012). The Arctic Ocean is projected to become nearly ice-free in summer within this century, whereas some models suggest it may happen within the next thirty to forty years (IPCC, 2013). According to AMAP (2011), the duration of snow-cover extent and snow depth are decreasing in North America while increasing in Eurasia. These changes have important implications for permafrost. Since the late 1970s, permafrost temperatures have increased typically between 0.5 and 2 °C, with warming rates being much smaller for warm, ice-rich permafrost at temperatures close to 0 °C than for colder permafrost or bedrock (Romanovsky et al., 2010a).

Changes in permafrost are exacerbated in areas of human activities (Anisimov et al., 2010; Raynolds et al., 2014; Streletskiy et al., 2012a). Around 370 settlements exist in the Arctic tundra zone. Although these settlements are relatively small in most parts of the Arctic, several cities in the Russian Arctic have populations of >100,000 people. Thus, permafrost warming can have severe socioeconomic consequences as most of the existing infrastructure will require expensive engineering solutions to stabilize foundations on permafrost. Intensification of coastal erosion, mass wasting, and thermokarst processes in low-lying areas are likely to reshape tundra landscapes, with negative effects on northern communities.

10.3 PERMAFROST OBSERVATIONS AND DATA

A major milestone in permafrost investigations was the creation of the Global Terrestrial Network on Permafrost (GTN-P). GTN-P was created in 1999 within the framework of the Global Climate Observing System/Global Terrestrial Observing System in support of the United Nations Framework Convention on Climate Change as a network of permafrost observatories to obtain a set of standardized temperature measurements in all permafrost regions of the planet to provide a baseline for temperature change assessments and data for validation of climatic models. Two components of GTN-P, the Circumpolar Active Layer Monitoring (CALM) program and the Thermal State of Permafrost currently serve as the major providers of permafrost and active-layer data (Figure 10.2) (Brown et al., 2000; Romanovsky et al., 2010a; Shiklomanov et al., 2012).

Despite the growing observational network and data rescue attempts, permafrost continues to be a data-limited science. Growing computational resources and available geospatial and remote sensing techniques allow permafrost scientists to scale limited local observations to large geographical areas. In order to do so, however, they first need to know what drives permafrost changes.

10.4 DRIVERS OF PERMAFROST AND ACTIVE LAYER CHANGE ACROSS SPACE AND TIME

Interactions between permafrost and the factors responsible for its dynamics are complex, and depend on many interrelated physical processes as well as on geographic scale. Climatic factors play a dominant role in explaining the spatial variability of the permafrost and active layer over large areas, whereas nonclimatic factors can be assumed to be represented by generalized vegetation and soil classes. Over small areas, the leading role in spatial variability is attributed to edaphic factors, such as soil−vegetation associations changing considerably over areas of similar climatic conditions. This explains why, at small geographical scales, the broad spatial pattern of permafrost parameters are closely related to climate: permafrost temperature and thickness, together with the active layer, vary systematically along the climatic gradient. In similar soil−vegetation units, active-layer thickness (ALT) increases from <15 cm in the High Arctic up to 1.5−2.0 m in the southern reaches of the permafrost zone. Permafrost thickness generally decreases from the north to the south, whereas its temperature increases following the same gradient.

These general geographic trends can be modified significantly by the influence of more localized factors such as topography, vegetation, soil type, moisture, and snow redistribution. For example, even at the southern fringes of permafrost distribution, taller vegetation with denser plant canopies and thicker organic horizons can counter the effects of warmer temperatures, explaining why there can be little correspondence between climatic factors and permafrost temperature and active layer even within small areas.

10.4.1 Role of Climate: Air Temperature and Precipitation

On an annual basis, the temperature at the top of permafrost and ALT will vary according to the particular climatic conditions of a given year. The majority of studies confirm that seasonal variations of ALT are controlled by the duration of the warm period, fluctuations in accumulated degree-days of thawing, and

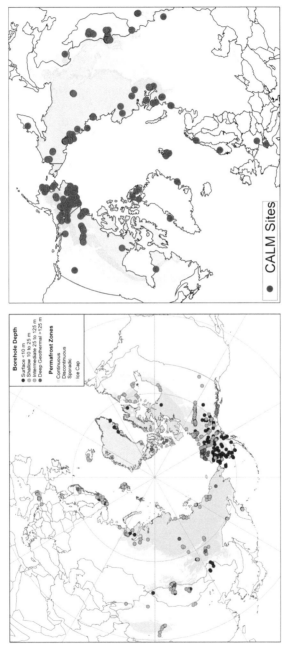

FIGURE 10.2   Location of GTN-P monitoring sites: Left—temperature boreholes, right—active-layer sites.

by landscape features (Brown et al., 2000). A strong relationship between thawing degree-days and ALT has been demonstrated by several studies showing that warm summers on average produce a deeper active layer and cold summers produce shallow depths of thaw (Hinkel and Nelson, 2003; Romanovsky and Osterkamp, 1997; Streletskiy et al., 2008). Heavy rain in midsummer is a warming factor that promotes thaw through the convective warming of permeable soils, whereas scattered showers in the fall are a cooling factor that suppresses active-layer thickening during the next summer by increasing the ice content at the base of the active-layer (Melnikov et al., 2004). However, the role of summer precipitation relative to temperature is less significant. Although summer temperatures can be quite variable from year to year explaining the large temporal variability of ALT, permafrost temperature is a function of mean annual temperature and its amplitude and hence a more conservative characteristic, since inter-annual variations are smaller than seasonal variations from year to year.

Over a period of several years, both ALT and permafrost temperature will approach the climatic mean. If climate change exceeds the established equilibrium, the permafrost will adjust accordingly to the new conditions, establishing a new equilibrium in the land—atmosphere system. Substantial evidence exists that the Arctic is experiencing an unprecedented degree of warming (IPCC, 2013; Serreze and Francis, 2006). Due to the attenuation of the climatic signal with depth and differences in ground conditions, particularly with the presence of ice and unfrozen water, the reaction of the permafrost to warming can be quite variable, which is reflected in the observed permafrost trends.

10.4.2 Role of Topography

Relief is an important factor that influences the temperature regime within the ground as it is a prime factor of heat-exchange differentiation in the atmosphere—ground system, especially in mountain environments. The amount of solar radiation reaching the Earth's surface depends on elevation, slope, and exposure. Normal environmental lapse rates lead to decreases of air temperature by about 6.5 °C per 1 km of altitude, which leads to a decreased heat flux, lowered permafrost temperature, and shallower ALT. Therefore, a general decrease in permafrost temperature and ALT with elevation is expected if other factors are constant. However, local wind circulation (e.g., foehn winds and cold-air drainage) can substantially change the altitudinal distribution of temperature. Moreover, temperature inversions are a common phenomenon in Arctic regions during the cold season, so actual environmental lapse rates can significantly differ from the normal ones. The difference between ground temperatures on north- and south-facing slopes vary, depending on the slope gradient. Slope aspect also influences the variation of the ground temperature amplitude. Terrain geometry also exerts an important control over ALT through its influence on the spatial variability of soil moisture and vegetation. Repelewska-Pekalowa and Pekala (2004) found that high values of ALT in Spitsbergen were associated with terrain units (primarily concave hillslope segments) in which the movement of surface or ground water produced convective heat transfers. The presence of microtopographic relief features at large geographical scales increases the variability of snow-cover distribution, leading to a delay of ground thawing in depressions as compared to thawing in blowouts (Voitkovskiy, 1999).

Leveling of topography prior to economic development leads to the alteration of heat and moisture exchange and can lead to the development of subsidence and thermokarst, particularly in areas with ice-rich permafrost. Thermal erosion and solifluction rates increase under warmer climatic conditions, which may further be exacerbated if measures are not taken to protect vegetation on slopes. Decreased slope stability resulting from the melting of ground ice, decreased soil cohesion, and deeper water infiltration under a warmer climate is a major concern in mountain environments and is discussed in detail in the corresponding chapter of this volume (Deline et al., 2014) and elsewhere (Bommer et al., 2010; Gruber and Haeberli, 2007; Harris and Isaksen, 2008; Kääb et al., 2007).

Substantial alterations to topography in permafrost regions are found in areas of mining. Mining activities on permafrost commonly result in the dumping of crushed rocks near the extraction sites, which have the potential to form technogenic rock glaciers. The largest technogenic rock glacier is located in the Norilsk area as a byproduct of mining activities that lasted from 1960 to 1984. It is located on the northern slope of Rudnaya Mountain, up to 120 m thick, and has a volume of 60 million cubic meters or about 110 million tons (Grebenets et al., 1998). Rain and snowmelt water percolation through the dump site and its subsequent refreezing, the overall increase in permafrost temperature, the decreased cohesion of the dump and bedrock slope, all resulted in the displacement of the frontal side of the dump body with a speed that has reached 60–80 mm/day. About 60 percent of the rock glacier collapsed into the nearby Medvezhy Ruchei River destroying the road and water pipeline supplying the city in 1996. The permafrost creep (Haeberli et al., 2006) still poses a danger to the city of Norilsk and numerous examples of other such technogenic rock glaciers exist, particularly on the Kola Peninsula. Smaller features called either "frozen debris lobes" or rock glaciers in the Brooks Range of Northern Alaska were recently found to have accelerated movement, which creates a potentially hazardous situation for the *Trans*-Alaskan Pipeline and Dalton Highway operations (Daanen et al., 2012).

10.4.3 Role of Vegetation and Mosses

Vegetation cover acts as a thermal insulator between the atmosphere and the ground and can play a substantial role in the redistribution of snow cover. One study involving 17 sites along a North Alaskan bioclimatic gradient has shown that increased plant biomass along the gradient acts as a negative feedback to increases in ALT, because of the insulative effect of vegetation and the highly organic soil horizons. This relationship underscores the importance of vegetation on the development of the active layer and near-surface permafrost (Walker et al., 2003).

A significant role in heat exchange at the surface in Arctic environments is attributed to moss cover. Generally, its presence leads to lower mean annual temperatures, as moss has a low thermal conductivity in summer and a high thermal conductivity of the frozen material in winter facilitating the effective cooling and storage of cold within the permafrost. The evaporative regime of the moss layer also promotes cooler summer temperatures (Riseborough and Burn, 1988). The addition of a 10-cm layer of moss resulted in almost a 3 °C decrease in the mean summer soil temperature and a 15 percent reduction in ALT in northern Alaska (Kade and Walker, 2008). Mosses dominate the surface cover in high northern latitudes and have the potential to play a key role in modifying the thermal and hydrologic regime of Arctic soils. These modifications in turn feed back into surface energy exchanges and hence may affect regional climate. Beringer et al. (2001) concluded that the addition of a surface layer of moss underlain by peat and loam had a substantial impact on modeled surface processes. They found that the thermal conductivity of the top layer consisting of moss, lichen, and peat in experiments was only one-quarter of that for sand and loam, because of the high air volume and lower water content. The addition of a surface moss layer resulted in higher simulated winter soil temperatures and lower summer temperatures. Decrease of organic layer due to forest fires may trigger long-term changes in permafrost stability as regrowth of the moss takes a considerable number of years (Burn, 1998; Jafarov et al., 2013). Removal of vegetation including the moss layer caused by economic development or forest fires leads to an increased heat flux in the ground during summer months, increased active layer and ground temperature, melting of ground ice and subsidence and may lead to the development of thermokarst terrain.

10.4.4 Role of Snow

It is difficult to overstress the influence of snow cover on the heat balance at the ground surface in cold environments. Variations in the duration, thickness, accumulation and melting processes, structure, density, and thermal properties have significant impacts on the insulating effect of seasonal snow cover (Zhang et al., 1996). Variations in permafrost temperatures are largely attributed to changes in snow-cover thickness. A recent study found that in Siberia changes in the mean annual ground temperature (MAGT) are more dependent on snow-cover thickness than on changes in air temperature (Sherstyukov, 2008). Late-lying snow cover in the summer increases albedo

and therefore decreases the surface temperature. In spring, melting snow delays the warming of the soil surface while it is close to $0\,°C$, even if the air
temperature is positive (Romanovsky and Osterkamp, 1997). Interception and
sublimation of snow strongly control snow accumulation in forest environments (Pfister and Schneebeli, 1999). For example, increases of 30−45 percent
in seasonal snow accumulation have been measured after the removal of
evergreen forest cover by clear-cutting at sites across Canada. Snowmelt under
forest canopies is very different from that in open environments because the
overlying canopies intercept radiation and suppress turbulent transfer. As a
result, melting rates are lower in forests than in equivalent open areas
(Woo, 2012). Tirtikov (1978) found that beneath trees and tall shrubs in the
forest tundra of West Siberia, where about 1 m of snow accumulates, ground
temperatures were 3−5 °C higher than in the surrounding tundra. However,
Smith (1975) found that in some locations in the Mackenzie Delta the
MAGT was raised above $0\,°C$ because of the higher amounts of snow accumulation in areas of low willow shrubs as compared to those with tall willow
shrubs and sites without vegetation. These examples show a complexity of
snow−vegetation interactions in modulating the ground thermal regime. Based
on the results from long-term monitoring of the insulating effect of different
landcover types in Central Yakutia, Varlamov (2003) concluded that the
greatest insulating effect of snow occurs on wet and oversaturated surfaces,
while it is minimized on dry sandy soils. This fact is attributed to the higher
rates of heat exchange associated with saturated soils. Snow also plays a role
in thaw depth development. Experimental data from Igarka and Yakutsk show
that in areas where snow was removed after each snowfall event during the
winter, ALT in the following summer was from 5 to 10 percent lower
compared to areas with undisturbed snow cover, since variations in ALT
depend on the MAGT, which in turn depends on snow-cover conditions.

   Redistribution of snow is a common practice in areas of human activities in
cold environments. Roads, entrances to residential houses, industrial facilities,
parking lots, and airports all require snow removal in order to be operational.
At the same time, certain areas are used to store the removed snow. Areas of
snow removal and storage occupy the same locations from year to year,
resulting in significant alterations to the ground thermal regime relative to
natural conditions. Piling snow in the same locations is likely to result in
warmer permafrost temperatures over a several-year period, but snow
compaction may offset the thermoinsulating effect of thicker snow. Moreover,
late-lying snow cover increases albedo and can decrease surface temperature,
further offsetting the warming impact of thicker snow. The areas of snow
removal lack this natural seasonal thermoinsulating layer and have winter
temperatures similar to those of the air. Not only does the removal of snow
decrease the temperature of permafrost resulting in permafrost aggradation
under the roads but it also increases the amplitude of temperature variations
leading to the intensification of frost cracking and cryogenic weathering. This

leads to a much faster destruction of construction materials and to the deterioration of road networks in permafrost-dominated areas.

A major part of the soil's moisture comes from melted snow (Sokratov and Barry, 2002). Areas of artificial snow accumulation, especially along roads, are subjected to water logging. Roads act as frozen dams with little or no infiltration capacity even by the end of the warm period of the year. Standing water along the roads in permafrost environments is one of the biggest problems requiring constant attention, as the standing water accumulating and heat retained during warm periods leads to the development of thermokarst topography.

10.4.5 Role of Soil Properties

The ability of permafrost to transfer and hold heat depends on thermal conductivity and heat capacity, which change depending on soil texture and water content (Romanovsky and Osterkamp, 1997). Under dry conditions, thermal conductivity increases along the textural sequence clay—loam—silt—sand—indurated rock, leading to much thicker permafrost in bedrock relative to fine-grained sediments under similar climatic conditions. Higher soil-moisture content requires more latent heat for water phase transitions than does low moisture content leading to a thinner active layer. On the other hand, increases in moisture content during the summer or winter (without phase transition) increases soil-thermal conductivity, because air with low thermal conductivity (0.023 W/m$^{\circ}$C) replaced by water with a higher conductivity (0.57 W/m$^{\circ}$C), or ice (2.29 W/m$^{\circ}$C) leads to the deeper propagation of freeze/thaw (Yershov and Williams, 1998). The latter case was supported by a hydrological manipulation experiment in the drained lake basin near Barrow, Alaska. ALT in the experimentally drained area was found to be 25 percent less than in the experimentally flooded area (Shiklomanov et al., 2010), which shows the importance of the moisture regime and surface hydrology in the variability of ALT.

Development of construction sites on permafrost begins with the construction of pads. Artificial pads commonly have soil characteristics that are different from natural landscape settings. Large logistic costs associated with the transportation of raw materials commonly result in the dragging of local streams in order to obtain the necessary sand and gravel. Corresponding changes in ground properties may result in changing permafrost thermal conditions and hydraulic conductivity resulting in increased water infiltration and warming permafrost during summer months.

Technogenic salinization and water logging is another problem facing some of the cities located on permafrost, particularly those with developed mining and metallurgy industries, such as Norilsk and Vorkuta. For example, the salinization of soils in Norilsk is reaching up to 21 mg/l near the nickel plant (Grebenets et al., 2012). Technogenic salinization is not only leading to decreases in the stability of infrastructure through increases in ALT and the decreased freezing temperature of the soil but it also directly affects foundations through the corrosion of metal and concrete within the active layer.

10.5 OBSERVED PERMAFROST AND ACTIVE-LAYER CHANGES

Changes in the air temperature and snow depth over the second half of the twentieth century have resulted in permafrost degradation, which is evident from the limited observations of permafrost temperatures summarized in Romanovsky et al. (2010a). It was concluded that following air temperature trends, permafrost has been experiencing warming in the majority of sites in the Arctic since the 1970s, but that this warming is not uniform (Figure 10.3). Large areas occupied by permafrost are still not monitored, so the conclusions are still based on relatively few and for most places rather short time series.

Romanovsky et al. (2010b), based on analysis of a large observational data set, estimated that permafrost has warmed from 0.5 to 2 °C over the last 20 to 30 years in the Russian Arctic. The European Russian North was characterized by warming trends of about 0.01–0.04 °C/year in the western portion (Malkova, 2010) and up to 0.08 °C/year in the eastern portion (Oberman, 2008). The northward retreat of permafrost by 30–40 km in the Pechora lowland and up to 70–100 km in the Foothills of the Ural Mountains occurred during 1970–2005 (Oberman and Shesler, 2009). The highest rates of permafrost warming occurred in peatlands, and the lowest rates were in sediments composed of loams. A similar situation has been characterized for the forest tundra of Western Siberia, where the highest observed warming trends were attributed to frost mounds covered by peat (0.04 °C/year). In locations where permafrost temperature is close to 0 °C, no changes were registered for the entire period of measurements (1975–2009), as important exchange of latent heat melting was already taking place in the ground. Permafrost thawing was also observed over the last 30 years in the southern forest tundra of the Urengoy gas-condensate field (Vasiliev et al., 2008). Both in West Siberia and in the European Russian North, temperature changes were observed to be more significant in colder permafrost sites. Northern Yakutia experienced permafrost warming up to 1.5 °C since 1980 in the eastern part; however, sites located in the west did not show a significant warming trend until recently. Smaller warming trends were observed near Chara (Baykal region), ranging from 0.025 to 0.04 °C/year.

Even higher permafrost changes are observed in some locations of North America (Smith et al., 2010). Since the 1980s, permafrost was recorded as having warmed 0.3–1.0 °C in Interior Alaska, 1–2 °C in the Arctic Foothills, and up to 3–4 °C in the Arctic Coastal Plain (Clow, 2008). Burn and Zhang (2009) estimated that the permafrost on Herschel Island warmed only 2 °C in the last 100 years, showing that permafrost warming has accelerated in recent

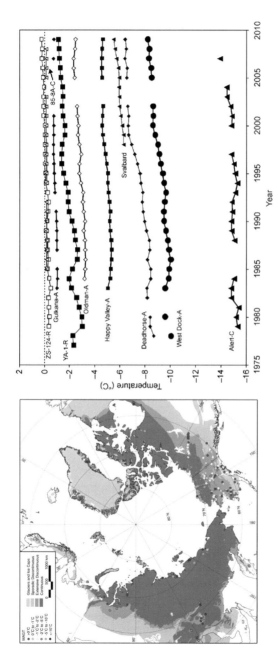

FIGURE 10.3   Left: Mean annual ground temperature (MAGT) snapshot. Right: time series of MAGTs at depths between 10 and 20 m for boreholes throughout the circumpolar northern permafrost regions. *From Romanovsky et al. (2010a).*

decades. Canadian High Arctic permafrost has warmed at a rate of 0.03−0.05 °C/year (Taylor et al., 2006) and up to 0.1 °C/year on Ellesmere Island. Warming in the western part of the North American Arctic has been underway since the 1970s, but it has slowed down in the last decade. Meanwhile, the eastern part is still undergoing warming trends that started in the early 1990s (Smith et al., 2010). A 130-km northward retreat of the southern permafrost boundary over the last 50 years was reported in the Quebec province of Canada (Thibault and Payette, 2009).

Permafrost temperature is significantly increasing in Scandinavia (Christiansen et al., 2010). Harris and Isaksen (2008) analyzed permafrost temperatures in the European Arctic and observed that permafrost temperature increase is characteristic for three deep boreholes located along a latitudinal transect through southern Norway, northern Sweden, and in Svalbard. At a 30-m depth warming rates were about 0.025−0.035 °C/year. Higher rates are characteristic for the top of the permafrost (0.04−0.07 °C/year) with the greatest changes attributed to permafrost on Svalbard. Monitoring sites located in Southern Norway experienced permafrost warming about 0.015−0.095 °C/year (Isaksen et al., 2011); permafrost temperature is also increasing in Sweden (Johansson et al., 2011).

Central Asia is also characterized by increased permafrost temperatures and ALT (Zhao et al., 2010). The highest rate of permafrost temperature increase in the region is characteristic for sites located on the Qinghai−Tibet Plateau and is 0.06 °C/year. Reported ground temperature increases in the Tian Shan Mountains were considerably less over the last 25 years (0.012−0.025 °C/year). Sharkhuu et al. (2008) reported that the increase of permafrost temperature in Mongolia was higher over the last 15−20 years compared to that in previous years, averaging at a rate of about 0.015 °C/year.

Examples from various regions show that near-surface permafrost is quite sensitive to observed climate change. In general, colder permafrost undergoes faster warming rates compared to warmer permafrost; however, large heterogeneity persists both spatially and temporarily, even within relatively small regions (Figure 10.3). Colder temperature permafrost has less unfrozen water, which allows for more effective heat conduction. As climate shifts to warmer temperatures, permafrost temperature increases accordingly, until a new climate−permafrost equilibrium is reached. Increasing the amount of unfrozen water decreases the ability of permafrost to transfer heat. When permafrost is close to the melting point of ice, a significant amount of latent heat is required to convert the water from frozen to liquid state, requiring substantial time to thaw permafrost rather than warm it. This explains why colder permafrost warms at a faster rate than permafrost that is close to the melting point. Other environmental and technogenic factors can add to the current permafrost warming trend, such as an increase in snow-cover depth or the removal of vegetation.

Permafrost degradation occurs differently in the continuous (cold) permafrost zone and the discontinuous permafrost zones (warm). Degradation of the continuous permafrost zone is expressed as an increase in permafrost temperature and a corresponding decrease in permafrost thickness in order to establish a new equilibrium. Although the adjustment of the permafrost temperature as measured at the depth of zero mean annual amplitude (10−20 m) is a relatively fast process (e.g., lags only several years behind air temperature changes), permafrost that thaws from the bottom takes much longer (hundreds of years).

Degradation of the discontinuous permafrost zone is expressed as a slow increase in temperature until the melting point, after which it may take decades to hundreds of years to thaw the permafrost depending on ground composition and ice content. Sporadic and island permafrost zones in the Arctic are largely controlled by nonclimatic factors, such as the presence of peat.

Increases in the ground temperature in the majority of permafrost regions were accompanied by increases in ALT (Shiklomanov et al., 2012); however, the magnitude was variable even within small regions (Figure 10.4). Significant increases in ALT were reported in Greenland and Svalbard, but are not spatially uniform (Christiansen et al., 2010). A progressive increase in ALT since 1970 resulted in the disappearance of permafrost in several mire landscapes in the Abisco area (Sweden) (Åkerman and Johansson, 2008). A similar situation occurred in the Russian European North, where a progressive increase of ALT was observed during 1996−2013 (Kaverin et al., 2012; Malkova, 2010), and in the Alaskan Interior. Central and Eastern Canada (Smith et al., 2010) and Central Asia (Zhao et al., 2010) also underwent increases in ALT. Sites located in West Siberia, on the North Slope of Alaska, and in Western Canada did not show long-term increasing trends but over the last several years have experienced slightly deeper depths of annual thaw (Smith et al., 2009; Streletskiy et al., 2008; Vasiliev et al., 2008). Spatially distributed long-term monitoring of thaw subsidence showed that an apparent lack of active-layer thickening may be attributed to melting of segregation ice at the bottom of the active layer due to consolidation of ice-rich permafrost (Shiklomanov et al., 2013).

10.6 PERMAFROST MODELING AND FORECAST

Geocryology remains a data-limited science with relatively small amounts of ground-truth data. Development of computational and modeling techniques have resulted in a dramatic increase in the number of permafrost models and in the variety of their applications (Anisimov and Reneva, 2006, 2011; Arzhanov et al., 2013; Goodrich, 1978; Koven et al., 2013; Malevsky-Malevich et al., 2001; Nelson and Outcalt, 1987; Shiklomanov and Nelson, 2002; Stendel and Christensen, 2002; Streletskiy et al., 2012b).

The simplest approaches are based on several variations of the analytical Stefan solution to the heat-conduction problem with phase change. Several variations of a more comprehensive model, based on the adaptation of the

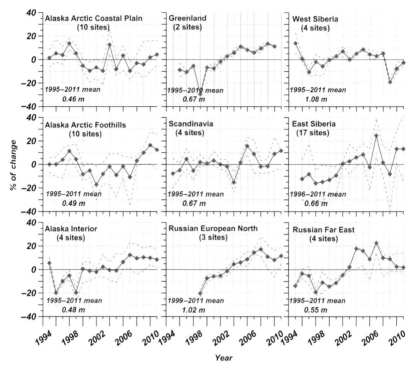

FIGURE 10.4 Active-layer change in nine different Arctic regions according to the Circumpolar Active Layer Monitoring program. The data are presented as annual percentage deviations from the mean value for the period of observations (indicated in each graph). *Figure from Shiklomanov et al. (2012).*

analytical solution by Kudryavtsev et al. (1974) to the geographic context, have been developed and used with the aid of a Geographic Information System (GIS) technology to calculate both ALT and MAGTs at regional and circumpolar scales (Anisimov and Reneva, 2006; Anisimov et al., 1997; Sazonova and Romanovsky, 2003; Shiklomanov and Nelson, 1999; Slater and Lawrence, 2013). The input data for these models include mean annual air temperature, and its annual amplitude, the thickness and thermal properties of ground cover (snow and vegetation), soil-thermal properties, and soil-moisture content (Figure 10.5). More complex models include dynamic interactions between permafrost and other components of the environment, but require more rigorous parameterization, which is largely unavailable.

Although the complexity of parameterization ranges from very simple, heat-conduction solutions to fully coupled transient Global Climate Models (GCMs), all models represent permafrost distribution depending on the nature of the climatic forcing. Climate inputs required to run these models come in the form of gridded data sets, which are created through the interpolation of point observations. Differences in data sources, interpolation, and validation

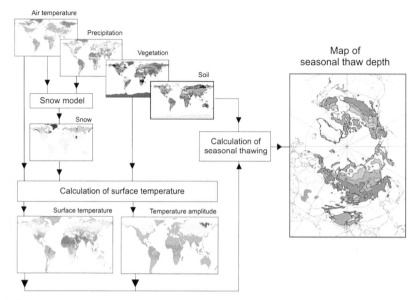

FIGURE 10.5　Computational design of the hemispheric-scale permafrost model. *From Anisimov and Reneva (2006).*

techniques used to construct the data sets result in differences in the resulting climatic fields. For example, the differences between commonly used observational data sets and reanalysis may reach about 1 °C over the Arctic and 10 times that within a relatively small area within a given grid (Streletskiy et al., 2012b). Projected climate changes have much higher uncertainty ranges, as some models produce warmer or colder temperatures relative to the others.

Slater and Lawrence (2013) compared the ability of numerous state-of-the art climate models (CMIP5) used by IPCC to portray present permafrost characteristics and tested their ability to predict future permafrost changes. The authors found that significant biases of temperature and snow depth representations lead to large deviations from the present permafrost extent (ranging from 4 to 25 million square kilometers, Figure 10.6). That is why scientists commonly apply the so-called ensemble approach that uses a combination of several climate models.

Despite large differences in the extent and rate of changes of permafrost, all models agree that projected warming and increases in snow thickness will result in near-surface permafrost degradation over large geographic areas (Koven et al., 2013). The sensitivity of permafrost to climate change is between 0.8 and 2.3 million square kilometers per degree centigrade meaning that an increase of global temperature by 1° corresponds to a decrease in near-surface permafrost area equal to the size of Mongolia or even Greenland. The combination of local factors such as snow redistribution, vegetation, and hydrological and soil properties may offset or increase the rate, but are unlikely

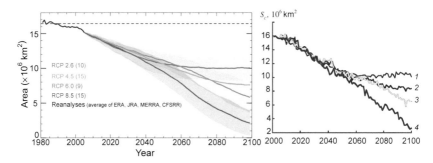

FIGURE 10.6   Total area of near-surface permafrost on the Northern Hemisphere land computed by Slater and Lawrence (2013) (left) and Arzhanov et al. (2013) (right) under the scenarios: (1) RCP 2.6, (2) RCP 4.5, (3) RCP 6.0, and (4) RCP 8.5.

to reverse the overall trend. Slater and Lawrence (2013) estimated that the reduction of the near-surface permafrost extent by the end of the century will be about 10 million square kilometers under RCP 2.6 and to 2.1 million square kilometers under RCP 8.5. Arzhanov et al. (2013) arrived at slightly larger estimates of 10.5 million square kilometers and 3 million square kilometers, respectively (Figure 10.6). Thawing of near-surface permafrost may result in differential subsidence of several decimeters or more in areas with ice-rich permafrost leading to the deterioration of overlying infrastructure, especially linear structures on permafrost.

10.7 PERMAFROST AND INFRASTRUCTURE

Arctic settlements range in size from small native villages in Alaska and Scandinavia to large industrial towns in Siberia. Depending on location, climatic and environmental conditions, proximity to the coastline, density, structure, and economic orientation of the settlements, the impact of climate change is expected to be diverse. Changing climatic conditions may benefit some aspects of arctic economies as heating costs are likely to decrease, and diminishing sea-ice extent has the potential to foster economic development along the coasts. However, these benefits may be outnumbered by the new challenges facing northern communities as coastal erosion is expected to intensify due to increasing wave activity, shorter winter road operational seasons lower the accessibility of remote settlements, and thawing permafrost may not be able to support existing infrastructure. A recent study of the economic effects of climate change on infrastructure in Alaska concluded that an additional 3.6−6.1 billion USD will be required to maintain infrastructure between 2010 and 2030 and up to 7.6 billion USD between 2010 and 2080 (Larsen et al., 2008). Despite no such comparison being conducted for other regions on permafrost, the direct cost of permafrost degradation is at least several billion dollars per year. For example, pipeline maintenance in permafrost regions of Russia requires more than 1.5 billion USD annually (Streletskiy et al., 2012a). The impacts of permafrost degradation are expected to be most pronounced in large population and mining centers, as substantial populations and infrastructure will be affected. Impacts of permafrost degradation in mountain regions are omitted as they are discussed in (Deline et al., 2014).

10.7.1 Buildings on Permafrost

A large number of structural deformations have been reported in settlements built on permafrost (Grebenets et al., 2012; Hinkel et al., 2003; Khrustalev et al., 2011; Kronic, 2001). Observed increases in near-surface permafrost temperatures are decreasing the ability of foundations to support structures at rates that may not have been anticipated at the time of construction (Khrustalev and Davidova, 2007; Streletskiy et al., 2012c). An increased ALT decreases the effective area of a foundation freezing within permafrost, intensifying the processes of frost heave and thaw subsidence, and increasing the corrosion of the foundation materials. The projected climate change in the Arctic has the potential to cause a further deterioration and deformation of structures on permafrost, which can have severe socioeconomic consequences, as most of the existing infrastructure will require expensive mitigation strategies to support them.

Russian permafrost regions account for half of the population in the Arctic, and at the same time, this is the most problematic region with respect to infrastructure on permafrost. The majority of Russian urban infrastructure consists of mass-produced standard design structures built in the 1960s—1970s, which were rarely designed to withstand the changes in climatic conditions beyond natural variability at the time of construction (Khrustalev et al., 2011). A survey of structures in a series of settlements across the Russian Arctic indicated that buildings with deformations account for 10 percent of their total numbers in Norilsk, 22 percent in Tiksi, 55 percent in Dudinka, 35 percent in Dikson, 50 percent in Pevek and Amderma, 60 percent in Chita, and 80 percent in Vorkuta (Kronic, 2001). During 1990—2000, the rate of building failures has increased by 42 percent in the city of Norilsk, 61 percent in Yakutsk, and 90 percent in Amderma. The number of observed deformations in structures in Norilsk during 2003—2013 was higher than the corresponding number of deformations during 1963—2013 (Grebenets et al., 2012).

Construction of houses on permafrost can be a tricky business, as heat and permafrost contradict each other. Through trial and error, Russia accumulated a tremendous amount of experience with construction on permafrost, which was later adopted by western engineers. Despite the fact that by 1906 Nezdanov had proposed the use of ventilated basements to protect permafrost from thawing and by 1912 Bogdanov had summarized much of the engineering problems in his book "Permafrost and construction upon it", houses without proper thermoinsulation of the ground floor were a common reality until the midtwentieth century. To avoid warming permafrost from buildings and structures and to redistribute the pressure associated with high structural loads, engineer Mikhail Kim promoted the idea of piling foundations, which were originally designed in 1956 in Norilsk. The piles were installed and frozen into the permafrost prior to the construction of the structure itself. The piles embedded in permafrost allowed for much greater structural weights relative to the shallow foundations since much of the structural load was carried by the side contact of piles with permafrost. It also decreased the disturbance of permafrost during construction as pile diameters were relatively small compared to other foundation types and allowed for the redistribution of pressure from pile to pile through the system of horizontal beams on which the buildings were erected. The piles, commonly 4−12 m in length and 0.2−0.5 m in diameter, were installed in predrilled boreholes, with the upper portion of the pile sticking 0.5−2 m above the ground. The system of horizontal beams was put on top of the piles redistributing the structural load. The elevated first floor with a crawl space basement was clearly advantageous, as the basement was ventilated during the winter and provided shading during the summer preventing the permafrost from thawing.

Presently, urban architecture in the Russian Arctic is predominantly represented by a mixture of prefabricated panel or standard design, brick, five- to nine-floor buildings. The majority of such structures on permafrost were built using piling foundations according to the ideas of Kim, which have become known as the passive principle of construction on permafrost (Shur and Goering, 2009). According to the passive principle of construction (also known as Principle I in the Russian literature), the permafrost is used as the base for foundation and is protected from thaw during construction and throughout the entire life span of the structure, which is expected to be between 30 and 50 years.

The ability of a pile foundation to carry a building structural load depends on many factors such as ground temperature, texture, density, salinity, the amount of ice in the ground, and the presence of unfrozen water. In coarse soils (gravels and denser), soil strength is primarily a function of internal friction. In frozen fine-grained materials, however, it is primarily a function of ice bonding. The strength and deformation characteristics of frozen soils are temperature dependent. Two parameters, namely, the thickness of the active layer (maximum thaw depth) and the maximum ground temperature, along the embedded pile length are incorporated into the site-specific design for a foundation in a permafrost region (Instanes and Anisimov, 2008). As previously mentioned, piles embedded in permafrost allow for much heavier loads compared to those in nonpermafrost regions, as the majority of the bearing capacity is gained by adfreezing of the sides of piles to the surrounding permafrost. The shear stress per side unit of a pile is much lower compared to the normal stress at the bottom of a pile, but this also means a much larger adfreezing area of the pile side relative to the bottom, allowing for the redistribution of up to 80 percent of the pile-bearing capacity to its sides.

The dependence of freezing strength on permafrost temperature along the sides of piles makes a foundation's bearing capacity highly dependent on permafrost temperature, which in turn depends on air temperature variations and characteristics of heat exchange at the ground surface, such as the presence and properties of snow and vegetation. Air temperature and snow-cover oscillations create conditions in which strength and bearing capacity vary with climatic conditions (Khrustalev and Davidova, 2007).

The previously described conventional permafrost engineering design is based on the analysis of historical variations of climatic and permafrost data and accounts for the frequency of extreme events, such as abnormally high temperatures resulting in deeper seasonal thawing of permafrost. Each foundation is designed with a construction-specific safety factor that depends on the probability of such extremes. The adfreeze bond strength between the pile surface and surrounding frozen soil is both temperature and time dependent. For a specific pile foundation design, this parameter should be determined from geotechnical field and laboratory investigations (Figure 10.7).

Engineers, however, were commonly using historical climatic averages available at the time of construction, in which case climate change was not accounted for. This fact raises questions about the stability of structures whose design is based on the purported climatic normal of past decades when climatic conditions were changing rapidly.

Assessing the stability of infrastructure located in permafrost regions Streletskiy et al. (2012a) found that the present deterioration of infrastructure can, to some degree, be attributed to climatic changes observed in permafrost regions over the last 50 years. Using a combination of permafrost-geotechnical modeling and climate data, these authors mapped changes in foundation bearing capacity for five Russian regions located on permafrost (Figure 10.8). These five regions, with a total population of almost two million people, experienced an overall increase of mean annual air temperature of 1 °C since the 1970s. Associated increases in near-surface permafrost temperatures and thickening of the active layer resulted in the decreased ability of foundations within the studied regions of Russia to support their structures. Combined with the low safety coefficients this raises a question about the safety of thousands of people living in houses built on permafrost in the Russian North. Projected climate change is likely to further decrease the stability of existing infrastructure as large areas of ice-rich permafrost are expected to experience differential subsidence (Anisimov and Reneva, 2006; Arzhanov et al., 2010; Nelson et al., 2001) and decreased foundation bearing capacity (Khrustalev et al., 2011; Streletskiy et al., 2012c). A comprehensive study of cumulative geoecological effects of infrastructure development and climate change in Prudhoe Bay Oil field (Northern Alaska) showed that significant

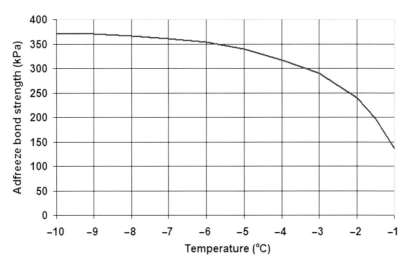

FIGURE 10.7   Typical diagram of adfreeze bond strength as a function of the ground temperature. *From Instanes and Anisimov (2008).*

intensification of thermokarst development occurred over the life span of the oil field (Raynolds et al., 2014). Although some of the thermokarst resulted directly from the infrastructure development (construction of roads, pipelines, gavel pads, and others) or indirect impacts associated with infrastructure (infrastructure-related flooding, roadside dust, and offroad vehicle traffic), numerous changes were attributed to warmer climatic conditions during 1994−2014, particularly during summer months. The increase of thermokarst extent may have serious negative implications on future economic development of the oil field, wildlife habitat, and sustainability of the native community of Nuiqsut located west of the development area.

10.7.2 Pipelines on Permafrost

Pipelines are a common means of transporting liquid and semiliquid products and are widely used in cold regions to transport heat, water, and by-products of oil and mineral resources. The latter ones are of greatest interest since they extend over large areas with drastically different environmental and climatic conditions. For example, the *Trans*-Alaskan pipeline extends more than 1,200 km through taiga and tundra and crosses through all permafrost zones. Extensive networks of oil and gas pipelines in Russia extend more than 71,000 km, with a considerable portion built on permafrost. Several gas pipelines are currently under consideration for construction in Alaska, Canada, and Russia. Deformation of pipelines and associated spills of petroleum products are particularly hazardous in regions on permafrost as the low biologic activity in tundra environments means that it would require years to restore these fragile

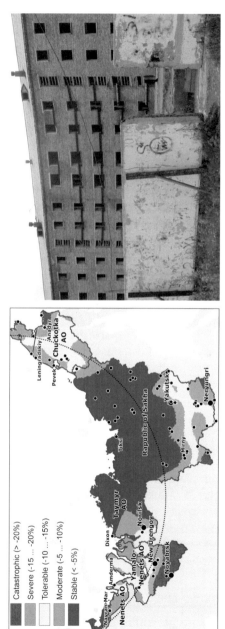

FIGURE 10.8   Left: Map of changes in the ability of foundations to support the structures in permafrost areas (from Streletskiy et al. (2012a)). Deformation of building constructed on permafrost using passive construction principle in Igarka. *Photograph by D. Streletskiy, July 2013.*

Arctic ecosystems. Despite the fact that microbiological transformation of hydrocarbon products in the Arctic occurs at a slow rate, it should not be ignored in restoration efforts (Kachinskii et al., 2014).

Deformations of pipelines in regions on permafrost are commonly associated with frost heave and subsidence processes, ground ice melting, water logging and thermokarst development, and thermoerosion and solifluction on slopes. Above-ground pipelines may experience deformation associated with intensified frost heave as the active-layer increases following the removal of vegetation and the organic layer during the construction. Increases in the ALT also lead to a decrease in the effective area of side freezing for supporting piles in permafrost, which eventually leads to pile jacking and the decreased ability to support structural weight. Redistribution of weight among neighboring piles creates additional mechanical stresses, which may result in the fracturing of the overhead pipe(s). For example, due to frost heave processes, about 8,000 pipeline supports had to be readjusted during 2007–2009 in the Yamburg region (West Siberia) alone. Subsurface pipelines suffer from intensive corrosion, heat loss associated with transportation of petroleum products, ground subsidence, and water logging. Subsidence can continue for more than a decade after the construction of trenches for underground cables and pipelines (Streever, 2012). Although it is rare in the North America, it is not uncommon to see floating pipelines in permafrost regions (Figure 10.9).

A study conducted in Yakutia analyzed 2,174 km of pipelines to classify deformations during 2000–2010. The authors found that permafrost conditions were responsible for only 3 percent of all deformations (Chuhareva and Tikhonova, 2012). However, if deformations due to operational maintenance are taken into account, this number reaches 17 percent, illustrating the indirect effects of permafrost changes. Other studies of pipeline deformations in West Siberia give a comparable value of 21 percent due to mechanical deformations, including those attributed to the decreased ability of piles to support structural weight. In West Siberia, there are 35,000 pipeline accidents occurring annually. The pipeline deformations commonly occur in locations where linear structures cross ice wedges and massive ice deposits (Stanilovskaya and Merzlyakov, 2013) (Figure 10.10). In Russia, 55 billion RUB are spent annually to maintain the operational stability of pipelines on permafrost (Streletskiy et al., 2012c). Considering that oil and gas development in the Arctic is likely to continue, aging infrastructure and projected climate change will increase the number of deformations and leaks from pipelines leading to the deterioration of ecosystems.

10.7.3 Railroads, Roads, and Utility on Permafrost

Permanent road networks on permafrost are likely to suffer from projected subsidence associated with melting ground ice. The projected increase in snow accumulation, and consequently larger volumes of redistributed snow off roads, will further enhance permafrost warming, resulting in the development

FIGURE 10.9 Thermokarst development around the subsurface pipeline, Kharp (Yamal-Nenents Autonomous Okrug, Russia). *Photograph by D. Streletskiy, July 2012.*

FIGURE 10.10 Above-ground pipeline in Northern Yakutia crossing the area with ice-rich permafrost. *Photograph by A.N. Fedorov.*

of thermokarst and waterlogged surfaces. Railroads can also suffer significant deformation due to differential frost heave and thaw settlement. Out of 3,539 km of Far East road in Siberia, 18 percent is deformed due to differential permafrost settlement. A similar situation is happening to the Baikal-Amur railroad where about 20 percent of the rail is deformed. The problem of railroad deformation in areas of ice-rich permafrost has been relevant not only in Russia but also in Alaska, Canada, and recently in China with the construction of the Qinghai-Tibetan railroad on warm permafrost (Kondratiev,

2013). About 550 out of 2,000 km of the Qinghai-Tibetan railroad was constructed on permafrost. A high operational speed of 100 km/h over the
permafrost sections required innovative designs and construction methods.
Some areas on warm permafrost required construction of rail track on elevated
piping foundation using an extensive network of thermosiphons, as in the case
of the *Trans*-Alaska pipeline. The presence of permafrost significantly
increased the cost of the project, which was estimated on the order of 30
billion Yuan. Thermosiphons are used quite extensively to decrease permafrost
temperature and to mitigate negative consequences of climate change or
technogenic disturbances. However, the high price of thermosiphons and
relatively small effective area (a few meters) can significantly increase the cost
of construction and maintenance of linear infrastructure, including that of
railroads and pipelines.

Large arctic countries heavily rely on winter roads and drivable ice
pavements to supply communities in remote areas. As a result, many areas of
the Canadian North and the Russian Arctic are land accessible only during
winter. Several studies indicate that changes to the operational length of winter
roads have already occurred. These changes, however, are not uniform across
space. For example, some regions of East Siberia show an increase in the
operational length of winter roads, whereas regions of intensive oil and gas
exploration and development in West Siberia have experienced a significant
decrease. Other economically vital regions with decreased overland accessibility include areas along the Yenesey River north of Igarka up to Dickson,
around Cherskiy in North-Eastern Yakutia, and Pevek and Anadyr in Chukotka
(Streletskiy et al., 2012a). A comprehensive study of the effects of climate
change on transportation in Northern Canada concluded that a shorter winter
road operational season will be compensated by longer navigation seasons on
lakes and rivers during summer months (Lonergan et al., 1993). A decrease in
winter road operation could also be offset by improved maritime transportation
conditions in the Arctic (Stephenson et al., 2011).

10.8 COASTAL EROSION AND PERMAFROST

The majority of Arctic settlements are located in the coastal zone. Increasing
air and permafrost temperature, sea-ice retreat, and accelerating rates of
coastal erosion can have highly detrimental impacts on arctic coastal communities. Arctic permafrost coasts are sensitive to climate changes due to the
presence of sediments with high ice content. Degradation of Arctic coasts is
one of the leading processes reflecting complex interactions between terrestrial
and marine systems. Primarily, the destruction of coasts occurs due to the
kinetic energy of waves and permafrost thawing. The entire suite of processes
leading to the destruction of Arctic coasts is known as coastal erosion. Coastal
erosion occurs only during the sea ice-free season. The formation of sea ice in
the beginning of a cold season temporarily stops coastal erosion processes for

the duration of the land-fast ice until it opens enough again along the coastal zone allowing for wave action.

The main factors determining rates of coastal erosion in the Arctic include wave energy applied to the base of the coastal cliffs (wave erosion), and the thawing and mass wasting of sediments to the beach zone (thermal denudation). Strong storms are considered to play a leading role in coastal erosion. However, this is not always the case, such as in western part of the Russian Arctic where regular wave action at the base of the cliffs is considered to be the primary erosional agent. Persistent wave activity was found to be more effective at eroding the coasts compared to strong, but short wave activity associated with storm surges. Storm surges account for less than a quarter of annual wave energy. The role of storm activity in coastal erosion is shown in Figure 10.11, which demonstrates a strong dependence of coastal retreat rate on the total wave energy compared to the weak dependence of the rate of retreat and storm energy. Coastal retreat rates have high internal variability. Long-term monitoring of coastal erosion in the Barents and Kara Sea coasts (Vasiliev et al., 2005) showed that periods of high retreat rates alternate with periods of low retreat rates (Figure 10.12).

Retreat rates are quite variable throughout the Arctic, depending on climatic and geologic conditions that range from just a few centimeters per year to 30 m per year, but usually fall within 0.5–2 m/year (Forbes, 2011). The highest retreat rates are found in ice-complex ("edoma") coasts composed of silty sediments with high ice content, up to 90 percent by volume. Minimal rates are characteristic for bedrock coasts with a low ice content. Spatial and temporal variabilities of coastal erosion rates in the Arctic are monitored under the umbrella of the Arctic Coastal Dynamics (ACD) project, initiated by the International Permafrost Association in 1999. Recently, the ACD group released the geomorphologic classification database for arctic coasts covering 101,500 km of coastline (Figure 10.13). Roughly 65 percent of arctic coasts are characterized by the presence of ground ice, such as the Beaufort, Laptev, and East Siberian coasts (Jones et al., 2009; Lantuit et al., 2012). The presence of various types of ground ice creates unique features of thermal–mechanical

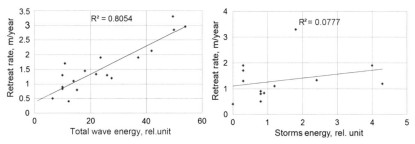

FIGURE 10.11   Dependence of coastal retreat rate at Western Yamal on total wave energy and storm energy.

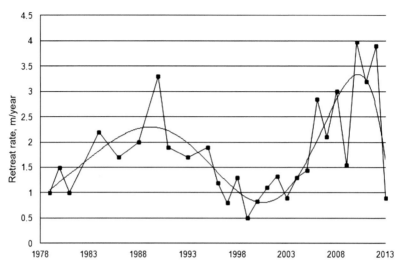

FIGURE 10.12   Temporal changes of retreat rates of Kara sea coasts (Western Yamal Peninsula example).

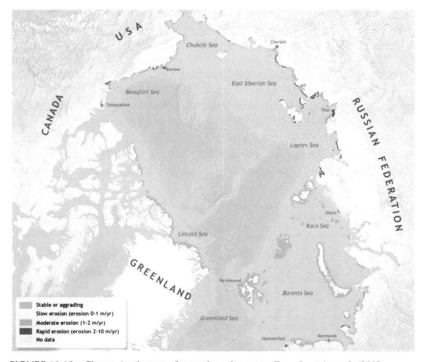

FIGURE 10.13   Circum-Arctic map of coastal erosion rates. *From Lantuit et al. (2012).*

erosion, making the coasts especially vulnerable to changes in land—ocean and land—atmosphere fluxes.

Climate change in the Arctic has resulted in increased air temperatures, decreased sea-ice extent, and a lengthened ice-free period. Although no direct effects between increasing air temperatures and coastal erosion rates have been found, the combined effect of climate change is likely to have caused the increased coastal erosion rates in the Arctic through the increased length of the ice-free period. The rough estimate of land loss along Arctic sea coasts can be computed using the total length of 101,500 km and the average retreat rate of 0.5 m/year (Lantuit et al., 2012). The resulting 51 km^2 gives an idea of the land lost annually, which is roughly equal to the size of a town such as the Swiss capital of Bern, or, over a two-year period, roughly equal to the size of Paris, the capital of France. Increased rates of coastal erosion have already resulted in a series of negative consequences for native settlements in Alaska and elsewhere. For example, Kivalina, a native settlement located in Northwest Alaska, is planning to be relocated due to coastal erosion. The estimated price for this relocation was 176 million USD (Wilson, 2007). Coastal erosion is also threatening oil terminals located in Varandei (Yamal, Russia). Industrial activity around Varandei Oil Field (Pechora Sea coast) has already led to intensified aeolian, mass wasting, and erosional processes along the coast, resulting in twice the rate of coastal retreat compared to a natural environment (Ogorodov, 2005). One of the major factors that played a role in increasing the rate of erosion was the excavation and removal of shoreline sediment for construction, which combined with vegetation degradation, led to increased ALT. Deeper thaw made sand dunes more susceptible to wind deflation and erosion.

Coastal erosion in permafrost regions supplies sediments and organic carbon to the sea in amounts comparable to organic carbon supplied by large rivers. Organic carbon is an important source of nutrients for sea biota in coastal zones (Lisitzin, 1999). One study estimated that the amount of material generated from coastal erosion dumped into the Kara Sea is about 35 million tons, including 0.4 million tons of organic carbon (Streletskaya et al., 2009).

Projected increases in air temperature and a subsequent decline in sea-ice extent will further promote coastal erosion as the duration and intensity of wave activity is likely to increase and the cohesion of permafrost coasts is likely to decrease. Intensified rates of coastal erosion will require additional costs to maintain existing and planned coastal infrastructure, and will lead to increased sedimentary and organic carbon input to the Arctic seas.

10.9 PERMAFROST AND THE CARBON CYCLE IN THE CONTEXT OF CLIMATE CHANGE

One of the relatively poorly studied natural hazards associated with changes in permafrost is the potential amplification of global warming through emission of greenhouse gases with implications for the climate-dependent systems and

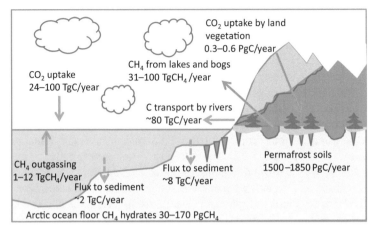

FIGURE 10.14   Major carbon pools and flows in the Arctic domain, including permafrost on land, continental shelves and ocean. *Adapted from McGuire et al. (2009) and Tarnocai et al. (2009).*

processes. The permafrost−climate feedback is described by the following conceptual model (Ciais and Sabine, 2013). Permafrost contains large amounts of carbon, of which the bulk was sequestered during the Holocene (Figure 10.14). Part of it is localized in the active layer and is involved in the annual carbon cycle, whereas the other part is trapped in frozen sediments on land and in the shelf of the Arctic seas. Climate change and rising temperature may have the twofold effect on terrestrial permafrost-carbon cycle by (1) thickening the active layer and increasing the portion of soil carbon stock involved in the seasonal cycle and (2) shifting the balance between the annual carbon sink and source. Thawing subaquatic permafrost may further enhance the climate feedback. It may lead to the increased gas permeability of the bottom sediments on the Arctic shelf, creating pathways for methane venting to the water column and to the atmosphere from deep layers, where it is currently stored in the form of gas hydrate. The strength of climate−permafrost feedback depends on the soil carbon pool; net fluxes, which are the balance between the sources and sinks of CO_2 and methane; and the sensitivity of climate to these two gases. Results from recent field studies and comprehensive carbon−permafrost models significantly reduced the uncertainty in quantifying the elements of this chain.

10.9.1 Permafrost Soil Carbon Pool

In the early 1990s, the carbon content in arctic soils was estimated to be 455 Pg, whereas the global content in the upper one meter of all terrestrial ecosystems totaled 1,400−1,600 Pg C (Oechel et al., 1993). More recent investigations showed that the upper 3 m of arctic soils hold about 750 Pg C

(Schuur et al., 2008). This estimate excluded arctic bogs, which according to Gorham (1991) store an additional 200–450 Pg C depending on the assumed thickness of average accumulated peat of 1 or 2.3 m McGuire et al. (2009), gave a slightly lower estimate of 270–370 Pg C in arctic bogs. Later work by Tarnocai et al. (2009) gave an estimate of 150 Pg C for Canadian Arctic bogs, which confirmed the estimates by Gorham (1991), as roughly one-third of arctic bogs are located in Canada.

Several studies of ice complex ("edoma") occupying large areas in East Siberia showed that significant amounts of carbon can be stored much deeper than 3 m as sediments of edoma were syngenetically frozen allowing for the accumulation of carbon during sedimentation (Walter et al., 2007; Zimov et al., 2006). These findings increase the estimated carbon pool by 400 Pg C in the upper 25 m of edoma. An additional 250 Pg C are potentially stored in seven deltas of major Arctic rivers (Schuur et al., 2008). Summarizing these findings gives about 1,400 to 1,850 Pg C, including 1,000 Pg C in arctic soils to a depth of 25 m, 200–450 in bogs and 200–400 Pg C in the edoma of Siberia. These numbers are consistent with the 1,672 Pg estimate by Tarnocai et al. (2009), and constitute about half of the global carbon stored in soils worldwide. The majority of the terrestrial permafrost soil carbon occurs within the upper 3-m layer, which is most susceptible to thawing under projected climate change.

10.9.2 Carbon Fluxes in Terrestrial Permafrost Regions

The net carbon flux between the atmosphere and permafrost depends on the interplay of sink (photosynthetic uptake of CO_2 by Arctic vegetation) and source (release of CO_2 due to soil respiration and/or emission of methane due to biogenic production and release from hydrates). Earlier observational studies suggested the near-zero average carbon balance in the circumpolar Arctic with large spatial and interannual variations (Corradi et al., 2005). In contrast, models showed a small sink of about 20 ± 40 g/cm^2/year, and a consensus was reached between models and observations that the Arctic acts as a net source of methane (Anisimov and Vaughan, 2007). Observational data summarized by Callaghan et al. (2011) confirmed these findings and provided more details suggesting that tundra regions currently act as sources of carbon in warm and dry years or in well-drained settings, and as sinks in cold and wet years. The most recent IPCC AR5 assessment concluded that permafrost regions are currently a net sink of CO_2 sequestering about 0.4 ± 0.4 PgC/year, and a modest source of methane estimated to be 15–50 Tg(CH_4)/year, which is emitted mostly from seasonally thawing wetlands (Ciais and Sabine, 2013). Methane (CH_4) has an approximately 20 times stronger greenhouse effect than CO_2 has. Deeper seasonal thawing of permafrost will lead to the involvement of the previously frozen organic material in the annual carbon cycle, and in combination with the higher soil temperatures and longer warm season will enhance the biogenic methane production, particularly in the organic-rich Arctic wetlands.

Although emissions from terrestrial permafrost are supposed to be gradual under conditions of climate change, the release of methane from hydrates underneath the subsea permafrost may potentially be abrupt. Higher-than average atmospheric methane concentrations occur around the continental shelves of East Siberian Seas (ESS), which could have been attributed to local methane sources, including those associated with decomposition of methane hydrates underneath the sea bottom (Shakhova et al., 2010). Because the ESS average depth is only 45 m, a significant portion of methane is able to reach the atmosphere without oxidation, which would be the case for a deeper water column. Total methane emission at ESS under current conditions is estimated at 7.9 Tg(CH_4)/year, which is relatively small compared to other sources (Shakhova et al., 2010).

10.9.3 The Effect on Global Climate

The impact of enhanced methane emissions from thawing permafrost on global temperature may be evaluated either by running the full-scale hydro-dynamic climate model, or by calculations using climate sensitivity to greenhouse gases, that are relatively well known. The latter approach was used in several studies to access the effects in terrestrial and subaquatic permafrost regions. All studies involved a comprehensive dynamic permafrost—carbon model (Anisimov et al., 2012b; Lavrov and Anisimov, 2011). A modeling study of carbon fluxes from Russian Arctic bogs found that increased methane emissions of the order of 0.008—0.01 Pg C per year could occur by mid-twenty-first century under the most extreme climatic scenario (Anisimov and Reneva, 2011). It would result in an average increase of atmospheric methane concentrations by 0.04 ppm, with a corresponding increase in the average global annual temperature by 0.012 °C, at most (Anisimov et al., 2012a; Dmitrenko et al., 2011). Modeling also indicated that in the following few centuries it is very unlikely that thawing of submarine permafrost in the East Siberian shelf will increase perforation and gas permeability of the frozen bottom sediments (Anisimov et al., 2012a; Dmitrenko et al., 2011). Except for selected permafrost-free locations near fault zones, gas hydrates in deep layers will remain trapped beneath the layer of the frozen sediments (Figure 10.15).

It would be appropriate to add that due to the absence of robust data some highly speculative sensational hypotheses have been suggested, as exemplified by the paper of Whiteman et al. (2013). These authors assessed the combined economic effect of an enhanced methane flux to the atmosphere. According to their estimate, the release of methane from thawing permafrost in the East Siberian Sea alone will cost 60 trillion USD to the global economy, which is comparable to the entire world economy (70 trillion USD in 2012). These authors concluded that 80 percent of economic consequences will occur in the

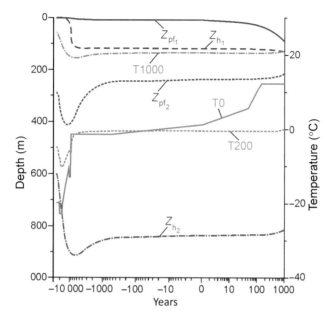

FIGURE 10.15 Modeled depths to upper (index 1) and lower (index 2) boundaries of permafrost (Zpf) and methane hydrate stability zone (Zh), and temperatures at 0, 200, and 1000 m depths on the time interval from the last glacial maximum 18,000 years ago to the year 3000. X-axis designates time in logarithmic scale from before and after the contemporary period, 0 corresponds to the year 2000. *From Anisimov et al. (2012a).*

developing countries of Africa, Asia, and South America. According to their model, a release of 50 Pg of methane from the shelf may increase the mean global temperature by 2 °C by 2035–2040. However, these dramatic projections are not supported by real data, and the model used in this study proves to be highly unrealistic. The recent IPCC assessment report analyzed the full set of available observational and model data and estimated the total amount of methane trapped in the hemispheric-scale Arctic shelves at 2–65 PgCH$_4$ (Ciais and Sabine, 2013); observations in the Arctic show that so far methane concentrations in the atmosphere are rising gradually and not abruptly (Figure 10.16); and comprehensive permafrost-carbon modeling results referenced above leave no space for such speculative judgments.

10.10 SUMMARY

Permafrost plays an important role in global climate, environmental systems, and human activities in the Arctic. Permafrost degradation has occurred many times throughout geological history, but the rate of climate change, presence of substantial population, and diverse economic and landuse activities in the Arctic, make contemporary permafrost degradation a unique process. The impacts of permafrost degradation are diverse and range from local to global,

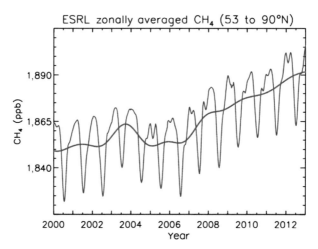

FIGURE 10.16   Methane mole fraction averaged over polar northern latitudes (blue). The red line is the deseasonalized trend. *From Bruhwiler and Dlugokencky (2012).*

such as the potential enhancement of climatic change through emission of greenhouse gases. Observational data on permafrost characteristics are limited, but show permafrost temperature is increasing, whereas the active layer is progressively thickening in the majority of regions. Although climate change is the main driver of permafrost changes, other environmental characteristics may significantly alter these general trends.

Natural landscapes, but more importantly areas of human activities, including large population and industrial centers, are facing numerous hazards associated with permafrost degradation, such as melting of ground ice, thaw subsidence, and thermokarst development; decrease of foundation stability and increase of corrosion of foundation materials; intensification of slope processes and coastal erosion. Although permafrost changes are already taking their toll on various types of infrastructure, numerous studies have showed that permafrost degradation will continue in the future. A series of planning activities and engineering solutions (e.g., thermosiphons) are capable of mitigating permafrost degradation to a certain extent and preventing immediate hazards to existing infrastructure. Climate change should be taken into account more seriously, when constructing on permafrost. More research is needed to develop informed and adequate adaptation and mitigation strategies with regard to land use and infrastructure. The limited permafrost observational network should be expanded to serve these needs.

ACKNOWLEDGMENTS

We thank Chris Burn (Carleton University, Canada), Hanne Christiansen (The University Centre in Svalbard, Norway) and the volume editors for their reviews of an earlier version of this paper. This work was supported by the Russian Science Foundation, project 14-17-00037.

REFERENCES

ACIA, 2005. Arctic Climate Impact Assessment. Cambridge University Press, Cambridge.

Åkerman, H.J., Johansson, M., 2008. Thawing permafrost and thicker active layers in sub-arctic Sweden. Permafrost Periglacial Process. 19 (3), 279−292.

AMAP, 2011. Snow, Water, Ice and Permafrost in the Arctic (SWIPA). Arctic Monitoring and Assessment Programme (AMAP), Oslo.

Anisimov, O.A., Reneva, S.A., 2006. Permafrost and changing climate: the Russian perspective. AMBIO J. Hum. Environ. 35 (4), 169−175.

Anisimov, O.A., Reneva, S.A., 2011. Carbon balance of Russian permafrost: current state and model-based projections. In: Kotliyakov, V.M. (Ed.), Results of the International Polar Year, Polar Cryosphere and Freshwater, vol. III. Paulsen, Moscow-St. Petersburg, pp. 122−140.

Anisimov, O.A., Vaughan, D.G., 2007. Polar Regions, in Climate Change 2007: Impacts, Adaptation, and Vulnerability. Contribution of Working Group II to the Fourth Assessment Report of the Intergovernmental Panel on Climate Change, 653−686 pp. Cambridge University Press, Cambridge.

Anisimov, O.A., Anokhin, Y.A., Lavrov, S.A., Malkova, G.V., Pavlov, A.V., Romanovsky, V.E., Streletskiy, D.A., Kholodov, A.L., Shiklomanov, N.I., 2012b. Terrestrial Permafrost. Roshydromet, Moscow.

Anisimov, O.A., Belolutskaya, M.A., Grigoriev, M.N., A., Instanes, Kokorev, V.A., Oberman, N.G., Reneva, S.A., Strelchenko, Y.G., Streletskiy, D., Shiklomanov, N.I., 2010. Major Natural and Social-economic Consequences of Climate Change in the Permafrost Region: Predictions Based on Observations and Modeling, 44 pp. Greenpeace, Moscow, Russia.

Anisimov, O.A., Borzenkova, I.I., Lavrov, S.A., Strelchenko, J.G., 2012a. Dynamics of sub-aquatic permafrost and methane emission at eastern Arctic sea shelf under past and future climatic changes. Ice Snow 2, 97−105.

Anisimov, O.A., Shiklomanov, N.I., Nelson, F.E., 1997. Global warming and active-layer thickness: results from transient general circulation models. Global Planet. Change 15 (3), 61−77.

Arzhanov, M.M., Demchenko, P.F., Eliseev, A.V., Mokhov, I.I., 2010. Modelling of subsidence of perennially frozen soil due to thaw for the Northern Hemisphere during the 21st century. Earth Cryosphere 14 (3), 37−42.

Arzhanov, M., Eliseev, A., Mokhov, I., 2013. Impact of climate changes over the extratropical land on permafrost dynamics under RCP scenarios in the 21st century as simulated by the IAP RAS climate model. Russian Meteorol. Hydrol. 38 (7), 456−464.

Beringer, J., Lynch, A.H., Chapin III, F.S., Mack, M., Bonan, G.B., 2001. The representation of arctic soils in the land surface model: the importance of mosses. J. Climate 14 (15), 3324−3335.

Bommer, C., Phillips, M., Arenson, L.U., 2010. Practical recommendations for planning, constructing and maintaining infrastructure in mountain permafrost. Permafrost Periglacial Process. 21 (1), 97−104.

Brown, J., Hinkel, K.M., Nelson, F.E., 2000. The circumpolar active layer monitoring (CALM) program: research designs and initial results. Polar Geogr. 24 (3), 166−258.

Bruhwiler, L., Dlugokencky, E., 2012. Carbon Dioxide (CO_2) and Methane (CH_4) Arctic Report Card 2012. http://www.arctic.noaa.gov/reportcard.

Burn, C., 1998. The response (1958−1997) of permafrost and near-surface ground temperatures to forest fire, Takhini River valley, southern Yukon Territory. Can. J. Earth Sci. 35 (2), 184−199.

Burn, C., Zhang, Y., 2009. Permafrost and climate change at Herschel Island (Qikiqtaruq), Yukon Territory, Canada. J. Geophys. Res. Earth Surf. (2003−2012) 114 (F2).

Callaghan, T., Johansson, M., Anisimov, O., Christiansen, H., Instanes, A., Romanovsky, V., Smith, S., 2011. Changing permafrost and its impacts. In: Snow, Water, Ice and Permafrost in the Arctic (SWIPA): Climate Change and the Cryosphere. Arctic Monitoring and Assessment Programme, Oslo, pp. 5-1−5-62.

Christiansen, H.H., Etzelmüller, B., Isaksen, K., Juliussen, H., Farbrot, H., Humlum, O., Johansson, M., Ingeman-Nielsen, T., Kristensen, L., Hjort, J., 2010. The thermal state of permafrost in the nordic area during the international polar year 2007−2009. Permafrost Periglacial Process. 21 (2), 156−181.

Chuhareva, N., Tikhonova, T., 2012. Accident and damage prediction system in gas trunk pipelines. Res. J. Int. Studies (5−2), 122−124.

Ciais, P., Sabine, C.L., 2013. Carbon and other biogeochemical cycles. In: Stocker, T.F., Qin, D., Plattner, G.K., Tignor, M. (Eds.), Climate Change 2013: The Physical Science Basis. Contribution of Working Group I to the Fifth Assessment Report of the Intergovernmental Panel on Climate Change. Cambridge University Press, Cambridge, pp. 465−570.

Clow, G.D., 2008. Continued permafrost warming in northwest Alaska as detected by the DOI/GTN-P borehole array. In: Kane, D.L., Hinkel, K.M. (Eds.), Proceedings of the Ninth International Conference on Permafrost, June 29−July 3. Institute of Northern Engineering, University of Alaska Fairbanks, Fairbanks, Alaska, pp. 47−48.

Corradi, C., Kolle, O., Walter, K., Zimov, S., Schulze, E.D., 2005. Carbon dioxide and methane exchange of a north-east Siberian tussock tundra. Global Change Biol. 11 (11), 1910−1925.

Daanen, R., Grosse, G., Darrow, M., Hamilton, T., Jones, B., 2012. Rapid movement of frozen debris-lobes: implications for permafrost degradation and slope instability in the south-central Brooks Range, Alaska. Nat. Hazards Earth Syst. Sci. 12 (5), 1521−1537.

Deline, P., Gruber, S., Delaloye, R., Fischer, L., Geertsema, M., Giardino, M., Hasler, A., Kirkbride, M., Krautblatter, M., Magnin, F., McColl, S., Ravanel, L., Schoeneich, P., 2014. Ice loss and slope stability in high-mountain regions. In: Haeberli, W., Whiteman, C. (Eds.), Snow and Ice-related Hazards, Risks and Disasters. Elsevier pp. 521−562.

Dmitrenko, I.A., Kirillov, S.A., Tremblay, L.B., Kassens, H., Anisimov, O.A., Lavrov, S.A., Razumov, S.O., Grigoriev, M.N., 2011. Recent changes in shelf hydrography in the Siberian Arctic: potential for subsea permafrost instability. J. Geophys. Res. Oceans (1978−2012) 116 (C10).

Forbes, D., 2011. State of the Arctic Coast 2010—Scientific Review and Outlook. International Arctic Science Committee, Land-Ocean Interactions in the Coastal Zone, Arctic Monitoring and Assessment Programme, vol. 178. International Permafrost Association, Helmholtz-Zentrum Geesthacht: Geesthacht, Germany. http://arcticcoasts.org.

French, H., Demitroff, M., Newell, W.L., 2009. Past permafrost on the Mid-Atlantic Coastal Plain, eastern United States. Permafrost Periglacial Process. 20 (3), 285−294.

Goodrich, L., 1978. Efficient numerical technique for one-dimensional thermal problems with phase change. Int. J. Heat Mass Transfer 21 (5), 615−621.

Gorham, E., 1991. Northern peatlands: role in the carbon cycle and probable responses to climatic warming. Ecol. Applications 1 (2), 182−195.

Grebenets, V.I., Kerimov, A.G., Titkov, S.N., 1998. Dangerous movement of an anthropogenic rock glacier, Norilsk region northern Siberia. In: Proc. 7th International Conference on Permafrost, Yellowknife, pp. 23−27.

Grebenets, V.I., Streletskiy, D.A., Shiklomanov, N.I., 2012. Geotechnical safety issues in the cities of Polar regions, geography. Environ. Sustain. 5 (3), 104−119.

Gruber, S., Haeberli, W., 2007. Permafrost in steep bedrock slopes and its temperature-related destabilization following climate change. J. Geophys. Res. Earth Surf. (2003−2012) 112 (F2).

Haeberli, W., 1992. Construction, environmental problems and natural hazards in periglacial mountain belts. Permafrost Periglacial Process. 3 (2), 111−124.

Haeberli, W., Hallet, B., Arenson, L., Elconin, R., Humlum, O., Kääb, A., Kaufmann, V., Ladanyi, B., Matsuoka, N., Springman, S., 2006. Permafrost creep and rock glacier dynamics. Permafrost Periglacial Process. 17 (3), 189−214.

Harris, C., Isaksen, K., 2008. Recent warming of European permafrost: evidence from borehole monitoring. In: Kane, D., Hinkel, K.M. (Eds.), Proceedings of the 9th International Conference on Permafrost, June 29−July 3. Institute of Northern Engineering, University of Alaska Fairbanks, Fairbanks, Alaska, pp. 655−661.

Hinkel, K., Nelson, F., 2003. Spatial and temporal patterns of active layer thickness at Circumpolar Active Layer Monitoring (CALM) sites in northern Alaska, 1995−2000. J. Geophys. Res. 108 (D2), 8168.

Hinkel, K., Nelson, F., Parker, W., Romanovsky, V., Smith, O., Tucker, W., Vinson, T., Brigham, L., 2003. Climate Change, Permafrost, and Impacts on Civil Infrastructure. US Arctic Research Commission, Permafrost Task Force. Report. 72 pp.

Hinzman, L.D., Bettez, N.D., Bolton, W.R., Chapin, F.S., Dyurgerov, M.B., Fastie, C.L., Griffith, B., Hollister, R.D., Hope, A., Huntington, H.P., 2005. Evidence and implications of recent climate change in northern Alaska and other arctic regions. Climatic Change 72 (3), 251−298.

Instanes, A., Anisimov, O., 2008. Climate change and Arctic infrastructure. In: Kane, D., Hinkel, K.M. (Eds.), Proceedings Ninth International Conference on Permafrost, June 29−July 3. Institute of Northern Engineering, University of Alaska Fairbanks, Fairbanks, Alaska, pp. 779−784.

IPCC, 2013. Climate Change 2013: The Physical Science Basis.

Isaksen, K., Ødegård, R.S., Etzelmüller, B., Hilbich, C., Hauck, C., Farbrot, H., Eiken, T., Hygen, H.O., Hipp, T.F., 2011. Degrading mountain permafrost in southern Norway: spatial and temporal variability of mean ground temperatures, 1999−2009. Permafrost Periglacial Process. 22 (4), 361−377.

Jafarov, E., Romanovsky, V., Genet, H., McGuire, A., Marchenko, S., 2013. The effects of fire on the thermal stability of permafrost in lowland and upland black spruce forests of interior Alaska in a changing climate. Environ. Res. Lett. 8 (3), 035030.

Jeffries, M.O., Richter-Menge, J.A., Overland, J.E., 2012. Arctic Report Card. http://www.arctic.noaa.gov/reportcard.

Johansson, M., Åkerman, J., Keuper, F., Christensen, T.R., Lantuit, H., Callaghan, T.V., 2011. Past and present permafrost temperatures in the Abisko area: redrilling of boreholes. AMBIO 40 (6), 558−565.

Jones, B., Arp, C., Jorgenson, M., Hinkel, K., Schmutz, J., Flint, P., 2009. Increase in the rate and uniformity of coastline erosion in Arctic Alaska. Geophys. Res. Lett. 36 (3).

Jorgenson, M.T., Harden, J., Kanevskiy, M., O'Donnell, J., Wickland, K., Ewing, S., Manies, K., Zhuang, Q., Shur, Y., Striegl, R., 2013. Reorganization of vegetation, hydrology and soil carbon after permafrost degradation across heterogeneous boreal landscapes. Environ. Res. Lett. 8 (3), 035017.

Kääb, A., Chiarle, M., Raup, B., Schneider, C., 2007. Climate change impacts on mountain glaciers and permafrost. Global Planet. Change 56 (1), vii−ix.

Kachinskii, V., Zavgorodnyaya, Y.A., Gennadiev, A., 2014. Hydrocarbon contamination of arctic tundra soils of the Bol'shoi Lyakhovskii Island (the Novosibirskie Islands). Euras. Soil Sci. 47 (2), 57−69.

Kade, A., Walker, D.A., 2008. Experimental alteration of vegetation on nonsorted circles: effects on cryogenic activity and implications for climate change in the Arctic. Arct. Antarct. Alp. Res. 40 (1), 96−103.

Kaverin, D., Mazhitova, G., Pastukhov, A., Rivkin, F., 2012. The transition layer in permafrost-affected soils, Northeast European Russia. In: 10th International Conference on Permafrost Salekhard, Russia. June 25−29, 2012, 2 (145−148).

Khrustalev, L.N., Davidova, I.V., 2007. Forecast of climate warming and account of it at estimation of foundation reliability for buildings in permafrost zone. Earth Cryos. XI (2), 68−75.

Khrustalev, L.N., Parmuzin, S.Y., Emelyanova, L.V., 2011. Reliability of Northern Infrastructure in Conditions of Changing Climate, 260 pp. University Book Press, Moscow.

Kondratiev, V.G., 2013. Geocryological problems of railroads on permafrost. Paper Presented at ISCORD 2013: Planning for Sustainable Cold Regions. In: Proceedings of the 10th International Symposium on Cold Regions Development, June 2−5, 2013. ASCE Publications, Anchorage, Alaska.

Koven, C.D., Riley, W.J., Stern, A., 2013. Analysis of permafrost thermal dynamics and response to climate change in the CMIP5 earth system models. J. Climate 26 (6), 1877−1900.

Kronic, Y.A., 2001. Accident rate and safety of natural-technogenic systems in cryolithozone. In: Proceedings of the Second Conference of Geocryologists of Russia. Russia, Moscow, pp. 138−146.

Kudryavtsev, V., Garagula, L., Kondrat'yeva, K., Melamed, V., 1974. Osnovy Merzlotnogo Prognoza. M: Izdatel'stvo MGU, 431.

Lantuit, H., Overduin, P.P., Couture, N., Wetterich, S., Aré, F., Atkinson, D., Brown, J., Cherkashov, G., Drozdov, D., Forbes, D.L., 2012. The Arctic Coastal Dynamics database: A new classification scheme and statistics on Arctic permafrost coastlines. Estuaries Coasts 35 (2), 383−400.

Larsen, P.H., Goldsmith, S., Smith, O., Wilson, M.L., Strzepek, K., Chinowsky, P., Saylor, B., 2008. Estimating future costs for Alaska public infrastructure at risk from climate change. Global Environ. Change 18 (3), 442−457.

Lavrov, S.A., Anisimov, O.A., 2011. Modelling of the Hydrothermal Regime of Soils: Description of the Dynamical Model and Comparison with Observations. Planeta, Moscow.

Lisitzin, A., 1999. The continental−ocean boundary as a marginal filter in the world oceans. In: Biogeochemical Cycling and Sediment Ecology. Springer, pp. 69−103.

Lonergan, S., Difrancesco, R., Woo, M.-K., 1993. Climate change and transportation in northern Canada: an integrated impact assessment. Climatic Change 24 (4), 331−351.

Ma, W., Mu, Y., Wu, Q., Sun, Z., Liu, Y., 2011. Characteristics and mechanisms of embankment deformation along the Qinghai−Tibet Railway in permafrost regions. Cold Reg. Sci. Technol. 67 (3), 178−186.

Malevsky-Malevich, S., Molkentin, E., Nadyozhina, E., Shklyarevich, O., 2001. Numerical simulation of permafrost parameters distribution in Russia. Cold Reg. Sci. Technol. 32 (1), 1−11.

Malkova, G.V., 2010. Mean-annual ground temperature monitoring on the steady-state-station "Bolvansky". Earth Cryos. XIV (3), 3−14.

McGuire, A.D., Anderson, L.G., Christensen, T.R., Dallimore, S., Guo, L., Hayes, D.J., Heimann, M., Lorenson, T.D., Macdonald, R.W., Roulet, N., 2009. Sensitivity of the carbon cycle in the Arctic to climate change. Ecol. Monogr. 79 (4), 523−555.

Melnikov, E., Leibman, M., Moskalenko, N., Vasiliev, A., 2004. Active-layer monitoring in the cryolithozone of west Siberia. Polar Geogr. 28 (4), 267−285.

Nelson, F.E., Outcalt, S.I., 1987. A computational method for prediction and regionalization of permafrost. Arct. Alp. Res., 279−288.

Nelson, F.E., Anisimov, O.A., Shiklomanov, N.I., 2001. Subsidence risk from thawing permafrost. Nature 410 (6831), 889−890.

Oberman, N.G., 2008. Contemporary permafrost degradation of the European north of Russia. In: Kane, D., Hinkel, K. (Eds.), Proceedings of the Ninth International Conference on Permafrost, June 29−July 3. Institute of Northern Engineering, University of Alaska Fairbanks, Fairbanks, Alaska, pp. 1305−1310.

Oberman, N.G., Shesler, I.G., 2009. Observed and projected changes in permafrost conditions within the European North-East of the Russian Federation. Problemy Severa i Arctiki Rossiiskoy Federacii 9, 96−106.

Oechel, W.C., Hastings, S.J., Vourlrtis, G., Jenkins, M., Riechers, G., Grulke, N., 1993. Recent change of Arctic tundra ecosystems from a net carbon dioxide sink to a source. Nature 361 (6412), 520−523.

Ogorodov, S., 2005. Human impacts on coastal stability in the Pechora sea. Geo-Marine Lett. 25 (2−3), 190−195.

Pfister, R., Schneebeli, M., 1999. Snow accumulation on boards of different sizes and shapes. Hydrol. Process. 13 (14−15), 2345−2355.

Raynolds, M.K., Walker, D.A., Ambrosius, K.J., Brown, J., Everett, K.R., Kanevskiy, M., Kofinas, G.P., Romanovsky, V.E., Shur, Y., Webber, P.J., 2014. Cumulative geoecological effects of 62 years of infrastructure and climate change in ice-rich permafrost landscapes, Prudhoe Bay Oilfield, Alaska. Global Change Biol. 20, 1211−1224.

Repelewska-Pekalowa, J., Pekala, K., 2004. Active-layer dynamics at the Calypsostranda CALM Site, Recherche Fiord Region, Spitsbergen. Polar Geogr. 28 (4), 326−343.

Riseborough, D., Burn, C., 1988. Influence of an organic mat on the active layer. In: Proc. Fifth Int. Conf. On Permafrost, pp. 633−638.

Romanovskii, N., Hubberten, H.W., 2001. Results of permafrost modelling of the lowlands and shelf of the Laptev Sea region, Russia. Permafrost Periglacial Process. 12 (2), 191−202.

Romanovsky, V., Osterkamp, T., 1997. Thawing of the active layer on the coastal plain of the Alaskan Arctic. Permafrost Periglacial Process. 8 (1), 1−22.

Romanovsky, V., Drozdov, D., Oberman, N., Malkova, G., Kholodov, A., Marchenko, S., Moskalenko, N., Sergeev, D., Ukraintseva, N., Abramov, A., 2010b. Thermal state of permafrost in Russia. Permafrost Periglacial Process. 21 (2), 136−155.

Romanovskii, N., Hubberten, H.-W., Gavrilov, A., Eliseeva, A., Tipenko, G., 2005. Offshore permafrost and gas hydrate stability zone on the shelf of East Siberian Seas. Geo-marine Lett. 25 (2−3), 167−182.

Romanovsky, V., Smith, S., Christiansen, H., 2010a. Permafrost thermal state in the polar Northern hemisphere during the international polar year 2007−2009: a synthesis. Permafrost Periglacial Process. 21 (2), 106−116.

Roshydromet, 2014. Report on 2013 Climate in Russia, p. 68.

Sazonova, T., Romanovsky, V., 2003. A model for regional-scale estimation of temporal and spatial variability of active layer thickness and mean annual ground temperatures. Permafrost Periglacial Process. 14 (2), 125−139.

Schuur, E.A., Bockheim, J., Canadell, J.G., Euskirchen, E., Field, C.B., Goryachkin, S.V., Hagemann, S., Kuhry, P., Lafleur, P.M., Lee, H., 2008. Vulnerability of permafrost carbon to climate change: implications for the global carbon cycle. BioScience 58 (8), 701−714.

Serreze, M.C., Francis, J.A., 2006. The Arctic amplification debate. Climatic Change 76 (3−4), 241−264.

Shakhova, N., Semiletov, I., Salyuk, A., Yusupov, V., Kosmach, D., Gustafsson, Ö., 2010. Extensive methane venting to the atmosphere from sediments of the East Siberian Arctic Shelf. Science 327 (5970), 1246−1250.

Sharkhuu, N., Sharkhuu, A., Romanovsky, V.E., Yoshikawa, K., Nelson, F.E., Shiklomanov, N.I., 2008. Thermal state of permafrost in Mongolia. In: Kane, D., Hinkel, K. (Eds.), Proceedings of the Ninth International Conference on Permafrost, June 29−July 3. Institute of Northern Engineering, University of Alaska Fairbanks, Fairbanks, Alaska, pp. 1633−1639.

Sherstyukov, A.B., 2008. Correlation of soil temperature with air temperature and snow cover depth in Russia. Earth Cryos. XII (1), 79−87.

Shiklomanov, N., Nelson, F., 1999. Analytic representation of the active layer thickness field, Kuparuk River Basin, Alaska. Ecol. Model. 123 (2), 105−125.

Shiklomanov, N., Nelson, F., 2002. Active-layer mapping at regional scales: a 13-year spatial time series for the Kuparuk region, north-central Alaska. Permafrost Periglacial Process. 13 (3), 219−230.

Shiklomanov, N., Nelson, F., 2013. 8.22 Thermokarst and civil infrastructure. In: Shroder, J.F. (Ed.), Treatise on Geomorphology. Academic Press, San Diego, pp. 354−373.

Shiklomanov, N., Streletskiy, D., Little, J., Nelson, F., 2013. Isotropic thaw subsidence in undisturbed permafrost landscapes. Geophys. Res. Lett. 40 (24), 6356−6361.

Shiklomanov, N., Streletskiy, D., Nelson, F., 2012. Northern hemisphere component of the Global Circumpolar Active Layer Monitoring (CALM) Program. In: Proceedings of the 10th International Conference on Permafrost, Salekhard, Russia, pp. 377−382.

Shiklomanov, N., Streletskiy, D.A., Nelson, F.E., Hollister, R.D., Romanovsky, V.E., Tweedie, C.E., Bockheim, J.G., Brown, J., 2010. Decadal variations of active-layer thickness in moisture-controlled landscapes, Barrow, Alaska. J. Geophys. Res. Biogeosci. (2005−2012) 115 (G4).

Shur, Y.L., Goering, D.J., 2009. Climate change and foundations of buildings in permafrost regions. In: Permafrost Soils. Springer, pp. 251−260.

Shur, Y.L., Hinkel, K.M., Nelson, F.E., 2005. The transient layer: implications for geocryology and climate-change science. Permafrost Periglacial Process. 16 (1), 5−17.

Slater, A.G., Lawrence, D.M., 2013. Diagnosing present and future permafrost from climate models. J. Climate 26, 5608−5623.

Smith, S., Romanovsky, V., Lewkowicz, A., Burn, C., Allard, M., Clow, G., Yoshikawa, K., Throop, J., 2010. Thermal state of permafrost in North America: a contribution to the International Polar Year. Permafrost Periglacial Process. 21 (2), 117−135.

Smith, S., Wolfe, S., Riseborough, D., Nixon, F., 2009. Active-layer characteristics and summer climatic indices, Mackenzie Valley, Northwest Territories, Canada. Permafrost Periglacial Process. 20 (2), 201−220.

Smith, M., 1975. Microclimatic influences on ground temperatures and permafrost distribution, Mackenzie Delta, Northwest Territories. Can. J. Earth Sci. 12 (8), 1421−1438.

Sokratov, S.A., Barry, R.G., 2002. Intraseasonal variation in the thermoinsulation effect of snow cover on soil temperatures and energy balance. J. Geophys. Res. 107 (D10), 4093.

Stanilovskaya, Y.V., Merzlyakov, V.P., 2013. The potential thermokarst hazard assessment of the wedge polygonal ice for the pipelines. Oil Oil Prod. Pipeline Transport. Sci. Technol. 3 (11), 48−54.

Stendel, M., Christensen, J., 2002. Impact of global warming on permafrost conditions in a coupled GCM. Geophys. Res. Lett. 29 (13), 1632.

Stephenson, S.R., Smith, L.C., Agnew, J.A., 2011. Divergent long-term trajectories of human access to the Arctic. Nat. Clim. Change 1 (3), 156−160.

Streever, B., 2012. Ice-rich permafrost and the rehabilitation of tundra on Alaska's North Slope: lessons learned from case studies, in. In: Hinkle, K. (Ed.), Proceedings of the Tenth International Conference on Permafrost, Salekhard, Russia. The Northern Publisher, Salekhard, Russia, pp. 573−574, 25−29 Jun 2012.

Streletskaya, I.D., Vasiliev, A.A., Vanstein, B.G., 2009. Erosion of sediment and organic carbon from the Kara Sea coast. Arct. Antarct. Alp. Res. 41 (1), 79−87.

Streletskiy, D.A., Shiklomanov, N.I., Hatleberg, E., 2012a. Infrastructure and a changing climate in the Russian Arctic: a geographic impact assessment. In: Proceedings of the 10th International Conference on Permafrost, vol. 1, pp. 407−412.

Streletskiy, D.A., Shiklomanov, N.I., Nelson, F.E., 2012b. Spatial variability of permafrost active-layer thickness under contemporary and projected climate in Northern Alaska. Polar Geogr. 35 (2), 95−116.

Streletskiy, D.A., Shiklomanov, N.I., Nelson, F.E., 2012c. Permafrost, infrastructure, and climate change: a GIS-based landscape approach to geotechnical modeling. Arct. Antarct. Alp. Res. 44 (3), 368−380.

Streletskiy, D.A., Shiklomanov, N.I., Nelson, F.E., Klene, A.E., 2008. 13 Years of observations at Alaskan CALM Sites: long-term active layer and ground surface temperature trends. In: Kane, D., Hinkel, K.M. (Eds.), Proceedings of the 9th International Conference on Permafrost, June 29−July 3. Institute of Northern Engineering, University of Alaska Fairbanks, Fairbanks, Alaska, pp. 1727−1732.

Tarnocai, C., Canadell, J., Schuur, E., Kuhry, P., Mazhitova, G., Zimov, S., 2009. Soil organic carbon pools in the northern circumpolar permafrost region. Global Biogeochem. Cycles 23 (2).

Taylor, A.E., Wang, K., Smith, S.L., Burgess, M.M., Judge, A.S., 2006. Canadian Arctic Permafrost Observatories: detecting contemporary climate change through inversion of subsurface temperature time series. J. Geophys. Res. Solid Earth (1978−2012) 111 (B2).

Thibault, S., Payette, S., 2009. Recent permafrost degradation in bogs of the James Bay area, northern Quebec, Canada. Permafrost Periglacial Process. 20 (4), 383−389.

Tirtikov, A.P., 1978. Permafrost and vegetation: heat exchange in the active layer. In: Proceedings of the Second International Conference on Permafrost, pp. 100−104.

Varlamov, S.P., 2003. Insulating Effect of Land Covers in Central Yakutia. Paper presented at Proceedings of the Eighth International Conference on Permafrost, 21−25 July 2003. A.A. Balkema Publishers, Zurich, Switzerland.

Vasiliev, A.A., Leibman, M.O., Moskalenko, N.G., 2008. Active layer monitoring in West Siberia under the CALM II Program. In: Kane, D., Hinkel, K. (Eds.), Proceedings of Ninth International Conference on Permafrost, June 29−July 3. Institute of Northern Engineering, University of Alaska Fairbanks, Fairbanks, Alaska, pp. 1815−1820.

Vasiliev, A.A., Kanevskiy, M., Cherkashov, G., Vanshtein, B., 2005. Coastal dynamics at the Barents and Kara Sea key sites. Geo-Marine Lett. 25 (2−3), 110−120.

Velichko, A., Faustova, M., 2009. Glaciation during the Late Pleistocene. GEOS, Moscow.

Velichko, A., Nechaev, V., 2009. Subaerial Cryolithozone of the Northern Hemisphere during the Late Pleistocene and Holocene. GEOS, Moscow.

Voitkovskiy, K.F. (Ed.), 1999. Fundamentals of Glaciology. Nauka, Moscow, p. 255.

Walker, D., Jia, G., Epstein, H., Raynolds, M., Chapin Iii, F., Copass, C., Hinzman, L., Knudson, J., Maier, H., Michaelson, G., 2003. Vegetation-soil-thaw-depth relationships along a low-arctic bioclimate gradient, Alaska: synthesis of information from the ATLAS studies. Permafrost Periglacial Process. 14 (2), 103−123.

Walter, K., Edwards, M., Grosse, G., Zimov, S., Chapin, F., 2007. Thermokarst lakes as a source of atmospheric CH_4 during the last deglaciation. Science 318 (5850), 633−636.

Whiteman, G., Hope, C., Wadhams, P., 2013. Climate science: vast costs of Arctic change. Nature 499 (7459), 401−403.

Wilson, K., 2007. Environmental Assessment and Finding of No Significant Impact. Section 117 Expedited Erosion Control Project Kivalina, Alaska. US Army Corps of Engineers, Alaska District. September 2007.

Wisser, D., Marchenko, S., Talbot, J., Treat, C., Frolking, S., 2011. Soil temperature response to 21st century global warming: the role of and some implications for peat carbon in thawing permafrost soils in North America. Earth Syst. Dyn. Dis. 2 (1), 161−210.

Woo, M.-K., 2012. Permafrost Hydrology, 563 pp. Springer-Verlag, Berlin.

Yershov, E.D., Williams, P.J., 1998. General Geocryology. Cambridge University Press.

Zhang, T., Osterkamp, T., Stamnes, K., 1996. Influence of the depth hoar layer of the seasonal snow cover on the ground thermal regime. Water Resour. Res. 32, 2075−2086.

Zhao, L., Wu, Q., Marchenko, S., Sharkhuu, N., 2010. Thermal state of permafrost and active layer in Central Asia during the International Polar Year. Permafrost Periglacial Process. 21 (2), 198−207.

Zimov, S.A., Schuur, E.A., Chapin III, F.S., 2006. Permafrost and the global carbon budget. Science (Washington) 312 (5780), 1612−1613.

Zoltai, S.C., 1995. Permafrost distribution in peatlands of west-central Canada during the Holocene warm period 6000 years BP. Géogr. Phys. Quat. 49 (1), 45−54.

Chapter 11

Radioactive Waste Under Conditions of Future Ice Ages

Urs H. Fischer [1], Anke Bebiolka [2], Jenny Brandefelt [3], Sven Follin [4],
Sarah Hirschorn [5], Mark Jensen [5], Siegfried Keller [2], Laura Kennell [5],
Jens-Ove Näslund [3], Stefano Normani [6], Jan-Olof Selroos [3] and
Patrik Vidstrand [3]

[1] *Nationale Genossenschaft für die Lagerung radioactiver Abfälle (Nagra), Wettingen,*
Switzerland, [2] *Bundesanstalt für Geowissenschaften und Rohstoffe (BGR), Stilleweg, Hannover,*
Germany, [3] *Svensk Kärnbränslehantering AB (SKB), Stockholm, Sweden,* [4] *SF GeoLogic AB,*
Täby, Sweden, [5] *Nuclear Waste Management Organization (NWMO), Toronto, ON, Canada,*
[6] *Civil and Environmental Engineering, University of Waterloo, ON, Canada*

ABSTRACT

The evolution of the landscape and the behavior of the groundwater system are key
concerns for the long-term management of radioactive waste in deep geological repositories. At time scales relevant for repository safety assessments (i.e., 1 Ma), future ice
ages are an important external perturbation to repository integrity in mid-latitudes and
northern latitudes. Ice ages result in the alteration of the landscape by glacial erosion and
sedimentation; lead to the formation of periglacial permafrost; and impose significant
transient hydraulic, mechanical, thermal, and chemical changes that can influence
groundwater flow and radionuclide mobility. Four case studies are presented: (1) discussing the processes involved in the formation of deeply incised troughs and overdeepened
valleys in the Swiss Plateau; (2) outlining the occurrence and dimensions of tunnel valleys
in the North German Plain and their impact on repository host rocks; (3) illustrating the
integration of techniques and geoscientific evidence, as applied in assessing the stability
and resilience of a deep-seated groundwater system in the Great Lakes Basin of North
America; and (4) detailing results of a numerical modeling study of the impact of
periglacial and glacial climate conditions on groundwater flow in the Fennoscandian
Shield.

11.1 INTRODUCTION

Radioactive waste has been produced since the 1940s through the use of
radioactive materials in power production, defense, industry, research, and
medicine. Many countries are developing plans for, or are proceeding with,

programs for the long-term management of short- and long-lived radioactive waste, generated by such activities, within deep geological repositories. Consistent with these plans, numerous countries support advanced technical programs (e.g., underground research laboratories; research, development, and engineering activities) that are focused on establishing a scientific basis and consensus for long-term safety of deep geological repositories.

The deep geological repository concept has been developed over many years, with emphasis placed on a multiple-barrier (both natural and engineered) system that is designed to safely contain and isolate radioactive waste over long time frames. Deep geological repositories are constructed at significant depth (hundreds of meters) below the ground surface, in rock formations characterized as having favorable properties for radionuclide retention, and consist of a network of placement rooms for radioactive waste. Because the primary purpose of the deep geological repository is to ensure that the radionuclides in the waste will remain contained and isolated from humans and the environment, a robust and comprehensive safety case must be developed that demonstrates, with confidence, that the deep geological repository can be safely implemented at a proposed site.

Due to the long half-lives of some of the radionuclides found in radioactive waste, deep geological repositories for spent nuclear fuel, and other high-level radioactive waste, need to provide safety on timescales of hundreds of thousands of years or longer (Fyfe, 1999), a time span during which future climate change will create significant ground surface perturbations (Talbot, 1999). In this regard, glaciation has been identified as one of the most intense perturbations associated with long-term climate change in northern latitudes. Much of northern Europe and northern North America have repeatedly been glaciated in the past two to three million years, and such cycles of glaciation and deglaciation are expected to continue into the future in response to variations in solar insulation due to the Earth's orbital dynamics. Climate-induced changes, such as the advance and retreat of continental-scale ice sheets and permafrost, may influence and alter both the ground surface and subsurface environment, including their hydrology, and such changes must be considered when assessing the expected performance of a deep geological repository (Sheppard et al., 1995; Talbot, 1999).

This chapter looks at some of the potential impacts of future ice ages on the long-term performance of deep geological repositories for radioactive waste. The chapter first addresses the timing of future glacial inception by Näslund and Brandefelt (Section 11.2), followed by four case studies in which glacial erosion and sedimentation processes, as well as impacts on groundwater hydrology resulting from glaciation and deglaciation events, are discussed.

The first case study by Fischer (Section 11.3) is from the midland areas in northern Switzerland. Here, the emphasis is on the processes that lead to deep glacial erosion beneath the frontal reaches of ice age glaciers far away from the Alps. The second case study by Bebiolka and Keller (Section 11.4) is from

the lowlands in northern Germany and addresses future tunnel valley formation and its potential impact on host rocks and deep geological repository
safety. The third case study by Kennell, Normani, Hirschorn, and Jensen
(Section 11.5) is from southern Ontario in Canada and touches on the various
parameters used to demonstrate deep groundwater system stability and
longevity in the context of glaciation. Finally, the fourth case study by Vidstrand, Selroos, and Follin (Section 11.6) is from the southwestern part of the
Fennoscandian Shield in Sweden. The numerical modeling study investigates
changes in groundwater flow at repository depth during advance and retreat of
an ice sheet.

This chapter provides, for the first time, a compilation of contributions that
summarize a variety of current and relevant topics related to the challenges
associated with the long-term management of radioactive waste under conditions of future ice ages. The case studies also reflect the considerable differences in geological and hydrological settings, as well as glaciation histories,
between four different nations in northern latitudes. This chapter illustrates the
necessity of applying integrated, site-specific approaches when assessing the
complex processes and interactions that act over large spatial and temporal
scales during climate-related events, such as glacial advances and retreats or
deep ground freezing.

11.2 TIMING OF FUTURE GLACIAL INCEPTION

11.2.1 Introduction

Geological records show that over the Quaternary period, that is, the past
approximately 2.6 Ma, the Earth's climate has repeatedly varied from warm
interglacial to cold glacial periods. The cold periods have been characterized
by the presence of extensive ice sheets in high northern latitudes. For the past
approximately 900 ka, interglacials have occurred every 80–120 ka (e.g.,
Lisiecki, 2010). To illustrate these variations, atmospheric CO_2 concentrations
as measured in Antarctic ice cores and stacked $\delta\ ^{18}O$ as measured in marine
sediment records, a proxy for global ice volume, are shown in Figure 11.1. In
the analyses of long-term safety for nuclear waste repositories, one obvious
question in this context is the timing of the next glacial inception, that is, the
initiation of the next phase of ice sheet build-up. To address this question, an
understanding of the dynamics of past climate evolution and variability is
essential. Further, known and projected future variations in forcing conditions,
specific for the coming tens and hundreds of thousands of years, must be
considered.

The latitudinal and seasonal distribution of incoming solar radiation
(insolation) is the major external driver of the Earth's climate. This distribution
changes over time due to variations in the Earth's orbit, as illustrated with the
June insolation at 60°N displayed in Figure 11.1. These variations are due to

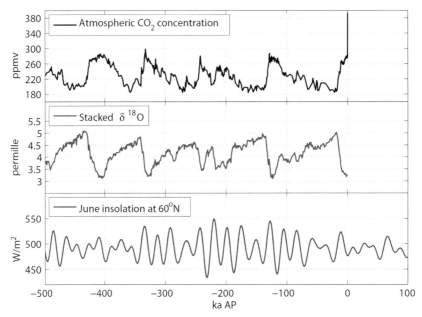

FIGURE 11.1 Upper panel: CO_2 variations over the past 500 ka. Composite record from Lüthi et al. (2008) complemented with the observed atmospheric CO_2 concentration in 2012 (394 ppmv; www.esrl.noaa.gov). Middle panel: Proxy data for global ice volume from a stack of benthic $\delta^{18}O$ records (Lisiecki and Raymo, 2005). High $\delta^{18}O$ values correspond to glacial periods (cold climate) and low values to interglacial states (warm climate). Lower panel: Summer insolation at high northern latitudes (June insolation at 60°N). *From Berger (1978) and Berger and Loutre (1991).*

variations in the eccentricity of the Earth's orbit, the obliquity (i.e., the tilt of the Earth's axis of rotation), and the precession of the equinoxes, with dominant frequencies at approximately 100 ka (eccentricity), 41 ka (obliquity), and 21—23 ka (precession). The approximately 100-ka time scale in the glacial/interglacial cycles of the last approximately 900 ka is commonly attributed to be controlled by these orbital variations. Astronomer Milutin Milankovitch proposed that summer insolation at high northern latitudes, with their large land masses, drives the glacial cycles, so that the Earth is in an interglacial state when its rotational axis both tilts to a high obliquity and precesses to align the Northern Hemisphere summer with the Earth's nearest approach to the Sun. Statistical analyses of long climate records support this theory (e.g., Hays et al., 1976; Huybers, 2011; Lisiecki and Raymo, 2007; Tzedakis et al., 2012). However, many questions remain on how orbital cycles in insolation produce the observed climate response (e.g., Huybers, 2011; Imbrie et al., 2011; Lisiecki, 2010).

The amplitude and the sawtooth shape of the variations in the glacial cycle climatic records (Figure 11.1) imply that non-linearities and amplifications exist, for example, through ice/snow albedo, atmosphere and ocean circulation, and the carbon cycle. A number of modeling studies have been performed to increase the understanding of the physical mechanisms associated with deglaciation and glaciation (see e.g., Mysak, 2008). Several climate model studies find that the combination of orbital insolation variations and glacial—interglacial atmospheric CO_2 variations gives a reasonable agreement between simulated glacial cycles and paleoclimate reconstructions (e.g., Loutre and Berger, 2000; Ganopolski et al., 2010).

The most contentious problem is why late Pleistocene climate records are dominated by the 100-ka cyclicity. The variations in insolation are dominated by the 41-ka obliquity and 23-ka precession cycles, whereas the 100-ka eccentricity cycle produces very small variations in seasonal or mean annual insolation. Recently, Abe-Ouchi et al. (2013) used comprehensive climate and ice sheet modeling to simulate the ice sheet variation for the past 400 ka forced by the insolation and atmospheric CO_2 content. Their model remarkably well simulates the sawtooth characteristic of glacial cycles (Figure 11.1), the timing of the terminations, and the amplitude of the Northern Hemisphere ice-volume variations, as well as the ice sheet configurations at the last glacial maximum (LGM) and the subsequent deglaciation. Their results suggest that the approximately 100-kyr cycle is essentially produced by the eccentricity modulation of precession amplitude through the changes in summer insolation with the support of obliquity for glacial terminations. Further, they conclude that insolation needs to be combined with internal feedbacks between the climate, the ice sheets and the lithosphere—asthenosphere system to explain the 100-ka periodicity.

11.2.2 Future Long-term Atmospheric Greenhouse-Gas Concentrations and Insolation

During previous interglacials, the atmospheric CO_2 concentration reached peak values of approximately 300 ppmv. The current (2013) atmospheric CO_2 concentration is approximately 396 ppmv (www.esrl.noaa.gov), and the Holocene peak value will be determined by human activities. The future CO_2 concentration is controlled by cumulative anthropogenic carbon emissions and the processes that act to reduce the atmospheric concentration and also by processes involved in the natural glacial—interglacial cycles (Lüthi et al., 2008). Under the assumption of continued anthropogenic carbon emissions, Archer et al. (2009) suggested that the atmospheric CO_2 concentration will remain above the preindustrial value of 280 ppmv for the next 10 ka and possibly even the next 100 ka. An improbable lower bound for future CO_2 concentration was obtained by Brandefelt et al. (2013) by assuming that human carbon emissions would be reduced to zero in the near future, resulting in an atmospheric CO_2 concentration below the preindustrial 280 ppmv around 10 ka after present (AP). The situation with high atmospheric greenhouse gas concentrations has no late Pleistocene analog, and several modeling studies indicate that the high CO_2 concentration will be of great importance for the Earth's climate evolution in the next 100–200 ka. Therefore, when analyzing the climate development over the coming tens and hundreds of thousands of years, including the timing of the next glacial inception, the influence of high atmospheric greenhouse gas concentrations needs to be taken into account.

Another thing to consider in this context is that over the next 100 ka, the amplitude of insolation variations will be exceptionally small, considerably smaller than, for example, during the last glacial cycle. For example, at latitude 60°N in June, insolation will vary by <25 W/m^2 over the next 25 ka, compared with 110 W/m^2 between 125 ka and 115 ka before present (BP) (Figure 11.1). Since, as previously described, glacial–interglacial cycles are believed to be driven by changes in insolation, this circumstance is also expected to influence the timing of the next glacial inception (e.g., Loutre and Berger, 2000; Berger and Loutre, 2002).

11.2.3 Modeling of Future Glacial Inception

The potential timing of future glacial inception may be analyzed by climate modeling. Complex climate models such as atmosphere-ocean general circulation models (AOGCMs) and Earth System Models have not yet been applied for studies of climate evolution beyond 10 ka into the future, since these models are too computationally expensive at present. However, several modeling studies of climate evolution over the coming 100 ka have been performed with simplified so-called simple climate models (SCMs) and earth system models of intermediate complexity (EMICs). These models include simplified descriptions of the main components of the climate system (i.e., atmosphere, ocean, sea ice, ice sheets, and sometimes vegetation).

Following Brandefelt et al. (2013), seven studies of the Earth's climate evolution in the coming 100–200 ka, performed with five different EMICs and one SCM, are briefly described here with a focus on the timing of glacial inception. In all the EMIC studies, the models are forced by the known future variations in orbital parameters and different scenarios for future atmospheric CO_2 concentration. The timing of glacial inception in these different studies is displayed as a function of the atmospheric CO_2 concentration in Figure 11.2.

Loutre and Berger performed simulations of Northern Hemisphere climate and ice sheet evolution for the period from 200 ka before present to 130 ka after present (Loutre and Berger, 2000; Berger and Loutre, 2002). They found that glacial inception occurs approximately 50 ka into the future in simulations with a constant CO_2 concentration lower than the preindustrial value (280 ppmv), whereas for CO_2 concentrations above this value, glaciation is

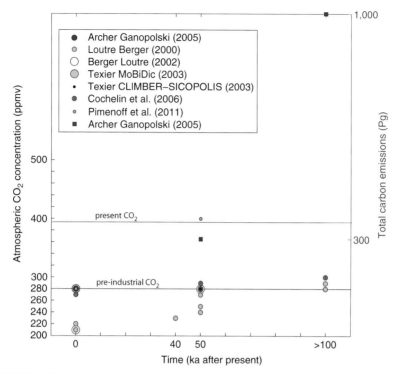

FIGURE 11.2   Approximate timing of glacial inception versus atmospheric CO_2 concentration (circles and asterisk) or total carbon emissions (squares) for the studies summarized in Section 11.2.3. In these model studies, glacial inception occurs during periods of minima in the incoming summer solar radiation at high northern latitudes, that is, around 0 ka AP, 54 ka AP, and 100 ka AP.

postponed beyond 50 ka into the future (Loutre and Berger, 2000). For CO_2 concentrations <220 ppmv, glaciation inception was found to be imminent.

Texier et al. (2003) simulated the climate over the coming 200 ka with two different models. They found that the models responded quite differently to the imposed forcing conditions; one model produced ice sheets over North America from the start of the simulation when CO_2 was kept constant at 280 ppmv, whereas the other produced no glaciation until after 50 ka after present. Under this scenario, neither of the models produced ice sheets over northern Europe during the next 100 ka.

Archer and Ganopolski (2005) performed climate simulations over the coming 500 ka. They found that glacial inception occurs approximately 50 ka into the future with a constant CO_2 concentration of 280 ppmv. They also performed simulations with future atmospheric CO_2 scenarios based on carbon cycle modeling following an instant release of 300, 1000, and 5000 Pg

(10^{15} g), respectively. Humankind has up to now released approximately 300 Pg carbon to the atmosphere, and the total cumulative emissions will surpass 1000 Pg carbon before the end of this century under business-as-usual scenarios (Archer et al., 2009). The remaining fossil fuel reserve that it might be reasonable to extract under present economic conditions totals approximately 5000 Pg carbon. For cumulative emissions of 300 and 1000 pg of carbon, glacial inception occurred around 50 ka and 130 ka, respectively. They also concluded that a carbon release from fossil fuels or methane hydrate deposits of 5000 Pg carbon could prevent glaciation for the next 500 ka.

Cochelin et al. (2006) simulated the climate of the coming 100 ka. Their results showed three types of evolution for the ice volume: an imminent glacial inception (low CO_2 levels), a glacial inception around 50 ka after present (CO_2 levels of 280 or 290 ppmv), or no glacial inception during the next 100 ka (CO_2 levels of 300 ppmv and higher).

Crucifix and Rougier (2009) utilized a three-degree-of-freedom stochastic model (SCM) to investigate the timing of the next glacial inception. Their provisional results indicated that without anthropogenic intervention (i.e., only natural processes) peak glacial conditions would occur in 60 ka after present, with glacial inception starting at 40 ka after present. Since atmospheric CO_2 concentration is a variable in the SCM used in this study, the results are not included in Figure 11.2.

Pimenoff et al. (2011) performed EMIC simulations for the coming 120 ka, coupled to a thermomechanical ice sheet model. Assuming a constant CO_2 concentration of 280 ppmv, glacial inception was immediate, whereas a constant CO_2 concentration of 400 ppmv resulted in glacial inception around 50 ka after present.

In addition to the above studies that were all conducted with simplified climate models (EMICs or in one case with an SCM), Vettoretti and Peltier (2011) investigated the impact of insolation, greenhouse-gas forcing and ocean circulation changes on glacial inception with a full AOGCM. They simulated the climate for orbital year 10 ka after present, when obliquity is at a minimum and eccentricity is low. Provided that atmospheric greenhouse-gas concentrations have reached preindustrial or lower levels at this time, they concluded that this period is favorable for a full onset of the next glacial cycle.

In this context, it should be mentioned that the results of these models may be sensitive to small disturbances and variations of model parameters, as noted by, for example, Archer and Ganopolski (2005). The sensitivity of the results to model parameters is studied by, for example, Charbit et al. (2013). They found large differences in the simulated ice sheet evolution depending on the chosen parameterization. Further, Crucifix (2011) shows that the timing of the next glacial inception, simulated with conceptual models designed to capture the gross dynamics of the climate system, is sensitive to small disturbances. He concludes, in agreement with Raymo and Huybers (2008), that the target of developing a dynamical system to convincingly model glacial cycles "is still elusive."

11.2.4 Timing of Future Glacial Inception and Concluding Remarks

In summary, the modeling studies reviewed in the previous section indicate two potential future timings of the next glacial inception, around 50 ka after present and around 100 ka after present (Figure 11.2). These timings occur in periods of low summer insolation at high northern latitudes. Low summer insolation is however not sufficient for glacial inception to take place in these models. A further requirement is that the atmospheric CO_2 concentration has decreased from the present high level (almost 400 ppmv) to the preindustrial (280 ppmv) level or below. As previously mentioned, the uncertainty is large concerning the future atmospheric CO_2 concentrations on these timescales. However, it is not unlikely that a preindustrial level is reached 50 ka after present, and the same is valid also for the period around 100 ka after present. In addition to the above results, the reviewed studies also indicate that glacial inception would be imminent if the present CO_2 concentration would have been at preindustrial levels.

The large uncertainty in the future evolution of the Earth's climate, illustrated by this review and discussed by, for example, Crucifix (2011), must be taken into account in safety assessments for nuclear waste repositories. Long-term future climate development cannot be predicted with enough confidence and detail for assessments of nuclear waste repository safety. Therefore, analyses of repository safety typically need to be based on a "range" of possible future climate scenarios to cover the uncertainty in future climate development (Näslund et al., 2013). The scenario range is determined from the current knowledge on reconstructed past and projected future climate evolution for the repository site, of which this review on the timing of the next glacial inception serves as an example. The character of the nuclear waste material to be deposited in the repository (radioactivity level and lifetime of key radionuclides), which influences the total time that needs to be covered by the safety assessment, as well as the repository concept (barrier material and the depth of geological repositories), also need to be taken into account in the construction of safety assessment climate scenarios (Näslund et al., 2013).

11.3 DEEP GLACIAL EROSION IN THE ALPINE FORELAND OF NORTHERN SWITZERLAND

11.3.1 Background

The National Cooperative for the Disposal of Radioactive Waste (Nagra) is in charge of developing deep geological repositories in Switzerland. This also includes proposing sites for such repositories and performing analyses of their long-term safety. As part of these safety analyses, the evolution of the landscape has to be evaluated for a time scale of one million years for the disposal of high-level radioactive waste (Nagra, 2008a). As the climate in the time period of concern (1 Ma) is expected to continue to oscillate between glacial and interglacial periods, a question comes to our mind as to why, how, where, and when future glaciations may affect the long-term safety of radioactive waste repositories in northern Switzerland. Of significance in this respect is that all of the proposed geological siting regions (Nagra, 2008b) are located within the ice extent during the Most Extensive Glaciation (MEG), some of them even within that at the LGM (Figure 11.3). One of the relevant aspects concerns the effects of "deep glacial erosion." This term refers to the origin of deeply incised troughs and overdeepened valleys (Nagra, 2008c, 2010; Jordan, 2010) beneath glaciers that extended from the Alps far into the Alpine foreland and covered the midland areas of northern Switzerland (Swiss Plateau) with ice hundreds of meters thick several times during the Quaternary (Figure 11.3; e.g., Schlüchter, 1979; Preusser et al., 2011).

Although the principles of glacial erosion are generally known, the processes that lead to deep glacial erosion beneath the frontal reaches of glaciers far away from the Alps (Figure 11.3) remain incompletely understood (Fischer, 2009; Haeberli, 2010; Fischer and Haeberli, 2010, 2012). Specifically, a lack in consensus exists about the factors that control processes and erosion rates beneath the ice. Uncertainty is also centered on the question whether the deeply incised troughs and overdeepened valleys in the Swiss Plateau were primarily carved out during a single glacial cycle or eroded and deepened in successive glaciations. Another important issue relates to the topographic steering of the flow of advancing ice at the onset of glaciations and thereby the focusing of future erosion in already existing valleys.

11.3.2 Ice-Age Conditions

Quantitative reconstructions together with model calculations have created a rather plausible view of the climatic and glaciological conditions that existed in northern Switzerland at the LGM (Haeberli, 2010; Haeberli et al., 1984). The strong and stable anticyclone that formed over the cold and high-elevation surface of the Laurentide Ice Sheet (LIS) resulted in the deviation of a branch of the Pacific jet stream via the Polar Ocean toward the Atlantic Ocean (e.g., Manabe and Broccoli, 1984; Hofer et al., 2012a,b) where the cold and dry air masses led to a penetration of winter sea ice to low latitudes (Frenzel et al., 1992). The corresponding closure of the Atlantic Ocean as a primary humidity source caused extremely cold and arid conditions in central Europe with mean annual air temperature values in northern Switzerland up to 15 °C lower than today's value (Vandenberghe and Pissart, 1993).

During the time period of lowest temperature and most extended area of surface ice, the frontal reaches of the glaciers spreading out over the Swiss

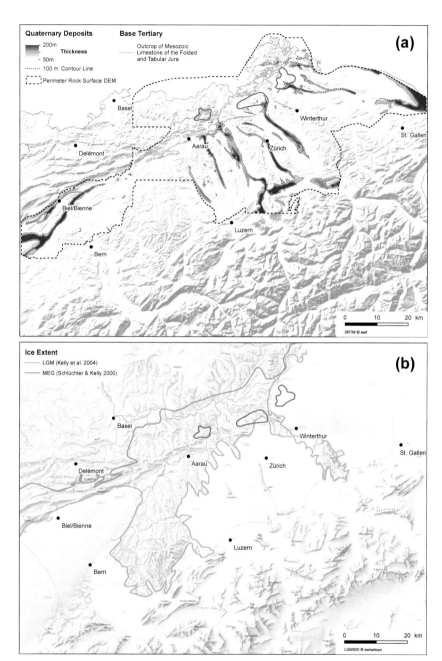

FIGURE 11.3   Map of northern Switzerland showing the proposed siting regions for a high-level radioactive waste repository (regions outlined in red): (a) Thickness of Quaternary sediments indicating a system of deep, overdeepened and buried valleys (Jordan, 2008, 2010) and outcrop of the Mesozoic limestone that separates the perialpine Molasse basin from the adjacent Jura. Mapping is based on information from several thousand boreholes and other sources *Shuttle Radar Topography Mission (SRTM) esri*; (b) Ice extent at the last glacial maximum (LGM) (Kelly et al., 2004; Bini et al., 2009) and during the Most Extensive Glaciation (MEG) (Schlüchter and Kelly, 2000) *LGM 500 swisstopo, reproduced by permission of swisstopo (BA14011).*

Plateau (Figure 11.3b) were predominantly polythermal to cold (Blatter and Haeberli, 1984), surrounded by continuous periglacial permafrost up to 150 m thick (Deslisle et al., 2003) and characterized by low driving stresses and correspondingly low mass balance gradients, mass turnover, and flow velocities (Haeberli and Penz, 1985; Benz-Meier, 2003). Subsurface temperatures and groundwater conditions were strongly influenced by the presence of extended surface and subsurface ice (Speck, 1994). More humid conditions with higher flow velocities, increased mass turnover, and meltwater effects must have prevailed during ice advance across the Swiss Plateau, and rapid downwasting or even collapse of the ice masses accompanied by violent outburst floods from temporarily ice-dammed lakes is likely to have taken place during the retreat phase into the Alpine valleys.

11.3.3 Processes of Glacial Erosion

The main processes of direct glacial erosion responsible for the landforms associated with glaciation are abrasion and quarrying and were identified over a century ago (e.g., Forbes, 1843; Tyndall, 1864). While abrasion involves the wearing down of rock surfaces by the grinding effect of glacier ice charged with basal debris, quarrying, or plucking, is the removal of well jointed or loosened blocks of bedrock by the overriding glacier. For abrasion and quarrying to take place, there must be some relative movement between the glacier sole and the bed, which therefore requires the basal ice to be at the pressure-melting point and sliding. Abrasion and quarrying, therefore, are dependent on the temperature and velocity of the basal ice, debris concentrations, ice thickness, and subglacial water pressure (e.g., Iverson, 2002; Bennett and Glasser, 2009; Burki, 2009; Swift et al., 2014).

Significant wear to the glacier bed can also be caused by subglacial fluvial action. Erosion of bedrock in subglacial meltwater channels proceeds in a manner similar to the bedrock incision by subaerial streams and is dependent on water velocity, discharge, and turbulence as well as the quantity of suspended sediments and bedload in traction (e.g., Drewry, 1986; Burki, 2009; Benn and Evans, 2010; Swift et al., 2014). Of specific importance in subglacial environments is that high water velocities and discharges require strong pressure gradients in relatively large, hydraulically efficient and well-connected subglacial channels.

Glacial erosion occurs where ice is warm based, which enables basal sliding and thus abrasion, and where surface meltwater is produced and is able to reach the bed because this promotes rapid sliding and stimulates quarrying. Concerning integrated effects over large areas, the processes of abrasion and quarrying may be largely responsible for the overall eroded volumes (e.g., Drewry, 1986; Burki, 2009). However, with respect to maximum possible erosion depths, subglacial fluvial action likely plays an important role (Fischer and Haeberli, 2010).

11.3.4 Glacial Overdeepening

Glacial erosion is closely linked to spatial patterns of ice flow and sediment transport and deposition (e.g., Alley et al., 1997). Maximum erosion is expected at a location close to the long-term average equilibrium line altitude (ELA) (Figure 11.4a) where ice flux and hence basal sliding is large (e.g., Hallet et al., 1996; Boulton, 1996; Anderson et al., 2006). However, without the evacuation of sediment by water from the ice-bed interface, the products of erosion may accumulate and inhibit further erosion (Alley et al., 2003a) (Figure 11.4b). Erosion may therefore be more efficient near the warm-based glacier terminus (Figure 11.4c) where abundant subglacial meltwater often flows in hydraulically efficient channels that have a high sediment transport capacity (Alley et al., 1997). Due to the ability of ice to move water and sediment along adverse bed slopes, continued erosion and deepening of the glacier bed will produce a closed topographic depression known as over-deepening (e.g., Linton, 1963; Harbor, 1989; Evans, 2008).

Cirque floors, trunk valleys, and fjords are commonly overdeepened or have overdeepened sections (Figure 11.5). Glacial erosion models that incorporate a simple relationship of erosion to glacier basal motion are able to generate overdeepenings in cirques and trunk valleys that appear realistic (e.g., MacGregor et al., 2000, 2009; Anderson et al., 2006). These models also predict confluence-type bedrock steps (hanging valleys) and over-deepenings, because ice convergence causes an increase in ice flux and velocity. However, observations of overdeepenings in areas of divergent flow where ice flux and hence erosional potential is low, such as those beneath former glacier termini in the Swiss Plateau (Figure 11.3), highlight the limitations of the simple velocity-based erosion rules. Moreover, localized bedrock quarrying may play a significant role in glacial overdeepening, whereby the formation of crevasses above initial perturbations caused by bed irregularities focuses the delivery of surface melt and, hence, water-pressure fluctuations to the head of incipient overdeepenings (Figure 11.4d; Hooke, 1991; Iverson, 1991). Constrictions in valley width or convexities over transitions from more to less resistant rock types (Jansson and Hooke, 1989) may provide the initial bed irregularities required for surface crevassing to initiate focused erosion (Hooke, 1991).

11.3.5 Water Flow in Overdeepenings

Overdeepenings have a fundamental influence on the nature of water flow and sediment transport within the glacial system. The most critical part of the water flow paths is the drainage along the adverse bed slope, where some of the heat generated by viscous dissipation must be used to keep the water at the pressure-melting point, such that less heat is available to maintain melting of the channel walls (e.g., Hooke, 1991; Alley et al., 2003a).

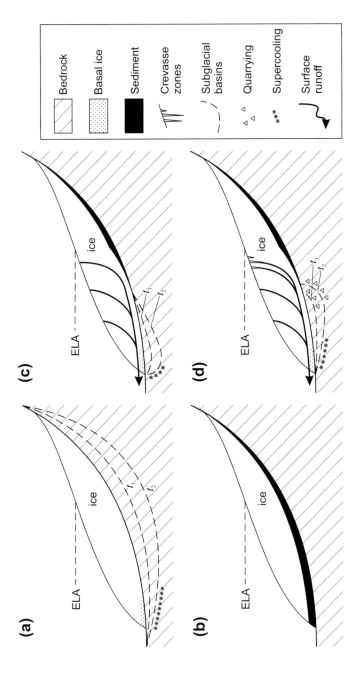

FIGURE 11.4 Schematic diagrams summarizing the processes associated with glacial erosion. *Adapted from Cook and Swift, 2012; with permission from Elsevier.* Dashed lines indicate glacier bed after erosion during two different time intervals. (a) Maximum erosion near the equilibrium line altitude (ELA) without hydrological forcing of basal sliding and sediment evacuation. (b) No erosion as a result of erosion products having accumulated to form a sediment layer that protects the bed. (c) Efficient erosion near the glacier terminus as a result of subglacial meltwater enhancing basal sliding and sediment evacuation. (d) Enhanced quarrying at the overdeepening head as result of focused delivery of surface meltwater through crevasses.

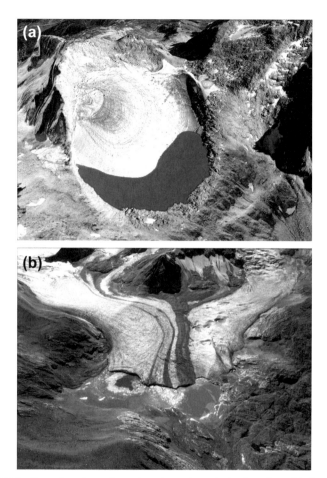

FIGURE 11.5   Examples of overdeepenings in modern glaciated settings. (a) Overdeepened cirque basin (Küehbodengletscher, Switzerland). *Image: Google.* (b) Confluence-type overdeepening (unnamed glacier in Peru (11°56'36S/76°04'10W)). *Image: Google, DigitalGlobe US Department of State Geographer.*

When the adverse slope exceeds 1.2−1.7 times the ice-surface slope, water becomes supercooled because the pressure-melting point rises faster than the water can be warmed, which causes some of the water to freeze (Röthlisberger, 1972; Röthlisberger and Lang, 1987; Hooke, 1991; Alley et al., 2003a; Cook et al., 2006; Creyts and Clarke, 2010; Creyts et al., 2013). Consequently, ice accretion and frazil ice formation (Alley et al., 1998; Lawson et al., 1998; Larson et al., 2006; Cook et al., 2007, 2010) will cause the transmissivity of the subglacial drainage system to reduce and basal water pressures to increase (Hanz and Lliboutry, 1983; Hooke and Pohjola, 1994). Water is thus forced to spread out across the bed in a hydraulically inefficient sheet or is diverted upward into englacial conduits or flows around the overdeepening in efficient channels along the valley sides (Hooke and Pohjola, 1994; Lawson et al., 1998; Fountain et al., 2005). By inhibiting the efficiency of the subglacial drainage system and thereby reducing the evacuation of sediment, glaciohydraulic super-cooling has been suggested to stabilize the geometry of overdeepenings (Alley et al., 1999, 2003b). As such, the depth of overdeepenings is likely to depend on the extent to which processes and feedbacks are able to focus erosion and maintain the evacuation of water and sediment (Cook and Swift, 2012).

11.3.6 Deep Glacial Erosion in the Swiss Plateau

Analysis of the deeply incised troughs and overdeepened valleys in the Swiss Plateau shows that the overdeepenings are mostly restricted to areas with a Molasse substratum (soft clastics) and end abruptly on reaching the Meso-zoic limestone of the Folded and Tabular Jura (Figure 11.3a; Jordan, 2010). This suggests a potential influence of bedrock lithology on the efficiency of erosion, for example, because large amounts of subglacial water may be lost into the underlying limestone karst (e.g., Grust, 2004; Gremaud and Gold-scheider, 2010) thereby preventing the build-up of high water pressures that are favorable for active erosion, and the energetic water flow necessary for sediment evacuation. Geological control on overdeepening formation is further supported by the observation of Preusser et al. (2010) that weaker lithologies correlate with the location of some of the overdeepenings within the European Alps.

The deeply incised troughs and overdeepened valleys in the Swiss Plateau (Figure 11.3) are likely to have formed either time transgressively during ice advance, when the steep surface slope of the ice margin provided the large hydraulic gradients necessary for a vigorous subglacial water flow and an efficient evacuation of erosion products from the ice-bed interface, or during the retreat phase by the outburst of subglacial meltwater dammed behind a cold-based ice margin or a marginal permafrost wedge (Piotrowski, 1994; Hooke and Jennings, 2006; Stumm, 2010; Kehew et al., 2012; Van der Vegt et al., 2012). Although the finding that pre-LGM sediments are preserved within many overdeepenings (e.g., Dehnert et al., 2012) suggests that not all glaciations have been able to excavate the postglacial sediment infill and produce renewed overdeepening, progressive deepening over successive glacial cycles may take place because ice is likely steered away from hills and ridges into already existing troughs and valleys (Kessler et al., 2008). This topographic steering causes the thicker and faster flowing ice to generate more basal melting that will further increase basal sliding and sediment evacuation and thus result in enhanced erosion. An interpretation of the observed maximum depth of the troughs and overdeepened valleys that can therefore arguably be defended is that they represent the greatest erosion that has occurred in the Swiss Plateau during the Quaternary, and that erosion at a similar pace can generally be expected over the next million years.

11.3.7 Future Research Focus

Despite the importance of subglacial water for basal sliding, erosion, and the requirement for erosion products to be evacuated from the glacial system, a weakness of many numerical modeling studies of glacial erosion is the exclusion of water and sediment transport processes. By incorporating a simplified representation of subglacial drainage into a glacial erosion model, Herman et al. (2011) have recently demonstrated the important influence of hydrology on spatial patterns of erosion and the formation of overdeepenings. Localized overdeepening occurs downglacier of the ELA where spatially focused surface meltwater contributions lead to a significant increase in sliding and erosion at the glacier bed. Modeling by Egholm et al. (2012) has further included processes of sediment transport by ice, basal sediment deformation, and subglacial water. With this approach, overdeepenings are formed even in regions of divergent ice flow beneath glacier termini, mainly because of the importance of sediment evacuation by subglacial water. Although the significance of erosion and evacuation of erosion products by glacial meltwater is being increasingly recognized, glaciofluvial processes remain to be fully investigated, accurately implemented in glacial erosion models and validated in realistic numerical simulations (Fischer, 2009; Fischer and Haeberli, 2010, 2012; Cook and Swift, 2012) when evaluating the hazard of a deep geological repository becoming exposed during future glaciations.

11.4 TUNNEL VALLEYS WITHIN THE NORTH GERMAN PLAIN AND THEIR RELEVANCE TO THE LONG-TERM SAFETY OF NUCLEAR WASTE REPOSITORIES

11.4.1 Background

The safety specifications in Germany for the disposal of heat-generating radioactive waste with long half-lives include the assessment of whether the waste can be safely contained for a period of over one million years. The comprehensive site-specific safety analysis takes into consideration and evaluates all the potential changes that could affect the geologic repository system, the containment-providing rock zone, and the cover rock and surrounding rock masses, in light of the expected geological, climate-related, and waste disposal-related processes.

In the German geologic repository and safety concepts, the geological barrier plays the key role in isolating the high-level radioactive substances. The rock types, halite and argillaceous rocks, are both seen as preferential barrier rocks in northern Germany because of their rock-mechanical properties (low permeability, thermal conductivity) and their occurrence at suitable depths and inadequate thicknesses.

A special phenomenon within the North German Plain is the formation of deep tunnel valleys during previous Quaternary glacial periods. These tunnel valleys, which were formed by extreme subglacial erosion processes, could diminish the protective effect of the cover rock and/or the surrounding rock, and in the case of unfavorably planned geologic repositories, also the host rock at a geologic repository site, if they were to form again under future climatic conditions. Estimating the evolution of the climate and the consequences of changes in climate on the potential formation of tunnel valleys in the future and at potential geologic repository sites are therefore important elements of safety analysis. The following looks at the geological situation in northern Germany and at the possible formation of deep tunnel valleys.

11.4.2 Tunnel Valleys in Northern Germany

Tunnel valleys are defined as elongate depressions with overdeepened areas along their floors cut into bedrock or unconsolidated sediment (Ó Cofaigh, 1996). They are also termed subglacial valleys or channels. The term tunnel valley infers a subglacial origin and follows the terminology of Jørgensen and Sandersen (2006).

The surface of the North German Plain is primarily covered by Quaternary sediments. These were deposited during three glacial periods (Elsterian, Saalian, and Weichselian) and corresponding interglacials, as well as in the Holocene. The maximum ice extent during the Elsterian glacial period (Figure 11.6) corresponds over most of northern Germany to the maximum ice cover during all three glacial periods. Positions of the ice-sheet margin during the Weichselian glacial period were located to the north of the Elbe River.

11.4.2.1 Tunnel Valley Distribution and Dimensions

The Base of the Quaternary in northern Germany (Figure 11.6) lies at a generally uniform depth of between 0 and 100 m below the sea level (Stackebrandt et al., 2001). Greater depths generally indicate the position of a tunnel valley system, mainly formed during the Elsterian glacial period. In northern Germany, the Elsterian tunnel valleys dominate with respect to their maximum depth. As shown by the depth map of the Base of the Quaternary, they cut down in some cases to up to 500–600 m below the sea level. The maximum depth of the tunnel valleys formed during the Saalian and Weichselian glacial periods reaches only around 200 m (e.g., Niedermayer, 1965). In northwestern Germany, tunnel valleys are oriented typically NNE–SSW, whereas those in northeastern Germany, typically run NE–SW. The tunnel valleys are therefore oriented almost orthogonally to the margin of the

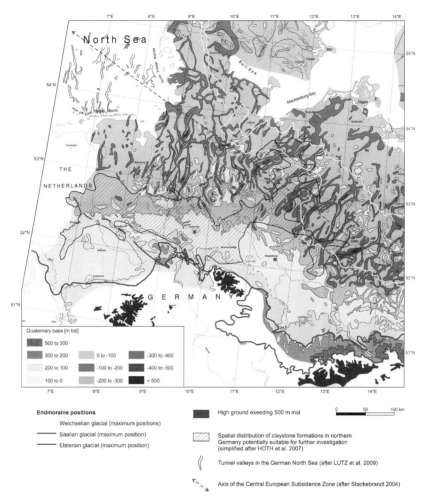

FIGURE 11.6   Base of Quaternary deposits and ice sheet margins in northern Germany. *Modified after Stackebrandt et al., 2001.*

Elsterian ice sheet. The deep Elsterian tunnel valleys end approximately along the Base Quaternary zero line. The tunnel valleys are located well to the north of the maximum extent of the Elsterian ice sheet and in areas whose current topographies do not exceed 100 m above the sea level. No obvious reason exists as to why no tunnel valleys are found in the remaining area toward the southern margin of the ice sheet. Perhaps this is because the ice sheet was not thick enough to generate the hydraulic conditions necessary for the formation of tunnel valleys.

The dimensions of the tunnel valleys formed during the Elsterian glacial period vary considerably. Onshore, the longest tunnel valley with depths

>200 m below sea level is 110 km long and has a maximum width of 12 km. Tunnel valleys individually have no linear courses but change direction in gentle curves. The width of a tunnel valley can change moderately within a very short distance. A general distinction can be made between tunnel valleys with narrow widths of up to 3 km and those whose widths exceed 5 km. The narrower type of tunnel valley is dominant in northwestern Germany, whereas the wider tunnel-valley type dominates in the east. The distance between neighboring tunnel valleys with depths >200 m lies between 10 km to >50 km. The tunnel valleys branch in net-like patterns, where the branches are in the form of shallow tunnel valleys with depths of <200 m. The parts of the tunnel valley systems with depths of >200 m are not branched. A tunnel valley can have several depth maxima, but this is not a general rule.

The position of the Pleistocene tunnel valleys in the German North Sea interpreted by Lutz et al. (2009) from seismic surveys supplements the work of Stackebrandt et al. (2001). The tunnel valleys are up to 66 km long and 8 km wide, and cut down 400 m and more into the Neogene sediments. The distribution of the tunnel valleys in the central German North Sea is irregular, and the directions of the tunnel valley longitudinal axes run NNW−SSE to NNE−SSW. The North Sea tunnel valleys are therefore similar to those found in the North German Plain (cf. Huuse and Lykke-Anderson, 2000).

Stackebrandt (2009), interpreted the distribution of the tunnel valleys as being related to the position of the neotectonically formed Central European Subsidence Zone. This runs in an SE−NW direction from Wroclaw via Berlin to Hamburg and on into the North Sea. The deepest parts of the tunnel valleys are located within this zone. The thick, poorly consolidated, and easily erodible sediments filling this subsidence zone are interpreted as a necessary condition for deep erosion by subglacial meltwater.

11.4.2.2 Tunnel Valley Development, Underlying Sediments, and Formation

The tunnel valleys are cut almost exclusively into unconsolidated Quaternary and Tertiary sediments because the base of the Cenozoic is deeper than 500 m over large swathes of northern Germany (Figure 12.1 in Doornenbal and Stevenson, 2010). When tunnel valleys cross anticlines in the vicinity of salt domes or structural highs, they also cut into solid Mesozoic rocks.

The course of tunnel valleys in northern Germany is not controlled by the position of the salt structures. Tunnel valleys can curve around a salt structure, for example, Boitzenburg, or cross it directly, for example, Gorleben tunnel valley (Figure 11.7). However, the hardness of the subsurface being eroded appears to have an influence on tunnel valley depths. The depth of erosion during the formation of tunnel valleys is therefore smaller in consolidated rocks. The dependency of tunnel valley formation on the rock hardness is also demonstrated by the "Kreuzbrücken fissure" (see Figure 5.4 in Cepek, 1993) near Rüdersdorf east of Berlin. According to Cepek (1993) and Schroeder (1995), this takes the form of an up to 50-m deep gorge cut into the Muschelkalk underlying the Quaternary (Anisian and Ladinian limestones). The cross-section of the gorge is dependent on the hardness of the rock: in the hard limestones its width is 35 m, which reduces to 5 m in the ravine-like narrowest zones. However, in the softer sediments, such as marl–limestone intercalations, and argillaceous rocks, the tunnel valley widens to a maximum of 160 m.

The formation of tunnel valleys during glacial periods is due to the outflow of large volumes of meltwater at the base of glaciers. According to Smed (1998) and Piotrowski (1994), the prerequisites for the formation of tunnel valleys are

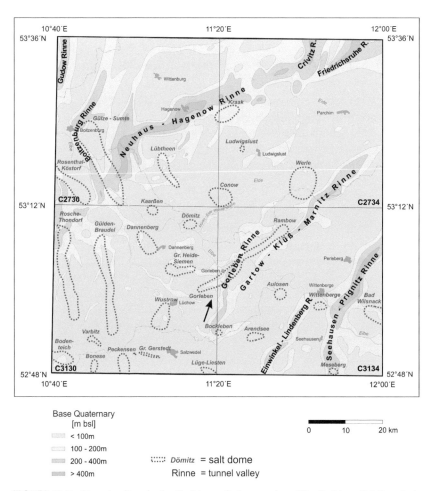

FIGURE 11.7   Tunnel valleys in the Gorleben salt dome region. *After Brückner-Röhling et al., 2002.*

pressurized meltwater and the presence of unconsolidated, water-saturated, and unfrozen sediments at the base of the glacier. Tunnel valleys form during the retreat phases of the glacier when melting gives rise to large volumes of water. According to Kehew et al. (2007), tunnel valleys are formed by a number of catastrophic outflow events from previously dammed volumes of water because the constant outflow volume resulting from meltwaters and groundwater is thought to have inadequate erosive power. The absence of a directed dip of the tunnel valley floor toward the end of the tunnel valley and the undulating longitudinal profile, with intense depth maxima, are typical features of formation beneath an ice sheet, and make these tunnel valleys very different from channels that have formed by subaerial fluviatile processes. Internal sedimentary structures point to multiple areas of erosion, and subsequent rapid refilling of the tunnel valleys. The concepts presented by the aforementioned authors on the formation of tunnel valleys are also adhered to in this article.

11.4.3 Impact of Tunnel Valley Formation on Geologic Repositories and Host Rocks

The safety concepts for geologic repositories discussed in Germany envisage the emplacement of high-level radioactive, heat-generating waste in deep geological formations. The host rocks with the potential to satisfy this concept in northern Germany are salt structures unaffected by earlier mining activities, or the thick argillaceous formations of the Mesozoic (Reinhold et al., 2008). Depending on the geologic repository concept pursued, the potential emplacement depths are between 800 and 1000 m in salt domes (e.g., Gorleben exploration mine: Bollingfehr et al., 2011), and around 400 m in argillaceous rocks, on the basis of petrophysical, engineering, and economic considerations (Amelung et al., 2007; Uhlig et al., 2007). The effect at depth of climate-related processes and the possible associated formation of tunnel valleys are therefore critical for the long-term safety of radioactive waste repositories.

The impact of the formation of tunnel valleys differs with respect to the type of the host rocks. In the case of salt host rocks, one not only has to take into consideration the hydromechanical aspects but also hydrochemical erosion (subrosion). In the case of argillaceous rocks, however, the only relevant parameter is hydromechanical erosion.

11.4.3.1 Rock Salt

A few salt structures have been identified in northern Germany, which have been crossed by tunnel valleys formed during the Elsterian glacial period, and which are eroded down as far as the cap rock or even to top salt (Kuster and Meyer, 1979). The northern German salt structures are not considered to be a prerequisite for the formation of tunnel valleys (Hinsch, 1979). However, the structures can influence the course of tunnel valleys: via the nature of the sediments; depressions overlying the salt, the presence of rim synclines, and the degree of hardness of the salt minerals and/or the cap rocks. The effects of tunnel valleys, formed during the Elsterian glacial period, on a host rock for a geologic radioactive waste repository is shown, for example, on the basis of the Gorleben salt dome that has been investigated in detail (Köthe et al., 2007). It can be concluded, from the results of the analysis, that the higher geomechanical strength of the anhydritic cap rock and the rock salt, compared to the Tertiary sediments, prevented deeper hydromechanical erosion of the top of the salt dome. Hard consolidated rock sequences are apparently an effective form of protection against deep erosion during the formation of tunnel valleys (see also Section 11.4.2.2 "Kreuzbrücken fissure").

In addition to the hydromechanical erosion, the contact between the rock salt and large volumes of meltwater during the formation of tunnel valleys also means that it was possible for the salt dome to be influenced by hydrochemical effects. The subrosion under the influence of meltwater that occurred at Gorleben during or shortly after the tunnel valley formation was more intense compared to the situation with salt-saturated groundwater and minor groundwater movement. Moreover, at certain special localities, where the protective subrosion-resistant evaporites are absent, deeper subrosion or alteration of the rock salt also occurs. In the case of the Gorleben salt dome, this gave rise to subrosion and alteration processes within the potash seam down to a maximum depth of approximately 170 m below today's top salt level (Bornemann et al., 2008).

Overall, the salt host rock at the investigated Gorleben site was affected hydromechanically and hydrochemically down to a maximum depth of around 470 m below sea level by the formation of the Elsterian tunnel valley. This value is the sum of the depth of the base of the deepest point of the tunnel valley derived from the hydromechanical erosion, and the maximum depth of subrosion and alteration processes within the rock salt. If the conditions in Gorleben are considered analogous to most northern German salt structures, and if the maximum tunnel valley influences during the past can be taken as a benchmark to assess the long-term safety, the construction of a safe geologic repository should be carried out at a depth considerably greater than the maximum depth of the range known from the hydromechanical erosion and subrosion processes. The only exceptions would be those salt structures whose tops are at appropriately greater depths, and which are located in regions outside of the zone of extreme tunnel valley formation. Such exceptions can be assumed to occur between the Lower Saxony Hills and areas where Base Quaternary lies at around 0 m below sea level (cf. Figure 11.6).

11.4.3.2 Argillaceous Rocks

The effects of the formation of tunnel valleys in argillaceous rocks can be similar to those in rock salt host rocks. In the North German Plain, the weakly consolidated sandy and argillaceous horizons of the younger Tertiary were primarily eroded during the formation of the tunnel valleys. Effects would also be possible on the underlying argillaceous rocks of Lower Cretaceous age that are considered as potential host rocks for a geologic repository in northern Germany (cf. Hoth et al., 2007), if they occur at shallow depths and no protective harder beds such as Upper Cretaceous limestones exist. The protective effect of chalk is indicated by studies on tunnel valleys in Denmark, which reach erosion depths of 50−350 m, but are only cut down a few tens of meters in chalk within the most deeply eroded rocks (Jørgensen and Sandersen, 2006). The geological investigations are interpreted in the case that greater erosion was prevented by the presence of hard and consolidated limestone.

To the south, toward the edge of the North German Plain, the thickness of the Tertiary horizons diminishes because of the Pliocene to Pleistocene uplift of the Earth's crust. The hard protective horizons of the Upper Cretaceous are absent here in places so that the Lower Cretaceous argillaceous rocks can crop out at the surface. Within the Lower Cretaceous sequence, sandy-marly, and in parts very hard siliceous horizons were formed in the southern part of the North German Plain in Upper Albian rocks, which also have a protective effect on underlying rocks with respect to erosion that takes place during the formation of tunnel valleys.

The information currently available on the distribution of tunnel valleys in northern Germany, their location with respect to special structural situations such as salt domes, and the morphology, reveal no rules with respect to the locations where they formed, making it difficult to draw conclusions about preferential locations for the formation of tunnel valleys in the future. The only suggestion is that the depth or width of tunnel valleys is limited by the presence of hard, resistant rock types. A safety analysis therefore has to take into consideration the impact of the formation of tunnel valleys on the cover rocks and surrounding rocks, as well as the host rocks, of potential geologic repository sites in northern Germany.

11.5 PALEOHYDROGEOLOGY AND GLACIAL SYSTEMS MODELING—CANADIAN PERSPECTIVE

11.5.1 Background

A key aspect in the implementation of the Deep Geologic Repository (DGR) concept for long-term management of radioactive waste concerns the evolution and behavior of the deep-seated groundwater system in which the repository is positioned. The Canadian Nuclear Waste Management Organization (NWMO) has dedicated technical programs focused on improving the overall understanding of dynamic processes affecting deep-seated groundwater systems in sedimentary and crystalline rock environments under ice age conditions, over the time frames relevant to establishing a DGR safety case, that is, 1 million years (Ma). The phenomena and coupled processes associated with glaciation, affecting a deep-seated groundwater system, are transient in nature and, in many cases, extreme. Among others, they include ice-sheet loading, permafrost, glacial melt-water recharge, crustal flexure, surface erosion, and glacial isostatic adjustment. Their evaluation in the context of developing a DGR safety case requires an integrated, multidisciplinary approach focused on deriving independent lines of geoscientific evidence to test and establish notions of long-term groundwater system stability, and natural formation barrier integrity.

To illustrate the integrated approach described above (i.e., a Geosynthesis), a case study describing the application of site-specific and regional-scale geoscientific investigations, and advanced numerical modeling methods, for a proposed Canadian low- and intermediate-level waste (L&ILW) DGR within a Paleozoic age sedimentary sequence at the Bruce nuclear site ("Bruce site") in southern Ontario (Figure 11.8) is presented.

11.5.2 Bruce Nuclear Site—Paleohydrogeologic Case Study

This section provides a summary of the glacial history, as well as geologic and hydrogeologic conditions observed at the Bruce site. Details of physical, chemical, and isotopic hydrogeologic studies, as well as numerical simulations of Laurentide glacial ice-sheet history and the Paleozoic bedrock groundwater system, are presented, all of which offer an insight into paleohydrogeologic evolution and stability as relevant to a DGR safety case.

11.5.2.1 Glacial History

Over the last 1 Ma, the North American continent has most probably experienced nine glacial events (Peltier, 2011), in which the Canadian landscape was

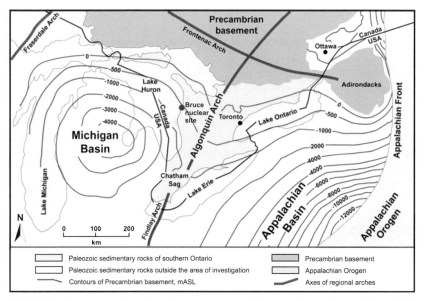

FIGURE 11.8 Location of the Bruce Site and geology of Southern Ontario. *From NWMO, 2011.*

subject to numerous cycles of glacial advance and retreat. During these cycles, the response (i.e., stability and resilience to change) of the geosphere is of relevance for a DGR safety case. The University of Toronto (UT) glacial systems model (GSM), developed by Peltier (2011), provides numerous constrained realizations of LIS and permafrost evolution over the North American continent for the last glacial event ($\sim$120,000 years).

During the last ice age, the LIS advanced over most of Canada, beginning approximately 120,000 years ago, and lasting over 100,000 years, reaching a total thickness at the LGM ($\sim$25,000 years ago) of over 2.5 km at the Bruce site (Figure 11.9; Peltier, 2011). The continental land mass was depressed significantly ($\sim$600 m) by the ice sheet, and rebound, initiated following glacial retreat, is still occurring today at rates of approximately 1 mm/year (Peltier, 2011). Glaciological reconstructions of the LIS suggest that the Bruce Site could be overlain by up to 3 km of ice during subsequent glaciations, in addition to periglacial and permafrost conditions (Peltier, 2011).

11.5.2.2 Glacial erosion

According to Hallet (2011), future glaciation events over southern Ontario should behave similarly to those that have affected the region over the last 1–2 Ma (including ice flow dynamics, erosion rates, and sediment deposition/accumulation). Based on the work of Peltier (2011), as well as the observed topography and observed evidence of glacial erosion in the vicinity of the Bruce site, Hallet (2011) suggested a conservative estimate of 100 m of erosion over a period of 1 Ma for the purposes of assessing long-term safety of a DGR at the Bruce site in the context of climate change.

11.5.2.3 Physical Geology and Hydrogeology

The Bruce site is underlain by approximately 860 m of Late Cambrian ($\sim$510 Ma) to middle Devonian ($\sim$350 Ma) age sedimentary rocks that rest unconformably on Precambrian ($\sim$1600–540 Ma) gneisses and metamorphic rocks of the Canadian Shield. The sedimentary sequence at the site has been characterized extensively by the drilling, logging, and testing of six deep boreholes (NWMO, 2011). The L&ILW DGR is proposed to be located within the argillaceous limestone of the Ordovician Cobourg Formation ($\sim$680 m depth; Figure 11.10).

Three groundwater regimes are defined at the Bruce site—shallow, intermediate, and deep—based on fluid chemistry and total dissolved solids (TDS) distribution (Figure 11.11a). The shallow regime contains waters with TDS values of $\leq$5.0 g/L. The intermediate regime is a transition zone from the relatively fresh shallow fluids above to the basin brines residing below. The deep regime contains sedimentary brine, with TDS values ranging between 200 and 275 g/L.

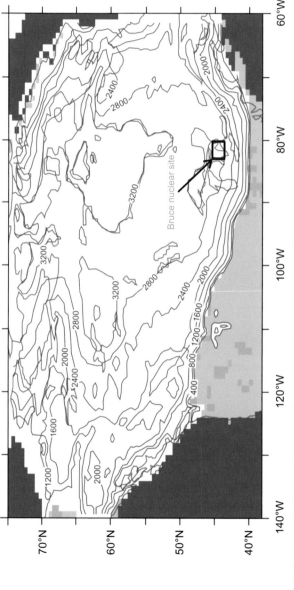

FIGURE 11.9   Laurentide ice sheet thickness at last glacial maximum. *From Peltier, 2011.*

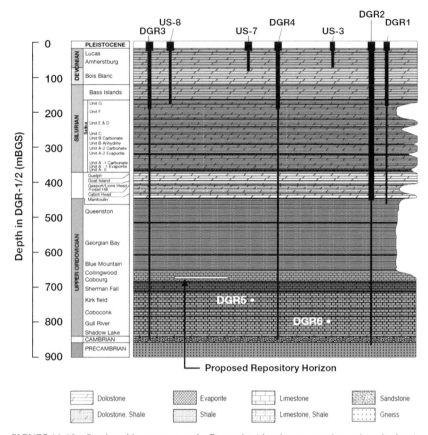

FIGURE 11.10 Stratigraphic sequence at the Bruce site (showing proposed repository horizon). *Modified from NWMO, 2011.*

The Ordovician shales and Trenton Group limestones at the Bruce Site possess very low rock mass horizontal hydraulic conductivity (K_h 10^{-12} to 10^{-15} m/s; Figure 11.11b) and low effective diffusion coefficients ($D_e \leq 10^{-11}$ m²/s) (NWMO, 2011). The hydraulic formation pressures within these sediments are significantly underpressured—by as much as 300 m below hydrostatic conditions—suggesting that the Ordovician shales have acted as a natural barrier to both hydrocarbon migration and solute transport (Engelder, 2011). In addition to the sealing capacity of the Ordovician shales, the presence of hydrostatic and overpressured conditions in the rock formations that bound the Ordovician sequence (Figure 11.11c) suggest that the Ordovician formations behave as a hydraulic sink both from above and below (Engelder, 2011). The observed abnormal hydraulic heads and significant vertical gradients (i.e., $\approx 1-2$) in the Ordovician and Silurian formations at

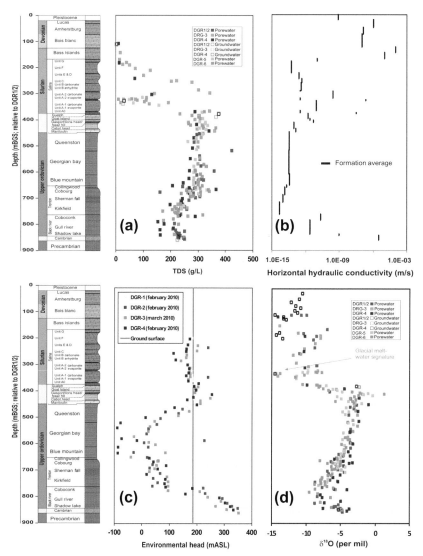

FIGURE 11.11   Vertical profiles of (a) TDS, (b) hydraulic conductivity (K_h), (c) environmental head, and (d) δ ^{18}O with depth at the Bruce site. *Modified from NWMO, 2011.*

the Bruce Site strongly suggest that significant vertical connectivity across the formation aquitards/aquicludes does not exist. Further, the under-pressures observed in the Odrovician strata (Figure 11.11c) could neither have developed, nor persisted, if the vertical formation-scale permeabilities were not as low as measured (NWMO, 2011).

11.5.2.4 Natural Tracers

Geochemical tools are used to assess groundwater/porewater age and the conditions under which they entered the groundwater system. If chemical signatures of recharge events are retained in the groundwater system, these tools allow for the interpretation of system dynamics and response to glacial perturbations. For the Bruce Site, stable water isotopes ($\delta\,^{18}O$, $\delta\,^{2}H$) are used to infer the climate under which groundwater recharged, and radioactive isotopes (e.g., ^{14}C, ^{129}I, and $^{3}He/^{4}He$) are used to infer groundwater age.

The $\delta\,^{18}O$ and $\delta\,^{2}H$ porewater profiles from the Bruce site (Figure 11.11d) show evidence for mixing of glacial meltwater and water from warmer conditions in the shallow Devonian and confined Silurian formations (up to ~330 m depth), suggesting that infiltration of these waters occurred during both glacial and interglacial periods (NWMO, 2011). Below the Silurian formations, the deep brines show no evidence of mixing with glacial meltwaters. This is consistent with findings that glacial melt and recharge waters occur in the shallow sediments of the Michigan basin (e.g., McIntosh and Walter, 2005, 2006), but high TDS groundwaters at depths indicate that it has not been possible for glacial meltwaters to infiltrate beyond approximately 350 m into the subsurface and displace deep basin brines. Chloride and bromide porewater chemistry at the Bruce site suggest an ancient seawater origin for the deep brines, consistent with other studies of Michigan Basin brines (NWMO, 2011; Clark et al., 2013; Wilson and Long, 1993a,b). Illustrative modeling suggests that the time frames required for the development of the vertical salinity and, in particular, $\delta\,^{18}O$ profiles within the Ordovician sediments are on the order of hundreds of millions of years, consistent with the occurrence of a long-lived diffusion-dominated solute transport regime. These vertical solute and environmental isotope profiles, derived from drill core porewater analysis within the Ordovician strata, are consistent with a stable groundwater system in which advective transport processes are not evident.

Shallow groundwater ages within the permeable Devonian strata at the Bruce site, using ^{14}C, range between 4500 and 7300 years before present, suggesting recharge in the mid-Holocene. The presence of tritium in the majority of shallow samples indicates mixing with modern meteoric water (INTERA, 2011). Below the Silurian, all porewater age estimates indicate an ancient origin. ^{129}I results indicate that porewaters at proposed repository depth are >80 Ma (upper age limit of the method; INTERA, 2011). $^{3}He/^{4}He$ indicates that minimum estimates of porewater age for the Ordovician shales at the Bruce Site are between 180 and 260 Ma (Clark et al., 2013). Consistent with observed solute and environmental isotope profiles, this interpretation suggests that the low permeability, Ordovician formations are part of a long-lived groundwater system that has remained stable in spite of multiple glaciations during the latter half of the Pleistocene.

11.5.2.5 Paleohydrogeologic Modeling

Multidisciplinary field investigations at the Bruce Site are complemented with glacial systems modeling and numerical simulations of coupled processes to better understand the spatial variability, magnitude, and time rate of change of groundwater system characteristics. Numerical models are based on representative geometry, physics, and parameters, and represent a physically-based means to link, in space and time, separate sources of data. Numerical models are also used to validate or test conceptual models based on regional or site-specific data. Data collected at different scales, such as core samples from the laboratory or in situ hydraulic testing in the field, can be integrated within a numerical model at one scale for use at a different scale.

Paleohydrogeological modeling can be used to assess groundwater system response to the temporally varying thermal, hydraulic, and mechanical loads on a geosphere imposed by continental-scale glaciations (Sykes et al., 2011). Simulation results from the UT GSM (Section 11.5.2.1) provided mechanical, hydraulic, and permafrost inputs to the paleohydrogeologic models developed for the Bruce Site (Sykes et al., 2011). The performance measures used by the model include mean life expectancy (MLE), which is a measure of the time required for a water particle to exit the model domain subject to both advective and diffusive processes. Peclet numbers of molecular diffusion are used to further quantify the dominant mass transport process at repository depth. A unit concentration tracer for recharge waters, applied at the top surface of the geosphere model, provides an indication of recharge water migration to depth—driven by temperate or cold-based glacial processes.

The combination of transient hydraulic boundary conditions imposed by subglacial fluid pressures and hydromechanical coupling govern vertical hydraulic gradients in glacially affected groundwater systems. Impacts of varying the surface hydraulic boundary condition and the degree of hydro-mechanical coupling in a paleohydrogeologic model can be examined through simulations that deliberately trace the migration of low saline recharge into the underlying groundwater system (Sykes et al., 2011). The temporally-integrated nature of the glacial recharge simulations accounts for the numerous vertical gradient reversals that occur during glaciation. Deep migration of tracers is more likely to occur when hydromechanical coupling is ignored, as vertically downward gradients are maximized. The presence of permafrost is integrated into the analyses as well, because hydraulic conductivities can be significantly reduced where permafrost is present, affecting tracer migration; greater tracer migration to depth is observed in scenarios where permafrost impacts the geosphere for a short period of time.

At the Bruce Site, paleohydrogeologic modeling was used to test the physical and geochemical conceptualizations of the site. Numerical modeling for the Bruce Site suggests that overpressured conditions in the deep Cambrian aquifer result from a combination of high TDS concentrations (i.e., high fluid density) and the basin hydrostratigraphy, which regionally confines the permeable Cambrian unit (Sykes et al., 2011). Modeling was used to investigate the effect of vertical hydraulic conductivities of the underlying formations on the evolution of the underpressures in the Ordovician shales and Trenton Group limestones. Base-case (i.e., steady-state) MLE values for the regional study area are shown in Figure 11.12a, and values for the deep system at the Bruce site are consistent with geochemical estimates of groundwater age (i.e., >100 Ma). Paleohydrogeologic simulations were performed using two bounding GSM scenarios (nn9930, nn9921; Figure 11.12b) of ice-sheet and permafrost evolution at the Bruce site (NWMO, 2011). These two GSM simulations provide constrained surface boundary conditions for ice-sheet advance/retreat and permafrost formation that were linked explicitly to the groundwater simulations, that is, transient surface hydraulic head boundaries and permafrost depth influenced the assignment of near surface hydraulic conductivities. Figure 11.12c illustrates the simulated distribution and maximum depth of penetration by glacial recharge after one glacial cycle (120,000 years). Two glacial cycles (240,000 years) were modeled as well, and the maximum tracer infiltration was the same at the Bruce Site—indicating that glacial meltwaters did not reach proposed repository depth. The modeling results support the interpretations of a shallow system impacted by glacial meltwater and a long-lived deep groundwater system that is stable and resilient to perturbations (NWMO, 2011).

In summary, the evaluation of phenomena and coupled processes associated with glaciation for a deep-seated groundwater system, in the context of developing a DGR safety case, requires an integrated multidisciplinary approach. Paleohydrogeologic investigations aim to explore the influence of glacial processes through field evidence, complemented with numerical simulations of coupled processes, to better understand spatial variability, as well as magnitude and time rate of change, of groundwater system characteristics. The case study presented illustrates the integration of techniques, or geosynthesis, as applied in assessing the stability and resilience of a deep-seated groundwater system over long (i.e., 1 Ma) time frames.

11.6 IMPACT OF GLACIAL AND PERIGLACIAL CLIMATE CONDITIONS ON GROUNDWATER FLOW AND TRANSPORT—EXAMPLES FROM A SAFETY ASSESSMENT OF A GEOLOGICAL REPOSITORY FOR SPENT FUEL IN FRACTURED CRYSTALLINE ROCK, SWEDEN

11.6.1 Introduction

The Forsmark site was proposed by the Swedish Nuclear Fuel and Waste Management Company (SKB) in 2009 to serve as the potential site for construction of a future geological repository for spent nuclear fuel in fractured

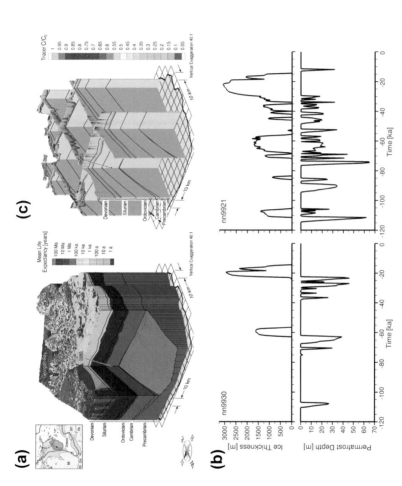

FIGURE 11.12 Numerical modeling for the Bruce site showing (a) base-case mean life expectancy (steady-state); (b) GSM simulations used—nn9930 and nn9921; and (c) tracer infiltration following one glacial cycle (transient boundary conditions based on GSM output) of 120,000 years. *Modified from Sykes et al., 2011.*

crystalline rock. The Forsmark site is very flat and low-lying, about 10 km^2 in size. The bedrock in the Forsmark area is situated inside the 1.9- to 1.8-Ga-old Svecokarelian orogen in the southwestern part of the Fennoscandian Shield (Figure 11.13), which forms one of the ancient continental parts on Earth (Koistinen et al., 2001).

A comprehensive description of the repository design utilized by SKB, the so-called KBS-3 system, is provided in the safety assessment safety report (SR)-Site (SKB, 2011). In short, the waste is placed in approximately 7,000 copper canisters surrounded by bentonite clay. The canisters are deposited in deposition holes drilled vertically from the tunnel floor in tunnels located at an approximately 470 m depth in the crystalline rock. The canisters have a steel insert, to increase mechanical strength, and contain the waste in the form of fuel elements. After the operational period of the repository, the tunnels are back filled with a low permeability material and pumping is stopped such that the repository becomes water filled.

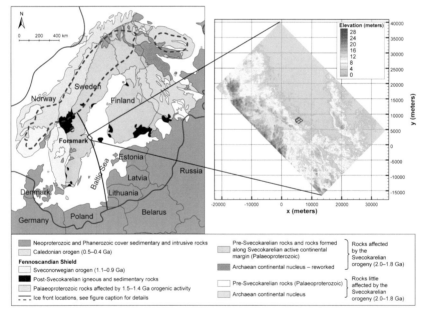

FIGURE 11.13   Left: Map showing major tectonic units in the northern part of Europe *(modified after Koistinen et al., 2001)* with the ice front location at the onset of glacial conditions (dotted line) and ice front location at the maximum ice sheet extent (solid line) indicated. Also shown is an ice flow line passing through the groundwater flow model domain location. The dotted line is based on Mangerud et al. (2011) and corresponds to time 38−35 ka BP, that is, MIS 3 (marine isotope stage 3). The solid line is based on Svendsen et al. (2004) and corresponds to a time approximately 20 ka BP. Right: Map showing the model domain, the present-day topography at Forsmark. The repository area is indicated by black lines.

The copper canister and bentonite clay are the engineered barriers of the KBS-3 concept. The fractured rock of the geosphere constitutes a natural barrier by providing conditions of low groundwater flow and high retention of potentially migrating radionuclides. The geosphere also provides stable mechanical and geochemical conditions for the repository, which ensures that the engineered barriers operate as intended.

The actual environmental hazard implied by the subsurface nuclear waste repository is possible leakage of radionuclides to the biosphere if the barriers of the repository fail. An ice sheet represents one case of adverse conditions with increased risks (SKB, 2011). The main adverse issues associated with an ice sheet are increased groundwater flow at repository depth, seismic hazard during deglaciation that can result in both canister failure and modified hydraulic properties of the fractured rock mass, and breaching of the repository by glacial erosion. Here, we only deal with the risks associated with an increased groundwater flow. In addition to an increased flow, also the groundwater chemistry affects the engineered barriers. Specifically, dilute water, such as glacial meltwater, increases the potential for bentonite erosion.

The formal procedure to assess hazards (i.e., radiation dose to humans and biota) or risks (i.e., additional life-time risk to develop cancer) in the context of nuclear waste repositories is through a safety assessment study (e.g., NEA, 2007, 2009, 2012). The exact contents of a safety assessment may vary depending on national conditions and legislation, but typically involve a description of the geological site and repository system, and an identification of features, events, and processes that affect the repository and geosphere. Different scenarios may be formulated that combine various features, events, and processes, and these typically result in modeling cases of radionuclide transport through the geosphere and biosphere.

In the Swedish regulatory framework, it is specified that the safety assessment should strive for more detail and realism for the first 1,000 years of the assessment. For longer assessment periods, a less detailed realism is allowed motivated by the fact that greater uncertainties prevail concerning future conditions.

It is emphasized that when future glacial conditions are modeled in the safety assessment, a number of simplifying assumptions are made. These assumptions are typically of a pessimistic nature in order not to underestimate the risks. The calculations performed do thus not necessarily represent the most likely future; they are intended to bound the potential risks.

11.6.2 Climate Conditions

The last glacial period (the Weichselian) started approximately 115,000 years ago. Interpretations of geological observations suggest several ice-free inter-stadial periods over large parts of Fennoscandia during the Weichselian (e.g.,

Svendsen et al., 2004; Helmens and Engels, 2010; Wohlfarth, 2010; Mangerud et al., 2011). When studying past influences of periglacial and glacial conditions on the hydrogeological and hydrogeochemical evolution of a site, boundary conditions derived from climate constraints and modeling are needed. This, in turn, needs to be based on a reconstruction of the paleoclimatic evolution. In the SR-site safety assessment (SKB, 2011), paleoclimatic and environmental conditions were reconstructed for the last glacial cycle (SKB, 2010a).

Periods with glacial climate conditions were in the SR-site subdivided into glacial advance, glacial maximum, and glacial retreat. It is likely that during the advance of the ice sheet margin over low-lying areas, such as the Forsmark site in Sweden, the proglacial area is frozen as a result of periglacial climate conditions. If this was the case, and if the ice sheet was warm based, subglacial permafrost may have melted due to the insulating effect of the ice sheet and due to the frictional and geothermal conditions at the ice-sheet bed. However, during an ice-sheet advance, subglacial cold-based, frozen, conditions could also prevail for some time, and stretch many kilometers under an advancing ice-sheet margin. When parts of an ice sheet—subsurface interface becomes warm based, groundwater recharge from basal melting may take place. In addition, surface melting during the warm seasons yields substantial meltwater rates that can reach the ice sheet—subsurface interface through crevasses or moulins, preferentially at low ice sheet altitudes. For warm-based conditions, the average hydrostatic pressure at the ice sheet—subsurface interface is typically found to be close to the floatation pressure, which equals approximately 90 percent of the local ice-sheet thickness (Cuffey and Paterson, 2010).

The glacial maximum represents a period with very thick ice. In the reconstruction of the last glacial cycle made for the SR-site safety assessment, the largest ice thickness over the Forsmark site is approximately 2,900 m. Ice-sheet simulations (SKB, 2010a) indicate an ice surface gradient of approximately 1.2 m per kilometer over Forsmark. For such gradients and local characteristics in site geology, a warm-based ice sheet will affect the groundwater flow pattern also at a depth (Vidstrand et al., 2012).

The final deglaciation of the Weichselian ice sheet was characterized by a calving ice-sheet margin over the Baltic Sea, with a water depth in the proglacial area of around 100—200 m at the Forsmark site after the ice-sheet margin passage. In the SR-site, it was considered important to analyze the effects of a steeper gradient at all stages. Therefore, a retreating ice-sheet margin similar to that of an advancing ice-sheet margin was pessimistically assumed. After the deglaciation, the Forsmark site underwent a long period when it was completely submerged by different phases of the evolving Baltic Sea. During this time, the hydraulic gradient at the site was almost zero, but some density-driven flow may have occurred due to changing salinity in the aquatic system in front of the ice-sheet margin. Once the Forsmark site became terrestrial as a result of the postglacial isostatic rebound, temperate climate conditions once again prevailed.

11.6.3 Groundwater Flow Model

The right-hand side of Figure 11.13 illustrates the location and horizontal dimensions of the flow model domain. The present-day topography is shown along with the outline of the main repository tunnels (located in the center of the illustration).

Figure 11.14 shows statistics of all Darcy fluxes at the deposition holes at repository depth for a few characteristic phases during a glacial cycle. The median value of the Darcy flux increases by approximately two orders of magnitude during glacial conditions when an advancing ice-sheet margin is located right above the repository. Significant differences in the Darcy flux changes may be due to site- and location-specific properties. The direction of the vertical flux may, for instance, change from upward to downward and vice versa many times during the period of a glacial cycle depending on the ice-front location. Such effects are not only due to the fact that the measurement location could be in front of, or behind the ice sheet margin but also due to local bedrock characteristics such as geometries and properties of the deformation zone. Vidstrand et al. (2012) showed how, during periods of

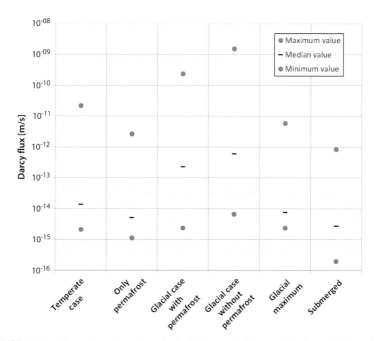

FIGURE 11.14   Darcy flux at repository depths for the main characteristic phases during a glacial cycle.

ice-sheet coverage, glacial meltwater typically recharges along deformation zones deep into the bedrock, whereas discharge occurs predominantly in between deformation zones.

Figure 11.14 further illustrates that the strongest influence on the Darcy flux occurs for an advancing ice-sheet margin without permafrost in the proglacial area. Further beneath the ice sheet where the permafrost has melted, no distinction between Darcy fluxes generated by the different simulations can be observed. The glacial maximum represents a reduced flow period compared to temperate climate conditions. When the site is subject to submerged conditions, the average value of all fluxes at repository depth is small as shown in Figure 11.14.

Figure 11.15 shows the Darcy flux evolution at three different measurement locations (occurring within 200 by 200 m of the modeled bedrock) along with the passage of the ice sheet margin. The differences in Darcy flux between the flux in the rock between deformation zones and measurement locations within deformation zones are larger than the increase in Darcy flux due to the passage of the ice sheet margin.

Figure 11.15 further shows that a deformation zone parallel to the general flow direction of the ice sheet is subject to a stronger impact than a deformation zone perpendicular to the ice-sheet flow direction. In such structures, the impact of the passage of the ice-sheet margin is almost three orders of magnitude, while a deformation zone orthogonal to the ice-sheet flow is impacted in a similar manner to the rock in between the deformation zones (approximately two orders of magnitude).

The salinity in the fracture water at a repository depth varies significantly during the passage of the ice-sheet margin. The salinity first increases and then decreases; this effect is the strongest for a case with no permafrost in front of the ice sheet. For such a system, the discharge primarily occurs along the ice sheet margin (e.g., Vidstrand et al., 2012; DeFoor et al., 2011). Figure 11.16 illustrates the development of fracture and rock matrix water salinity. The salinity is monitored at a location that represents an area of rock between deformation zones. This location undergoes a small, upward flux during the ice-sheet coverage. In the simulated exchange of salt between the fracture water and the matrix porewater, the exchange is from the matrix porewater to the fracture water for a limited period of time only coinciding with the passages of the ice-sheet margin, that is, when the fracture water salinity gets flushed out by glacial meltwater. During the long period of complete ice coverage, the conditions are the opposite.

11.6.4 Radionuclide Transport Model

In the radionuclide transport analysis of SR-Site (SKB, 2010b), the effect of varying flow conditions during glacial cycles is analyzed and illustrated. A

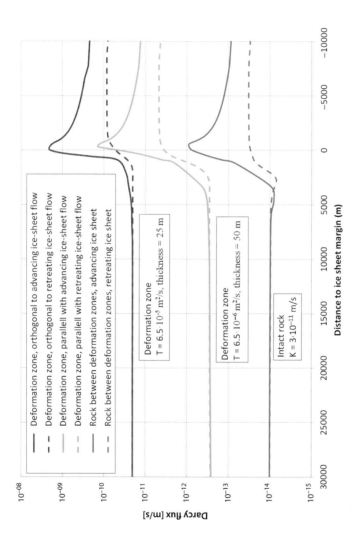

FIGURE 11.15 Illustration of the effect on Darcy flux at three different measurement locations at a −470-m elevation at the repository volume at the Forsmark site for an advancing (solid lines) and a retreating (dashed lines) ice sheet margin. On the horizontal axis, a positive distance value represents the distance from the measurement location to an approaching ice sheet margin, whereas a negative value represents the distance from the measurement location to the ice sheet margin when the margin has passed the measurement location.

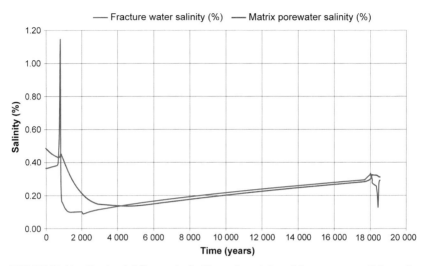

FIGURE 11.16   Simulated difference in flushing and regaining of fracture water salinity at the measurement location situated in rock between deformation zones close to the center of the repository at the Forsmark site during the period of ice coverage. Results are presented for mobile fracture water and immobile matrix porewater.

glacial cycle is repeated multiple times such that a 1 million years assessment is performed.

The resulting mean annual effective dose is shown in Figure 11.17. Also shown is the resulting dose if temperate climate conditions are assumed to prevail during the whole assessment period. It is observed that only the peaks of the case with a varying climate exceed the case with a constant climate. The main reason for this is that the dose conversion factor (quantifying the dose in the biosphere for a given flux of nuclides from the geosphere) for glacial conditions is much smaller than the dose factor for temperate conditions. The glacial climate is dominated by very large dilution effects due to, for example, glacial melting and submerged conditions in front of the ice sheet. Thus, the regulatory criterion is only met due to the changing biosphere conditions; the actual flux of radionuclides from the geosphere is increased during periods with high groundwater flow. As seen in Figure 11.14, the highest flow rates are observed for glacial conditions when the ice sheet is right above the repository (either advancing or retreating); these high flow conditions yield the peaks in the dose curve in Figure 11.17.

The results presented in Figure 11.17 were obtained by scaling the groundwater velocity along flow paths obtained for temperate climate conditions with flow factors representing the simulated change in groundwater flux at repository depths between temperate conditions and the other climate conditions (flow factors correspond to the median values presented in Figure 11.14). The results clearly show that the calculated dose is well below the regulatory criterion.

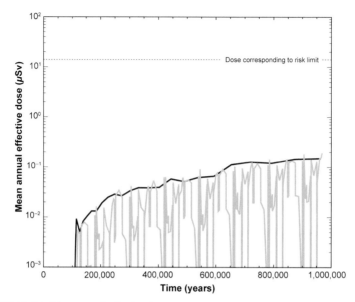

FIGURE 11.17   Mean annual effective dose (green curve) assuming varying climate conditions. For comparison the resulting dose assuming no flow changes is also given (black curve) as well as the dose corresponding to the regulatory risk criterion.

REFERENCES

Abe-Ouchi, A., Saito, F., Kawamura, K., Raymo, M.E., Okuno, J., Takahashi, K., Blatter, H., 2013. Insolation-driven 100,000-year glacial cycles and hysteresis of ice sheet volume. Nature 500 (7461), 190−193. http://dx.doi.org/10.1038/nature12374.

Alley, R.B., Cuffey, K.M., Evenson, E.B., Strasser, J.C., Lawson, D.E., Larson, G.J., 1997. How glaciers entrain and transport basal sediment: physical constraints. Quat. Sci. Rev. 16, 1017−1038.

Alley, R.B., Lawson, D.E., Evenson, E.B., Larson, G.J., 2003a. Sediment, glaciohydraulic supercooling, and fast glacier flow. Ann. Glaciol. 36, 135−141.

Alley, R.B., Lawson, D.E., Evenson, E.B., Strasser, J.C., Larson, G.J., 1998. Glaciohydraulic supercooling: a freeze-on mechanism to create stratified, debris-rich basal ice: II. Theory. J. Glaciol. 44, 563−569.

Alley, R.B., Lawson, D.E., Larson, G.J., Evenson, E.B., Baker, G.S., 2003b. Stabilizing feedbacks in glacier-bed erosion. Nature 424, 758−760.

Alley, R.B., Strasser, J.C., Lawson, D.E., Evenson, E.B., Larson, G.J., 1999. Glaciological and geological implications of basal-ice accretion in overdeepenings. In: Mickelson, D.M., Attig, J.W. (Eds.), Glacial Processes Past and Present, Geological Society of America Special Paper, vol. 337, pp. 1−9.

Amelung, P., Jobmann, M., Uhlig, L., 2007. Untersuchungen zur sicherheitstechnischen Auslegung eines generischen Endlagers im Tonstein in Deutschland—GENESIS—(Anlagenband Geologie der Referenzregionen im Tonstein). DBETec-Bericht, Internet: 04/2009, 67 pp.

Anderson, R.S., Molnar, P., Kessler, M.A., 2006. Features of glacial valley profiles simply explained. J. Geophys. Res.: Earth Surf. 111, F01004.

Archer, D., Ganopolski, A., 2005. A movable trigger: Fossil fuel CO_2 and the onset of the next glaciation. Geochem. Geophys. Geosyst. 6 (5), 1−7. http://dx.doi.org/10.1029/2004GC000891.

Archer, D., Eby, M., Brovkin, V., Ridgwell, A., Cao, L., Mikolajewicz, U., Caldeira, K., Matsumoto, K., Munhoven, G., Montenegro, A., Tokos, K., 2009. Atmospheric lifetime of fossil fuel carbon dioxide. Annu. Rev. Earth Planet. Sci. 37 (1), 117−134. http://dx.doi.org/10.1146/annurev.earth.031208.100206.

Benn, D.I., Evans, D.J.A., 2010. Glaciers and Glaciation, second ed. Hodder Education, London.

Bennett, M.R., Glasser, N.F., 2009. Glacial Geology. Ice Sheets and Landforms, second ed. John Wiley & Sons Ltd, Chichester.

Benz-Meier, C., 2003. Der würmeiszeitliche Rheingletscher—Maximalstand. Digitale Rekonstruktion, Modellierung und Analyse mit einem Geographischen Informations System (Ph.D. thesis). Universität Zürich.

Berger, A., 1978. Long-term variations of daily insolation and Quaternary climatic changes. J. Atmos. Sci. 35, 2362−2367.

Berger, A., Loutre, M.F., 1991. Insolation values for the climate of the last 10 million years. Quat. Sci. Rev. 10 (1988), 297−317.

Berger, A., Loutre, M.F., 2002. An exceptionally long interglacial ahead? Science 297 (5585), 1287.

Bini, A., Buoncristiani, J.-F., Couterrand, S., Ellwanger, D., Felber, M., Florineth, D., Graf, H.R., Keller, O., Kelly, M., Schlüchter, C., Schoeneich, P., 2009. Die Schweiz während des letzteiszeitlichen Maximums (LGM) 1:500'000. Bundesamt für Landestopografie swisstopo, Wabern.

Blatter, H., Haeberli, W., 1984. Modelling temperature distribution in Alpine glaciers. Ann. Glaciol. 5, 18−22.

Bollingfehr, W., Filbert, W., Lerch, C., Tholen, M., 2011. Endlagerkonzepte. Bericht zum Arbeitspaket 5. Vorläufige Sicherheitsanalyse für den Standort Gorleben. Gesellschaft für Anlagen- und Reaktorsicherheit (GRS), GRS-272, 198 pp.

Bornemann, O., Behlau, J., Fischbeck, R., Hammer, J., Jaritz, W., Keller, S., Mingerzahn, G., Schramm, M., 2008. Standortbeschreibung Gorleben—Teil 3: Ergebnisse der über- und untertägigen Erkundung des Salinars. Geol. Jb., C 73, 211.

Boulton, G.S., 1996. Theory of glacial erosion, transport and deposition as a consequence of subglacial sediment deformation. J. Glaciol. 42, 43−62.

Brandefelt, J., Zhang, Q., Hartikainen, J., Näslund, J.O., 2013. Potential for Cold Climate Conditions and Permafrost in Forsmark in the Next 60,000 Years. Report TR-13−04. Svensk Kärnbränslehantering AB, Sweden.

Brückner-Röhling, S., Espig, M., Fischer, M., Fleig, S., Forsbach, H., Kockel, F., Krull, P., Stiewe, H., Wirth, H., 2002. Standsicherheitsnachweise Nachbetriebsphase: Seismische Gefährdung—Teil 1: Strukturgeologie. Bundesanstalt für Geowissenschaften und Rohstoffe (BGR), 253 pp.

Burki, V., 2009. Glaziale Erosion: Prozesse und ihre Kapazitäten. Nagra Arbeitsbericht NAB 09−06, Wettingen, Switzerland.

Cepek, A.G., 1993. Die Schichtenfolge: Pleistozän. Ablagerungen und Erosionserscheinungen. In: Schroeder, J.H. (Ed.), Führer zur Geologie von Berlin und Brandenburg. No. 1: Die Struktur Rüdersdorf. 2. Erweiterte Auflage, pp. 118−130. Selbstverlag.

Charbit, S., Dumas, C., Kageyama, M., Roche, D.M., Ritz, C., 2013. Influence of ablation-related processes in the build-up of simulated Northern Hemisphere ice sheets during the last glacial cycle. Cryosphere 7, 681−698. http://dx.doi.org/10.5194/tc-7-681-2013.

Clark, I.D., Al, T., Jensen, M., Kennell, L., Mazurek, M., Mohapatra, R., Raven, K.G., 2013. Paleozoic-aged brine and authigenic helium preserved in an Ordovician shale aquiclude. Geology 41 (9), 951−954.

Cochelin, A.-S.B., Mysak, L.A., Wang, Z., 2006. Simulation of long-term future climate changes with the green McGill paleoclimate model: the next glacial inception. Clim. Change 79 (3−4), 381−401. http://dx.doi.org/10.1007/s10584-006-9099-1.

Cook, S.J., Swift, D.A., 2012. Subglacial basins: their origin and importance in glacial systems and landscapes. Earth Sci. Rev. 115 (4), 332−372.

Cook, S.J., Knight, P.G., Waller, R.I., Robinson, Z.P., Adam, W.G., 2007. The geography of basal ice and its relationship to glaciohydraulic supercooling: Svínafellsjökull, southeast Iceland. Quat. Sci. Rev. 26, 2309−2315.

Cook, S.J., Robinson, Z.P., Fairchild, I.J., Knight, P.G., Waller, R.I., Boomer, I., 2010. Role of glaciohydraulic supercooling in the formation of stratified facies basal ice: Svínafellsjökull and Skaftafellsjökull, southeast Iceland. Boreas 39, 24−38.

Cook, S.J., Waller, R.I., Knight, P.G., 2006. Glaciohydraulic supercooling: the process and its significance. Progr. Phys. Geogr. 30 (5), 577−588.

Creyts, T.T., Clarke, G.K.C., 2010. Hydraulics of subglacial supercooling: theory and simulations for clear water flows. J. Geophys. Res.: Earth Surf. 115, F03021.

Creyts, T.T., Clarke, G.K.C., Church, M., 2013. Evolution of subglacial overdeepings in response to sediment redistribution and glaciohydraulic supercoiling. J. Geophys. Res.: Earth Surf. 118, 1−24. http://dx.doi.org/10.1002/jgrf.20033.

Crucifix, M., 2011. How can a glacial inception be predicted? Holocene 21 (5), 831−842. http://dx.doi.org/10.1177/0959683610394883.

Crucifix, M., Rougier, J., 2009. On the use of simple dynamical systems for climate predictions. Eur. Phys. J. Special Topics 174 (1), 11−31. http://dx.doi.org/10.1140/epjst/e2009-01087-5.

Cuffey, K.M., Paterson, W.S.B., 2010. The Physics of Glaciers, fourth ed. Elsevier, Oxford, Pergamon.

DeFoor, W., Pearson, M., Larsen, H.C., Lizarralde, D., Cohen, D., Dugan, B., 2011. Ice sheet-derived submarine groundwater discharge on Greenland's continental shelf. Water Resour. Res. 47, W07549. http://dx.doi.org/10.1029/2011WR010536.

Dehnert, A., Lowick, S.E., Preusser, F., Anselmetti, F.S., Drescher-Schneider, R., Graf, H.R., Heller, F., Horstmeyer, H., Kemna, H.A., Nowaczyk, N.R., Züger, A., Furrer, H., 2012. Evolution of an overdeepened trough in the northern Alpine Foreland at Niederweningen, Switzerland. Quat. Sci. Rev. 34, 127−145.

Deslisle, G., Caspers, G., Freund, H., 2003. Permafrost in north-central Europe during the Weichselian: how deep?. In: ICOP 2003 Permafrost: Proceedings of the Eighth International Conference on Permafrost, 21−25 July 2003, Zurich, Switzerland, vol. 1. A.A. Balkema Publishers, pp. 187−191.

Doornenbal, J.C., Stevenson, A.G. (Eds.), 2010. Petroleum Geological Atlas of the Southern Permian Basin Area. EAGE Publications, 342 pp.

Drewry, D., 1986. Glacial Geologic Processes. Edward Arnold, London.

Egholm, D.L., Pedersen, V.K., Knudsen, M.F., Larsen, N.K., 2012. Coupling the flow of ice, water, and sediment in a glacial landscape evolution model. Geomorphology 141−142, 47−66.

Engelder, T., 2011. Analogue Study of Shale Cap Rock Barrier Integrity. Nuclear Waste Management Organization, Toronto, Canada. Report NWMO DGR-TR-2011-23. http://www.nwmo.ca/dgrgeoscientificsitecharacterization.

Evans, I.S., 2008. Glacial erosional processes and forms: mountain glaciation and glacier geography. In: The History of the Study of Landforms or the Development of Geomorphology:

Quaternary and Recent Processes and Forms (1890−1965) and the Mid-century Revolutions. Geological Society, London, p. 413.

Fischer, U.H., 2009. Glacial Erosion: A Review of Its Modelling. Nagra Arbeitsbericht NAB 09−23, Wettingen, Switzerland.

Fischer, U.H., Haeberli, W., 2010. Glacial Erosion Modelling—Results of a Workshop Held in Unterägeri, Switzerland, 29 April−1 May 2010. Nagra Arbeitsbericht NAB 10−34, Wettingen, Switzerland.

Fischer, U.H., Haeberli, W., 2012. Glacial Overdeepening—Results of a Workshop Held in Zürich, Switzerland, 20−21 April 2012. Nagra Arbeitsbericht NAB 12-48, Wettingen, Switzerland.

Forbes, J.D., 1843. Travels through the Alps of Savoy and Other Parts of the Pennine Chain, with Observations on the Phenomena of Glaciers. Adam and Charles Black, Edinburgh.

Fountain, A.G., Jacobel, R.W., Schlichting, R., Jansson, P., 2005. Fractures as the main pathways of water flow in temperate glaciers. Nature 433 (7026), 618−621.

Frenzel, B., Pécsi, M., Velichko, A.A., 1992. Atlas of Paleoclimates and Paleoenvironments of the Northern Hemisphere—Late Pleistocene/Holocene. Hungarian Academy of Sciences and Gustav Fischer Verlag, Budapest and Stuttgart, 153 pp.

Fyfe, W.S., 1999. Nuclear waste isolation: an urgent international responsibility. Eng. Geol. 52, 159−161.

Ganopolski, A., Calov, R., Claussen, M., 2010. Simulation of the last glacial cycle with a coupled climate ice sheet model of intermediate complexity. Clim. Past 6, 229−244.

Gremaud, V., Goldscheider, N., 2010. Geometry and drainage of a retreating glacier overlying and recharging a karst aquifer, Tsanfleuron-Sanetsch, Swiss Alps. Acta Carsol. 39 (2), 289−300.

Grust, K., 2004. The hydrology and dynamics of a glacier overlying a linked-cavity drainage system (Ph.D. thesis). University of Glasgow.

Haeberli, W., 2010. Glaciological Conditions in Northern Switzerland during Recent Ice Ages. Nagra Arbeitsbericht NAB 10−18, Wettingen, Switzerland.

Haeberli, W., Penz, U., 1985. An attempt to reconstruct glaciological and climatological characteristics of 18 ka BP ice age glaciers in and around the Swiss Alps. Zeitschrift für Gletscherkunde und Glazialgeologie 21, 351−361.

Haeberli, W., Rellstab, W., Harrison, W.D., 1984. Geothermal effects of 18 ka BP ice conditions in the Swiss Plateau. Ann. Glaciol. 5, 56−60.

Hallet, B., 2011. Glacial Erosion Assessment. Nuclear Waste Management Organization, Toronto, Canada. Report NWMO DGR-TR-2011-18. http://www.nwmo.ca/dgrgeoscientificsitecharacterization.

Hallet, B., Hunter, L., Bogen, J., 1996. Rates of erosion and sediment evacuation by glaciers: a review of field data and their implications. Global and Planet. Change 12, 213−235.

Hantz, D., Lliboutry, L., 1983. Waterways, ice permeability at depth, and water pressures at glacier d'Argentière, French Alps. J. Glaciol. 29 (102), 227−239.

Harbor, J., 1989. Early discoverers 36: W. J. McGee on glacial erosion laws and the development of glacial valleys. J. Glaciol. 35, 419−425.

Hays, J., Imbrie, J., Shackleton, N., 1976. Variations in the Earth's orbit: pacemaker of the ice ages. Science 194 (4270), 1121−1132.

Helmens, K.F., Engels, S., 2010. Ice-free conditions in eastern Fennoscandia during early Marine Isotope Stage 3: lacustrine records. Boreas 39, 399−409.

Herman, F., Beaud, F., Champagnac, J.-D., Lemieux, J.-M., Sternai, P., 2011. Glacial hydrology and erosion patterns: a mechanism for carving glacial valleys. Earth Planet. Sci. Lett. 310, 498−508.

Hinsch, W., 1979. Rinnen an der Basis des glaziären Pleistozäns in Schleswig-Holstein. Eiszeitalter und Gegenwart 29, 173–178.

Hofer, D., Raible, C.C., Dehnert, A., Kuhlemann, J., 2012a. The impact of different glacial boundary conditions on atmospheric dynamics and precipitation in the North Atlantic region. Clim. Past 8 (3), 935–945. http://dx.doi.org/10.5194/cp-8-935-2012.

Hofer, D., Raible, C.C., Merz, N., Dehnert, A., Kuhlemann, J., 2012b. Simulated winter circulation types in the North Atlantic and European region for preindustrial and glacial conditions. Geophys. Res. Lett. 39, L15805. http://dx.doi.org/10.1029/2012GL052296.

Hooke, R.LeB., 1991. Positive feedbacks associated with erosion of glacial cirques and overdeepenings. Geol. Soc. Am. Bull. 103, 1104–1108.

Hooke, R.LeB., Jennings, C.E., 2006. On the formation of the tunnel valleys of the southern Larentide ice sheet. Quat. Sci. Rev. 25, 1364–1372.

Hooke, R.LeB., Pohjola, V.A., 1994. Hydrology of a segment of a glacier situated in an overdeepening, Storglaciären, Sweden. J. Glaciol. 40, 140–148.

Hoth, P., Wirth, H., Reinhold, K., Bräuer, V., Krull, P., Feldrappe, H., 2007. Endlagerung radioaktiver Abfälle in tiefen geologischen Formationen Deutschlands. Untersuchung und Bewertung von Tongesteinsformationen. Bundesanstalt für Geowissenschaften und Rohstoffe, 118 pp.

Huuse, M., Lykke-Anderson, H., 2000. Overdeepend Quaternary valleys in the eastern Danish North Sea: morphology and origin. Quat. Sci. Rev. 19 (12), 1233–1253.

Huybers, P., 2011. Combined obliquity and precession pacing of late Pleistocene deglaciations. Nature 480 (7376), 229–232. http://dx.doi.org/10.1038/nature10626.

Imbrie, J.Z., Imbrie-Moore, A., Lisiecki, L.E., 2011. A phase-space model for Pleistocene ice volume. Earth Planet. Sci. Lett. 307, 94–102. http://dx.doi.org/10.1016/j.epsl.2011.04.018.20.

INTERA, 2011. Descriptive Geosphere Site Model. Intera Engineering Ltd, Toronto, Canada. Report for Nuclear Waste Management Organization NWMO DGR-TR-2011-24. http://www.nwmo.ca/dgrgeoscientificsitecharacterization.

Iverson, N.R., 1991. Potential effects of subglacial water pressure fluctuations on quarrying. J. Glaciol. 37 (125), 27–36.

Iverson, N.R., 2002. Processes of glacial erosion. In: Menzies, J. (Ed.), Modern and Past Glacial Environments, revised student edition. Butterworth/Heinemann, Oxford, pp. 131–146.

Jansson, P., Hooke, R.LeB., 1989. Short-term variations in strain and surface tilt on Storglaciären, Kebnekaise, northern Sweden. J. Glaciol. 35, 201–208.

Jordan, P., 2008. Designing the DEM of the base of the Swiss Plateau Quaternary sediments. In: Proceedings of the 6th International Cartographic Association (ICA) Mountain Cartography Workshop "Mountain Mapping and Visualisation", pp. 107–113.

Jordan, P., 2010. Analysis of overdeepened valleys using the digital elevation model of the bedrock surface of Northern Switzerland. Swiss J. Geosci. 103 (3), 375–384.

Jørgensen, F., Sandersen, P.B.E., 2006. Buried and open tunnel valleys in Denmark—erosion beneath multiple ice sheets. Quat. Sci. Rev. 25, 1339–1363.

Kehew, A.E., Lord, M.L., Koslowski, A.L., 2007. Glacifluvial landforms of erosion. In: Elias, S.A. (Ed.), Encyclopedia of Quaternary Science, 1. Aufl, vol. 1. Elsevier, pp. 818–831.

Kehew, A.E., Piotrowski, J.A., Jørgensen, F., 2012. Tunnel valleys: concepts and controversies—a review. J. Earth Sci. Rev. 113, 33–58.

Kelly, M.A., Buoncristiani, J.-F., Schlüchter, C., 2004. A reconstruction of the last glacial maximum (LGM) ice-surface geometry in the western Swiss Alps and contiguous Alpine regions in Italy and France. Eclogae Geol. Helv. 97, 57–75.

Kessler, M.A., Anderson, R.S., Briner, J.P., 2008. Fjord insertion into continental margins driven by topographic steering of ice. Nat. Geosci. 1, 365–369.

Koistinen, T., Stephens, M.B., Bogatchev, V., Nordgulen, Ø., Wennerström, M., Korhonen, J., 2001. Geological Map of the Fennoscandian Shield, Scale 1:2,000,000. Geological Surveys of Finland, Norway and Sweden and the North-West Department of Natural Resources of Russia.

Köthe, A., Hoffmann, N., Krull, P., Zirngast, M., Zwirner, R., 2007. Standortbeschreibung Gorleben. Teil 2: die Geologie des Deck- und Nebengebirges des Salzstocks Gorleben. Geol. Jb., C 72, 201.

Kuster, H., Meyer, K.-D., 1979. Glaziäre Rinnen im mittleren und nordöstlichen Niedersachsen. Eiszeitalter und Gegenwart 29, 135−156.

Larson, G.J., Lawson, D.E., Evenson, E.B., Alley, R.B., Knudsen, O., Lachniet, M.S., Goetz, S.L., 2006. Glaciohydraulic supercooling in former ice sheets? Geomorphology 75, 20−32.

Lawson, D.E., Strasser, J.C., Evenson, E.B., Alley, R.B., Larson, G.J., Arcone, S.A., 1998. Glaciohydraulic supercooling: a freeze-on mechanism to create stratified, debris-rich basal ice: I. Field evidence. J. Glaciol. 44, 547−562.

Linton, D.L., 1963. The forms of glacial erosion. Trans. Inst. Brit. Geogr. 33, 1−28.

Lisiecki, L.E., 2010. Links between eccentricity forcing and the 100,000-year glacial cycle. Nat. Geosci. 3 (5), 349−352. http://dx.doi.org/10.1038/ngeo828.

Lisiecki, L.E., Raymo, M.E., 2005. A Plio−Pleistocene stack of 57 globally distributed benthic $\delta^{18}O$ records. Paleoceanography 20 (1), 1−17. http://dx.doi.org/10.1029/2004PA001071.

Lisiecki, L.E., Raymo, M.E., 2007. Plio−Pleistocene climate evolution: trends and transitions in glacial cycle dynamics. Quat. Sci. Rev. 26 (1−2), 56−69. http://dx.doi.org/10.1016/j.quascirev.2006.09.005.

Loutre, M.F., Berger, A., 2000. Future climatic changes: are we entering an exceptionally long interglacial? Clim. Change 46 (1), 61−90.

Lüthi, D., Le Floch, M., Bereiter, B., Blunier, T., Barnola, J.-M., Siegenthaler, U., Raynaud, D., Jouzel, J., Fischer, H., Kawamura, K., Stocker, T.F., 2008. High-resolution carbon dioxide concentration record 650,000−800,000 years before present. Nature 453 (7193), 379−382. http://dx.doi.org/10.1038/nature06949.

Lutz, R., Kalka, S., Gaedicke, C., Reinhardt, L., Winsemann, J., 2009. Pleistocene tunnel valleys in the German North Sea: spatial distribution and morphology. Z. dt. Ges. Geowiss. 160, 225−235.

MacGregor, K.R., Anderson, R.S., Anderson, S.P., Waddington, E.D., 2000. Numerical simulations of glacial-valley longitudinal profile evolution. Geology 28, 1031−1034.

MacGregor, K.R., Anderson, R.S., Waddington, E.D., 2009. Numerical modeling of glacial erosion and headwall processes in alpine valleys. Geomorphology 103, 189−204.

Manabe, S., Broccoli, A.J., 1984. Ice-Age climate and continental ice sheets: some experiments with a general circulation model. Ann. Glaciol. 5, 100−105.

Mangerud, J., Gyllencreutz, R., Svendsen, J.I., 2011. Glacial history of Norway. In: Ehlers, J., Gibbard, P.L., Hughes, P.D. (Eds.), Developments in Quaternary Science, vol. 15. Elsevier Amsterdam, the Netherlands, ISBN 978-0-444-53447-7, pp. 279−298. http://dx.doi.org/10.1016/B978-0-444-53447-7.00022-2.

McIntosh, J.C., Walter, L.M., 2005. Volumetrically significant recharge of Pleistocene glacial meltwaters into epicratonic basins: constraints imposed by solute mass balances. Chem. Geol. 222, 292−309.

McIntosh, J.C., Walter, L.M., 2006. Paleowaters in the Silurian−Devonian carbonate aquifers: geochemical evolution of groundwater in the Great Lakes region since the late Pleistocene. Geochim. Cosmochim. Acta 70, 2454−2479.

Mysak, L.A., 2008. Glacial inceptions: past and future. Atmos.-ocean 46 (3), 317−341. http://dx.doi.org/10.3137/ao.460303.

Nagra, 2008a. Vorschlag geologischer Standortgebiete für das SMA- und das HAA-Lager. Begründung der Abfallzuteilung, der Barrierensysteme und der Anforderungen an die Geologie. Bericht zur Sicherheit und technischen Machbarkeit. Nagra Technischer Bericht NTB 08−05, Wettingen, Switzerland.

Nagra, 2008b. Vorschlag geologischer Standortgebiete für das SMA- und das HAA-Lager. Darlegung der Anforderungen, des Vorgehens und der Ergebnisse. Nagra Technischer Bericht NTB 08−03, Wettingen, Switzerland.

Nagra, 2008c. Vorschlag geologischer Standortgebiete für das SMA- und das HAA-Lager. Geologische Grundlagen. Nagra Technischer Bericht NTB 08−04, Wettingen, Switzerland.

Nagra, 2010. Beurteilung der geologischen Unterlagen für die provisorischen Sicherheitsanalysen in SGT Etappe 2. Klärung der Notwendigkeit ergänzender geologischer Untersuchungen. Nagra Technischer Bericht NTB 10−01, Wettingen, Switzerland.

Näslund, J.O., Brandefelt, J., Claesson Liljedahl, L., 2013. Climate considerations in long-term safety assessments for nuclear waste repositories. Ambio 42 (4), 393−401. http://dx.doi.org/10.1007/s13280-013-0406-6.

NEA, 2007. Regulating the Long-term Safety of Geological Disposal. NEA No. 6182. Organisation for Economic Co-operation and Development.

NEA, 2009. International Experiences in Safety Cases for Geological Repositories (INTESC). NEA No. 6251. Organisation for Economic Co-operation and Development.

NEA, 2012. Methods for Safety Assessment of Geological Disposal Facilities for Radioactive Waste. NEA No. 6923. Organisation for Economic Co-operation and Development.

Niedermayer, J., 1965. Gliederung und Ausbildung des Quartärs im Niederelbegebiet. Max Richter-Festschrift, 73−82.

NWMO, 2011. Geosynthesis. Nuclear Waste Management Organization, Toronto, Canada. Report NWMO DGR-TR-2011-11. http://www.nwmo.ca/dgrgeoscientificsitecharacterization.

Ó Cofaigh, C., 1996. Tunnel valley genesis. Progr. Phys. Geogr. 20, 1−19.

Peltier, W.R., 2011. Long-term Climate Change. Nuclear Waste Management Organization, Toronto, Canada. Report NWMO DGR-TR-2011-14. http://www.nwmo.ca/dgrgeoscientificsitecharacterization.

Pimenoff, N., Venäläinen, A., Järvinen, H., 2011. Posiva 2011−04. Climate Scenarios for Olkiluoto on a Time-scale of 120,000 Years, vol. 31. Posiva, OY, Finland.

Piotrowski, J.A., 1994. Tunnel-valley formation in northwestern Germany—geology, mechanisms of formation, and subglacial bed conditions for the Bornhöved tunnel valley. Sediment. Geol. 89, 107−141.

Preusser, F., Graf, H.R., Keller, O., Krayss, E., Schlüchter, C., 2011. Quaternary glaciation history of northern Switzerland. Quat. Sci. J. 60 (2−3), 282−305.

Preusser, F., Reitner, J.M., Schlüchter, C., 2010. Distribution, geometry, age and origin of overdeepened valleys and basins in the Alps and their foreland. Swiss J. Geosci. 103 (3), 407−426.

Raymo, M.E., Huybers, P., 2008. Unlocking the mysteries of the ice ages. Nature 451 (7176), 284−285. http://dx.doi.org/10.1038/nature06589.

Reinhold, K., Krull, P., Kockel, F., 2008. Salzstrukturen Norddeutschlands, Geologische Karte 1: 500 000. Bundesanstalt für Geowissenschaften und Rohstoffe, Hannover.

Röthlisberger, H., 1972. Water pressure in intra- and subglacial channels. J. Glaciol. 11, 177−202.

Röthlisberger, H., Lang, H., 1987. Glacial hydrology. In: Gurnell, A.M., Clark, M.J. (Eds.), Glaciofluvial Sediment Transfer: An Alpine Perspective. John Wiley, Chichester.

Schlüchter, C., 1979. The quaternary glaciations of Switzerland, with special reference to the Northern Alpine Forland. Quat. Sci. Rev. 5, 413−419.

Schlüchter, C., Kelly, M., 2000. Das Eiszeitalter in der Schweiz. Eine schematische Zusammen-fasssung. Überarbeiteter Neudruck. Stiftung Landschaft und Kies, Uttigen.

Schroeder, J.H., 1995. Die Kreuzbrückenspalte von Rüdersdorf—subglaziale Erosion im Wellenkalk. Berliner Geowiss. Abh., Reihe A, Bd 168, 177—189.

Sheppard, M.I., Amiro, B.D., Davis, P.A., Zach, R., 1995. Continental glaciations and nuclear fuel waste disposal: Canada's approach and assessment of the impact on nuclide transport through the biosphere. Ecol. Modell. 78 (3), 249—265.

SKB, 2010a. Climate and Climate Related Issues for the Safety Assessment SR-site. SKB TR-10—49. Swedish Nuclear Fuel and Waste Management Company, Stockholm, Sweden.

SKB, 2010b. Radionuclide Transport Report for the Safety Assessment SR-site. SKB TR-10—50. Swedish Nuclear Fuel and Waste Management Company, Stockholm, Sweden.

SKB, 2011. Long-term Safety for the Final Repository for Spent Nuclear Fuel at Forsmark. Main Report of the SR-site Project. SKB TR-11—01. Swedish Nuclear Fuel and Waste Management Company, Stockholm, Sweden.

Smed, P., 1998. Die Entstehung der dänischen und norddeutschen Rinnentäler (Tunneltäler)—Glaziologische Gesichtspunkte. Eiszeitalter U. Gegenwart 48, 1—18.

Speck, C., 1994. Änderung des Grundwasserregimes unter dem Einfluss von Gletschern und Permafrost. Mitteilung der ETHZ/VAW 134.

Stackebrandt, W., 2009. Subglacial channels of northern Germany—a brief review. Z. dt. Ges. Geowiss. 160 (3), 203—210.

Stackebrandt, W., Ludwig, A.O., Ostaficzuk, S., 2001. Base of Quaternary deposits of the Baltic Sea depression and adjacent areas (map 2). In: Garetsky, R.G., et al. (Eds.), Neogeodynamics of the Baltic Sea Depression and Adjacent Areas. Results of IGCP Project 346. Brandenburgische Geowiss. Beitr., vol. 8 (1), pp. 13—19.

Stumm, D., 2010. Deep Glacial Erosion—Review with Focus on Tunnel Valleys in Northern Europe. Nagra Arbeitsbericht NAB 10—33, Wettingen, Switzerland.

Svendsen, J.I., Alexanderson, H., Astakhov, V.I., Demidov, I., Dowdeswell, J.A., Funder, S., Gataullin, V., Henriksen, M., Hjort, C., Houmark-Nielsen, M., Hubberten, H.W., Ingólfsson, O., Jakobsson, M., Kjær, K.H., Larsen, E., Lokrantz, H., Lunkka, J.P., Lyså, A., Mangerud, J., Matiouchkov, A., Murray, A., Moller, P., Niessen, F., Nikolskaya, O., Polyak, L., Saarnisto, M., Siegert, C., Siegert, M.J., Spielhagen, R.F., Stein, R., 2004. Late Quaternary ice sheet history of northern Eurasia. Quat. Sci. Rev. 23, 1229—1271.

Swift, D.A., Cook, S., Heckmann, T., Moore, J., Gärther-Roer, I., Korup, O., 2014. Ice and snow as landforming agents. In: Haeberli, W., Whiteman, C. (Eds.), Snow and Ice-Related Hazards, Risks and Disasters. Elsevier, pp. 167—199.

Sykes, J.F., Normani, S.D., Yin, Y., 2011. Hydrogeologic Modelling. Nuclear Waste Management Organization, Toronto, Canada. Report NWMO DGR-TR-2011-16. http://www.nwmo.ca/dgrgeoscientificsitecharacterization.

Talbot, C.J., 1999. Ice ages and nuclear waste isolation. Eng. Geol. 52, 177—192.

Texier, D., Degnan, P., Loutre, M., Paillard, D., Thorne, M., 2003. Modelling sequential BIOsphere systems under CLIMate change for radioactive waste disposal-Project BIOCLIM. In: Proceedings of the International High-level Waste Management Conference (30.3-2.4. 2003, Las Vegas, USA).

Tyndall, J., 1864. On the conformation of the Alps. Philos. Mag. 28, 255—271.

Tzedakis, P.C., Channell, J.E.T., Hodell, D.A., Kleiven, H.F., Skinner, L.C., 2012. Determining the natural length of the current interglacial. Nat. Geosci. 5 (1), 1—4. http://dx.doi.org/10.1038/ngeo1358.

Uhlig, L., Amelung, P., Billaux, D., Polster, M., Schmidt, H., 2007. Untersuchungen zur sicherheitstechnischen Auslegung eines generischen Endlagers im Tonstein in Deutschland—GENESIS—(Abschlussbericht). DBETec-Bericht, Peine. Internet, 04/2009, 84 pp.

Vandenberghe, J., Pissart, A., 1993. Permafrost changes in Europe during the last glacial. Permafrost Periglac. 4, 121—135.

Van der Vegt, P., Janszen, A., Moscariello, A., 2012. Tunnel valleys: current knowledge and future perspectives. In: Huuse, M., Redfern, J., Le Heron, D.P., Dixon, R.J., Moscariello, A., Craig, J. (Eds.), Glaciogenic Reservoirs and Hydrocrabon Systems, vol. 368. Geological Society, London, pp. 75—97. Special Publication.

Vettoretti, G., Peltier, W.R., 2011. The impact of insolation, greenhouse gas forcing and ocean circulation changes on glacial inception. Holocene 21 (5), 803—817. http://dx.doi.org/10.1177/0959683610394885.

Vidstrand, P., Follin, S., Selroos, J.-O., Näslund, J.-O., Rhén, I., 2012. Modeling of groundwater flow at depth in crystalline rock beneath a moving ice-sheet margin, exemplified by the Fennoscandian Shield, Sweden. Hydrogeol. J.. http://dx.doi.org/10.1007/s10040-012-0921-8.

Wilson, T.P., Long, D.T., 1993a. Geochemistry and isotope chemistry Ca-Na-Cl brines in Silurian strata, Michigan Basin, U.S.A. Appl. Geochem. 8, 507—524.

Wilson, T.P., Long, D.T., 1993b. Geochemistry and isotope chemistry of Michigan Basin brines: Devonian formations. Appl. Geochem. 8, 81—100.

Wohlfarth, B., 2010. Ice free conditions in Sweden during marine Oxygen isotope stage 3? Boreas 39, 377—398.

Chapter 12

Snow Avalanches

Jürg Schweizer, Perry Bartelt and Alec van Herwijnen

WSL Institute for Snow and Avalanche Research SLF, Davos, Switzerland

ABSTRACT

Snow avalanches are a major natural hazard in most snow-covered mountain areas of the world. They are rapid, gravity-driven mass movements and are considered a meteorologically induced hazard. Snow avalanches are one of the few hazards that can be forecast, and in situ measurements of instability are feasible. Advanced hazard-mitigation measures exist, such as land-use planning based on modeling avalanche dynamics. The most dangerous snow avalanches start as a dry-snow, slab avalanche that is best described with a fracture mechanical approach. How fast and how far an avalanche flows is the fundamental question in avalanche engineering. Models of different levels of physical complexity enable the prediction of avalanche motion. Although the avalanche danger (probability of occurrence) for a given region can be forecast—in most countries with significant avalanche hazard, avalanche warnings are issued on a regular basis—the prediction of a single event in time and space is not (yet) possible.

12.1 INTRODUCTION

Snow avalanches occur in snow-covered mountain regions throughout the world and have caused natural disasters as long as mountainous areas have been inhabited or traveled. One of the oldest records dates back to 218 BC when Roman historian Livius described that Hannibal, while crossing the Alps, lost 12,000 soldiers and 2000 horses due to avalanches. Large disasters have often been associated with military operations such as the crossing of the Alps by Napoleon in 1800, the fighting in the Dolomites in 1916 during World War I, and most recently the conflict between India and Pakistan where, for example, an avalanche killed about 130 soldiers in April 2012.

The number of avalanche fatalities per year due to snow avalanches is estimated to be about 250 worldwide. In fact, in Europe and North America alone, avalanches claimed the lives of about 1,900 people during the 10-year period of 2000–2001 to 2009–2010. In addition to these well-established

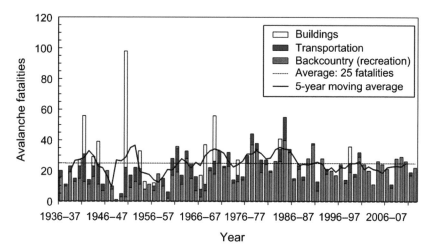

FIGURE 12.1   Avalanche fatalities in the Swiss Alps for the period of 1936−1937 to 2012−2013 (77 years). Most victims were caught during recreational activities such as skiing ("Backcountry"); victims on roads, etc. ("Transportation") and in villages ("Buildings") are less frequent.

statistics, occasional large disasters occur in mountainous countries in Asia. In Europe and North America, most of the fatalities involve personal recreation on public land. Avalanche fatalities on roads or in houses have become less frequent during the twentieth century due to extensive mitigation measures.

In Switzerland, for example, the number of avalanche fatalities on roads or in settlements was about 11 per year until the mid-1970s and has now decreased to less than three, with a long-term total average of 25 victims per year (Figure 12.1). Since the disastrous avalanche winter in 1950−1951 when 98 people in Switzerland (and 135 people in Austria) were killed (Figure 12.2), Switzerland has constructed avalanche defense works worth about $1.5 billion. The effect of these mitigation measures was clearly shown during the winter of 1998−1999 when a similar number of avalanches to that in the winter of 1950−1951 released, but only 17 fatalities occurred (on roads or in buildings), despite the obvious increase in land use and mobility in the Swiss Alps (Figure 12.3). The total damages amounted to $800 million (Wilhelm et al., 2001). In Canada, the yearly average direct and indirect costs are estimated to be more than $5 million (CAA, 2002b).

Even if avalanches are a major threat to people living and recreating in mountain communities, their contribution to the overall risk due to natural hazards in a country such as Switzerland is only about 3 percent—though they contribute more than one-third of all injuries and deaths. The risks due to an earthquake and flooding are estimated to be considerably higher in Switzerland primarily due to the larger area that is affected by these hazards so that damage to people, property and infrastructure is expected to be higher than in the case of snow avalanches (BABS, 2003).

FIGURE 12.2   In Airolo (Switzerland), the Vallascia avalanche destroyed 23 buildings and killed 10 residents on February 12, 1951. *(Photograph: SLF archive.)*

Avalanche risk analysis involves the determination of an avalanche return period or frequency and some measure of consequences that describe the destructive potential (CAA, 2002a). To reduce avalanche risk, protective measures are characteristically used in combination. Avalanche mitigation includes temporary measures (forecasting, road closure) and permanent measures (land-use planning, protective means such as snow sheds or tunnels, reforestation). By combining temporary and permanent measures in a cost-efficient way, otherwise known as integral risk management (Bründl and Margreth, 2014), the avalanche risk can be reduced to an acceptable level. Because snow avalanches are still relatively rare events, personal experience is limited and expertise is generally not readily available. Therefore, it is

FIGURE 12.3   In Evolène (Switzerland), a large avalanche destroyed or damaged several chalets and killed 12 people on February 21, 1999. *(Photograph: M. Phillips.)*

essential for hazard mitigation to increase the awareness of land managers, consultants, governmental agencies, and individual recreationists about snow avalanches.

12.2 THE AVALANCHE PHENOMENON

Snow avalanches are a type of fast-moving mass movement. They can additionally contain rocks, soil, vegetation, or ice. Avalanche size is classified according to its destructive power (Table 12.1). A medium-sized slab

TABLE 12.1 Avalanche Size Classification

Size	Description	Destructive Potential (Definition)	Typical Mass (Tons)	Typical Path Length (m)	Typical Impact Pressure (kPa)
1	Small	Relatively harmless to people	<10	10–30	1
2	Medium	Could bury, injure, or kill a person	100	50–250	10
3	Large	Could bury a car, destroy a small building (e.g., a wood frame house), or break a few trees	1,000	500–1,000	100
4	Very large	Could destroy a railway car, large truck, several buildings, or a forest with an area up to 4 ha	10,000	1,000–2,000	500
5	Extreme	Largest snow avalanches known; could destroy a village or forest of 40 ha	100,000	>2,000	1,000

Adapted from McClung and Schaerer (2006).

avalanche may already involve $10,000 \text{ m}^3$ of snow, equivalent to a mass of about 2000 tn (snow density 200 kg/m^3). Avalanche speeds vary between 50 and 200 km/h for large dry-snow slides, whereas wet-snow avalanches are denser and slower ($20-100$ km/h). If the avalanche path is steep, dry-snow avalanches generate a powder cloud.

Snow avalanches come in many different types (e.g., wet or dry) and sizes. The morphological classification published by the former International Commission on Snow and Ice (UNESCO, 1981) takes into account the three principal zones of an avalanche: origin (or starting zone), transition (or track), and runout (Table 12.2). It helps one to classify the type of avalanche based on observable features such as the manner of starting or the form of movement.

A snow avalanche path consists of a starting zone, a track, and a runout zone where the avalanche decelerates and the snow is deposited (Figure 12.4). The starting zone, or in analogy to hydrology, the catchment area, is where the initial snow mass releases and generally consists of terrain steeper than 30°. Only a low percentage of dry-snow avalanches start on terrain under 30°. Wet-snow slides, on the other hand, can occur on slopes under 25°. Slope angle is the most important terrain factor influencing avalanche release. A snow avalanche will then flow downstream from the starting zone along the track, which often consists of creek beds and gullies. If the track is steep and a powder cloud develops, the powder snow avalanche may run straight down, regardless of the topography, that is, not follow, for example, any bends in the creek bed. Although small avalanches may stop in the track (typically $15-30°$ steep), large ones move with an approximately constant speed to the runout zone where they slow down and stop. On large avalanche paths, the slope angle in the runout zone is generally $<15°$ (Jamieson, 2001). Runout zones for large avalanche paths are common on alluvial fans—a preferred area for infrastructure, including businesses and residences, in mountain areas.

12.3 AVALANCHE RELEASE

A snow avalanche may release in two distinctly different ways: as a loose snow avalanche or as a snow slab avalanche (Figure 12.5). Loose snow avalanches start from a point, in a relatively cohesionless surface layer of either dry or wet snow. The initial failure originates in one location when a small mass of snow fails and begins to move and entrain additional snow. The process is analogous to the rotational slip of cohesionless sands or soil, but occurs within a small volume ($<1 \text{ m}^3$) in comparison to much larger initiation volumes in soil slides (Perla, 1980). As the snow mass descends, the avalanche spreads outward in an inverted V shape. Most loose snow avalanches are relatively small and harmless since only a cohesionless surface layer is involved. However, when the entire snow cover is saturated with water, loose snow avalanches can entrain large volumes of snow and cause damage.

TABLE 12.2 International Morphological Avalanche Classification (UNESCO, 1981)

Zone	Criterion	Characteristics	Denomination
Origin (starting zone)	Manner of starting	From a point	Loose snow avalanche
		From a line	Slab avalanche
	Position of failure layer	Within the snowpack	Surface-layer avalanche
		On the ground	Full-depth avalanche
	Liquid water in snow	Absent	Dry-snow avalanche
		Present	Wet-snow avalanche
Transition (track)	Form of path	Open slope	Unconfined avalanche
		Gully or channel	Channeled avalanche
	Form of movement	Snow dust cloud	Powder snow avalanche
		Flow along ground	Flowing snow avalanche
Deposition	Surface roughness of deposit	Coarse	Coarse deposit
		Fine	Fine deposit
	Liquid water in snow	Absent	Dry avalanche deposit
		Present	Wet avalanche deposit
	Contamination of deposit	No apparent contamination	Clean avalanche
		Rock debris, soil, branches, trees	Contaminated avalanche

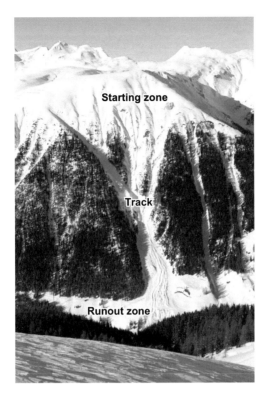

FIGURE 12.4   Large avalanche path showing the starting zone where the avalanche initiates, the track and the runout zone where the avalanche decelerates and the snow is deposited. *(Breitzug, Davos; photograph: J. Schweizer.)*

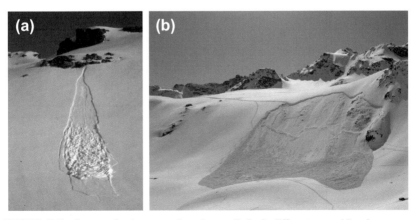

FIGURE 12.5   Snow avalanches may release in two distinctly different ways: (a) as loose snow avalanche or (b) as snow slab avalanche. *(Photograph: J. Schweizer.)*

Snow slab avalanches behave quite differently. They involve the release of a cohesive snow slab over an extended plane of weakness, analogous to the planar failure of rock slopes and landslides rather than to the rotational failure of soil slopes (Perla, 1980). The observed ratio between width and thickness of the slab varies between 10 and 10^3, and is typically about 10^2. Slab thickness is generally <1 m, typically about 0.5 m, but can reach several meters in the case of large disastrous avalanches (Schweizer et al., 2003). Slab avalanches are the more hazardous of the two types and represent the vast majority of fatal avalanches. Slab avalanches are more harmful as they typically involve more snow and are harder to predict than loose snow avalanches. Slab avalanches are the focus of most avalanche-related studies.

Predicting snow slab-avalanche release can be approached either by (1) exploring the complex interaction between three main contributing factors: terrain, weather and snowpack or by (2) studying the physical and mechanical processes of avalanche formation (Schweizer et al., 2003). We first discuss the former approach that is applied by most avalanche forecasting services. It involves empirically weighting the influence of the contributory factors in a specific situation.

Terrain is an essential factor and the only factor that is constant in time. A slope angle of about 30° is required for a slab avalanches to release. However, other topographic parameters such as curvature, aspect, distance to a ridge, and forest cover are also important. In general, the identification of potential avalanche release areas is a difficult task requiring considerable expertise, but it is a prerequisite for large-scale, hazard mapping, numerical avalanche simulations, and planning of hazard-mitigation measures. Today, starting zones can automatically be identified within a geographic information system (GIS), provided that a high-resolution digital terrain model (DTM) is available. However, for detailed planning, including various release scenarios, manual adjustment of the starting zone perimeter is generally required (Bühler et al., 2013). Forests inhibit avalanche formation; in particular, in dense forests, the snow cover is too irregular to produce avalanches. The main effects that alter the snow cover characteristics in forests compared to open unfrosted terrain are (1) the interception of falling snow by trees; (2) the reduction of near-surface wind speeds; (3) the modification of the radiation and temperature regimes; and (4) the direct support of the snowpack by stems, remnant stumps, and dead wood (Schneebeli and Bebi, 2004; Teich et al., 2012).

The main meteorological conditions contributing to avalanche formation are precipitation (new snow or rain), wind, air temperature, and solar radiation. For large, catastrophic avalanches, precipitation is the strongest forecasting parameter. Although the total amount of precipitation plays an important role, the precipitation rate can also strongly influence avalanche release. Wind contributes to loading and is often considered the most active contributing factor after precipitation. Loading by wind-transported snow can be fast and produce irregular deposits, increasing the probability of avalanching in certain areas. Snow deposition by wind is strongly influenced by terrain so that snow-drift accumulations commonly occur at the same location year after year. The persistence of accumulation patterns has been revealed by terrestrial layer scanning (Prokop, 2008) and offers the possibility for modeling these patterns, based simply on topography and mean wind direction (Schirmer et al., 2011). Temperature, and in general, the energy balance at the snow–atmosphere interface, can be decisive factors contributing to avalanche formation, especially in the absence of loading. Its effects on snow stability are complex and commonly subtle, but it is often assumed that a rapid increase in air temperature and/or solar radiation promotes instability. In any case, instability always stems from changes in slab properties (Reuter and Schweizer, 2012; Schweizer and Jamieson, 2010a). Increased deformation due to reduced stiffness of the surface layers increases the strain rate in the weak layer, increases the energy release rate, or increases the skier stress at depth. Although these effects occur rapidly and promote instability, delayed effects, such as snow metamorphism and settlement, tend to promote stability (McClung and Schweizer, 1999). Surface warming (of a dry snowpack) is most efficient with warming by solar radiation as radiation penetrates the surface layers where the energy is released. Surface warming due to warm (relative to the snow surface) air temperatures is a secondary effect—except in the case when a moderate or strong wind blows (Schweizer and Jamieson, 2010a).

Finally, snow cover stratigraphy is recognized as the key contributing factor for snow slab avalanche formation (Schweizer et al., 2003). The mountain snowpack can contain many different snow layers with distinctive properties. Each layer is the result of a snowfall, wind transport or energy exchange between the snow surface, and the atmosphere. Each interface between two layers was once the snow surface and was influenced by the atmosphere before it was buried. Snow layers are generally characterized according to grain type, grain size, hardness, and density following the International Classification for Seasonal Snow on the Ground (Fierz et al., 2009). Some layers are softer so that the strain is concentrated within these layers (Reiweger and Schweizer, 2010); they have lower strength than the layers above and below, and are more often associated with slab avalanching and are hence termed weak layers. Some weak layers are very discernible within the snowpack and are several centimeters or even tens of centimeters thick. Other weak layers can be very thin (few millimeters) and hard to identify, but equally important. This weakness can either be within the old snow (typically a weak layer composed of facets, depth hoar, or surface hoar) or at the old snow surface underlying the new snow. Weak layers differ distinctly in grain size and hardness from the adjacent layers (Schweizer and Jamieson, 2003). They can be grouped into non-persistent and persistent weak layers (Jamieson, 1995).

Non-persistent weak layers, also called storm-snow instabilities, generally consist of precipitation particles that may remain weaker and lower in density than the adjacent layers during the initial stages of rounding. These layers tend to stabilize within a few days after burial, hence the name non-persistent. Persistent weak layers can remain weak for extended periods of time, sometimes months. They consist of surface hoar, faceted crystals, or depth hoar; these layers are more prone to failure in shear than in compression (Reiweger and Schweizer, 2013b). Although wet layers on the snow surface that freeze and become melt-freeze crusts form the bed surface for many slab avalanches, they are not considered weak layers. Failure often occurs in a layer of facets above the crust, so thin it is hard to identify. These so-called weakly bonded crusts form when snow falls on a wet snow surface so that a weak layer of faceted crystals develops while the underlying wet layer freezes into a crust, often within a day (Jamieson, 2006). With regard to avalanche accidents, persistent weak layers are the main concern for skiers as the majority of fatal avalanches occur on persistent weak layers (Schweizer and Jamieson, 2001).

Any loading by new or wind-blown snow or any temperature increase has no effect on snow stability if no weakness exists within the snowpack. The presence of a weak layer is a necessary, but not sufficient condition for slab-avalanche formation. Apart from the weak layer, the properties of the overlying slab are equally important for avalanche formation, in particular because the slab provides parts of the energy for crack propagation.

The slab-avalanche nomenclature (Perla, 1977) reflects the fact that a slab avalanche is the result of a fracture process involving at least four fracture surfaces. The first failure is within the weak layer, and the bed surface is defined as the surface over which the slab slides. The bed surface can be the ground or older snow. The weak layer is always just above the bed surface and just under the slab. The breakaway wall at the top periphery of the slab is called the crown (fracture), and is approximately perpendicular to the bed surface reflecting the fact that the initial failure is in the weak layer below the slab. The flanks are the left and right sides of the slab. The flanks are generally smooth surfaces, as is the crown. The lowest down-slope fracture surface is termed the stauchwall (Figure 12.6) (Schweizer et al., 2003).

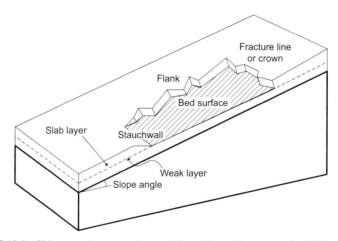

FIGURE 12.6 Slab-avalanche nomenclature. *(Adapted from Schweizer et al. (2003).)*

Depending on the processes leading to slab release, three types of slab avalanches occur: dry-snow, wet-snow, and glide-snow avalanches.

12.3.1 Dry-Snow Avalanches

The release of a dry-snow slab avalanche is due to the overloading of an existing weakness in the snowpack. Most dry-snow slab avalanches start naturally during or soon after snow storms. High precipitation rates favor snowpack instability. In general, about 50 cm of new snow within 24 h (equivalent to about 50 mm of precipitation) is critical for avalanche initiation. Large disastrous avalanches usually follow storms that deposit >1 m of snow. Therefore, for large new snow avalanches, the 3-day sum of precipitation is the strongest forecasting parameter (Schweizer et al., 2009) and closely related to avalanche danger (Schweizer et al., 2003). The triggering of a dry-snow slab avalanche can also occur artificially by localized, rapid, near-surface loading by, for example, people (usually unintentionally) or intentionally by explosives used as part of avalanche control programs. In general, naturally released avalanches mainly threaten residents and infrastructure, whereas human-triggered avalanches are the main threat to recreationists (Schweizer et al., 2003).

For a dry-snow slab avalanche to release, an initial crack in a weak layer has to propagate below the slab. For natural slab avalanches, it is believed that the initial failure is caused by a gradual damage process at the micro-scale leading to failure localization within the weak layer (Figure 12.7). For artificially triggered avalanches (e.g., skier-triggered avalanches), the external trigger induces localized deformations that are large enough to initiate a crack within the weak layer (van Herwijnen and Jamieson, 2005). In any case, if the initial crack in the weak layer reaches a critical size (length)—of the order of several tens of centimeters—it will propagate through the weak layer below the slab. Typically, weak layers are extremely porous and therefore the fracture

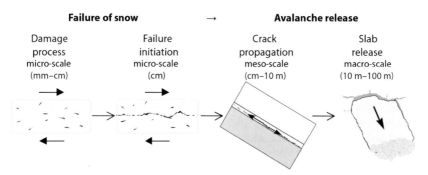

FIGURE 12.7   Conceptual model of dry-snow slab-avalanche release. *(After Schweizer et al. (2003).)*

process is associated with the collapse of the weak layer (van Herwijnen et al., 2010). Once the weak layer has fractured, slab and bed surface come into contact. If the gravitational pull on the detached snow slab is large enough to overcome friction, that is, the slope is steep enough ($>30°$), a snow slab avalanche releases (van Herwijnen and Heierli, 2009).

Considering avalanche release as a fracture process, and describing it accordingly, goes back to McClung (1979, 1981a) who adapted a failure model by Palmer and Rice (1973). Two decades later, Kirchner et al. (2000) performed the first fracture mechanical measurements needed to eventually apply models based on fracture mechanics. Based on laboratory measurements (Schweizer et al., 2004), a field test has been developed (Gauthier and Jamieson, 2006; Sigrist and Schweizer, 2007) that now enables the determination of two crucial properties: the specific fracture energy of the weak layer and the stiffness of the slab (Schweizer et al., 2011; van Herwijnen and Heierli, 2010).

12.3.2 Wet-Snow Avalanches

Wet-snow avalanches release due to the percolation of liquid water within the snow cover and primarily endanger communication lines and infrastructure. Wet-snow avalanches mostly release spontaneously and characteristically cannot be triggered artificially—in contrast to dry-snow avalanches. Although dry-snow avalanches cause most avalanche fatalities, mainly among winter recreationists, wet-snow avalanches may occasionally cause severe damage. Analysis of a 10-year record of avalanche victims in the Swiss Alps showed that about 50 percent of the fatalities caused by naturally released snow avalanches were due to wet-snow avalanches (either slab or loose snow avalanche) (Schweizer and Lütschg, 2001). When only considering human-triggered avalanches, however, fatalities due to wet-snow avalanches drop to 1 percent. Hence, spontaneous releases of wet-snow avalanches are as lethal as naturally released dry-slab avalanches but wet-snow avalanches are seldom triggered by recreationists themselves. Our understanding of the triggering conditions for wet-snow avalanches is still somewhat limited. This is partly due to a lack of observations, and the fact that wet-snow instability is a highly transient and spatially variable phenomenon related to the water transport in snow (Schneebeli, 2004).

Two prerequisites exist for wet-snow avalanche formation: (1) the presence of liquid water within the snowpack; and (2) a (large) part of the snowpack must be isothermal ($0°C$). Water production at the snow surface is determined by the energy balance at the snow—air interface and/or the amount of water delivered through rain (Mitterer and Schweizer, 2013). Based on experience and observations, three possible triggering mechanisms (Baggi and Schweizer, 2009) occur: (1) loss of strength due to water infiltration and storage at a capillary barrier; (2) overloading of a partially wet and weak snowpack due to precipitation; and (3) gradual weakening of the snowpack due to warming to 0 °C and eventual failure of basal layers. Clearly, combinations of these three mechanisms may exist. Overall, it is still not entirely clear as to how water infiltration influences wet-snow instability. Snow stratigraphy is key as it controls the rate of infiltration, the pattern of infiltration and the concentration of water at a given location—which then ultimately will affect the mechanical strength (Peitzsch, 2008). Yamanoi and Endo (2002) observed a continuous decrease in shear strength with increasing liquid water content (up to 8 percent). Recent preliminary results of shear measurements on different substrates indicate that the decrease is highly nonlinear.

Wet-snow avalanches are particularly difficult to forecast. Once the snowpack becomes partly wet, the release probability rapidly increases, but determining the peak and end of a period of high wet-snow avalanche activity is particularly difficult (Techel and Pielmeier, 2009).

Air temperature is commonly related to days with wet-snow instability (Kattelmann, 1985), but is not suitable for forecasting wet-snow avalanches because many false alarms are produced (Mitterer and Schweizer, 2013; Trautman, 2008). By introducing a combination of air and snow surface temperature, predictive performance for days with high wet-snow avalanche activity improved (Mitterer and Schweizer, 2013). In addition, Mitterer and Schweizer (2013) showed that when modeling the entire energy balance for virtual slopes, avalanche and non-avalanche days could be classified with a good accuracy. However, modeling and interpreting the energy balance in terms of wet-snow avalanche release probability is complex and sometimes not feasible for operational avalanche forecasting. Therefore, Mitterer et al. (2013) introduced an index of liquid water content describing the amount of liquid water in the entire snowpack. The index related well with observed wet-snow avalanche activity and indicated spatial and temporal patterns of wet-snow avalanche activity. Onset and peak of wet-snow avalanche activity were mostly well detected, particularly when high temperatures and high values of shortwave radiation caused the percolating melt water. However, determining the end of a period with wet-snow avalanche activity was not possible, because the index showed no distinctive pattern at the end of such periods.

12.3.3 Glide-Snow Avalanches

Glide-snow avalanches can represent a serious challenge to avalanche programs protecting roads, towns, ski lifts, and other operations. These avalanches are notoriously difficult to forecast and can be very destructive, as large volumes of dense snow are mobilized at once. Mitterer and Schweizer (2012) have recently summarized what is currently understood about glide-snow avalanches. Glide-snow avalanches occur when the entire snowpack glides over the ground until an avalanche releases. Glide cracks, that is, full-depth

FIGURE 12.8   Glide-snow avalanches are often preceded by the appearance of a glide crack through the snow cover. Avalanche release can occur within minutes, hours, weeks, or not at all. *(Photograph: R. Meister.)*

tensile cracks exposing the ground, are often observed before glide-snow avalanches (Figure 12.8).

Glide-snow avalanches mostly release from specific and well-known starting zones, and their location is highly dependent on topography (Lackinger, 1987; Leitinger et al., 2008). Glide-snow avalanches occur mostly on steep terrain, that is, 30−40° steep slopes (Leitinger et al., 2008; Newesely et al., 2000), covered with smooth rock (e.g., Stimberis and Rubin, 2011), grass (in der Gand and Zupancic, 1966), or tipped-over bamboo bushes (Endo, 1985). Newesely et al. (2000) observed increased snow gliding on abandoned pastures compared to slopes with short grass. Leitinger et al. (2008) and Höller (2001) observed that the lack of dense forest stands contributed to glide-snow activity, in particular if the distance between surrounding anchor points is >20 m. Observations suggest that most glide-snow avalanches release on convex rolls (e.g., in der Gand and Zupancic, 1966).

Snow gliding processes and glide-snow avalanches are conceptually well understood, and it is widely accepted that a reduction in friction at the base of the snow cover due to the presence of liquid water is the main driver (e.g., Lackinger, 1987; McClung, 1981b). Once friction is reduced and a glide crack has opened, the peripheral strength, in particular of the stauchwall, seems to be crucial for stability (Bartelt et al., 2012a). Since various processes are involved in reducing friction at the base of the snow cover, the relationship between meteorological conditions and glide-snow avalanching is complex. Thus, relying on weather data to forecast glide-snow avalanches is still difficult and relatively inaccurate (Peitzsch et al., 2012; Simenhois and Birkeland, 2010; Stimberis and Rubin, 2005). As the presence of water seems to be decisive for the formation of glide-snow avalanches, it is paramount to know the processes that are responsible for the presence of water at the snow−soil interface. Three different processes may deliver liquid water to the snow−soil interface (McClung and Clarke, 1987): (1) Water percolating through the snow cover; (2) heat released from the warm ground melting the snow at the base of the snow cover after the first major snowfall; and (3) water, produced at terrain features with strong energy release (e.g., bare rocks), running downward along the snow−soil interface or originating from springs (ground-water outflow). In addition, it seems possible that the lowermost snowpack layer becomes wet due to capillary rise caused by different hydraulic pressures along the snow−soil interface (Mitterer and Schweizer, 2012).

The triggering case (1) is very similar to the triggering process related to wet-snow avalanches: the less permeable substrate below the snowpack often acts as a capillary barrier to infiltrating water. Thus, determining the arrival time of water at the base of the snow cover is crucial for predicting avalanche events (Mitterer et al., 2011). The main processes associated with producing the water are due to melting at the snow surface and rain-on-snow events. Many glide-snow avalanches are therefore observed during warm periods or rain events (Clarke and McClung, 1999; Lackinger, 1987). However, the so-called cold temperature events also release when the snow cover is still mostly below freezing. in der Gand and Zupančič (1966) stated that for these events the existence of a lowermost moist snow layer is especially important as a dry boundary layer would not cause glide motion on a grass surface. Moreover, they suggested that liquid water is produced due to warm ground temperatures. Snow layers with low temperatures ($<0\,°C$) may exist above the wet layer. These observations have been confirmed by several later studies (Höller, 2001; Newesely et al., 2000). In addition, the release of glide-snow avalanches, in particular cold-snow events, is often observed during snow loading as the additional load increases creep and glide (Dreier et al., 2013).

The reduction of friction at the base of the snow cover results in increased snow glide rates, and snow glide rates are closely related to glide-snow avalanche release (e.g., in der Gand and Zupančič, 1966). Clarke and McClung (1999) and Stimberis and Rubin (2011) suggested that glide-snow avalanche release may best correlate with periods of rapid increases in glide rates. Thus, measuring glide rates could improve glide-snow avalanche forecasting. Several methods have been used to measure glide rates in snow, including sprung probes (Wilson et al., 1997), seismic sensors (Lackinger, 1987; Stimberis and Rubin, 2005), accelerometers (Rice et al., 1997), and, most frequently, glide shoes (in der Gand and Zupančič, 1966; Lackinger, 1987; Clarke and McClung, 1999; Stimberis and Rubin, 2005). While all these methods can be used to measure glide rates, they are currently costly, somewhat unreliable, and difficult to conduct in multiple paths.

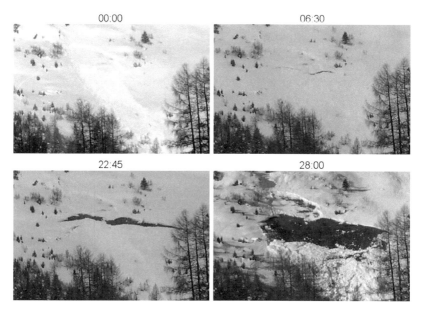

FIGURE 12.9   Sequence of images showing the gradual expansion of a glide crack, followed by the release of a glide-snow avalanche after 28 h. *(Photograph: A. van Herwijnen.)*

More recently, a different approach for measuring glide rates was proposed by tracking the expansion of glide cracks with time-lapse photography (van Herwijnen et al., 2013; van Herwijnen and Simenhois, 2012) (Figure 12.9). When a glide crack appears, the ground below the snow cover is exposed. Since the ground is much darker than snow, it can clearly be identified on the time-lapse images. Using a simple method based on dark pixel counting, glide rates can be derived. van Herwijnen and Simenhois (2012) showed that for glide cracks that resulted in avalanche release, the number of dark pixels rapidly increased a few hours before avalanche release, in line with previously published glide rate measurements. Although such new developments are encouraging with regard to glide-snow avalanche forecasting, it is still not clear how increases in glide rates relate to avalanche release.

12.4 AVALANCHE FLOW

Avalanche flow begins after snow is released and set in motion. How fast and how far an avalanche flows is one of the essential questions in avalanche engineering. Hazard maps and the planning of mitigation measures require an estimation of avalanche speed and the extent of the inundated area as a function of the release zone location and characteristics of the mountain terrain (steepness, roughness, gullies, vegetation, etc.). Hazard scenarios are typically based on the concept of avalanche return period, which is linked to snow cover fracture heights (e.g., Ancey et al., 2004; Bründl et al., 2010). Information regarding snowfall history and past avalanche events is necessary to predict reasonable starting masses, which greatly affect avalanche runout distances. Climatic conditions need to be considered to assess the possibility of extreme wet-snow avalanches or powder snow avalanches. To predict the motion of avalanches including these effects, models with different levels of physical complexity are employed in engineering practice. These include empirical (e.g., Lied and Bakkehoi, 1980), statistical (e.g., McClung and Mears, 1991), and physical and semiphysical analytical models (e.g., Salm et al., 1990). Increasingly physics-based numerical models (e.g., Christen et al., 2010) are being used because of the availability of high-resolution digital elevation models that allow an accurate representation of mountain terrain (Bühler et al., 2013). The newest generation numerical models also account for snow entrainment processes that allow the treatment of avalanche growth.

12.4.1 The Transition to Flow: Slab Break-up and Snow Granularization

When a dry-snow slab avalanche releases, it may first slide as a rigid block. For dry-snow avalanches, the sliding surface is typically a harder (older) snow layer; for wet-snow avalanches, the sliding surface can be the ground. Friction is low because the slab often slides on a hard layer lubricated by the poorly bonded remnants of the fractured weak layer; wet-snow avalanches often slide on wetted surfaces. Because of the low friction of the sliding surface, accelerations can be large and sliding snow can quickly reach velocities of >10 m/s. Terrain undulations or variations in surface roughness contribute to the quick break-up of the slab (Figure 12.10). First, large snow cover fragments form, but as the slab continues to displace, the fragments disintegrate into smaller granules of various shapes and sizes (Bozhinskiy and Losyev, 1998). By the

FIGURE 12.10 The break-up of the slab and the start of a large flowing avalanche. *(Photograph: T. Feistl.)*

FIGURE 12.11   The granular deposits of (a) dry-snow flowing avalanche and (b) wet-snow avalanche. *(Photograph: P. Bartelt.)*

time the entire slab has exited the release zone and passed the stauchwall (which can also enhance the break-up process), the slab avalanche has transformed into a granular flow.

The bulk flow density of the avalanche is given by the distribution of granules within an avalanche flow volume. The break-up process, granule collisions, and other interparticle interactions lead to hardened granules and snow fragments. Typical granule densities range between 300 and 500 kg/m³ (Bozhinskiy and Losyev, 1998). It is also possible to find hard ice granules in avalanche deposits as well as in rocks and woody debris (Figure 12.11).

The bulk flow densities of the avalanche can vary strongly (Gauer et al., 2007, 2008). For example, highly fluidized avalanches—a dry-snow flowing avalanche or powder snow avalanche—can have bulk flow densities at the front of <100 kg/m³. Such regions in the avalanche are commonly termed "saltation" layers because of their low solid mass content (Gauer et al., 2008). Wet-snow avalanches or the dense core of flowing avalanches have higher bulk flow densities, varying between 200 and 400 kg/m³. That is, with bulk densities near the granule densities suggesting less interstitial air space. Avalanches, depending on the snow properties and the granule formation, can exhibit both "collisional" and "frictional" flow regimes. It is entirely possible that with one single avalanche a transition will occur between flow regimes (Bartelt et al., 2011). For example, at the head of a dry-snow flowing avalanche or powder avalanche, we are often confronted with a collisional regime or dispersed, dilute flow regime, whereas toward the tail of the avalanche, the flow densifies, producing the dense core of the avalanche. An understanding of the avalanche flow density is necessary as it determines both the mobility of the flow and the magnitude of the avalanche impact pressures.

The second salient feature of the granularization process is that it modifies the internal energy fluxes of the avalanche (Figure 12.12) (Bartelt et al., 2006). When a slab releases, potential energy P is transformed into kinetic energy K and heat E. This kinetic energy is associated with the mean slope-parallel

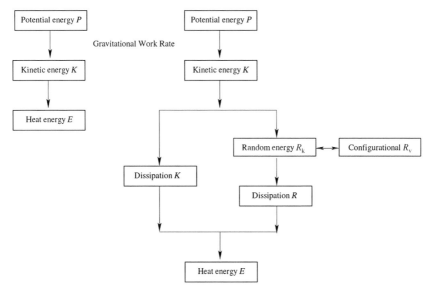

FIGURE 12.12   Two models of avalanche flow. (Left) Potential energy is transformed into kinetic energy and heat (block models, simple hydrodynamic models). (Right) Potential energy is transformed into kinetic energy, random kinetic energy R_k, configurational energy R_v and heat E (granular models).

movement of the avalanche, the velocity U. Frictional processes dissipate the kinetic energy, raising the internal heat energy E of the avalanche. In this model of avalanche flow (Figure 12.12), the motion of the avalanche can be described completely, once the frictional processes have been defined. This flow model is the basis of many (useful) block-type avalanche models (Perla et al., 1980; Salm et al., 1990). However, the granularization of the snow slab at release changes this simple energy model dramatically. As the slab displaces, space opens up between the granules. Movements in the slope-perpendicular direction are possible, implying an increase in the avalanche flow volume and a decrease in the avalanche flow density. This energy, denoted R_v, is termed "configurational" as it is associated with volume changes and therefore with flow densities and flow regimes. Importantly, the granules no longer all move with the same speed or direction of the mean flow. That is, the avalanche motion is no longer a rigid block, but a highly variable and deformable mass of particles. This energy has been termed R_k for "random kinetic" as it is associated with granule movements that vary from the mean translational velocity (Bartelt et al., 2006; Buser and Bartelt, 2009).

12.4.2 Avalanche Flow Regimes

After the transition phase of release and slab break-up in an avalanche, the flow phase begins. The avalanche at first accelerates and reaches a relatively

FIGURE 12.13 The muddy deposits of a large wet-snow avalanche near Klosters (Switzerland), April 2008. *(Photograph: P. Bartelt.)*

steady velocity (transition zone) before decelerating on a flatter slope (runout zone). The length of the transition and runout phases depends strongly on the geometry of the avalanche path and also on the avalanche type and the frictional properties of the flowing snow (Figure 12.13).

Mitigation of snow avalanche hazard requires determining the propagation speed of the leading edge, the avalanche flow density and the dimensions of the avalanche body (depth, width, and length). These parameters vary over the length of the avalanche track, from the point of release to the point of maximum runout. The total volume of avalanche deposits is often an additional parameter that is required to assess the height of avalanche dams and avalanche deflecting structures.

The problem of predicting avalanche flow in mountain terrain is difficult. Many different models have been developed to estimate the parameters of avalanche flow. The models are based on relatively simple observations of avalanche drop height versus runout, or on detailed flow measurements (Harbitz, 1998). The models can be simple, for example, empirical formulas that average terrain features, or more intricate numerical models that require digital elevation models to represent the complexities of real mountain terrain. To date, however, both simple and complex models divide avalanche flow into one of two flow distinct and general categories: "dense flowing" and "powder" (suspension). This division essentially splits avalanche flow into a regime (or model) with relatively dense snow mass ($\rho = 100-500$ kg/m^3) and a regime (or model) of highly dispersed snow and air mass ($\rho = 3-50$ kg/m^3). The dispersed phase can be further subdivided into a "saltation layer" ($\rho = 10-50$ kg/m^3) and a "suspension layer" ($\rho = 3-10$ kg/m^3). Dry-mixed avalanches consist of a combination of both dense and powder parts and

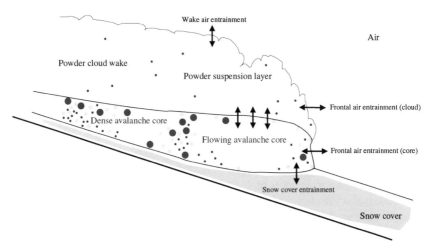

FIGURE 12.14   Avalanche cross-section. Avalanches consist of a dense avalanche core and suspension layer. The density of the avalanche core varies from 50 to 100 kg/m³ ("saltation" like flows) to 500 kg/m³ (heavy, wet-snow avalanches). The snow cover and ambient air can be entrained by the avalanche.

therefore consist of a dense core with saltation and suspension layers (Figure 12.14). See Gauer et al. (2008) for an overview of inferred density measurements in snow avalanches.

Here, it is important to reiterate that avalanches are composed of flowing snow debris mixed with air and the very fact that avalanche density varies is a result of the granularization and break-up process that allows density variations. Therefore, flow regime and flow regime transitions are intricately related to the granular properties of flowing snow, which are strongly temperature and moisture dependent (Gauer et al., 2008).

12.4.3 The Avalanche Core

The dense and saltation layers can be grouped together to form the "avalanche core." This is the destructive center of the avalanche. The core contains mass in granular form, that is, both large and small aggregates of ice grains. Mean granule sizes for dry-snow avalanches are in the range of 5–10 cm; wet-snow granules are larger, a result of the cohesive properties of moist snow (Bartelt and McArdell, 2009). Of course, there are many smaller and larger particles within the flow. The particles exist in a continual state of flux; they can break or they can combine to form particle agglomerates, especially in the runout zone where the terrain flattens and the flow becomes slower. Air mixed with ice dust is blown out of the core to form the suspension layer. Both small granules and ice dust are blown out of the avalanche core as well. However, the smaller granular aggregates are not suspended, but once ejected from the core, they will return rapidly to flow. The granular aggregates can be found high above the ground (10−20 m) increasing the height of the zone where large, but local, impact pressures can be exerted on tall structures, including trees. Because heavier particles fall out of the dust cloud, this region of the avalanche core is sometimes termed the "segregation layer." These aspects of the flow core are a result of the fact that the upper surface of the avalanche is essentially a free surface.

The denser regions within the avalanche core also exhibit significant variations in bulk flow density (Gauer et al., 2008). Because of several factors (ground roughness, terrain undulations, large overburden pressures, as well as the braking effects of snow cover deconstruction and entrainment), granules at the running surface move slower than granules in the upper region of the flow. Measured velocity profiles have been reported in Kern et al. (2009). Velocity gradients indicate not only granular collisions and thus frictional dissipation but also strong dilatative movements within the avalanche core (Buser and Bartelt, 2011). This causes an expansion of the avalanche flow volume and therefore a decrease in bulk avalanche flow density. The degree of volume expansion depends on the avalanche flow height. It is harder to change the flow volume of larger flow heights where the overburden pressures are larger. At the front of the avalanche, where flow heights are small and frictional forces are large, strong expansion movements can occur within the avalanche core.

12.4.4 The Suspension Cloud

If the avalanche snow is dry, ice grains (ice dust) become suspended in the surrounding air to form a "powder" or "suspension" cloud. The dust can arise from a layer of weakly bonded fresh snow, or the abrasive granular interactions between dry-snow granules within the avalanche. Mean grain sizes are 0.1 mm (Rastello et al., 2011). The ice dust is mixed with air in the avalanche core and blown out of the avalanche, often in violent vertical movements to form irregular plume-like structures. The plumes are formed at different locations and rise at different speeds; the stronger plumes will often rise and spread more quickly, spreading over and consuming slower rising plumes. The powder cloud surface is therefore uneven and composed of billows and clefts. This gives powder avalanches their distinctive turbulent appearance.

Japanese measurements of powder cloud densities from pressure measurements reveal bulk densities between 3 and 5 kg/m^3 (Nishimura et al., 1993). Although these densities appear low, they are more than enough to give powder clouds their opaque grey-white color. A cubic meter of powder cloud dust will contain some 10,000,000 ice particles. This is enough to obstruct the view of the avalanche core flowing below. It hinders a clear understanding of the formation process of powder clouds.

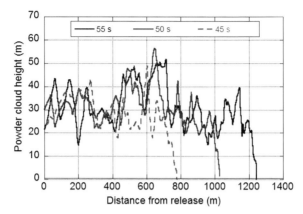

FIGURE 12.15   Cross-section of measured powder cloud heights of a Vallée de la Sionne avalanche for three times $t = 45$, 50 and 55 s. The heights reach over 50 m. The cloud is created at the front of the avalanche. The plumes initially move forward but are almost stationary 100 m behind the avalanche front.

Using stereogrammetric, georeferenced photographic images of the powder cloud at the Swiss Vallée de la Sionne test site, it has been possible to measure the total volume of powder clouds. These are immense. For avalanches consisting of 50,000−100,000 m³ of flowing snow debris, the powder cloud volumes can exceed 10,000,000 m³. The powder cloud volume can be well over 100 times the volume of the core. The suspension ratio (the ratio of powder mass to core mass) is estimated to be between 5 percent and 20 percent. This is an indication of the large amount of air entrained by the powder cloud.

Another use of stereogrammetric imaging is to trace the movement of plumes (Figure 12.15). This reveals that the plumes are primarily created at the front and edges of the flow. The plumes quickly decelerate and become stationary, sometimes only a 100 m behind the leading edge of the avalanche. This is an indication of large drag forces acting on the cloud. These drag forces arise from air entrainment and drag as the particles are expelled into the surrounding air. The front of the cloud is moving faster than the tail, or wake of the cloud. The movement of the powder cloud resembles the smoke blown out by a steam engine, that is, there is a continual creation of new powder at the avalanche front that becomes stationary as the avalanche moves forward. The motion of the cloud does not resemble a rigid body movement (in which the front and rear move at the same speed) as some powder snow avalanches models would suggest (Ancey, 2004; Beghin and Olagne, 1991).

12.4.5 Snow Entrainment

Observations reveal that avalanches can entrain snow layers, typically a layer of fresh snow. Wet-snow avalanches frequently remove the entire snowpack exposing the ground creating muddied depositions (Jomelli and Bertran, 2001). In fact, avalanches play an important role in transporting debris (soil, rocks, and dead wood) from the starting zone to the runout zone (Bozhinskiy and Losyev, 1998; Swift et al., 2014).

Entrainment processes are usually divided into three phenomenological categories (Cherepanov and Esparragoza, 2008; Gauer and Issler, 2004; Sovilla, 2004): (1) frontal ploughing; (2) basal erosion; and (3) snow layer fracture entrainment.

Snow entrainment affects avalanche motion in four ways: (1) It changes the overall mass balance of the avalanche; (2) It can change the temperature and moisture content of the flowing snow and therefore snow entrainment can change the avalanche flow regime from dry to wet, or from dilute to dense; (3) As the avalanche runs over the snow cover, it produces internal shear gradients in the avalanche core, and therefore, entrainment can enhance the production of chaotic and vertical motions (granular fluctuations and rotations), especially at the front of the avalanche; (4) Dry snow covers are a good source of ice dust that is subsequently mixed with the air blown out of the avalanche core to create the suspension cloud of powder snow avalanches.

Avalanche flow volume and avalanche mobility are related: in general, observations reveal that larger avalanches have longer runout distances (Bozhinskiy and Losyev, 1998). This fact is reflected in calculation guidelines where larger (extreme) avalanches are assigned lower friction values. Theoretically, this indicates that snow entrainment should positively affect avalanche mobility. At present, however, no physical explanation exists as to why this should occur. In fact, many experts maintain that entrainment will slow down the avalanche and reduce runout distances because accelerating the entrained snow to the avalanche velocity, breaking the snow cover, or snow cover ploughing consume avalanche flow energy. In practice, the effect of avalanche entrainment is handled simply by using extreme (and therefore low) friction values for large avalanches, without actual consideration of the entrained mass.

More advanced avalanche dynamics models not only consider the entrained mass but also consider the entrained heat energy (temperature) and moisture content (Vera and Bartelt, 2013). This modifies the internal energy fluxes in the avalanche core. That is, even after the energy losses of entrainment are accounted for, an increase in mobility occurs due to flow regime transitions induced by the production of random energy (more dilute flows), or melt water lubrication (temperature effects), or enhanced moisture content (reduced shear resistance) (Bartelt et al., 2012c). Clearly, entrainment has a much greater effect on avalanche flow than simply increasing the flow mass.

12.4.6 Stopping and Depositional Features

A better understanding of how avalanches stop is being driven by new methods to measure avalanche deposits, particularly by accurate terrestrial and airborne laser scanning techniques (Bartelt et al., 2012b; Bühler et al., 2009). Increasingly accurate and hand-held global positioning system devices are useful to trace the extent of avalanche deposits in field investigations. Because modern avalanche dynamics models predict the distribution of mass in the runout zone, an increased interest exists to compare measured and calculated runout distances, as well as calculated and measured deposition heights (Christen et al., 2010). Avalanche deposition patterns are strongly dependent on three-dimensional terrain characteristics (roughness, steepness of runout zone, counterslope, terrain undulations, gullies, etc.) and therefore are ideal to test the stopping mechanics of avalanche dynamics models. Commonly, information gathered from avalanche deposits is the only hard evidence available to reconstruct an avalanche event.

Stopping is driven by the imbalance between the driving gravitational forces G and shear friction S. The point where avalanche deceleration begins is generally estimated to be given at the point P where the tangent of the slope angle Φ_P is equal to the value of the Coulomb friction μ:

$$\tan(\phi_P) = \mu.$$

This is a useful formula as most avalanche dynamics models rely on some kind of Coulomb friction to describe when an avalanche stops, that is,

$$\frac{S}{N} = \mu,$$

where S is shear and N is the normal pressure exerted by the avalanche on the basal running surface. For far reaching (extreme) avalanches, different investigators have found minimum values between $\mu \approx 0.12$ and 0.15 (e.g., Bozhinskiy and Losyev, 1998; Salm et al., 1990). The formula is well tested and facilitates a simple, direct approximation of runout distance.

Actual shear and normal force measurements in snow chute experiments (Platzer et al., 2007) reveal that avalanche friction cannot be well described by a constant μ. It appears that avalanche friction varies over the length of the avalanche (it is lowest at the front of the avalanche and increases toward the tail). Other factors, including snow temperature and moisture content, can also influence μ. Interestingly, snow chute experiments show that the shear S can have different values for the same normal pressure N. The fact that μ can vary—it is a frictional process, rather than a frictional constant—explains the tremendous variation in "calibrated" friction parameters used to model snow avalanches. Because this frictional process is mass dependent, smaller avalanches can stop on steeper slopes, giving the impression that smaller avalanches have higher friction values.

How the friction μ evolves over time determines the distribution of depositions in the runout zone. If the terrain permits, avalanche deposits can be very regular. A tendency exists for higher deposition heights at lower slope

FIGURE 12.16   Different avalanche deposition types: (a) Levee with sidewalls. (b) Levee. (c) Avalanche flow finger with pile up. (d) Avalanche flow finger. *(From Bartelt et al. (2012b).)*

angles. However, avalanche deposits can be highly irregular and exhibit four unique features (Bartelt et al., 2012b):

1. *Levees and sidewall-type constructions* (Figure 12.16(a) and (b)). Levees commonly occur in wet-snow avalanche deposits. They arise when avalanche snow stops flowing at the outer boundaries of an avalanche. The interior flow continues to move forward, and internal shear planes are created several meters inward from the stopped outer boundary. As the interior flow drains, it leaves the sidewalls exposed, creating channel-like structures in the depositions. Levees are an indication of strong internal friction gradients within the flow. Although these friction gradients can be induced by terrain undulations, levees also occur on flat and relatively homogeneous runout zones, indicating that levee formation is triggered by internal variations of avalanche velocity (the velocity is smaller at the outer edges of the flow). Levees can form at the front of slow moving (wet-snow) avalanches, or at the slow moving tail of dry-snow flowing avalanches. The importance of levee sidewalls is that they can be large (>5 m in height) and thus form structures that can deflect the flowing mass of the same avalanche, or subsequent avalanches on the same avalanche track (Figure 12.16(b)).

2. *Avalanche flow fingers* (Figure 12.16(c) and (d)). Avalanche flow fingers commonly extend beyond the reach of the bulk of the deposited avalanche

mass. Typically, they are narrow structures (5—10 m in width, see Figure 12.16(d)), and they can also be rather wide depositional lobes (up to say 50 m). Flow fingers typically arise in moist or wet-type avalanches and are an indication of strong preferential flow behavior. Flow fingers can follow roads or other terrain features (gullies). The sides of flow fingers are steep, signifying strong cohesive properties of the flowing snow. In a sense, they are similar to the interior flow of a self-formed levee channel, except that the mass at the outer boundaries does not stop and builds no sidewalls. Flow fingers are a problem for avalanche runout calculations as they are difficult to predict. The flow fingers can bend $\geq 90°$.

3. *Shear planes and en-echelon shear faults.* Both vertical and horizontal shear planes occur in avalanche deposits. Vertical planes, parallel to the flow direction, occur on levee sidewalls. Horizontal (or basal) shear planes are formed as the avalanche moves in a plug-like motion over terrain undulations. Snow fills in the undulations and acts to smoothen the terrain. Commonly, the basal shear planes are formed in the avalanche interior between levee sidewalls. Frictional rubbing can produce heat which, depending on the initial temperature of the snow, can melt the frictional surfaces of avalanching snow. The melt water characteristically refreezes. This hardens the sheared surface, giving it a shiny appearance that is visible from a considerable distance. Because of the evolution of flow friction, avalanches do not stop at once: the avalanche tail can stop first, behind the front. This leads to the formation of en-echelon type fault structures in which the front is "pulled" away from the tail.

4. *Pile-ups* (Figure 12.16(c)). Avalanche pile-ups can occur at the end of levee-type channels, or they can occur at steep slope transitions, as, for example, at the exit of a gully at the beginning of a flat runout zone. Avalanche snow can also pile up at the transition between a slope and steep counterslope. Avalanche pile-ups are dangerous depositional features because they can be very high (>20 m) and therefore change the topography of the avalanche track.

12.4.7 Avalanche Interaction with Obstacles

The problem of how to adequately design buildings and other structures that stand in the path of an avalanche is central to hazard mapping and the planning of mitigation measures. Real-scale experiments in Switzerland, Russia, and Canada show that avalanches can exert immense pressures on obstacles (e.g., Bozhinskiy and Losyev, 1998; McClung and Schaerer, 1985; Sovilla et al., 2008). The obstacles (measurement pylons) are regarded as stationary, rigid bodies in comparison to the fluid behavior of the avalanche body, which can flow around or completely immerse the obstacle. Obstacle interaction is therefore not only an impact phenomenon but it also lasts significantly longer as the avalanche flows past the obstacle. Engineers are

FIGURE 12.17   Pressure measurements of a large dry-mixed flowing avalanche at the Vallée de la Sionne test site (Switzerland). The measurement pylon pressure cells with various diameters measure avalanche impact pressures at different heights. *(Photograph: F. Dufour.)*

mostly interested in the maximum pressure and pressure distribution, as this determines both the dynamic forces and overturning moments exerted on the structure.

Maximum pressures of 100 tn/m^2 have been measured at the Swiss Vallée de la Sionne test site (Figure 12.17); Russian researchers have reported pressures of up to 200 tn/m^2 (Bozhinskiy and Losyev, 1998). Several dry-mixed flowing avalanches recorded at Vallée de la Sionne site have exerted peak pressures between 20 and 50 tn/m^2 when traveling at speeds between 30 and 40 m/s. Pressure signals can be both intermittent and continuous functions in time, depending on the avalanche flow regime (dispersed, dense) and type (wet, dry). A problem with the analysis of pressure signals is that different measurement techniques can measure different pressures at different periods during the passage of the avalanche.

Pressure models of avalanche—obstacle interaction assume that the avalanche body is a continuum. Pressures p can be evaluated using the hydrodynamic formula

$$p = c \frac{1}{2} \rho U^2,$$

where c is the dimensionless coefficient of resistance, dependent on the shape of the object ($c = 1$ for slender objects; $c = 2$ for wider, wall-like structures). A prerequisite for a pressure analysis is an estimate of the avalanche speed U and bulk flow density ρ. It is necessary to have an idea of the avalanche flow height to determine the height of the applied pressure.

As we have seen, snow avalanches consist of hard snow granules (as well as rocks and dead wood) and will undergo significant variations in the bulk flow density. The application of the continuum formula is therefore questionable since it does not take into account the granular nature of the flow. The formula

$$p_g = \frac{4}{3}\rho_g U^2$$

can be used to predict local, granule impact pressures. The formula assumes that the time of a granule collision is $t = d/2U$, where d is the diameter of the granule. It assumes a plastic collision in which the granule is completely destroyed at impact. Our experience with this formula is that it well approximates measured impact pressures, which are often performed with pressure cells of size d. The granule density ρ_g varies between 400 and 500 kg/m^3 indicating significant densification of the snow during flow.

12.5 AVALANCHE MITIGATION

To avoid avalanche disasters, first of all, potential avalanche terrain and accordingly a potential avalanche problem needs to be recognized. Indications include oral and written history of previous avalanche events, vegetation clues and snow depth records, and most importantly, terrain. A terrain analysis can be done by identifying all areas steeper than about 25° by using a GIS and a DTM. Obviously, any infrastructure in these potential starting zones is endangered. The next step, the risk analysis, involves assessing the frequency, magnitude, and runout of avalanches initiating from the identified potential starting zones (e.g., Schweizer, 2004).

Depending on the object endangered and the frequency and size of potential avalanches, the resulting risk is too high, and mitigation measures are planned. If residential areas or areas where development is planned, are endangered, land-use planning measures should be established based on the hazard map so that buildings in hazard zones are avoided, restricted, or designed to withstand potential avalanche pressures. In Switzerland, for example, hazard zones are defined based on avalanche frequency and impact pressure. For residential areas large avalanches with return periods as high as 300 years need to be taken into account. For roads, return periods are typically much shorter, up to several times a year. If a road is endangered by several potentially large avalanches with return periods of <10 years, an active avalanche control program (including the use of explosives) and/or the construction of snow sheds should be considered (Figure 12.18). If the avalanches potentially hitting the road are rather small and or infrequent, temporary closures based on an avalanche forecast may be the adequate protection measure (CAA, 2002b). For highly endangered infrastructure (highways, railways, mining operations, ski areas, etc.) local forecasting services need to be established that also run the avalanche control program (Schweizer et al., 1998).

FIGURE 12.18 Avalanche mitigation measures include snow sheds (protecting the railway line), deflecting dams (protecting the Trans-Canada Highway) and warning signs that reduce the risk by reducing the exposure. *(Field BC, Canada; photograph: J. Schweizer.)*

Various mitigation measures should be applied, usually in combination, in a coordinated manner to reduce the avalanche risk to an acceptable level. This approach is known as integral risk management (Wilhelm et al., 2001). In general, mitigation measures are grouped into temporary (or short-term) and permanent (or long-term) measures, both of which can be subdivided into either active (e.g., preventing avalanche release by snow supporting structures) or passive measures (e.g., establishing hazard maps). The various mitigation measures are described in more detail by, for example, Bründl and Margreth (2014), McClung and Schaerer (2006) and Rudolf-Miklau and Sauermoser (2009).

12.6 AVALANCHE FORECASTING

Predicting snow avalanches is generally called avalanche forecasting, which McClung (2000) defined as predicting snowpack instability in space and time relative to a given triggering level. In the framework of integral risk management avalanche forecasting is considered a short-term, passive mitigation measure. Snow avalanches are the only natural hazard—apart from some purely meteorological hazards such as storms and heavy precipitation—that can be forecast.

Snow avalanche prediction (forecasting snowpack instability) can be made for various scales (McClung and Schaerer, 2006). We focus here on the regional scale (1,000 km²), the local scale (100 km²), and the scale of an individual avalanche path (1 km²). Typically, the prediction is about the danger level, the avalanche activity (or occurrence), and the probability of the single event for these three scales, respectively. In avalanche forecasting, a mismatch often occurs between the scale of the forecast and the scale of the underlying data (Hägeli and McClung, 2004).

Today, most countries with areas where people and infrastructure are significantly endangered by avalanches operate an avalanche forecasting service, similar to a weather service, which issues warnings, or so-called avalanche bulletins. These services primarily forecast the regional avalanche danger by describing the hazard situation with one of five degrees of danger: very high (or extreme), high, considerable, moderate, and low (Table 12.3).

Operational avalanche forecasting mainly relies on meteorological observations and forecasts in combination with observations of snow cover instability, ideally direct observations of avalanches (McClung and Schaerer, 2006). It involves the assimilation of multiple data sources to make predictions over complex interacting processes. For many decades now, avalanche professionals have developed successful decision-making strategies to deal with this complexity (LaChapelle, 1980). These rule-based empirical strategies, sometimes applied intuitively, are able to deal with many of the numerous processes and scales involved in avalanche formation. However, the employed methods are largely based on the experience of the forecaster and are therefore prone to subjectivity, unable to deal with unusual situations and difficult to

TABLE 12.3 Example of Avalanche Danger Scale as Used by Most Avalanche Forecasting Services. The Five Degree Danger Scale was Introduced by the European Warning Services in 1994

North American public avalanche danger scale
Avalanche danger is determined by the likelihood, size and distribution of avalanches.

Danger Level		Travel Advice	Likelihood of Avalanches	Avalanche Size and Distribution
5 Extreme		Avoid all avalanche terrain.	Natural and human-triggered avalanches certain.	Large to very large avalanches in many areas.
4 High		Very dangerous avalanche conditions. Travel in avalanche terrain not recommended.	Natural avalanches likely; human-triggered avalanches very likely.	Large avalanches in many areas; or very large avalanches in specific areas.
3 Considerable		Dangerous avalanche conditions. Careful snowpack evaluation, cautious route-finding and conservative decision-making essential.	Natural avalanches possible; human-triggered avalanches likely.	Small avalanches in many areas; or large avalanches in specific areas; or very large avalanches in isolated areas.
2 Moderate		Heightened avalanche conditions on specific terrain features. Evaluate snow and terrain carefully; identify features of concern.	Natural avalanches unlikely; human-triggered avalanches possible.	Small avalanches in specific areas; or large avalanches in isolated areas.
1 Low		Generally safe avalanche conditions. Watch for unstable snow on isolated terrain features.	Natural and human-triggered avalanches unlikely.	Small avalanches in isolated areas or extreme terrain.

Safe backcountry travel requires training and experience. You control your own risk by choosing where, when and how you travel.

transfer to new personnel. Recently, rule-based decision support tools have been developed for recreationists that relate terrain to the danger level (McCammon and Hägeli, 2007).

Numerous attempts have been made to develop objective techniques for avalanche forecasting. Such efforts predominantly consist of statistically relating local weather observations to avalanche occurrence data or estimated avalanche danger. Statistical forecasting models are based on the idea that similar weather conditions lead to comparable avalanche situations. Various methods have been used, including linear regression analysis, multivariate discriminant analysis, time series analysis, nearest neighbors, or pattern recognition techniques (e.g., Buser, 1983; Pozdnoukhov et al., 2011; Schweizer and Föhn, 1996). Depending on the scale of the study area and the quality of the input data, the accuracy of the statistical models can vary greatly. A major drawback is the poor temporal resolution of avalanche observations or estimated avalanche danger. Therefore, coarse scale meteorological parameters, such as the amount of new snow in the last 24, 48, or 72 h, are typically used as input parameters (McClung and Tweedy, 1994). High-frequency meteorological data (e.g., recorded at 10-min intervals) or forecasts, which nowadays are widely available, can therefore not be exploited to their fullest extent.

Another shortcoming of most statistical forecasting models is the failure to include quantitative snowpack stratigraphy data (e.g., Schweizer and Föhn, 1996). Such snowpack data are important for avalanche forecasting, especially during periods of low avalanche activity and stable weather conditions. In fact, snow avalanches are the only hazard where in situ tests exist that can provide information on the state of instability. Of course these tests provide point information only, which needs to be extrapolated—a task hampered by the inherently variable nature of the mountain snowpack (Schweizer et al., 2008). A first step toward a more deterministic approach to avalanche forecasting (which would include simulating avalanche release) therefore consists of modeling snowpack characteristics and deriving stability information.

Essentially two types of numerical models exist: one-dimensional snow cover models and three-dimensional model systems. One-dimensional models, such as CROCUS (Brun et al., 1992) and SNOWPACK (Bartelt and Lehning, 2002; Lehning et al., 2002a, 2002b), compute the snow cover stratigraphy by solving the one-dimensional mass and energy balance equations using meteorological data as input. The modeled stratigraphy can then be interpreted with regard to stability (e.g., Monti et al., 2012; Schweizer et al., 2006). Alternatively, modeled snow stratigraphy variables can be used as input variables to improve the performance of statistical avalanche forecasting models (Schirmer et al., 2009). A limitation of one-dimensional models is the lack of spatial snow cover information.

Such information is provided by three-dimensional model systems, such as SAFRAN—CROCUS—MEPRA (SCM) (Durand et al., 1999) and Alpine3D, which includes SNOWPACK (Lehning et al., 2006). These model systems use spatial meteorological input data and account for three-dimensional processes such as radiation redistribution and snow transport by wind. A large part of the complexity was removed in the French SCM model chain by providing automated avalanche danger prediction for virtual slopes. It is the only real operational forecasting model on the basis of physical modeling. However, given the overall complexity of the interactions between the snowpack and meteorological parameters and the high spatial resolution required, surface process model systems are very computationally intensive and many uncertainties still remain. Furthermore, interpretation of spatial snow cover data with regard to snow-slope stability remains unclear.

Field observations on snow stratigraphy are widely used to investigate snow cover instability for avalanche forecasting (e.g., Schweizer and Jamieson, 2010b). Over the last decade, new in situ measurement techniques have been developed to objectively describe the stratigraphy of a natural snowpack, including near-infrared photography (Matzl and Schneebeli, 2006), the snow micro-penetrometer (SMP) (Schneebeli and Johnson, 1998), microcomputer tomography (Schneebeli and Sokratov, 2004), contact spectrometry (Painter et al., 2007), and upward-looking ground-penetrating radars (Schmid et al., 2014). While these methods primarily provide new insight into the microstructure of snow and its physical processes, interpretation of the data remains complicated, in particular with regards to snow stability (Bellaire et al., 2009; van Herwijnen et al., 2009). Recently, promising advances have been made by combining micro-tomography and SMP measurements (Reuter et al., 2013). Nevertheless, manual snow cover measurements, consisting of snow profiles and snow stability tests, presently remain the method of choice for avalanche forecasting. Snow cover observations are typically compared to observed avalanche activity or estimated avalanche danger to obtain empirical relationships (e.g., Schweizer and Jamieson, 2007; van Herwijnen and Jamieson, 2007). Although these in situ tests have a number of deficiencies, including an error rate of at least 5–10 percent, correlations between manual snow cover measurements and snow-slope stability have been identified (Schweizer and Jamieson, 2010b).

In general, prediction requires predictability, that is, some sort of a precursor, or observational variable that announces the event. In the case of snow avalanches, avalanche prediction during storms is mainly based on precipitation amounts; new snow is considered as precursor. Obviously, precursors more related to the state of the snowpack would be useful. Pioneering work in the 1970s tried to relate acoustic emissions from a natural snow cover to avalanche formation and slope stability (e.g., Sommerfeld, 1977; St. Lawrence and Bradley, 1977; Gubler, 1979). Given the differences in instrumental setup and the lack of a thorough description of the field experiments, in particular with regard to the signal processing and the treatment of environmental background noise, results from early microseismic studies remain ambiguous.

Whereas laboratory measurements on snow failure have indicated that monitoring acoustic emissions may reveal imminent failure, applying the technique in the field is difficult due to strong attenuation of high-frequency signals in snow (Reiweger and Schweizer, 2013a). Alternatively, seismic sensors can be used, but van Herwijnen and Schweizer (2011b) were unable to detect precursors to slab-avalanche release. However, the avalanche occurrence data (van Herwijnen and Schweizer, 2011a) suggest that avalanches could be used as precursors because often at the beginning of periods of significant avalanche activity the number of events increases (Schweizer and van Herwijnen, 2013). Obviously, direct observations of avalanches are the most reliable evidence of unstable snow conditions and of fundamental importance for avalanche forecasting.

12.7 CONCLUDING REMARKS

Disasters due to snow avalanches have caused the loss of life and property damage in most populated, snow-covered mountain areas. Although in the past people had primarily to rely on experience, which is known to be of limited value when dealing with rare and extreme events such as large snow avalanches, today advanced methods exist that enable recognition of the risk as well as its mitigation. Still, this does not mean that disasters due to snow avalanches will no longer occur. Even with modern monitoring and modeling methods it is not possible to predict the exact location and time of an avalanche release—primarily due to the lack of reliable precursors and the inherently variable nature of the mountain snowpack. Clearly, physics-based numerical models are helpful tools to predict avalanche motion and impact, but uncertainties remain despite major advances—also in view of the uncertain consequences of climate change. Even after applying mitigation measures, a residual risk remains because completely reducing the risk is definitely not cost-efficient. Therefore, combining permanent and temporary protection measures is most promising, but requires that the risk is actively managed. By doing so the risk to people—living, traveling, or recreating in the mountains—can effectively be reduced to an acceptable level.

REFERENCES

Ancey, C., 2004. Powder snow avalanches: approximation as non-Boussinesq clouds with a Richardson number-dependent entrainment function. J. Geophys. Res. 109, F01005. http://dx.doi.org/10.1029/2003JF000052.

Ancey, C., Gervasoni, C., Meunier, M., 2004. Computing extreme avalanches. Cold Reg. Sci. Technol. 39 (2−3), 161−180.

BABS, 2003. KATARISK. Katastrophen und Notlagen in der Schweiz. Eine Risikobeurteilung aus der Sicht des Bevölkerungsschutzes. Bundesamt für Bevölkerungsschutz BABS (Swiss Federal Office for Civil Protection FOCP), Bern, Switzerland, 83 pp.

Baggi, S., Schweizer, J., 2009. Characteristics of wet snow avalanche activity: 20 years of ob-servations from a high alpine valley (Dischma, Switzerland). Nat. Hazards 50 (1), 97–108.

Bartelt, P., Lehning, M., 2002. A physical SNOWPACK model for the Swiss avalanche warning; part I: numerical model. Cold Reg. Sci. Technol. 35 (3), 123–145.

Bartelt, P., McArdell, B.W., 2009. Granulometric investigations of snow avalanches. J. Glaciol. 55 (193), 829–833.

Bartelt, P., Buser, O., Platzer, K., 2006. Fluctuation-dissipation relations for granular snow ava-lanches. J. Glaciol. 52 (179), 631–643.

Bartelt, P., Feistl, T., Buhler, Y., Buser, O., 2012a. Overcoming the stauchwall: viscoelastic stress redistribution and the start of full-depth gliding snow avalanches. Geophys. Res. Lett. 39.

Bartelt, P., Glover, J., Feistl, T., Buhler, Y., Buser, O., 2012b. Formation of levees and en-echelon shear planes during snow avalanche run-out. J. Glaciol. 58 (211), 980–992.

Bartelt, P., Meier, L., Buser, O., 2011. Snow avalanche flow-regime transitions induced by mass and random kinetic energy fluxes. Ann. Glaciol. 52 (58), 159–164.

Bartelt, P., Vera, C., Steinkogler, W., Feistl, T., Buser, O., 2012c. The role of thermal temperature in avalanche flow. In: International Snow Science Workshop ISSW 2012, Anchorage AK, U.S.A., 16–21 September 2012, pp. 32–37.

Beghin, P., Olagne, X., 1991. Experimental and theoretical study of the dynamics of powder snow avalanches. Cold Reg. Sci. Technol. 19 (3), 317–326.

Bellaire, S., Pielmeier, C., Schneebeli, M., Schweizer, J., 2009. Stability algorithm for snow micro-penetrometer measurements. J. Glaciol. 55 (193), 805–813.

Bozhinskiy, A.N., Losyev, K.S., 1998. The Fundamentals of Avalanche Science. Mitteilungen des Eidg. Instituts für Schnee- und Lawinenforschung SLF, 55, Davos, Switzerland, 280 pp.

Brun, E., David, P., Sudul, M., Brunot, G., 1992. A numerical model to simulate snow-cover stratigraphy for operational avalanche forecasting. J. Glaciol. 38 (128), 13–22.

Bründl, M., Margreth, S., 2014. Integrative risk management: the example of snow avalanches. In: Haeberli, W., Whiteman, C. (Eds.), Snow and Ice-Related Hazards, Risks and Disasters. Elsevier, pp. 263–301.

Bründl, M., Bartelt, P., Schweizer, J., Keiler, M., Glade, T., 2010. Review and future challenges in snow avalanche risk analysis. In: Alcantara-Ayala, I., Goudie, A.S. (Eds.), Geomorpho-logical Hazards and Disaster Prevention. Cambridge University Press, New York, USA, pp. 49–61.

Bühler, Y., Hüni, A., Christen, M., Meister, R., Kellenberger, T., 2009. Automated detection and mapping of avalanche deposits using airborne optical remote sensing data. Cold Reg. Sci. Technol. 57 (2–3), 99–106.

Bühler, Y., Kumar, S., Veitinger, J., Christen, M., Stoffel, A., Snehmani, 2013. Automated iden-tification of potential snow avalanche release areas based on digital elevation models. Nat. Hazards Earth Syst. Sci. 13 (5), 1321–1335.

Buser, O., 1983. Avalanche forecast with the method of nearest neighbours: an interactive approach. Cold Reg. Sci. Technol. 8 (2), 155–163.

Buser, O., Bartelt, P., 2009. Production and decay of random kinetic energy in granular snow avalanches. J. Glaciol. 55 (189), 3–12.

Buser, O., Bartelt, P., 2011. Dispersive pressure and density variations in snow avalanches. J. Glaciol. 57 (205), 857–860.

CAA, 2002a. In: McClung, D.M., Stethem, C.J., Schaerer, P., Jamieson, J.B. (Eds.), Guidelines for Snow Avalanche Risk Determination and Mapping in Canada. Canadian Avalanche Associ-ation, Revelstoke, BC, Canada, p. 24.

CAA, 2002b. In: Jamieson, J.B., Stethem, C.J., Schaerer, P., McClung, D.M. (Eds.), Land Managers Guide to Snow Avalanche Hazards in Canada. Canadian Avalanche Association, Revelstoke, BC, Canada, p. 28.

Cherepanov, G.P., Esparragoza, I.E., 2008. A fracture-entrainment model for snow avalanches. J. Glaciol. 54 (184), 182−188.

Christen, M., Kowalski, J., Bartelt, P., 2010. RAMMS: numerical simulation of dense snow avalanches in three-dimensional terrain. Cold Reg. Sci. Technol. 63 (1−2), 1−14.

Clarke, J.A., McClung, D.M., 1999. Full-depth avalanche occurrences caused by snow gliding. Coquihalla, B.C., Canada. J. Glaciol. 45 (151), 539−546.

Dreier, L., Mitterer, C., Feick, S., Harvey, S., 2013. The influence of weather on glide-snow avalanches. In: Naaim-Bouvet, F., Durand, Y., Lambert, R. (Eds.), Proceedings ISSW 2013. International Snow Science Workshop, Grenoble, France, 7−11 October 2013. ANENA, IRSTEA, Météo-France, Grenoble, France, pp. 247−252.

Durand, Y., Giraud, G., Brun, E., Mérindol, L., Martin, E., 1999. A computer-based system simulating snowpack structures as a tool for regional avalanche forecasting. J. Glaciol. 45 (151), 469−484.

Endo, Y., 1985. Release mechanism of an avalanche on a slope covered with bamboo bushes. Ann. Glaciol. 6, 256−257.

Fierz, C., Armstrong, R.L., Durand, Y., Etchevers, P., Greene, E., McClung, D.M., Nishimura, K., Satyawali, P.K., Sokratov, S.A., 2009. The International Classification for Seasonal Snow on the Ground. HP-VII Technical Documents in Hydrology, 83. UNESCO-IHP, Paris, France, 90 pp.

Gauer, P., Issler, D., 2004. Possible erosion mechanisms in snow avalanches. Ann. Glaciol. 38, 384−392.

Gauer, P., Issler, D., Lied, K., Kristensen, K., Iwe, H., Lied, E., Rammer, L., Schreiber, H., 2007. On full-scale avalanche measurements at the Ryggfonn test site, Norway. Cold Reg. Sci. Technol. 49 (1), 39−53.

Gauer, P., Lied, K., Kristensen, K., 2008. On avalanche measurements at the Norwegian full-scale test-site Rvggfonn. Cold Reg. Sci. Technol. 51 (2−3), 138−155.

Gauthier, D., Jamieson, J.B., 2006. Evaluating a prototype field test for weak layer fracture and failure propagation. In: Gleason, J.A. (Ed.), Proceedings ISSW 2006. International Snow Science Workshop, Telluride CO, U.S.A., 1−6 October 2006, pp. 107−116.

Gubler, H., 1979. Acoustic emission as an indication of stability decrease in fracture zones of avalanches. J. Glaciol. 22 (86), 186−188.

Hägeli, P., McClung, D.M., 2004. Hierarchy theory as a conceptual framework for scale issues in avalanche forecasting modeling. Ann. Glaciol. 38, 209−214.

Harbitz, C.B. (Ed.), 1998. EU Programme SAME—A Survey of Computational Models on Snow Avalanche Motion, p. 126.

Höller, P., 2001. Snow gliding and avalanches in a south-facing larch stand. In: Dolman, A.J., Hall, A.J., Kavvas, M.L., Oki, T., Pomeroy, J.W. (Eds.), Soil-Vegetation-Atmosphere Transfer Schemes and Large-scale Hydrological Models. IAHS Publ. 270. International Association of Hydrological Sciences, Wallingford, Oxfordshire, UK, pp. 355−358.

in der Gand, H.R., Zupancic, M., 1966. Snow Gliding and Avalanches, Symposium at Davos 1965—Scientific Aspects of Snow and Ice Avalanches. IAHS Publication, 69. Int. Assoc. Hydrol. Sci., Wallingford, U.K, pp. 230−242.

Jamieson, J.B., 1995. Avalanche Prediction for Persistent Snow Slabs (Ph.D. thesis). University of Calgary, Calgary, AB, Canada, 258 pp.

Jamieson, B., 2001. Snow avalanches. In: Brooks, G.R. (Ed.), A Synthesis of Geological Hazards in Canada. Bulletin 548. Geological Survey of Canada, pp. 81−100.

Jamieson, J.B., 2006. Formation of refrozen snowpack layers and their role in slab avalanche release. Rev. Geophys. 44 (2), RG2001. http://dx.doi.org/10.1029/2005RG000176.

Jomelli, V., Bertran, P., 2001. Wet snow avalanche deposits in the French Alps: structure and sedimentology. Geogr. Ann. Ser. A Phys. Geogr. 83A (1−2), 15−28.

Kattelmann, R., 1985. Wet slab instability. In: Proceedings International Snow Science Workshop, Aspen, Colorado, U.S.A., 24−27 October 1984. ISSW 1984 Workshop Committee, Aspen CO, USA, pp. 102−108.

Kern, M., Bartelt, P., Sovilla, B., Buser, O., 2009. Measured shear rates in large dry and wet snow avalanches. J. Glaciol. 55 (190), 327−338.

Kirchner, H.O.K., Michot, G., Suzuki, T., 2000. Fracture toughness of snow in tension. Phil. Mag. A 80 (5), 1265−1272.

LaChapelle, E.R., 1980. The fundamental process in conventional avalanche forecasting. J. Glaciol. 26 (94), 75−84.

Lackinger, B., 1987. Stability and fracture of the snow pack for glide avalanches. In: Salm, B., Gubler, H. (Eds.), Symposium at Davos 1986-Avalanche Formation, Movement and Effects. IAHS Publ., 162. International Association of Hydrological Sciences, Wallingford, Oxfordshire, UK, pp. 229−241.

Lehning, M., Bartelt, P., Brown, R.L., Fierz, C., 2002a. A physical SNOWPACK model for the Swiss avalanche warning; part III: meteorological forcing, thin layer formation and evaluation. Cold Reg. Sci. Technol. 35 (3), 169−184.

Lehning, M., Bartelt, P., Brown, R.L., Fierz, C., Satyawali, P.K., 2002b. A physical SNOWPACK model for the Swiss avalanche warning; part II. Snow microstructure. Cold Reg. Sci. Technol. 35 (3), 147−167.

Lehning, M., Völksch, I., Gustafsson, D., Nguyen, T.A., Stähli, M., Zappa, M., 2006. ALPINE3D: a detailed model of mountain surface processes and its application to snow hydrology. Hydrol. Process. 20 (10), 2111−2128.

Leitinger, G., Holler, P., Tasser, E., Walde, J., Tappeiner, U., 2008. Development and validation of a spatial snow-glide model. Ecol. Model. 211 (3−4), 363−374.

Lied, K., Bakkehoi, S., 1980. Empirical calculations of snow-avalanche run-out distance based on topographic parameters. J. Glaciol. 26 (94), 165−177.

Matzl, M., Schneebeli, M., 2006. Measuring specific surface area of snow by near infrared photography. J. Glaciol. 42 (179), 558−564.

McCammon, I., Hägeli, P., 2007. Comparing avalanche decision frameworks using accident data from the United States. Cold Reg. Sci. Technol. 47 (1−2), 193−206.

McClung, D.M., 1979. Shear fracture precipitated by strain softening as a mechanism of dry slab avalanche release. J. Geophys. Res. 84 (87), 3519−3526.

McClung, D.M., 1981a. Fracture mechanical models of dry slab avalanche release. J. Geophys. Res. 86 (B11), 10783−10790.

McClung, D.M., 1981b. A physical theory of snow gliding. Can. Geotech. J. 18 (1), 86−94.

McClung, D.M., 2000. Predictions in avalanche forecasting. Ann. Glaciol. 31, 377−381.

McClung, D.M., Clarke, G.K.C., 1987. The effects of free water on snow gliding. J. Geophys. Res. 92 (B7), 6301−6309.

McClung, D.M., Mears, A.I., 1991. Extreme value prediction of snow avalanche runout. Cold Reg. Sci. Technol. 19, 163−175.

McClung, D.M., Schaerer, P.A., 1985. Characteristics of flowing snow and avalanche impact pressures. Ann. Glaciol. 6, 9−14.

McClung, D.M., Schaerer, P., 2006. The Avalanche Handbook. The Mountaineers Books, Seattle, WA, USA, 342 pp.

McClung, D.M., Schweizer, J., 1999. Skier triggering, snow temperatures and the stability index for dry slab avalanche initiation. J. Glaciol. 45 (150), 190–200.

McClung, D.M., Tweedy, J., 1994. Numerical avalanche prediction. Kootenay Pass, British Columbia, Canada. J. Glaciol. 40 (135), 350–358.

Mitterer, C., Schweizer, J., 2012. Towards a better understanding of glide-Snow avalanche formation. In: International Snow Science Workshop ISSW 2012, Anchorage AK, U.S.A., 16–21 September 2012, pp. 610–616.

Mitterer, C., Schweizer, J., 2013. Analysis of the snow-atmosphere energy balance during wet-snow instabilities and implications for avalanche prediction. Cryosphere 7 (1), 205–216.

Mitterer, C., Hirashima, H., Schweizer, J., 2011. Wet-snow instabilities: comparison of measured and modelled liquid water content and snow stratigraphy. Ann. Glaciol. 52 (58), 201–208.

Mitterer, C., Techel, F., Fierz, C., Schweizer, J., 2013. An operational supporting tool for assessing wet-snow avalanche danger. In: Naaim-Bouvet, F., Durand, Y., Lambert, R. (Eds.), Proceedings ISSW 2013. International Snow Science Workshop, Grenoble, France, 7–11 October 2013. ANENA, IRSTEA, Météo-France, Grenoble, France, pp. 334–338.

Monti, F., Cagnati, A., Valt, M., Schweizer, J., 2012. A new method for visualizing snow stability profiles. Cold Reg. Sci. Technol. 78, 64–72.

Newesely, C., Tasser, E., Spadinger, P., Cernusca, A., 2000. Effects of land-use changes on snow gliding processes in alpine ecosystems. Basic Appl. Ecol. 1 (1), 61–67.

Nishimura, K., Maeno, N., Kawada, K., Izumi, K., 1993. Structures of snow cloud in dry-snow avalanches. Ann. Glaciol. 18, 173–178.

Painter, T.H., Molotch, N.P., Cassidy, M., Flanner, M., Steffen, K., 2007. Contact spectroscopy for determination of stratigraphy of snow optical grain size. J. Glaciol. 53 (180), 121–127.

Palmer, A.C., Rice, J.R., 1973. The growth of slip surfaces in the progressive failure of over-consolidated clay. Proc. Roy. Soc. Lond. Ser. A 332 (1591), 527–548.

Peitzsch, E., 2008. Wet slabs: what do we really know about them. Avalanche Rev. 26 (4), 20–21.

Peitzsch, E.H., Hendrikx, J., Fagre, D.B., Reardon, B., 2012. Examining spring wet slab and glide avalanche occurrence along the going-to-the-Sun Road corridor, Glacier National Park, Montana, USA. Cold Reg. Sci. Technol. 78, 73–81.

Perla, R., 1977. Slab avalanche measurements. Can. Geotech. J. 14 (2), 206–213.

Perla, R.I., 1980. Avalanche release, motion, and impact. In: Colbeck, S.C. (Ed.), Dynamics of Snow and Ice Masses. Academic Press, New York, pp. 397–462.

Perla, R., Cheng, T.T., McClung, D.M., 1980. A 2-parameter model of snow-avalanche motion. J. Glaciol. 26 (94), 197–207.

Platzer, K., Bartelt, P., Kern, M., 2007. Measurements of dense snow avalanche basal shear to normal stress ratios (S/N). Geophys. Res. Lett. 34 (7).

Pozdnoukhov, A., Matasci, G., Kanevski, M., Purves, R.S., 2011. Spatio-temporal avalanche forecasting with support vector machines. Nat. Hazards Earth Syst. Sci. 11 (2), 367–382.

Prokop, A., 2008. Assessing the applicability of terrestrial laser scanning for spatial snow depth measurements. Cold Reg. Sci. Technol. 54 (3), 155–163.

Rastello, M., Rastello, F., Bellot, H., Ousset, F., Dufour, F., Meier, L., 2011. Size of snow particles in a powder-snow avalanche. J. Glaciol. 57 (201), 151–156.

Reiweger, I., Schweizer, J., 2010. Failure of a layer of buried surface hoar. Geophys. Res. Lett. 37, L24501. http://dx.doi.org/10.1029/2010GL045433.

Reiweger, I., Schweizer, J., 2013a. Measuring acoustic emissions in an avalanche starting zone to monitor snow stability. In: Naaim-Bouvet, F., Durand, Y., Lambert, R. (Eds.), Proceedings ISSW 2013. International Snow Science Workshop, Grenoble, France, 7−11 October 2013. ANENA, IRSTEA, Météo-France, Grenoble, France, pp. 942−944.

Reiweger, I., Schweizer, J., 2013b. Weak layer fracture: facets and depth hoar. Cryosphere 7 (5), 1447−1453.

Reuter, B., Schweizer, J., 2012. The effect of surface warming on slab stiffness and the fracture behavior of snow. Cold Reg. Sci. Technol. 83−84, 30−36.

Reuter, B., Proksch, M., Loewe, H., van Herwijnen, A., Schweizer, J., 2013. On how to measure snow mechanical properties relevant to slab avalanche release. In: Naaim-Bouvet, F., Durand, Y., Lambert, R. (Eds.), Proceedings ISSW 2013. International Snow Science Workshop, Grenoble, France, 7−11 October 2013. ANENA, IRSTEA, Météo-France, Grenoble, France, pp. 7−11.

Rice, B., Howlett, D., Decker, R., 1997. Preliminary investigations of glide/creep motion sensors in Alta, Utah. In: Proceedings International Snow Science Workshop, Banff, Alberta, Canada, 6−10 October 1996. Canadian Avalanche Association, Revelstoke, BC, Canada, pp. 189−194.

Rudolf-Miklau, F., Sauermoser, S. (Eds.), 2009. Handbuch Technischer Lawinenschutz. Ernst & Sohn, Berlin, Germany, 464 pp.

Salm, B., Burkard, A., Gubler, H., 1990. Berechnung von Fliesslawinen. Eine Anleitung für Praktiker mit Beispielen. Mitteilungen des Eidg. Instituts für Schnee- und Lawinenforschung, 47. Swiss Federal Institute for Snow and Avalanche Research SLF, Weissfluhjoch/Davos, Switzerland, 37 pp.

Schirmer, M., Lehning, M., Schweizer, J., 2009. Statistical forecasting of regional avalanche danger using simulated snow cover data. J. Glaciol. 55 (193), 761−768.

Schirmer, M., Wirz, V., Clifton, A., Lehning, M., 2011. Persistence in intra-annual snow depth distribution: 1. Measurements and topographic control. Water Resour. Res. 47, W09516.

Schmid, L., Heilig, A., Mitterer, C., Schweizer, J., Maurer, H., Okorn, R., Eisen, O., 2014. Continuous snowpack monitoring using upward-looking ground-penetrating radar technology. J. Glaciol. 60 (221), 509−525.

Schneebeli, M., 2004. Mechanisms in wet snow avalanche release. In: Proceedings ISSMA-2004, International Symposium on Snow Monitoring and Avalanches. Snow and Avalanche Study Establishment, India, Manali, India, 12−16 April 2004, pp. 75−77.

Schneebeli, M., Bebi, P., 2004. Snow and avalanche control. In: Evans, J., Burley, J., Youngquist, J. (Eds.), Encyclopedia of Forest Sciences. Elsevier, Oxford, pp. 397−402.

Schneebeli, M., Johnson, J.B., 1998. A constant-speed penetrometer for high-resolution snow stratigraphy. Ann. Glaciol. 26, 107−111.

Schneebeli, M., Sokratov, S.A., 2004. Tomography of temperature gradient metamorphism of snow and associated changes in heat conductivity. Hydrol. Process. 18 (18), 3655−3665.

Schweizer, J., 2004. Snow avalanches. Water Resour. Impact 6 (1), 12−18.

Schweizer, J., Föhn, P.M.B., 1996. Avalanche forecasting—an expert system approach. J. Glaciol. 42 (141), 318−332.

Schweizer, J., Jamieson, J.B., 2001. Snow cover properties for skier triggering of avalanches. Cold Reg. Sci. Technol. 33 (2−3), 207−221.

Schweizer, J., Jamieson, J.B., 2003. Snowpack properties for snow profile analysis. Cold Reg. Sci. Technol. 37 (3), 233−241.

Schweizer, J., Jamieson, J.B., 2007. A threshold sum approach to stability evaluation of manual snow profiles. Cold Reg. Sci. Technol. 47 (1−2), 50−59.

Schweizer, J., Jamieson, B., 2010a. On surface warming and snow instability. In: International Snow Science Workshop ISSW, Lake Tahoe CA, U.S.A., 17−22 October 2010, pp. 619−622.

Schweizer, J., Jamieson, J.B., 2010b. Snowpack tests for assessing snow-slope instability. Ann. Glaciol. 51 (54), 187−194.

Schweizer, J., Lütschg, M., 2001. Characteristics of human-triggered avalanches. Cold Reg. Sci. Technol. 33 (2−3), 147−162.

Schweizer, J., van Herwijnen, A., 2013. Can near real-time avalanche occurrence data improve avalanche forecasting? In: Naaim-Bouvet, F., Durand, Y., Lambert, R. (Eds.), Proceedings ISSW 2013. International Snow Science Workshop, Grenoble, France, 7−11 October 2013. ANENA, IRSTEA, Météo-France, Grenoble, France, pp. 195−198.

Schweizer, J., Bellaire, S., Fierz, C., Lehning, M., Pielmeier, C., 2006. Evaluating and improving the stability predictions of the snow cover model SNOWPACK. Cold Reg. Sci. Technol. 46 (1), 52−59.

Schweizer, J., Jamieson, J.B., Schneebeli, M., 2003. Snow avalanche formation. Rev. Geophys. 41 (4), 1016.

Schweizer, J., Jamieson, J.B., Skjonsberg, D., 1998. Avalanche forecasting for transportation corridor and backcountry in Glacier National Park (BC, Canada). In: Hestnes, E. (Ed.), 25 Years of Snow Avalanche Research, Voss, Norway, 12−16 May 1998. NGI Publication. Norwegian Geotechnical Institute, Oslo, Norway, pp. 238−243.

Schweizer, J., Kronholm, K., Jamieson, J.B., Birkeland, K.W., 2008. Review of spatial variability of snowpack properties and its importance for avalanche formation. Cold Reg. Sci. Technol. 51 (2−3), 253−272.

Schweizer, J., Michot, G., Kirchner, H.O.K., 2004. On the fracture toughness of snow. Ann. Glaciol. 38, 1−8.

Schweizer, J., Mitterer, C., Stoffel, L., 2009. On forecasting large and infrequent snow avalanches. Cold Reg. Sci. Technol. 59 (2−3), 234−241.

Schweizer, J., van Herwijnen, A., Reuter, B., 2011. Measurements of weak layer fracture energy. Cold Reg. Sci. Technol. 69 (2−3), 139−144.

Sigrist, C., Schweizer, J., 2007. Critical energy release rates of weak snowpack layers determined in field experiments. Geophys. Res. Lett. 34 (3), L03502. http://dx.doi.org/10.1029/2006GL028576.

Simenhois, R., Birkeland, K., 2010. Meteorological and environmental observations from three glide avalanche cycles and the resulting hazard management technique. In: International Snow Science Workshop ISSW, Lake Tahoe CA, U.S.A., 17−22 October 2010, pp. 846−853.

Sommerfeld, R.A., 1977. Preliminary observations of acoustic emissions preceding avalanches. J. Glaciol. 19 (81), 399−409.

Sovilla, B., 2004. Field Experiments and Numerical Modelling of Mass Entrainment and Deposition Processes in Snow Avalanches (Ph.D. thesis). ETH Zurich, Zurich, Switzerland, 190 pp.

Sovilla, B., Schaer, M., Rammer, L., 2008. Measurements and analysis of full-scale avalanche impact pressure at the Vallée de la Sionne test site. Cold Reg. Sci. Technol. 51 (2−3), 122−137.

St. Lawrence, W.F., Bradley, C.C., 1977. Spontaneous fracture initiation in mountain snow-packs. J. Glaciol. 19 (81), 411−417.

Stimberis, J., Rubin, C., 2005. Glide avalanche detection on a smooth rock slope, Snoqualmie Pass, Washington. In: Elder, K. (Ed.), Proceedings ISSW 2004. International Snow Science Workshop, Jackson Hole WY, U.S.A., 19−24 September 2004, pp. 608−610.

Stimberis, J., Rubin, C.M., 2011. Glide avalanche response to an extreme rain-on-snow event, Snoqualmie Pass, Washington, USA. J. Glaciol. 57 (203), 468−474.

Swift, D.A., Cook, S., Heckmann, T., Moore, J., Gärther-Roer, I., Korup, O., 2014. Ice and snow as landforming agents. In: Haeberli, W., Whiteman, C. (Eds.), Snow and Ice-Related Hazards, Risks and Disasters. Elsevier, pp. 167−199.

Techel, F., Pielmeier, C., 2009. Wet snow diurnal evolution and stability assessment. In: Schweizer, J., van Herwijnen, A. (Eds.), International Snow Science Workshop ISSW, Davos, Switzerland, 27 September−2 October 2009. Swiss Federal Institute for Forest, Snow and Landscape Research WSL, pp. 256−261.

Teich, M., Marty, C., Gollut, C., Grêt-Regamey, A., Bebi, P., 2012. Snow and weather conditions associated with avalanche releases in forests: rare situations with decreasing trends during the last 41 years. Cold Reg. Sci. Technol. 83−84, 77−88.

Trautman, S., 2008. Investigations into wet snow. Avalanche Rev. 26 (4), 16−17, 21.

UNESCO, 1981. Avalanche Atlas—Illustrated International Avalanche Classification. Natural Hazard. UNESCO, Paris, France, 265 pp.

van Herwijnen, A., Heierli, J., 2009. Measurement of crack-face friction in collapsed weak snow layers. Geophys. Res. Lett. 36 (23), L23502. http://dx.doi.org/10.1029/2009GL040389.

van Herwijnen, A., Heierli, J., 2010. A field method for measuring slab stiffness and weak layer fracture energy. In: International Snow Science Workshop ISSW, Lake Tahoe CA, U.S.A., 17−22 October 2010, pp. 232−237.

van Herwijnen, A., Jamieson, B., 2005. High-speed photography of fractures in weak snowpack layers. Cold Reg. Sci. Technol. 43 (1−2), 71−82.

van Herwijnen, A., Jamieson, J.B., 2007. Fracture character in compression tests. Cold Reg. Sci. Technol. 47 (1−2), 60−68.

van Herwijnen, A., Schweizer, J., 2011a. Monitoring avalanche activity using a seismic sensor. Cold Reg. Sci. Technol. 69 (2−3), 165−176.

van Herwijnen, A., Schweizer, J., 2011b. Seismic sensor array for monitoring an avalanche start zone: design, deployment and preliminary results. J. Glaciol. 57 (202), 267−276.

van Herwijnen, A., Simenhois, R., 2012. Monitoring glide avalanches using time-lapse photography. In: International Snow Science Workshop ISSW 2012, Anchorage AK, U.S.A., 16−21 September 2012, pp. 899−903.

van Herwijnen, A., Bellaire, S., Schweizer, J., 2009. Comparison of micro-structural snowpack parameters derived from penetration resistance measurements with fracture character observations from compression tests. Cold Reg. Sci. Technol. 59 (2−3), 193−201.

van Herwijnen, A., Failletaz, J., Berthod, N., Mitterer, C., 2013. Investigating glide snow avalanche release using seismic monitoring in combination with time-lapse photography. Geophys. Res. Abstr. 15. EGU2013−14045.

van Herwijnen, A., Schweizer, J., Heierli, J., 2010. Measurement of the deformation field associated with fracture propagation in weak snowpack layers. J. Geophys. Res. 115, F03042. http://dx.doi.org/10.1029/2009JF001515.

Vera, C., Bartelt, P., 2013. Modelling wet snow avalanche flow with a temperature dependent Coulomb friction function. In: Naaim-Bouvet, F., Durand, Y., Lambert, R. (Eds.), Proceedings ISSW 2013. International Snow Science Workshop, Grenoble, France, 7−11 October 2013. ANENA, IRSTEA, Météo-France, Grenoble, France, pp. 691−696.

Wilhelm, C., Wiesinger, T., Bründl, M., Ammann, W.J., 2001. The avalanche winter 1999 in Switzerland—an overview. In: Proceedings International Snow Science Workshop, Big Sky, Montana, U.S.A., 1−6 October 2000. Montana State University, Bozeman, MT, USA, pp. 487−494.

Wilson, A., Statham, G., Bilak, R., Allen, B., 1997. Glide avalanche forecasting. In: Proceedings International Snow Science Workshop, Banff, Alberta, Canada, 6–10 October 1996. Canadian Avalanche Association, Revelstoke, BC, Canada, pp. 200–202.

Yamanoi, K., Endo, Y., 2002. Dependence of shear strength of snow cover on density and water content (in Japanese with English Abstract). Seppyo J. Jpn. Soc. Snow Ice 64 (4), 443–451.

Chapter 13

Glacier Surges

William D. Harrison[1], **Galina B. Osipova**[2], **Gennady A. Nosenko**[2], **Lydia Espizua**[3], **Andreas Kääb**[4], **Luzia Fischer**[5], **Christian Huggel**[6], **Patty A. Craw Burns**[7], **Martin Truffer**[8] and **Alexandre W. Lai**[9]

[1] *Geophysical Institute, University of Alaska, Fairbanks, AK, USA,* [2] *Institute of Geography, Russian Academy of Sciences, Moscow, Russia,* [3] *Instituto Argentino de Nivología, Glaciología y Ciencias Ambientales (IANIGLA), Mendoza, Argentina,* [4] *Department of Geosciences, University of Oslo, Norway,* [5] *Norwegian Geological Survey, Trondheim, Norway,* [6] *Department of Geography, University of Zurich, Switzerland,* [7] *Department of Natural Resources, Division of Mining, Land & Water, Lands Section, Fairbanks, AK, USA,* [8] *Geophysical Institute and the Department of Physics, University of Alaska, Fairbanks, AK, USA,* [9] *Alyeska Pipeline Service Company, Integrity Management Department, Fairbanks, AK, USA*

ABSTRACT

Surge-type glaciers periodically undergo large flow acceleration after extended quiescent phases of slow movement, usually accompanied by terminus advance. Such glaciers are relatively rare but occur in many of the world's glacierized areas. High water pressures and extreme basal sliding are obvious characteristics but key questions concerning this, usually spectacular phenomenon, remain open. Why are glaciers in some regions surge-type but not in others, what sort of "memory" lets glaciers surge again and again, what is the influence of climate, geology, and topography? Besides their scientific interest, glacier surges can also be a threat to humans, especially in connection with rapidly forming lakes and their sudden outbursts. Cases of hazard- and disaster-related glacier surges are described from the Pamirs, the Andes, the Italian Alps, and Alaska.

13.1 INTRODUCTION

Surge-type glaciers are those subject to episodes of rapid motion. This chapter addresses some of the problems, which they and their surroundings pose. It has five parts, beginning with a discussion by Harrison of surges and their causes. Then follow four case histories. The first is the central Pamirs in Russia, by Osipova and Nosenko; the second is from the central Andes in South America, by Espizua; the third is from the Alps in Northern Italy, by Kääb, Fischer, and Huggel; and the fourth is from the central Alaska Range, by Burns, Truffer,

Snow and Ice-Related Hazards, Risks, and Disasters. http://dx.doi.org/10.1016/B978-0-12-394849-6.00013-5

437

and Lai. The authors have different stories to tell, but their themes are the same: the prediction and mitigation of the hazards.

Historically, the principal cause of damage from surges has usually been the outbursts of water from lakes dammed by advance of the glacier terminus across a side valley, from water within the glacier valley, or possibly from water stored on or within the glacier itself. But there are other problems, such as the over-riding of proglacial sediments and iceberg discharge. The lake problem is especially complicated. The downstream damage is largely determined by the volume of the lake and the rate at which it drains, but the latter, in particular, is not understood. For this reason the authors examine the historical record both of the damage and of the lake drainages, an approach that is particularly valuable since surging tends to be a quasi-periodic phenomenon. One might be tempted to count on a decrease in surge magnitude with time, but this is not universal even though in most areas the number of surging-type glaciers and the magnitude of surges have decreased with climate change. Also, climate change can unleash a new suite of problems, such as debris flows.

Mitigation measures are discussed by all the authors. Loss of life, damage to infrastructure, and mitigation (particularly the first two) have been important in Russia and the Andes, but the case history in northern Italy is especially complicated and the most detailed. The case history from Alaska is different from the others because it details how potential problems have been avoided by careful routing, design, and monitoring of a major oil pipeline system.

13.2 PROPERTIES AND CAUSES OF GLACIER SURGES

13.2.1 What Are Surge-Type Glaciers?

Surge-type glaciers are those subject to periods of relatively rapid motion, but because glaciers seldom if ever flow at a constant speed, the definition is difficult and remains to be clarified. Speeds on most glaciers typically fluctuate on timescales of hours (or less) to days, seasons, years, or centuries. "Temperate" glaciers (those near 0°C throughout) or almost temperate glaciers, undergo significant seasonal variations in speed, often punctuated by events lasting several hours or days. In these glaciers most of the variations in speed, including variations in the subset of those that surge, are primarily due to liquid water input (or lack of it as we shall see) and storage. However a key effect in some colder glaciers is change of thermal regime at the bed from cold to temperate, and it has received considerable attention (e.g., Frappé and Clarke, 2007; Murray et al., 2000 and 2003). Also, more than one surge mechanism may be involved in some cases.

In what follows our discussion focuses mainly on the water-induced surges of the temperate or nearly temperate glaciers that are common in relatively temperate places like Alaska. Nevertheless, because these glaciers share some surge properties with colder ones in such places as Svalbard, reference to studies on the latter is sometimes included. We focus on the warmer glaciers because, unlike the colder, they are known to be capable of fast and strong surges, which makes them particularly hazardous to activities of men. In one of the case studies below (Black Rapids Glacier, BRG), direct measurements show that the glacier is essentially temperate (Harrison et al., 1975) but in other cases we base the inference of water-driving on a glacier's large and rapid changes during surge. A more general approach is followed in summaries by Jiskoot (2011) and Cuffey and Paterson (2010). A recent reference on seasonal variation is Burgess et al. (2013).

Depending upon surface slope, typical temperate or nearly temperate mountain glaciers flow at speeds of less than a meter per day. At the "fast" end of what is probably a spectrum, a surging glacier is one whose speed rapidly increases by an order of magnitude or more, and does so quasi-periodically, usually on the timescale of decades. The classic photo pair taken before and after a surge is shown in Figure 13.1. Surge speeds can reach tens of meters or even more than a hundred meters per day. The rapid motion lasts from less than a year to a few years, may be interrupted by brief periods of relative quiescence (e.g., Clarke, 1987; Meier and Post, 1969), and ends suddenly. Crevassing becomes intense, and commonly propagating topographic waves can be identified (e.g., Raymond et al., 1987). Surge initiation is generally but not always local (Murray et al., 2003), sometimes where different channels join, and propagates up and down from there. Only part of a glacier may be affected. The most important common factor is that surging is caused by enhanced basal motion, by which we mean sliding over the glacier bed or deformation of any sediments beneath it; deformation of the ice alone cannot account for the speeds attained (Kamb et al., 1985; Raymond, 1987). "Enhanced" is the key word here because basal motion occurs in both non-surge- and surge-type glaciers (e.g., Boulton and Hindmarsh, 1987; Truffer et al., 2000). Also, surface and probably basal speed increase between surges before the surge instability is reached (Dolgushin and Osipova, 1975; Raymond, 1987).

Finally, several other characteristics of surges are interesting and point to possible physical mechanisms (Meier and Post, 1969). Glaciers with different sizes and shapes and in different climate settings can surge, and only part of a glacier may be involved. At the conclusion of a surge one can identify a hinge point above which mass has been lost (the "reservoir" area), and below which (the "receiving" area) the mass has been received. Elevation changes can be 100 m or more. This implies a flattening of the surface topography, which is recovered as the glacier steepens between surges, an important process discussed below. In addition to the crevassing, faulting may occur along the sides. The hinge point generally does not coincide with the equilibrium line, the line above which the glacier receives annual net accumulation of mass, mainly from snow. Surges do not necessarily imply an advance of the terminus because a surge may peter out in stagnant ice left by a previous surge (e.g.,

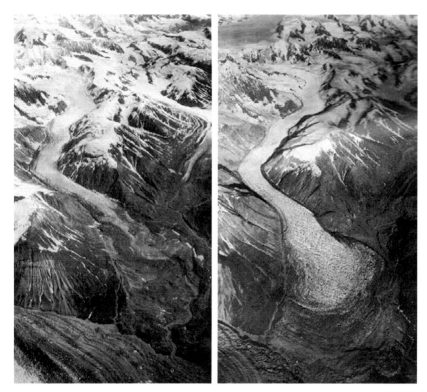

FIGURE 13.1   Variegated Glacier, Alaska. Left: before surge or soon after its commencement (August 29, 1964). Right: after surge (August 22, 1965). *Photographs by A. Post, U.S. Geological Survey.*

Harrison et al., 2008; Murray et al., 1998). Nevertheless, terminus advances of several kilometers can occur as noted in the case histories below.

13.2.2 Geographic Distribution of Surge-Type Glaciers

Surge-type glaciers are relatively rare, but they occur throughout many, but not all, of the world's glacierized areas. When they are surging, their identification is usually obvious because of the enhanced surface crevassing. When not surging, they can be identified, at least tentatively, by several indirect methods, an example of which is the study of looped moraine patterns where tributaries enter; an extreme case is shown in Figure 13.2 (Post, 1972). Surge-type glaciers have a highly nonrandom and therefore interesting geographic distribution. For example, in certain parts of the St. Elias Mountains of Alaska and the Yukon Territory, it is only a slight exaggeration to say that all of the glaciers surge, while in the eastern Chugach Mountains of Alaska, none do (Post, 1969). These and subsequent studies (e.g., Jiskoot et al., 2000) have suggested that the most common property of surge-type glaciers is their

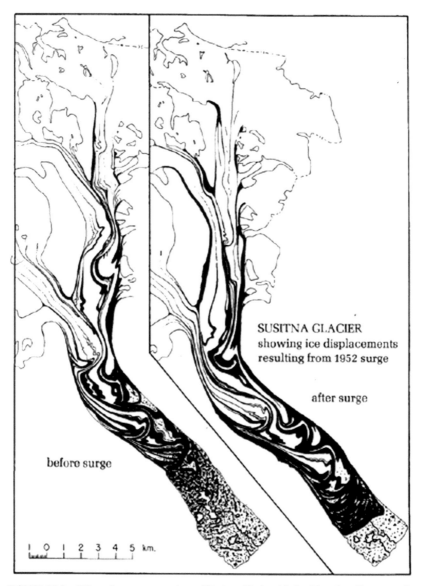

FIGURE 13.2   Effect of surge on moraines of Susitna Glacier, Alaska. Note the tributary on the left pushing into the quiescent main glacier. It was subsequently cut off and transported several kilometers by the 1952 surge. *From Meier and Post (1969).*

geologic setting on faults or easily erodible beds. This suggests, but does not prove, that significant parts of the beds of surge-type glaciers consist of till rather than bedrock—by till we mean unconsolidated sediment. However, this has been confirmed by the study of exposed tills, by drilling programs where the bed has been sampled, or by geophysical measurements (e.g., Fischer and Clarke, 1994; Harrison and Post, 2003; Nolan and Echelmeyer, 1999a; Porter et al., 1997; Truffer et al., 1999). By common usage, a bed consisting of unconsolidated sediment is sometimes called a "soft" bed, while bedrock is a "hard" bed. Obviously beds can be mixed. Although soft beds seem to be necessary for surging, one should keep in mind that surges can sometimes advance over bedrock (Björnsson et al., 2003; Harrison and Post, 2003).

Although the presence of till seems to be a key ingredient of surging, the picture is not simple because till's presence is a necessary but not a sufficient condition. This is illustrated by a climate connection. Some formerly surge-type glaciers no longer surge in this time of general global warming and glacier shrinking, although in the Karakoram new surge-types have been reported (Copland et al., 2011). In some but not all cases a simple connection occurs between the interval between surges and the time to fill the reservoir area (e.g., Dolgushin and Osipova, 1978; Harrison et al., 2008), which is an obvious example of the climate connection. But when the surge behavior stops, the reservoir area does not fill to the critical level, and the glacier is able to transport the ice accumulating in the reservoir by normal flow processes. Before this happens, however, the surges may weaken (Figure 13.3) or become ill-defined, even though significant speed variations may still occur (e.g., Frappé and Clarke, 2007; Heinrichs et al., 1996).

13.2.3 Liquid Water

One of the most important factors in surge behavior of temperate or nearly temperate glaciers, or even in the seasonal behavior of the subset of most nonsurge glaciers, is the role of liquid water. However, the dependence is the opposite of what one might guess, because the sustained input of surface water tends to inhibit rather than promote motion. This is indicated by several observations, the first related to seasonality. In Alaska, at least, surges usually (not always) start in late fall or early winter, when the supply of surface water has been shut off, and terminate early in the melt season (Eisen et al., 2005; Raymond, 1987). A late spring can delay a surge termination and may result in an anomalously strong surge (Figure 13.3). Also, anomalously hot weather resulting in high melt rates may cause early termination of a surge. Despite these complications, the surge period at least sometimes can be thought of as the time required to refill the reservoir area after a surge, although the amount of water storage within a glacier must also be a factor (e.g., Lingle and Fatland, 2003). Storage is important in nonsurge glaciers also, but the effects in surging are more obvious, because termination or even temporary slowdown events are accompanied by large floods of water and sediment exiting the glacier (e.g., Humphrey and Raymond, 1994; Kamb et al., 1985). In fact, one of the more interesting properties of glacier drainage systems, as observed with dye tracing and other methods, is the presence of two distinct types. One

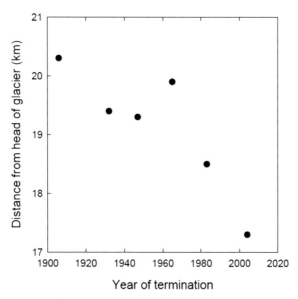

FIGURE 13.3   Length of Variegated Glacier, Alaska, affected by surges since 1900. A failed surge in 1995, caused by hot weather, is off scale and not shown. Note the tendency of the surges to weaken with time. The exception is the 1964–1965 surge, probably caused by a late-starting melt season. *Modified from Eisen et al. (2005) and Harrison et al. (2008).*

is a "discreet" or "channelized" system, which means the presence of tunnels or "conduits," and the other "distributed" in the sense that water pursues tortuous slow paths to the terminus in which the pressures are relatively high. During surge the latter dominates (or subglacial drainage is completely shut off (Harrison et al., 1994)), but at surge termination, as defined by an abrupt decrease in speed, there is a rapid shift to the discreet system (Björnsson, 1998; Kamb et al., 1985). Finally, there seems to be no fundamental difference in the drainage systems of surge-type glaciers in their quiescent state and of nonsurge glaciers; the peculiar drainage system of the surge-type glaciers seems to be organized near the onset of the surge (Raymond et al., 1995).

In summary, the available observations indicate the importance of three factors for the surging of temperate or nearly temperate glaciers: water and its storage within the glacier, till under large parts of the glacier, and the climatic environment. In addition to these three we need to consider the effect of a glacier's evolving shape, which we illustrate below. We emphasize that surging is a basal phenomenon, involving not the enhanced flow of ice, but of the basal motion over or within the underlying till.

13.2.4 A Simple Model

Some insight into the surging mechanism can be gained from a model, which, although probably overly simple, is consistent with observation. First, because of the importance of water, we need to consider how water might be transported in a glacier, especially because such large amounts are generated at the surface by melt (or sometimes by rain) and enter the glacier. We follow the reasoning of Röthlisberger (1972) and Shreve (1972). As the melt season progresses, evidence exists that this water eventually (but not always) ends up in a system of a few conduits near the bed. These conduits conduct water rapidly to the terminus. An interesting property of this "plumbing system" is that unlike conventional systems through fixed diameter pipes, an inverse relationship exists between pressure and the amount of flow. In other words, low pressure in a conduit is associated with high flow, although this is expected only under more or less, steady-state conditions. This inverse relationship holds because the heat generation by water flow causes melting and enlarges the "pipes." If the conduit is near a soft bed, the situation is less clear, but if the bed slope is not too small, the same discussion may apply (Walder and Fowler, 1994). The simplest approximation for the relation between water flux, Q and conduit pressure, p_c is then

$$P - p_c \approx AQ^b \tag{13.1}$$

The overburden pressure, P also enters. A is a quantity, probably spatially variable, which depends upon hydraulic, ice, and bed parameters (including bed slope), and $b \approx 1/12$ (simplified from Paterson, 1994, p. 117). Because of the negative sign, for a given P the conduit pressure, p_c indeed decreases for increasing water flux, Q. Equation (13.1) lacks direct observational justification and has other limitations, but it is simple and will illustrate our key points.

To see how this water-pressure dependence influences the motion of a surge-type glacier, we invoke the mechanical properties of the ubiquitous subglacial till. Under shear, till has a reasonably well-defined yield stress, τ_y approximated by

$$\tau_y \approx \varsigma(P - p) \tag{13.2}$$

where ς is a constant depending upon the composition of the till and $P - p$ is the "effective pressure," in which P is the pressure due to overlying ice as in Eqn (13.1), and p is the pore water pressure in the till (e.g., Tulaczyk et al., 2000). Equation (13.2) is a key result from soil mechanics. Laboratory measurements (e.g., Iverson et al., 1998) and order of magnitude calculations (Cuffey and Paterson, 2010) confirm that yielding in subglacial sediment is common, a conclusion that field geologists would not find surprising. An important step is the identification of the pore water pressure, p with the pressure, p_c in the conduit with which it is more or less in contact if the conduit is near the bed. With this approximation Eqns (13.1) and (13.2) can be combined to describe the shear stress τ_y at which the sediment begins to yield:

$$\tau_y \approx \varsigma AQ^b \tag{13.3}$$

which depends upon both the properties of the sediment and the water flux in a conduit.

Under nonsurging conditions, the shear stress, τ acting at the bed due to the downslope component of gravity is generally approximated by

$$\tau \approx \rho g H \sin \alpha \qquad (13.4)$$

where ρg is the weight density of ice, H is its thickness, and α is the surface slope. This would apply before the onset of the surge. If $\tau < \tau_y$, the till remains relatively rigid and relatively little motion occurs at the base of the glacier. However, if τ increases and approaches τ_y as the thickness, H and slope of the ice, α evolve between surges, the till will flow and maintain τ at the τ_y yield value. Thus on a till bed τ can never exceed τ_y. Other effects, such as longitudinal forces and shear along the valley walls, would then become important in preventing the glacier from running away downhill.

The inverse of Eqn (13.3) is

$$Q_{\text{crit}} \approx \left(\frac{\tau}{\varsigma A} \right)^{\frac{1}{b}} \qquad (13.5)$$

Equation (13.5) says that given a shear stress τ (approximated by Eqn (13.4) before the surge), a critical amount of water flux Q_{crit} is required to keep a conduit open. As long as this is satisfied, the till bed is more or less rigid and there will be little basal motion. If the requirement is not met, the bed yields, which means that it cannot support a shear stress greater than τ_y. Deformation of the till results, which may destroy the conduit and lead to the different, distributed drainage system noted above. If such a disturbance spreads, a surge would be expected. Given the evolving shear stress between surges, also noted above, it is possible that this critical state is achieved late in the melt season as the water input decreases, and indeed most surges in Alaska begin soon after the end of the melt season as noted above. Water would then be trapped in the glacier, and the high pressures characteristic of the distributed drainage system would prevail. We know that stored water is an important factor in surging, because surge termination is usually accompanied by flooding from under the glacier as noted above. This behavior is plausible because the magnitude of the exponent, $1/b$ in Eqn (13.4) is about 12. Thus the water flux required to keep a conduit open is an extremely sensitive function of shear stress, τ, which usually increases as the slope and thickness of a glacier evolve between surges (Raymond, 1987). It is interesting that some of this discussion also applies to the seasonal dependence of the speed of most nonsurge glaciers.

Several points need to be made. First, this calculation applies to temperate or almost temperate glaciers. It is not meant to be quantitative but to illustrate key points: that the drainage system of a surge-type glacier is extremely sensitive to the evolving surface geometry between surges, and that this affects the basal water pressure, water storage, and basal motion. Second, there may be more than one mechanism operating during surges, as noted above. Third, we have considered conditions at one locality in the bed whereas in fact the yielding of the till there may be compensated by increased drag elsewhere such as at the valley walls or via longitudinal forces, as noted above. This may be common even in nonsurge glaciers. Therefore before a full surge is nucleated, yielding needs to occur over some finite region. Still, it is easy to imagine how the blocking of a conduit could cause enlargement of the zone, which initially yields (Raymond, 1987). Fourth, one might ask whether the yielding of the till occurs throughout its thickness or in a relatively thin layer, the latter of which would effectively amount to sliding. It seems likely that both can occur, although the huge amount of sediment released in some "fast" surges, at least, suggests a major mobilization of the till for them (Truffer et al., 2000).

Fifth, it does not follow that this model by itself supports the requirement of a soft bed for a surge-type glacier, and in fact one can develop a theory for basal motion over a hard bed and include the effect of water (Kamb, 1987). If one considers a simplified version of this, a conduit flowing down a bedrock washboard, the condition for the leakage of the conduit into the lee sides of the washboard bumps, and therefore its destruction, is formally the same as Eqn (13.1) (Eisen et al., 2005). This suggests that the fundamental conclusion, that the drainage system is extremely sensitive to the evolving surface geometry of the glacier, does not depend upon the bed morphology, even though the constants in the above equations would have a different meaning. However, the observations strongly favor the importance of soft beds for surge-type glaciers, even though some surging glaciers can advance over bedrock, as noted above.

13.2.5 Conclusions

Much remains to be learned about surging and even its definition, but we can draw some tentative conclusions. There is a strong enhancement of basal motion of temperate or nearly temperate glaciers during a surge, in which liquid water plays a key role. The behavior of the water is largely controlled by the surface geometry of the glacier, which besides its dependence on the geometry of the valley, evolves between surges and determines the basal shear stress. This implies a sensitivity to climate, because climate and weather determine the rate of accumulation of snow on the glacier, how much ice has to be transported, and as a result, the basal shear stress and the recurrence interval of the surges. Climate and weather also determine the water availability. A soft bed appears to be a necessary condition for surging, but not a sufficient one because of the climate connection and the effect of valley geometry. A high-stress requirement is reminiscent of Budd's (1975) idea that surge-type glaciers are those which transport relatively large ice flux per unit width, and therefore require a relatively large basal shear stress. Finally, our discussion has focused mainly on temperate mountain glaciers, some of which undergo fast surges. We have not considered glaciers terminating in water.

The presence of basal till and its mobilization during surges raises the fundamental but unanswered question of its mass balance (see Walder and Fowler, 1994). For example where does till originate, how is it metamorphosed and sustained, and what determines its active thickness?

13.3 MEDVEZHIY AND GEOGRAPHICAL SOCIETY GLACIERS, CENTRAL PAMIRS, TAJIKISTAN

In 2006, 55 glaciers with a record of one or multiple surges were identified in the Pamirs (Kotlyakov et al., 2008b; cf. Osipova et al., 1998). Many of these surges go unnoticed by the public because of their low power, short duration, and occurrence in sparsely populated areas. Some however, can advance across river valleys, damming temporary glacial lakes that eventually drain in sudden, catastrophic outburst floods and mudflows. Two potentially dangerous glaciers exist in the upper reaches of the Vanj River valley with its rather dense population (Figure 13.4). Despite the large number of long-term observations of these glaciers and a global tendency of continued glacier retreat, the existing uncertainties about surge mechanisms and possible future surge extents as well as the poorly known influence of large rockfalls onto glaciers make reliable scenario construction and long-term forecasting difficult. As a consequence, the possibility of future ice-dammed lake formation and outbursts with disastrous consequences not only in the Vanj valley, but also on the Panj River further downstream, cannot be safely excluded, and systematic ground and space monitoring of the development must continue.

13.3.1 Surges and Lake Outbursts at Medvezhiy Glacier

Medvezhiy Glacier, one of the most famous surging glaciers with disastrous implications, is located in the upper reaches of the Vanj River, Central Pamirs (Figure 13.4). This valley glacier has a large accumulation area (≈ 20 km^2) and a long narrow tongue (up to 5 km^2), separated from the firn field by a steep icefall.

Studies of the dynamics and regime of Medvezhiy Glacier were initiated after a powerful, catastrophic surge in 1963, which began in early spring and ended in late June. Within a few months, the glacier snout advanced by 1.7 km (at a rate of up to 100 m per day) destroyed the local electric power station and geology field station, and blocked the Abdukagor River, a major tributary to the Vanj River. The 180 m-high ice dam caused a lake to form with a volume up to 14.5 million m^3 and a maximum depth of 80 m (Figure 13.5). The lake finally burst out across the glacier along englacial openings at some sort of a shear plane separating the upper quickly moving, heavily crevassed part of the glacier from a lower, possibly dead part of the ice enriched by moraine material. The outburst produced a discharge (calculated as a function of decreasing lake volume) of up to 1,000 m^3/s and caused devastating floods in

FIGURE 13.4   Part of ETM+ image of the Central Pamirs, September 16, 2000: (1) Medvezhiy Glacier, (2) Geographical Society Glacier. Scale is in kilometers.

the Vanj River valley, where the bridges were demolished, the motorway and power lines destroyed, and—even at a distance of 80 km from the glacier—the airfield near the Vanj settlement was flooded and a plane whirled away (Dolgushin et al., 1964).

Subsequent studies revealed the general surge pattern in the behavior of Medvezhiy glacier and formed the basis for the rapid development of advance

FIGURE 13.5   Medvezhiy Glacier and the lake in the Abdukagor valley at maximal filling in 1963. *Photograph by K. P. Rototaev.*

forecasting. The prediction of the next surge, which happened in 1973, had been made in advance (Dolgushin and Osipova, 1973, 1975).

The surge of Medvezhiy Glacier in 1973 was more powerful than the previous one. Front advance was recorded early in spring and the surge ended in August. The glacier advanced by 1.9 km, the ice surface in the advancing part of the tongue rose by 200 m, and in the upper part of the tongue it dropped by 110 m (Figure 13.6).

The ice-dammed lake in the Abdukagor valley formed again, this time with a volume up to 16.4 million m^3 and a maximum depth of 110 m. Two powerful

FIGURE 13.6   Medvezhiy Glacier tongue just after the surge in June 1973. *Photograph by G. Osipova.*

lake outbursts were observed at an interval of about a fortnight. The outbursts occurred, as in 1963, across the glacier with a discharge this time up to 1,400 m³/s (Figure 13.7). In contrast to the surge in 1963, the effects of the lake-outburst floods were smaller—several bridges were demolished, some parts of the motorway and of power lines were damaged. The power plant had been protected with a dam beforehand, and some bridges had been dismantled as a preventive measure (Dolgushin and Osipova, 1982).

The next glacier surge (1988–1989) was documented by repeated aerial surveys. Its character was similar to the previous one, but the scale was less significant—the glacier advanced by only 1.1 km. The lake in the Abdukagor River valley formed again, but its volume was less considerable and its water was released along the left side of the glacier without catastrophic consequences (Kotlyakov et al., 1997; Osipova and Tsvetkov, 1991).

The next surge was in 2001. The glacier advanced only by 450–500 m and did not reach the Abdukagor River; no catastrophic consequences occurred (Desinov et al., 2001).

During the following years the glacier observations made from the International Space Station (ISS) revealed that the activated bulge of the surge front propagated down the tongue with a speed of 250 m per year. In August 2010, the shape of the glacier showed signs of an upcoming surge, and in March 2011 the ISS reported on the beginning of fast movement—dark bands of longitudinal disruptions appeared along the glacier margins although no signs of tongue advance were visible. In late May, as a result of its rapid advance, the front of the glacier tongue was already close to the Abdukagor River. The surge terminated in July. A picture from the ISS on August 22 shows that the tongue had advanced by 1.2 km and blocked the Abdukagor River. However, traces of a lake were not found—the Abdukagor River had probably found an outlet between the left side of the glacier and the valley wall (Kotlyakov and Desinov, 2012).

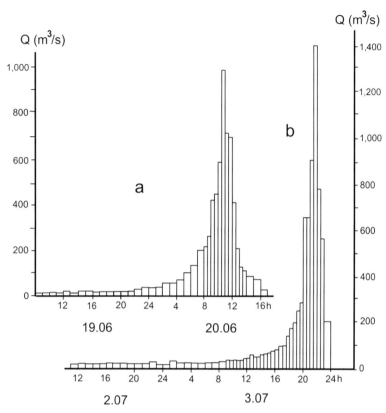

FIGURE 13.7   Hydrographs of the lake outbursts in 1973 at Medvezhiy Glacier as calculated from the measured lake level and volume reduction (a) 19–20 June 1973; (b) 2–3 July 1973.

According to historical records (Dolgushin and Osipova, 1982) and direct observations, surges of Medvezhiy Glacier occurred in 1913–1915, 1937, 1951, 1963, 1973, 1989, 2001, and 2011 (cf. Figure 13.8). The surges in 1963 and 1973 were accompanied by the formation and outbursts of an ice-dammed lake with catastrophic consequences. Quantitative data collected during these extraordinary events provide important empirical information in view of estimating potential peak discharge from ice-dammed lakes in connection with hazard assessment, risk reduction, and disaster prevention in comparable cases. The volume/peak-flow ratio is at least at the upper limit if not beyond values known from approaches using principles of progressively enlarging subglacial channels (Clague and Mathews, 1973) but rather characteristic for sudden break mechanisms (Haeberli, 1983; Walder and Costa, 1996), which are known to sometimes take place in heavily broken ice dams from ice avalanches and glacier surges. A combination of both processes, triggered first by hydraulic uplift of the ice dam on the lakeside, may be a plausible

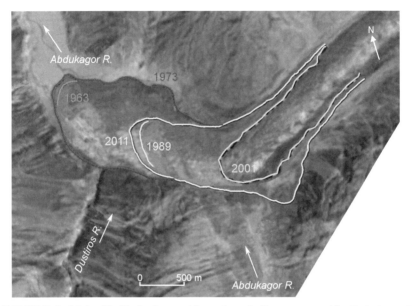

FIGURE 13.8   Medvezhiy Glacier front position at the end of known surges. *After Kotlyakov and Desinov (2012).*

interpretation of the measured hydrographs and visual observations. This indicates that limiting assessments of potential lake outbursts to the classical theory of progressively enlarging subglacial channels may lead to an underestimation of real peak discharge and an overestimation of outburst time-duration, which means an overestimation of prewarning times. Both effects may cause a serious underestimation of the hazard potential.

13.3.2 Advancing Geographical Society Glacier

Geographical Society Glacier (≈ 65 km^2, Figure 13.9) is the sixth largest glacier in the Pamirs, located 5 km downvalley from Medvezhiy Glacier. At the beginning of the last century, the glacier advanced into the Abdukagor River valley and blocked it, forming an ice-dammed lake of about 50 million m^3. The lake-outburst floods were disastrous for the inhabitants of the Vanj valley. By the middle of the last century, the glacier had retreated by 2.6 km, and its 6.5 km-long lower part became a debris-covered stagnant ice body. Ice-cored moraines were still preserved in recently deglaciated areas, and the traces of at least two lake shorelines occurred on the slopes of the Abdukagor River valley above the glacier. In 1965, an activation of the glacier was noticed. Its thickness had increased, although the glacier front remained in a stationary position (Dolgushin and Osipova, 1971).

ASTER images in May 2001 confirmed the reactivation of the glacier—its tongue became convex with fresh crevasses. The front had started to advance

FIGURE 13.9   The Geographical Society Glacier. Space image ASTER August 8, 2003. The rock avalanche/mudflow deposits are seen at the glacier surface (dotted outline). The detachment place of this event is indicated by an arrow.

and by August 2003 parts of the strongly thickened tongue had advanced by 150−200 m. In 2004−2006, the glacier advance continued at a slow rate of 100−125 m for this period. By July 2007 the tongue had overridden the ancient moraine and closely approached the Abdukagor River bank, advancing in one single year by 500 m (Figure 13.10). The ice thickness on the right side of the glacier tongue was at least 200 m. By March 2008 the glacier tongue advanced a further 70−100 m and started to be eroded by the Abdukagor River waters, which seems to have stopped the advance (Kotlyakov et al., 2008a).

In 2001−2002, a large rock avalanche ran out over the middle part of the glacier tongue covering an area of about 5 km^2 (Figure 13.9; Kotlyakov et al., 2003). The influence of this event on the mass balance and flow dynamics of the glacier is not yet clear.

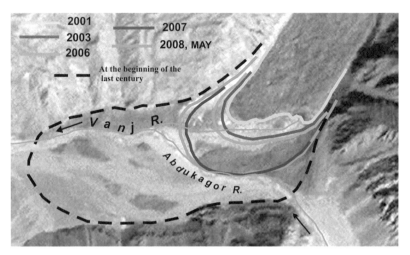

FIGURE 13.10   Margin positions of the Geographical Society Glacier in 2000–2008.

13.4 SURGES OF GLACIAR GRANDE DEL NEVADO DEL PLOMO, CENTRAL ANDES, ARGENTINA, AND RELATED DISASTERS/HAZARDS

Glacier surges have been observed on both sides of the Central Andes. In Chile, surge events were reported by Lliboutry (1956, 1958) for Glaciar Nieves Negras (33°49′ S; southwest of Volcán San José, 5,785 m) in 1927, for Glaciar Universidad (34°39′ S; in the Tinguiririca basin at the transition between the Dry and Wet Andes) in 1943, and for Glaciar Juncal Sur (33°05′ S; on the southern flank of Nevados del Juncal in the upper part of the Río Maipo basin and only 7 km to the northwest of the Glaciar Grande del Nevado del Plomo described in detail below) in 1947. In Argentina, surges are known for Glaciar de la Laguna (34°29′ S; in the Atuel basin) between 1970 and 1982 (Cobos and Boninsegna, 1983) and for Glaciar Horcones Inferior (32°40′ S; below the south wall of Cerro Aconcagua (6,959 masl; Figure 13.11), the highest peak in the Western Hemisphere) in spring 1984 and even more powerful in 2004 (Espizua et al., 2008; Pitte et al., 2009); no surge of this glacier had been reported before since the beginning of the twentieth century (Happoldt and Scrott, 1993; Unger et al., 2000). In the Río Tunuyán basin at 33°23′ S, Williams and Ferrigno (1998) reported looped moraines visible on satellite imagery as an indication of surge behavior for Glaciar Unnamed (33°27′ S) and for the Glaciar Marmolejo (33°43′ S)—a photograph taken by Luis Krahl in 1946 for this last glacier shows that it was indeed surging at that time—and 24 km to the north of this glacier, Glaciar Nevado de los Piuquenes (33°31′ S) surged in January 1997 (A. Aristarain, oral communication).

The most important hazard-related events happened in the Río del Plomo sub-basin (Argentina) at 33°S. Glaciar Grande del Juncal (33°03′ S) surged

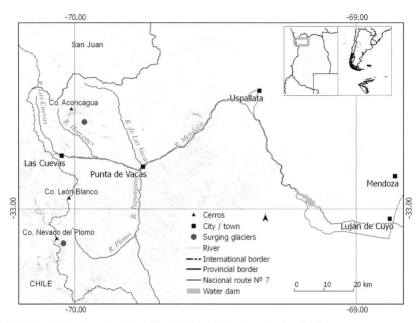

FIGURE 13.11   Location of the Rio del Plomo in the Rio Mendoza basin.

between March and November 1910 (Helbling, 1919) and another rapid advance can be deduced from Helbling's observations in 1934 and the 1955 aerial photographs (Espizua, 1986). During its surge of 1933, nearby Glaciar Grande del Nevado del Plomo crossed the Rio del Plomo valley, damming a large temporary lake, the outburst of which in 1934 caused catastrophic consequences—historic sources indicate the occurrence of a similar event in 1786 (Prieto, 1986). Later surges occurred in 1984 and 2007 creating hazardous situations but no damage (Espizua, 1986; Espizua and Bengochea, 1990; Ferri Hidalgo and Espizua, 2010; Helbling, 1919, 1935, 1940).

13.4.1 The Surges of Glaciar Grande Del Nevado Del Plomo

The Rio del Plomo sub-basin, the most important glacierized part of the Rio Mendoza basin, is located at 32°57′ - 33°12′ S latitude and 69°57′ - 70°06′ W longitude (Figures 13.11−13.12). Large glaciers in the sub-basin covered an area of 711 km^2 in the early 1970s (Corte and Espizua, 1981). Using aerial photographs and the information provided by Helbling (1919, 1935) and Reichert (1929), Espizua (1986) reconstructed the glacier fluctuations from 1909 to 1974, a period of general and important glacier retreat in this sub-basin. The Glaciar Grande del Nevado del Plomo flows down to the east from the slopes of the Cerro Nevado del Plomo (6,050 m), in the border range between Chile and Argentina. The repeated surges and related hazards are described in the following.

FIGURE 13.12   ASTER image in 2010 showing the glaciers in the Rio del Plomo valley: (1) Bajo del Plomo; (2) Alto del Plomo; (3) Oriental del Juncal; (4) Grande del Juncal; (5, 6, and 7) Alfa, Beta, and Gamma; (8) Grande del Nevado del Plomo. Red line = Chile−Argentina border.

13.4.1.1 Surge and Flood Disaster of Glaciar Grande Del Nevado Del Plomo in 1933/1934

On January 10, 1934, a catastrophic flood affected the Rio Mendoza. The volume of the water involved was estimated about 60,000,000 m^3/s. At Luján de Cuyo village located about 170 km to the northeast from the Glaciar del Nevado del Plomo, peak discharge was around 3,000 m^3/s and arrived in the form of a frontal wave 5−6 m high propagating at a velocity of about 3.3 m/s. The flood lasted less than 5 h, caused many deaths and destroyed a power plant, seven bridges, and 12 km of the Transandino railway (King, 1935). According to Helbling (1935) this flood was produced by the sudden outburst of a lake, 3 km long and 75 m deep, formed in the Rio del Plomo valley by the advance or surge of Glaciar Grande del Nevado del Plomo across the valley river in 1933 (Figure 13.13).

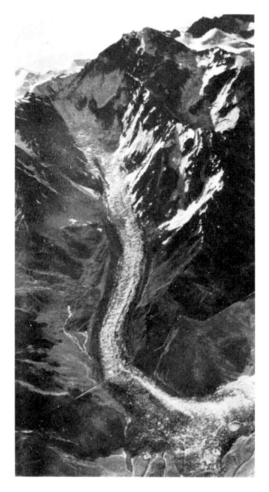

FIGURE 13.13   Surge of the Glaciar Grande del Nevado del Plomo across the valley of the Río del Plomo in 1934. *Helbling (1935).*

The damage and especially the number of people killed were related to the massive and sudden peak discharge superimposed onto an already elevated runoff. This indicates mechanical rupture mechanisms (Haeberli, 1983) as part of the lake drainage. Figure 13.13 indeed shows a striking collapse structure in the central part of the heavily fractured ice dam directly above the subglacial drainage tunnel. This collapse structure may have been related to an intermittent blockage of water flow followed by mechanical rupture of the obstacle. Higher up on the glacier, Helbling (1935) reported marked surface lowering in the central part of the firn area, which indicates an onset of rapid mass displacements above the equilibrium line. Helbling considered that the ice

FIGURE 13.14 Sixty meters thick ice cover with pinnacles over the former moraine at the outcrop (right) known as "Roca Pulida" (glacial-polished rock). *Photograph of the Argentine Transandino Railway Company taken in 1934 by W.D. King.*

advance must have occurred over a short time period of days, weeks, or months. A mass of "white ice," 60 m thick spread over the former moraine reaching an outcrop known as "Roca Pulida" on the left side of the Río del Plomo (Figure 13.14). The glacier advanced 900 m beyond the 1909−1912 position. In the middle sector of the glacier the ice was chaotically faulted and broken up. Beyond the valley mouth, the heavily crevassed glacier overrode its former moraine.

13.4.1.2 Surge of Glaciar Grande Del Nevado Del Plomo in 1984

Between 1934 and aerial photographs of 1955 and 1963 the glacier front had retreated by 4.1 km and between 1963 and 1974 advanced by 1.05 km (Espizua, 1986). In 1984, Glaciar Grande del Nevado del Plomo surged again (Figure 13.15). On Landsat MSS and TM images of the March 15, 1976−February 16, 1984 time period, no significant changes in the position of the debris-covered terminus are recognizable. The glacier front was approximately 2.7 km away from the Roca Pulida outcrop (Figure 13.16). On the image of April 4, 1984 the glacier terminus was free of debris and had advanced 500 m sometime since February 16, 1984. On the November 14, 1984 image, the glacier front had already reached the Roca Pulida outcrop. Nine days later, on the November 23, 1984 image, a lake 1.65 km long and 600 m wide is visible. This lake reached a length of 2.8 km and a width of 1.1 km on January 9, 1985 (Figure 13.16). On the February 27, 1985 image,

FIGURE 13.15   Southern view of the Glaciar Grande del Nevado del Plomo surge in February 1985. The surge had come to an end and the river drained through the dam of broken ice. *Photograph by W. Haeberli.*

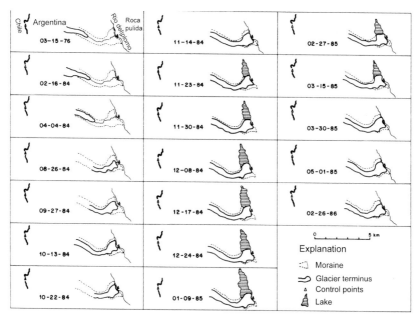

FIGURE 13.16   The evolution of the Glaciar Grande del Nevado del Plomo surge and the ice-dammed lake in 1984 through satellite images. *Espizua and Bengochea (1990).*

the lake dimensions are reduced and on the March 30, 1985 it had completely disappeared (Espizua and Bengochea, 1990, Figure 13.16). On February 28, 1985, Bruce et al. (1987) measured in the field a lake length of 1,494 m, a width of the lake against the glacier ice-dam of 703 m, and estimated the lake volume at $12.1 \times 10^6 \, \text{m}^3$.

Downstream people became aware of the surge event through the occurrence of three unexpected peaks in the flow of Rio Tupungato and Rio Mendoza that occurred on February 14, 22, and March 13, 1985 (Figure 13.17(a)). Maximum peak flow of 293 m³/s in Río Tupungato occurred on February 14 and was followed by two lesser peaks on February 22 and March 13. Similar peaks with the same timing occurred on the Río Mendoza. These peak flows were up to 147 m³/s and 155 m³/s higher than normal flow in these rivers (Figure 13.17(b)).

No damage was reported downvalley during the occurrence of the Rio Mendoza flood peaks. It is nevertheless interesting to note that the discharge peaks were again extremely sharp, especially the February 22 and March 13 events, which show very steep ascending limbs and much less steep descending limbs (Figure 13.17(b)). Such hydrograph shapes are atypical for lake outbursts by progressive enlargement of subglacial channels and indicate that sudden, mechanical rupture mechanisms may at least have been part of the outburst process (Haeberli, 1983).

13.4.1.3 Surge of Glaciar Grande Del Nevado Del Plomo in 2007

Glaciar Grande del Nevado had a new surge in 2007. After 23 years of quiescence, the glacier advanced by some 400 m between September 25, 2006 and March 20, 2007. On September 28, 2007, around or just after surge termination, it had advanced over a total of 3.8 km. The glacier thereby again reached the Roca Pulida outcrop but the width of its terminal part was less important than in the 1933 and 1984 events. On the November 14, 2007 image a small lake (about 190 m long and 45 m wide) can be observed upstream of the glacier front. The ice-dammed lake had completely disappeared by November 23 (Ferri Hidalgo and Espizua, 2010). Later images (March 30, 2008 and January 12, 2009) show the glacier front covered by debris and with reduced volume. On an image of March 2010 the glacier front was still in contact with Roca Pulida and the Rio del Plomo (cf. photograph of April 2013 in Figure 13.18) but the ice is becoming more and more debris-covered. The 2007 surge of Glaciar Grande del Nevado del Plomo was less dangerous than the 1984 surge because only a small temporary lake formed.

13.4.2 Hazard Mitigation and Monitoring

In relation to the 1984/1985 surge, an emergency expert mission by the United Nations Disaster Relief Organisation (UNDRO) was carried out on the demand

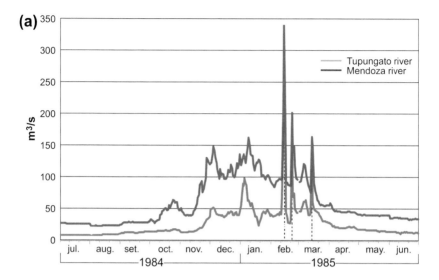

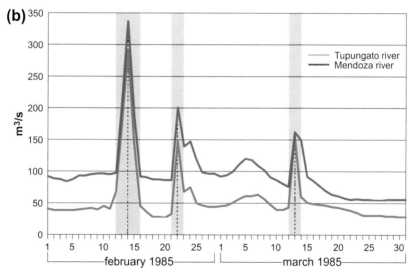

FIGURE 13.17   (a). Mean daily discharge (m³/s) record of the Rio Tupungato and the Rio Mendoza during July–December 1984 and January–June 1985. (b) Mean daily discharge (m³/s) record of the Rio Tupungato and the Rio Mendoza on February 14, 22 and March 13, 1985.

of the Governor of Mendoza to assess the situation and reduce the risks for the population and for the important infrastructure in the reaches of Rio Mendoza. During a helicopter-supported field visit it could be recognized that the surge had come to an end and that the river was no longer dammed. As there was no more imminent hazard, the primary task was to organize adequate long-term

FIGURE 13.18 Glacier observations made from a helicopter in April 2013. The volume of the ice has diminished and the surge is being covered partially by debris. *Photograph by R. Villa Real, Departamento General de Irrigación de Mendoza.*

observation and to develop ideas for hazard mitigation and risk reduction for the case of expected future surges. Three corresponding options were discussed: (1) excavation of a drainage tunnel through the Roca Pulida outcrop to suppress lake formation; (2) construction of a dam with coarse blocky material on the lakeside of the ice dam to control dam permeability and outflow of a possible future lake; and (3) construction of a retention dam/basin in a suitable downvalley river section. Option (3) would be expensive but could enable multipurpose use for irrigation, tourism, and/or hydropower production during the quiescent time between surges. Glacier monitoring by satellite imagery was assumed to provide enough warning time for hazard mitigation measures in case of a new surge. This task is being carried out by IANIGLA-CONICET in Mendoza. Corresponding analyses of available imagery and other sources of information point to some important facts: The active phase of Glaciar Grande del Nevado del Plomo surges seems to start during wintertime when the ice reaches its maximum advance rate. During the different surges the ice advanced within a period of months over a distance of a few kilometers. Accelerated flow seems to start in the upper part and propagates downglacier as a wave; during and after the surge, the glacier surface is characterized by deep crevasses and jagged ice pinnacles. In the central part of the dam consisting of heavily broken ice, a collapse structure may have been related to an intermittent blockage of water flow followed by sudden rupture and extremely sharp discharge peaks.

13.5 A SURGE-LIKE FLOW INSTABILITY OF BELVEDERE GLACIER, ITALIAN ALPS, AND ASSOCIATED HAZARDS 2001—2003

Changes in glacier and permafrost conditions are currently shifting beyond empirical historical knowledge (Kääb et al., 2004). The type and magnitude of processes and hazards involved at Belvedere Glacier (Ghiacciaio del Belvedere, Macugnaga, Italian Alps) during 2001—2003 confronted the responsible authorities with complex problems without precedent in the European Alps. The following case study is mainly based on Kääb et al. (2004), Kääb (2005), and Mortara and Tamburini (2009).

13.5.1 Preceding Events

Belvedere Glacier is a humid-temperate glacier with a flat, heavily debris-covered tongue and fed by steep glaciers, ice and snow avalanches, as well as rockfalls from the large east face of Monte Rosa (Figure 13.19). It has been investigated for many decades and is known as a classic example of a glacier with an elevated sediment bed (Mazza, 1998). At least seven outburst floods

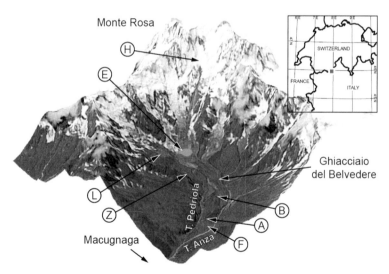

FIGURE 13.19 Oblique perspective view of Ghiacciaio del Belvedere and Monte Rosa from the Northeast. The synthetic view is computed from ASTER false color data of July 19, 2002, and a digital terrain model from photogrammetry and ASTER stereo data. Section size is approximately 8 × 8 km. Note the existence of the supraglacial Lago Effimero at the foot of the Monte Rosa east face. A: Alpe Burki, B: Belvedere, E: Lago Effimero, F: Fontanone, H: Hanging glaciers, L: Lago delle Locce, Z: Rifugio Zamboni.

are known to have originated from the glacier before 2001, threatening and partly damaging the village of Macugnaga and affecting the original moraine geometry (Haeberli et al., 2002):

- August 1868: water pocket outburst and moraine collapse at the left tongue
- August 1896: moraine breaching after heavy rainfall
- 1904: water pocket outburst and cutting through the right lateral moraine
- September 1922: water outburst from the glacier after a rainfall period
- August 1970, August 1978, July 1979: outburst floods from the Lago delle Locce.

In July 1979, Lago delle Locce (L in Figure 13.19) broke out through/ underneath its ice dam toward Ghiacciaio Nord delle Locce. The flood wave traveled through the natural channel between the Belvedere Glacier and its right lateral moraine. At a slight counterslope near the hut Rifugio Zamboni (Z in Figure 13.19), the flood cuts through the moraine progressively eroding a large breach. The subsequent debris flow destroyed the lower section of the Belvedere chairlift and flooded the valley bottom for a length of 1 km and a mean width of 150 m, almost reaching the hamlet of Pecetto near Macugnaga. Following the 1979 flood, prevention work such as artificial lowering of the lake level at Lago delle Locce and dam construction in the main torrent below the glacier became necessary (Haeberli and Epifani, 1986; Mazza, 1998).

Since mid-1980, increasing numbers of rockfalls, rock/ice avalanches, and debris flows have been observed in the east face of Monte Rosa (Fischer et al., 2006, 2011, 2013; cf. Deline et al., 2014), where the rapid vanishing of glaciers and smooth ice covers left large parts of the underlying permafrost rock unprotected against warming and destabilization (cf. Davies et al., 2001; Noetzli et al., 2003). Concern increased that these phenomena may possibly be precursors of large events with potential horizontal travel distances of several kilometers, comparable to the rock avalanche at Brenva, Mont Blanc, in 1997 (Barla et al., 2000; Deline, 2001; Giani et al., 2001), or the 2002 Kolka/ Karmadon rock/ice avalanche, Caucasus (Haeberli et al., 2003; Huggel et al., 2005; Kääb et al., 2003; cf. also Huggel et al., 2010).

In summer 2001 heavy crevassing of the glacier surface, a dramatic increase in ice thickness, dirty ponds between the glacier and its lateral moraines, and enhanced ice- and rockfall activity from the glacier margins were observed. The processes were interpreted as indicating a surge-like glacier movement and the local authorities were informed (Haeberli et al., 2002). As a consequence monitoring of the area was intensified and hazard-prevention measures taken. The latter were especially necessary in connection with the rapid formation of a large supraglacial lake at the very foot of the Monte Rosa east face and the possibility of its sudden outburst or of an impact/ flood wave triggered by large rock/ice avalanches.

13.5.2   A Surge-Like Movement of Ghiacciaio Del Belvedere

Since spring 2001 and possibly already starting in autumn 1999 or spring 2000 (Mazza, 2003; Kääb et al., 2004), Belvedere Glacier experienced drastic changes in flow regime and related features. Numerous and large crevasses formed on the glacier, first at the foot of the Monte Rosa east face and later downwards along the entire glacier. The almost completely debris-covered surface of the glacier tongue with its smooth topography suddenly became heavily crevassed and disrupted with debris cover only surviving in a few places (Figure 13.20). In the mid-1980s and during 1995−1999, average surface speeds on the lower part of Belvedere Glacier had been in the order of 35−45 m/a (VAW, 1985; cf. Mazza, 2003). During 1999−2001, average speeds of up to 110 m/a and, during autumn 2001, of up to 200 m/a were observed photogrammetrically (Figure 13.21; Kääb, 2005). Terrestrial surveying and photogrammetry in summer 2002 yielded somewhat lower speeds of up to 80 m/a.

At most parts of the glacier, speedup was accompanied by an increase in ice thickness of 20 m and more (Figures 13.22 and 13.23). The corresponding mass shift traveled down the glacier reaching its maximum at Belvedere (B in Figure 13.19) around summer 2002 (cf. Mazza, 2003). In August 2003, ice thickness began to diminish again on most parts of Belvedere Glacier. Photogrammetric analyses suggest that the enhanced movement was mainly restricted to the flat lower part of the glacier (Kääb, 2005). The large depression at the foot of the Monte Rosa east face, filled by a supraglacial lake (E in Figure 13.19), might partially originate from strongly extending flow and, thus, mark the upper limit of the primary surge-like movement. At this site, a total loss in ice thickness of about 20 m had already occurred during 1995−1999 (Figure 13.23; Kääb, 2005).

The exact mechanism of the surge-like movement since 2001 still remains unclear (cf. Mazza, 2003). Enhanced basal water pressure seems to have been

FIGURE 13.20 Right lateral moraine and ice margin of Ghiacciaio del Belvedere (view upglacier; left: summer 1996; right: end of June 2002). The ice wall near the middle of the right image is approximately 20 m high. *Photographs by W. Haeberli (left) and A. Kääb (right).*

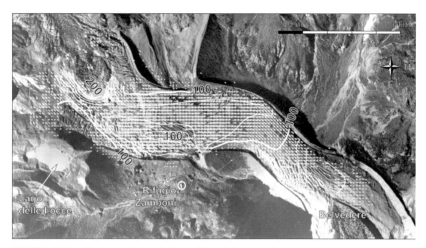

FIGURE 13.21 Surface velocity field on Ghiacciaio del Belvedere between September 6 and
October 11, 2001. Underlying is the orthoimage of October 11, 2001 (air-photo CNR-IRPI). The
black numbers and white isolines indicate the glacier speeds of the autumn 2001 period. Speeds are
given in meters per year.

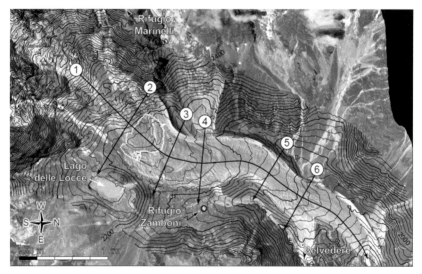

FIGURE 13.22 Orthoimage of Ghiacciaio del Belvedere, Italian Alps, from October 12, 1995
(original aerial photo by swisstopo, permission BA057212). Numbers 1—6 and corresponding lines
indicate the elevation profiles of Figure 13.23; arrows give the respective profile direction. The
dotted line marks the shoreline of the supraglacial lake at the beginning of July 2012.

involved as suggested by dirty ice-marginal ponds in 2001 (Haeberli et al.,
2002) and the evolution of the huge supraglacial lake in 2002 and 2003. An
ice-marginal band of strong shearing combined with an undisturbed snow
cover on the glacier surface itself, as observed in winter 2002/2003, indicated

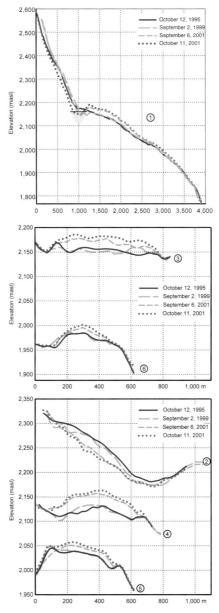

FIGURE 13.23   Repeat longitudinal and transverse profiles over Ghiacciaio del Belvedere from stereo-photogrammetry. For profile numbers see Figure 13.22. In profile no.1 the position of the summer 2002 supraglacial lake is indicated schematically.

that the glacier moved as a block by sliding, sediment deformation or internal shearing rather than by enhanced deformation throughout the entire ice column (e.g., Harrison and Post, 2003; Raymond, 1987).

13.5.3 Consequences of Sudden Changes

Hazards related to the enhanced glacier speed and mass transport included increased rock- and icefall activity at its margins, overrunning of infrastructure and installations, and destabilization of moraines.

Due to its debris cover and elevated moraine bed the surface of Belvedere Glacier had remained close to the lateral moraines of the Little Ice Age. As a consequence of the surge-like movement, however, the glacier surface vertically exceeded the level of lateral moraines by up to 20 m. At some locations the glacier overran its moraines producing frequent ice- and rockfalls over the outer moraine slope. Parts of the access trails to the Rifugio Zamboni (Z in Figure 13.19) and further to the Lago delle Locce (L in Figure 13.19) had to be closed in 2002 and a detour constructed. In some obvious places, the enhanced lateral pressure from the surging and advancing glacier led to weakening or even first cutting through the moraines. Breaching of lateral moraines was of major concern in view of enhanced runoff from lake-outburst floods (cf. the experience with the 1979 flood).

Already in summer 2001, but most noticeably since winter 2001/2002, the glacier started to overrun its lateral moraines along the entire left side of the right tongue near the location "Belvedere" (B in Figure 13.19) at the divergence between the two tongues of Belvedere Glacier (Figure 13.24). The material cable-lift from Belvedere to the Rifugio Zamboni across the glacier had to be closed in autumn 2001 and the trail crossing the glacier toward the Rifugio was closed in 2002 and had to be repeatedly rebuilt after its reopening. Extensive rock- and icefalls reached the ski run (O in Figure 13.24) and increasingly buried it as well as the forest and the hiking trail along the outer moraine slope. In autumn 2002, a new ski run (N in Figure 13.24) was constructed to avoid the direct danger zone. A blocky retention dam of approximately 100 m length (location at D in Figure 13.24, built after image date) was built for holding back rock- and icefalls toward the ski run and toward the middle station of the chairlift at Alpe Burki (A in Figure 13.19). A further point of concern was the very terminus of the advancing right tongue. It descends into the steep flanks of the river Torrente Pedriola. Heavy erosion in the riverbed (e.g., due to a glacier flood) or destabilization within the ice mass itself could have led to damming of the river.

13.5.4 Evolution of a Supraglacial Lake

A supraglacial lake of about 3,500 m^2 developed in September 2001 on the Ghiacciaio del Belvedere at the foot of the Monte Rosa east face (Haeberli et al., 2002). In October 2001 the lake reached an area of about 20,000 m^2.

FIGURE 13.24   The right tongue of Ghiacciaio del Belvedere, between Torrento Pedriola (left) and Belvedere (B, end of June 2002). The glacier overran its historic moraines and caused frequent rockfalls onto the ski run (O). Later in 2002, after the date of this photograph, a new ski run (N) and an ice- and rock-retention dam (D) were constructed. *Photograph by A. Kääb, June 2006.*

By the end of May 2002, the lake had an area of roughly 20,000−40,000 m^2 as reconstructed from ASTER satellite imagery (Kääb et al., 2003). The first annual control visit in mid-June 2002 encountered an exceptionally large lake of nearly 150,000 m^2 with an estimated volume of 3 million m^3 (Figure 13.25). The lake level was rising by up to 1 m per day and had only a few meters of freeboard left. At the right margin of the lake, a small ice dam only a few tens of meters wide above lake level held the water at a level up to 25 m higher than the right lateral moraine. The associated hydraulic gradient within the ice dam (hydraulic head) must have been far in excess of 10 percent. The exact volume of the lake was measured by bathymetric soundings at the beginning of July 2002 in order to plan for the capacity and location of emergency pumping (Mortara and Tamburini, 2009; Tamburini et al., 2003). The highest lake level was reached on June 26/27, 2002. As a consequence of intense ice melting by the lake water (cf. concerning involved processes Benn et al., 2000; Chikita et al., 1999; Kääb and Haeberli, 2001; Reynolds, 2000), the lake depth had reached a maximum of about 60 m. A cold spell in early July 2002 significantly reduced meltwater input. Together with pumping (Figure 13.26) and naturally occurring subglacial drainage, this helped to stabilize and then to slowly lower the lake. By the end of October 2002, the lake area had decreased to a size comparable to that of autumn 2001.

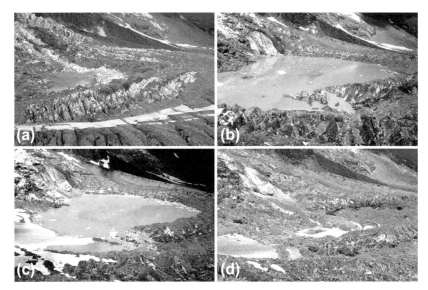

FIGURE 13.25   Location of Lago Effimero as seen from the moraine threshold near Lago delle Locce: July 17, 2001 (a), June 26, 2002 (b), June 18, 2003 (c), August 5, 2003 (d). Width of the section shown is approximately 400 m. *Photographs by A. Kääb.*

FIGURE 13.26   Installation of pumps on a raft on Lago Effimero, July 2002. *Photograph by A. Kääb.*

Tracer experiments performed in mid-October 2002 (Tamburini et al., 2003) indicated average flow velocities of sub- or englacial water in the order of a few centimeters per second between the lake and the resurgence sites at the moraine sources between the two glacier tongues (Fontanone; F in Figure 13.19) and the orographic left glacier outlet (Torrente Anza). No tracer was found at the sources in the moraine breach near Rifugio Zamboni and the outlet of the right glacier tongue (Torrente Pedriola). From ground-based seismic and radar soundings in the mid-1980s (VAW, 1985), helicopter-based radar soundings in 2002 (Tamburini et al., 2003), and again ground-based radar soundings in spring 2003, Belvedere Glacier is known to be up to 200 m thick and possibly underlain by bedrock along the section between the foot of the Monte Rosa east face and Rifugio Zamboni. Below this point, a sediment bed several tens of meters thick forms a marked bed-overdeepening with an adverse slope and decreasing glacier thickness. The probable existence of a transverse sediment riegel and the observed low velocities of water flow indicate that lake drainage takes place through subglacial sediments.

In spring 2003, Lago Effimero rapidly refilled with meltwater, reaching a volume similar to the one found at the end of June 2002. In fact, the 2003 Lago Effimero was similar to the one of 2002 with respect to surface shape as well (Figure 13.25). Its water level was about 10 m lower compared to 2002, presumably as an effect of continued ice thickness loss. Between June 18 and June 20, 2003 Lago Effimero burst out. On June 18, 2003, a day of fine weather but without especially enhanced melting conditions, increasing discharge of very dirty water from the left glacier tongue (Torrente Anza) was observed. At the same time the automatic monitoring systems for the water level of the lake and for the flow height of the Torrente Anza at Pecetto showed unusually de- and increasing values. Torrente Anza experienced above-average flow-heights from June 18 to 20, with peak discharge values in the order of $15-20$ m^3/s on June 19. The maximum lake level before the outburst had been reached on June 16. Within a few days the lake level dropped by about 20 m and about 2.3 million m^3 were released. The outburst path occurred en- and/or subglacially toward the left glacier gate (Torrente Anza). During the outburst, along a short section of roughly 100 m length, a dirty stream between the ice margin and the left lateral moraine became visible.

Supraglacial lakes like the one on Belvedere Glacier have a potential for catastrophic floods. A volume of 3 million m^3 can cause peak discharges in the order of up to $100-200$ m^3/s in the case of a hydraulic ice-dam break (i.e., progressive enlargement of englacial flow channels; Clague and Mathews, 1973; Walder and Costa, 1996). Mechanical breaking of an ice dam (very unlikely under the 2003 topographic conditions) or temporary runoff blockage through ice collapses at the glacier margins or under the glacier could have resulted in even higher peak discharge (Haeberli, 1983). Moreover, the reach of Torrente Pedriola on the right side of the glacier contains an amount of erodible debris, which is apparently large enough for a debris flow to be triggered by a massive glacier flood. Accordingly, a serious hazard scenario for Lago Effimero was an impact wave from a large rock/ice avalanche with subsequent breaching of a lateral moraine and formation of a large debris flow reaching the densely populated area of the tourist station Macugnaga.

13.5.5 Hazard Management

The rapid evolution of a large supraglacial lake has not been previously observed in the European Alps at this scale. In view of the unusual and fast developing threat, which the 2002 Lago Effimero posed to the populated downvalley areas, the Italian National Department for Civil Defence (Dipartimento di Protezione Civile) took over by decree the responsibility at beginning of July 2002. In such a case, command over the involved parts of the armed forces, fire departments, police departments, regional and local authorities, etc. is transferred by law to the operation leader within the Protezione Civile (Legge 225, 1992). As a consequence, the actions necessary in July 2002 could be performed with high efficiency and with a considerable effort in personnel and material (Figure 13.26). Emergency actions included continuous monitoring of the level at Lago Effimero, surveying of the moraine stability carried out twice a day by alpine guides and volunteers of the Macugnaga Alpine Rescue, continuous observations with video cameras at the lake and the glacier rivers, an automatic alarm system, evacuation of certain parts of the village of Macugnaga during critical time periods, and the installation of a pump system at the lake (Mortara and Tamburini, 2009).

After these high-level emergency actions and substantial hazard reduction due to the decrease in lake volume, the responsibility was transferred back to the technical and civil defense authorities of the Regione Piemonte in mid-July 2002. Over winter 2002/2003 mitigation plans were elaborated for the case of a new formation of Lago Effimero in 2003. Planned measures included pumps mounted on a cable crossing the glacier and artificial outlet channels through ice and moraines. Due to uncertainties in the lake development, and in technical and financial issues it was decided to prepare but not to perform emergency actions in advance and to put into service again the monitoring and alarm system if needed. During the lake outburst in mid-June 2003, warning and alarm thresholds for river discharge and changes in lake level were set up, and additional monitoring personnel placed at critical points. The local authorities prepared evacuations. The Belvedere chairlift and endangered trails were, as already in 2002, temporarily closed.

The drastic loss in ice cover on the Monte Rosa east face and the associated mechanical and thermal adjustments of ice and rock continued to express themselves in several ways, including substantial rock-ice avalanches in August 2005, April 2007, and September 2010 into the now empty Lago Effimero basin (Fischer et al. 2013; Huggel et al., 2010; Mortara and Tamburini, 2009). Between 2001 and 2005, the surface elevation of Belvedere

Glacier first increased by several meters per year, and then decreased over most of the tongue except near the terminus (Mortara and Tamburini, 2009). Ice speeds during 2006, 2007, and 2008 above Belvedere were up to 30−35 m/a, i.e., similar to speeds before the surge-like instability. The massive ice thickness loss at and around the location of Lago Effimero triggered, presumably through debuttressing, instability of the steep right-internal moraine flank between Lago delle Locce and Belvedere Glacier/Lago Effimero, which produced a landslide spreading over a horizontal surface area of roughly 5,000 m^2 and moving toward Belvedere Glacier/Lago Effimero. Since 2003, Lago Effimero is visible on satellite imagery but never exceeds its September 2001 and August 2003 extent of a few thousand square meters.

13.6 SURGING GLACIERS AND THE TRANS ALASKA PIPELINE SYSTEM: POTENTIAL HAZARDS AND MONITORING

On its way from the northern coast of Alaska to the marine terminal in southern Alaska the Trans Alaska Pipeline System (TAPS) crosses three mountain ranges, numerous rivers and climate zones that range from the sparse northern tundra with continuous permafrost (cf. Figure 13.14 in Haeberli and Whiteman, 2014) to a lush temperate marine climate (Figure 13.27). With this large variety of terrain comes a multitude of potential natural hazards, some of

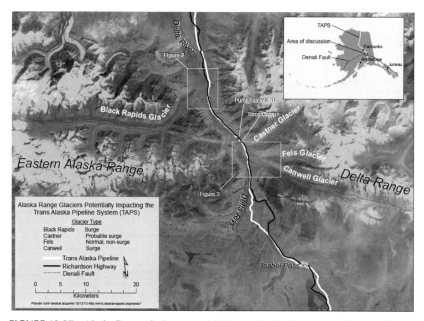

FIGURE 13.27   Alaska Range glaciers potentially impacting the Trans Alaska Pipeline System.

which are related to glaciers. TAPS comes to close proximity of glaciers as it crosses the Eastern Alaska Range. With this proximity comes a range of potential hazards including direct impact from advancing glaciers, glacier-dammed lakes, and glacial outburst floods. The Alaska Range poses a particularly interesting case, because several of its glaciers are known to be of the surge-type.

Here, we provide a summary of assessments concerning potential hazards to TAPS posed by surging glaciers and Alyeska Pipeline Service Company's (APSC) monitoring strategy related to surging glaciers.

13.6.1 Glacier Descriptions

The Delta River crosses the Alaska Range from south to north and separates the Eastern Alaska Range (or Hayes Range) from the Delta Range (Figure 13.27). The TAPS takes advantage of this natural corridor to cross this major topographic hurdle. Approximately 20 km north of its crossing of Isabel Pass, TAPS passes within a few kilometers downstream of four glaciers, at least one of which is known to have surged. Located at approximately 63.5° latitude in a cold and dry continental climate, these glaciers are polythermal with large temperate parts and some thin layer of cold, near-surface ice in the upper ablation area (Harrison et al., 1975). The area has been characterized by ice loss for at least the past half century (e.g., Arendt et al., 2002) and none of the four glaciers considered here have shown signs of advance in at least 50 years.

The 40 km-long Black Rapids Glacier (BRG) is the biggest and best-studied glacier of the area. It is the only glacier there with a recorded surge in 1936/1937 (Hance, 1937). This was perhaps the first surge that caught widespread public attention, at least in the United States, and the glacier has since been known as the "galloping glacier." Evidence for several prior surges exists in a multitude of surge moraines in the glacier forefield (Reger et al., 1993). Cutoff looped moraines on the glacier document at least the two past surges and indicate a surge period of several decades, which is thought to be typical of the Alaska Range. Like many of the surge-type Alaska Range glaciers, BRG directly overlies the Denali Fault, a major tectonic boundary that runs along the entire length of the range.

BRG studies include point mass balance and surface elevations and velocities since the early 1970s (Heinrichs et al., 1996), studies of basal processes through geophysical methods (radar and active seismicity; Gades et al., 2012; Nolan and Echelmeyer, 1999b), and, directly through borehole drilling and sampling (Harrison et al., 2004; Truffer et al., 1999), and studies of glacier hydrology (Raymond and Nolan, 2000; Raymond et al., 1995).

The 2002 M7.9 Denali Fault earthquake triggered several large rockslides that covered more than $10 \, km^2$ of the lower glacier (Shugar et al., 2012; Truffer et al., 2002). The protective effect of the rock cover is clearly visible, and the rock-protected ice stands in excess of 10 m above bare ice and leads to water pooling in the summer (Shugar et al., 2012). The landsat image showing BRG in Figure 13.27 was acquired prior to the Denali Fault earthquake; therefore, 2002 landslides are not visible. The glacier has several marginal lakes along the northern side of its ablation area, which are known to drain catastrophically and lead to temporary increases in ice velocities (Cochran, 1995; Nolan and Echelmeyer, 1999a).

Canwell Glacier overlies the eastern continuation of the Denali Fault across the Delta River from Black Rapids. The presence of looped moraines indicates a possible surge history, although no surges have been observed. It is possible that the observed moraine loop is a consequence of pulsing flow, rather than a full surge. The upper Canwell Glacier was also affected by a rockslide on the upper glacier caused by the 2002 Denali Fault earthquake.

Castner Glacier is located north of Canwell. Wilbur (1988) classified it as probable surge-type based on topographical and geometrical evidence, but no surges have been observed.

Fels Glacier is located between Canwell and Castner Glaciers and was confluent with Canwell Glacier during the oldest Holocene advance. Its current appearance and overall morphology do not suggest that it is of surge-type. Raymond et al. (1995) studied the hydrology of Fels and BRG with the goal of finding differences between surge-type and nonsurge-type glacier hydraulics, but all the recorded differences could be attributed to the different glacier size.

13.6.2 Glacier Impacts on TAPS

13.6.2.1 Direct Effects of Glacial Advance on TAPS

BRG would need to advance several kilometers and cross the Delta River to make direct contact with TAPS. This has happened twice in the past 3,400 years, last about 1,700 years ago (Reger et al., 1993; Figure 13.28). Under current climate the glacier does not appear to build up to surge geometry (Heinrichs et al., 1995; Truffer et al., 2005), and the great amount of ice wastage at low elevation appears to preclude a vigorous surge with a large advance. This is also evidenced by other recent surges in the Alaska Range, where the surge front has not reached the terminus. A potential caveat is given by the effects of the landslides that protect ice from melting. Some preliminary model results suggest that this will eventually lead to a modest glacier advance, but not one strong enough to reach the Delta River. However, the scenario of advancing debris-protected ice that is then activated by a surge has not been explored.

The Castner and Canwell Glaciers have not surged in historic times but advances to the TAPS corridor have occurred during the Holocene. Currently a large and persistent reversal in climate trends would be necessary to generate a sufficiently large advance for direct impact.

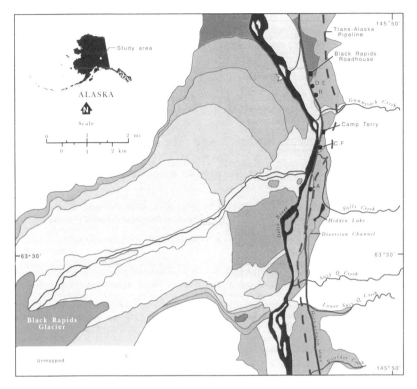

EXPLANATION

FIGURE 13.28   Geologic map of the terminal area of Black Rapids Glacier (BRG). *(From Reger et al. (1993)).* Used with permission.

13.6.2.2 Indirect Effects of Glacial Advance on TAPS

Seismic effects: Ground vibration was reported during the 1936–1937 surge of the BRG (Giddings, 1988). The owner of Black Rapids Roadhouse at the Richardson Highway, which runs parallel to TAPS, thought she was detecting earthquakes for a month before she realized the source was coming from the glacier (Moffit, 1939). While these tremors must have been sufficiently strong to be felt and to have caused concern, it is unlikely that they would have exceeded those of the 2002 M7.9 Denali Fault Earthquake, which occurred directly underneath the pipeline and resulted in relatively minor damage with no oil spill (Ebarhart-Phillips et al., 2003).

Drainage change: All glaciers in the area have significant outwash plains that are associated with changing drainage paths and unconsolidated sediments. BRG is unlikely to pose significant immediate hazards as such changes are mostly restricted to the west side of the Delta River. Even the oldest mapped Holocene outwash alluvium is located well below the TAPS corridor.

Potential hazards from drainage change are higher for the three glaciers to the east of TAPS (Figure 13.29). If Canwell or Castner Glaciers were to readvance, a potential exists for adverse impacts through changes in outflow channels. Reactivation of currently abandoned drainage channels may lead to potential flood hazards to the pipeline. If Miller Creek, the currently active channel draining Canwell Glacier, were to shift to its adjacent abandoned channel, Lower Miller Creek, which drains Fels Glacier, the potential for a significantly increased flow in Lower Miller Creek also exists. This could lead to bank and/or bed erosion requiring increased maintenance to protect the pipeline. Another abandoned outwash channel, Chrome Creek, is associated with Canwell glacier. It was probably formed as an outlet of a glacier-dammed lake. Its reactivation would require a substantial advance of Canwell Glacier, which is unlikely under current climatic conditions.

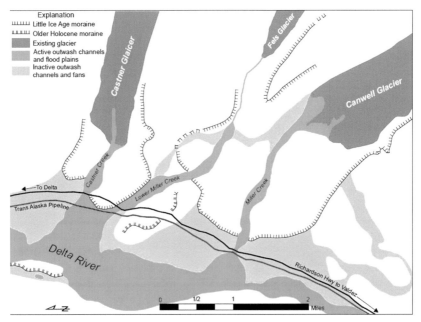

FIGURE 13.29   Glacial deposits and drainage features of the Castner, Fels, and Canwell Glaciers.

Glacier-dammed lake formation: Black Rapids and Castner glaciers both advanced across the Delta River and created large, glacier-dammed lakes in the geologic past (Péwé and Reger, 1983). Holocene BRG advances led to the formation of lakes on at least two occasions between 1,700 and 3,400 years ago (Reger et al., 1993). If such a lake formed at present, it would have the potential for inundating TAPS but not its nearby pump station. A surge of BRG is currently not expected, and even if one were to occur, it is highly unlikely that it would advance across the Delta River.

The proglacial morphology of Castner Glacier does indicate a damming of the Delta River in the past. The extent and timing of the lake are unknown. A sufficiently large advance of the glacier to dam the river is not likely under current climate.

Outburst floods: Glacial outburst floods originate from trapped englacial, supraglacial, or ice-marginal water. Such floods routinely occur on many glaciers, but surge-type glaciers are known to release large volumes of water at surge termination or even during the surge itself (for example, Kamb et al., 1985). These floods can occur during winter. It is difficult to estimate maximum flood discharge, because the potential volume of englacially trapped water cannot be reliably estimated. Outburst floods pose particular hazards when they engage unconsolidated debris, which can have large erosive power.

BRG has several marginal lakes on the northern side of the upper ablation area. These lakes drain annually, but do not affect the pipeline corridor. A flood was also associated with the surge termination in 1937, but neither the size nor the nature of the event is known. If an outburst flood were to occur as a result of the Castner or Canwell glaciers surging, increased flow resulting in more erosion may impact TAPS. Inactive outwash channels could be reactivated and increased flow could increase scour on pipeline supports. Lake-outburst floods from Castner and Canwell Glaciers have been known to cause major road damage on the Richardson Highway but have not had an impact on the pipeline.

13.6.3 APSC's Monitoring Strategy

APSC conducts weekly helicopter surveillance of the TAPS corridor and ground surveillance as needed. During weekly surveillance, APSC staff observe stream flow and any unusual activities, such as ice or debris flows toward the pipeline corridor. In late spring and early fall glaciers are monitored in greater detail for signs of terminus advance, surge development, landslides, levels of glacier-dammed lakes, and changes in channel patterns.

Since operation of the pipeline in 1977, APSC has been using aerial imagery of glaciers of concern to monitor changes in their movement. These images are taken at a frequency of 5 years or less. In more recent years, APSC began using orthorectified aerial photography and high-resolution digital elevation models generated from LiDAR data to monitor glacier termini and drainage patterns with greater accuracy and precision. Starting in 2014, APSC will integrate satellite imagery into its monitoring routine in addition to the current protocol utilizing fixed wing imagery and LiDAR elevation data.

13.6.4 Conclusion

APSC has planned the routing of the pipeline to minimize potential glacier hazards, particularly with regards to the well-known surge-type BRG. Potential hazards do remain, and APSC is regularly monitoring nearby glaciers. The highest damage potential comes from outburst floods, debris flows, and possible changes in drainage channels.

The current regime of widespread glacier recession in this area of the Alaska Range generally serves to reduce the potential glacier hazards, but with the possibility of glacier-dammed lakes forming the existing monitoring at APSC is essential in order to detect potential hazards and mitigate risks to the pipeline. Also, effects of the 2002 landslides on BRG have not been fully assessed; however, it is highly unlikely that the landslides will create a change in glacier behavior large enough to damage the pipeline. Finally, should glaciers in the area revert to an extended period of positive mass balance, many of the risks outlined here will need to be reassessed.

ACKNOWLEDGMENTS

The authors thank Hester Jiskoot and Luke Copland for their useful comments.

REFERENCES

Arendt, A., Echelmeyer, K.A., Harrison, W.D., Lingle, C.S., Valentine, V., 2002. Rapid wastage of Alaska glaciers and their contribution to rising sea level. Science 297 (5580), 382−386.

Barla, G., Dutto, F., Mortara, G., 2000. Brenva glacier rock avalanche of 18 January 1997 on the Mont Blanc range, NW Italy. Landslide News 13, 2−5.

Björnsson, H., 1998. Hydrological characteristics of the drainage system beneath a surging glacier. Nature 395 (6704), 771−774.

Björnsson, H., Pálsson, F., Sigurðsson, O., Flowers, G.E., 2003. Surges of glaciers in Iceland. Ann. Glaciol. 36, 82−90.

Benn, D.I., Wiseman, S., Warren, C.R., 2000. Rapid growth of a supraglacial lake, Ngozumpa glacier, Khumbu Himal, Nepal. In: Nakawo, M., Raymond, C.F., Fountain, A. (Eds.), Debris-Covered Glaciers, vol. 265. IAHS publications, pp. 177−185.

Boulton, G.S., Hindmarsh, R.C.A., 1987. Sediment deformation beneath glaciers: rheology and geologic consequences. J. Geophys. Res. 92 (B9), 9059−9082.

Bruce, R.H., Cabrera, G.A., Leiva, J.C., Lenzano, L.E., 1987. The 1985 surge and ice dam of Glaciar Grande del Nevado del Plomo, Argentina. J. Glaciol. 113 (33), 131−132.

Budd, W.F., 1975. A first simple model for periodically self-surging glaciers. J. Glaciol. 14 (70), 3−21.

Burgess, E.W., Larsen, C.F., Forster, R.R., 2013. Summer melt regulates winter glacier flow speeds throughout Alaska. Geophys. Res. Lett. 40, 6160−6164 http://dx.doi.org/10.002/2013GL058228.

Chikita, K., Jha, J., Yamada, T., 1999. Hydrodynamics of a supraglacial lake and its effect on the basin expansion: Tsho Rolpa, Rolwaling valley, Nepal Himalaya. Arct. Antarct. Alp. Res. 31, 58−70.

Clague, J.J., Mathews, W.H., 1973. The magnitude of jökulhlaups. J. Glaciol. 12/66, 501−504.

Clarke, G.K.C., 1987. Fast glacier flow: ice streams, surging, and tidewater glaciers. J. Geophys. Res. 92 (B9), 8835−8841.

Cobos, D.R., Boninsegna, J.A., 1983. Fluctuations of Some Glaciers in the Upper Atuel River Basin, Mendoza, Argentina. IANIGLA-CONICET. Quaternary of South America and Antarctic Peninsula (1), 61−82.

Copland, L., Sylvestre, T., Bishop, M.P., Shroder, J.F., Seong, Y.B., Owen, L.A., Bush, A., Kamp, U., 2011. Expanded and recently increased glacier surging in the Karakoram. J. Arct. Antarct. Alp. Res. 43 (4), 503−516.

Cochran, O., 1995. The Subglacial Hydraulics of the Surge-Type Black Rapids Glacier, Alaska: A schematic model (M.S. thesis). University of Alaska Fairbanks.

Corte, A.E., Espizua, L.E., 1981. Inventario de Glaciares de la Cuenca del Rio Mendoza. IANIGLA-CONICET, Mendoza, Argentina, 62 pp, 17 maps.

Cuffey, K.M., Paterson, W.S.B., 2010. The Physics of Glaciers, Fourth ed. Elsevier. Amsterdam, etc.

Davies, M.C.R., Hamza, O., Harris, C., 2001. The effect of rise in mean annual temperature on the stability of rock slopes containing ice-filled discontinuities. Permafrost Periglac. Process. 12 (1), 137−144.

Deline, P., 2001. Recent Brenva rock avalanches (Valley of Aosta): new chapter in an old story? Supplementi Geogr. Fis. Din. Quatern. 5, 5−63.

Deline, P., Gruber, S., Delaloye, R., Fischer, L., Geertsema, M., Giardino, M., Hasler, A., Kirkbride, M., Krautblatter, M., Magnin, F., McColl, S., Ravanel, L., Schoeneich, P., 2014. Ice Loss and Slope Stability in High-mountain Regions. Elsevier, pp. 303−344.

Desinov, L.V., Kotlyakov, V.M., Osipova, G.B., Tsvetkov, D.G., 2001. Medvezhiy glacier again commemorated itself. Data Glaciol. Stud. 91, 249−253 (in Russian).

Dolgushin, L.D., Osipova, G.B., 1971. New data on the recent glacier surges. Data Glaciol. Stud. 18, 191−218 (in Russian).

Dolgushin, L.D., Osipova, G.B., 1973. The regime of a surging glacier between advances. In: International Symposium on the Role of Snow and Ice in Hydrology. Symposium on Measurements and Forecasting. IAHS publication No. 107, pp. 212−217.

Dolgushin, L.D., Osipova, G.B., 1975. Glacier Surges and the Problem of Their Forecast. In: Symposium on Snow and Ice in Polar Regions, IUGG, IAHS Commission of Snow and Ice, XVth General Assembly, Moscow, 1971, vol. 104. IAHS Pub, 292−304.

Dolgushin, L.D., Osipova, G.B., 1978. Balance of a surging glacier as a basis for forecasting its periodic advances. Mater. Glyatsiologicheskikh Issled. Khronica Obsuzhdeniya 32, 260−265.

Dolgushin, L.D., Osipova, G.B., 1982. Surging Glaciers. Gidrometeoizdat, Leningrad, 192 pp (in Russian).

Dolgushin, L.D., Evteev, S.A., Krenke, A.N., Rototaev, K.P., Svatkov, N.M., 1964. Periodic rapid glacier movement and the recent advancement of Medvezhiy glacier in the Pamirs. Izv. Acad. Nauk USSR Ser. Geogr. 5, 30−39 (in Russian).

Ebarhart-Phillips, D., Haeussler, P.J., Freymueller, J.T., Frankel, A.D., Rubin, C.M., Craw, P.A., Ratchkovski, N.A., Anderson, G., Crone, A.J., Dawson, T.E., Fletcher, H., Hansen, R., Harp, E.L., Harris, R.A., Hill, D.P., Hreinsdottir, S., Jibson, R.W., Jones, L.M., Keefer, D.K., Larsen, C.F., Moran, S.C., Personius, S.F., Plafker, G., Sherrod, B., Sieh, K., Wallace, W.K.,

2003. The 2002 denali fault earthquake, Alaska: a large magnitude, slip-partitioned event. Science, 1113—1118. May 2003.

Eisen, O., Harrison, W.D., Raymond, C.F., Echelmeyer, K.A., Bender, G.A., Gorda, J.L.D., 2005. Variegated Glacier, Alaska, USA: a century of surges. J. Glaciol. 51 (174), 399—406.

Espizua, L.E., 1986. Fluctuations of the Rio del Plomo glaciers. Geogr. Ann. 68A (4), 317—327.

Espizua, L.E., Bengochea, J.D., 1990. Surge of Grande del Nevado glacier (Mendoza, Argentina) in 1984: its evolution through satellite images. Geogr. Ann. 72A (3—4), 255—259.

Espizua, L.E., Pitte, P., Hidalgo, L., 2008. Horcones inferior glacier surge. fluctuations of glaciers 2000—2005. In: Haeberli, W., Zemp, M., Kääb, A., Paul, F., Hoelzle, M. (Eds.), World Glacier Monitoring Service, Vol IX. ICSU (FAGS) — IUGG (IACS) — UNEP — UNESCO — WMO, Zürich, pp. 43—44.

Ferri Hidalgo, L., Espizua, L.E., 2010. A new surge event of Grande del Nevado Glacier, Mendoza, Argentina. In: International Glaciological Conference. Ice and Climate Change: A view from the South. Valdivia, Chile, 37.

Fischer, L., Eisenbeiss, H., Kääb, A., Huggel, C., Haeberli, W., 2011. Monitoring topographic changes in a periglacial high-mountain face using high-resolution DTMs, Monte Rosa East Face, Italian Alps. Permafrost Periglac. Process. 22 (2), 140—152.

Fischer, L., Huggel, C., Kääb, A., Haeberli, W., 2013. Slope failures and erosion rates on a glacierized high-mountain face under climatic changes. Earth Surf. Process. Land. 38 (8), 836—846.

Fischer, L., Kääb, A., Huggel, C., Noetzli, J., 2006. Geology, glacier retreat and permafrost degradation as controlling factors of slope instabilities in a high-mountain rock wall: the Monte Rosa east face. Nat. Hazards Earth Syst. Sci. 6 (5), 761—772.

Fischer, U.H., Clarke, G.K.C., 1994. Plowing of subglacial sediment. J. Glaciol. 40 (134), 97—106.

Frappé, T., Clarke, G.K.C., 2007. Slow surge of Trapridge glacier, Yukon territory, Canada. J. Geophys. Res. 112, F03S32. http://dx.doi.org/10.1029/2006JF000607.

Gades, A.M., Raymond, C.F., Conway, H., 2012. Radio-echo probing of Black Rapids Glacier, Alaska, USA, during onset of melting and spring speed-up. J. Glaciol. 58 (210), 713—724. http://dx.doi.org/10.3189/2012JoG11J145.

Giani, G.P., Silvano, S., Zanon, G., 2001. Avalanche of 18 January 1997 on brenva glacier, Mont Blanc Group, western Italian alps: an unusual process of formation. Ann. Glaciol. 32, 333—338.

Giddings, J.L., 1988. Thunder from below — 1937 University of Alaska Black Rapids expeditions. J. Northern Sci. 2, 33—39.

Haeberli, W., 1983. Frequency and characteristics of glacier floods in the Swiss Alps. Ann. Glaciol. 4, 85—90.

Haeberli, W., Epifani, F., 1986. Mapping the distribution of buried glacier ice - an example from Lago Delle Locce, Monte Rosa, Italian Alps. Ann. Glaciol. 8, 78—81.

Haeberli, W., Whiteman, C., 2014. Snow and Ice-related Hazards, Risks and Disasters — a General Framework. Elsevier, pp. 437—485.

Haeberli, W., Huggel, C., Kääb, A., Polkvoj, A., Zotikov, I., Osokin, N., 2003. Permafrost conditions in the starting zone of the Kolka-Karmadon rock/ice slide of 20th September 2002 in North Osetia (Russian Caucasus). In: Extended Abstracts, Eighth International Conference on Permafrost, pp. 49—50.

Haeberli, W., Kääb, A., Paul, F., Chiarle, M., Mortara, G., Mazza, A., Richardson, S., 2002. A surge-type movement at Ghiacciaio del Belvedere and a developing slope instability in the east face of Monte Rosa, Macugnaga, Italian Alps. Norw. J. Geogr. 56 (2), 104—111.

Hance, J.H., 1937. The recent advance of black rapids glacier, Alaska. J. Geol. 45 (7), 775—783.

Happoldt, H., Scrott, L., 1993. Horcones inferior glacier surge. (Abstract). Fluctuations of glaciers 1985-1990. In: Haeberli, W., Hoelzle, M. (Eds.), World Glacier Monitoring Service, Vol VI. IAHS (ICSI) — UNEP — UNESCO, Zürich, p. 70.

Harrison, W.D., Post, A.S., 2003. How much do we really know about glacier surging? Ann. Glaciol. 36, 1−6.

Harrison, W.D., Echelmeyer, K.A., Chacho, E.F., Raymond, C.F., Benedict, R.J., 1994. The 1987-88 surge of West Fork glacier, Susitna Basin, Alaska. U.S.A. J. Glaciol. 40 (135), 241−254.

Harrison, W.D., Eisen, O., Fahnestock, M.A., Moran, M.T., Motyka, R.J., Nolan, M., Raymond, C.F., 2008. Another surge of Variegated Glacier, Alaska, USA, 2004/04. J. Glaciol. 54 (184), 192−194.

Harrison, W.D., Mayo, L.R., Trabant, D.C., 1975. Temperature measurements on Black Rapids Glacier, Alaska, 1973. Clim. Arct., 9−11.

Harrison, W.D., Truffer, M., Echelmeyer, K.A., Pomraning, D.A., Abnett, K.A., Ruhkick, R.H., 2004. Probing the till beneath Black Rapids Glacier, Alaska, USA. J. Glaciol. 50 (171), 608−614. http://dx.doi.org/10.3189/172756504781829693.

Heinrichs, T.A., Mayo, L.R., Echelmeyer, K.A., Harrison, W.D., 1996. Quiescent-phase evolution of a surge-type glacier: Black Rapids Glacier, Alaska, U.S.A. J. Glaciol. 42 (140), 110−122.

Heinrichs, T.A., Mayo, L.R., March, R.S., Trabant, D.C., 1995. Observations of surge-type Black Rapids Glacier, Alaska, during a quiescent period, 1972-92. U.S. Geol. Surv. Open File Rep., 94−512.

Helbling, R., 1919. Beiträge zur Topographischen Erschliessung der Cordilleren der Anden zwischen Aconcagua und Tupungato. XXIII Jahresbericht des Akademischen Alpenklubs, Zürich, 1−8.

Helbling, R., 1935. The origin of the Rio del Plomo Ice-Dam. Geogr. J. 85 (1), 41−49.

Helbling, R., 1940. Ausbruch eines Gletschersees in den argentininschen Anden und ausserge-wöhnliche Gletscherschwankungen im Allgemeinen. Schweiz. Bauztg. 115, 121−128.

Huggel, C., Salzmann, N., Allen, S., Caplan-Auerbach, J., Fischer, L., Haeberli, W., Larsen, C., Schneider, D., Wessels, R., 2010. Recent and future warm extreme events and high-mountain slope stability. Philos. Trans. R. Soc. A Math. Phys. Eng. Sci. 368 (1919), 2435−2459.

Huggel, C., Zgraggen-Oswald, S., Haeberli, W., Kääb, A., Polkvoj, A., Galushkin, I., Evans, S.G., 2005. The 2002 rock/ice avalanche at Kolka/Karmadon, Russian Caucasus: assessment of extraordinary avalanche formation and mobility, and application of QuickBird satellite im-agery. Nat. Hazards Earth Syst. Sci. 5 (2), 173−187.

Humphrey, N.F., Raymond, C.F., 1994. Hydrology, erosion and sediment production in a surging glacier: Variegated Glacier, Alaska, 1982-83. J. Glaciol. 40 (136), 539−552.

Iverson, N.R., Hooyer, T.S., Baker, R.W., 1998. Ring-shear studies of till deformation: Coulomb-plastic behavior and distributed strain on glacier beds. J. Glaciol. 44 (148), 634−642.

Jiskoot, H., 2011. Glacier surging. In: Singh, V.P., Singh, P., Haritashya, U.K. (Eds.), Encyclopedia of Snow, Ice and Glaciers, 1300 pp. Encyclopedia of Earth Sciences Series. Springer, pp. 415−428.

Jiskoot, H., Murray, T., Boyle, P., 2000. Controls on the distribution of surge-type glaciers in Svalbard. J. Glaciol. 46 (154), 412−422.

Kääb, A., 2005. Remote Sensing of Mountain Glaciers and Permafrost, Chapter 8.5. Series in Physical Geography, 48, 266 pages. University of Zurich.

Kääb, A., Haeberli, W., 2001. Evolution of a high-mountain thermokarst lake in the Swiss Alps. Arct. Antarct. Alp. Res. 33 (4), 385−390.

Kääb, A., Huggel, C., Barbero, S., Chiarle, M., Cordola, M., Epifani, F., Haeberli, W., Mortara, G., Semino, P., Tamburini, A., Viazzo, G., 2004. Glacier hazards at Belvedere Glacier and the Monte Rosa east face, Italian Alps: processes and mitigation. Interpraevent 1, 67−78.

Kääb, A., Reynolds, J.M., Haeberli, W., 2005. Glacier and permafrost hazards in high mountains. In: Huber, U.M., Reasoner, M.A., Bugmann, B. (Eds.), Global Change and Mountain Regions: A State of Knowledge Overview. Advances in Global Change Research. Kluwer Academic Publishers, Dordrecht.

Kääb, A., Wessels, R., Haeberli, W., Huggel, C., Kargel, J., Khalsa, S.J.S., 2003. Rapid ASTER imaging facilitates timely assessment of glacier hazards and disasters. EOS Trans. Am. Geophys. Union 84 (13), 117,121.

Kamb, B., 1987. Glacier surge mechanism based on linked cavity configuration of the basal water conduit system. J. Geophys. Res. 92 (B9), 9083−9100.

Kamb, B., Raymond, C.F., Harrison, W.D., Engelhardt, H., Echelmeyer, K.A., Humphrey, N., Brugman, M.M., Pfeffer, T., 1985. Glacier surge mechanism: 1982-83 surge of Variegated Glacier. Science 227 (4686), 469−479.

King, W.D., 1935. El aluvión del Rio Mendoza de Enero de 1934. In: Conferencia dada en el Centro Nacional de Ingenieros, 26 de Septiembre de 1934. La Ingeniería, Buenos Aires, pp. 309−400.

Kotlyakov, V.M., Desinov, L.V., 2012. Medvezhiy Glacier surge in 2011. Ice Snow 1 (117), 129−132 (in Russian).

Kotlyakov, V.M., Desinov, L.V., Osipova, G.B., Hauser, M., Tsvetkov, D.G., Schneider, J.F., 2003. Events in 2002 on Russian Geographical Society glacier (RGO), pamirs. Data Glaciol. Stud. 95, 221−230 (in Russian).

Kotlyakov, V.M., Nosenko, G.A., Osipova, G.B., Homidov, A.Sh, Tsvetkov, D.G., 2008a. Space monitoring of the Geographical Society glacier surge in Pamirs. Data Glaciol. Stud. 105, 145−148 (in Russian).

Kotlyakov, V.M., Osipova, G.B., Tsvetkov, D.G., 1997. Fluctuations of unstable glaciers: scale and character. Ann. Glaciol. 24, 338−343.

Kotlyakov, V.M., Osipova, G.B., Tsvetkov, D.G., 2008b. Monitoring surging glaciers of the Pamirs, central Asia, from space. Ann. Glaciol. 48, 125−134.

Legge 225, 1992. Legge 24 febbraio 1992, n. 225. Istituzione del Servizio Nazionale della Protezione Civile.

Lingle, C.S., Fatland, D.R., 2003. Does englacial water storage drive temperate glacier surges? Ann. Glaciol. 36, 14−20.

Lliboutry, L., 1956. Nieves y glaciares de Chile. Fundamentos de Glaciología: Ediciones de la Universidad de Chile. Santiago, Chile. 471 pp.

Lliboutry, L., 1958. Studies of the shrinkage after a sudden advance, blue bands and wave ogives on Glaciar Universidad (Central Chilean Andes). J. Glaciol. 3 (24), 262−270.

Mazza, A., 1998. Evolution and dynamics of Ghiacciaio Nord delle Locce (Valle Anzasca,Western Alps) from 1854 to the present. Geogr. Fis. Din. Quatern. 21, 233−243.

Mazza, A., 2003. The kinematics wave theory: a possible application to "Ghiacciaio del Belvedere" (Valle Anzasca, Italian Alps). Preliminary hypothesis. Terra Glacialis 6, 23−36.

Meier, M.F., Post, A., 1969. What are glacier surges? Can. J. Earth Sci. 6, 807−817.

Moffit, F.H., 1939. Geology of the Gerstle River District, Alaska, with a Report on the Black Rapids Glacier, 936-B. United States Geological Survey Bulletin.

Mortara, G., Tamburini, A., 2009. Il Ghiacciaio del Belvedere e l'emergenza del Lago Effimero. Societa Meteorologica Subalpina, Bussolena (Italia), 192 pp.

Murray, Dowdeswell, J.A., Drewry, D.J., Frearson, I., 1998. Geometric evolution and ice dynamics during a surge of Bakaninbreen, Svalbard. J. Glaciol. 44 (147), 263−272.

Murray, T., Strozzi, T., Luckman, A., Jiskoot, H., Christakos, P., 2003. Is there a single surge mechanism? contrasts in dynamics between glacier surges in Svalbard and other regions. J. Geophys. Res. 108 (B5), 2237. http://dx.doi.org/10.1029/2002JB001906.

Murray, T., Stuart, G.W., Miller, P.J., Woodward, J., Smith, A.M., Porter, P.R., Jiskoot, H., 2000. Glacier surge propagation by thermal evolution at the bed. J. Geophys. Res. 135 (B6), 13491−13507.

Noetzli, J., Hoelzle, M., Haeberli, W., 2003. Mountain permafrost and recent Alpine rock-fall events: a GIS-based approach to determine critical factors. In: Proceedings, 8th International Conference on Permafrost, Zurich. Balkema, pp. 827−832.

Nolan, M., Echelmeyer, K.A., 1999a. Seismic detection of transient changes beneath Black Rapids Glacier, Alaska, U.S.A.: I. Techniques and observations. J. Glaciol. 45 (149), 119−131.

Nolan, M., Echelmeyer, K., 1999b. Seismic detection of transient changes beneath Black Rapids Glacier, Alaska, U.S.A.: II. Basal morphology and processes. J. Glaciol. 45 (149), 132−146.

Osipova, G.B., Tsvetkov, D.G., 1991. Kinematics of the Surface of a Surging Glacier (Comparison of the Medvezhiy and Variegated Glaciers). IAHS Publication 208, 345−357.

Osipova, G.B., Tsvetkov, D.G., Schetinnikov, A.S., Rudak, M.S., 1998. Inventory of surging glaciers of the Pamirs. Data Glaciol. Stud. 105, 3−136 (in Russian).

Paterson, W.S.B., 1994. The Physics of Glaciers, third ed. Elsevier, Oxford, etc.

Péwé, T.L., Reger, R.D., 1983. Delta river area, Alaska range. In: Péwé, T.L., Reger, R.D. (Eds.), Guidebook to Permafrost and Quaternary Geology along the Richardson and Glenn Highways between Fairbanks and Anchorage, Alaska-Guidebook 1, Fourth International Conference on Permafrost: Alaska Division of Geological and Geophysical Surveys, pp. 47−135.

Pitte, P., Ferri Hidalgo, L., Espizua, L.E., 2009. Aplicación de sensores remotos al estudio de glaciares en el Cerro Aconcagua. In: Anais XIV Simpósio Brasileiro de Sensoriamento Remoto. Natal, Brasil, pp. 1473−1480.

Porter, P.R., Murray, T., Dowdeswell, J.A., 1997. Sediment deformation and basal dynamics beneath a glacier surge front: Bakaninbreen, Svalbard. Ann. Glaciol. 24, 21−26.

Post, A., 1969. Distribution of surging glaciers in western North America. J. Glaciol. 8 (53), 229−240.

Post, A., 1972. Periodic surge origin of folded medial moraines on Bering piedmont glacier, Alaska. J. Glaciol. 11 (62), 219−226.

Prieto, M. del R., 1986. The glacier dam on the Rio Plomo: a cyclic phenomenon? Z. Gletscherkd. Glazialgeol. Innsbruck 22 (H. 1), 73−78.

Raymond, C.F., 1987. How do glaciers surge? A review. J. Geophys. Res. 92 (B9), 9121−9134.

Raymond, C.F., Nolan, M., 2000. Drainage of a Glacial Lake through an Ice Spillway. IAHS PUBLICATION (264), 199−210.

Raymond, C.F., Benedict, R.J., Harrison, W.D., Echelmeyer, K.A., Sturm, M., 1995. Hydrological discharges and motion of fels and black rapids glaciers, Alaska, U.S.A.: implications for the structure of their drainage systems. J. Glaciol. 41 (138), 290−304.

Raymond, C., Jóhannesson, T., Pfeffer, T., Sharp, M., 1987. Propagation of a glacier surge into stagnant ice. J. Geophys. Res. 92 (B9), 9037−9049.

Reger, R.D., Sturmann, A.G., Beget, J.E., 1993. Dating holocene moraines of black rapids glacier, Delta river Valley, central Alaska range. In: Solie, D.N., Tannian, Fran (Eds.), Short Notes on Alaskan Geology 1993: Alaska Division of Geological & Geophysical Surveys Professional Report 113F, pp. 51−59.

Reichert, F., 1929. La exploración de la Alta Cordillera de Mendoza. Círculo Militar. Biblioteca Oficial. 91 rev, Buenos Aires, Argentina, 401 pp.

Reynolds, J.M., 2000. On the formation of supraglacial lakes on debris-covered glaciers. In: Nakawo, M., Raymond, C.F., Fountain, A. (Eds.), Debris-Covered Glaciers. IAHS publications, 264, pp. 153−161.

Röthlisberger, H., 1972. Water pressure in intra- and subglacial channels. J. Glaciol. 11 (62), 177–203.

Shreve, R.I., 1972. Movement of water in glaciers. J. Glaciol. 11 (62), 205–214.

Shugar, D.H., Rabus, B.T., Clague, J.J., Capps, D.M., 2012. The response of Black Rapids Glacier, Alaska, to the Denali earthquake rock avalanches. J. Geophys. Res. 117, F01006. http://dx.doi.org/10.1029/2011JF002011.

Tamburini, A., Mortara, G., Belotti, M., Federici, P., 2003. The emergency caused by the "Short-lived Lake" of the Belevdere Glacier in the summer 2002 (Macugnaga, Monte Rosa, Italy). Studies, survey techniques and main results. Terra Glacialis 6, 37–54.

Truffer, M., Craw, P., Trabant, D.C., March, R.S., 2002. Effects of the M7.9 Denali Fault earthquake on glaciers in the Alaska Range. EOS Trans. AGU 83 (47), S72F–S1334.

Truffer, M., Harrison, W.D., Echelmeyer, K., 2000. Glacier motion dominated by processes deep in underlying till. J. Glaciol. 46 (153), 213–221.

Truffer, M., Harrison, W.D., March, R.S., 2005. Record negative glacier balances and low velocities during the 2004 heatwave in Alaska, USA: implications for the interpretations of observations by Zwally and others in Greenland. Correspondence. J. Glaciol. 51 (175), 663–664. http://dx.doi.org/10.3189/172756505781829016.

Truffer, M., Motyka, R.J., Harrison, W.D., Echelmeyer, K.A., Fisk, B., Tulaczyk, S., 1999. Subglacial drilling at Black Rapids Glacier, Alaska, U.S.A.: drilling method and sample descriptions. J. Glaciol. 45 (151), 495–505.

Tulaczyk, S., Kamb, W.B., Engelhardt, H.F., 2000. Basal mechanics of ice stream B, west Antarctica: I. Till mechanics. J. Geophys. Res. 105 (B1), 463–481.

Unger, C., Espizua, L.E., Bottero, R., 2000. Untersuchung von Gletscherständen im Tal des Río Mendoza (Zentralargentinishe Anden) – Kartierung eines surge-vorstosses des Horcones Inferior. Z. Gletscherkd. Glazialgeol. 36, 151–157.

VAW, 1985. Studi sul comportamento del Ghicacciaio del Belvedere, Macugnaga, Italia. Relazione, Versuchsanstalt für Wasserbau. Hydrol. Glaziol. der ETH Zürich 97 (3), 157.

Walder, J.S., Costa, J.E., 1996. Outburst floods from glacier-dammed lakes: the effect of mode of lake drainage on flood magnitude. Earth Surf. Process. Land. 21, 701–723.

Walder, J.S., Fowler, A., 1994. Channelized subglacial drainage over a deformable bed. J. Glaciol. 40 (134), 3–15.

Wilbur, S.W., 1988. Surging versus Non-Surging Glaciers: A Comparison using Morphometry and Balance (M.Sc. thesis). University of Alaska Fairbanks.

Williams Jr., R.S., Ferrigno, J.G., 1998. Satellite image Atlas of glaciers of the world. South America. In: Williams Jr., R.S., Ferrigno, J.G. (Eds.), Unites States geological survey professional Paper 1386 – I. Washington.

Chapter 14

Glacier-Related Outburst Floods

John J. Clague [1] and Jim E. O'Connor [2]

[1] *Centre for Natural Hazard Research, Simon Fraser University, Burnaby, B C, Canada,*
[2] *U.S. Geological Survey, Oregon Water Science Center, Portland, Oregon, USA*

ABSTRACT

Water bodies impounded by glaciers and moraines can drain suddenly, with disastrous downstream consequences. Lakes can form at the margins of an alpine glacier or ice cap, on its surface, or at its base. Smaller pockets of water may also be present within some glaciers. In all cases, these ice-dammed water bodies might drain either via enlarging subglacial tunnels or by overtopping or mechanical collapse of the glacier dam. Most formerly stable glacier lakes failed over the past century, in many cases repeatedly, as glaciers downwasted and receded. The peak discharge, duration, and volume of an outburst flood resulting from subglacial tunnel enlargement depend mainly on the: (1) geometry and rate of development of the tunnel at the base of the glacier; and (2) size and geometry of the impounded water body. Discharge commonly increases exponentially during the outburst, but then decreases when either the drainage tunnel is plugged by collapse of the tunnel roof or closes due to plastic ice flow.

Most lakes dammed by lateral and end moraines formed in the twentieth century when valley and cirque glaciers retreated from advanced positions reached during the Little Ice Age. Moraine dams are susceptible to failure because they are steep and relatively narrow, they comprise loose poorly sorted sediment, and they may contain ice cores or interstitial ice. These dams generally fail by overtopping and incision. The triggering event might be a heavy rainstorm, strong winds, or an avalanche or landslide into the lake that generates waves that overtop the dam. Melting of moraine ice cores and piping are other possible failure mechanisms. Outflow from a moraine-dammed lake increases as the breach enlarges and then decreases as the level of the lake falls. The moraine breach may become armored, preventing further incision, or the hydraulic gradient at the breach may decrease to a point at which erosion ceases.

Outburst floods from glacier- and moraine-dammed lakes typically entrain, transport, and deposit large amounts of sediment. If the channel is steeper than about 6–9° and contains abundant loose sediment, the flood likely will transform into a debris flow. Such debris flows may be larger and more destructive than the flood from which they

Snow and Ice-Related Hazards, Risks, and Disasters. http://dx.doi.org/10.1016/B978-0-12-394849-6.00014-7
487

formed. A period of protracted atmospheric warming is required to trap lakes behind moraines and create conditions that lead to dam failure. The warming can also force glaciers to retreat, prompting ice avalanches, landslides, and jökulhlaups that have destroyed many moraine dams.

14.1 INTRODUCTION

About 10 percent of our planet is covered by ice sheets and glaciers, and about 99 percent of this ice is in Greenland and Antarctica. The other one percent forms ice fields, ice caps, and cirque, valley, and piedmont glaciers, mainly in mountains of northwest North America, Arctic Canada, and Asia.

Glaciers provide many benefits; for example, melt water from alpine glaciers augments run off during summer, which is important for agriculture, municipal water supply, and hydroelectric power generation. However, processes associated with glaciers can also be hazardous, and some of these processes may be amplified by climate change. This chapter explores hazards and risks associated with outburst floods caused by failures of lakes impounded by glaciers and their moraines. We review, in sequence, flood sources, dam failure mechanisms and flood magnitude, and downstream flood behavior. Next, we discuss outburst flood scenarios in a warming world. A final section deals with risk assessment and risk reduction. Our contribution builds on an earlier paper on these topics by Haeberli et al. (2010).

During the Pleistocene, outbursts from water bodies impounded by glaciers produced some of the largest floods on Earth (Figure 14.1; Bretz, 1923a,b; O'Connor et al., 2002; O'Connor and Costa, 2004; Herget, 2005; Baker, 2009). These floods inundated large tracts of land and produced long-lasting geomorphic effects. Outburst floods that occur in the modern world, although much smaller than those of the Pleistocene, have caused much damage, injury, and death (Lliboutry et al., 1977; Yesenov and Degovets, 1979; Hewitt, 1982; Haeberli, 1983; Eisbacher and Clague, 1984; Ruren and Deji, 1986; Vuichard and Zimmerman, 1987; Costa and Schuster, 1988; Haeberli et al., 1989; Reynolds, 1992; O'Connor and Costa, 1993; Evans and Clague, 1994; Watanabe and Rothacher, 1996; Cenderelli, 2000; Richardson and Reynolds, 2000; Cenderelli and Wohl, 2003; Vilimek et al., 2005; Bajracharya and Mool, 2009; ICIMOD, 2010, 2011; Narama et al., 2010). Most mountain ranges where these floods occur were formerly sparsely populated, but in recent times have experienced explosive population growth accompanied by increased tourism. The European Alps, for example, are presently home to about 14 million people, and about 120 million people visit the Alps every year. As a result, the risk from outburst floods and other natural hazards has greatly increased.

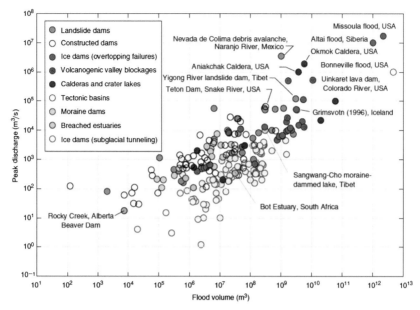

FIGURE 14.1 Outburst floods for which flood volume and peak discharge are known. The largest floods of each type are labeled. *Adapted from Figure 13.1 of O'Connor et al. (2013); data from Walder and Costa (1996); and O'Connor and Beebee (2009).*

14.2 FLOOD SOURCES

14.2.1 Glacier-Dammed Lakes

A glacier can impound a water body at its margins, on its surface, at its base, or within it (Figure 14.2). Some lakes at the margins of alpine glaciers contain more than 100 million m^3 of water, although most are smaller. These lakes are small compared to the more than 400 lakes that currently exist at the base of the East Antarctic Ice Sheet—the largest of these lakes is Lake Vostok, which holds about 5400 km^3 of water, more than three times the amount of water in Lake Ontario, one of the five Great Lakes in North America. Most lakes at the base of the Antarctic Ice Sheet, however, are stable, and in any case they do not pose a risk to people. In contrast, lakes impounded by alpine glaciers and ice caps are vulnerable to sudden emptying, and consequently are a threat to residents and infrastructure in valleys below.

Relations between valley-blocking ice and the surrounding landscape are complex, leading to a variety of unstable situations. Lakes can exist where glaciers block a tributary or trunk valley (Figures 14.3 and 14.4). They also can form during glacier retreat when formerly confluent glaciers separate, leaving space between the glaciers (Costa and Schuster, 1988; Tweed and Russell, 1999). Lakes of these types have been described in mountainous areas around

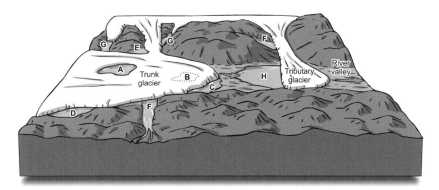

FIGURE 14.2   Schematic diagram showing locations where a water body can be trapped by a glacier: (A) supraglacial, (B) subglacial, (C) proglacial, (D) embayment in slope at glacier margin, (E) area of coalescence of two glaciers, (F) tributary valley adjacent to a trunk or tributary glacier, (G) same as F except that glaciers dam both ends of the lake, and (H) main valley adjacent to a tributary valley. Toned areas are land; unpatterned areas are ice. *Clague and Evans (1994).*

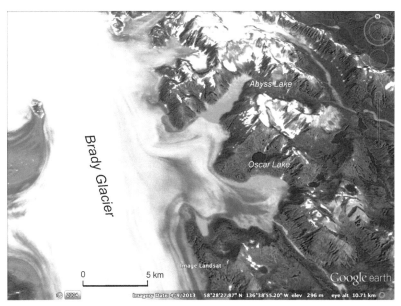

FIGURE 14.3   Lakes dammed by Brady Glacier in southeast Alaska. The largest of the lakes is Abyss Lake, which when full is over 250 m deep at the glacier dam. *Google Earth image.*

the world, including Alaska (Post and Mayo, 1971), Western Canada (Marcus, 1960; Clague and Mathews, 1973; Clague and Evans, 1994, 2000; Geertsema and Clague, 2005), Scandinavia (Liestol, 1956), Iceland (Roberts, 2005; Björnsson, 2009), Europe (Haeberli, 1983; Eisbacher and Clague, 1984; Haeberli et al., 1989, 2001; Kääb et al., 2004), South Asia (Hewitt, 1968, 1982; Ives, 1986; Ding and Liu, 1992; Richardson and Reynolds, 2000;

FIGURE 14.4   Glacial Lake Alsek formed repeatedly during the Little Ice Age when Lowell Glacier, a large valley glacier in the St. Elias Mountains in southwest Yukon Territory, advanced across the Alsek River valley and blocked the flow of the river. This figure depicts the lake in the mid-nineteenth century, when it reached to the present site of Haines Junction on the Alaska Highway, and earlier when the site of Haines Junction was inundated by water about 70 m deep. *Courtesy of Jeff Bond, Yukon Geological Survey.*

Xin et al., 2008; Hewitt and Liu, 2010), and South America (Lliboutry et al., 1977; Dussaillant et al., 2010).

Unstable glacial lakes also exist beneath ice caps in Iceland. The lakes are located in geothermal areas where melting of ice takes place continuously, ultimately leading to sudden draining and outburst floods known as "jökulhlaups" (Björnsson, 1974, 1992, 2002, 2009, 2010; Tómasson, 2002; Waitt, 2002; Alho et al., 2005; Roberts, 2005; Carrivick, 2007, 2009; Russell et al., 2010). Occasional volcanic eruptions add to the melting and thereby influence the outburst process. The 1918 Katla jökulhlaup had a peak discharge of about 300,000 m^3/s (Tómasson, 1996), and was one of the largest floods on Earth during the historic period. An eruption of Grímsvötn volcano beneath Vatnajökull in 1996 had a peak discharge of about 50,000 m^3/s and released 3.3 km^3 of water (Björnsson, 2002, 2009).

Small, but potentially damaging outburst floods and debris flows also result from the sudden draining of "water pockets" beneath and within cirque and valley glaciers (Richardson, 1968; Haeberli, 1983; Driedger and Fountain, 1989; Walder and Driedger, 1994). Haeberli (1983) documented more than 26 outburst floods from water pockets in glaciers in the Swiss Alps. Dozens of similar outburst floods have happened on Mount Rainier, Washington, in the twentieth century (Driedger and Fountain, 1989; Walder and Driedger, 1994). Some of the Mount Rainier floods happened during rainstorms, suggesting a possible trigger, but many occurred during dry warm periods. Driedger and Fountain (1989) inferred that the floods on Mount Rainier resulted from sudden emptying of one or a few water-filled cavities into lower pressure, subglacial drainage channels. Pressurized water-filled cavities can arise from glacier flow over steep or stepped beds (Haeberli, 1983; Driedger and Fountain, 1989), and such outbursts can happen without warning in warm weather on clear days, and pose a particular hazard to tourists visiting the termini of glaciers, for example in New Zealand and Europe.

Catastrophic outburst floods and debris flows occasionally occur without failure of the glacier dam. A recent example is the coupled flood and debris flow in the Cordillera Blanca of Peru in April 2010. An avalanche of rock and ice from the top of Hualcán Mountain northeast of the town of Carhuaz entered a glacial lake (Laguna 513). The avalanche displaced water from the lake, triggering a flood that transformed into a debris flow that caused considerable property damage up to 11 km downstream (Carey et al., 2012; Valderrama and Vilca, 2012; Schneider et al., 2014).

14.2.2 Moraine-Dammed Lakes

Water was trapped behind many Little Ice Age moraines when glaciers retreated in the late nineteenth and early twentieth centuries (Figure 14.5; Costa and Schuster, 1988; O'Connor and Costa, 1993; Clague and Evans, 1994, 2000). Moraine dams typically are steep-sided, consist of loose sediment, and are sparsely vegetated. As a consequence, they are potentially unstable and vulnerable to failure. Rapid incision of moraine dams may be caused by a large overflow triggered by an ice avalanche or a rockfall (Figure 14.6; Blown and Church, 1985; Costa and Schuster, 1988; Clague and Evans, 2000; Kershaw et al., 2005). Thawing of permafrost on steep rock slopes bordering these lakes may increase the likelihood of landslides and ice avalanches into both moraine- and ice-dammed lakes in the future (Haeberli, 2013). Breaching can also occur during periods of rapid snowmelt or intense rainfall when large amounts of water flow over the dam, initiating outlet incision. Other failure mechanisms include earthquakes, slow melt of buried ice (Reynolds, 1992), and removal of fine sediment from the dam by groundwater ("piping"). Outbursts from these lakes are sudden and rapid; they can produce impressive floods far from their sources. Maximum historical breakout volumes have approached 50 million m^3 with breach depths of up to 40 m (O'Connor and Beebee, 2009).

Outburst floods from moraine-dammed lakes were first documented comprehensively in the Cordillera Blanca of Peru, where the phenomenon is known by the name "aluvión" (Lliboutry et al., 1977). By far the most deadly

FIGURE 14.5   Moraine-dammed Nostetuko Lake in the southern Coast Mountains of British Columbia *(Photo taken in July 1977 by J.M. Ryder)*. The large moraine impounding the lake was built during the Little Ice Age by Cumberland Glacier, which has subsequently thinned and retreated. Compare with Figure 14.6, a Photograph taken after the outburst flood of July 19, 1983.

aluvión destroyed much of the city of Huaraz in 1941, killing about 5,000 people (Figure 14.7). Cenderelli (2000), Clague and Evans (2000), Richardson and Reynolds (2000), and Kattelmann (2003) provide recent summaries of floods from moraine-dammed lakes.

Because moraine-dammed lakes form in the wake of retreating glaciers, the relation of the hazard to the atmospheric warming over the past century is a subject of considerable interest (Lliboutry et al., 1977; Liu and Sharma, 1988; O'Connor and Costa, 1993; Clague and Evans, 1994; O'Connor et al., 2001; Kattelmann, 2003). Most moraine dams that have breached in recent decades formed during the Little Ice Age. Terminal and lateral moraines constructed during this period are up to 100 m high and thus are formidable barriers. During the twentieth century, valley glaciers thinned and retreated from their maximum Little Ice Age positions, allowing lakes with volumes up to 100 million m^3 and depths of nearly 100 m to form in the basins between the moraines and the retreating ice (Yamada and Sharma, 1993). Hundreds of these lakes are present in the mountains of Tibet (Liu and Sharma, 1988), Nepal and Bhutan (Yamada and Sharma, 1993; Richardson and Reynolds, 2000), Peru (Lliboutry et al., 1977; Reynolds, 1992), the European Alps (Haeberli, 1983), British Columbia (Clague and Evans, 1994; McKillop and

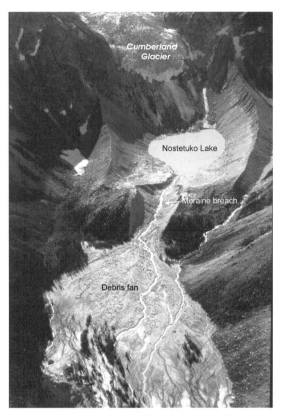

FIGURE 14.6 Breached Little Ice Age moraine of Cumberland Glacier and the remains of Nostetuko Lake in the southern Coast Mountains of British Columbia. About 6.5 million m^3 of water flowed out of Nostetuko Lake on July 19, 1983, after an ice avalanche from the toe of Cumberland Glacier entered the lake and produced a displacement wave that overtopped and incised the moraine. The level of the lake fell 28 m during the outburst. *Photograph by J.J. Clague.*

Clague, 2007a,b), and the Cascade Range of Oregon and Washington in the United States (O'Connor et al., 2001).

In some cases, a moraine dam breaches soon after the lake forms behind it (O'Connor et al., 2001), but most dams fail years or decades later. As a general rule, the likelihood of breaching increases as lakes increase in size and then decreases as glaciers continue to retreat, reducing the likelihood that ice avalanches will trigger overtopping and breaching (Clague and Evans, 2000). Moraine dams that are most likely to fail have already done so, thus there is a reduced population. For example, 8 of the 13 moraine-dammed lakes in the Oregon Cascade Range that formed between 1924 and 1956 partially or completely drained between 1934 and 1981. Only one breach occurred after 1981, indicating that breaches in the Oregon Cascades are becoming rarer with time. The incidence of moraine breaches in the Himalaya, however, has not

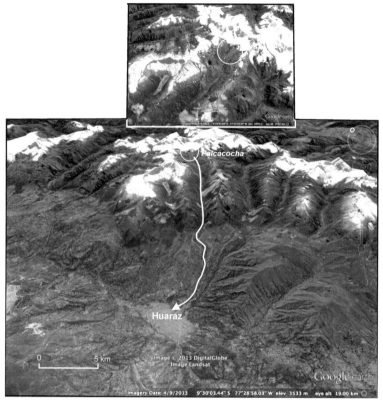

FIGURE 14.7   Path of the 1941 outburst flood (aluvión) that killed about 5,000 people in the city of Huaraz, Peru. An ice avalanche into Palcacocha (Lake Palca) (circled) in December 1941 triggered waves that overtopped and breached the moraine impounding the lake. The path of the resulting aluvión is shown by the white line. *Google Earth image.*

slowed; many lakes have continued to grow behind moraines and remain in contact with glacier ice (Liu and Sharma, 1988; Mool, 1995; Richardson and Reynolds, 2000; Kattelmann, 2003).

Not all moraine dams fail. The large number of existing lakes impounded by Pleistocene moraines attests to long-term stability of some moraine dams. In addition, many Little Ice Age moraine dams have persisted for more than a century without breaching. McKillop and Clague (2007a) found that only 10 of 175 lakes in their inventory of moraine-dammed lakes in southern British Columbia had partly or completely drained. Armoring of outlet channels with boulders can stabilize a moraine dam, and some dams are stable because they are broad and have low-gradient outlet channels (Clague and Evans, 2000). Nevertheless, even a wide moraine dam can be breached if overtopped by a wave caused by a large landslide or ice avalanche (Kershaw et al., 2005). Moraine dams formed on stratovolcanoes are particularly susceptible to breaching because they are narrow and steep-sided, are situated on steep slopes, and consist of easily erodible volcanic rock debris. Over 60 percent of the moraine-dammed lakes on stratovolcanoes in Oregon have failed (O'Connor et al., 2001), far surpassing the 6 percent that have failed in the mostly crystalline rocks of the British Columbia Coast Mountains (McKillop and Clague, 2007a).

Some large outburst floods occur without breaching of the moraine. When a landslide or ice avalanche enters a lake, the displaced water may overtop the dam without breaching it. An example is the 2002 Laguna Safuna Alta outburst in the Cordillera Blanca, Peru (Hubbard et al., 2005). During that event, a large landslide fell onto Glaciar Pucajirca and then entered moraine-dammed Laguna Safuna Alta, creating a displacement wave that increased in height as it traveled down the lake. The wave was over 100 m high when it reached and overtopped the moraine dam. It did not, however, breach the moraine. The overtopping wave flowed into Laguna Safuna Baja, another moraine-dammed lake about 350 m downvalley from Laguna Safuna Alta. The moraine impounding the lower lake had about 15 m of freeboard, sufficient to completely contain the inflow of water from the upper lake.

Another example of an overtopping flood, but in this case accompanied by breaching, is the 1997 Queen Bess event in the British Columbia Coast Mountains (Kershaw et al., 2005). In July 1997, about 2 million m^3 of ice detached from the toe of Diadem Glacier located on a steep rock slope about 200 m above Queen Bess Lake. The ice plunged into the lake and produced a large surge wave over 30 m high that overtopped the moraine dam and ran down the valley below the lake (Figure 14.8). The overtopping wave initiated a breach that lowered the level of the lake by 8 m and produced a second flood peak. Floodwaters surged 20 km to Homathko River and thence west to tidewater at Bute Inlet. The flood attenuated as it moved downvalley, but it still generated a marked spike on a hydrograph 100 km from the source.

14.3 FAILURE MECHANISMS AND FLOOD MAGNITUDE

14.3.1 Glacier Dams

Glacier-dammed lakes drain by overtopping or flow through subglacial tunnels; the latter mechanism is the more common of the two. Initial outflow via a subglacial tunnel may be triggered by flotation of a part of the ice dam that is in contact with the lake, but flotation does not appear to be a requirement for the initiation of drainage (Fowler, 1999; Flowers et al., 2004; Ng and Liu, 2009). As water begins to flow along a subglacial channel, the channel walls enlarge by both thermal and mechanical erosion (Liestol, 1956; Nye, 1976; Clarke, 1982; Roberts, 2005; Björnsson, 2010). The channels can, however, narrow or close during or after the flood by ice deformation caused by glacier flow, especially if the ice is thick.

FIGURE 14.8   Queen Bess Lake and its breached Little Ice Age moraine. This photo was taken one day after the outburst flood, which happened on August 13, 1997. An ice avalanche from the toe of Diadem Glacier triggered a large displacement wave that overtopped a 600 m length of the end moraine by 15—25 m and breached it. The level of Queen Bess Lake fell 8 m during the breach phase of the outburst. *Photograph courtesy of Interfor Forest Products.*

Much of our understanding of glacial outburst floods has come from analyses of Icelandic jökulhlaups, beginning with Thórarinsson (1939) who recognized that a glacier dam might float due to hydrostatic stresses once the depth of the impounded water reaches a threshold value. Floating of the edge of the glacier dam may explain why few dams fail by overtopping (Björnsson, 1974, 1976; Walder and Costa, 1996), but other processes must operate to explain hydrographs of observed jökulhlaups. Glen (1954) and Liestol (1956) proposed that subglacial tunneling may be an important process, and Mathews (1973) inferred that tunnels enlarge by thermal erosion, with the energy supplied by the waters escaping from the impounded lake.

Nye (1976) first explained the physics of nonsteady state, glacier tunnel enlargement. Although his model has been refined (Spring and Hutter, 1981, 1982; Clarke, 1982, 2003), the basic principles remain the same: (1) both sensible and frictional heat derived from water flowing through a tunnel within or at the base of a glacier is transferred to the tunnel walls; (2) the tunnel enlarges, increasing the flow of water through it; and (3) more heat further enlarges the tunnel. This positive feedback process produces an exponential rise in water discharge (Clarke, 1982; Björnsson, 1992, 2010). Outbursts produced by this process generally develop more slowly than those caused by breaching of moraine and landslide dams involving similar impoundment volumes (Figure 14.1).

The tunnel enlargement process has solid theoretical underpinnings, but the initiation of outflow is not well understood. Outflow can begin at much lower lake levels than required for hydrostatic flotation, indicating that other factors, for example, englacial water pressure or the presence of a network of sub-glacial channels downstream of the dam, are important (Roberts, 2005; Björnsson, 2010). Similarly, the end of a jökulhlaup is not entirely predictable or understood. In some instances, a jökulhlaup ends when the water supply in the lake is exhausted (Clarke, 1982), but in many others ice deformation may seal the outlet before the lake completely empties (Nye, 1976; Ng and Björnsson, 2003; Roberts, 2005). Furthermore, in the case of thin glacier dams, the roof of a tunnel may collapse, ending outflow (Mathews, 1973; Sturm and Benson, 1985).

Some englacial or subglacial water bodies drain too rapidly to be the result of simple tunnel enlargement (Björnsson, 1977, 1992, 2009, 2010; Haeberli, 1983; Walder and Driedger, 1995; Walder and Fountain, 1997; Roberts, 2005). These floods are not well understood but may involve sudden rupture of pressurized water-filled englacial cavities (Walder and Fountain, 1997), fracturing of the base of the glacier due to high hydrostatic pressures (Glen, 1954; Fowler, 1999) or rapid changes in flow (Roberts, 2005; Björnsson, 2009). Some jökulhlaups, notably those resulting from subglacial eruptions, may involve sheet flow over a large area of the glacier bed rather than channelized flow (Björnsson et al. 2001; Björnsson, 2002; Johannesson, 2002; Flowers et al., 2004).

Some glacier-dammed lakes fail by overtopping, followed by incision and occasionally mechanical collapse of the dam (Haeberli, 1983; Walder and Costa, 1996; Harrison et al., 2014). These floods typically have much larger peak discharges than jökulhlaups resulting from thermal erosion of subglacial channels. Their character is similar to floods associated with failures of con-structed dams (Figure 14.1). The similarity is intuitive because peak discharge from an overtopped ice dam approximates critical flow through a breach that is enlarging by mechanical erosion. Historical examples of this type of flood are the 1986 and 2002 Russell Lake, Alaska, outbursts. The earlier of the two floods had a peak flow of about 10,000 m^3/s (Mayo, 1989). Lake Alsek in Yukon Territory (Figure 14.4) also may have drained in this way (Clague and Rampton, 1982).

A relation has been noted between the size of glacier-dammed lakes and the peak discharge of the floods they produce—the largest lakes generally produce the largest floods (Figure 14.1; Clague and Mathews, 1973). This observation has led researchers to develop empirical regression equations based on variables such as impoundment volume and impoundment depth to estimate peak discharge from glacier-dam failures (Clague and Mathews, 1973; Haeberli, 1983; Evans, 1986; Costa and Schuster, 1988; Walder and Costa, 1996; Walder and O'Connor, 1997; Cenderelli, 2000). However, the regression equations can only provide rough estimates of peak discharge because glacier dams differ markedly in length, thickness, and other characteristics, because outflow is strongly influenced by hydraulic factors and breach erosion rates, and because the lake may only partially drain (Walder and O'Connor, 1997; Ng and Björnsson, 2003). Another complication is the difficulty of accurately estimating peak discharges of the floods on which these regression equations are based. Flood discharges at the breach are rarely known; typically discharges used to derive empirical equations are estimated at different distances downstream of the breach and are underestimates of peak discharge.

14.3.2 Moraine Dams

Once a lake forms behind a moraine, several processes facilitate breaching of the dam. First, a hydraulic gradient is established across the dam, promoting groundwater flow through the sediments forming the moraine. Second, seepage of groundwater from the downstream face of the moraine may increase the likelihood of mass movements that destabilize the dam (Massey et al., 2010). Third, retrograde erosion of sapping channels in permeable sediment may incise the barrier and trigger overflow. Fourth, a rapid inflow of water or the generation of high waves by wind, landslides, or ice avalanches may increase water flow through the outlet channel and initiate incision.

Most outbursts from moraine-dammed lakes, like those from landslide-dammed lakes and artificial reservoirs, result from overtopping and breaching of the barriers. Outflow from the lake increases once a stream begins to incise a moraine dam. The increased outflow further incises the channel, leading to increased outflow in a self-enhancing process. The breach commonly grows through lake-ward migration of one or more knick points on the downstream face of the dam. At the same time, the breach widens by slumping and sliding from the steep banks of the deepening outlet channel (Lee and Duncan, 1975; Plaza-Nieto and Zevallos, 1994; Dwivedi et al., 2000). The breach continues to enlarge until either the level of the lake drops to a critical level with an associated reduction in the hydraulic gradient or until the outlet channel becomes sufficiently armored to halt erosion. In some instances, incision ceases when the channel is lowered onto bedrock. Once the outlet stabilizes, outflow continues at a diminishing rate until the lake drops to the level of the stable outlet.

Peak discharge at the dam during a breach event is controlled by the interplay between erosion and enlargement of the outflow channel and drawdown of the impounded lake (Figure 14.9). The important factors are: (1) the speed at which the breach grows; (2) the final depth of the breach; and (3) downstream water and sediment interactions that affect the volume, peak discharge, and type of flow. The rate of breach growth depends on the critical flow, or velocity, at some cross-section within the breach. Critical flow is defined as $v = (gd)^{0.5}$, where v is critical flow velocity, g is gravitational

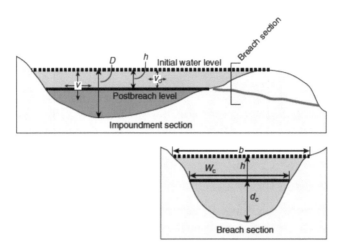

FIGURE 14.9 Definition diagram for impoundment and breach geometry. V is the total impoundment volume; V_o is the volume released during the breach event; D is the maximum impoundment depth; h is the change in impoundment level during the breach event and is approximately equal to the maximum possible specific energy of flow through the breach; b is the breach width at the maximum impoundment elevation; W_c is the breach width at flow stage d_c associated with critical flow $(V = (gd)^{0.5})$ through the breach section with specific energy h. *O'Connor et al. (2013).*

acceleration, and d is flow depth. The specific energy of the flow at any point is the sum of the flow depth, d, and $0.5(v^2/g)$.

Parameterized and physical models used to estimate outflow and peak discharges during breaching of natural and constructed dams are based on coupling between breach development and the drawdown of the impounded water body (Ponce and Tsivoglou, 1981; Fread, 1987; Froehlich, 1987, 1995, 2008; Singh et al., 1998; Webby and Jennings, 1994; Webby, 1996; Walder and O'Connor, 1997; Manville, 2001; Marche et al., 2006; Worni et al., 2012). The models assume critical outflow, and breach growth is parameterized using either shape and time functions (e.g., the DAMBRK model of Fread, 1988), or physically based sediment transport and mass movement rules (e.g., the BREACH model of Fread, 1987; the BEED model of Singh et al., 1998; and the ERODE model of Marche et al., 2006). The modeler iteratively fits parameters until the measured and calculated hydrographs are in agreement, although the large number of parameters rarely permits a unique fit.

Recent experimental studies have provided new insights into earthen dam failure that are pertinent to failure of moraine dams (Coleman et al., 2002; Hanson and Cook, 2005; Morris et al., 2007; Zhang et al., 2009; Pickert et al., 2011). The studies show that parameterized erosion laws and assumptions about critical flow poorly replicate breach evolution in observed events. As a result, several investigators (e.g., Wang and Bowles, 2006; Faeh, 2007; Wu et al., 2009) have developed models that differ from the parametric models mentioned above. The new models assume that embankment breaching can be replicated by calculating fluid flow along an erodible channel, where only a small part of the flow passes through the embankment. From this perspective, there is little difference between dam-breach modeling and flood routing.

When considering breach formation, it is useful to discriminate between large and small impoundments (Walder and O'Connor, 1997). In the case of a "large impoundment," as the term is used by Walder and O'Connor (1997), the breach develops fully before the level of the lake drops significantly, either because the lake is large relative to the breach or because the breach develops rapidly. In either case, peak flow is approximately equivalent to the critical flow through the fully developed breach. If we assume that the breach cross-section is about as wide as it is deep, $Q \approx g^{0.5}h^{2.5}$, where Q is the peak outflow discharge, g is gravitational acceleration, and h is the height difference between the level of the impounded water body and the bottom of the breach at the point of outflow. If the breach is significantly wider than it is deep, peak discharge will be larger by a factor of approximately b/h, where b is the breach width. The strong dependence of peak discharge on breach depth (h) explains why deep lakes can produce such large outflows. In the case of "small impoundments," the surface of the lake falls significantly as the breach develops and thus the evolving breach geometry controls peak discharge. The peak in discharge generally occurs before the breach fully develops and is dependent largely on the vertical erosion rate.

Final breach depth is an important factor for both peak flow and the total volume of water discharged (Webby, 1996). Escaping waters may erode only partway through a moraine dam, to the base of the dam, or, in rare cases, even deeper. Most breaches, however, extend only part way through the dam (O'Connor and Beebee, 2009). The character of the dam materials is important in controlling the depth of breaching. The interplay between outflow rates and breach deepening, widening, and armoring will dictate the rate of breach growth and the ultimate geometry of the breach (Clague and Evans, 1992). These aspects make it difficult to accurately predict outburst flood magnitude because: (1) materials forming moraines are heterogeneous; (2) a variety of processes contribute to breach enlargement (Walder and O'Connor, 1997); and (3) predictions are sensitive to the bed-load transport formulas that are used to parameterize models (Cencetti et al., 2006).

14.4 DOWNSTREAM FLOOD BEHAVIOR

Outburst floods from glacier- and moraine-dammed lakes entrain, transport, and deposit huge amounts of sediment over distances up to tens of kilometers from the flood source (Figure 14.10). Erosion and deposition by the flood-waters can themselves significantly affect flow behavior. Outburst floods commonly broaden floodplains, destroy preflood channels, and alter river planform. The changes can persist for years or decades after the flood,

FIGURE 14.10   Valley of a tributary of the Nostetuko River about 1 year after the outburst flood from moraine-dammed Queen Bess Lake, British Columbia, in August 1997. The flood removed forest from the valley floor and left a sheet of gravel and boulders in its wake. Arrows show direction of flow. *Photograph by John J. Clague.*

although rivers quickly reestablish their preflood gradients by incising flood deposits.

Many outburst floods from moraine-dammed lakes change into debris flows by entrainment of sediment from the breached moraine and by incorporation of downstream bank and bed materials (Figure 14.11; Lliboutry et al., 1977; Eisbacher and Clague, 1984; Clague et al., 1985; Clague and Evans, 1994; Gallino and Pierson, 1985; Schuster, 2000; Huggel et al., 2004; McKillop and Clague, 2007b; Procter et al., 2010). Such transformations also have occurred during glacier outburst floods, particularly on volcanoes (Walder and Driedger, 1994, 1995). Downvalley sediment bulking can increase peak discharge by an order of magnitude (Figure 14.12). An extreme example is the 1985 Dig Tsho outburst in Nepal, where about 4 million m^3 of water redeposited 3.3 million m^3 of sediment over a distance of 40 km downstream from the dam (Vuichard and Zimmerman, 1987). Analysis of the flow transformation process is difficult because there are few observations of flow rheology. Nevertheless, it is clear that steep channel slopes ($>6-9°$) are required to sustain debris flows and hyperconcentrated flows (Haeberli, 1983; Clague and Evans, 1994; O'Connor et al., 2001; McKillop and Clague, 2007b; Procter et al., 2010). Most flow bulking happens within the first 10 km of the breach, because entrainment requires steep channels that are most common in watershed headwaters where many moraine dams are located.

FIGURE 14.11   An outburst flood from moraine-dammed Klattasine Lake in the British Columbia Coast Mountains transformed into a debris flow that traveled 8 km to the fan into the foreground of this photo. Up to 20 m of debris accumulated on the fan, temporarily stemming the flow of the river flowing toward the bottom of the photo. The event occurred sometime between 1971 and 1973. *Photograph by John J. Clague.*

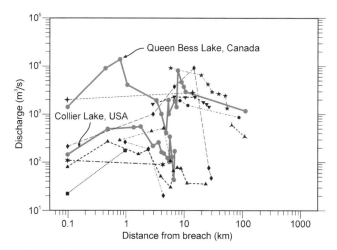

FIGURE 14.12   Plot of peak discharge vs distance from the breach for outburst floods from moraine-dammed lakes. Note the rapid increase in peak discharge, followed by rapid flow attenuation. *Adapted from O'Connor et al. (2013); data mainly from O'Connor et al. (2001).*

Debris flows or hyperconcentrated flows will deposit sediment in unconfined reaches of a valley or where the gradient of the valley floor decreases to less than $\sim 9°$ (Clague and Evans, 2000). Entrainment and deposition of sediment and woody plant debris by floodwaters have important implications for hazard appraisal, because debris flows are more destructive than floods of the same size.

Downstream attenuation of outburst floods controls the distribution and size of the resulting flood features. In general, a given flood will attenuate more rapidly if there is substantial downstream space for temporary storage of floodwaters or where the valley gradient is low. Floods of small volume and short duration attenuate more rapidly than floods of large volume and long duration. The latter do not attenuate as rapidly because the first floodwaters fill available space in the valley prior to the passage of the peak.

The characteristics of flood-produced landforms and the particle size of the associated sediments depend on the flow type and size, flow attenuation rates, and channel morphology and materials (O'Connor, 1993; Cenderelli and Wohl, 1998, 2003; Kershaw et al., 2005). Landform size scales with cross-sectional flow geometry, whereas maximum clast size scales with flow strength. Assessments of the relation between flood landforms and flow are based on measures of flow strength, notably shear stress and stream power. Both the caliber of entrained sediment and the distribution of erosional features are governed by the distribution of shear stresses and stream power magnitude (Baker, 1973; O'Connor, 1993; Benito, 1997; Cenderelli and Wohl, 2003; Carrivick, 2009; Denlinger and O'Connell, 2010). Shear stress, τ, is the tangential force applied to the bed per unit area—for hydrostatic conditions $\tau = \rho g d_f S$, where ρ is the fluid density, d_f is the flow depth, and S is the local flow energy gradient, which for steady and uniform flow corresponds to channel slope. Unit stream power, ω (Bagnold, 1966), can be expressed as $\omega = \rho g d_f v S = \tau v$, where v is the flow velocity. These formulas are indices of geomorphic work, although only a portion of available mechanical energy is expended as geomorphic work, with the rest dissipated in other forms of energy. The formulas do not consider vertical and horizontal accelerations that can produce stresses of the same magnitude as the hydrostatic forces (Iverson, 2006).

Outburst floods generate shear stresses and stream powers greatly exceeding those of "normal" meteorological floods because of their large flow depths and locally high energy gradients (Baker and Costa, 1987). Cenderelli and Wohl (2003) concluded that floods generated by breaching of moraine dams produce stream power values several times greater than those from the largest meteorological floods affecting the same reaches. Montgomery et al. (2004) suggested that stream power values up to 5×10^6 W/m^2 might have been attained during Holocene floods from breached glacier dams in the Tsangpo River gorge, Tibet. These values are two orders of magnitude higher than those reconstructed from discharge measurements of flash floods in steep basins (Baker and Costa, 1987).

FIGURE 14.13   Large boulder transported in the hyperconcentrated phase of the outburst flood from Queen Bess Lake (British Columbia) in August 1997. *Photograph by John J. Clague.*

Outburst floods commonly transport huge boulders, some more than 10 m across (Figure 14.13). Theory and observations (Baker and Ritter, 1975; Costa, 1983; Williams, 1983; Komar, 1987; O'Connor, 1993) demonstrate that entrainment and tractive transport of boulders increase with local flow strength. Empirical analyses by Costa (1983) and Williams (1983) indicate that a boulder, 1 m in diameter, can be moved tractively when local stream power reaches about $500 \ W/m^2$. Although subject to many uncertainties (Jacobson et al., 2003), such flow competence relations have been used in many paleohydrologic analyses of megafloods (Lord and Kehew, 1987; Waythomas et al., 1996; Manville et al., 1999; Hodgson and Nairn, 2005; Carrivick, 2009). It is unclear, however, whether the commonly used flow competence relations apply in the case of hyperconcentrated flows, in which large clasts are transported on a mobile basal carpet of sediment. Such flows may be responsible for many extremely coarse outburst flood deposits.

14.5 OUTBURST FLOODS AND CLIMATE CHANGE

Glaciers around the world repeatedly grew and shrank in response to climate change during the Holocene. Most alpine glaciers achieved their maximum Holocene size during the latest stage of the Little Ice Age, in the eighteenth and nineteenth centuries (Grove, 1988). Since the end of the nineteenth century, alpine glaciers have thinned and receded in response to atmospheric warming, and today the area of ice cover in most mountain ranges is one-half to two-thirds of what it was in the middle of the nineteenth century.

Climate change affects the stability of glacier dams. Fewer glacier-dammed lakes drain when climate is stable than when climate changes markedly, causing glaciers to advance or retreat. During the Little Ice Age, new lakes formed where glaciers advanced across streams and blocked drainage. When these lakes first formed, the glacier dams may have been weak and many of the lakes perhaps drained repeatedly (Clague and Evans, 1994). As these glaciers continued to advance, both the dams and the impounded lakes stabilized. Most Little Ice Age lakes drained one or more times in the twentieth century when their dams weakened due to glacier thinning and retreat (Costa and Schuster, 1988; Clague and Evans, 1994; Bolch et al., 2012).

A "jökulhlaup cycle" can develop in a period when the atmosphere is warming. For an individual lake, the cycle is characterized by recurrent outburst floods separated by times when the lake refills (Evans and Clague, 1994). Outbursts may occur annually or on shorter or longer timescales depending on interactions between the blocking glacier and filling of the reservoir (Hulsing, 1981; Mathews and Clague, 1993; Depretris and Pasquini, 2000; Walder et al., 2006). Flood character changes with overall glacier conditions—outbursts commonly decrease in peak discharge and volume as the glacier thins and re-treats (Clague and Evans, 1994; Evans and Clague, 1994). Many former glacier-dammed lakes no longer exist because the glaciers that dammed them have thinned and retreated so much that they no longer impound water. However, lakes have formed in new locations at the margins of some receding glaciers, typically at higher elevations than former lakes, and pose new risks to down-valley areas (Figure 14.14; Geertsema and Clague, 2005).

A relation also exists between climate change and the stability of moraine dams. Most existing moraine-dammed lakes were formed in the twentieth century when glaciers retreated from bulky Little Ice Age end moraines. Soon thereafter, the lakes began to fail. If warming and glacier retreat continue, the supply of moraine-dammed lakes susceptible to failure in most mountain ranges will decrease and the threat they pose will diminish (Clague and Evans, 2000). An exception may be moraine-dammed lakes in the high mountains of South Asia. In parts of the Himalaya, for example, moraine-dammed lakes are still evolving due to downwasting and fragmentation of ice tongues behind moraines.

14.6 RISK ASSESSMENT AND REDUCTION

Assessment and mitigation of an outburst flood hazard requires knowledge of both the hazard itself and physical and socioeconomic vulnerability down-stream (ICIMOD, 2011). The potential for damage and loss of life can only be reduced, and resilience enhanced, after the possible outburst has been char-acterized and its impact reliably assessed.

Elements at risk from outburst floods include people and their property, infrastructure, systems such as tourism and trade that support people's liveli-hood, and environmental resources such as forest, pasture and grazing land, and fisheries. The capacity of people to cope with an outburst flood, as any

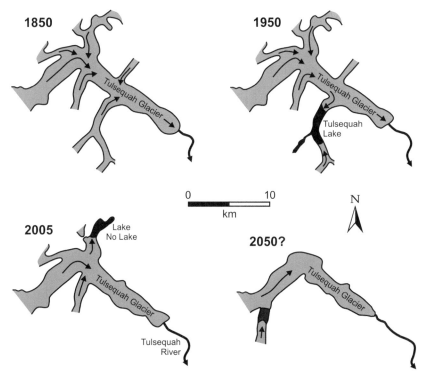

FIGURE 14.14   Evolution of Tulsequah Glacier (British Columbia) and its ice-marginal lakes from the end of the Little Ice Age to c. AD 2050. During the first half of the twentieth century, Tulsequah Lake produced large annual outburst floods (upper right). By the early years of the twenty-first century, Tulsequah Glacier had downwasted and receded, and Tulsequah Lake no longer held enough water to generate significant outburst floods. During the same period, however, Lake-No-Lake formed in a tributary valley about 5 km farther upvalley and is now the primary source of large outburst floods. If Tulsequah Glacier continues to recede, Lake-No-Lake itself will cease to produce outburst floods and the locus of outburst activity might shift farther up the glacier. *Geertsema and Clague (2005).*

disaster, depends largely on individual and community assets and access to information, technology, services, and institutions.

    An assessment of outburst flood hazard and risk requires: (1) mapping and classification of glacier- and moraine-dammed lakes using satellite imagery and aerial photographs; (2) field inspection of lakes that are hazardous; (3) assessment of likelihood and magnitude of an outburst flood, which includes mapping of downstream areas that would be impacted, commonly coupled with numerical modeling of scenario events; and (4) assessment of socio-economic and environmental vulnerability in the hazard zone (Hegglin and Huggel, 2008). In the case of high-risk lakes, it is advisable to monitor lakes to document changes that might signal an imminent outburst. Monitoring could involve repeat mapping of lakes using remote sensing, field monitoring of seepage, degradation of ice cores of moraines, slope instability in moraine dams, measurements of water inflow into lakes, and inspection of adjacent glaciers and slopes for signs of instability.

Much research has been done on identifying, mapping, characterizing, and monitoring glacier-dammed lakes using aerial photographs and satellite images (Buchroithner, 1996; Huggel et al., 2002, 2003; Iwata et al., 2002; Kääb et al., 2005; Quincey et al., 2005, 2007; Randhawa et al., 2005; McKillop and Clague, 2007a,b; Bolch et al., 2008, 2011; Fujita et al., 2008, 2009; Babu-Govinda-Raj, 2010; Mergili and Schneider, 2011; Wang et al., 2012a, 2012b; Mergili et al., 2013; Xie et al., 2013). The marked increase in the number and variety of active satellite sensors and improvements in image resolution now allow researchers to monitor hazardous lakes on a continuous basis.

In recent years, a major advance in predicting the downstream impacts of outburst floods and debris flows has been the use of coupling different physically based numerical models to simulate process cascades. Schneider et al. (2014), for example, used a physically based mass movement model (RAMMS) to model an ice avalanche into Laguna 513 in the Cordillera Blanca, Peru, in 2010, and a hydrodynamic model (IBER) to simulate the resulting impact wave that formed in the lake and overtopped the moraine dam. They tested the model results against an observational dataset for the event. Numerical models have been applied, for example to outbursts in the Indian Himalaya (Worni et al., 2013) and the Argentine Andes (Worni et al., 2012).

Another noteworthy development in outburst flood hazard assessment is the use of glacier hypsometry to predict subglacial closed basins where lakes will develop if valley and cirque glaciers continue to thin and retreat through the remainder of this century, as is expected (Paul et al., 2007; Frey et al., 2010; Linsbauer et al., 2012). These lakes could be sources of future outburst floods if impacted by landslides or ice avalanches from adjacent slopes.

High-risk situations also require preparedness and precautionary measures to minimize damage and loss of life should the lake empty. One element of preparedness is land-use planning and structural mitigation. Another, equally important element of preparedness is early warning—the provision of information on a potential threat in sufficient time to allow an appropriate response. Early warning systems include video cameras or water-level recorders that detect changes in lake level. Data are transmitted in real time to communities that might be impacted by outburst floods, and sirens or other devices provide warning of floods or debris flows. Early warning systems must be simple to operate, easy to maintain or replace, and reliable. The linked communication network must be capable of relaying the warning to the appropriate authorities and is effective only if placed in the hands of the local communities with appropriate response systems.

Early warning systems have met with mixed success. For example, an automatic early warning system was set up at and below Tsho Rolpa, a moraine-dammed lake in Nepal, in 1998 before mitigation work was undertaken to reduce the risk of an outburst (Reynolds, 1999). However, by 2002 the system was no longer operating, in part because local residents assumed the lake had been lowered to a safe level.

An important structural mitigation measure for reducing the chance of an outburst flood is to lower the level of the lake by pumping or siphoning water, or tunneling through the moraine barrier or under an ice dam (Lliboutry et al., 1977; Kääb et al., 2004; Vincent et al., 2010). The hazard can also be mitigated by constructing an outlet control structure, armoring the outlet of a moraine-dammed lake to prevent breaching, or through a controlled breaching of the moraine dam. Downstream infrastructure can be protected against floods, for example, by designing bridges with appropriate flow capacity at elevations higher than levels likely to be reached by an outburst flood. Land-use zoning should also be considered, as it reduces the number of structures at risk.

14.7 SUMMARY

Outburst floods from glacier- and moraine-dammed lakes can be large and dangerous, with effects ranging tens to hundreds of kilometers from the source. The largest floods have been caused by failures of glacier dams during the Pleistocene Epoch, but great floods also have occurred during the Holocene and future floods pose a serious threat to downvalley communities and infrastructure in many mountain ranges.

The peak discharge, duration, and volume of outburst floods are interrelated and depend on complex outburst mechanisms, the geometry and rate of development of breaches (moraines) and subglacial channels (glaciers), and on the size and geometry of the impounded water body. The largest floods derive from large water bodies. In the case of moraine dams, flows from large lakes through rapidly developing breaches have peak discharges that approximate critical flow through the final breach. Floods from small lakes have peak discharges that are strongly dependent on the breach rate.

The characteristics of outburst floods, and the deposits and landforms they produce, depend on the dam failure process, the rate and duration of flow out of the lake, and downstream interactions with sediment and the valley floor. Outbursts that incorporate large amounts of sediment, either from a moraine dam or from downstream channels, may evolve into debris flows. Some debris flows have peak discharges several times the peak outflow at the breach and substantially increase the flow volume. However, debris flows attenuate rapidly downstream by deposition, and they commonly come to rest where channel gradients fall below 9°. Jökulhlaups attenuate because of channel and valley storage. Large floods with relatively low peak discharges attenuate at a slower rate than small, high-discharge floods. Erosional and depositional features left by large floods reflect the depth and breadth of inundation and the large forces applicd by the deep, fast-moving floodwaters. Floods from natural dam failures can achieve local stream power values one to two orders of magnitude greater than those generated by the large meteorological floods.

Climate is an important determinant of the long-term stability of moraine and glacier dams. Most moraine-dammed lakes formed and drained in the past century as glaciers retreated from bulky end moraines constructed during the late Holocene. With continued warming and glacier retreat, the supply of moraine-dammed lakes that are susceptible to failure will decrease and the threat they pose will diminish. Over the past century, many glacier-dammed lakes have gone through a period of cyclic or sporadic outburst activity lasting up to several decades. The outburst floods from any one lake ended when the glacier dam weakened to the point that it could no longer trap water behind it. However, with continued glacier retreat, the locus of outburst activity, in some cases, shifted upglacier to sites where new lakes developed in areas that became newly ice-free.

ACKNOWLEDGMENTS

We thank Helgi Björnsson, Wilfried Haeberli, Glen Hess, Martin Mergili, and Colin Whiteman for helpful reviews that improved the quality of the manuscript.

REFERENCES

Alho, P., Russell, A.J., Carrivick, J.L., Kâyhkö, J., 2005. Reconstruction of the largest Holocene jökhulhlaup within Jökulsá á Fjöllum, NE Iceland. Quat. Sci. Rev. 24, 2319—2334.

Babu-Govindha-Raj, K., 2010. Remote sensing based hazard assessment of glacial lakes: a case study in Zanskar basin, Jammu and Kashmir, India. Geo. Nat. Hazards Risk 1, 339—347.

Bagnold, R.A., 1966. An Approach to the Sediment Transport Problem from General Physics. U.S. Geol. Surv. Prof. Pap. 422-I, 37 pp.

Bajracharya, S.R., Mool, P.K., 2009. Glaciers, glacial lakes and glacial lake outburst floods in the Mount Everest region, Nepal. Ann. Glaciol. 50, 81—86.

Baker, V.R., 1973. Paleohydrology and Sedimentology of Lake Missoula Flooding in Eastern Washington. Geol. Soc. Am. Spec. Pap. 144, 79 pp.

Baker, V.R., 2009. Overview of megaflooding: Earth and Mars. In: Burr, D.M., Carling, P.A., Baker, V.R. (Eds.), Megaflooding on Earth and Mars. Cambridge University Press, Cambridge, UK, pp. 1—12.

Baker, V.R., Costa, J.E., 1987. Floodpower. In: Mayer, L., Nash, D. (Eds.), Catastrophic Flooding. Allen and Unwin, Boston, MA, pp. 1—24.

Baker, V.R., Ritter, D.F., 1975. Competence of rivers to transport coarse bedload material. Geol. Soc. Am. Bull. 86, 975—978.

Benito, G., 1997. Energy expenditure and geomorphic work of the cataclysmic Missoula flooding in the Columbia River Gorge, USA. Earth Surf. Process. Land. 22, 457—472.

Björnsson, H., 1974. Explanation of jökulhlaups from Grimsvötn, Vatnajökull, Iceland. Jökull 24, 1—26.

Björnsson, H., 1976. Marginal and supraglacial lakes in Iceland. Jökull 26, 40—51.

Björnsson, H., 1977. The cause of jökulhlaups in the Skaftá river, Vatnajökull. Jökull 27, 71—78.

Björnsson, H., 1992. Jökulhlaups in Iceland: prediction, characteristics and simulation. Ann. Glaciol. 15, 95−106.

Björnsson, H., 2002. Subglacial lakes and jökulhlaups in Iceland. Global Planet. Change 35, 255−271.

Björnsson, H., 2009. Jökulhlaups in Iceland: sources, release and drainage. In: Burr, D.M., Carling, P.A., Baker, V.R. (Eds.), Megaflooding on Earth and Mars. Cambridge University Press, Cambridge, UK, pp. 50−64.

Björnsson, H., 2010. Understanding jökulhlaups: from tale to theory. J. Glaciol. 56, 1002−1010.

Björnsson, H., Rott, H., Gudmundsson, S., Fisher, A., Slegel, A., Gudmundsson, A.T., 2001. Glacier-volcano interactions deduced by SAR interferometry. J. Glaciol. 47, 58−70.

Blown, I., Church, M., 1985. Catastrophic lake drainage within the Homathko River basin, British Columbia. Can. Geotech. J. 22, 551−563.

Bolch, T., Buchroitner, M., Peters, J., Baessler, M., Bajarcharya, S., 2008. Identification of glacier motion and potentially dangerous glacial lakes in the Mt. Everest region/Nepal using space-borne imagery. Nat. Hazards Earth Syst. Sci. 8, 1329−1340.

Bolch, T., Kulkarni, A., Kääb, A., Huggel, C., Paul, F., Cogley, J.G., Frey, H., Kargel, J.S., Fujita, K., Scheel, M., Bajracharya, S., Stoffel, M., 2012. The state and fate of Himalayan glaciers. Science 336, 310−314.

Bolch, T., Peters, J., Yegrorov, A., Pradhan, B., Buchroithner, M., Blagoveshchensky, V., 2011. Identification of potentially dangerous glacial lakes in the northern Tien Shan. Nat. Hazards 59, 1691−1714.

Bretz, J.H., 1923a. The Channeled Scabland of the Columbia plateau. J. Geol. 31, 617−649.

Bretz, J.H., 1923b. Glacial drainage on the Columbia plateau. Geol. Soc. Am. Bull. 34, 573−608.

Buchroithner, M., 1996. Jökulhlaup mapping in the Himalaya by means of remote sensing. Kartographische Bausteine 12, 75−86.

Carey, M., Huggel, C., Bury, J., Portocarrero, C., Haeberli, W., 2012. An integrated socio-environmental framework for climate change adaptation and glacier hazard management: lessons from Lake 513, Cordillera Blanca, Peru. Clim. Change 112, 733−767.

Carrivick, J.L., 2007. Hydrodynamics and geomorphic work of jökulhlaups (glacial outburst floods) from Kverkfjöll volcano, Iceland. Hydrol. Process. 21, 725−740.

Carrivick, J.L., 2009. Jökulhlaups from Kverkfjöll volcano, Iceland: modelling transient hydraulic phenomena. In: Burr, D.M., Carling, P.A., Baker, V.R. (Eds.), Megaflooding on Earth and Mars. Cambridge University Press, Cambridge, UK, pp. 273−289.

Cencetti, C., Fredduzzi, A., Marchesini, I., Naccini, M., Tacconi, P., 2006. Some considerations about the simulation of breach channel erosion on landslide dams. Comput. Geosci. 10, 201−219.

Cenderelli, D.A., 2000. Floods from natural and artificial dam failures. In: Wohl, E.E. (Ed.), Inland Flood Hazards: Human, Riparian, and Aquatic Communities. Cambridge University Press, New York, pp. 73−103.

Cenderelli, D.A., Wohl, E.E., 1998. Sedimentology and clast orientation of deposits produced by glacial-lake outburst floods in the Mount Everest regions, Nepal. In: Kalvoda, J., Rosenfield, C.L. (Eds.), Geomorphological Hazards in High Mountain Areas. Kluwer Academic Publishers, Dordrecht, The Netherlands, pp. 1−26.

Cenderelli, D.A., Wohl, E.E., 2003. Flow hydraulics and geomorphic effects of glacial-lake outburst floods in the Mount Everest region, Nepal. Earth Surf. Process. Land. 28, 385−407.

Clague, J.J., Evans, S.G., 1992. A self-arresting moraine dam failure, St. Elias Mountains, British Columbia. In: Current Research, Part A. Geol. Surv/ Can. Pap. 92-1A, pp. 185−188.

Clague, J.J., Evans, S.G., 1994. Formation and Failure of Natural Dams in the Canadian Cordillera. Geol. Surv. Can. Bull. 464, 35 pp.

Clague, J.J., Evans, S.G., 2000. A review of catastrophic drainage of moraine-dammed lakes in British Columbia. Quat. Sci. Rev. 19, 1763−1783.

Clague, J.J., Mathews, W.H., 1973. The magnitude of jökulhlaups. J. Glaciol. 12, 501−504.

Clague, J.J., Rampton, V.N., 1982. Neoglacial Lake Alsek. Can. J. Earth Sci. 19, 94−117.

Clague, J.J., Evans, S.G., Blown, I., 1985. A debris flow triggered by the breaching of a moraine-dammed lake, Klattesine Creek, British Columbia. Can. J. Earth Sci. 22, 1492−1502.

Clarke, G.K.C., 1982. Glacier outburst floods from 'Hazard Lake', Yukon Territory, and the problem of flood magnitude prediction. J. Glaciol. 28, 3−21.

Clarke, G.K.C., 2003. Hydraulics of subglacial outburst floods: new insights from the Spring-Hutter formulation. J. Glaciol. 49, 299−313.

Coleman, S.E., Andrews, D.P., Webby, M.G., 2002. Overtopping breaching of noncohesive homogenous embankments. J. Hydraul. Eng. 128, 829−838.

Costa, J.E., 1983. Paleohydraulic reconstruction of flash-flood peaks from boulder deposits in the Colorado Front Range. Geol. Soc. Am. Bull. 94, 986−1004.

Costa, J.E., Schuster, R.L., 1988. The formation and failure of natural dams. Geol. Soc. Am. Bull. 100, 1054−1068.

Denlinger, R.P., O'Connell, D.R.H., 2010. Simulations of cataclysmic outburst floods from Pleistocene Glacial Lake Missoula. Geol. Soc. Am. Bull. 122, 678−689.

Depretris, P.J., Pasquini, A.I., 2000. The hydrological signal of the Perito Moreno Glacier damming of Lake Argentino (southern Andean Patagonia): the connection to climate anomalies. Global Planet. Change 26, 367−374.

Ding, Y., Liu, J., 1992. Glacier lake outburst flood disasters in China. Ann. Glaciol. 16, 180−184.

Driedger, C.L., Fountain, A.G., 1989. Glacier outburst floods at Mount Rainier, Washington state, USA. Ann. Glaciol. 13, 51−55.

Dussaillant, A., Benito, G.I., Buytaert, W., Carling, P., Meier, C., Espinoza, F., 2010. Repeated glacial-lake outburst floods in Patagonia: an increasing hazard? Nat. Hazards 54, 469−481.

Dwivedi, S.K., Archarya, M.D., Simard, D., 2000. The Tam Pokhari glacier lake outburst flood of 3 September 1998. J. Nepal Geol. Soc. 22, 539−546.

Eisbacher, G.H., Clague, J.J., 1984. Destructive Mass Movements in High Mountains − Hazard and Management. Geol. Surv. Can. Pap. 84-16, 230 pp.

Evans, S.G., 1986. The maximum discharge of outburst floods caused by the breaching of man-made and natural dams. Can. Geotech. J. 23, 385−387.

Evans, S.G., Clague, J.J., 1994. Recent climatic change and catastrophic geomorphic processes in mountain environments. Geomorphology 10, 107−128.

Faeh, R., 2007. Numerical modeling of breach erosion of river embankments. J. Hydraul. Eng. 133, 1000−1009.

Flowers, G.E., Björnsson, H., Pálsson, F., Clarke, G.K.C., 2004. A coupled sheet-conduit mechanism for jökulhlaup propagation. Geophys. Res. Lett. 31, L05401. http://dx.doi.org/10.1029/2003GL019088.

Fowler, A.C., 1999. Breaking the seal at Grímsvötn. J. Glaciol. 45, 506−516.

Fread, D.L., 1987. BREACH: An Erosion Model for Earthen Dam Failures. National Oceanic and Atmospheric Association, National Weather Service. Hydrologic Research Laboratory, Silver Spring, MD, 35 pp.

Fread, D.L., 1988. The NWS DAMBRK Model: Theoretical Background and User Documentation, HRL-258. National Oceanic and Atmospheric Association, National Weather Service. Hydrologic Research Laboratory, Silver Spring, MD, 325 pp.

Frey, H., Haeberli, W., Linsbauer, A., Huggel, C., Paul, F., 2010. A multi-level strategy for anticipating future glacial-lake formation and associated hazard potentials. Nat. Hazards Earth Syst. Sci. 10, 339–352.

Froehlich, D.C., 1987. Embankment-dam breach parameters. In: Proceedings, 1987 National Conference on Hydraulic Engineering. American Society of Civil Engineers, New York, pp. 570–575.

Froehlich, D.C., 1995. Peak outflow from breached embankment dams. J. Water Resour. Plann. Manage. 121, 90–97.

Froehlich, D.C., 2008. Embankment dam breach parameters and their uncertainties. J. Hydraul. Eng. 134, 1708–1721.

Fujita, K., Sakai, A., Nuimura, T., Yamaguchi, S., Sharma, R.R., 2009. Recent changes in Imja glacial lake and its damming moraine in the Nepal Himalaya revealed by in situ surveys and multi-temporal ASTER imagery. Environ. Res. Lett. 4 (4). http://dx.doi.org/10.1088/1748-9326/4/4/045205.

Fujita, K., Suzuki, R., Nuimura, T., Sakai, A., 2008. Performance of ASTER and SRTM DEMs, and their potential for assessing glacial lakes in the Lunana region, Bhutan Himalaya. J. Glaciol. 54, 220–228.

Gallino, G.L., Pierson, T.C., 1985. Polallie Creek Debris Flow and Subsequent Dam-break Flood of 1980, East Fork Hood River basin, Oregon. U.S. Geol. Surv. Water Supply Pap. 2273, 22 pp.

Geertsema, M., Clague, J.J., 2005. Jökulhlaups at Tulsequah Glacier, northwestern British Columbia, Canada. Holocene 15, 310–316.

Glen, J.W., 1954. The stability of ice-dammed lakes and other waterfilled holes in glaciers. J. Glaciol. 2, 316–318.

Grove, J.M., 1988. The Little Ice Age. Methuen Press, London, UK.

Haeberli, W., 1983. Frequency and characteristics of glacier floods in the Swiss Alps. Ann. Glaciol. 4, 85–90.

Haeberli, W., 2013. Mountain permafrost — research frontiers and a special challenge. Cold Reg. Sci. Technol. 96, 71–76.

Haeberli, W., Allen, J.C., Müller, P., Funk, M., 1989. Assessing the risks from glacier hazards in high mountain regions: some experiences in the Swiss Alps. Ann. Glaciol. 13, 77–101.

Haeberli, W., Kääb, A., Vonder, D., Teysseire, P., 2001. Prevention of outburst floods from periglacial lakes at Grubengetscher, Valais, Swiss Alps. J. Glaciol. 47, 111–122.

Haeberli, W., Clague, J.J., Huggel, C., Kääb, A., 2010. Hazards from lakes in high-mountain glacier and permafrost regions: climate change effects and process interactions. In: Avances de la Geomorphología en España, 2008–2010, vol. XI. Reunión Nacional de Geomorphología, Solsona, pp. 439–446.

Hanson, G.J., Cook, K.R., 2005. Physical modeling of overtopping erosion and beach formation of cohesive embankments. Trans. Am. Soc. Agric. Eng. 48, 1783–1794.

Hegglin, E., Huggel, C., 2008. An integrated assessment of vulnerability to glacier hazards: a case study in the Cordillera Blanca, Peru. Mt. Res. Dev. 28, 310–317.

Herget, J., 2005. Reconstruction of Pleistocene Ice-dammed Lake Outburst Floods in the Altai Mountains, Siberia. Geol. Soc. Am. Spec. Pap. 386, 118 pp.

Hewitt, K., 1968. Record of natural damming and related floods in the Upper Indus Basin. Indus 10 (3), 11–19.

Hewitt, K., 1982. Natural dams and outburst floods of the Karakoram Himalaya. In: Hydrological Aspects of Alpine and High-Mountain Areas, vol. 138. International Association of Hydrological Sciences Publication, pp. 259–269.

Hewitt, K., Liu, J., 2010. Ice-dammed lakes and outburst floods, Karakoram Himalaya: historical perspectives on emerging threats. Phys. Geogr. 31, 528–551.

Hodgson, K.Λ., Nairn, I.Λ., 2005. The c. AD 1315 syn-eruption and AD 1904 post-eruption breakout floods from lake Tarawera, Haroharo crater, North Island, New Zealand. N. Z. J. Geol. Geophys. 48, 491−506.

Hubbard, B., Heald, A., Reynolds, J.M., Quincey, D., Richardson, S.D., Zapata Luyo, M., Santillan Portilla, N., Hambrey, M.J., 2005. Impact of a rock avalanche on a moraine-dammed lake: Laguna Safuna Alta, Cordillera Blanca, Peru. Earth Surf. Process. Land. 30, 1251−1264.

Huggel, C., Haeberli, W., Kääb, A., Bieri, D., Richardson, S., 2004. An assessment procedure for glacial hazards in the Swiss Alps. Can. Geotech. J. 41, 1068−1083.

Huggel, C., Kääb, A., Haeberli, W., Krummenacher, B., 2003. Regional-scale GIS-models for assessment of hazards from glacier lake outbursts: evaluation and application in the Swiss Alps. Nat. Hazards Earth Syst. Sci. 3, 647−662.

Huggel, C., Kääb, A., Haeberli, W., Teysseire, P., Paul, F., 2002. Remote sensing based assessment of hazards from glacier lake outbursts: a case study in the Swiss Alps. Can. Geotech. J. 39, 316−330.

Hulsing, H., 1981. The Breakout of Alaska's Lake George. U.S. Government Printing Office, Washington, DC, 15 pp.

ICIMOD, 2010. Formation of Glacial Lakes in the Hindu Kush-Himalayas and GLOF Risk Assessment. International Centre for Integrated Mountain Development, Kathmandu, Nepal, 56 pp.

ICIMOD, 2011. Glacial Lakes and Glacial Lake Outburst Floods in Nepal. International Centre for Integrated Mountain Development, Kathmandu, Nepal, 109 pp.

Iverson, R.M., 2006. Langbein lecture: shallow flows that shape earth's surface. EOS Trans. Am. Geophys. Union 87 (52). Abstract H22A-01.

Ives, J.D., 1986. Glacial Lake Outburst Floods and Risk Engineering in the Himalaya − a Review of the Langmoche Disaster, Khumbu Himal, 4 August 1985. International Centre for Integrated Mountain Development, Kathmandu, Nepal. Occasional Paper 5, 42 pp.

Iwata, S., Ageta, Y., Naito, N., Sakai, A., Narama, C., Karma, C., 2002. Glacial lakes and their outburst flood potential assessment in the Bhutan Himalaya. Global Environ. Res. 6, 3−17.

Jacobson, R.B., O'Connor, J.E., Oguchi, T., 2003. Surficial geologic tools in fluvial geomorphology. In: Kondolf, G.M., Piégay, H. (Eds.), Tools in Fluvial Geomorphology. John Willey and Sons, Chichester, UK, pp. 25−57.

Johannesson, T., 2002. Propagation of a subglacial flood wave during the initial of a jökulhlaup. J. Hydrol. Sci. 47, 417−434.

Kääb, A., Huggel, C., Barbero, S., Chiarle, M., Cordola, M., Epifani, F., Haeberli, W., Mortara, G., Semino, P., Tamburini, A., Viazzo, G., 2004. Glacier hazards at Belvedere Glacier and the Monte Rosa east face, Italian Alps: processes and mitigation. In: Internationales Symposion Interpraevent 2004 − Riva/Trient, pp. 1-67−1-78.

Kääb, A., Huggel, C., Fischer, L., Guex, S., Paul, F., Roer, I., Salzmann, N., Schlaefli, S., Schmutz, K., Schneider, D., Strozzi, T., Weidmann, Y., 2005. Remote sensing of glacier- and permafrost-related hazards in high mountains: an overview. Nat. Hazards Earth Syst. Sci. 5, 527−554.

Kattelmann, R., 2003. Glacial lake outburst floods in the Nepal Himalaya: a manageable hazard? Nat. Hazards 28, 145−154.

Kershaw, J.A., Clague, J.J., Evans, S.G., 2005. Geomorphic and sedimentological signature of a two-phase outburst flood from moraine-dammed Queen Bess Lake, British Columbia, Canada. Earth Surf. Process. Land. 30, 1−25.

Komar, P.D., 1987. Selective gravel entrainment and the empirical evaluation of flow competence. Sedimentology 34, 1165−1176.

Lee, K.L., Duncan, J.M., 1975. Landslide of Aril 25, 1974, on the Mantaro River, Peru. National Research Council, Commission on Sociotechnical Systems, Committee on Natural Disasters, Washington, DC, 72 pp.

Liestol, O., 1956. Glacier dammed lakes in Norway. Norsk Geogr. Tidsskr. Norw. J. Geogr. 15, 122−149.

Linsbauer, A., Paul, F., Haeberli, W., 2012. Modeling glacier thickness distribution and bed topography over entire mountain ranges with GlabTop: application of a fast and robust technique. J. Geophy. Res. 117, F03007. http://dx.doi.org/10.1029/2011JF002313.

Liu, C., Sharma, C.K., 1988. Report on First Expedition to Glaciers and Glacier Lakes in the Pumqu (Arun) and Poiqu (Bhote-Sun Kosi) River Basins, Xizang (Tibet), China: Sino-Nepalese Investigation of Glacier Lake Outburst Floods in the Himalayas. Science Press, Beijing, China, 192 pp.

Lliboutry, L., Arnao, V.M., Pautre, A., Schneidter, B., 1977. Glaciological problems set by the control of dangerous lakes in Cordillera Blanca, Peru: 1. Historical failures of morainic dams, their causes and prevention. J. Glaciol. 18, 239−254.

Lord, M.L., Kehew, A.E., 1987. Sedimentology and paleohydrology of glacial-lake outburst deposits in southeastern Saskatchewan and northwestern North Dakota. Geol. Soc. Am. Bull. 99, 663−673.

Manville, V., 2001. Techniques for Evaluating the Size of Potential Dam-break Floods from Natural Dams. New Zealand Institute of Geological and Nuclear Sciences, Science Report 2001/28, 72 pp.

Manville, V., White, J.D.L., Houghton, B.F., Wilson, C.J.N., 1999. Paleohydrology and sedimentology of a post-1.8 ka breakout flood from intracaldera Lake Taupo, North Island, New Zealand. Geol. Soc. Am. Bull. 111, 1435−1447.

Marche, C., Mahdi, T., Quach, T., 2006. ERODE: Une methode fiable pour etablir l'hydrogramme de rupture potentielle par surverse de chaque digue en terre. Trans. Int. Congr. Large Dams 22, 337−360.

Marcus, M.G., 1960. Periodic drainage of glacier-dammed Tulsequah Lake, British Columbia. Geogr. Rev. 50, 89−106.

Massey, C., Manville, V., Hancox, G., Keys, H., Lawrence, C., McSaveney, M., 2010. Outburst flood (lahar) triggered by retrogressive landsliding, 18 March 2007 at Mt Ruapehu, New Zealand − a successful early warning. Landslides 7, 303−315.

Mathews, W.H., 1973. Record of two jökulhlaups. Int. Assoc. Hydrol/ Sci. Publ. 95, 99−110.

Mathews, W.H., Clague, J.J., 1993. The record of jökulhlaups from Summit Lake, northwestern British Columbia. Can. J. Earth Sci. 30, 499−508.

Mayo, L.R., 1989. Advance of Hubbard Glacier and 1986 outburst of Russell Fiord, Alaska. J. Glaciol. 13, 189−194.

McKillop, R., Clague, J.J., 2007a. A procedure for making objective preliminary assessments of outburst flood hazard form moraine-dammed lakes in southwestern British Columbia. Nat. Hazards 41, 131−157.

McKillop, R., Clague, J.J., 2007b. Statistical, remote sensing-based approach for estimating the probability of catastrophic drainage from moraine-dammed lakes in southwestern British Columbia. Global Planet. Change 56, 153−171.

Mergili, M., Schneider, J.F., 2011. Regional-scale analysis of lake outburst hazards in the southwestern Pamir, Tajikistan, based on remote sensing and GIS. Nat. Hazards Earth Syst. Sci. 11, 1447−1462.

Mergili, M., Müller, J.P., Schneider, J.F., 2013. Spatial-temporal development of high-mountain lakes in the headwaters of the Amu Darya River (Central Asia). Global Planet. Change 107, 13—24.

Montgomery, D.R., Hallet, B., Liu, Y., Finnegan, N., Anders, A., Gillespie, A., Greenberg, H.M., 2004. Evidence for Holocene megafloods down the Tsangpo River gorge, southeastern Tibet. Quat. Res. 62, 201—207.

Mool, P.K., 1995. Glacier lake outburst floods in Nepal. J. Nepal Geol. Soc. 11, 273—280.

Morris, M.K., Hassan, M.A.A.M., Vaskinn, K.A., 2007. Breach formation: field test and laboratory experiments. J. Hydraul. Res. 45 (Suppl. 1), 9—17.

Narama, C., Duishonakunov, M., Kääb, A., Daiyrov, M., Abdrakhmatov, K., 2010. The 24 July 2008 outburst flood at the western Zyndan glacier lake and recent regional changes in glacier lakes of the Teskey Ala-Too range, Tien Shan, Kyrgyzstan. Nat. Hazards Earth Syst. Sci. 10, 647—659.

Ng, F., Björnsson, H., 2003. On the Clague-Mathews relation for jökulhlaups. J. Glaciol. 49, 161—172.

Ng, F., Liu, S., 2009. Temporal dynamics of a jökulhlaup system. J. Glaciol. 55, 651—655.

Nye, J.F., 1976. Water flow in glaciers: Jökulhlaups, tunnels and veins. J. Glaciol. 17, 181—207.

O'Connor, J.E., 1993. Hydrology, Hydraulics, and Geomorphology of the Bonneville Flood. Geol. Soc. Am. Spec. Pap. 274, 83 pp.

O'Connor, J.E., Beebee, R.A., 2009. Floods from natural rock-material dams. In: Burr, D.M., Carling, P.A., Baker, V.R. (Eds.), Megafloods on Earth and Mars. Cambridge University Press, Cambridge, UK, pp. 128—171.

O'Connor, J.E., Costa, J.E., 1993. Geologic and hydrologic hazards in glacierized basins in North America resulting from 19th and 20th century global warming. Nat. Hazards 8, 121—140.

O'Connor, J.E., Costa, J.E., 2004. The World's Largest Floods, Past and Present — Their Causes and Magnitudes. U.S. Geol. Surv. Circ. 1254, 19 pp.

O'Connor, J.E., Clague, J.J., Walder, J.S., Manville, V., Beebee, R.A., 2013. Outburst floods. In: Shroder, J., Wohl, E. (Eds.), Treatise on Geomorphology. Fluvial Geomorphology, vol. 9. Academic Press, San Diego, CA, pp. 475—510.

O'Connor, J.E., Grant, G.E., Costa, J.E., 2002. The geology of geography of floods. In: House, P.K., Webb, R.H., Levish, D.R. (Eds.), Ancient Floods, Modern Hazards; Principles and Applications of Paleoflood Hydrology. American Geophysical Union, Washington, DC, pp. 191—215.

O'Connor, J.E., Hardison II, J.H., Costa, J.E., 2001. Debris Flows from Failures of Neoglacial Moraine Dams in the Three Sisters and Mt. Jefferson Wilderness Areas, Oregon. U.S. Geol. Surv. Prof. Pap. 1608, 93 pp.

Paul, F., Maisch, M., Rothenbuehler, C., Hoelzle, M., Haeberli, W., 2007. Calculation and visualization of future glacier extent in the Swiss Alps by means of hypsographic modeling. Global Planet. Change 55, 343—357.

Pickert, G., Weitbrecht, V., Bieberstein, A., 2011. Breaching of overtopped river embankments controlled by apparent cohesion. J. Hydraul. Res. 49, 143—156.

Plaza-Nieto, G., Zevallos, O., 1994. The 1993 La Josefina rockslide and Rio Paute landslide dam, Ecuador. Landslide News 8, 4—6.

Ponce, V.M., Tsivoglou, A.M., 1981. Modeling gradual dam breaches. J. Hydraul. Div. Proc. Am. Soc. Civ. Eng. 107, 829—838.

Post, A., Mayo, L.R., 1971. Glacial-dammed Lakes and Outburst Floods in Alaska. U.S. Geol. Surv. Hydrol. Atlas HA-455, 10 pp., 3 plates.

Procter, J.N., Cronin, S.J., Fuller, I.C., Lube, G., Manville, V., 2010. Quantifying the geomorphic impacts of a lake break-out flood, Mt. Ruapehu, New Zealand. Geology 39, 67−70.

Quincey, D.J., Lucas, R.M., Richardson, S.D., Glasser, N.F., Hambrey, M.J., Reynolds, J.M., 2005. Optical remote sensing techniques in high-mountain environments: application to glacial hazards. Prog. Phys. Geogr. 29, 475−505.

Quincey, D.J., Richardson, S.D., Luckman, A., Lucas, R.M., Reynolds, J.M., Hambrey, M.J., Glasser, N.F., 2007. Early recognition of glacial lake hazards in the Himalaya using remote sensing datasets. Global Planet. Change 56, 137−152.

Randhawa, S.S., Sood, R.K., Rathore, B.P., Kulkarni, A.V., 2005. Moraine-dammed lake study in the Chenab and the Satluj River basins using IRS data. J. Indian Soc. Remote Sens. 33, 285−290.

Reynolds, J.M., 1992. The identification and mitigation of glacier-related hazards: examples from the Cordillera Blanca, Peru. In: McCall, G.J.H., Laming, D.J.C., Scott, S.C. (Eds.), L Geohazards − Natural and Man-made. Chapman and Hall, London, pp. 143−157.

Reynolds, J.M., 1999. Glacial hazard assessment at Tsho Rolpa, Rolwaling, central Nepal. Quart. J. Eng. Geol. 32, 209−214.

Richardson, D., 1968. Glacier Outburst Floods in the Pacific Northwest. U.S. Geol. Surv. Prof. Pap. 600-D, D79−D86.

Richardson, S.D., Reynolds, J.M., 2000. An overview of glacial hazards in the Himalayas. Quat. Int. 65−66, 31−47.

Roberts, M.J., 2005. Jökulhlaups: a reassessment of floodwater flow through glaciers. Rev. Geophys. 43, RG1002.

Ruren, L., Deji, L., 1986. Debris flow induced by ice lake burst in the Tangbulang gully, Gangbujlangda, Xizang (Tibet). J. Glaciol. Cryopedol. 8, 61−71.

Russell, A.J., Tweed, F.S., Roberts, M.J., Harris, T.D., Gudmundsson, M.T., Knudsen, O., Marren, P.M., 2010. An unusual jökulhlaup resulting from subglacial volcanism, Sölheimajökull, Iceland. Quat. Sci. Rev. 29, 1363−1381.

Schneider, D., Huggel, C., Cochachin, A., Gullén, S., García, J., 2014. Mapping hazards from glacier lake outburst floods based on modelling of process cascades at Lake 513, Cuaraz, Peru. Adv. Geosci. 35, 145−155.

Schuster, R.L., 2000. Outburst debris flows from failure of natural dams. In: Wieczorek, G.F., Naeser, N.D. (Eds.), Debris-flow Hazards Mitigation: Mechanics, Prediction, and Assessment. Proceedings, 2nd International Conference on Debris Flow Hazard Mitigation, Taipei, Taiwan, pp. 29−42.

Singh, V.P., Scarlatos, P.D., Collins, J.G., Jourdan, M.R., 1998. Breach erosion of earthfill dams (BEED) model. Nat. Hazards 1, 161−180.

Spring, U., Hutter, K., 1981. Numerical studies of jökulhlaups. Cold Reg. Sci. Technol. 4, 227−244.

Spring, U., Hutter, K., 1982. Conduit flow of a fluid through its solid phase and its application to intraglacial channel flow. Int. J. Eng. Sci. 20, 327−363.

Sturm, M., Benson, C.S., 1985. A history of jökulhlaups from Strandline Lake, Alaska. J. Glaciol. 31, 272−280.

Thórarinsson, S., 1939. The ice dammed lakes of Iceland with particular reference to their values as indicators of glacier oscillations. Geogr. Ann. 21, 216−242.

Tómasson, H., 1996. The jökulhlaup from Katla in 1918. Ann. Glaciol. 22, 249−254.

Tómasson, H., 2002. Catastrophic floods in Iceland. In: Proceedings of Extremes of the Extremes: Extraordinary Floods Symposium. Int. Assoc. Hydrol. Sci. Pub. 271, pp. 121−126.

Tweed, F.S., Russell, A.J., 1999. Controls on the formation and sudden drainage of glacier-impounded lakes: implications for jökulhlaup characteristics. Prog. Phys. Geogr. 23, 79−110.

Valderrama, P., Vilca, O., 2012. Dinámica e implicancia del aluvión de la laguna 513, Cordillera Blanca, Ancash, Perú. Rev. Asoc. Geol. Argent. 69, 400−406.

Vilimek, V., Zapata Luyo, M., Klimes, J., Patzelt, Z., Santillan, N., 2005. Influence of glacial retreat on natural hazards of the Palcacocha Lake area, Peru. Landslides 2, 107−115.

Vincent, C., Auclair, S., LeMeur, E., 2010. Outburst flood hazard for glacier-dammed Lac de Rochemelon, France. J. Glaciol. 56, 91−100.

Vuichard, D., Zimmerman, M., 1987. The 1985 catastrophic drainage of a moraine-dammed lake, Khumbu Himal, Nepal: cause and consequences. Mt. Res. Dev. 7, 91−110.

Waitt, R.B., 2002. Great Holocene floods along Jökulsá á Fjöllum, north Iceland. In: Martini, P.I., Baker, V.R., Garzon, G. (Eds.), Flood and Megaflood Processes and Deposits: Recent and Ancient Examples. Blackwell Science, Oxford, UK, pp. 37−51.

Walder, J.S., Costa, J.E., 1996. Outburst floods from glacier-dammed lakes: the effect of mode of lake drainages on flood magnitude. Earth Surf. Process. Land. 21, 701−723.

Walder, J.S., Driedger, C.L., 1994. Rapid geomorphic change caused by glacial outburst floods and debris flows along Tahoma Creek, Mount Rainier, Washington, U.S.A. Arct. Alp. Res. 26, 319−327.

Walder, J.S., Driedger, C.L., 1995. Frequent outburst floods from South Tahoma Glacier, Mount Rainier, U.S.A.: relation to debris flows, meteorological origin and implications for subglacial hydrology. J. Glaciol. 41, 1−10.

Walder, J.S., Fountain, A.G., 1997. Glacier generated floods. In: Leavesley, G.H., Lins, H.F., Nobilis, F. (Eds.), Destructive Water: Water-Caused Natural Disasters − Their Abatement and Control. Int. Assoc. Hydrol. Sci. Pub. 239, pp. 107−113.

Walder, J.S., O'Connor, J.E., 1997. Methods for predicting peak discharge of floods caused by failure of natural and constructed earthen dams. Water Resour. Res. 33, 2337−2348.

Walder, J.S., Trabant, D.C., Fountain, A.G., Anderson, S.P., Anderson, R.S., Malm, A., 2006. Local response of a glacier to annual filling and drainage of an ice-marginal lake. J. Glaciol. 52, 440−450.

Wang, W., Yang, X., Yao, T., 2012a. Evaluation of ASTR GDEM and SRTM and their suitability in hydraulic modeling of a glacial lake outburst flood in southeast Tibet. Hydrol. Process. 26, 213−225.

Wang, X., Liu, S., Ding, Y., Guo, W., Jiang, Z., Lin, J., Han, Y., 2012b. An approach for estimating breach probabilities of moraine-dammed lakes in the Chinese Himalayas using remote-sensing data. Nat. Hazards Earth Syst. Sci. 12, 3109−3122.

Wang, Z., Bowles, D.S., 2006. Three-dimensional non-cohesive earthen dam breach model. Part 1: theory and methodology. Adv. Water Resour. 29, 1528−1545.

Watanabe, T., Rothacher, D., 1996. The 1994 Lugge Tsho glacial lake outburst flood, Bhutan Himalaya. Mt. Res. Dev. 16, 77−81.

Waythomas, C.F., Walder, J.S., McGimsey, R.G., Neal, C.A., 1996. A catastrophic flood caused by drainage of a caldera lake at Aniakchak Volcano, Alaska, and implications for volcanic hazards assessment. Geol. Soc. Am. Bull. 108, 861−871.

Webby, M.G., 1996. Discussion [of Froehlich (1995)]. J. Water Resour. Plann. Manage. 122, 316−317.

Webby, M.G., Jennings, D.N., 1994. Analysis of dam-break flood caused by failure of Tunawaea landslide dam. In: Proceedings of International Conference on Hydraulics in Civil Engineering 1994. University of Brisbane, Queensland, Australia, pp. 163−168.

Williams, G.P., 1983. Paleohydrological methods and some examples from Swedish fluvial environments, 1 − cobble and boulder deposits. Geogr. Ann. 65A, 227−243.

Worni, R., Huggel, C., Stoffel, M., 2013. Glacial lakes in the Indian Himalayas − from an area-wide glacial lake inventory to on-site and modeling based risk assessment of critical glacial lakes. Sci. Total Environ. 468−469, (Suppl. 1 December 2013), S71−S84.

Worni, R., Stoffel, M., Huggel, C., Volz, C., Casteller, A., Luckman, B.H., 2012. Analysis and dynamic modeling of a moraine failure and glacier lake outburst flood at Ventisquero Negro, Patagonia Andes (Argentina). J. Hydrol. 444−445, 134−145.

Wu, W., He, Z., Wang, S.S.Y., 2009. A depth-averaged 2-D model of non-cohesive dam/levee breach processes. In: Proceedings of World Environmental and Water Resources Congress 2009: Great Rivers. American Society of Civil Engineers Conference Proceedings 342, 3341−3350.

Xie, Z., ShangGuan, D., Zhang, S., Ding, Y., Liu, S., 2013. Index for hazard of glacial lake outburst flood of Lake Merzbacher by satellite-based monitoring of lake area and ice cover. Global Planet. Change 107, 229−237.

Xin, W., Shiyin, L., Wanqin, G., Junli, X., 2008. Assessment and simulation of glacier lake outburst floods for Longbasaba and Pida Lakes, China. Mt. Res. Dev. 28, 310−317.

Yamada, T., Sharma, C.K., 1993. Glacier lakes and outburst floods in the Nepal Himalaya. In: Young, G.J. (Ed.), Snow and Glacier Hydrology. Int. Assoc. Hydrol. Sci. Pub. 218, pp. 319−330.

Yesenov, U.Y., Degovets, A.S., 1979. Catastrophic mudflow on the Bol'shaya River in 1977. Sov. Hydrol. 18, 158−160.

Zhang, J.Y., Li, Y., Xuan, G.X., Wang, X.G., Li, J., 2009. Overtopping breaching of cohesive homogeneous earth dam with different cohesive strength. Sci. China Ser. E Technol. Sci. 52, 3024−3029.

Ice Loss and Slope Stability in High-Mountain Regions

Philip Deline [1], Stephan Gruber [2], Reynald Delaloye [3], Luzia Fischer [4], Marten Geertsema [5], Marco Giardino [6], Andreas Hasler [7], Martin Kirkbride [8], Michael Krautblatter [9], Florence Magnin [1], Samuel McColl [10], Ludovic Ravanel [1] and Philippe Schoeneich [11]

[1] *EDYTEM Lab, Université de Savoie, CNRS, Le Bourget-du-Lac Cedex, France,* [2] *Department of Geography and Environmental Studies, Carleton University, Ottawa, Canada,* [3] *Department of Geosciences, Geography, University of Fribourg, Fribourg, Switzerland,* [4] *Norwegian Geological Survey, Trondheim, Norway,* [5] *Ministry of Forests, Lands, and Natural Resource Operations, Prince George, BC, Canada,* [6] *GeoSitLab, Dipartimento di Scienze della Terra, Università di Torino, Italy,* [7] *Department of Geography, University of Zurich, Switzerland,* [8] *Geography, School of the Environment, University of Dundee, United Kingdom,* [9] *Technische Universität München, Germany,* [10] *Physical Geography Group, Institute of Agriculture and Environment, Massey University, Palmerston North, Australia,* [11] *Institut de Géographie Alpine, Université de Grenoble, CNRS, Grenoble, France*

ABSTRACT

The present time is one significant stage in the adjustment of mountain slopes to climate change, and specifically atmospheric warming. This review examines the state of understanding of the responses of mid-latitude alpine landscapes to recent cryospheric change, and summarizes the variety and complexity of documented landscape responses involving glaciers, moraines, rock and debris slopes, and rock glaciers. These indicate how a common general forcing translates into varied site-specific slope responses according to material structures and properties, thermal and hydrological environments, process rates, and prior slope histories. Warming of permafrost in rock and debris slopes has demonstrably increased instability, manifest as rock glacier acceleration, rock falls, debris flows, and related phenomena. Changes in glacier geometry influence stress fields in rock and debris slopes, and some failures appear to be accelerating toward catastrophic failure. Several sites now require expensive monitoring and modeling to design effective risk-reduction strategies, especially where new lakes as multipliers of hazard potential form, and new activities and infrastructure are developed.

15.1 INTRODUCTION

Especially in the presence of climatic changes, processes resulting from steep slope instability are compounded by the effects of a changing cryosphere: High-mountain slopes are commonly characterized by ice and firn cover. Less visible, but not less common is the presence of permafrost, that is, ground with a temperature of $\leq 0\,°C$ during at least two consecutive years, that likely contains ice in the subsurface. In contrast to the well-studied relationship of rock-slope instability and glacier retreat (cf. McColl, 2012), research on the effects of permafrost and its dynamics on bedrock slope stability is more recent (cf. Gruber and Haeberli, 2007).

Changes to the cryosphere have been documented for both historic and Holocene times. Since the termination of the Little Ice Age (LIA), the extent of mountain glaciers has decreased worldwide, with an acceleration of their shrinkage during recent decades in relation to the increase in mean air temperature. Glaciers in the European Alps have lost 35 percent of their total area from 1850 until 1975, and almost 50 percent by 2000 (Zemp et al., 2006), leaving newly exposing rock and debris slopes at the terrain surface.

Monitoring of temperature in debris slopes started in the 1970s (cf. Marchenko et al., 2007), and in the 1990s it was extended to bedrock slopes. These measurements allow investigating multiyear trends and seasonal variations in the permafrost (PERMOS, 2013). Similarly, strong seasonal and multiannual fluctuations and longer-term acceleration of rock glacier creep were observed during the 1990s and 2000s (Delaloye et al., 2008). An apparent increase in slope failures in steep rock walls observed over recent decades in the European Alps (Allen and Huggel, 2013) also suggests current degradation of the permafrost.

Due to the high potential energy inherent in steep environments and the possibility of compound events, consequences of slope instability following ice loss can be far reaching. Especially in densely populated mountain areas such as the Himalaya or the European Alps, such events can have devastating consequences, as illustrated by the dramatic Kolka-Karmadon ice-rock avalanche triggered on September 2002 in the Caucasus (Haeberli et al., 2004; Evans et al., 2009). Lake formation due to glacier shrinkage and expanding human activities are increasing the risks (cf. Schaub et al., 2013; Haeberli, 2013).

Slope instability and mass movements, in general and when related to ice loss, occur at a wide range of magnitudes and frequencies. This chapter is focused on less frequent events that often have larger magnitudes and that are—or can be—tied to the unprecedented ice loss during recent decades. Generally, these events are also less understood than the more frequent ones.

This chapter describes the consequences of the loss of perennial ice, both at the surface and in the subsurface, rather than processes related to seasonal change. It has been compiled and edited by P. Deline and S. Gruber and represents a joint effort of 13 authors contributing their specific expertise. It is structured in two parts: the first revisits the mechanisms of cryosphere control on slope stability, dealing with instabilities in hanging glaciers (L. Fischer), debuttressing and ice unloading (S. McColl), and thermal and hydrologic changes in bedrock (A. Hasler and M. Krautblatter) and sediments (R. Delaloye). The second part illustrates the relationship between ice loss and slope instability with case studies, organized partly by mechanism and partly by morphology. Concerning bedrock slopes, case studies are presented on the Monte Rosa East face (L. Fischer), rock falls from permafrost-affected rock walls in the Mont Blanc massif (L. Ravanel), where this permafrost is modeled (F. Magnin); relations between debuttressing and rock fall, rock avalanche, rock slide, and deep-seated gravitational slope deformation (DSGSD) are illustrated by cases in the Mont Blanc massif (L. Ravanel), British Columbia (BC) and Alaska ranges (M. Geertsema), Mueller Glacier, New Zealand (S. McColl), and at Alp Bäregg, Switzerland (P. Deline), and Mont de la Saxe, Italy (M. Giardino), respectively. Concerning debris slopes, the processes (sliding, gullying, and breaching) that impact the moraine stability because of glacier shrinkage are developed (P. Deline and M. Kirkbride); rock–glacier displacement changes, ground-ice melting, and debris supply from permafrost areas are exposed through several Alpine case studies (P. Schoeneich and R. Delaloye). The last section deals with the interactions of differing processes that can translate cryospheric change into far-reaching and difficult-to-forecast events (A. Hasler and M. Geertsema). The chapter concludes with a synopsis and an outlook on likely future challenges.

15.2 MECHANISMS OF CRYOSPHERE CONTROL ON SLOPE STABILITY

15.2.1 Unstable Ice

Changes in glacier geometry and slope topography as well as in the thermal regime of steep ice can have a severe impact on the stability of hanging glaciers and eventually lead to ice avalanches (Figure 15.1). Such ice avalanches can pose a considerable threat to society in the case of process chains (e.g., impact waves in lakes) or new infrastructure (cf. Evans and Delaney, 2014). A periodic or occasional breakoff of small to medium-volume ice avalanches is common in the mass balance cycle of many hanging glaciers, where snow accumulation is compensated by a break-off of ice mass at the glacier front (Pralong and Funk, 2006). This chapter focuses mainly on instabilities that are more unusual, and highlights the influence of temperature and topographic changes on ice stability.

Firn and ice temperatures and the topography are the most important factors for the stability of steep ice and hanging glaciers (Alean, 1985; Haeberli et al., 1999; Wagner, 1996). Additional factors are adhesion of cold

FIGURE 15.1 The east face of Monte Rosa above the Belvedere Glacier, Monte Rosa massif, Italy. Several ice and ice-rock avalanches have affected this face since the 1990s. *Photograph by S. Gruber, 2003.*

and polythermal ice on bedrock, cohesion with more stable upslope ice, supporting effects from flatter downslope glacier parts and lateral bedrock abutments, and the englacial and subglacial hydrology (Röthlisberger, 1981; Haeberli et al., 1999; Pralong, 2005; Faillettaz et al., 2012). The thermal state of the ice is related to the glacier location, aspect, and elevation with three types of glacier (Röthlisberger, 1981; Alean, 1985; Faillettaz et al., 2012): (1) cold-based glaciers frozen to the bedrock; (2) polythermal glaciers stabilized by the cold ice frozen to the bedrock in the frontal and lateral parts; and (3) temperate glaciers sliding on the bedrock.

Climatic changes may introduce complex feedback mechanisms involving changes in englacial and subglacial temperatures, ice accumulation rate and surface geometry, melt water infiltration, and stress distribution. Such changes may reduce the stability of parts of, or the entire, hanging glacier (Wegmann et al., 1998; Pralong and Funk, 2006; Fischer et al., 2013). Atmospheric warming affects the hydraulic conditions of the ice and increases the proportion of temperate ice in steep high-mountain faces (Gruber and Haeberli, 2007). A rise in ice temperature lowers the viscosity of the ice mass and decreases the basal cohesion of the previously cold ice frozen to the bedrock; it increases flow velocities and sliding processes within the ice mass and at the glacier bed, causing a changed stress field in the glacier. Furthermore, the tensile strength of the warmed and water-containing ice is reduced (Haeberli et al., 1997, 1999). Melting can occur at higher elevations, and increased infiltration of melt water through temperate firn and ice layers, and crevasses may lead to a warming of the basal ice and cause a decrease in the effective basal pressure and friction at the glacier bed (Pralong, 2005; Faillettaz et al., 2012).

Changes in glacier geometry and the topography around the glacier can influence the ice stability. A retreat of the glacier terminus from a moderate into a steep slope (e.g., due to an ice avalanche from the frontal part or a decoupling from a valley glacier) can strongly change the stress field within the glacier, possibly leading to its instability. Rock falls or debris flows can lead to a destabilization of hanging glaciers when changing the topography around steep glaciers by decreasing the lateral/downslope abutments (Fischer et al., 2011, 2013). Finally, the thickness and mass of a steep glacier are limited for a given slope gradient and thermal regime, as they affect the shear stresses determining stability (Pralong, 2005); additional load from mass movement deposits can trigger an ice failure when the basal shear resistance or tensile strength is exceeded (Fischer et al., 2013).

Due to the ongoing climate change, these complex factors and processes might lead to the development of potentially large instabilities in steep ice at locations with known events (e.g., Alean, 1985; Pralong, 2005) and also without precedence.

15.2.2 Debuttressing and Ice Unloading

Aside from creating a void into which slope failures can freely move, the thinning and retreat of glaciers is thought to influence slope instability through three main stress mechanisms: (1) glacial debuttressing; (2) stress-release fracturing; and (3) crustal rebound (McColl, 2012).

The term "glacial debuttressing" refers to the loss of slope support provided by glacial ice; it is considered to be a preparatory factor, or trigger, for slope failure in deglaciated terrain. Slopes are supported by normal stresses imposed by the glacier ice through the static self-weight loading or the active flow of the ice against the slope, that vary through space and time with ice thickness, density and flow, and groundwater and thermal regimes (McColl and Davies, 2012). For rock slopes, where glacial erosion permanently changes the geometry and stress distribution in the slope (e.g., glacial over-steepening, sensu Augustinus, 1995), the slope stability upon debuttressing will likely be lower. Failure of the slope may occur long after or synchronously with ice retreat, possibly leading to failure of the remaining buttressing ice (e.g., the 1975 Devastation Glacier landslide; Evans and Clague, 1988). For moraine or drift-covered slopes, removal of the ice support will always shift the slope out of equilibrium, likely resulting in collapse (Blair, 1994). Failures may be large and deep seated, or consist of numerous smaller slumps, rock falls, and debris flows (e.g., Hugenholtz et al., 2008; Blair, 1994; Ballantyne and Benn, 1994). Both deep-seated and shallow failures, along with fluvial erosion, have been observed in moraine exposed by the gradually thinning Tasman Glacier in New Zealand (Blair, 1994, Figure 15.2). Failures will

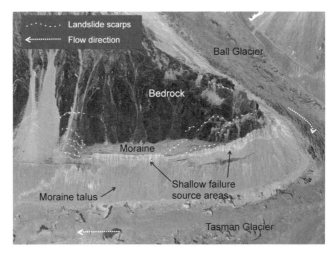

FIGURE 15.2   Tasman Glacier, Southern Alps, New Zealand. Downwasting of the glacier since the Little Ice Age has exposed and unloaded the lateral moraine wall. The moraine has responded by a combination of shallow failures and slumping, which has damaged and threatened huts constructed on it. The rock wall behind the moraine is also unstable. *Photograph: S. Winkler.*

continue until the slopes reach a new strength—equilibrium profile independent of ice support.

"Stress-release fracturing" that develops as a result of ice unloading is another preparatory factor for rock—slope failure. The removal of confining glacial ice can cause instantaneous or time-dependent rebound of crustal material (Nichols, 1980). One consequence of this is the initiation and propagation of fractures in the near-surface rock parallel to its surface, referred to as stress-release fractures, or sheeting, exfoliation, or unloading joints. Their development creates potential failure surfaces for mass movements. There appears to be a spatial coincidence between the occurrence of stress-release joints and rock slope failures (e.g., Cossart et al., 2008). A lag time between maximum ice unloading and fracturing may in part explain the timing of paraglacial rock slope failures. However, this lag time is complicated by the additional (and possibly greater) effects of unloading of rock overburden by glacial erosion and modification of slope form. Further, it appears that the response of the rock slope to these stress changes can depend on the erosional history of slopes (Jarman, 2006), and tectonic, geomechanical, and topographic processes (Leith et al., 2010; Leith, 2012).

The unloading of glacial ice can cause larger scale "crustal rebound", that is, the isostatic, and elastic flexural adjustment of the crustal material. Its time scale depends on the rheological properties of the upper mantle and strength of the crust; it may last tens of thousands of years (Stewart et al., 2000). In some locations, isostatic adjustments (i.e., uplift) in the order of hundreds of meters have occurred since deglaciation, and differential uplift and changes to stress fields around faults are sufficient to cause coseismic faulting (Stewart et al., 2000; Hampel et al., 2010). Links between glacial unloading, fault activity, high magnitude seismicity, and landsliding have been documented (e.g., Arvidsson, 1996; Firth and Stewart, 2000; Muir-Wood, 2000; Hetzel and Hampel, 2005; Hampel et al., 2010; Sanchez et al., 2010; Cossart et al., 2013). Isostatic adjustment is one mechanism that will influence paraglacial landsliding, explaining any lag time between maximum ice retreat and maximum landslide activity.

The influence of each of these mechanisms on spatial and temporal distributions of slope failures is uncertain, as they also operate alongside a number of other environmental factors discussed in this chapter. Statistical testing provides an approach for unraveling this uncertainty (e.g., Cossart et al., 2013). The power of such approaches relies on good landslide distribution data, supported by both sound geomorphological interpretations and knowledge of the timing of glacial, seismic, rock damage, and landslide processes.

15.2.3 Bedrock Permafrost: Thermal and Hydrologic Impact

Permafrost as a thermal phenomenon affects the hydrological and mechanical behavior of the naturally fractured bedrock: Water will freeze in cavities, fissures, and fractures at $0\,°C$, whereas the freezing point in the pore system is significantly lower (-0.1 to $-1.5\,°C$; Krautblatter, 2009) in water-saturated, tough (low-porosity) alpine and arctic bedrock (Krautblatter et al., 2010; Draebing and Krautblatter, 2012). Rock temperature changes and water percolation into cold fractured rock with corresponding phase changes are accompanied by latent heat exchange, strong changes in hydraulic permeability and mechanical strength, and an alteration of the stress field.

In the pore system of compact rock (interjoint rock mass), the liquid water content and hydraulic permeability changes gradually below the freezing point (Kleinberg and Griffin, 2005). In this temperature range, crack initiation and ice lens formation are expected to occur where the cryogenic stresses exceed rock strength and overburden. For porous chalk, ice segregation has been illustrated in the laboratory (Murton et al., 2006), whereas acoustic emissions measured in metamorphic alpine bedrock indicate frost cracking activity significantly below $0\,°C$ (Girard et al., 2013).

On a larger scale in low-porosity rock, the hydrology is mainly controlled by the permeability of the fracture system. Pogrebiskiy and Chernyshev (1977) found that the permeability of frozen fissured granite is one to three orders of magnitude lower than the permeability of identical thawed rock. Hence, saturated fractured bedrock permafrost acts as an aquiclude and is important for the development of perched water levels and corresponding hydrostatic pressures. Even though several observations of ice saturated fractures in alpine bedrock exist (cf. Gruber and Haeberli, 2007), reverse observations of "dry" bedrock permafrost from borehole pictures (oral. comm. V. Mair) and deep-reaching unfrozen fracture systems (microtaliks) percolating into railway tunnels (Tang and Wang, 2006; Haeberli and Gruber, 2008) exist as well. Water flowing into permafrost bedrock is a highly efficient transport medium of latent energy that may instantaneously warm the rock fractures at depth, and is unlikely to freeze in subsurface channels if not impounded (Hasler et al., 2011b). For the erosion (melt) of the ice within rock fractures, the heat uptake of percolation water at the surface is important. These hydrothermal processes play a crucial role for both the hydrostatic and the thermal conditions at potential failure planes in fractured bedrock. From a mechanical point of view, both the strength of bedrock permafrost and the subsurface stress field changes with temperature and hydraulic conditions. The stress change is due to additional cryostatic and hydrostatic pressures and thermomechanical forcing. Increased cryostatic pressures up to a few megapascals can derive from ice, whereas volume expansion of (fast-) freezing, trapped water may cause larger stresses that decline rapidly due to ice extrusion (Matsuoka and Murton, 2008). Hydrostatic pressure can derive from perched groundwater blocked by impermeable permafrost bedrock (Fischer et al., 2010) or from melt water input into fractured bedrock with limited permeability. It reduces the effective normal stress and as a consequence the friction of a potential failure plane or simply increases gravitational downslope forces. Temperature variation of the rock mass leads to thermal expansion/contraction and related stress changes and movements within fractured bedrock (Gischig et al., 2011). The change in rock- and ice-mechanical strength inside the bedrock affects the strength of intact bedrock, the strength of rock−ice interfaces and the deformation of ice in fractures (Krautblatter et al., 2013). In the intact rock, fracture initiation thresholds and friction within rough fractures decrease when thawing (Mellor, 1973; Dwivedi et al., 2000). Along rock−ice interfaces, fracturing is facilitated by increasing temperatures. The deformability and fracturing of ice are also temperature-dependent and strongly react to warming (Sanderson, 1988; Budd and Jacka, 1989). All these processes work effectively for different rock instability magnitudes (stress levels) and deformation velocities (strain levels) (Krautblatter et al., 2013). Lower magnitudes are more prone to be affected by ice-mechanical changes, whereas slow deformation in high magnitudes is more likely to be affected by rock-mechanical changes. Although the stress changes tend to be more short-term reactions (except ice segregation) to particular meteorological events and annual extremes, the strength changes all predict a stability reduction with increased rock temperature.

Different explanations of the role of liquid water in rock instability currently exist; one hypothesis is that hydrostatic pressures due to the sealing of rock surfaces by ice may play a vital part in the destabilization of rock slopes (Terzaghi, 1962), as illustrated by coupled hydromechanical modeling of the Tschierva rock avalanche by Fischer and Huggel (2008). Another hypothesis is that hydrothermal heat transport causes rapid warming at depths and corresponding strength reductions (Hasler et al., 2012).

Permafrost degradation can occur on different scales in time and space and is highly sensitive to the surface conditions of steep bedrock. For example, extreme warm summers or disappearance of thin ice covers lead to active-layer thickening or new active-layer formation with rather fast reactions and small magnitude events. On the other hand, the signal of current atmospheric warming is delayed by decades at deep-seated potential failure planes even in steep topography where signal propagation is faster due to multilateral warming (Noetzli and Gruber, 2009).

15.2.4 Debris: Hydrologic and Thermal Change

Thermal and hydrological conditions in frozen accumulations of mainly coarse debris are primarily responding to shifts in temperature occurring at the ground surface. Increased subsurface temperatures induce a thickening of the active layer. The thickening rate is mainly dependent on the ice content in the upper permafrost layer (Scherler et al., 2013). In the Swiss Alps, for permafrost conditions close to 0 °C, the active-layer thickness in debris terrain remains almost constant or has increased by up to 10 cm/year over the last 10−15 years (Zenklusen Mutter and Philips, 2012a; PERMOS, 2013). A deeper active layer induces changes in the suprapermafrost water circulation: drainage through the active layer increases in autumn due to a thaw depth larger than in early summer (Buchli et al., 2013). On steep terrain mass movements could also start from the active layer. On the one hand, a larger thaw depth increases the availability of loose material for larger mass movement; on the other hand, frozen ground at depth limits the regressive erosion during an event—but also the infiltration of water, which could favor the triggering of the mass movement (Figure 15.3).

The warming of permafrost favors the infiltration of water at depth and the formation of talik, as observed, for instance, in a borehole on the Ritigraben rock glacier (Zenklusen Mutter and Philips, 2012b). During warming, the viscosity of ice decreases, and the permafrost creep rate increases nonlinearly, accelerating significantly when the temperature is approaching 0 °C (Kääb et al., 2007); under cooling conditions, the response is reversed. Interannual variations in the displacement rate of rock glaciers are almost synchronous and similar in range in a given region (Delaloye et al., 2008; PERMOS, 2013). Rock glaciers with temperatures close to the melting point may also display short-term reactions to water infiltration (recharge of intrapermafrost or sub-permafrost aquifer by snowmelt or a heavy rainfall event). Destabilization of rock glaciers with dramatic acceleration of the creep velocity has been reported (see the Section 15.3.4), the onset of which appears to be, at least for some cases, a direct response to warmer ice conditions.

FIGURE 15.3   Left: ground ice exposure (dark gray) in the starting zone of a 4,000-m^3 rock fall at the front of a deep-seated landslide in permafrost conditions, Grabengufer, Swiss Alps, September 2010; Right: on-going regressive erosion scar produced by a series of debris flows triggered at the front of an active rock glacier, likely due to the limitation of snowmelt percolation by the frozen ground, Gugla-Bielzug, Swiss Alps, June 2013 *(Photographs: R. Delaloye)*. The exposed ground ice started to melt for weeks to years after both events.

When a coarse debris layer is not or no longer completely sealed with ice, air convection and advection are favored (Delaloye and Lambiel, 2005). The process could be limited to the active layer but may also develop beneath the frozen layer. The so-called chimney effect (air advection) occurs in talus slopes and coarse-grained rock glaciers (Morard et al., 2008). It contributes to preserving cold ground conditions, particularly in the lower and deeper parts of the affected landforms, especially by the aspiration of cold air deep into the ventilated terrain in wintertime, while heat is expelled toward the upper part of the landform. The efficiency of air circulation increases with decreasing sealing by ice. Ground ice could thus be formed or preserved in inactive rock glaciers and talus slopes under climatic conditions unfavorable for permafrost occurrence.

15.3 CASE STUDIES

15.3.1 Bedrock Slopes: Hanging Glaciers and Permafrost

15.3.1.1 Hanging Glaciers on the Monte Rosa East Face

The Monte Rosa east face (Figure 15.1) is a prominent example of recent dramatic changes in hanging glaciers, including large ice avalanches. It is among the highest rock walls in the European Alps extending from 2,200 m to >4,600 masl (45°56′N, 7°53′E). Large sections are covered by hanging glaciers, firn fields, as well as steep glaciers connected downslope to more gently sloping valley glaciers. Between the last glacial maximum at the end of the LIA around 1850 and the 1980s, the hanging glaciers and firn fields on the face changed only slightly (Fischer et al., 2011). Recently, however, the ice cover has experienced an accelerated loss in extent and thickness (Haeberli et al., 2002; Kääb et al., 2004; Fischer et al., 2006, 2011), leaving large parts of the underlying rock unprotected against mechanical and thermal erosion (cf. Davies et al., 2001; Noetzli et al., 2003). Since around 1990, new slope instabilities have developed in hanging glaciers and in bedrock areas. Frequent small-volume as well as several large-volume ice and rock avalanches and debris flow events have led to significant reduction of the ice-covered area. The total material loss in the face was around 25 Mm^3 from 1988 to 2007, from both steep glaciers and subjacent bedrock.

Measured englacial temperatures on the Colle Gnifetti high plateau, located directly above the east face, show cold thermal regime throughout, with a mean annual surface temperature near $-14\,^\circ$C and a basal temperature at 124 m ice depth slightly below $-12\,^\circ$C (Haeberli and Funk, 1991). However, the terminus of the largest steep glacier is at 3,300 masl, an elevation close to the modeled lower boundary of permafrost conditions (Zgraggen, 2005; Fischer et al., 2006). This implies changing thermal conditions from cold glaciers that are frozen to the bedrock in the uppermost part of the face to polythermal glaciers in the middle elevations and temperate glaciers at the foot of the slope. The three-dimensional permafrost distribution and its spatiotemporal evolution are closely linked to the ice and snow cover of the face (and vice-versa). Thus, the obvious decrease in glaciation is most likely accompanied by a less visible but, nevertheless, also marked change in the permafrost thermal regime of the Monte Rosa east face (Fischer et al., 2006, 2013).

The increased ice avalanche activity caused a reduction in the size and disappearance of several hanging glaciers. The corresponding topographic change was investigated based on a time series of high-resolution digital terrain models (DTMs) and terrestrial photographs (Fischer et al., 2011, 2013). The investigations suggest a strong connection between failures in steep glaciers and the underlying bedrock. Failure zones are spatially correlated and commonly proceed from lower elevation upward.

Around 1990, an entire hanging glacier, which is assumed to be located in the cold ice regime, and large parts of the underlying bedrock disappeared. Thereby, a steep rock wall became newly exposed in the uppermost part of the face. The instability in the hanging glacier was most likely influenced by previous rock fall and debris flow activity right below it. The changed bedrock geometry at the terminus of the hanging glacier and additional mass movement activity below the glacier were subsequently inducing a stepwise failure of the glacier front and retreat of the ice mass.

A second phase of increased ice avalanche occurred around 1999–2000 in the central part of the face, where the lowest part of the hanging glacier reaching down to 3,300 masl disappeared (Figure 15.4, glacier part below marked H). The ice failures started in the lowest part of the glacier and proceeded progressively upward in a combination of one major ice avalanche and several small-volume events. As the lowest part of this steep glacier is likely polythermal or temperate, mean annual air temperatures rising by about 0.5–1.5 °C from 1955 to 2007 might have affected stability (Fischer et al., 2013).

In the following years, especially heavy ice and rock fall originated from the newly ice-free zone right below the retreated glacier front (Figure 15.4). Mass wasting frequently occurred also by debris flows, which were presumably triggered by melt water. The continuation of rock-fall activity during the winter months pointed to a substantial change in glacier and rock conditions rather than to seasonal melt effects alone. These mass wasting processes culminated in August 2005 in a major ice avalanche event, where a large part of the remaining hanging glacier failed in a single event with a maximum detachment thickness of 40 m and a total volume of around 1.2 Mm^3 (Figure 15.4, at location H). Also here, changing ice temperature might have locally increased water percolation, indicated by strong water outflow at the front of the remaining glacier at an elevation of 3,700 masl, and hence reduced ice stability. Additional destabilizing factors for this large ice failure were supposedly the lack of the supporting glacier parts below, which detached in 1999, and the accumulation of ongoing deposition of rock fall debris on the hanging glacier from the wall above.

The topographic analysis has also revealed very rapid ice dynamics in such a steep face, with both large mass loss and mass accumulation within a short time period. From 2005 to 2007, considerable ice volume gain with up to 20-m thickness took place in the area of the former detachment zone of the 2005 ice avalanche in the Imseng Channel. This also shows that rapid ice accumulation and buildup of steep glaciers in eroded areas following a rock or ice avalanche are possible (Fischer et al., 2011).

FIGURE 15.4   Upper part of the Monta Rosa east face in the mid-1980s (left) and on August 6, 2003 (right). Some steep glaciers have totally disappeared; some have significantly lost mass. Zones with highest slope failure activity and mass waste are located in the center of the face (H). *From Kääb et al., 2004.*

15.3.1.2 Modeling the Temperature of High-Elevation Steep Slopes in the Mont Blanc Massif

Steep alpine faces have reduced snow and debris covers. Thus, surface temperature is mainly, and more simply than in other areas, controlled by atmospheric variables: the spatial distribution of rock surface temperature mostly results from sun exposure, whereas changes over time are dominated by air temperature (Gruber et al., 2004a). Most simulations of subsurface temperature consider that heat is transported from the surface only by conduction in compact rock (Noetzli et al., 2007). However, recent studies highlight temperature offsets (i.e., temperature difference between rock surface and permafrost body below the active layer) in steep and compact bedrock due to heterogeneous snow accumulation on microtopography and air ventilation in clefts (Hasler et al., 2011a).

The linear regression formulated for a statistical model of mean annual rock surface temperature (MARST) of the entire European Alps (Boeckli et al., 2012) has been used for simulating MARST on a high-resolution digital elevation model (DEM) of the Mont Blanc massif. Explanatory variables are the air temperature, and the potential incoming solar radiation, which depends on exposure, on slope angle, and on the surrounding topography controlling horizon shading.

Results show that the 0 °C isotherm of MARST extends as far down as 2,600 masl in the most shaded faces and rises up to 3,800 m in the most sun-exposed rock surfaces (Figure 15.5). As the modeling and measurements described above concern near-surface temperatures only, their relationship with temperature and permafrost occurrence at depths has been investigated

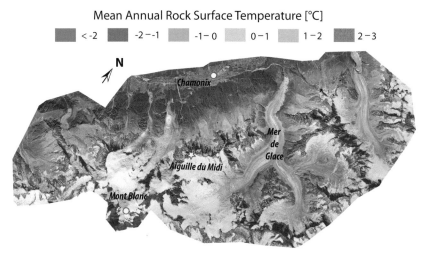

FIGURE 15.5 Simulated MARST of high-elevation rock walls on the French side of the Mont Blanc massif for the 1961−1990 period. Orthophoto draped on 4-m DEM (RGD 73−74).

with boreholes and geophysical soundings. The temperature offset can reach up to 3 °C in the south-facing 10-m-deep borehole at Aiguille du Midi (3,842 masl), whereas a cleft locally cooled the rock by 1 °C at the NW borehole. Electric resistivity tomography on five rock walls show high resistivity bodies at a depth of approximately 20−30 m below simulated MARST up to +2−3 °C. Validated with specific resistivity−temperature calibration, these results reinforce the hypothesis that permafrost can exist below MARST up to +3 °C in steep alpine rock walls.

15.3.1.3 Rock Fall from Permafrost Slopes in the European Alps

An increase of rock falls with volume from a few hundreds to tens of thousands of cubic meters is today largely recognized as one probable impact of warming in permafrost-affected mountain regions (Gruber and Haeberli, 2007). Consideration has recently been given to the potential direct role of extremely high air temperatures in triggering rock fall. Ravanel and Deline (2008, 2011) showed an important increase in the frequency of rock falls and a strong correlation between the warmest periods since 1860 and the occurrence of 58 rock falls in the Mont Blanc massif (Figure 15.6). In the Swiss Alps, a high occurrence of extremely warm days in the week leading up to rock falls has been observed over recent decades (Allen and Huggel, 2013).

Due to the lack of measurements (thermal in particular) in rock walls prior to their collapse, it is difficult to be categorical about the role of permafrost in the triggering of rock fall. However, the analysis of nearly 400 rock falls that occurred in the Mont Blanc massif since 2003 (Ravanel et al., 2010, 2011) shows (1) a close relationship between rock falls and air temperature; (2) massive ice in at least 40 detachment scars; (3) the mean scar elevation is higher than the mean rock wall elevation; (4) the most affected altitudinal belt is characterized by modeled warm permafrost (i.e., −2 to 0 °C); (5) the hotter

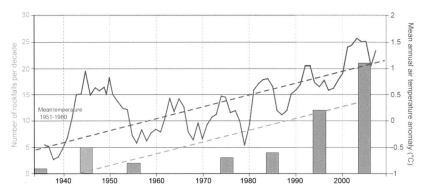

FIGURE 15.6   Mean annual air temperature in Chamonix (1,040 masl) since 1934 and number of rock falls per decade in the West face of the Drus and on the North side of the Aiguilles de Chamonix, Mont Blanc massif, France. Dashed lines: linear regressions.

the summer, the higher the scar elevation; (6) a sharp contrast in scar elevation between north and south faces, consistent with modeled permafrost distribution; and finally (7) rock fall especially affects topography prone to rapid permafrost degradation such as pillars, spurs, and ridges.

The most prominent episode of extreme rock fall from permafrost slopes in the European Alps has been the hot summer of 2003, which set the stage for much of the current research into the topic. Although backanalyses of rock temperature were consistent with the expected extraordinary depth of thaw during 2003, the timing of exceeding previous maxima was suspect: observed rock fall occurred almost simultaneously with highest air temperatures, whereas full active-layer development would have occurred much later (Gruber et al., 2004b). Fast and linear thaw by water percolation into fractures has been suggested to accommodate this fast reaction (Hasler et al., 2011b). Furthermore, warming on convex topography is faster than in flat terrain due to geometric effects (Noetzli and Gruber, 2009).

15.3.2 Bedrock Slopes and Debuttressing: Rock Fall

Rock falls and rock avalanches can acutely threaten human infrastructure and activities in mountain areas. One factor that could increase their frequency is the impact of the current glacier shrinkage. An accurate observation of active rock falls is possible in places that are well frequented by people, whereas only large events such as rock avalanches are usually observed in remote areas. The possible control that glacial debuttressing exerts on rock falls and rock avalanches in areas likely containing warm permafrost can be explored by jointly interpreting case studies from these different areas.

15.3.2.1 Rock Falls in the Mont Blanc Massif

The evolution of the ice masses since the end of the LIA—especially during the 1990s and 2000s—quite clearly influences the triggering of rock fall close to the glacier tongues and also at higher elevations. During two decades, the Mer de Glace lost >65 m of ice thickness below the Montenvers, and up to tens of meters above 3,000 masl. This could lead to the trigger of rock falls due to the removal of the ice pressure combined with gravity: 20 percent of the 139 collapses recorded between 2007 and 2009 in the massif occurred at the base of recently exposed rock slopes.

Ravanel et al. (2012) have examined rock falls at the lower Arête des Cosmiques over 15 years. Its slopes differ strongly in terms of height ($\sim$350 m for the NW face and 50 m for the SE face) and structure (no particular structure visible in the NW face, more massive but much fractured SE face). The vulnerability of this 400-m-long ridge results from a mountain hut with 140 beds situated on its crest (3,613 masl). After the 600-m^3 rock fall that destabilized a part of the refuge in 1998 (Figure 15.7(a)), the SE face of the ridge was observed, and surveyed each year by terrestrial laser scanning

FIGURE 15.7   Instability of the lower Arête des Cosmiques, Mont Blanc massif, France. (a) The rock fall of August 1998 (600 m³); a portion of the refuge became unsettled *(Photograph A. Sage)*. (b) Comparison of the high-resolution 3D models of the SE face of the ridge of October 2009 and October 2010, and position of the four identified rock detachments between the two dates. Scale on the right is in meters (Ravanel et al., 2012).

since 2009 (Figure 15.7(b)): 16 rock falls (20–256 m³) occurred between 2003 and 2011.

Topographic maps from the 1950s and 1970s show an ice cover on the area east and SE of the refuge, which is not the case today. Ice thickness reduced by 40 m between 1979 and 2003 because of the interplay of reduced annual mass balance and the submergence velocity. The evolution of the glacier could thus have interfered with the stability of the ridge. For example, the 1998 rock fall affected a slab that had an ice-covered base until that year, whereas the 2010 rock fall in the couloir located east of the refuge (Figure 15.7(b)) was possibly related to the recent lowering of the glacier. The observed presence of permafrost within the rock mass and the concentration of rock falls during or at the end of warm periods suggest that permafrost degradation could have contributed to their occurrence.

The Cosmiques case highlights how the monitoring of rock slopes, permafrost, and glaciers is necessary for the sustainability of infrastructure and the management of risks in high-mountain areas (Bommer et al., 2010).

15.3.2.2 Rock Avalanches in Canada and Alaska

Rock avalanches onto glaciers are generally characterized by their large-volume, long runout distance, and low travel angle (Geertsema and Cruden, 2008). Many post-LIA rock avalanches in BC occurred on rock slopes above glaciers (Evans and Clague, 1994; Holm et al., 2004). For example, two-thirds of the 18 rock avalanches that occurred in northern BC between 1973 and 2003 (Geertsema et al., 2006a) affected cirque walls where glaciers thinned up to several hundred meters (Geertsema and Chiarle, 2013), as, for example, the 1999 Howson rock avalanche (Figure 15.8). Here some 0.9 Mm^3 of fractured and jointed granodiorite toppled at 1,900 masl, fell and slid 150 m over a 48° slope onto a glacier (Schwab et al., 2003); the rubble transformed into a rock avalanche (2.7 km runout), incorporated some 0.6 Mm^3 of colluvium and till along the slide path, severed a natural gas pipeline, and dammed a stream.

More recent examples of rock/ice avalanches onto glaciers include the events at Mount Steele (Yukon) in 2007, and at Mount Lituya (Alaska) in 2012, with horizontal travel distances of 7 and 9 km (Lipovsky et al., 2008; Geertsema, 2012). Other examples are more modest (Figure 15.9), with travel distances of 2–3 km, but there is a clear association with either cirque walls, or zonation within LIA trim lines.

FIGURE 15.8   The September 11, 1999, Howson rock fall avalanche, Coast Mountains, British Columbia. Note main scarp (yellow line) 150 m above the glacier, and vertical joints (arrows). The rock fall triggered a 2.7-km-long landslide that severed a natural gas pipeline. *Photograph: J.W. Schwab, 16/09/1999.*

FIGURE 15.9   Recent Alaskan rock falls that transformed into rock avalanches near Nunatak Fjord (1.8-km runout) and Tsirku River (3-km runout). The landslides initiated in recently deglaciated zones. *Photographs: M. Geertsema.*

15.3.3 Bedrock Slopes and Debuttressing: Rock Sliding and Deep-Seated Gravitational Slope Deformation

Active over long periods (decades to millennia), many rock slides and DSGSDs in formerly glaciated valleys—characterized by a slow deformation evolving along planes or shear zones at depth—are considered as paraglacial (Ballantyne, 2002). However, lack of dating makes the testing of this hypothesis difficult as the landslides often undergo multiple different phases in their development. Some slides have been studied because they are clearly related to the post-LIA shrinkage of glaciers, whereas others have been monitored because of their strong hazard potential. As bedrock and debris-cover stability are strongly affected by these landslides, they form a source for other mass movements such as earth flows or rock avalanches.

15.3.3.1 Mueller Rock Slide

The Mueller Rock slide (~ 1.1 Mm^2; $100-200$ Mm^3) is a very slow moving, glaciated translational rock slide in the Sealy Range in the central Southern Alps of New Zealand (Figure 15.10). Its crown is near the crest of the slope ($\sim 1,800$ masl) and its toe is some $600-800$ m further down the slope, hidden beneath the surface of the Mueller Glacier. The bedrock is unweathered to

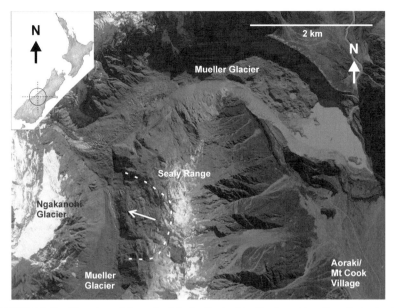

FIGURE 15.10   Plan view of the active Mueller Rock slide, Southern Alps, New Zealand, with the debris-covered Mueller Glacier at the base of the slope. The dashed line delineates the main scarp and subaerial rock slide boundary. Movement is thought to have been initiated or increased by downwasting of the glacier since the Little Ice Age. *Imagery source: GoogleEarth, 5/04/2006.*

slightly weathered very strong greywacke sandstone interbedded with strong argillaceous mudstone. Bedding, which is inferred to be the primary failure surface, dips steeply down the slope on a limb of an anticline (Hancox, 1994; Lillie and Gunn, 1964). The instability of the rock slope was primed by undercutting by the Mueller Glacier, and movement was probably initiated by post-LIA debuttressing from glacier downwasting of 150 m. The magnitude of cumulative annual movement (1−4 m) appears to depend on external factors such as rainfall; however, because the rock slide pushes into and deforms the side of the glacier, it is probably moderated by the ductile behavior of the ice (McColl and Davies, 2012). If the glacier thins further, the stability of the slope will decrease; this will lower the triggering thresholds for rainfall events and increase the rate of movement. Eventually, rapid, catastrophic failure could occur, but not necessarily before the slope is completely debuttressed. This rock slide thus provides an example of a geologically controlled rock slope failure primed for failure by glacial erosion but has its movement rate moderated by glacial debuttressing.

15.3.3.2 Schlossplatte/Alp Bäregg

The instability and then collapse of a 2 Mm^3, 250-m-high, compact limestone mass in 2006−2009 at Schlossplatte, Bernese Oberland, Switzerland, was caused by the Lower Grindelwald Glacier shrinkage, the LIA surface of which was 300 m above its 2006 elevation (Figure 15.11). The collapsed area was subject to a high glacial compression due to its position on a topographic ridge (Figure 15.12).

After the opening of two 250-m-long valley-parallel cracks, and one month of small rock falls, 0.17 Mm^3 toppled in July 2006 (Oppikofer et al., 2008). During one year, the total subsidence of the rear block was 52 m, whereas the front block slid forward by approximately 20 m (Figure 15.11). Dismantling of the rear block was achieved in June 2008, whereas the front block collapsed in August 2008; two residual rock columns partly toppled in October, and disappeared during the summer of 2009.

15.3.3.3 Mont de la Saxe

As a result of the fluctuations of the Pleistocene glaciers (Porter and Orombelli, 1982), the NW side of the ridge of Mont de la Saxe, Aosta Valley, Italy, underwent a DSGSD, forming aligned trenches and closed depressions, scarps,

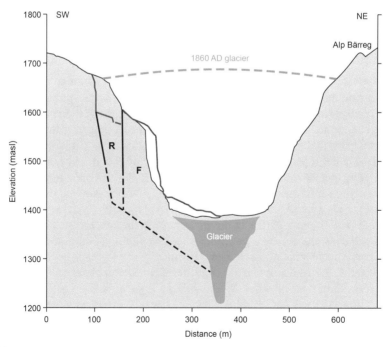

FIGURE 15.11 Cross-section of Schlossplatte/Alp Bäregg trough, Bernese Oberland, Switzerland, with Lower Grindelwald Glacier elevation at the end of the Little Ice Age (Oppikofer et al., 2008; modified). Black profile: prefailure state; red profile: Summer 2007 profile, with deposit at the glacier surface; R: rear block; F: front block.

and counterslope scarps in intensely deformed metasedimentary rocks over an area >4 km². The DSGSD induced block-flexural toppling on rock masses acting on the subvertical tectonic and metamorphic discontinuities. An extremely active and shallow complex landslide with an estimated volume of 8 Mm³ affects the SW end of the ridge, with velocities of 1.5 mm per day during the winter and 15 mm per day during the spring. Since it poses a high risk to the village of Courmayeur, the highway, national road, and the access to the Mont Blanc tunnel, an intensive monitoring system has been initiated since 2002 for understanding and modeling this complex rock slide (Crosta et al., 2013).

15.3.3.4 Northeastern British Columbia

Slope reactions to glacial debuttressing can be complex. Many catastrophic rock avalanches in BC are preceded by long periods of DSGSD, probably with early postglacial onsets. The volume of the catastrophic movements, though often several millions of cubic meters, is only a fraction of the DSGSD volume. Two prominent examples include landslides at Pink Mountain and Turnoff Creek, both 2 km long. Here tension fractures and antiscarps are prominent above the crowns of the rock slides. Strong lichen covers on some of the surfaces corroborate long periods of slope deformation.

Debuttressing effects can result in both rotational and translational rock slides. Geertsema and Chiarle (2013) identified a previously glaciated slope near Neves Creek that responded to glacial debuttressing by two undated prehistoric rock avalanche and rotational rock slide.

FIGURE 15.12   Schlossplatte/Alp Bäregg area in September 2006. A small proglacial lake was dammed by a part of the rock slide deposit; the distal area of the Lower Grindelwald Glacier developed upstream, with a supraglacial lake. *Photograph: S. Coutterand.*

15.3.4 Moraine Stability and Glacier Shrinkage

Glacier surface lowering and retreat have exposed an increasing number of moraine slopes to denudation. Glacial debuttressing of proximal moraine slopes is commonly associated with deep-seated slope failure or smaller slumps. Once adjusted to the glacier withdrawal, moraines continue to be affected by gullying for one to two centuries (Curry et al., 2006), and by breaching.

15.3.4.1 Slides

Many studies have documented translational sliding in moraines (Table 15.1), although debris is often released through a combination of sliding, slumping, debris flow, and blockfall (Ballantyne, 2002). Mattson and Gardner (1991) documented 25 landslides from ice-cored moraines at Boundary Glacier, Alberta Rockies, Canada, which initiated at the ice−debris interface. Most were triggered by rainfall, with some resulting from ice melt that reduced the strength of overlying sediment.

Large deformation of lateral moraines appears to be less common. The sharp-crested right-lateral moraine of the Athabasca Glacier, Alberta Rockies, Canada, rises 150 m above the glacier foreland. Hugenholtz et al. (2008) found that deformation along a 540-m-long section began in the early 1950s, and the (possibly ice-cored) moraine underwent progressive bulging and gravitational deformation from the late 1960s, to create a gap in the moraine crest and a displacement of 55 m toward the glacier with a vertical lowering of 41 m. A network of fractures up to 40 m long has transformed the sharp crest into a series of discontinuous crests. Why the moraine deformed only in this section (a quarter of its length) may be due to both the interaction with adjacent rock glaciers impinging on the distal side of the moraine, and to vibrations caused by daily heavy vehicle traffic used since 1961 for summer glacier tourism.

At the Belvedere Glacier, the crest of a curved section of the right-lateral moraine has been sliding since at least 1889 on a 40° plane, generating a 300-m-long double-crested moraine with a vertical lowering of 10−20 m (Chiarle and Mortara, 2001; Mortara et al., 2009); this has survived three glacier advances in the twentieth Century (Figure 15.13). Sometime before 1951, a translational slide detached several meters below the crest of the proximal side of another section of the same moraine several hundred meters upstream; this 250-m-long section was probably eroded by the 1960s advance. Adjacent to the Belvedere Glacier, the LIA terminal moraine of the Locce Glacier has been affected by a rotational slide starting in early 2005, resulting from the lowering of the surface of the Belvedere after its surge-type flow acceleration of 1999−2004 (Haeberli et al., 2002); by the end of the summer of 2007, the sliding mass had entirely collapsed (Tamburini, 2009).

Lateral moraines at several large valley glaciers in New Zealand show a range of large-scale failure types. At the Tasman Glacier, the right-lateral

TABLE 15.1 Records of Deformation of Alpine Moraines

Glacier	Number, Magnitude, and Mode of Failures	Preparatory Factors	Triggering Factors	Reference
Boundary (Alaska, USA)	25 Landslides	Ice-cored moraine degradation	Rainfall	Mattson and Gardner (1991)
Athabasca (Canada)	Post-1960 gravity deformation with multiple failure surfaces	Glacier thinning >65 m	Traffic vibration? Rock glacier push from upslope?	Hugenholtz et al. (2008)
Tortin (Switzerland)	7 Slides of 110–1,140 m³	Glacier thinning >11 m (2007–2011); warming of permafrost	Contact between base of summer active layer and buried glacier ice	Ravanel and Lambiel (unpublished)
Lower Grindelwald (Switzerland)	2009 Slide of 0.3 Mm³	Glacier thinning >80 m (1985–2000)	Not specified	Stoffel and Huggel (2012)
Belvedere (Italy)	Two slides of 300-m (post-1889) and 250-m length (post 1954)	Glacier thinning	Not specified	Chiarle and Mortara (2001); Mortara et al. (2009)
Locce (Italy)	Rapid rotation slide in 2005, collapsing in 2007.	Glacier thinning; melting of buried ice	Infiltration from moraine-dammed lake	Tamburini (2009)
Tasman (New Zealand)	Progressive down-glacier failure >5 km since 1960s. Multiple slow slides of moraine and valley-side bedrock	Glacier thinning >150 m (1890–2000)	Rainfall; snowmelt	Blair (1994)

FIGURE 15.13 Double-crested moraine on the right margin of the Belvedere Glacier, Monte Rosa massif, Italy. Yellow arrows indicate the slide direction. *Photograph: G. Mortara.*

moraine displays a complex suite of mass movement styles (Figure 15.2). Failure commenced 10 km upstream of the contemporary terminus, where post-1890 glacier thinning was greatest above the insulated, debris-covered tongue. Gullying of the enlarging proximal moraine face preceded the initiation of deep-seated failure in the 1960s. By 1986, this had propagated 4.5 km toward the terminus, at a rate apparently controlled by the rate of glacier surface lowering (Blair, 1994). Subsequently, failure surfaces activated by 1986 have continued to develop. Slumping comprises translational sliding of rigid slabs down the proximal moraine face, rotational slumps incorporating blocks of the entire superposed moraine ridge, slumps incorporating the alluvial fill of the adjacent lateral morainic trough and valley-side debris cones, and deep-seated rock slope failure extending approximately 250 m above the trough. The entire valley side now appears to be at residual strength, perhaps posing the risk of catastrophic failure in coming decades after the supporting glacier commenced rapid calving retreat in 2008 (Dykes et al., 2011).

Melting of internal ice has also caused strong surface deformation in some moraines, for example, at the terminal moraine of Locce Glacier, which caused the hut Rifugio Paradisio to be abandoned by 1975 because of toppling (Mortara et al., 2009).

15.3.4.2 Gullying

Gullying is the fretting of moraine faces by weathering and erosion, to form geometrically regular suites of gullies, most commonly on oversteepened proximal lateral moraine slopes, where gullied faces can extend for several kilometers up valley. Gullying results from frost weathering, surface wash, wind ablation, debris fall, and avalanche (Ballantyne, 2002; Curry et al., 2006).

The upper proximal faces of large moraines (>120-m-height, >30°) are characteristically covered by an array of parallel gullies separated by sharp

FIGURE 15.14   Active erosion on the right-lateral moraine of Mer de Glace, Mont Blanc massif, France, is evident in comparing these two pictures separated by one century. Left: photograph by E. Spelterini from his balloon, August 1909 (Kramer and Stadler, 2007). Right: 2008 orthophoto draped on 4-m DEM. *Courtesy M. Le Roy and RGD 73–74.*

ridges (Figure 15.14), with a gully density up to 60 per km in the central Swiss Alps (Curry et al., 2006) or 110 per km in western Norway (Curry, 1999). Gullies are wider and deeper when incised by streams flowing along the mountain side; debris flows along poorly consolidated and sometimes ice-cored moraines are initiated by rainstorms or melting of snowbeds at gully heads (Ballantyne, 2002). The lower moraine faces comprise debris cones or debris aprons derived from the gullies above.

Glacier thinning exceeds 200 m at the present terminus of the Mer de Glace since the end of the LIA, and 71 m 3.5 km upstream between 1979 and 2008 (Berthier and Vincent, 2012). The right-lateral moraine is incised by many streams coming from the upper glacier basins (Figure 15.14). Terrestrial laser scanning of the moraine's distal face above the two recent proglacial lakes shows that 50- to 200-m-high gullies average 10–20 m deep (up to 64 m), with a longitudinal profile slope angle in the range 21–41°. The rate of gully erosion of this moraine averages 50–100 mm/year with a maximum of 300 mm/year, comparable to the highest values measured on LIA moraines in Norway (Curry, 1999) and Switzerland (Curry et al., 2006).

However, the sediment release is short: maximum dimensions of gullies in the central Swiss Alps are obtained within approximately 50 years of deglaciation, and stabilization occurs within approximately 80–140 years (Curry et al., 2006). Once stabilized, the paraglacial gully system in Norway comprises an upper bedrock-floored source area, a midslope area of gullies with approximately 25° sidewalls, and a slope-foot debris talus (Curry, 1999).

15.3.4.3 Breaching

Moraine breaching is the complete destruction of a section of moraine, and may occur when the glacier is still present or a long time after paraglacial sediment supply from upslope has been exhausted. Breaching represents the terminal grade of linear incision in this environment, and can result from either internal processes (ice core melting, saturation collapse) or the impact of external erosive events (e.g., ice avalanching, lake outburst flood). Where the breached moraine dams a lake, downstream flooding forms a serious hazard due to its magnitude, rapid onset, and unpredictability (e.g., Nostetuko Lake in 1983: Clague and Evans, 2000).

Moraine water saturation due to precipitation was responsible for the September 1993 failure of the LIA moraine of the Mulinet Glacier, Levanne massif, Italy, at 2,525 masl. A heavy and prolonged rainstorm eroded a 50-m-deep and 450-m-long breach, facilitated by buried glacier ice. A 0.8-Mm^3 debris flow traveled >5.6 km and flooded the village of Forno Alpi Graie (Mortara et al., 1995).

Breaching usually affects terminal moraines (e.g., Dolent Glacier in 1990: Lugon et al., 2000), because they are located in more vulnerable positions with respect to the causative processes, but lateral moraines may be affected too. The left lateral moraine of the Belvedere Glacier was progressively buried by an aggrading debris cone from a lateral proglacial stream. The cone reached the moraine crest by the 1940s, triggering incision and removal of a >100-m-long section of this moraine by the stream between 1950 and 1968 (Mortara et al., 2009).

15.3.5 Debris Slopes

Hazards related to permafrost degradation on debris slopes can be subdivided into (1) direct hazards due to the movements of periglacial landforms, especially rock glaciers; (2) indirect hazards induced by the debris supply from active or degrading permafrost; and (3) thermokarst features due to the melting of ground ice.

15.3.5.1 Rock Glacier Movements

Monitoring and observations of rock glaciers throughout the European Alps permit several types of behavior and/or evolution to be distinguished in a continuum (Figure 15.15; Schoeneich et al., submitted for publication):

- Type 1: Moderate multi-annual velocity fluctuations in the range of a few centimeters per year to >2 m/year (Delaloye et al., 2008, 2010).
- Type 2a: Accelerated displacement of a rock glacier with opening of crevasses or scarps on the surface (Roer et al., 2008; Lambiel et al., 2008), ranging from a few meters per year to >10 m/year.
- Type 2b: Very strong acceleration with speed up to >80 m/year (Delaloye et al., 2013).
- Type 3: Acceleration and dislocation of the lower part of a rock glacier, with the formation of scarps (Roer et al., 2008).

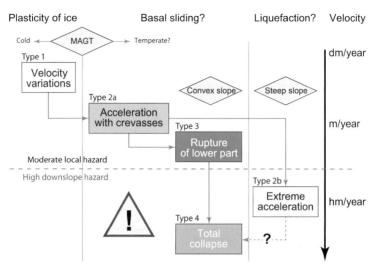

FIGURE 15.15   Evolution trajectories of rock glacier behavior with increasing temperature. *From Schoeneich et al. (2011).*

- Type 4: Collapse of the lower part of the rock glacier, which breaks down as a debris flow; a new front develops from the scarp (Krysiecki et al., submitted for publication).

Type 1 Laurichard rock glacier (Ecrins massif, France) provides one of the longest available displacement measurement series (Bodin et al., 2009). Velocity increased until 2004, then abruptly decreased, and finally showed variations with a frequency of 1–2 years, in relation to changes of mean annual ground surface temperature with a time lag of one to two years (Delaloye et al., 2008; Bodin et al., 2009)—which is not the case with the mean annual air temperature (Buck and Kaufmann, 2008). A warm summer after or before an early snow cover and snowy winters induce acceleration, whereas a late and/or thin snow cover allows a strong cooling of the ground and a deceleration. Velocity fluctuations without a long-term trend represent the response of rock glaciers to multi-annual climate variability; intra-annual variations of the displacement are specific to each rock glacier, in response to its seasonal recharge with liquid water and to temperature shifts in the upper frozen layer (Delaloye et al., 2010).

Recently discovered types 2–4 are often called destabilized rock glaciers. Many of the destabilization phenomena started in the post-1980s, warm decades, but at the Derochoir rock glacier, Mont Blanc massif, France, archives document acceleration between 1888 and 1895, which triggered large debris flows in the downstream torrent; during the whole twentieth century, its mean velocity remained very low (10–20 cm/year) before a very moderate acceleration since 2000.

The internal composition of rock glaciers plays an important role in the destabilization: type 4 Bérard rock glacier (Southern French Alps), for instance, is a "pebbly" or "fine-grained rock glacier" according to Ikeda and Matsuoka (2006), composed of fine schist debris prone to sliding when water-saturated. The initiation of a destabilization phase results from the combined influence of thermal, geometrical/topographical, and mechanical factors over different time scales (Delaloye et al., 2013). Rock glacier creep rates are dependent on the thermal state of permafrost (Kääb et al., 2007; Delaloye et al., 2008). The relationship between atmospheric warming, extreme heat waves, and acceleration or destabilization of many rock glaciers in the Alps during recent decades can be explained by (1) the lower viscosity of ice closer to melting point, (2) possibly the onset of basal sliding in some cases, and (3) an increased presence of unfrozen pore water. For instance the destabilization of type 3 Petit-Vélan and type 2 Tsaté rock glaciers (Valais Alps) between 1988 and 1995 succeeded the strong increase in permafrost temperature that occurred around 1990 (Delaloye and Morard, 2011; Lambiel, 2011), and the collapse of the Bérard rock glacier was triggered by the summer heat waves of 2003 and 2006 (Krysiecki, 2009; Krysiecki et al., submitted for publication).

The topography of the surface over which a rock glacier is moving is a major parameter for destabilization (Avian et al., 2009). A steep slope angle favors higher shear stress, and a convex long profile topography favors extending flow and hence thinning and splitting of the rock glacier, as shown by all reported destabilizations/ruptures of the lower part of rock glaciers.

Slides, rock avalanches, or rock falls may partly overload a rock glacier, initiating a destabilization phase. If the overloading affects the rooting zone, a longer time is necessary for the induced compressive wave to reach the terminal part of the rock glacier: 25 years to achieve the recent "mechanical surge" of the 400-m-long Grabengufer rock glacier (Delaloye et al., 2013).

15.3.5.2 Debris Supply from Permafrost Areas

The connections between permafrost degradation and debris flows have received increased attention in the aftermath of the catastrophic rain and flooding in the Swiss Alps during the summer of 1987, which triggered numerous debris flows on steep, till-covered slopes deglaciated since the end of the LIA (Zimmermann and Haeberli, 1992). Rock glacier fronts also provide significant amounts of debris to torrential systems (Zischg et al., 2011), but the link is not necessarily direct. In the case of active rock glacier fronts, debris supply to the torrential system is a regular process; modulated by the velocity variations of the rock glacier, the rate of debris supply is generally in the range of 10 to a few 100 m^3/year. Episodic acceleration or destabilization phases can drastically increase the debris supply, from 10^3 to more than several 10^4 m^3/year. Melting of ground ice leaves a large amount of loose debris available to erosion, but in many cases, this erosion is limited by the very coarse size of debris.

The effects that this variation of debris supply have on the torrential activity will largely depend on the characteristics of the torrential system. In debris-limited systems, variations of the debris supply due to permafrost degradation could lead to an increase of the magnitude and frequency of debris flows, and a strong destabilization could even trigger a torrential crisis.

15.3.5.3 Ground Ice Melting and Thermokarst

In the European Alps, thermokarst phenomena are commonly due to the presence of buried ice originating from glaciers or avalanche deposits, and more rarely due to periglacial ice lenses. The long-term conservation of buried ice is favored by permafrost conditions. The most striking phenomena are thermokarst lakes and associated outburst floods (e.g., Gruben Glacier area: Haeberli et al., 2001). In the southern French Alps, the Lac Chauvet (2,800 masl) is an ephemeral lake on debris-covered dead ice in a proglacial area that grows for a couple of years before draining through a glacial tunnel, triggering debris flows that can dam the main river (Assier, 1996). The phenomenon repeated at least six times since the 1930s, once in 2008. Geophysical investigations show that the ice and frozen debris are still >40 m thick.

In recent years, several cases of ground ice revealed by thermokarst features were reported in the Alps in places where no presence of ground ice was suspected. This could point to an increase in the melting rate of buried ice bodies, representing an emerging hazard in periglacial mountain areas. Mapping of proglacial areas combined with permafrost distribution modeling could help in identifying potential areas of thermokarst development.

Ice-cemented debris layers can occur on aggrading talus slopes where debris and avalanche snow are deposited in an interleaved fashion. The Dents Blanches (Valais Alps) rock fall in 2006 provided a rare exposure of such a permafrost debris slope by eroding it (Gruber and Haeberli, 2009). The thaw of the pore ice and the interwoven ice layers make sediment hitherto preserved from erosion available for transport. As a consequence, debris flows of unexpected magnitude have to be expected from talus slopes that contain ice.

15.3.6 Interactions and Complex Landslides

Large landslides in mountain environments typically are complex landslides (Cruden and Varnes, 1996), that is, events during which one mode of movement transforms into another. The following cases (mainly from BC) all involve transitions between mass movement types, and illustrate how complex landslides can lead to unexpected consequences, and thus may cause events related to ice loss to have consequences reaching much further than commonly anticipated (cf. Evans and Delaney, 2014).

15.3.6.1 Rock Slide—Debris Avalanche

In 2002 and 2005, landslides at Pink Mountain and Sutherland River traveled approximately 2 and 1.5 km, respectively. Rock rubble slid less than one-third of the total runout on respective sedimentary and volcanic dip slopes; undrained loading of cohesive soil triggered debris avalanches. These landslides are typically modeled with two rheologies: after a frictional rock slide, the rheology resulted in a Bingham flow (Geertsema et al., 2006b).

15.3.6.2 Rock Slide—Debris Flow

Besides the huge rock slide—debris flow that occurred at Mount Meager in 2010 (Guthrie et al., 2012; Evans and Delaney, 2014), a large rock slide at Zymoetz River was initiated in 2002 in a steep cirque basin, but traveled >3 km as a channelized debris flow (Boultbee et al., 2006; Geertsema et al., 2006a; McDougall et al., 2006). The landslide ruptured a natural gas pipeline triggering a forest fire, and dammed Zymoetz River, flooding an important industrial road for more than one year.

15.3.6.3 Rock Slide—Debris Avalanche—Debris Flow

In 2002 ice-cored rubble in a cirque near Harold Price Creek failed and triggered a 4-km-long landslide (Geertsema et al., 2006a). The volcanic rock rubble traveled some 500 m before it loaded the till and colluvium lower on the slope, transforming the moving mass into a debris avalanche that traveled 1.5 km before it channelized into a debris flow.

15.3.6.4 Rock Fall—Debris Storage—Debris Flow (Chain Reaction)

Another type of process interaction is the time-lagged remobilization of material that has been deposited by an ice-loss-related mass movement. For example, the Ritzlihorn rock fall, Switzerland, in July 2009 led to a series of large debris flows during intense rainfall in 2009 and 2010; it damaged a highway and a transnational gas pipeline in the valley floor near Guttannen (Huggel et al., 2012). A large rock fall in the north face of Piz Cengalo, Bergell valley, Switzerland, occurred in December 2011 and deposited 2–3 Mm^3 debris in the upper Bondasca Valley; during a storm in August 2012, several large debris flows traveled 4 km and reached the valley near the village of Bondo. The travel angle of rock fall was exceptionally steep (33°) for an event of this size, whereas debris flows extended the reach of the combined event (Figure 15.16); such debris flows are expected to persist for several years because much erodible material remains in the upper Bondasca Valley.

15.3.6.5 Landslide-Generated Displacement Waves

When large rock slides (often originating from steep and recently deglaciated areas) enter water bodies, they create local displacement tsunami. A wave

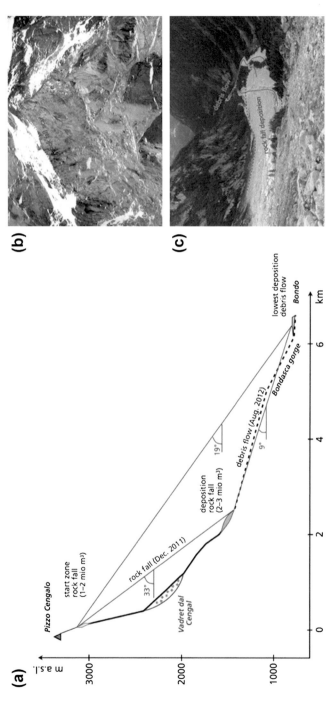

FIGURE 15.16   Piz Cengalo-Bondasca events, Bergell valley, Switzerland: (a) Profile of Piz Cengalo rock fall and Bondasca debris flow; (b) Scar of the rock fall from December 27, 2011 with massive ice in the right section; (c) Rock fall deposit with debris flow channels. *Photographs by S. Salzmann, 25 February 2012 (b) and J- Noetzli, 20 September 2012 (c).*

>50 m high resulted in 1946 from a landslide at the cirque wall at Mount Colonel Foster, Vancouver Island (Evans, 1989). A 2- to 6-m-high displacement wave in Knight Inlet some 500 years ago destroyed a First Nation's Village; it was generated by a 4-Mm^3 landslide detached at a distance of 5 km on an 840-m-high cliff oversteepened during the Last Glacial Maximum/Fraser Glaciation (Bornhold et al., 2007).

15.4 CONCLUSION AND OUTLOOK

The mountain cryosphere has been changing rapidly in recent decades, with strong consequences for morphodynamics. Glaciers are shrinking at a rate never observed since the termination of the LIA. This influences slope instability at different spatial and temporal scales through glacial debuttressing, stress-release fracturing, and crustal rebound. Therefore, large rock slides and DSGSD are triggered or reactivated. Denudation of moraine slopes due to glacier surface lowering allows very active sliding, gullying, and breaching of moraines due to rainfall, infiltration, or outburst flood. Change in the geometry of temperate glaciers reduces abutment and can generate ice avalanches. Warming of cold-based glaciers changes rheology, basal cohesion, water content, and tensile strength of ice, which enables glacier sliding.

At the same time, mountain permafrost is strongly affected by atmospheric warming. In steep rock walls, rock temperature increases and water percolation can lead to strong changes in hydraulic permeability and mechanical strength that alter the stress field. As a consequence, rock fall frequency is increasing in mountain ranges worldwide. In frozen accumulations of coarse debris, thermal changes thicken the active layer with consequences at depths on water infiltration and creep velocity. This can lead to rock glacier destabilization, dislocation, or even collapse, whereas ice-cemented talus slopes and rock glaciers become a source for debris flows and rock falls. This illustrates how processes specific to a high-elevation geomorphic belt can affect lower areas and generate cascading processes, with catastrophic consequences where long runout phenomena (e.g., lake impacted by an ice-rock avalanche) threaten downvalley populations.

Many scientific advances concerning the relationship between ice loss and slope stability in high-mountain regions have been realized in recent decades, often stimulated by critical moments such as the 1987 summer flooding in the Swiss Alps, or the very hot summer of 2003 in the European Alps. As impacts of a changing climate on the cryosphere and the related morphodynamics will be reinforced during the twenty-first century, a race has begun between this acceleration of processes and the understanding of these processes, their sensitivity, and response times. To do this, coordinated observations and monitoring are needed, as illustrated, for instance, by the WGMS (World Glacier Monitoring Service) for glaciers or PERMOS (Swiss Permafrost Monitoring Network) for Swiss mountain permafrost, supporting long-term scientific research as well as physical and numerical modeling. These future research steps will be based on an increasing international scientific collaboration—of which this chapter is in its own way an illustration.

ACKNOWLEDGMENTS

We acknowledge the two editors Wilfried Haeberli and Colin Whiteman, and the two external reviewers Norikazu Matsuoka and Tristram Hales, whose remarks and suggestions helped to improve the chapter.

REFERENCES

Alean, J., 1985. Ice avalanches: some empirical information about their formation and reach. J. Glaciol. 31 (109), 324−333.

Allen, S., Huggel, C., 2013. Extremely warm temperatures as a potential cause of recent high mountain rockfall. Glob. Planet. Change 107, 59−69.

Arvidsson, R., 1996. Fennoscandian earthquakes: whole crustal rupturing related to postglacial rebound. Science 274 (5288), 744−746.

Assier, A., 1996. Glaciers et glaciers rocheux de l'Ubaye. Association Sabença de la Valeia, Barcelonnette, 32 pp.

Augustinus, P.C., 1995. Rock mass strength and the stability of some glacial valley slopes. Z. Geomorphol. 39, 55−68.

Avian, M., Kellerer-Pirklbauer, A., Bauer, A., 2009. LiDAR for monitoring mass movements in permafrost environments at the cirque Hinteres Langtal, Austria, between 2000 and 2008. Nat. Hazards Earth Syst. Sci. 9, 1087−1094.

Ballantyne, C.K., 2002. Paraglacial geomorphology. Quat. Sci. Rev. 21, 1935−2017 http://dx.doi.org/10.1016/S0277-3791(02)00005-7.

Ballantyne, C.K., Benn, D.I., 1994. Paraglacial slope adjustment and resedimentation following recent glacier retreat, Fåbergstølsdalen, Norway. Arct. Alp. Res. 26 (3), 255−269.

Berthier, E., Vincent, V., 2012. Relative contribution of surface mass balance and ice flux changes to 1979−2008 thinning of the Mer de Glace (Alps). J. Glaciol. 58 (209), 501−512.

Blair, R.W., 1994. Moraine and valley wall collapse due to rapid deglaciation in Mount Cook National Park, New Zealand. Mt. Res. Dev. 14, 347−358.

Bodin, X., Thibert, E., Fabre, D., Ribolini, A., Schoeneich, P., Francou, B., Reynaud, L., Fort, M., 2009. Two decades of responses (1986−2006) to climate by the Laurichard rock glacier, French Alps. Permafrost Periglacial Processes 20, 331−344.

Boeckli, L., Brenning, A., Gruber, S., Noetzli, J., 2012. A statistical approach to modelling permafrost distribution in the European Alps or similar mountain ranges. Cryosphere 6, 125−140. http://dx.doi.org/10.5194/tc-6-125-2012.

Bommer, C., Phillips, M., Arenson, L.U., 2010. Practical recommendations for planning, constructing and maintaining infrastructure in mountain permafrost. Permafrost Periglacial Processes 21, 97−104.

Bornhold, B.D., Harper, J.R., McLaren, D., Thomson, R.E., 2007. Destruction of the First Nations village of Kwalate by a rock avalanche—generated tsunami. Atmos. Ocean 45, 123−128.

Boultbee, N., Stead, D., Schwab, J.W., Geertsema, M., 2006. The 2002 Zymoetz River rock slide—debris flow, northwestern British Columbia. Eng. Geol. 83, 76−93.

Buchli, T., Merz, K., Zhou, X., Kinzelbach, W., Springman, S., 2013. Characterization and monitoring of the Furggwanghorn rock glacier, Turtmann Valley, Switzerland: results from 2010 to 2012. Vadose Zone J. 12 (1). http://dx.doi.org/10.2136/vzj2012.0067.

Buck, S., Kaufmann, V., 2008. The influence of air temperature on the creep behaviour of three rockglaciers in the Hohe Tauern. In: Proceedings of the 10th International Symposium on High Mountain Remote Sensing Cartography, Kathmandu.

Budd, W.F., Jacka, T.H., 1989. A review of ice rheology for ice-sheet modeling. Cold Reg. Sci. Technol. 16, 107−144.

Chiarle, M., Mortara, G., 2001. Esempi di rimodellamento di apparati morenici nell'arco alpino italiano. Geografia Fisica e Dinamica Quaternaria (Supplemento V), 41−54.

Clague, J.J., Evans, S.G., 2000. A review of catastrophic drainage of moraine-dammed lakes in British Columbia. Quat. Sci. Rev. 19, 1763−1783.

Cossart, E., Braucher, R., Fort, M., Bourlés, D.L., Carcaillet, J., 2008. Slope instability in relation to glacial debuttressing in alpine areas (Upper Durance catchment, southeastern France): evidence from field data and 10Be cosmic ray exposure ages. Geomorphology 95, 3−26.

Cossart, E., Mercier, D., Decaulne, A., Feuillet, T., Jonsson, H.A., Saemundsson, P., 2013. Impacts of post-glacial rebound on landslide spatial distribution at a regional scale in Northern Iceland (Skagafjörður). Earth Surf. Processes Landforms 39, 336−350. http://dx.doi.org/10.1002/esp. 3450.

Crosta, G.B., di Prisco, C., Frattini, P., Frigerio, G., Castellanza, R., Agliardi, F., 2013. Chasing a complete understanding of the triggering mechanisms of a large rapidly evolving rockslide. Landslides. http://dx.doi.org/10.1007/s10346-013-0433-1.

Cruden, D.M., Varnes, D.J., 1996. Landslide types and processes. In: Turner, A.K., Schuster, R.L. (Eds.), Special Report 247: Landslides Investigation, Mitigation. National Research Council, Transportation Research Board, Washington, DC, pp. 36−75.

Curry, A.M., 1999. Paraglacial modification of slope form. Earth Surf. Processes Landforms 24, 1213−1228.

Curry, A.M., Cleasby, V., Zukowskyj, P., 2006. Paraglacial response of steep, sediment-mantled slopes to post-'Little Ice Age' glacier recession in the central Swiss Alps. J Quat. Sci. 21, 211−225.

Davies, M.C.R., Hamza, O., Harris, C., 2001. The effect of rise in mean annual temperature on the stability of rock slopes containing ice-filled discontinuities. Permafrost Periglacial Processes 12, 137−144.

Delaloye, R., Lambiel, C., 2005. Evidence of winter ascending air circulation throughout talus slopes and rock glaciers situated in the lower belt of alpine discontinuous permafrost (Swiss Alps). Nor. J. Geogr. 59 (2), 194−203.

Delaloye, R., Morard, S., 2011. Le glacier rocheux déstabilisé du Petit-Vélan (Val d'Entremont, Valais): morphologie de surface, vitesses de déplacement et structure interne. In: Lambiel, C., Reynard, E., Scapozza, C. (Eds.), La géomorphologie alpine: entre patrimoine et contrainte. Actes du colloque de la Société Suisse de Géomorphologie, Olivone, Géovisions 36. IGUL, Lausanne, pp. 195−210.

Delaloye, R., Lambiel, C., Roer, I., 2010. Overview of rock glacier kinematics research in the Swiss Alps: seasonal rhythm, interannual variations and trends over several decades. Geogr. Helv. 65 (2), 135−145.

Delaloye, R., Morard, S., Barboux, C., Abbet, D., Gruber, V., Riedo, M., Gachet, S., 2013. Rapidly moving rock glaciers in Mattertal. In: Jahrestagung der Schweizerischen Geomorphologischen Gesellschaft, pp. 21−31.

Delaloye, R., Perruchoud, E., Avian, M., Kaufmann, V., Bodin, X., Hausmann, H., Ikeda, A., Kääb, A., Kellerer-Pirklbauer, A., Krainer, K., Lambiel, C., Mihajlovic, D., Staub, B., Roer, I., Thibert, E., 2008. Recent interannual variations of rock glacier creep in the European Alps. In: Kane, D.L., Hinkel, K.M. (Eds.), Proceeding of the 9th International Conference on Permafrost. University of Alaska, Fairbanks, pp. 343−348.

Draebing, D., Krautblatter, M., 2012. P-wave velocity changes in freezing hard low-porosity rocks: a laboratory-based time-average model. Cryosphere 6, 1163−1174. http://dx.doi.org/10.5194/tc-6-1163-2012.

Dwivedi, R.D., Soni, A.K., Goel, R.K., Dube, A.K., 2000. Fracture toughness of rocks under subzero temperature conditions. Int. J. Rock Mech. Mining Sci. 37, 1267−1275.

Dykes, R., Brook, M., Roberston, C., Fuller, I., 2011. Twenty-first century calving retreat of Tasman Glacier, New Zealand. Arct. Antarct. Alp. Res. 43, 1−10.

Evans, S.G., 1989. The 1946 Mount Colonel Foster rock avalanche and associated displacement wave, Vancouver Island, British Columbia. Can. Geotech. J. 26, 447−452.

Evans, S.G., Clague, J.J., 1988. Catastrophic rock avalanches in glacial environments. In: Bonnard, C. (Ed.), Proceedings of 5th International Symposium on Landslides, vol. 2. Balkema, Rotterdam, pp. 1153−1158.

Evans, S.G., Clague, J.J., 1994. Recent climatic change and catastrophic geomorphic processes in mountain environments. Geomorphology 10 (1−4), 107−128.

Evans, S.G., Delaney, K.B., 2014. Catastrophic Mass Flows in the Mountain Glacial Environment. Elsevier, pp. 563−606.

Evans, S.G., Tutubalina, O.V., Drobyshev, V.N., Chernomorets, S.S., McDougall, S., Petrakov, D.A., Hungr, O., 2009. Catastrophic detachment and high-velocity long-runout flow of Kolka Glacier, Caucasus Mountains, Russia in 2002. Geomorphology 105, 314−321.

Faillettaz, J., Funk, M., Sornette, D., 2012. Instabilities on Alpine temperate glaciers: new insights arising from the numerical modelling of Allalingletscher (Valais, Switzerland). Nat. Hazards Earth Syst. Sci. 12, 2977−2991.

Firth, C.R., Stewart, I.S., 2000. Postglacial tectonics of the Scottish glacio-isostatic uplift centre. Quat. Sci. Rev. 19 (14−15), 1469−1493.

Fischer, L., Huggel, C., 2008. Methodical design for stability assessments of permafrost-affected high-mountain rock walls. In: Kane, D.L., Hinkel, K.M. (Eds.), Proceeding of the 9th International Conference on Permafrost. University of Alaska, Fairbanks, pp. 439−444.

Fischer, L., Amann, F., Moore, J., Huggel, C., 2010. Assessment of periglacial slope stability for the 1988 Tschierva rock avalanche (Piz Morteratsch, Switzerland). Eng. Geol. 116, 32−43. http://dx.doi.org/10.1016/j.enggeo.2010.07.005.

Fischer, L., Eisenbeiss, H., Kääb, A., Huggel, C., Haeberli, W., 2011. Monitoring topographic changes in a periglacial high-mountain face using high-resolution DTMs, Monte Rosa East Face, Italian Alps. Permafrost Periglacial Processes 22, 140−152. http://dx.doi.org/10.1002/ppp.717.

Fischer, L., Kääb, A., Huggel, C., Haeberli, W., 2013. Slope failures and erosion rates on a glacierized high-mountain face under climatic changes. Earth Surf. Processes Landforms 38 (8), 836−846. http://dx.doi.org/10.1002/esp.3355.

Fischer, L., Kääb, A., Huggel, C., Noetzli, J., 2006. Geology, glacier retreat and permafrost degradation as controlling factor of slope instabilities in a high-mountain rock wall: the Monte Rosa east face. Nat. Hazards Earth Syst. Sci. 6, 761−772. http://dx.doi.org/10.5194/nhess-6-761-2006.

Geertsema, M., 2012. Initial observations of the 11 June 2012 rock/ice avalanche, Lituya mountain, Alaska. In: The First Meeting of ICL Cold Region Landslides Network, Harbin, pp. 49−54.

Geertsema, M., Chiarle, M., 2013. Mass-movement causes: glacier thinning. In: Shroder, J.F. (Editor-in-chief), Marston, R.A., Stoffel, M. (Eds.), Treatise on Geomorphology, vol. 7, Mountain and Hillslope Geomorphology. Academic Press, San Diego, pp. 217−222.

Geertsema, M., Clague, J.J., Schwab, J.W., Evans, S.G., 2006a. An overview of recent large landslides in northern British Columbia, Canada. Eng. Geol. 83, 120−143.

Geertsema, M., Cruden, D.M., October 2008. Travels in the Canadian Cordillera. In: Locat, J., Perret, D., Demers, D., Lerouil, S. (Eds.), Proceedings 4th Canadian Conference on Geo-hazards: From Causes to Management. Quebec PQ.

Geertsema, M., Hungr, O., Evans, S.G., Schwab, J.W., 2006b. A large rock slide−debris avalanche at Pink Mountain, northeastern British Columbia, Canada. Eng. Geol. 83, 64−75.

Girard, L., Gruber, S., Weber, S., Beutel, J., 2013. Environmental controls of frost cracking revealed through in situ acoustic emission measurements in steep bedrock. Geophys. Res. Lett. 40 (9), 1748−1753. http://dx.doi.org/10.1002/grl.50384.

Gischig, V.S., Moore, J.R., Evans, K.F., Amann, F., Loew, S., 2011. Thermomechanical forcing of deep rock slope deformation: 1. Conceptual study of a simplified slope. J. Geophys. Res. 116 (F4), F04010.

Gruber, S., Haeberli, W., 2007. Permafrost in steep bedrock slopes and its temperature-related destabilization following climate change. J. Geophys. Res. 112, F02S18. http://dx.doi.org/10.1029/2006JF000547.

Gruber, S., Haeberli, W., 2009. Mountain permafrost. In: Margesin, R. (Ed.), Permafrost Soils, Biology Series, vol. 16. Springer, pp. 33−44. http://dx.doi.org/10.1007/978-3-540-69371-0_3.

Gruber, S., Hoelzle, M., Haeberli, W., 2004a. Rock-wall temperatures in the Alps: modelling their topographic distribution and regional differences. Permafrost Periglacial Processes 15, 299−307. http://dx.doi.org/10.1002/ppp.501.

Gruber, S., Hoelzle, M., Haeberli, W., 2004b. Permafrost thaw and destabilization of Alpine rock walls in the hot summer of 2003. Geophys. Res. Lett. 31, L13504. http://dx.doi.org/10.1029/2004GL020051.

Guthrie, R.H., Friele, P., Allstadt, K., Roberts, N., Evans, S.G., Delaney, K.B., Roche, D., Clague, J.J., Jakob, M., 2012. The 6 August 2010 Mount Meager rock slide-debris flow, Coast Mountains, British Columbia: characteristics, dynamics, and implications for hazard and risk assessment. Nat. Hazards Earth Syst. Sci. 12 (5), 1277−1294.

Haeberli, W., 2013. Mountain permafrost—research frontiers and a special long-term challenge. Cold Reg. Sci. Technol. 96, 71−76. http://dx.doi.org/10.1016/j.coldregions.2013.02.004.

Haeberli, W., Funk, M., 1991. Borehole temperatures at the Colle Gnifetti core-drilling site (Monte Rosa, Swiss Alps). J. Glaciol. 37 (125), 37−46.

Haeberli, W., Gruber, S., 2008. Research challenges for steep and cold terrain. In: Kane, D.L., Hinkel, K.M. (Eds.), Proceeding of the 9th International Conference on Permafrost, Fairbanks, pp. 597−605.

Haeberli, W., Huggel, C., Kääb, A., Zgraggen-Oswald, S., Polkvoj, A., Galushkin, I., Zotikov, I., Osokin, N., 2004. The Kolka−Karmadon rock/ice slide of 20 September 2002: an extraor-dinary event of historical dimensions in North Ossetia, Russian Caucasus. J. Glaciol. 50, 533−546.

Haeberli, W., Kääb, A., Hoelzle, M., Bösch, H., Funk, M., Vonder Mühll, D., Keller, F., 1999. Eisschwund und Naturkatastrophen im Hochgebirge. vdf Hochschulverlag an der ETH Zürich, Zürich, 190pp.

Haeberli, W., Kääb, A., Paul, F., Chiarle, M., Mortara, G., Mazza, A., Deline, P., Richardson, S., 2002. A surge-type movement at Ghiacciaio del Belvedere and a developing slope instability in the east face of Monte Rosa, Macugnaga, Italian Alps. Nor. Geogr. Tidsskr. — Nor. J. Geogr. 56, 104—111.

Haeberli, W., Kääb, A., Vonder Mühll, D., Teysseire, P., 2001. Prevention of outburst floods from periglacial lakes at Grubengletscher, Valais, Swiss Alps. J. Glaciol. 47 (156), 111—122.

Haeberli, W., Wegmann, M., Vonder Mühll, D., 1997. Slope stability problems related to glacier shrinkage and permafrost degradation in the Alps. Eclogae Geol. Helv. 90, 407—414.

Hampel, A., Hetzel, R., Maniatis, G., 2010. Response of faults to climate-driven changes in ice and water volumes on Earth's surface. Philos. Trans. R. Soc. A Math. Phys. Eng. Sci. 368 (1919), 2501—2517.

Hancox, G.T., April 1994. Report on Mt Cook National Park Hut Site Inspections and Establishment of Monitoring Lines. Institute of Geological and Nuclear Sciences. Client Report 353902.20, prepared for Department of Conservation, Mt Cook National Park.

Hasler, A., Gruber, S., Beutel, J., 2012. Kinematics in steep bedrock permafrost. J. Geophys. Res. 117, F01016. http://dx.doi.org/10.1029/2011JF001981.

Hasler, A., Gruber, S., Font, M., Dubois, A., 2011b. Advective heat transport in frozen rock clefts—conceptual model, laboratory experiments and numerical simulation. Permafrost Periglacial Processes 22, 387—398. http://dx.doi.org/10.1002/ppp.737.

Hasler, A., Gruber, S., Haeberli, W., 2011a. Temperature variability and offset in steep alpine rock and ice faces. Cryosphere 5, 977—988. http://dx.doi.org/10.5194/tc-5-977-2011.

Hetzel, R., Hampel, A., 2005. Slip rate variations on normal faults during glacial—interglacial changes in surface loads. Nature 435 (7038), 81—84.

Holm, K., Bovis, M., Jakob, M., 2004. The landslide response of alpine basins to Little Ice Age glacial thinning and retreat in southwestern British Columbia. Geomorphology 57, 201—216.

Hugenholtz, C., Moorman, B., Barlow, J., Wainstein, P., 2008. Large-scale moraine deformation at the Athabasca Glacier, Jasper National Park, Alberta, Canada. Landslides 5 (3), 251—260.

Huggel, C., Clague, J.J., Korup, O., 2012. Is climate change responsible for changing landslide activity in high mountains? Earth Surf. Processes Landforms 37 (1), S.77—S.91.

Ikeda, A., Matsuoka, N., 2006. Pebbly versus bouldery rock glaciers: morphology, structure and processes. Geomorphology 73, 279—296.

Jarman, D., 2006. Large rock slope failures in the Highlands of Scotland: characterisation, causes and spatial distribution. Eng. Geol. 83, 161—182.

Kääb, A., Frauenfelder, R., Roer, I., 2007. On the response of rock glacier creep to surface temperature increase. Glob. Planet. Change 56 (1—2), 172—187.

Kääb, A., Huggel, C., Barbero, S., Chiarle, M., Cordola, M., Epifani, F., Haeberli, W., Mortara, G., Semino, P., Tamburini, A., Viazzo, G., 2004. Glacier hazards at Belvedere Glacier and the Monte Rosa east face, Italian Alps: processes and mitigation. In: Proceedings of the Interpraevent 2004, Riva/Trient, pp. 67—78.

Kleinberg, R.L., Griffin, D.D., 2005. NMR measurements of permafrost: unfrozen water assay, pore-scale distribution of ice, and hydraulic permeability of sediments. Cold Reg. Sci. Technol. 42, 63—77.

Kramer, T., Stadler, H., 2007. Eduard Spelterini. Photographs of a Pioneer Balloonist. Scheidegger & Spiess, Zürich, 147 pp.

Krautblatter, M., 2009. Detection and Quantification of Permafrost Change in Alpine Rock Walls and Implications for Rock Instability (Ph.D. thesis). University of Bonn, Bonn, Germany.

Krautblatter, M., Funk, D., Günzel, F.K., 2013. Why permafrost rocks become unstable: a rock—ice-mechanical model in time and space. Earth Surf. Processes Landforms 38 (8), 876—887. http://dx.doi.org/10.1002/esp.3374.

Krautblatter, M., Verleysdonk, S., Flores-Orozco, A., Kemna, A., 2010. Temperature-calibrated imaging of seasonal changes in permafrost rock walls by quantitative electrical resistivity tomography (Zugspitze, German/Austrian Alps). J. Geophys. Res. Earth Surf. 115, F02003.

Krysiecki, J.-M., 2009. Rupture du glacier rocheux du Bérard (Alpes de Haute Provence): analyses géomorphologiques et premiers résultats du suivi mis en place sur le site. Environnements Périglaciaires 16, 65−78.

Krysiecki, J.-M., Bodin, X., Schoeneich, P., Le Roux, O., Lorier, L., Echelard, T., Peyron, M. The collapse of the Bérard rock glacier (Southern French Alps) in 2006 and its post-event evolution: insights from geodetic, thermal and geophysical datasets. Permafrost Periglacial Processes, submitted for publication.

Lambiel, C., 2011. Le glacier rocheux déstabilisé de Tsaté-Moiry (VS) : caractéristiques morphologiques et vitesses de déplacement. In: Lambiel, C., Reynard, E., Scapozza, C. (Eds.), La géomorphologie alpine: entre patrimoine et contrainte. Actes du colloque de la Société Suisse de Géomorphologie, Olivone, Géovisions 36. IGUL, Lausanne, pp. 211−224.

Lambiel, C., Delaloye, R., Strozzi, T., Lugon, R., Raetzo, H., 2008. ERS InSAR for detecting the rock glacier activity. In: Kane, D.L., Hinkel, K.M. (Eds.), Proceeding of the 9th International Conference on Permafrost, Fairbanks, pp. 1019−1024.

Leith, K.J., 2012. Stress Development and Geomechanical Controls on the Geomorphic Evolution of Alpine Valleys (Ph.D. thesis), ETH Zurich, 170 pp.

Leith, K., Amann, F., Moore, J.R., Kos, A., Loew, S., 2010. Conceptual modelling of near-surface extensional fracture in the Matter and Saas Valleys, Switzerland. In: Williams, A.L., Pinches, G.M., Chin, C.Y., McMorran, T.J., Massey, C.I. (Eds.), Geologically Active: Delegate Papers 11th Congress of the International Association for Engineering Geology and the Environment, Auckland, Aotearoa. CRC Press, Auckland, pp. 363−371.

Lillie, A.R., Gunn, B.M., 1964. Steeply plunging folds in the sealy range, southern Alps. N.Z. J. Geol. Geophys. 7, 403−423.

Lipovsky, P.S., Evans, S.G., Clague, J.J., Hopkinson, C., Couture, R., Bobrowsky, P., Ekström, G., Demuth, M.N., Delaney, K.B., Roberts, N.J., Clarke, G., Schaeffer, A., 2008. The July 2007 rock and ice avalanches at Mount Steele, St. Elias Mountains, Yukon, Canada. Landslides 5 (4), 445−455. http://dx.doi.org/10.1007/s10346-008-0133-4.

Lugon, R., Gardaz, J.-M., Vonder Mühll, D., 2000. The partial collapse of the Dolent glacier moraine (Mont Blanc range, Swiss Alps). Z. Geomorphol. Supplement-Band 122, 191−208.

Marchenko, S.S., Gorbunov, A.P., Romanovsky, V.E., 2007. Permafrost warming in the Tien Shan mountains, Central Asia. Glob. Planet. Change 56, 311−327. http://dx.doi.org/10.1016/j.gloplacha.2006.07.02.

Matsuoka, N., Murton, J., 2008. Frost weathering: recent advances and future directions. Permafrost Periglacial Processes 19 (2), 195−210.

Mattson, L.E., Gardner, J.S., 1991. Mass wasting on valley-side ice-cored moraines, Boundary Glacier, Alberta, Canada. Geogr. Ann. 73A, 123−128.

McColl, S.T., 2012. Paraglacial rock-slope stability. Geomorphology 153−154, 1−16.

McColl, S.T., Davies, T.R.H., 2012. Large ice-contact slope movements; glacial buttressing, deformation and erosion. Earth Surf. Processes Landforms 38 (10), 1102−1115. http://dx.doi.org/10.1002/esp.3346.

McDougall, S., Boultbee, N., Hungr, O., Stead, D., Schwab, J.W., 2006. The Zymoetz River landslide, British Columbia, Canada: description and dynamic analysis of a rock slide−debris flow. Landslides 3, 195−204.

Mellor, M., 1973. Mechanical properties of rocks at low temperatures. In: 2nd Int. Conference on Permafrost, Yakutsk, pp. 334−344.

Morard, S., Delaloye, R., Dorthe, J., 2008. Seasonal thermal regime of a mid-latitude ventilated debris accumulation. In: Kane, D.L., Hinkel, K.M. (Eds.), Proceeding of the 9th International Conference on Permafrost, Fairbanks, pp. 1233−1238.

Mortara, G., Dutto, F., Godone, F., 1995. Effetti degli eventi alluvionali nell'ambiente proglaciale. La sovraincisione della morena del Ghiacciaio del Mulinet (Stura di Valgrande, Alpi Graie). Geogr. Fis. Din. Quat. 18, 295−304.

Mortara, G., Tamburini, A., Chiarle, M., Haeberli, W., Epifani, F., 2009. Le morene: sull'orlo del collasso. In: Mortara, G., Tamburini, A. (Eds.), Il ghiaccao del Belvedere e l'emergenza del lago Effimero. Regione Piemonte − Edizioni Società Meteorologica Subalpina, pp. 123−134.

Muir-Wood, R., 2000. Deglaciation seismotectonics: a principal influence on intraplate seismo-genesis at high latitudes. Quat. Sci. Rev. 19 (14−15), 1399−1411.

Murton, J.B., Peterson, R., Ozouf, J.-C., 2006. Bedrock fracture by ice segregation in cold regions. Science 314, 1127−1129. http://dx.doi.org/10.1126/science.1132127.

Nichols Jr., T.C., 1980. Rebound, its nature and effect on engineering works. Q. J. Eng. Geol. Hydrogeol. 13 (3), 133−152.

Noetzli, J., Gruber, S., 2009. Transient thermal effects in Alpine permafrost. Cryosphere 3 (1), 85−99. http://dx.doi.org/10.5194/tc-3-85-2009.

Noetzli, J., Gruber, S., Kohl, T., Salzmann, N., Haeberli, W., 2007. Three-dimensional distribution and evolution of permafrost temperatures in idealized high-mountain topography. J. Geophys. Res. Earth Surf. 112 (F2), F02S13. http://dx.doi.org/10.1029/2006JF000545.

Noetzli, J., Hoelzle, M., Haeberli, W., 2003. Mountain permafrost and recent Alpine rock-fall events: a GIS-based approach to determine critical factors. In: Proceedings of the 8th International Conference on Permafrost, Zürich, pp. 827−832.

Oppikofer, T., Jaboyedoff, M., Keusen, H.-R., 2008. Collapse at the eastern Eiger flank in the Swiss Alps. Nat. Geosci. 1, 531−535. http://dx.doi.org/10.1038/ngeo258.

PERMOS, 2013. Permafrost in Switzerland 2008/2009 and 2009/2010. In: Noetzli, J. (Ed.), Glaciological Report (Permafrost) No. 10/11 of the Cryospheric Commission of the Swiss Academy of Sciences.

Pogrebiskiy, M.I., Chernyshev, S.N., 1977. Determination of the permeability of the frozen fissured rock massif in the vicinity of the Kolyma Hydroelectric Power Station. Cold Reg. Res. Eng. Lab. − Draft Translation 634, 1−13.

Porter, S.C., Orombelli, G., 1982. Late-glacial ice advances in the western Italian Alps. Boreas 11, 125−140.

Pralong, A., 2005. On the Instability of Hanging Glaciers, vol. 189. Mitteilungen, Laboratory of Hydraulics Hydrology and Glaciology, ETH Zürich, Switzerland, 156 pp.

Pralong, A., Funk, M., 2006. On the instability of avalanching glaciers. J. Glaciol. 52 (176), 31−48.

Ravanel, L., Deline, P., 2008. La face ouest des Drus (massif du Mont-Blanc): évolution de l'in-stabilité d'une paroi rocheuse dans la haute montagne alpine depuis la fin du Petit Age Glaciaire. Géomorphologie 4, 261−272.

Ravanel, L., Deline, P., 2011. Climate influence on rockfalls in high-Alpine steep rockwalls: the North side of the Aiguilles de Chamonix (Mont Blanc massif) since the end of the Little Ice Age. Holocene 21, 357−365.

Ravanel, L., Allignol, F., Deline, P., Bruno, G., 2011. Les écroulements rocheux dans le massif du Mont-Blanc pendant l'été caniculaire de 2003. In: Lambiel, C., Reynard, E., Scapozza, C.

(Eds.), La géomorphologie alpine : entre patrimoine et contrainte. Actes du colloque de la Société Suisse de Géomorphologie, Olivone, Géovisions 36. IGUL, Lausanne, pp. 245−261.

Ravanel, L., Allignol, F., Deline, P., Gruber, S., Ravello, M., 2010. Rock falls in the Mont Blanc massif in 2007 and 2008. Landslides 7, 493−501.

Ravanel, L., Deline, P., Lambiel, C., Vincent, C., 2012. Instability of a highly vulnerable high alpine rock ridge: the lower Arête des Cosmiques (Mont Blanc massif, France). Geogr. Ann. Ser. A 95 (1), 51−66. http://dx.doi.org/10.1111/geoa.12000.

Roer, I., Haeberli, W., Avian, M., Kaufmann, V., Delaloye, R., Lambiel, C., Kääb, A., 2008. Observations and considerations on destabilizing active rockglaciers in the European Alps. In: Kane, D.L., Hinkel, K.M. (Eds.), Proceeding of the 9th International Conference on Permafrost, Fairbanks, pp. 1505−1510.

Röthlisberger, H., 1981. Eislawinen und Ausbrüche von Gletscherseen. In: Kasser, P. (Ed.), Gletscher und Klima—glaciers et climat, Jahrbuch der Schweizerischen Naturforschenden, Gesellschaft, wissenschaftlicher Teil 1978, pp. 170−212.

Sanchez, G., Rolland, Y., Corsini, M., Braucher, R., Bourlès, D., Arnold, M., Aumaître, G., 2010. Relationships between tectonics, slope instability and climate change: cosmic ray exposure dating of active faults, landslides and glacial surfaces in the SW Alps. Geomorphology 117 (1−2), 1−13.

Sanderson, T., 1988. Ice Mechanics and Risks to Offshore Structures. Springer, Amsterdam, 270 pp.

Schaub, Y., Haeberli, W., Huggel, C., Künzler, M., Bründl, M., 2013. Landslides and new lakes in deglaciating areas: a risk management framework. In: Margottini, C., Canuti, P., Sassa, K. (Eds.), Landslide Science and Practice (Proceedings of the 2nd World Landslide Forum, Rome), Social and Economic Impacts and Policies, vol. 7. Springer, pp. 31−38. http://dx.doi.org/10.1007/978-3-642-31313-4_5.

Scherler, M., Hauck, C., Hoelzle, M., Salzmann, N., 2013. Modeled sensitivity of two alpine permafrost sites to RCM-based climate scenarios. J. Geophys. Res. Earth Surf. 118, 780−794. http://dx.doi.org/10.1002/jgrf.20069.

Schoeneich, P., Bodin, X., Echelard, T., Kellerer-Pirklbauer, A., Krysiecki, J.-M., Lieb, G.K., 2014. Velocity changes of rock glaciers and induced hazards. In: Proceedings IAEG XII Congress, 15−19 September 2014, Torino, Italy.

Schoeneich, P., Bodin, X., Kellerer-Pirklbauer, A., Krysiecki, J.-M., Lieb, G.K., 2011. Chapter 1: Rockglaciers. In: Schoeneich, P., Dall'Amico, M., Deline, P., Zischg, A. (Eds.), Hazards Related to Permafrost and to Permafrost Degradation. PermaNET project, state-of-the-art report 6.2. On-line publication. ISBN: 978-2-903095-59-8, 27 pp.

Schwab, J.W., Geertsema, M., Evans, S.G., 2003. Catastrophic rock avalanches, west-central B.C., Canada. In: 3rd Canadian Conference on Geotechnique and Natural Hazards, Edmonton, pp. 252−259.

Stewart, I.S., Sauber, J., Rose, J., 2000. Glacio-seismotectonics: ice sheets, crustal deformation and seismicity. Quat. Sci. Rev. 19 (14−15), 1367−1389.

Stoffel, M., Huggel, C., 2012. Effects of climate change on mass movements in mountain environments. Prog. Phys. Geogr. 36 (3), 421−439. http://dx.doi.org/10.1177/0309133312441010.

Tamburini, A., 2009. L'evoluzione della morena del Ghiacciao delle Locce. In: Mortara, G., Tamburini, A. (Eds.), Il Ghiacciao del Belvedere e l'emergenza del lago Effimero. Regione Piemonte − Edizioni Società Meteorologica Subalpina, pp. 140−144.

Tang, G.Z., Wang, X.H., 2006. Modeling the thaw boundary in broken rock zones in permafrost in the presence of surface water flows. Tunn. Undergr. Space Technol. 21, 684−689.

Terzaghi, K., 1962. Stability of steep slopes in hard unweathered rock. Geotechnique 12, 251−270.

Wagner, S., 1996. Dreidimensionale Modellierung zweier Gletscher und Deformationsanalyse von eisreichem Permafrost, vol. 146. Mitteilungen, Laboratory of Hydraulics Hydrology and Glaciology, ETH Zürich, Switzerland.

Wegmann, M., Gudmundsson, G.H., Haeberli, W., 1998. Permafrost changes in rock walls and the retreat of Alpine glaciers: a thermal modelling approach. Permafrost Periglacial Processes 9, 23–33.

Zemp, M., Haeberli, W., Hoelzle, M., Paul, F., 2006. Alpine glaciers to disappear within decades? Geophys. Res. Lett. 33, L13504. http://dx.doi.org/10.1029/2006GL026319.

Zenklusen Mutter, E., Phillips, M., 2012a. Active layer characteristics at ten borehole sites in Alpine permafrost terrain, Switzerland. Permafrost Periglacial Processes 23, 138–151. http://dx.doi.org/10.1002/ppp.1738.

Zenklusen Mutter, E., Phillips, M., 2012b. Thermal evidence of recent talik formation in Ritigraben rock glacier: Swiss Alps. In: Hinkel, K.M. (Ed.), Proceedings of the 10th International Conference on Permafrost, Salekhard. The Northern Publisher, Salekhard, pp. 479–483.

Zgraggen, A., 2005. Measuring and Modelling Rock Surface Temperatures in the Monte Rosa East Face (Diploma thesis), ETH Zurich.

Zimmermann, M., Haeberli, W., 1992. Climatic change and debris flows activity in high mountain areas: a case study in the Swiss Alps. Catena Supplement 22, 59–72.

Zischg, A., Curtaz, M., Galuppo, A., Lang, K., Mayr, V., Riedl, C., Schoeneich, P., 2011. Chapter 2: permafrost and debris-flows. In: Schoeneich, P., Dall'Amico, M., Deline, P., Zischg, A. (Eds.), Hazards Related to Permafrost and to Permafrost Degradation. PermaNET project, state-of-the-art report 6.2. On-line publication, pp. 29–66. ISBN: 978-2-903095-59-8.

Catastrophic Mass Flows in the Mountain Glacial Environment

Stephen G. Evans and Keith B. Delaney
Natural Disaster Systems, Department of Earth and Environmental Sciences,
University of Waterloo, Waterloo, Ontario, Canada

ABSTRACT

Catastrophic mass flows (CMFs) in the glacial environment are an important geomorphic process that may pose significant hazard to communities and infrastructure in glacierized mountains. CMFs form a broad range of glacial hazards and include mass movements of glacial ice, rock avalanches, ice—rock avalanches, glacial debris flows, and outburst-generated flows. Broadly, CMFs in the glacial environment are characterized by a sudden onset, high mean velocity (≥ 5 m/s), high mobility (i.e., long runout in relation to volume), and generally involve a mixture of earth materials, water, snow, and ice. In some cases, CMF runout may exceed 100 km from source. CMFs commonly undergo dramatic process transformation during movement in response to melting of entrained ice and snow, entrainment of additional materials along its path, river-damming effects, and incorporation or displacement of water in the periglacial environment; process complexity thus represents a challenge to quantitative hazard assessment. CMFs initiate in uninhabited glacial environments and frequently descend into denser populated areas where they have an impact on mountain communities and infrastructure. CMFs have been responsible for several notable mountain disasters since 1940, resulting in the death of >15,000 people worldwide. Our focus on an examination of process illuminates an assessment of CMF hazard in glacierized mountain regions and forms the basis for the development of mitigation strategies based on detection, warning, engineering techniques in source and run-out areas, and land-use controls. The precise relationship between the magnitude/frequency of CMFs and cryospheric change in the mountain glacial environment since ca. 1900 remains uncertain.

Snow and Ice-Related Hazards, Risks, and Disasters. http://dx.doi.org/10.1016/B978-0-12-394849-6.00016-0

16.1 INTRODUCTION

Catastrophic mass flows (CMFs) in the glacial environment[1] are an important geomorphic process that may pose a significant hazard to communities and infrastructure in glacierized mountains. They form a broad range of glacial hazards and include mass movements of glacial ice, ice—rock avalanches, rock avalanches (Figure 16.1), glacial debris flows, and outburst-generated flows (e.g., Hanke, 1966; Röthlisberger, 1978; Tufnell, 1984; Grove, 1987; Haeberli et al., 1989; Dutto and Mortara, 1992; Evans and Clague, 1993, 1994; McSaveney, 2002; Huggel et al., 2004a,b; Kääb et al., 2005; Petrakov et al., 2008; Allen et al., 2009; Stoffel and Huggel, 2012; Portocarrero, 2014; Delaney and Evans, 2014; Deline et al., 2014; Clague and O'Connor, 2014).

FIGURE 16.1   Mount Cook rock avalanche of December 14, 1991, Mount Cook National Park, Southern Alps, New Zealand (McSaveney, 2002). The CMF was initiated when a mass (volume ~11 Mm^3) of rock and ice detached from the peak of Mount Cook (3,754 m asl), New Zealand's highest mountain. Debris traveled over the Hochstetter Glacier and came to rest on debris-covered Tasman Glacier (vertical fall distance = 2,720 m) in the foreground. The average velocity was estimated as 60 m/s (McSaveney, 2002). *Photograph by L. Homer, GNS Science CN21418/12.*

---

1. For the purposes of this chapter, we define the glacial environment in a broad sense to include the glacier body itself, terrain, materials, and water bodies directly adjacent to the glacier, or within the glacier foreland marked by Holocene moraines (e.g., periglacial rock slopes, degraded morainic complexes, and proglacial lakes).

FIGURE 16.2 Oblique aerial photograph of Huaraz, Peru, showing the swath of destruction associated with the December 13, 1941, aluvión. Approximately 4,000 persons lost their lives in the 1941 event. The Rio Santa, visible in the foreground, flows to the left (north). The view is to the east. Compare with Figure 16.21. *Photograph taken by A. Heim on June 24, 1947 (reproduced from Heim, 1948).*

Broadly, CMFs in the glacial environment are characterized by a sudden onset, high mean velocity (≥ 5 m/s), high mobility (i.e., long runout in relation to volume), and generally involve a mixture of earth materials, water, snow, and ice. As we will detail, CMFs commonly undergo dramatic process transformation during movement in response to the melting of entrained ice and snow, entrainment of additional materials along their path, river-damming effects, and incorporation or displacement of water (cf. Scott et al., 2001).

CMFs in the glacier environment have been responsible for a number of major natural disasters in mountain regions since 1940. These events include the 1941 Huaraz disaster, Peru (Figure 16.2; Oppenheim, 1946; Heim, 1948; Carey, 2005, 2010), the Huascarán mass flows of 1962 and 1970, Peru (Evans et al., 2009a), the Kolka Glacier event of 2002, Russia (Haeberli et al., 2004; Huggel et al., 2005; Evans et al., 2009b), the 2012 Gayari ice—debris avalanche, Pakistan (Schneider et al., 2013), the 2012 Seti debris flows (Hanisch et al., 2013), Nepal, and the Kedarnath debris floods, India, in 2013.

The objective of this chapter is to undertake a global review of the range of processes involved in CMFs in the glacial environment and to examine how an understanding of process illuminates an assessment of CMF hazard in glacierized mountains.

16.2 CATASTROPHIC MASS FLOWS IN THE MOUNTAIN GLACIAL ENVIRONMENT—GENERAL CHARACTERISTICS

As noted above, CMFs in the glacial environment can consist of a mixture of (1) initially displaced earth materials (rock and soil), ice blocks, and snow;

(2) water (derived from melted snow and/or ice or incorporated/released from water bodies); and (3) material entrained from the path of the flow. The volumetric concentration of these components may vary at flow inception and can change dramatically during flow development (e.g., Niyazov and Degovets, 1975; Casassa and Marangunic, 1993; Evans et al., 2009a). Their complexity reflects the complexity of a source glacial environment that is characterized by high relative relief, steep slopes, fractured rock masses, glacial ice (either as part of an active glacier or as stagnant buried ice), available unconsolidated materials (including a variety of morainic materials), thermal sensitivity, mountain permafrost, marked seasonal changes in temperature, meltwater within source materials, and the availability of potentially large water volumes in englacial, supraglacial, and proglacial lakes. Superimposed on this complexity is an unstable climate-sensitive cryospheric environment in which rapid decadal/subdecadal-scale changes occur in glacier ice limits, glacial lake volume, and source-slope thermal regimes (e.g., Ravanel et al., 2010).

A significant feature of CMFs in the glacial environment that complicates accurate hazard assessment is that they are frequently transformed during movement by the addition of entrained materials (Seinova et al., 2003, 2007; Casassa and Marangunic, 1993; Evans et al., 2009a) and/or the melting of included ice and/or snow. These processes result in significant volume variations during movement, changes in mass flow behavior, and high fluidity-related mobility, the net effect of which is to increase mass and decrease frictional resistance that in turn leads to an increase in the specific energy (Joules per kilogram) of the flow event (Delaney and Evans, 2014). CMFs can be significant hazards to human activity in mountainous regions when they descend into inhabited areas at lower elevations; the destructive potential of CMFs arises from their substantial mass and high velocity that combine to produce high impact forces on community and infrastructural elements (e.g., Jakob et al., 2012) that may be located at some distance from the source glacial environment.

As a framework for discussion, we distinguish five groups of CMFs in the mountain glacial environment. The first group consists of mass flows mainly involving glacier ice (glacier avalanches and large-scale glacier detachments). The second group consists of mass flows (rock avalanches) that mostly involve, if not exclusively, fragmented rock originating in the failure of a periglacial rock slope and involving, at least in part, travel over a glacier surface. Rockfall-generated snow avalanches (e.g., Barla et al., 2000; Giani et al., 2001) are excluded from our discussion. A third group is a hybrid of the first two groups and involves mass flows containing significant proportions of ice and rock, so-called rock−ice avalanches. In all three cases, initial flows may be sequentially transformed into more fluid distal debris flows, debris floods, hyperconcentrated flows, and/or true water floods, particularly following interaction with water bodies (Kjartansson, 1967; Wiles and Calkin,

1992; McSaveney, 2002), in which the flow undergoes a reduction in sediment concentration.

Groups four and five consist of two types of proglacial debris flows that originate in the immediate proglacial environment in the vicinity of glacier limits, particularly where recent glacier retreat has exposed readily mobilized morainic material. Group four consists of glacial debris flows (not related to outburst processes) that initially mobilize morainic material from the proximity of glacier margins, and group five consists of debris flows (and related flow processes) generated by catastrophic outbursts involving the sudden release of water from glacial lakes and subsequent entrainment of material downstream (e.g., Clague et al., 1985; Clague and Evans, 1994; Petrakov et al., 2007, 2011; Portocarrero, 2014).

It is noted that this framework is not a rigorous typology, or classification, of mass flows in the glacial environment; rather we recognize that, most commonly, CMFs exhibit a range of processes during movement. Our framework is thus based on the identification of CMF source mechanisms and materials and the acknowledgment that the direction and type of process development during the flow (after initiation) may be multiple and complex.

16.3 MASS FLOWS INVOLVING MAINLY GLACIER ICE (GLACIER AVALANCHES AND LARGE-SCALE GLACIER DETACHMENTS)

Mass flows involving mainly, or exclusively, glacial ice originate in two settings: (1) as glacier avalanches, mainly on steep valley sides (from small "hanging glaciers"), or low-order tributary valleys; and (2) what we will term large-scale glacier detachments, originating in valley floor locations within higher order drainage basins.

Glacier avalanches (Alean, 1984, 1985a,b; Röthlisberger, 1977, 1978; van der Woerd et al., 2004) generally occur on steep valley-side slopes, steep tributary valleys involving "hanging glaciers" (Alean, 1985b), or steep glacier basin head walls; they involve a mass of glacier ice that breaks off from the toe region of the main body of the glacier (or segments of the glacier on steep head walls). As the mass moves downslope, it disintegrates and forms a high-velocity flow of fragmented ice.

Glacier avalanches have been well documented in the European Alps since the final decades of the nineteenth century (e.g., Heim, 1896; Capello, 1959; Röthlisberger, 1978; Alean, 1985a; Huggel et al., 2004a,b). Particularly well-documented examples are the September 11, 1895 Altels (Switzerland), and the August 30, 1965 Allalin (Switzerland), events.

In the 1895 Altels case (Heim, 1896; Forel, 1895; Du Pasquier, 1896; Röthlisberger, 1977, 1978; Faillettaz et al., 1896), approximately 4.5 Mm^3 of glacier ice detached from the toe of the Altels Glacier and descended a vertical distance of 1,440 m to the valley floor, running up a vertical distance of 320 m on the opposite valley side. Velocities in the valley bottom were estimated at 60 m/s (Röthlisberger, 1978). Detachment took place at the boundary of the Altels glacier and the underlying bedrock sloping at a 30−35° downslope (Alean, 1985a). The current understanding of the Altels event (Faillettaz et al., 2011) centers on the hypothesis that warming of the glacier bed and the redistribution of glacier mass, possibly linked to the degradation of an ice bonding to its bed after the warm summer of 1895, were key factors in failure. The Altels event is the largest glacier avalanche recorded in the European Alps; a similar glacier avalanche had occurred at the site in August 1782 (Forel, 1895; Röthlisberger, 1977).

In 1965, the Allalin glacier avalanche resulted in the deaths of 88 construction workers at the Mattmark dam site (Figure 16.3; Röthlisberger, 1978; Vivian, 1966; Faillettaz et al., 2012). The event involved the detachment of between 1 and 2 Mm^3 from the toe region of the Allalin Glacier, and an avalanche of fragmented ice into the glacier foreland marked by Little Ice Age lateral moraines (Figure 16.3). The vertical displacement in the center of gravity was about 450 m (Röthlisberger, 1978), indicating a specific energy for the event of 4.4×10^3 J/kg. The Allalin Glacier had undergone rapid retreat since the 1920s (Figure 16.3; Röthlisberger, 1978) and the retreat of the glacier up onto the steep valley-side rock slope (average slope of basal sliding plane $= 27°$) provided the conditions for catastrophic sliding in the 1965 event (Faillettaz et al., 2012). A further glacier avalanche involving about 1 Mm^3 broke off the toe of the glacier in 2000 (Faillettaz et al., 2012). It is noted by Röthlisberger (1978) that the tongue of the Allalin glacier showed marked signs of movement before the catastrophic avalanche of August 1965; similar prefailure movements were noted in 2000.

Hazard assessment for glacier avalanches is complex (e.g., Röthlisberger, 1978; Margreth and Funk, 1999; Salzmann et al., 2004; Margreth et al., 2011). Attempts have been made to quantify the initial detachment conditions for glacier avalanches involving the break-off of steep sections of hanging glaciers (Pralong and Funk, 2006; Faillettaz et al., 2011, 2012). Several factors determine the occurrence of catastrophic break-off and its magnitude. These include the shear strength at the base of the glacier ice (determined by freezing (adhesion) at the glacier ice/glacier bed boundary, and pore pressures at the base of the ice), the inclination and shape of the basal slope, and the tensile strength of the glacier body itself (determined by tension cracks). Some of these factors are altitude sensitive, for example, slope steepness and basal freezing increase with altitude (Alean 1985 a,b).

Monitoring of prefailure movements has developed time histories of displacement that show power law relationships in time to failure similar to that for rock slopes. Empirical data show that run-out distance is broadly related to volume (Alean, 1985a, Figure 7).

An example of a community exposed to glacier avalanche risk is the village of Randa (Valais) in the Swiss Alps. The village is subject to glacier avalanche

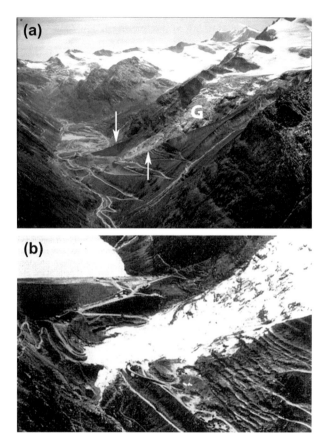

FIGURE 16.3   Aerial views of the site of the Allalin Glacier avalanche (Swiss Alps) before (a) and after (b) the August 30, 1965, event. (a) view upstream (to the south) of the Saas Valley during the initial stages of the Mattmark dam construction. Neoglacial lateral moraines and steep frontal slope (average slope 26°) in the vicinity of the terminus of the Allalin Glacier (G) are evident (vertical arrows). As noted by Röthlisberger (1978), the Allalin Glacier reached down into the Saas valley during the Little Ice Age from the end of the sixteenth century to the beginning of the twentieth Century and occasionally blocked it during this period. (ETH archive photograph http://dx.doi:10.3932/ethz-a-000040318). (b) View of glacier avalanche debris. The avalanche (volume ~2 Mm³) covered part of the Mattmark Dam construction camp causing the death of 88 workers at the construction site. Mattmark Dam visible in the left background. (cf. Röthlisberger 1978, Evans and Clague 1993).

hazard from a hanging glacier on the slopes of the Weisshorn, as well as possible instability of the steeply sloping tongue of the Bis glacier itself directly above the village (Röthlisberger, 1978). Randa was struck several times by glacier avalanches during the period 1636–1819, a period corresponding to the Little Ice Age; some of these events dammed the Vispa River that flows through the valley. Glaciers in the Randa area have been subject to dramatic retreat since the beginning of the twentieth century creating favorable conditions for further glacier ice avalanching (Röthlisberger, 1978).

Glacier avalanches can be triggered by earthquakes (Post, 1967; Slingerland and Voight, 1979; van der Woerd et al., 2004) and thus in these cases are not a passive response to changing thermal conditions and/or glacier ice loadings. van der Woerd et al. (2004) documented numerous large glacier avalanches in the Kunlun Shan mountains of northern Tibet triggered by an ~ M8.0 earthquake in 2001. Two glacier avalanches were in the volume range of $1-4$ Mm^3, and it is noted that they did not transport significant amounts of debris.

Glacier avalanches may generate other catastrophic processes in the glacial environment. First, they may cause potentially destructive displacement waves in confined water bodies such as fiords or proglacial lakes (Slingerland and Voight, 1979; Kershaw et al., 2005, Figure 16.4). Second, glacier avalanches may dam rivers and form temporary natural dams that can cause flooding upstream during damming, and outburst floods downstream upon failure (e.g., Chernomorets et al., 2007; Röthlisberger, 1978). Third, a glacier avalanche may be transformed into a debris avalanche with sufficient entrainment of morainal materials along its path (e.g., Dutto et al., 1991). In the April 2012 Gayari ice–debris avalanche (Karakoram Himalaya, Pakistan), for example, a glacier avalanche within a steep glacier watershed entrained a significant volume (>1 Mm^3) of morainal material, exposed during recent glacier retreat,

FIGURE 16.4   Aerial view of Queen Bess Lake, British Columbia, one day after the overtopping and subsequent breach of its moraine dam on August 12, 1997. The overtopping wave was generated by a glacier avalanche (volume ~2.5 Mm^3) from the toe of Diadem Glacier (on the right). Note icebergs floating in the lake and ice block at the foot of the downstream side of the terminal moraine (circled at lower left). Approximately 6.5 Mm^3 of water was lost in the outburst from Queen Bess Lake (Clague and Evans, 2000; Kershaw et al., 2005). *Photograph provided by Interfor Forest Products.*

FIGURE 16.5 LANDSAT-8 satellite image (acquired 2013) of the upper Saltoro Valley, Karakoram Himalayas, Pakistan, the showing the setting of the April 7, 2012, Gayari ice–debris avalanche. Debris is visible on Gayari fan (arrowed at G) and small lake formed by debris-dammed Saltoro River is shown by arrow (L). The tongue-shaped terminus of the Bilafond Glacier is visible at the upper right. The ice–debris avalanche buried a Pakistan army base (white circle) in the Saltoro Valley resulting in 140 fatalities. (Schneider et al. 2013).

to generate a mass flow that overwhelmed a Pakistan army camp (3,740 m asl) in the upper Saltoro valley, near the terminus of the Bilafond Glacier (Figure 16.5; Schneider et al., 2013). The debris avalanche dammed the headwaters of the Saltoro River, thus forming a small lake; a spillway was excavated through the debris to drawdown the impoundment (Schneider et al., 2013).

*Large-scale glacier detachment*s: This process originates within valley bottom locations and involves the decoupling of a glacier ice mass from its bed and catastrophic detachment of a large volume of a valley glacier. Large-scale glacier detachments have been documented from the Caucasus (Heybrock, 1935; Haeberli et al., 2004; Huggel et al., 2005; Huggel, 2009; Petrakov et al.,

2008; Evans et al., 2009b) and inferred from the Andes (Milana, 2007; Fauque et al., 2009). The detachment process may be related to a range of surging mechanisms in which rapid extension and thinning of the glacier mass occurs but is distinguished from conventional glacier surging (cf. Kotlyakov et al., 2004) on the basis of complete detachment, and subsequent high-velocity travel, of the glacier mass.

The largest CMF documented in the mountain glacial environment in the historical era occurred on September 20, 2002, on the northern slope of the Caucasus Mountains, Russian Federation (Figure 16.6). In an extraordinary event, approximately 130 Mm^3 of the Kolka Glacier detached from its bed

FIGURE 16.6   CMF resulting from large-scale glacier detachment—Quickbird image (acquired September 25, 2002) of glacier ice deposited in the Karmadon River valley, 19 km from its source, as a result of the September 20, 2002, Kolka Glacier detachment (average velocity ~50 m/s; Haeberli et al., 2004; Huggel et al., 2005; Evans et al., 2009b). Note the damming of a tributary stream by the glacier debris and the formation of a small lake.

(Evans et al., 2009b) between 3,100 and 3,300 m asl, evacuated its basin and moved downvalley, accelerating to 65 m/s in under 6 km of travel, and then traveled a further 13 km downstream as a high-velocity (average velocity $\sim$50 m/s) glacier mass flow (Haeberli et al., 2004; Huggel et al., 2005; Evans et al., 2009b). The glacier mass flow deposited approximately 110 Mm^3 of the detached Kolka Glacier ice mass 19 km from its source (Figure 16.6) having traveled a vertical distance of 1,480 m on an average slope of only 6°. During its travel, the flow expended approximately 1.5×10^{19} J of potential energy. The specific energy of the event is roughly estimated at 1.45×10^4 J/kg, comparable to other CMFs in the glacial environment (cf. Delaney and Evans, 2014). Most of the Kolka Glacier ice stopped in the Karmadon basin upstream of the Karmadon Gates, a narrow gorge through the Skalistyi Range (Figure 16.6), but a significant volume of the flow (including debris, water, and glacial ice) continued a further 15-km downstream as a debris flow—debris flood, for a total travel distance of approximately 35 km.

A number of questions remain about the initiation and behavior of the 2002 Kolka event including the actual conditions and trigger(s) of glacier detachment (cf. Haeberli et al., 2004; Huggel et al., 2005; Petrakov et al., 2008; Huggel, 2009; Evans et al., 2009b). Two causal mechanisms have been advanced for the catastrophic glacier detachment: (1) response to impact forces generated by a large-scale rockslide from a steep periglacial rock slope above the Kolka Glacier (Haeberli et al., 2004; Huggel et al., 2005; Huggel, 2009); (2) response to loss of shear resistance at the base of the glacier mass due to high pore pressures developed in stagnating and debris covered ice (Evans et al., 2009b); this process has been termed "catastrophic flotation" by Krenke and Kotlyakov (1985). Similar events have affected the Kolka Glacier (e.g., in 1902 when 50—70 Mm^3 of the Kolka Glacier detached and traveled at a high velocity down the Genaldon River) and other glaciers in the Caucasus in the last two centuries (Heybrock, 1935). In 1832, for example, a 15 Mm^3 detachment of Devdorak Glacier, on the eastern slope of Mt Kazbek, traveled 6.5 km and dammed the Terek River (Heybrock, 1935; Chernomorets et al., 2007).

16.4 MASS FLOWS INVOLVING MAINLY FRAGMENTED ROCK (ROCK AVALANCHES)

Rock avalanches generated by the large-scale failure of periglacial rock slopes and involving the rapid movement of a mass of disintegrated source rock are common occurrences in the glacial environment (Post, 1967; Evans and Clague, 1988; Evans et al., 1989; Hewitt, 2009; Schneider et al., 2011a; Huggel et al., 2010; Uhlmann et al., 2013; Ekstrom and Stark, 2013; Delaney and Evans, 2014; De Blasio, 2014) and contribute to the primary denudation of glacierized basins (Arsenault and Meigs, 2005; Uhlmann et al., 2013). We note that 8 (23.5 percent—five were seismically triggered) of 34 rock avalanches, with volumes in excess of 20 Mm^3, known to have occurred globally between

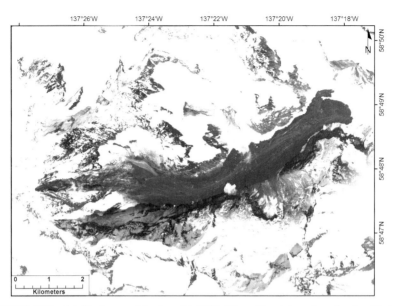

FIGURE 16.7   Satellite image of the June 11, 2012, Lituya Mountain rock avalanche, Alaska. The rock avalanche had an estimated volume of 20 Mm³ and ran out approximately 8 km on the surface of Johns Hopkins Glacier (Pleiades satellite image acquired June 23, 2012).

1945 and 2013, occurred on slopes directly above or adjacent to glacier surfaces (Evans, unpublished data). A recent example of this type of CMF is the 2012 Lituya Mountain rock avalanche (Figure 16.7; estimated volume ∼20 Mm³), one of 56 rock avalanches with volumes >1 Mm³, to have occurred in the glacial environments of NW North America in the period 1945−2014.

As these numbers indicate, periglacial rock slopes are particularly prone to catastrophic failure (for example, http://earthobservatory.nasa.gov/IOTD/view.php?id=81943) as a result of high relative relief, slope steepness, changing stresses due to glacier loading and unloading, the effects of water pressure, thermal effects, rock mass quality, the presence of warming mountain permafrost, and episodic seismic loading (e.g., Post, 1967; Matsuoka and Sakai, 1999; Oppikoffer et al., 2008; Fischer et al., 2010; McColl and Davies, 2013; Krautblatter et al., 2013). The resulting rock avalanche may be contained on the glacier surface itself (e.g., Shreve, 1966; Post, 1967; McSaveney, 1978; Gordon et al., 1978; Evans and Clague, 1990, 1998, 1999; Mauthner, 1996; Lipovsky et al., 2008; Cox and Allen, 2009; Sosio et al., 2012; Delaney and Evans, 2014, Figures 16.7 and 16.8) or it may run out into the glacier foreland (e.g., Crandell and Fahnestock, 1965) and beyond (Figure 16.9; Evans and Clague, 1988; Evans et al., 1989; Porter and Orombelli, 1980; McSaveney, 2002; Geertsema et al., 2006; Deline and Kirkbride, 2009; Akçar et al., 2012; Fauqué et al., 2009).

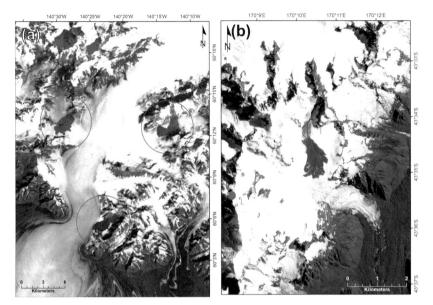

FIGURE 16.8 Satellite imagery of rock avalanches on glaciers. (a) LANDSAT-5 image of Seward Glacier area, St Elias Mountains, on the Alaska–Yukon border showing three rock avalanches (circled) triggered by M7.5 February 28, 1979, St Elias Earthquake (Lahr et al., 1979; Stover et al., 1980; Delaney and Evans, 2014; Evans and Delaney, unpublished data). Image acquired on August 25, 1979. (b) EO-1 image of Mount Dixon area, Mount Cook National Park, Southern Alps, New Zealand showing debris of January 21, 2013, Mount Dixon rock avalanche on the surface of the glacier forming the Grand Plateau. The source of the 2013 rock avalanche is 5 km ENE of the source of the 1991 Mount Cook rock avalanche (red dot at left) illustrated in Figure 16.1.

In the context of recent glacier ice loss, an important factor in initial rock slope failure is the debutressing of periglacial rock slopes by recent glacier thinning (e.g., Bovis, 1990; Evans and Clague, 1994; McSaveney, 1993, 2002; Holm et al., 2004; Oppikoffer et al., 2008; Cossart et al., 2008; Deline, 2009; Fischer et al., 2010, 2012, 2013; Strozzi et al., 2010; Stoffel and Huggel, 2012; McColl and Davies, 2013).

Rock avalanches in the glacial environment are frequently triggered by earthquakes (Post, 1967; McSaveney, 1978; Jibson et al., 2006; Delaney and Evans, unpublished data; Figures 16.8 and 16.10). In NW North America, for example, earthquakes in 1964, 1979, and 2002 triggered over 20 rock avalanches (with volumes in excess of 1 Mm^3) in Alaska and NW British Columbia (Delaney and Evans, unpublished data) including the iconic Sherman Glacier rock avalanche (estimated volume ~ 12 Mm^3; Figure 16.10; Shreve, 1966; Post, 1967; McSaveney, 1978).

Debris sheets of rock avalanches involving volumes ≥1 Mm^3 that run out on glaciers have distinctive geometries in that they are (1) longer in relation to volume and (2) more extensive in area in relation to volume than non-glacial

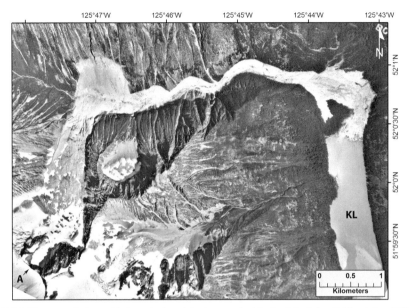

FIGURE 16.9 Orthorectified vertical aerial photograph (obtained in July 1965) of the 1959 Pandemonium Creek rock avalanche, Coast Mountains, British Columbia. The rock avalanche (est. volume 5 Mm^3) originated at A, traveled over a glacier surface below and ran into the main valley of Pandemonium Creek beyond the glacier limits. The debris ran up 335 m on the opposite side of the valley before turning east and running about 4 km to the mouth of Pandemonium Creek. The debris showed dramatic superelevation in the bends of Pandemonium Creek as it traveled down the valley (Evans et al., 1989). A significant volume of debris was deposited on the fan at the mouth of Pandemonium Creek, but some debris turned toward the south and entered Knot Lakes (KL); this generated a wave that swept down Knot lakes destroying trees along its shoreline. The peak velocity and mean velocity for the Pandemonium event are estimated at 81−100 and 30 m/s, respectively (Evans et al., 1989). *British Columbia aerial photograph BC5145-165.*

rock avalanches (Evans and Clague, 1988; Schneider et al., 2011a; De Blasio, 2014). These two effects result in a very thin supraglacial debris sheet. Delaney and Evans (2014) argue that debris sheet thinning (and an associated higher degree of fragmentation) is driven by the low friction of the debris— glacier surface interface.

In some cases, mass flows involving fragmented rock may be sequentially transformed into more fluid distal debris flows, debris floods, hyper-concentrated flows, and/or true water floods, by incorporation of snow and ice during travel (e.g., Casassa and Marangunic, 1993) and/or by displacing significant water volumes from water bodies (e.g., Wiles and Calkin, 1992: Kjartansson, 1967; McSaveney, 2002) in its path, in which the flow undergoes a reduction in sediment concentration. In both cases, travel distance is significantly enhanced, and the impact area is increased beyond the boundaries of "normal" rock avalanche deposition.

FIGURE 16.10   Sherman Glacier (Alaska) rock avalanche triggered by the March 27, 1964, M9.2 Great Alaska Earthquake (Shreve, 1966; Post, 1967; McSaveney, 1978). The rock avalanche (estimated volume = 12 Mm^3) originated in the mountain peak in distance (arrowed), well above the glacier surface, at an epicentral distance of 152 km. *Photograph by A. Post, USGS, in August 1965 http://libraryphoto.cr.usgs.gov/htmllib/batch81/batch81j/batch81z/ake00237.jpg.*

In 1967, a large rock avalanche (estimated volume 15 Mm^3; Kjartansson, 1967) fell from Steinsholt, a sector of the northern flank of Eyjafjallajökull, the Icelandic volcano that erupted in 2010, onto the Innstihaus glacier. Most of the debris was deposited on the surface of the glacier, but a fast moving part of the rockslide entrained about 1 Mm^3 of glacial ice from the surface of the glacier and continued its travel beyond the glacier snout into the lake Steinholtslon displacing a significant volume of water. As described by Kjartansson (1967, p. 254) "...augmented with this water, the whole mass traveled at great velocity down the valley of the Steinholtsi (river)." Some boulders of the original rockslide were transported 5 km from their source by a debris flood. Further downstream, the flow resembled a conventional water flood that reached up to 25 km downstream. Estimates of the flood volume calculated from a stream hydrograph are in the range 1.5−2.5 Mm^3 (Kjartansson, 1967).

In the 1987 Parraguirre case (Cassasa and Marangunic, 1993; Hauser, 1993, 2002), a rock avalanche (estimated vol. 5.5 Mm^3 including some ice masses) occurred on a steep 60° slope above Glacier 24 and incorporated a small amount of glacier ice as it traveled over the glacier surface before turning south and following the course of the Parraguirre river. At the turn, superelevation data suggested a minimum velocity of 24 m/s (Casassa and Marangunic, 1993). In its 15 km travel down the Parraguirre, the movement was transformed into a debris flow that entered the main channel of the Rio Colorado. The debris flow, augmented by Rio Colorado channel flow, then traveled downstream destroying the Los Maitenes hydroelectric plant located 41 km from the source of the movement. The flood wave continued downstream and was recorded at the Camimbao gauging station as a water flood, 220 km from the initial rock avalanche. The source of the water required to initiate and sustain the Parraguirre–Colorado flow (measured as 7 Mm^3 at Cambimbao, 220 km downstream) is problematical as discussed by Casassa and Marangunic (1993) who could not reconcile a large water deficit in their calculations. The flow downstream from the Rio Colorado confluence may have resulted from the temporary damming of the Rio Colorado channel by the Parraguirre debris and its subsequent outburst (cf. Evans et al., 2009a).

16.5 MASS FLOWS INVOLVING A MIXTURE OF GLACIER ICE AND ROCK (ICE–ROCK AVALANCHES AND FLOWS)

CMFs in the glacial environment may also originate in hybrid movements involving a mixture of rock and ice in significant proportions. The run-out, velocity, and depositional characteristics of ice–rock avalanches differ from those of true rock avalanches because of the rheological effects of contained ice, the melting of contained ice, and related water content (e.g., Schneider et al., 2011b; Turnbull, 2011) in transforming the ice–rock avalanche into an ice–rock debris flow.

The archetypical events of catastrophic flows of this nature are the Huascarán events of 1962 and 1970 (Morales, 1966; Plafker and Ericksen, 1978; Evans et al., 2009a) in the Cordillera Blanca of Peru. Both Huascarán mass flows, originated as rock/ice falls from the mountain's North Peak (Figure 16.11) between el. 6,400 and 5,600 m asl; the smaller 1962 event (Figure 16.12) did not have a definable trigger, whereas the 1970 flow (Figure 16.13) resulted from an offshore M7.8 earthquake at an epicentral distance of 150 km. Both initial falls were transformed into higher-volume, high-velocity mud-rich debris flows (Figures 16.12 and 16.13) by incorporation of snow from the surface of a glacier below Huascarán and the substantial entrainment of morainal and colluvial material from slopes below the glacier terminus between el. 4,000 and 2,900 m asl (Figure 16.11; Evans et al., 2009a). The flows were confined within the valley of the Rio Ranrahirca and continued to the Rio Santa at el. 2,400 m asl, 15.5 km from its source on the north peak of Huascarán (Figures 16.12 and 16.13). In 1970, a segment of the debris flow ($\sim$3.6 Mm^3 in volume) overtopped the Rio Ranrahirca valley side and overwhelmed the town of Yungay (Evans et al., 2009a) with the loss of

FIGURE 16.11   The upper part of the path of the 1962 and 1970 catastrophic mass flows that originated in a fall from the North Peak of Huascarán (6,654 m asl.). Note summit ice cap, source rock slope, and Glacier 511 below the rock slope. The former site of Yungay is in the foreground.

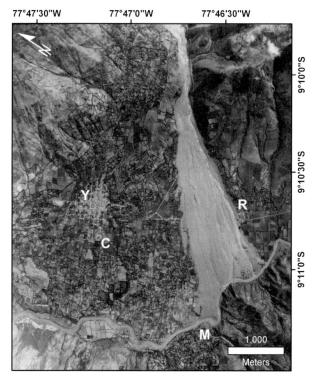

FIGURE 16.12   1962 Huascarán catastrophic mass flow; vertical aerial photograph of the runout area of the January 10, 1962, debris flow on the Ranrahirca fan taken the day after the event. Note location of the part of Ranrahirca (R) undamaged by flow, and the location of Yungay (Y), Cemetery Hill (C), and Matacoto (M). Rio Santa runs from the right to the left (*Servició Aerofotografíco Nacional de Perú 8600-62-3; January 11, 1962*).

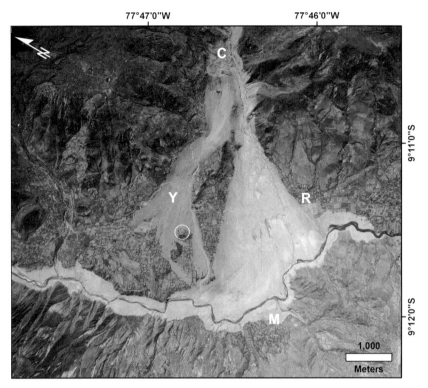

FIGURE 16.13   Aerial photograph of deposits of the 1970 Huascarán CMF (Evans et al., 2009a). Debris flow originated in the transformation of an earthquake-triggered ice/rock fall from Huascarán through the entrainment of material in its path. Debris flow surmounted Cerro de Aira at C. Y = Yungay Lobe; R = Ranrahica; M = Matacoto. Cemetery Hill is circled. Note the trace of significant downstream debris flood (aluvión) moving down the Rio Santa to the left. (*NASA aerial photograph, July 14, 1970*).

many lives (Figure 16.13; see further discussion below). Initial failure volumes were 3 and 7.5 Mm^3 in 1962 and 1970, respectively, which became increased to 16 and 58 Mm^3 (volume of solids) in 1962 and 1970, respectively, through bulking and entrainment (Evans et al., 2009a). Water for fluidization (estimated at 4 and 18 Mm^3 in 1962 and 1970, respectively) of the entrained material originated in the melting of incorporated snow and the liberation of soil moisture contained within the entrained materials. Eyewitness reports (Evans et al., 2009a) indicate very high mean velocities for the events; 17−35 m/s (1962) and 50−85 m/s (1970). Both mass movements continued downstream in the Rio Santa (Figures 16.12 and 16.13) as debris floods ("aluviones") that in 1970 reached the Pacific Ocean at a distance of 180 km from its source. From a glacier hazard assessment perspective, the Huascarán case highlights (1) the lesson of precedence, since the 1962 event presaged to a remarkable degree the larger and more destructive 1970 event; (2) the importance of entrainment in augmenting the volume of an initial failure; (3) the role of water in fluidization of a CMF; and (4) the need for extensive and comprehensive field studies in the immediate aftermath of a CMF-related disaster as a basis for the accurate reconstruction of the event. We note, however, that the mechanism and geometry of the 1970 Huascarán event was reported with remarkable accuracy by Peruvian engineer Luis Ghiglino Antunez de Mayolo in a Spanish language publication in 1971 (Ghiglino Antunez, 1971).

A similar hybrid event occurred on May 5, 2012, in the extreme relief of the Nepalese Himalaya in the vicinity of Pokhara (Figure 16.14; Hanisch et al., 2013). The flow originated in a rock-ice avalanche (estimated volume $10-15$ Mm^3) from Annapurna IV (7,524 m asl) that, in its transit through the

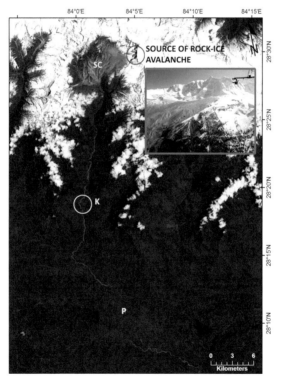

FIGURE 16.14   The May 5, 2012, Seti catastrophic mass flow, Nepal Himalaya. ASTER satellite image (acquired May 22, 2012) showing the location of source rock-ice avalanche (blue circle) on Annapurna IV, and the path of the debris flow down Seti Khola. Debris flow caused fatalities in the village of Kharapani (K) and reached Pokhara (P) some 45 km from the source on Annapurna IV. SC is Sabche Cirque discussed in the text. Lakes visible to the west and east of Pokhara are remnants of larger lakes dammed by deposits (Pokhara Gravels) of CMF(s) originating in the Annapurna massif about 500 years ago (Yamanaka et al., 1982; Fort, 1987, 2010). Inset photograph shows the source area of the initial ice-rock avalanche and debris in Sabche Cirque. *Nepali Times photograph by A. Maximov.*

Sabche Cirque (Figure 16.14), and its outlet into the Seti Gorge, became transformed into a debris flow/debris flood in the Seti Khola river that continued downstream for a distance of >40 km through an elevation difference of 6,594 m. Lives were lost in the debris flow at the village of Kharapani, 26 km from the site of the initial rock/ice avalanche (Figure 16.14). Uncertainty remains concerning the precise volumes involved and the mechanism of transformation of the Seti flow, particularly with respect to the source of water necessary for transformation. Possible mechanisms include hydraulic ponding in the confined Seti Gorge, transient river damming by debris, incorporation of snow from the upper parts of the path in the Sabche Cirque, and entrainment of an unknown volume of glacial ice. We note that massive CMF deposits (the Pokhara Gravels—volume estimated at >4 km^3 (Fort, 1987, 2010)) with their source in the Annapurna Range in the vicinity of the Sabche Cirque, fill the Pokhara valley damming major rivers to form the lakes visible in Figure 16.14 (Yamanaka et al., 1982; Fort, 1987, 2010). The most recent debris flow/CMF event occurred about 500 years ago during the early part of the Little Ice Age (Fort, 1987, 2010).

CMFs involving rock and ice also occur on active glacier-clad volcanoes and glacierized dissected Quaternary volcanic centers (Major and Newhall, 1989; Pierson and Janda, 1994; Delgado Granados et al., 2014). In eruptions of glacier-clad volcanoes water generated by the melting of ice may mobilize large-scale lahars as in the cases of the 1985 eruption of Nevado del Ruiz, Colombia (Pierson et al., 1990; Pierson and Janda, 1994), the 1989 and 2009 eruptions of Redoubt Volcano, Alaska (Pierson and Janda, 1994; Waythomas et al., 2013), and nineteenth century eruptions of Cotopaxi, Ecuador (Pistolesi et al., 2013). In eruption-related flank collapse of glacier-clad volcanoes, glacier ice may make up a significant proportion of the collapse volume and contributes to high debris avalanche mobility. In the case of the 1980 flank collapse of Mt St Helens, for example, it is estimated that 100 Mm^3 of snow and ice was involved in the 2.5-km^3 (4 percent) flank collapse (Brugman and Meier, 1981). Large-scale mass movements, involving the failure of periglacial rock slopes, can take place on active volcanoes unrelated to eruptive activity; examples include the rock and ice avalanches on Iliamna volcano, Alaska (Waythomas et al., 2000; Huggel et al., 2007; Schneider et al., 2010).

CMFs also occur on rock slopes in glacierized dissected Quaternary volcanic centers, such as the Mount Meager volcanic complex in southwestern British Columbia. On July 22, 1975, a complex series of landslide events took place at Devastation Glacier near Pemberton, British Columbia when approximately 13 Mm^3 of altered Quaternary volcanic rock and glacier ice was lost from the west flank of Pylon Peak within the Mount Meager volcanic complex (Figure 16.15). The rockslide was quickly transformed into a high-velocity debris flow, consisting of glacial ice and disaggregated pyroclastic rocks, which continued down Devastation Creek valley. The overall length of the flow path was 7 km, and its vertical height was 1,220 m,

FIGURE 16.15   Vertical aerial photograph of the source area of July 22, 1975, Devastation Glacier mass flow, Mount Meager Volcanic Complex, Coast Mountains, British Columbia. Mass flow originated in altered Quaternary pyroclastic rocks, which had undergone significant prefailure subglacial deformation as a result of downwasting of Devastation Glacier at the foot of the slope (note crevasse patterns around the head of the landslide scar). The mass involved in the initial failure (volume ~ 13 Mm^3 including rock and ice) disintegrated during movement forming a high-velocity debris flow that traveled 7 km downvalley (to the right of photograph). (*National Air Photo Library A37245-113*).

yielding a fahrböschung of 10°. Stability analysis of the initial failure (Evans, unpublished data) suggests that the 1975 rockslide was the result of a complex history of glacial erosion, loading and unloading of the toe of the slide mass caused by the Little Ice Age advance and subsequent retreat of Devastation Glacier (Figure 16.15). The shearing resistance along the base of the rockslide mass was reduced prior to 1975 by substantial previous slope displacements. Some of this displacement is likely to have occurred as subglacial slope deformation since ice fall and crevasse patterns suggest the presence of slide-like shearing displacements below the base of adjacent glacier ice. Increased fluid pressures acting along the base of the slide and on internal shear planes, which no doubt accompanied ice melting during a period of warm summer weather, probably reduced the overall shearing resistance sufficiently to trigger the initial slide.

In early August 2010, another mass flow occurred within the Meager Creek watershed (Guthrie et al., 2012). It involved the initial failure of the western flank of Mount Meager in Pleistocene rhyodacitic volcanic rock. Undrained loading of the sloping flank caused rapid evacuation of the entire flank, as the initial rock slope failure transformed into a massive ice–rock debris flow that exhibited dramatic high-velocity behavior (Figure 16.16). The disintegrating mass traveled down Capricorn Creek at an average velocity of 64 m/s, exhibiting marked superelevation in bends, to the intersection of Meager Creek, 7.8 km from the source. At Meager Creek, the debris impacted the south side of Meager valley causing a runup of 270 m above the valley floor and the deflection of the landslide debris both upstream (for 3.7 km), and downstream into the Lillooet River valley (for 4.9 km) where it blocked the Lillooet River (Figure 16.16) for some hours, approximately 13 km from the landslide source (Figure 16.16). Deposition at the Capricorn–Meager

FIGURE 16.16 Oblique aerial photograph of August 6, 2010, mass flow (view to the north) originating on the west flank of Mount Meager (arrowed) in the Mount Meager Volcanic Complex, Coast Mountains, British Columbia (Guthrie et al., 2012). The initial rockslide (volume ~48.5 Mm^3) developed into a high-velocity mass flow, containing glacier ice, in Capricorn Creek and traveled 7.8 km downvalley at an average velocity of 64 m/s, to enter Meager Creek. At the confluence, some debris ran 3.7 km upstream, and most of the debris ran 4.9 km downstream to enter the Lillooet River valley (foreground), temporarily blocking its braided channels. The remainder of the debris formed a transient blockage of Meager Creek (Guthrie et al., 2012), forming a 1.5-km-long lake. *Photograph taken on August 29, 2010.*

confluence also dammed Meager Creek for about 19 h creating a lake 1.5 km long. The overtopping of the dam and the predicted outburst flood was the basis for a night-time evacuation of 1,500 residents in the town of Pemberton, 65 km downstream. The volume of the initial displaced mass from the flank of Mount Meager was estimated to be to be 48.5 Mm^3, the height of the path (H) to be 2,183 m and the total length of the path (L) to be 12.7 km. This yields $H/L = 0.172$ and a fahrböschung of 9.75° (Guthrie et al., 2012). The initial failure involved a slope that had been debutressed by twentieth century glacial retreat (Holm et al., 2004).

16.6 GLACIAL DEBRIS FLOWS I; NON-OUTBURST RELATED

Debris flows are common in the mountain glacial environment (Jomelli et al., 2007; Chiarle et al., 2007; Huggel et al., 2012) due to a combination of favorable conditioning factors including steep slopes, thermal sensitivity, convective summer rainstorms, and availability of readily mobilized unconsolidated debris (including moraine, colluvium, and stream sediments; Broscoe and Thomson, 1969) for both initial movement volumes and entrainment downstream from its origin. Non-outburst related glacial debris flows are mobilized in steep proglacial and periglacial terrain exposed by recent glacial retreat and includes material in (1) ice-cored moraines or moraines adjacent to glaciers; and (2) along periglacial/proglacial stream channels. They are a common response to high summer temperatures and/or heavy summer rains that augment thermally induced summer run-off (cf. Chiarle et al., 2007; Jomelli et al., 2007). Initial failure volumes are commonly increased by entrainment of materials from the path of the flow resulting in very large deposit volumes that in many cases exceed 1 Mm^3. These types of glacial debris flow processes have been documented in the Andes, the European Alps, the Caucasus, and the Tien Shan mountains of Central Asia.

On February 27, 1998, a massive glacial debris flow occurred in the southern Andes of Peru approximately 3 km southwest of Machu Picchu (Huggel et al., 2012); the flow was triggered by heavy summer rains associated with an El Niño. Initial mobilization occurred in a Little Ice Age terminal moraine complex at el. 4,100 m asl within a glacier foreland on the north slope of Nevado Salcantay (Figure 16.17). Moving downstream, the high-velocity debris flow entrained material from the channel margins and valley sides of the Ahobamba Valley and entered the main valley of the Vilcanota River, 17.5 km downstream at el. 1,783 m asl. The debris temporarily dammed the Vilcanota River forming a lake (Figure 16.17) that flooded a hydroelectric power plant just upstream. The volume is reported as 25 Mm^3 (Huggel et al., 2012), which is probably an overestimate; we estimate a volume closer to 10 Mm^3 (Figure 16.17).

In the Alps (Haeberli and Naef, 1988; Jomelli et al., 2007; Chiarle et al., 2007; Huggel et al., 2012), glacial debris flows originate in periglacial moraine

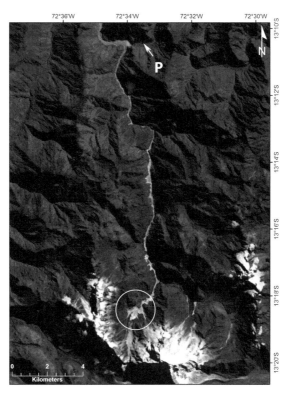

FIGURE 16.17   LANDSAT-5 image of the February 27, 1998, Salcantay-Ahobamba glacial debris flow, just to the southwest of Machu Picchu, Peruvian Andes. The debris flow, triggered by El Niño rains, originated in a Little Ice Age moraine complex (circled). The developing mass flow traveled downstream in the Ahobamba valley entraining material from the valley sides (note erosive scars). At 17.5 km, the debris flow entered the main Vilcanota River valley and temporarily blocked the channel forming a lake upstream. The lake waters flooded the Machu Piccu hydro-electric plant at P.

complexes in the alpine zone at 2,000—3,000 m asl and are triggered by heavy rains and/or high alpine summer temperatures (Figure 16.18; Zimmermann and Haeberli, 1992; Chiarle et al., 2007). Volumes range up to a maximum of 1 Mm^3 (Chiarle et al., 2007). In 1987, approximately 100 debris flows, with volumes >1,000 m^3, were triggered by heavy rains in Switzerland (Haeberli and Naef, 1988; Zimmermann and Haeberli, 1992; Rickenmann and Zimmerman, 1993); many of these debris flows originated in the periglacial zone (>2,400 m asl) and >50 percent originated in terrain that was glacier covered in the nineteenth century.

Detailed work on glacial debris flows in the Gerkhozan River, a tributary to the Baksan River in the Central Caucasus Mountains of southern Russia (Seinova, 1991; Seinova et al., 2003, 2007), has established initial mobilizing conditions in the proglacial environment and quantified the importance of

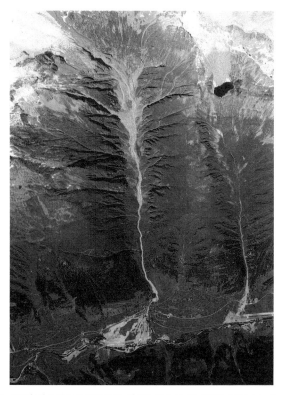

FIGURE 16.18   Vertical aerial photograph of the August 27, 1987, Münster glacial debris flow, Swiss Alps (from Haeberli, 1992). The debris flow originated below the tongue of the Ministigertal Glacier (upper left) in ice-rich moraine exposed by glacial retreat (Haeberli and Naef, 1988; Zimmermann and Haeberli, 1992) and traveled down Minstigertal impacting the village of Münster located on a fan at the mouth of the valley (Haeberli and Naef, 1988). Aerial photograph acquired August 29, 1987.

entrainment in augmenting debris flow volume during downstream travel. As in the Alps, initial conditions of non-outburst-related glacial debris flows in the Central Caucasus are associated with recent rapid deglaciation (Stokes et al., 2006, 2007), which has exposed large areas of unconsolidated morainal material containing buried glacial ice. The topography of source areas is often characterized by thermokarst features, which during rainfall in the summer ablation season can become saturated with excess meltwater and surface runoff (Seinova et al., 2011). Under these conditions, thermoerosional processes at the boundaries of buried ice, combined with the development of an active layer in melting permafrost, leads to regressive erosion, retrogressive collapse of morainal masses, and the initial mobilization of debris flows at elevations around 3,240 m asl (Seinova et al., 2003). Debris flows thus enter the headwaters of the fluvial system and travel downstream; as they do so, additional material is entrained from channel margins resulting in the further increase in the flow volume. This process has resulted in a number of debris flows with total volumes in excess of $1 \, Mm^3$ since the 1930s. Recent historical events in the Gerkhozan River have included rainfall-triggered debris flows in August 1977 ($\sim 3 \, Mm^3$), August 1999 ($\sim 4 \, Mm^3$) and July 2000 ($\sim 6 \, Mm^3$) (Seinova et al., 2003, 2007). The massive glacial debris flow of 2000 began in the collapse of headwater slopes (Figure 16.19) in an area of buried glacier ice and reached a maximum discharge of 2,000 m^3/s; the debris blocked the main channel of the Baksan river (at el. 1,270 m asl) forming an impoundment that flooded parts of Tyrnyauz City (Seinova et al., 2003, Figure 16.19).

In the Zailiiskiy Alatau (northern Tien Shan Mountains) of Kazakhstan, many glacial debris flows (both related and unrelated to outburst events) have occurred since the 1920s (Niyazov and Degovets, 1975; Khegai and Popov,

FIGURE 16.19 LANDSAT-7 image of Tyrnyauz debris flow of the path in the Gerkhozhan River, Caucasus Mountains, Russian Federation. The July 18, 2000, debris flow initiated by the thawing of dead ice in a moraine complex at A. The volume of the debris flow was increased by the incorporation of colluvium from valley sides in the mid to lower path of the flow (B), which struck Tyrnyauz City at C. The debris flow also blocked the channel of the Baksan River forming a small lake upstream (image obtained August 2, 2000).

1989; Passmore et al., 2008). As in other glacierized mountain regions examined here, heavy summer rainfalls are triggers for debris flows that originate in the degradation of moraine complexes exposed by recent glacial retreat. In 1921, multiple debris flows occurred in all of the tributaries of the Lesser and Greater Almatinka Rivers that coalesced to form massive debris flows at the mouths of the drainage basins; these covered a large area of Alma-Ata city resulting in many deaths. In the Lesser Almatinka, deposited debris volume was >3 Mm^3; the average velocity of the debris flows was estimated at 4 m/s (Niyazov and Degovets, 1975).

16.7 GLACIAL DEBRIS FLOWS II; LAKE OUTBURST-RELATED FLOWS

Outbursts from glacial lakes may generate significant CMFs ranging in characteristics from debris flows, debris floods, hyperconcentrated flows, to water floods with very low sediment concentrations (e.g., Lliboutry et al., 1977; Haeberli, 1983; Clague and Evans, 1994; Chiarle et al., 2007; Kershaw et al., 2005; Petrakov et al., 2007, 2011; Portocarrero, 2014; Clague and O'Connor, 2014). Glacial lakes include (1) moraine-dammed lakes (Lliboutry et al., 1977; Clague et al., 1985); (2) glacier-dammed lakes (englacial, sub-glacial, or proglacial) (Rabot, 1905); and (3) rock basin lakes directly adjacent to glaciers (Lliboutry et al., 1977). The mass flow initially develops as a result of entrainment of the moraine dam and/or downstream materials by the outburst flood. Sediment concentration may approach that of a true debris flow. Haeberli (1983) and Clague and Evans (1994) have pointed out that the debris flow—water flood transition is dependent on channel slope, with debris flows developing in steeper channels as a result of the mobilization of channel-margin material.

In the Zailiiskiy Alatau (northern Tien Shan mountains) of Kazakhstan, many debris flows generated by the breaching of moraine-dammed lakes have occurred since the 1950s (e.g., Niyazov and Degovets, 1975; Yesenov and Degovets, 1979; Keremkulov and Tsukerman, 1988; Passmore et al., 2008). Glacial retreat in the Zailiiskiy Alatau has created many moraine-dammed lakes, the conditions for moraine dam failure, and exposed large volumes of entrainable material (Bolch, 2007; Bolch et al., 2011). A number of debris flows generated by moraine-dammed lake breaching have taken place in the Bolshoya (Greater) Almatinka and the Malaya (Lesser) Almatinka Rivers since the 1920s and have caused destruction and life loss in the former Kazakh capital, Alma-Ata (Figure 16.20; Niyazov and Degovets, 1975). Major debris flows with a similar origin have taken place in other drainage basins to the east (e.g., Issyk River in 1963; Popov, 1991). As in the Caucasus, the debris flows of the Zailiiskiy Alatau are notable in that debris flow volumes are massive relative to the volumes of small glacial lake outbursts associated with their initiation.

FIGURE 16.20   LANDSAT-8 image of a segment of the northern slope of the Zailiiskiy Alatau, Tien Shan Mountains, Kazakhastan showing the location of Malaya (Lesser) Almatinka River basin. Debris flows (outburst and non-outburst related) in the Malaya Almatinka River, originating in the glacier foreland (red rectangle), reached the southeast limits of Alma-Ata (el. 1,200 m asl) on a number of occasions since the 1920s causing substantial damage and loss of life, before construction of mitigation structures in the 1960s (Niyazov and Degovets, 1975). Circled are two debris flow retention dams constructed to reduce glacial debris flow hazard: ME—Medeo Dam and MI—Mynzhilki Dam.

The debris flow of July 15, 1973, which formed in the channel of the Malaya Almatinka, had a total volume of 4.5 Mm^3 (Figure 16.20) and was the largest of the twentieth century in the Alma-Ata region. The failure mechanism of the 1973 event involved the thermal erosion of an ice core in the moraine dams of Lakes 2 and 3 (at 3,400 m asl) and the subsequent rapid progressive collapse of the thermokarst cavities (Niyazov and Degovets, 1975). Water (180,000 m^3) was released from the lakes in about 30 min with a peak discharge of 200—250 m^3/s. The initial velocity was 9 m/s, and it reached the Mynzhilki gabion debris flow control dam (at 3,031 m asl; Figure 16.20), 2 km downstream in about 9 min. The 36,000-m^3 debris storage capacity Mynzhilki dam filled up (Niyazov and Degovets, 1975; Zinevich, 1981) and collapsed releasing a debris flow pulse into the upper reaches of the Malaya

Almatinka. Reaching the Medeo debris flow retention dam (at 1,900 m asl; Figure 16.20), 10 km downstream from its source, maximum discharges had reached 5,180 m^3/s due to vertical and lateral channel erosion and subsequent entrainment; entrainment rates in the 8 km between the Mynzhilki and the Medeo Dam amounted to 340 m^3/m (equivalent to a total downstream entrainment of 2.72 Mm^3) at an average velocity of 10−12 m/s. The debris flow filled the Medeo Dam retention structure, and thus, precise estimates could be made of debris flow characteristics (see review by Evans et al. (2011)); the total volume of water in the debris flow was 580,000 m^3 (moraine dam release plus water incorporated from the channel at an estimated rate of 50 m^3/m); the volume of sediments was 2.31 Mm^3 and the volume of pore water in these sediments was 1.1 Mm^3. These data suggest a CMF sediment concentration of 67 percent (Niyazov and Degovets, 1975).

Thus, the Almatinka glacial debris flows, generated by the breaching of moraine-dammed lakes in the Zailiiskiy Alatau, suggest that the initial outburst volume constitutes only 4 percent of the total volume of a debris flow and only 31 percent of the total volume of free water in a flow with an average sediment concentration of about 67 percent. Taking into account the contribution of pore water in the entrained material to the total volume of water in the debris flow, the total volume of water is 1.68 Mm^3 of which the volume of initial release from the moraine-dammed lakes constituted only 10 percent. The imbalance between the initial volume of water (whether in ice or free water) and the much greater final volume of water thus emerges as a key characteristic of CMFs in the glacial environment (cf. Casassa and Maragunic, 1993; Evans et al., 2009a) and indicates the importance of incorporation of additional water during movement, in the form of channel water and pore water in entrained materials, in CMF development.

The outburst of glacier-dammed lakes (subglacial, englacial, and proglacial) may also generate destructive CMFs in the mountain glacial environment (e.g., El'manov and Mamdzhanov, 1975; Evans, 1987; Clague and Evans, 1994; Chiarle et al., 2007; Petrakov et al., 2007).

The presence of rock basin lakes adjacent to unstable rock slopes and glaciers may also present a CMF hazard in the mountain environment. CMFs are generated in this situation by high-velocity avalanches of glacier ice (or ice−rock avalanches) entering the rock basin lake and displacing a large volume of water over the lip of the lake, which then generates a debris flow downstream through entrainment of alluvial and glacial materials. We note that the rock basin lake is formed by glacial retreat. An example of this type of CMF is the 2010 event at Laguna 513, Cordillera Blanca, Peru (Carey et al., 2012). A rock-ice avalanche (estimated vol. 200,000−400,000 m^3) occurred on the SW slope of Nevado Hualcán at an altitude of about 5,400 m asl. It traveled over Glacier 513 and into Laguna 513 at 4,428 m asl generating a major displacement wave that overtopped the lip of the bedrock basin. At the time of the event, the bedrock dam had a freeboard of 20 m, and it is estimated that the total wave height reached 25 m above the lake surface (Haeberli et al., 2010). The volume of the overtopping was estimated at 1 Mm^3. This water entrained considerable moraine material downstream generating a significant debris flow in the Chuchun River, which reached the outskirts of the town of Carhuaz, approximately 13 km downstream from Laguna 513. As discussed below, the water level of Laguna 513 had been lowered in the 1990s by the construction of drainage tunnels through the bedrock dam (Reynolds et al., 1998; Carey et al., 2012).

16.8 CATASTROPHIC MASS FLOWS IN THE MOUNTAIN GLACIAL ENVIRONMENT: DISCUSSION

Global occurrence and climate change: as detailed above, CMFs in the glacial environment involve the mobilization of materials at high altitudes that include glacier ice and snow. Thus, it seems self-evident that elements of climate change involving atmospheric warming, and their cryospheric response including altitude-sensitive thermal instability, glacier ice loss, and an increase of meltwater in periglacial slope systems, would influence both the occurrence and characteristics of CMFs in the mountain glacier environment. Indeed, the broad paradigm that has developed in the last 20 years linking climate change (and associated cryospheric response) to CMFs involves the creation of conditions more favorable for their occurrence and an increase in their frequency (e.g., Evans and Clague, 1993, 1994; O'Connor and Costa, 1993; Haeberli et al., 1997; Abele, 1997; Huggel, 2009; Huggel et al., 2010, 2012; Stoffel and Huggel, 2012; Allen and Huggel, 2013). However, as Huggel et al. (2012) recently pointed out, no study has been published to date that shows an unambiguous link between atmospheric warming and an increase in the magnitude and/or frequency of CMF in the mountain glacial environment. This suggests that linkages in the CMF-climate change paradigm are complex (cf. Jomelli et al., 2007) and can be complicated by at least two factors. First, superimposed on the climate change effect is a background seismic signal that may produce CMFs unrelated to thermal changes in the slope environment (e.g., Tarr and Martin, 1912; van der Woerd et al., 2004; Post, 1967; Jibson et al., 2006; Allen et al., 2011). As Allen et al. (2011) and Uhlmann et al. (2013) have demonstrated, it is not a simple task to partition seismic and climate change effects in a record of CMF occurrences in a tectonically active glacierized region. A second complicating factor is the precise nature (timing and response) of the passage of the twentieth–twenty first century climate change perturbation through the mountain slope system. The perturbation is not a steady-state phenomenon in the last century or so. Rather, it is more likely that the effect is (1) pulse-like; (2) will exhibit (possibly multiple and offset) lag times; (3) will exhibit a peak at some point; and (4) will then decay as susceptible sites are successively mobilized. These issues await further exploration (cf. Huggel et al., 2012).

Hazard Assessment: the complexity of CMFs in the glacial environment has important implications for hazard assessment (Huggel et al., 2004a,b). The first implication relates to the interpretation of deposits in valley fills in attempting to reconstruct historic and prehistoric CMF events in a given mountain region. A number of studies have demonstrated the rock avalanche/ mass flow origin of coarse rubbly deposits in valley fills (Porter and Orombelli, 1980; Fort, 1987, 2000; Fort and Peulvast, 1995; Hewitt, 1999; Fauqué et al., 2009) in a reinterpretation of deposits previously mapped as morainal material. However, even in circumstances when knowledge about a historical event is well constrained, the interpretation of coarse rubbly deposits may not be straightforward (cf. Orombelli and Porter, 1988; Deline and Kirkbride, 2009; Akçar et al., 2012).

A second implication of the complexity relates to the modeling of both initial failure and runout for hazard assessment (Huggel et al., 2004a,b). In the analysis of initial failure, in addition to the geological and geotechnical characterization of the source rock mass, quantification of the role of the degradation of high-altitude permafrost in rock slope stability presents an analytical challenge (e.g., Gruber and Haeberli, 2007; Fischer et al., 2010; Krautblatter et al., 2013). With regard to runout, uncertainty about the volume and flow behavior of initially included and entrained ice and snow presents a major difficulty in the analysis of mobility (Evans et al., 2009a; Schneider et al., 2010). An additional issue regarding runout and mobility is the volume of entrainment of glacial and alluvial materials from the path of a given CMF, which is a strong determinant on modeling outcome (Evans et al., 2009a).

As examples noted above have demonstrated, a third implication arises from the fact that CMFs sometimes generate secondary processes that extend the impact limit of the CMF by some tens of kilometers beyond the boundaries of deposited debris. These include the formation and failure of natural dams and the generation of displacement waves, which in turn may generate a CMF. We have shown that the formation of debris dams on main river channels is an important secondary process of CMF and that these themselves may fail creating catastrophic downstream flows. In addition, transient river damming may be an important process of major CMFs as hypothesized in the Parraguire case described by Casassa and Marangunic (1993). A further secondary process is the catastrophic displacement of significant volumes of water by CMF from proglacial lakes, which in turn may generate destructive debris flows, debris floods, or water floods downstream from the impacted lakes (Haeberli, 2013; Schneider et al., 2014). Both scenarios present major challenges to the modeling of CMF hazard where the potential for river damming and lake water displacement exists.

Hazard Mitigation: approaches to CMF hazard mitigation are well developed in some national jurisdictions (e.g., Switzerland) and with reference to selected CMF processes in sub-national regions (e.g., glacial lakes in the Cordillera Blanca, Peru; Portocarrero, 2014). However, in many glacierized mountain regions, mitigation is piecemeal, informal, or non-systematized.

Mitigation can include observation and monitoring (and related warning systems), land-use planning (based on some spatial delimitation of hazard zones), hazard reduction by engineering intervention in the source area (e.g., removal of potentially unstable masses, slope stabilization, lake drainage and/or spillway reinforcement; e.g., Portocarrero, 2014), and hazard reduction by engineering intervention in the run-out area (e.g., defensive structures including debris retention dams (noted above and summarized by Evans et al., 2011), berms, channelization and levees). In the case of the 2010 Laguna 513 event, described above, drainage tunnels constructed in the early 1990s had lowered the lake level, reduced the volume of water that could be displaced in a displacement event (Reynolds et al., 1998; Carey et al., 2012), and thus effectively reduced the downstream debris flow hazard.

Socioeconomic Impact: as we note above, CMF in the glacier environment have been responsible for a number of major natural disasters in mountain regions since 1940. Three events in the Cordillera Blanca of Peru resulted in the deaths of about 12,000 people between 1941 and 1970. The 1941 Huaraz disaster resulted in the deaths of approximately 4,000 persons when a mass flow cut through the northern segment of the city (Figures 16.2 and 16.21)

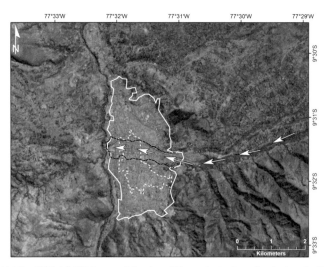

FIGURE 16.21   2013 ASTER image of the city of Huaraz in the Peruvian Andes. Outlined are the limits of the city in 1941 (dashed white line), present outline of city (heavy white line) and extent of swath of destruction resulting from the 1941 mass flow (black dashed line). The debris flood/debris flow originated in the outburst of Lake Palcacocha 22 km upstream, which developed into a destructive high-velocity flow (direction of flow indicated by arrows). Census data indicates that the 1940 population of Huaraz was 11,054 and the 1941 flow is thought to have resulted in 3,750 fatalities, one-third of the city's population. The 2013 population of Huaraz exceeds 100,000.

destroying about 1 km² of densely populated urban terrain (Oppenheim, 1946; Heim, 1948; Carey, 2005, 2010; Evans and Delaney, unpublished data). In the Huascarán mass flows of 1962 and 1970, Peru, it was established (Evans et al., 2009a) that previously reported death tolls (~25,000 in the case of the 1970 event) were overestimated by a misinterpretation of the 1961 Peru Census, a possibility first raised by Clapperton and Hamilton (1971). After consulting the original 1961 Peru Census documents, Evans et al. (2009a) concluded that the death toll of the 1970 event was approximately 6,000 and that total life loss in the two events did not exceed 7,000 people (Figure 16.22). The Huascarán events illustrate the need for a careful and objective analysis of the human impact of destructive CMFs in the glacial environment and highlights the risk to which a local population is exposed when CMF debris is reoccupied for housing and agriculture despite the fact that conditions appear to exist for a recurrence of a CMF originating once more on the North Peak of Huascarán (Evans et al., 2009a).

In the Kolka Glacier event of 2002, Russia (Haeberli et al., 2004; Huggel et al., 2005; Evans et al., 2009b), the glacier debris flow overwhelmed the village of Nizhniy Karmadon located in the valley floor, resulting in 125 fatalities (Figure 16.6). More recently, the 2012 Gayari ice−debris avalanche, Pakistan (Schneider et al., 2013) buried a Pakistan army camp with the loss of 128 soldiers (Figure 16.5). Lastly, in the Kedarnath floods, India, of 2013, >1,000 people lost their lives when a monsoon-triggered glacial debris flood overran the pilgrim town of Kedernath (Evans and Delaney, unpublished data) (Figure 16.23).

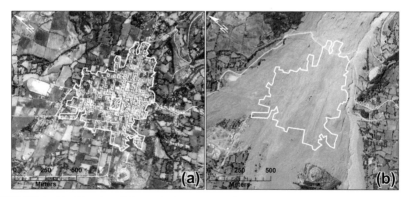

FIGURE 16.22   The site of Yungay before (a) and after (b) the May 31, 1970, debris flow in georeferenced aerial photographs (Evans et al., 2009a). (a) the urban area of Yungay is outlined by a white line. The outlined area is 0.36 km². In the 1961 Peru Census, the population of Yungay was recorded as being 3,543, a population density of 9,980 p/km². The Plaza de Armas is visible in the town center. (*Servicío Aerofotografíco Nacional de Perú photograph; January 9, 1962*). (b) the urban area of Yungay superimposed on the debris of the Yungay lobe deposited in the Huascarán mass flow of May 31, 1970. The 1970 population of Yungay has been estimated at 4,100 (Evans et al., 2009a), a population density of 11,549 p/km² (*NASA aerial photograph; July 14, 1970*).

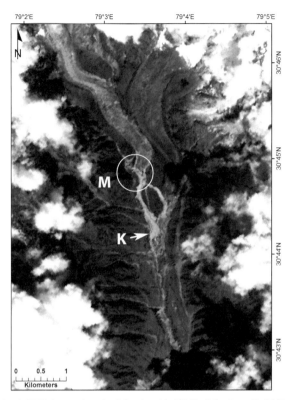

FIGURE 16.23   ASTER image (acquired October 21, 2013) of the June 17, 2013, mass flow that destroyed much of the pilgrim town of Kedernath, Uttarakhand, India (el. 3,544 m asl) arrowed at K, located in the glacier foreland of the Chorabari Glacier. The flow in the Mandakini River was triggered by heavy monsoon rains and involved the breach of a monsoon rain-filled glacial lake impounded by a lateral moraine of the Chorabari Glacier circled at M. More than 100 people lost their lives in the Kedernath event.

CMFs also impact engineering infrastructure in the mountain environment, in particular hydroelectric power facilities. High-velocity, distal debris-rich flows have destroyed diversion dams, blocked water intakes to turbines, as well as penetrating turbine machinery in a number of mountain regions, namely, Caucasus (Russia), the Tien Shan (Kazakhstan) (Niyazov and Degovets, 1975), Pamirs (Tajikistan) (El'manov and Mamadzhanov, 1975), and the Andes (Peru) (Hauser, 2002; Evans et al., 2009a; Huggel et al., 2012).

16.9 CONCLUSIONS

The occurrence of CMFs in the mountain glacial environment is relatively common. We recognize five broad groups as follows: mass flows involving mainly glacier ice (glacier avalanches and large-scale glacier detachments), rock avalanches, hybrid rock–ice avalanches, and glacial debris flows (both outburst and non-outburst related).

CMFs exhibit high-velocity behavior (mean velocity >5 m/s; maximum velocity in some cases >50 m/s), and high mobility represented by long run-out distance, which may exceed 100 km. A key characteristic of CMFs is that their behavior frequently involves one or more phases of process trans-formation during movement from initiation to cessation. The transformation reflects (1) the type of solid materials (fragmented rock, surficial materials, and glacial ice) involved in the flow, which may show dramatic changes in volumetric composition during movement through entrainment of surficial materials along its path and the melting of transported ice blocks; and (2) the fluid component of the flow that is determined by the melting of transported ice and snow, the water content of entrained materials, and the incorporation of entrained and/or displaced water from periglacial rivers and glacial lakes. Long runout is dependent on the decrease of the solid concentration in the flow relative to the fluid content, that is, an increase in fluidity with distance. However, many studies have been unable to provide credible water budgets to account for the large volumes of water required to initiate and maintain long runout fluid flows that commonly characterize the distal behavior of CMFs. These characteristics present analytical challenges and complicate the characterization of CMFs for hazard assessment.

Glacier ice loss, resulting from glacier retreat and associated glacier thinning, together with the degradation of permafrost in the mountain environment, have created favorable conditions for the initiation of CMF through such processes as unstable glacier tongues, decrease in source material shear strength, development of glacial lakes, glacial debutressing of periglacial rock slopes, and the exposure and degradation of periglacial moraine complexes. Yet several authors have reported the absence of unambiguous studies that clearly show that climate change has resulted in an increase in the frequency and magnitude of CMFs in the glacial environment. This may reflect a type of exhaustion effect in the response of the periglacial environment; we conclude that the precise role of cryospheric change (including glacier ice loss) in CMF magnitude and frequency requires additional detailed investigation. Our review also shows that CMFs are generally triggered by heavy rains, high summer temperatures, and earthquakes, creating an event record reflecting both climate and seismic forcing.

Significant disasters have been caused by CMF in mountain regions since 1940 leading to a loss of life in excess of 15,000 people worldwide; however, in some cases mitigation measures, both in the source area and run-out region of potential flows have been effective in reducing the hazard. We suggest that a process-oriented approach to understanding CMF hazard in the mountain glacial environment, and its relationship to cryospheric change, will lead to the development of enhanced mitigation strategies and thus reduced risk to mountain communities and infrastructure.

REFERENCES

Abele, G., 1997. Influence of glacier and climate variation on rockslide activity in the Alps. Paläoklimaforschung 19, 1–6.

Akçar, N., Deline, P., Ivy-Ochs, S., Alfimov, V., Hajdas, I., Kubik, P.W., Christl, M., Schlűchter, C., 2012. The AD 1717 rock avalanche deposits in the upper Ferret Valley (Italy): a dating approach with cosmogenic [10]Be. J. Quat. Sci. 27, 383–392.

Alean, J., 1984. Ice avalanches and a landslide on Grosser Aletschgletscher. Zeits. für Gletsch. u. Glaziol 20, 9–25.

Alean, J., 1985a. Ice avalanches: some empirical information about their formation and reach. J. Glaciol. 31, 324–333.

Alean, J., 1985b. Ice avalanche activity and mass balance of a high-altitude hanging glacier in the Swiss Alps. Ann. Glaciol. 6, 248–249.

Allen, S., Huggel, C., 2013. Extremely warm temperatures as a potential cause of recent high mountain rockfall. Global Planet. Change 107, 59–69.

Allen, S.K., Cox, S.C., Owens, I.F., 2011. Rock avalanches and other landslides in the central Southern Alps of New Zealand: a regional study considering possible climate change impacts. Landslides 8, 33–48.

Allen, S.K., Schneider, D., Owens, I.F., 2009. First approaches towards modelling glacial hazards in the Mount Cook region of New Zealand's Southern Alps. Nat. Hazards Earth Sys. Sci. 9, 481–499.

Arsenault, A.A., Meigs, A.J., 2005. Contribution of deep-seated bedrock landslides to erosion of a glaciated basin in southern Alaska. Earth Surf. Process. Landf. 30, 1111–1125.

Barla, G., Dutto, F., Mortara, G., 2000. Brenva glacier rock avalanche of 18 January 1997 on the mount Blanc range, northwest Italy. Landslide News 13, 2–5.

Bolch, T., 2007. Climate change and glacier retreat in northern Tien Shan (Kazakhstan/Kyrgyzstan) using remote sensing data. Global Planet. Change 56, 1–12.

Bolch, T., Peters, J., Yegorov, A., Pradhan, B., Buchroitner, M., Blagoveshchensky, V., 2011. Identification of potentially dangerous lakes in the northern Tien Shan. Nat. Hazards 59, 1691–1714.

Bovis, M.J., 1990. Rock-slope deformation at affliction creek, southern coast mountains, British Columbia. Can. J. Earth Sci. 27, 243–254.

Broscoe, A.J., Thomson, S., 1969. Observations on an alpine mudflow, Steele Creek, Yukon. Can. J. Earth Sci. 6, 219–229.

Brugman, M.M., Meier, M.F., 1981. Response of glaciers to the eruptions of Mount St. Helens. In: Lipman, P.W., Mullineaux, D.R. (Eds.), The 1980 eruptions of Mount St. Helens, Washington. U.S. Geological Survey Professional Paper, 1250, pp. 743–756.

Capello, C.F., 1959. Frane-valanghe di ghiaccio nel gruppo del Monte Bianco. Boll. Comitato Ital. 9 (Ser. II), 125–138.

Carey, M., 2005. Living and dying with glaciers: people's historical vulnerability to avalanches and outburst floods in Peru. Global Planet. Change 47, 122–134.

Carey, M., 2010. In the Shadow of Melting Glaciers: Climate Change and Andean Society. Oxford University Press, 273 pp.

Carey, M., Huggel, C., Bury, J., Portocarrero, C., Haeberli, W., 2012. An integrated socio-economical framework for glacier hazard management and climate change adaptation: lessons from Lake 513, Cordillera Blanca, Peru. Climate Change 112, 733–767.

Casassa, G., Marangunic, C., 1993. The 1987 Río Colorado rockslide and debris flow, central Andes, Chile. Bull. Assoc. Eng. Geologists 30, 321–330.

Chernomorets, S.S., Tutubalina, O.V., Seinova, I.B., Petrakov, D.A., Nososv, K.N., Zaporozhchenko, E.V., 2007. Glacier and debris flow disasters around Mt. Kazbek, Russia/ Georgia. In: Chen, C.-L., Major, J.J. (Eds.), Debris-flow Hazards, Mitigation, Mechanics, Prediction, and Assessment. Millpress, Netherlands, pp. 691−702.

Chiarle, M., Iannotti, S., Mortara, G., Deline, P., 2007. Recent debris flow occurrences associated with glaciers in the Alps. Global Planet. Change 56, 123−136.

Clague, J.J., Evans, S.G., 2000. A review of catastrophic drainage of moraine-dammed lakes in British Columbia. Quaternary Science Reviews 19, 763−1783.

Clague, J.J., Evans, S.G., 1994. Formation and failure of natural dams in the Canadian Cordillera. Geol. Surv. Canada, Bull. 464, 35 pp.

Clague, J.J., O'Connor, J.E. Glacier-related outburst floods. In: Haeberli, W., and Whiteman, C.A. (Eds.), Snow and Ice-related Hazards, Risks and Disasters, Elsevier, pp. 487−519.

Clague, J.J., Evans, S.G., Blown, I.G., 1985. A debris flow triggered by the breaching of a moraine-dammed lake, Klattasine Creek, British Columbia. Can. J. Earth Sci. 22, 1492−1502.

Clapperton, C.M., Hamilton, P., 1971. Peru beneath its eternal threat. Geogr. Mag. 43, 632−639.

Cossart, E., Braucher, R., Fort, M., Bourles, D.L., Carcaillet, J., 2008. Slope instability in relation to glacial debuttressing in alpine areas (upper Durance catchment, southeastern France): evidence from field data and [10]Be cosmic ray exposure ages. Geomorphology 95, 3−26.

Cox, S.C., Allen, S.K., 2009. Vampire rock avalanches of January 2008 and 2003, Southern Alps, New Zealand. Landslides 6, 161−166.

Crandell, D.R., Fahnestock, R.K., 1965. Rockfalls and Avalanches from Little Tahoma Peak on Mount Rainier. United States Geological Survey, Bulletin, Washington. 1221-A, 30 pp.

De Blasio, F.V., 2014. Friction and dynamics of rock avalanches travelling on glaciers. Geomorphology 213, 88−98. http://dx.doi.org/10.1016/j.geomorph.2014.01.001.

Delaney, K.B., Evans, S.G., 2014. The 1997 Mount Munday landslide (British Columbia) and the behaviour rock of avalanches on glacier surfaces. Landslides. http://dx.doi.org/10.1007/s10346-013-0456-7.

Delgado Granados, H., Julio Miranda, P., Ontiveros Gonzáles, G., Cortés Ramos, J., Carrasco Núñez, G., Pulgarín Alzate, B., Mothes, P., Moreno Roa, H., and Cáceres Correa, B.E. Hazards at ice-clad volcanoes. In: Haeberli, W., and Whiteman, C.A. (Eds.), Snow and Ice-related Hazards, Risks and Disasters, Elsevier, pp. 607−646.

Deline, P., 2009. Interactions between rock avalanches and glaciers in the Mount Blanc massif during the late Holocene. Quat. Sci. Rev. 28, 1070−1083.

Deline, P., Kirkbride, M.P., 2009. Rock avalanches on a glacier and morainic complex in Haut Val Ferret (Mont Blanc Massif, Italy). Geomorphology 103, 80−92.

Deline, P., Gruber, S., Delaloye, R., Fischer, L., Geertsema, M., Giardino, M., Hasler, A., Kirkbride, M., Krautblatter, M., Magnin, F., McColl, S., Ravanel, L., and Schoeneich, P. Ice loss and slope stability in high-mountain regions. In: Haeberli, W., and Whiteman, C.A. (Eds.). Snow and Ice-related Hazards, Risks and Disasters, Elsevier, pp. 521−561.

Du Pasquier, L., 1896. L'avalanche de l'Altels le 11 Septembre 1895. Bull. Soc. Sci. Nat. Neuchâtel 24, 149−176.

Dutto, F., Mortara, G., 1992. Rischi connessi con la dinamica glaciale nelle Alpi italiane. Geogr. Fis. Dinamica Quat. 15, 85−99.

Dutto, F., Godone, F., Mortara, G., 1991. L'écroulement due glacier supérieur de Coolidge (Paroi nord du Mont Viso, Alpes occidentales). Revue Géographie Alpine 79, 7−18.

Ekstrom, G., Stark, C., 2013. Simple scaling of catastrophic landslide dynamics. Science 339, 1416−1419.

El'manov, B.A., Mamadzhanov, D.R., 1975. Protection of structures of a diversion-canal hydroelectric station from a catastrophic mudflow. Hydrotechnical Construction 9, 870−873.

Evans, S.G., 1987. The breaching of moraine dammed-lakes in the southern Canadian Cordillera. In: Proceedings, International Symposium on Engineering Geological Environment in Mountain Areas, Beijing, vol. 2, 141−150.

Evans, S.G., Clague, J.J., 1988. Catastrophic rock avalanches in glacial environments. In: Proceedings, 5th International Symposium on Landslides, vol. 2, 1153−1158.

Evans, S.G., Clague, J.J., 1990. Reconnaissance observations on the Tim Williams Glacier rock avalanche, near Stewart, B.C. In: Current Research, Part E. Geological Survey of Canada, pp. 351−354. Paper 90-1E.

Evans, S.G., Clague, J.J., 1993. Glacier-related hazards and climatic change. In: Bras, R. (Ed.), The World at Risk: Natural Hazards and Climatic Change. American Institute of Physics Conference Proceedings 277, pp. 48−60.

Evans, S.G., Clague, J.J., 1994. Recent climatic change and catastrophic geomorphic processes in mountain environments. Geomorphology 10, 107−128.

Evans, S.G., Clague, J.J., 1998. Rock avalanche from mount Munday, Waddington range, British Columbia, Canada. Landslide News 11, 23−25.

Evans, S.G., Clague, J.J., 1999. Rock avalanches on glaciers in the coast and St. Elias Mountains, British Columbia. In: Proc. 13th Ann. Vancouver Geot. Soc. Symposium, pp. 115−123.

Evans, S.G., Bishop, N.F., Smoll, L.F., Valderrama Murillo, P., Delaney, K.B., Oliver-Smith, O., 2009a. A re-examination of the mechanism and human impact of catastrophic mass flows originating on Nevado Huascarán, Cordillera Blanca, Peru in 1962 and 1970. Eng. Geol. 108, 96−118.

Evans, S.G., Clague, J.J., Woodsworth, G.J., Hungr, O., 1989. The Pandemonium Creek rock avalanche, British Columbia. Can. Geotech. J. 26, 427−446.

Evans, S.G., Delaney, K.B., Hermanns, R.L., Strom, A.L., Scarascia-Mugnozza, G., 2011. The formation and behaviour of natural and artificial rockslide dams; implications for engineering performance and hazard management. In: Evans, S.G., et al. (Eds.), Natural and Artificial Rockslide Dams, Lecture Notes in the Earth Sciences, vol. 133. Springer, Heidelburg, pp. 1−75.

Evans, S.G., Tutubalina, O.V., Drobyshev, V.N., Chernomorets, S.S., McDougall, S., Petrakov, D.A., Hungr, O., 2009b. Catastrophic detachment and high-velocity long-runout flow of Kolka Glacier, Caucasus Mountains, Russia in 2002. Geomorphology 105, 314−321.

Faillettaz, J., Funk, M., Sornette, D., 2012. Instabilities on Alpine temperate glaciers: new insights arising from the numerical modelling of Allalingletscher (Valais, Switzerland). Nat. Hazards Earth Sys. Sci. 12, 2977−2991.

Faillettaz, J., Sornette, D., Funk, M., 2011. Numerical modeling of a gravity-driven instability of a cold hanging glacier: reanalysis of the 1895 break-off of Altelsgletscher, Switzerland. J. Glaciol. 57, 817−831.

Fauqué, L., Hermanns, R., Hewitt, K., Rosas, M., Wilson, C., Baumann, V., Lagorio, S., Di Tommaso, I., 2009. Mega-deslizamientos de la pared sur del cerro Aconcagua y su relación con depositos asignados a la glaciación Pleistocena. Revista de la Associatión Geólogica Argentina 65, 691−712.

Fischer, L., Amann, F., Moore, J.R., Huggel, C., 2010. Assessment of periglacial slope stability for the 1988 Tschierva rock avalanche (Piz Morteratsch, Switzerland). Eng. Geol. 116, 32−43.

Fischer, L., Huggel, C., Kääb, A., Haeberli, W., 2013. Slope failures and erosion rates on a glacierized high-mountain face under climatic changes. Earth Surf. Process. Landf. 38, 836−846.

Fischer, L., Purves, R.S., Huggel, C., Noetzli, J., Haeberli, W., 2012. On the influence of topographic, geological and cryospheric factors on rock avalanches and rockfalls in high mountain areas. Nat. Hazards Earth Sys. Sci. 12, 241−254.

Forel, F.A., 1895. L'éboulement du glacier de l'Altels. Archives des Sciences physiques et naturelles, Genève 34, 513−543.

Fort, M., 1987. Sporadic morphogenesis in a continental subduction setting: an example from the Annapurna Range, Nepal Himalaya. Zeitschrift für Geomorphologie, Supplement Band 63, 9−36.

Fort, M., 2000. Glaciers and mass wasting processes: their influence on the shaping of the Kali Gandaki valley (Higher Himalaya of Nepal). Quat. Int. 65/66, 101−119.

Fort, M., 2010. The Pokhara valley: a product of a natural catastrophe. In: Migori, P. (Ed.), Geomorphological Landscapes of the World. Springer, Berlin, pp. 265−274.

Fort, M., Peulvast, J.-P., 1995. Catastrophic mass-movements and morphogenesis in the Peri-Tibetan ranges: examples from West Kunlun, East Pamir, and Ladakh. In: Slaymaker, O. (Ed.), Steepland Geomorphology. J.Wiley & Sons, New York, pp. 171−198.

Ghiglino Antunez, L., 1971. Alud de Yungay y Ranrahirca del 31 de Mayo de 1970. Revista Peruana de Andinismo y Glaciologia 9, 84−88.

Giani, G.P., Silvano, S., Zanon, G., 2001. Avalanche of 18 January 1997 on Brenva glacier, Mont Blanc group, western Italian alps: an unusual process of formation. Ann. Glaciol. 32, 333−338.

Geertsema, M., Clague, J.J., Schwab, J., Evans, S.G., 2006. An overview of recent large catastrophic landslides in northern British Columbia. Eng. Geol. 83, 120−143.

Gordon, J.E., Birnie, R., Timmis, R., 1978. A major rockfall and debris slide on the Lyell Glacier, South Georgia. Arctic Alpine Res. 10, 49−60.

Grove, J.M., 1987. Glacier fluctuations and hazards. Geogr. J. 153, 351−369.

Gruber, S., Haeberli, W., 2007. Permafrost in steep bedrock slopes and its temperature-related destabilisation following climate change. J. Geophys. Res. 112, F02S18.

Guthrie, R.H., Friele, P., Allstadt, K., Roberts, N., Evans, S.G., Delaney, K.B., Roche, D., Clague, J.J., Jakob, M., 2012. The 6 August 2010 Mount Meager rock slide-debris flow, Coast Mountains, British Columbia: characteristics, dynamics, and implications for hazard and risk assessment. Nat. Hazards. Earth Sys. Sci. 12, 1277−1294.

Haeberli, W., 1983. Frequency and characteristics of glacier floods in the Swiss Alps. Ann. Glaciol. 4, 85−90.

Haeberli, W., 1992. Ursachen und Ablauf des Murgangereignisses von Münster 1987. Gemeine Münster VS: Als der Bach kam, Gommer Satz. Fiesch, 61−77.

Haeberli, W., 2013. Mountain permafrost—research frontiers and a special long-term challenge. Cold Reg. Sci. Technol. 96, 71−76.

Haeberli, W., Naef, F., 1988. Murgänge im hochgebirge. Die Alpen 64, 331−343.

Haeberli, W., Alean, J.-C., Müller, P., Funk, M., 1989. Assessing risks from glacier hazards in high mountain regions: some experiences in the Swiss Alps. Ann. Glaciol. 13, 96−102.

Haeberli, W., Huggel, C., Kääb, A., Zgraggen-Oswald, S., Polkvoj, A., Galushkin, I., Zotikov, I., Osokin, N., 2004. The Kolka-Karmadon rock/ice slide of 20 September 2002: an extraordinary event of historical dimensions in North Ossetia, Russian Caucasus. J. Glaciol. 50, 533−546.

Haeberli, W., Portocarrero, C., Evans, S.G., 2010. Nevado Hualcán, Laguna 513 y Carhuaz 2010—Observaciones, evaluación y recomendaciones (un corto informe técnico luego de las reuniones y visita de campo en Julio, 2010). Unpublished report, Huaraz.

Haeberli, W., Wegmann, M., Vonder Mühll, D., 1997. Slope stability problems related to glacier shrinkage and permafrost degradation in the Alps. Eclogae Geol. Helvetiae 90, 407−414.

Hanisch, J., Koirala, A., Bhandary, N.P., 2013. The Pokhara May 5th flood disaster: a last warning sign sent by nature? J. Nepal Geol. Soc. 46, 1−10.

Hanke, H., 1966. Gletscherkatastrophen. Der Bergsteiger 6, 433−463, 540−556.

Hauser, A., 1993. Remociones en masa en Chile. Servicio Nacional de Geologia y Mineria-Chile. Boletin 45, 75.

Hauser, A., 2002. Rock avalanche and resulting debris flow in Estero Parraguirre and Rió Colorado, Region Metroplitana, Chile. In: Evans, S.G., DeGraff, J.V. (Eds.), Catastrophic Landslides: Effects, Occurrence and Mechanisms, Geological Society of America Reviews in Engineering Geology, vol. 15, pp. 135−148.

Heim, A., 1896. Die Gletscherlawine an der Altels am 11 September 1895. Neujahrsblatt Naturforschenden Gesellschaft in Zurich auf das Jahr 1896. No. 98.

Heim, A., 1948. Wunderland Peru. Verlag Hans Huber, Bern, 301 pp.

Hewitt, K., 1999. Quaternary moraines vs catastrophic rock avalanches in the Karakoram Himalaya, northern Pakistan. Quat. Res. 51, 220−237.

Hewitt, K., 2009. Rock avalanches that travel onto glaciers and related developments, Karakoram Himalaya, Inner Asia. Geomorphology 103, 66−79.

Heybrock, W., 1935. Earthquakes as a cause of glacier avalanches in the Caucasus. Geogr. Rev. 25, 423−429.

Holm, K., Bovis, M., Jakob, M., 2004. The landslide response of alpine basins to post-Little Ice Age glacial thinning and retreat in southwestern British Columbia. Geomorphology 57, 201−216.

Huggel, C., 2009. Recent extreme slope failures in glacial environments: effects of thermal perturbation. Quat. Sci. Rev. 28, 1119−1130.

Huggel, C., Caplan-Auerbach, J., Waythomas, C., Wessels, R.L., 2007. Monitoring and modeling ice-rock avalanches from ice-capped volcanoes: a case study of frequent large avalanches on Iliamna volcano, Alaska. J. Volcanol. Geothermal Res. 168, 114−136.

Huggel, C., Caplan-Auerbach, J., Wessels, R., 2004a. Recent extreme avalanches: triggered by climate change? EOS 89, 469−470.

Huggel, C., Clague, J.J., Korup, O., 2012. Is climate change responsible for changing landslide activity in high mountains? Earth Surf. Process. Landf. 37, 77−91.

Huggel, C., Haeberli, W., Kääb, A., Bieri, D., Richardson, S., 2004b. An assessment procedure for glacial hazards in the Swiss Alps. Can. Geotech. J. 41, 1068−1083.

Huggel, C., Salzmann, N., Allen, S., Caplan-Auerbach, J., Fischer, L., Haeberli, W., Larsen, C., Schneider, D., Wessels, R., 2010. Recent and future warm extreme events and high-mountain slope stability. Phil. Trans. R. Soc. Ser. A. 368, 2435−2459.

Huggel, C., Zgraggen-Oswald, S., Haeberli, W., Kääb, A., Polkvoj, A., Galushkin, I., Evans, S.G., 2005. The 2002 rock/ice avalanche at Kolka/Karmadon, Russian Caucasus: assessment of extraordinary avalanche formation and mobility, and application of QuickBird satellite imagery. Nat. Hazards Earth Sys. Sci. 5, 173−187.

Jakob, M., Stein, D., Ulmi, M., 2012. Vulnerability of buildings to debris flow impact. Nat. Hazards 60, 241−261.

Jibson, R.W., Harp, E.L., Schulz, W., Keefer, D.K., 2006. Large rock avalanches triggered by the M 7.9 Denali Fault, Alaska, earthquake of 3 November 2002. Eng. Geol. 83, 144−160.

Jomelli, V., Brunstein, D., Grancher, D., Pech, P., 2007. Is the response of hill slope debris flows to recent climate change univocal? A case study in the Massif des Écrins (French Alps). Climate Change 85, 119−137.

Kääb, A., Huggel, C., Fischer, L., Guex, S., Pual, F., Roer, I., Saltzmann, N., Schlaefi, S., Schmutz, K., Schneider, D., Strozzi, T., Weidmann, Y., 2005. Remote sensing of glacier- and permafrost-related hazards in high mountains: an overview. Nat. Hazards. Earth Sys. Sci. 5, 527−554.

Keremlukov, V.A., Tsukerman, I.G., 1988. A review of information on breaching of moraine lakes of the Zailiyskii Alatau. Selevyye Potoki 10, 62−77.

Kershaw, J.A., Clague, J.J., Evans, S.G., 2005. Geomorphic and sedimentological signature of a two-phase outburst flood from moraine-dammed Queen Bess Lake, British Columbia, Canada. Earth Surf. Process. Landf. 30, 1−25.

Khegai, A.Y., Popov, N.V., 1989. The extent and economic significance of the debris flow and landslide problem in Kazakhstan, in the Soviet Union. In: Brabb, E.E., Harrod, B.L. (Eds.), Landslides: Extent and Economic Significance. Balkema, Rotterdam, pp. 221−225.

Kjartansson, G., 1967. The Steinholtslaup, Central-south Iceland on January 15th, 1967. Jokull 17, 249−262.

Kotlyakov, V.M., Rototaeva, O.V., Nosenko, G.A., 2004. The september 2002 Kolka glacier catastrophe in North Ossetia, Russian Federation: evidence and analysis. Mountain Res. Dev. 24, 78−83.

Krautblatter, M., Funk, D., Gunzel, F.K., 2013. Why permafrost rocks become unstable: a rock-ice-mechanical model in time and space. Earth Surf. Process. Landf. 38, 876−887.

Krenke, A.N., Kotlyakov, V.M., 1985. USSR Case Study: Catastrophic Floods, 149. IASH Publication, 115−125.

Lahr, J.C., Plafker, G., Stephens, C.D., Fogleman, K.A., Bleckford, M.E., 1979. Interim Report on the St. Elias, Alaska Earthquake of 28 February 1979. United States Geological Survey. Open-File Report 79−670, 35 pp.

Lipovsky, P.S., Evans, S.G., Clague, J.J., Hopkinson, C., Couture, R., Bobrowsky, P., Ekstrom, G., Demuth, M.N., Delaney, K.B., Roberts, N.J., Clarke, G., Schaeffer, A., 2008. The July 2007 rock and ice avalanches at Mount Steele, St. Elias mountains, Yukon, Canada. Landslides 5, 445−455.

Llibroutry, L., Morales Arnao, B., Pautre, A., Schneider, B., 1977. Glaciological problems set by the control of dangerous lakes in Cordillera Blanca Peru. I. Historical failures of morainic dams, their causes and prevention. J. Glaciol. 18, 239−254.

Major, J.J., Newhall, C.G., 1989. Snow and ice perturbation during historical volcanic eruptions and the formation of lahars and floods. Bull. Volcanol. 52, 1−27.

Margreth, S., Funk, M., 1999. Hazard mapping for ice and combined snow/ice avalanches—two case studies from the Swiss and Italian Alps. Cold Reg. Sci. Technol. 30, 159−173.

Margreth, S., Faillettaz, J., Funk, M., Vagliasindi, M., Diotri, F., Broccolato, M., 2011. Safety concept for hazards caused by ice avalanches from the Whymper hanging glacier in the Mont Blanc Massif. Cold Reg. Sci. Technol. 69, 194−201.

Matsuoka, N., Sakai, H., 1999. Rockfall activity from an alpine cliff during thawing periods. Geomorphology 28, 309−328.

Mauthner, T.E., 1996. Kshwan glacier rock avalanche, southeast of Stewart, British Columbia; in current research 1996−A. Geol. Surv. Can., 37−44.

McColl, S.T., Davies, T.R.H., 2013. Large ice-contact slope movements: glacial buttressing, deformation and erosion. Earth Surf. Process. Landf. 38, 1102−1115.

McSaveney, M.J., 1978. Sherman glacier rock avalanche, Alaska, U.S.A. In: Voight, B. (Ed.), Rockslides and Avalanches, v. 1. Elsevier Scientific Publishing Co., Amsterdam, pp. 197−258.

McSaveney, M.J., 1993. Rock avalanches of 2 May and 16 september 1992, Mount Fletcher, New Zealand. Landslide News 7, 2−4.

McSaveney, M.J., 2002. Recent rockfalls and rock avalanches in Mount Cook National Park, New Zealand. In: Evans, S.G., DeGraff, J.V. (Eds.), Catastrophic Landslides: Effects, Occurrence and Mechanisms, Geological Society of America Reviews in Engineering Geology, vol. 15. Geological Society of America, Boulder, CO., pp. 35−70.

Milana, J.P., 2007. A model of the Glacier Horcones Inferior surge, Aconcagua region, Argentina. J. Glaciol. 53, 565−572.

Morales, B., 1966. The Huascarán avalanche in the Santa Valley, Peru. International Association Scientific Hydrology, Publication 69, 304—315.

Niyazov, B.S., Degovets, A.S., 1975. Estimation of the parameters of catastrophic mudflows in the basins of the Lesser and Greater Almatinka Rivers. Sov. Hydrol. Sel. Papers 2, 75—80.

O'Connor, J.E., Costa, J.E., 1993. Geologic and hydrological hazards in glacierized basins in North America resulting from 19th and 20th Century global warming. Nat. Hazards 8, 121—140.

Oppenheim, V., 1946. Sobre las lagunas de Huaraz. Boletin de la Sociedad Geologica del Peru 19, 68—79.

Oppikofer, T., Jaboyedoff, M., Keusen, H.-R., 2008. Collapse at the eastern Eiger flank in the Swiss Alps. Nature Geosci. 1, 531—535.

Orombelli, G., Porter, S.C., 1988. Boulder deposit of upper Val Ferret (Courmayeur, Aosta valley): deposit of a historic giant rockfall and debris avalanche or a late-glacial moraine? Eclogae Geol. Helvetiae 81, 365—371.

Passmore, D.G., Harrison, S., Winchester, V., Rae, A., Severskiy, I., Pimankina, N.V., 2008. Late Holocene debris flows and valley floor development in the northern Zaliiskiy Alatau, Tien Shan Mountains, Kazakhstan. Arctic and Alpine Res. 40, 548—560.

Petrakov, D.A., Chernomorets, S.S., Evans, S.G., Tutubalina, O.V., 2008. Catastrophic glacial multi-phase mass movements: a special type of glacial hazard. Adv. Geosci. 14, 211—218.

Petrakov, D.A., Krylenko, I.V., Chernomorets, S.S., Tutubalina, O.V., Krylenko, I.N., Shakminha, M.S., 2007. Debris flow hazard of glacial lakes in the Central Caucasus. In: Chen, C.-L., Major, J.J. (Eds.), Debris-flow Hazards, Mitigation, Mechanics, Prediction, and Assessment, pp. 703—714.

Petrakov, D.A., Tutubalina, O.V., Aleinikov, A.A., Chernomorets, S.S., Evans, S.G., Kidyaeva, V.M., Krylenko, I.N., Norin, S.V., Shakhmina, M.S., Seynova, I.B., 2011. Monitoring of Bashkara glacier lakes (Central Caucasus, Russia) and modelling of their potential outburst. Natural Hazards 61, 1293—1316.

Pierson, T.C., Janda, R.J., 1994. Volcanic mixed avalanches: a distinct eruption-triggered mass flow process at snow-clad volcanoes. Geol. Soc. Am. Bull. 106, 1351—1358.

Pierson, T.C., Janda, R.J., Thouret, J.C., Borrero, C.A., 1990. Perturbation and melting of snow and ice by the 13 November 1985 eruption of Nevado del Ruiz, Colombia, and the consequent mobilization, flow, and deposition of lahars. J. Volcanol. Geothermal Res. 41, 17—66.

Pistolesi, M., Cioni, R., Rosi, M., Cashman, K.V., Rossotti, A., Aguilera, E., 2013. Evidence for lahar-triggering mechanisms in complex stratigraphic sequences: the post-twelfth century eruptive activity of Cotopaxi Volcano, Ecuador. Bull. Volcanol. 75, 698.

Plafker, G., Ericksen, G.E., 1978. Nevados Huascarán avalanches, Peru. In: Voight, B. (Ed.), Rockslides and Avalanches, vol. 1. Elsevier Scientific Publishing Co., Amsterdam, pp. 277—314.

Popov, N., 1991. Assessment of glacier debris flow hazard in the North Tien-Shan. In: Proceedings Soviet—China—Japan Symposium and Field Workshop on Natural Disasters, pp. 384—391.

Porter, S.C., Orombelli, G., 1980. Catastrophic rockfall of September 12, 1717 on the Italian flank of the Mont Blanc Massif. Zeitschrift für Geomorphologie 24, 200—218.

Portocarrero, C., 2014. The glacial lake handbook; reducing risk from dangerous glacial lakes in the Cordillera Blanca, Peru. Technical Report, US AID, Washington, D.C. 68 pp.

Post, A., 1967. Effects of the March 1964 Alaska Earthquake on Glaciers. United States Geological Survey. Professional Paper 544-D, 42 pp.

Pralong, A., Funk, M., 2006. On the instability of avalanching glaciers. J. Glaciol. 52, 31—48.

Rabot, C., 1905. Glacial reservoirs and their outburst. Geogr. J. 25, 534—548.

Ravanel, L., Allignol, F., Deline, P., Gruber, S., Ravello, M., 2010. Rock falls in the Mont Blanc Massif in 2007 and 2008. Landslides 7, 493−501.

Reynolds, J.M., Dolecki, A., Portocarrero, C., 1998. The construction of a drainage tunnel as part of glacial lake hazard mitigation at Hualcán, Cordillera Blanca, Peru. In: Maund, J.G., Eddleston, M. (Eds.), Geohazards in Engineering Geology, 15. Geological Society of London Engineering Geology Special Publications, pp. 41−48.

Rickenmann, D., Zimmerman, M., 1993. The 1987 debris flows in Switzerland: documentation and analysis. Geomorphology 8, 175−189.

Röthlisberger, H., 1977. Ice avalanches. J. Glaciol. 19, 669−671.

Röthlisberger, H., 1978. Eislawinen und Ausbrüche von Gletscherseen. Sonderdruck aus dem Jahrbuch der Schweizerischen Naturforschenden Gesellschaft. Wissenschaftlicher Teil 1978, 170−212.

Salzmann, N., Kääb, A., Huggel, C., Allgöwer, B., Haeberli, W., 2004. Assessment of the hazard potential of ice avalanches using remote sensing and GIS-modelling. Norsk Geografisk Tidskrift-Norwegian J. Geogr. 58, 74−84.

Schneider, D., Bartelt, P., Caplan-Auerbach, J., Christen, M., Huggel, C., McArdell, B.W., 2010. Insights into rock-ice avalanche dynamics by combined analysis of seismic recordings and a numerical avalanche model. J. Geophys. Res. 115, F04026.

Schneider, D., Huggel, C., Haeberli, W., Kaitna, R., 2011a. Unraveling driving factors for large rock-ice avalanche mobility. Earth Surf. Process. Landf. 36, 1948−1966.

Schneider, D., Huggel, C., Cochachin, A., Guillén, S., Garcia, J., 2014. Mapping hazards from glacier lake outburst floods based on modelling of process cascades at Lake 513, Carhuaz, Peru. Adv. Geosci. 35, 145−155.

Schneider, D., Kaitna, R., Dietrich, W.E., Hsu, L., Huggel, C., McArdell, B.W., 2011b. Frictional behaviour of granular gravel-ice mixtures in vertically rotating drum experiments and implications for rock−ice avalanches. Cold Reg. Sci. Technol. 69, 70−90.

Schneider, J.F., Gruber, F.E., Mergili, M., 2013. Impact of large landslides, mitigation measures. In: Proceedings, International Conference on Vajont−1963−2013. Sapienza Universitá Editrice, Padova, pp. 73−84.

Scott, K.M., Macías, J.L., Naranjo, J.A., Rodriguez, S., McGeehin, J., 2001. Catastrophic Debris Flows Transformed from Landslides in Volcanic Terrains: Mobility Hazard Assessment, and Mitigation Strategies. US Geological Survey. Professional Paper 1630, 59 pp.

Seinova, I.B., 1991. Past and present changes of mudflow intensity in the Central Caucasus. Mountain Res. Dev. 11, 13−17.

Seinova, I.B., Andreev, Y.B., Krylenko, I.N., Chernomorets, S.S., 2011. Regional short-term forecast of debris flow initiation for glaciated high mountain zone of the Caucasus. In: Proceedings, 5th International Conference on Debris−flow Hazards, Mitigation, Mechanics, Prediction, and Assessment, pp. 1003−1011. Padova, Italy.

Seinova, I.B., Popovnin, V.V., Zolotarov, Y.A., 2003. Intensification of glacial debris flows in the Gerkhozhan River basin, Caucasus, in the late 20th Century. Landslide News 14/15, 39−43.

Seinova, I.B., Sidorova, T.L., Chernomorets, S.S., 2007. Processes of debris flow formation and the dynamics of glaciers in the central Caucasus. In: Chen, C.-L., Major, J.J. (Eds.), Debris−flow Hazards, Mitigation, Mechanics, Prediction, and Assessment, pp. 75−85.

Shreve, R., 1966. Sherman landslide, Alaska. Science 154, 1639−1643.

Slingerland, R., Voight, B., 1979. Occurrences, properties, and predictive models of landslide-generated water waves. In: Voight, B. (Ed.), Rockslides and Avalanches, vol. 2. Elsevier Scientific Publishing Co., Amsterdam, pp. 317−397.

Sosio, R., Crosta, G.B., Chen, J.H., Hungr, O., 2012. Modelling rock avalanche propagation onto glaciers. Quate. Sci. Rev. 47, 23−40.

Stoffel, M., Huggel, C., 2012. Effects of climate change on mass movements in mountain environments. Prog. Phys. Geogr. 36, 421−439.

Stokes, C.R., Gurney, S.D., Shahgedanova, M., Popovnin, V., 2006. Late 20th-century changes in glacier extent in the Caucasus Mountains, Russia/Georgia. J. Glaciol. 52, 99−109.

Stokes, C.R., Popovnin, V., Aleynikov, A., Gurney, S.D., Shahgedanova, M., 2007. Recent glacier retreat in the Caucasus mountains, Russia, and associated increase in supraglacial debris cover and supra-/proglacial lake development. Ann. Glaciol. 46, 195−203.

Stover, C.W., Reagor, B.G., Wetmiller, R.J., 1980. Intensities and isoseismal map for the St. Elias earthquake of February 28, 1979. Bull. Seismol. Soc. Am. 70, 1635−1649.

Strozzi, T., Delaloye, R., Kääb, A., Ambrosi, C., Perruchoud, E., Wegmüller, U., 2010. Combined observations of rock mass movements using satellite SAR interferometry, differential GPS, airborne digital photogrammetry, and airborne photography interpretation. J. Geophys. Res. 115, 11. FO1014.

Tarr, R.S., Martin, L., 1912. The Earthquakes at Yakutat Bay, Alaska, in September 1899. United States Geological Survey. Professional Paper 69, 135 pp.

Tufnell, L., 1984. Glacier Hazards. Longman, London, 97 pp.

Turnbull, B., 2011. Scaling laws for melting ice avalanches. Phys. Rev. Lett. 107, 258001.

Uhlmann, M., Korup, O., Huggel, C., Fischer, L., Kargel, J.S., 2013. Supra-glacial deposition and flux of catastrophic rock-slope failure debris, south-central Alaska. Earth Surf. Process. Landf. 38, 675−682.

van der Woerd, J., Owen, L.A., Tapponnier, P., Xiwei, X., Kervyn, F., Finkel, R.C., Barnard, P.L., 2004. Giant, M∼8 earthquake-triggered ice avalanches in the eastern Kunlun Shan, northern Tibet: characteristics, nature and dynamics. GSA Bull. 116, 394−406.

Vivian, R., 1966. La catastrophe du glacier Allalin. Revue de géographie alpine 54, 97−112.

Waythomas, C.F., Miller, T.P., Beget, J.E., 2000. Record of Late Holocene debris avalanches and lahars at Iliamna volcano, Alaska. J. Volcanol. Geothermal Res. 104, 97−130.

Waythomas, C.F., Pierson, T.C., Major, J.J., Scott, W.E., 2013. Voluminous ice-rich and water-rich lahars generated during the 2009 eruption of Redoubt Volcano, Alaska. J. Volcanol. Geothermal Res. 259, 389−413.

Wiles, G.C., Calkin, P.E., 1992. Reconstruction of a debris-slide-initiated flood in the southern Kenai Mountains, Alaska. Geomorphology 5, 535−546.

Yamanaka, H., Yoshida, M., Arita, K., 1982. Terrace landforms and Quaternary deposits around the Pokhara valley, central Nepal. J. Nepal Geol. Soc. 2, 113−142.

Yesenov, U.Y., Degovets, A.S., 1979. Catastrophic mudflow on the Bolshaya Almatinka River in 1977. Sov. Hydrol. Sel. Papers 18, 158−160.

Zimmermann, M., Haeberli, W., 1992. Climate change and debris flow activity in high-mountain areas - a case study in the Swiss Alps. Catena Suppl. 22, 59−72.

Zinevich, Y.N., 1981. Mudflow detention dam at Mynzhilki. Hydrotech. Constr. 15, 762−767.

Chapter 17

Hazards at Ice-Clad Volcanoes: Phenomena, Processes, and Examples From Mexico, Colombia, Ecuador, and Chile

Hugo Delgado Granados [1], Patricia Julio Miranda [2],
Gerardo Carrasco Núñez [3], Bernardo Pulgarín Alzate [4],
Patricia Mothes [5], Hugo Moreno Roa [6], Bolívar E. Cáceres Correa [7]
and Jorge Cortés Ramos [1]

[1] *Departamento de Vulcanología, Instituto de Geofísica, Universidad Nacional Autónoma de México, México,* [2] *Escuela de Ciencias Sociales y Humanidades, Universidad Autónoma de San Luis Potosí, Frac. Talleres, México,* [3] *Centro de Geociencias, Campus UNAM Juriquilla, Querétaro, Qro,* [4] *Servicio Geológico Colombiano, Observatorio Vulcanológico y Sismológico de Popayán, Barrio Loma de Cartagena, Popayán, Colombia,* [5] *Instituto Geofísico, Escuela Politécnica Nacional, Quito, Ecuador,* [6] *Servicio Nacional de Geología y Minería (SERNAGEOMIN), Observatorio Volcanológico de los Andes del Sur (OVDAS), Rudecindo, Temuco, Chile,* [7] *Instituto Nacional de Meteorología (INAMHI), Corea, Quito, Ecuador*

ABSTRACT

The interaction of volcanic activity with snow and ice bodies can cause serious hazards and risks. These interactions relate to enhanced heat flow, tephra in contact with the ice and snow resulting in the alteration of surficial ablation, and interaction with pyroclastic flows and incandescent materials. Such interactions can result in the formation of differential ablation, in tephra remobilization and, especially, in the generation of sometimes far-reaching lahars. Disasters resulted from recent events at Mount St Helens (1980–1986) and Nevado del Ruiz (1985) illustrate the impact these interactions can have. Case studies from Mexico, Colombia, Ecuador, and Chile are described. These descriptions depict the way in which the volcanic activity has interacted with ice bodies in recent volcanic crises (Popocatépetl, Mexico; Nevado del Huila, Colombia; Llaima and Villarica, Chile) and how the lahar processes have been generated. Reconstruction of historical events (Cotopaxi, Ecuador) or interpretation of events from the geological remains (Citlatépetl, Mexico) help to document past events that today could be disastrous for people and infrastructure now existing at the

Snow and Ice-Related Hazards, Risks, and Disasters. http://dx.doi.org/10.1016/B978-0-12-394849-6.00017-2

corresponding sites. A primary challenge for hazard prevention and risk reduction is the difficulty of making decisions based on imperfect information and a large degree of uncertainty. Successful assessments have resulted in the protection of lives in recent cases such as that at Nevado del Huila (Colombia).

17.1 INTRODUCTION

Dealing with volcanic hazards is a difficult task not only because of the complexity of the natural hazard but also because societies are becoming more complex and population growth increases vulnerability. Assessing hazards related to volcano—ice interactions is of particular importance because the second largest volcano-related disaster in recent times was of this type. The destruction caused by the eruption of ice-clad Mount St Helens in May 1980 (Brantley and Myers, 1997), snow-covered Tokachi Dake in 1926 (Murai, 1960), and especially the eruption of Nevado del Ruiz in Colombia in November 1985 with a toll of >25,000 lives (Naranjo et al., 1986; Parra and Cepeda, 1990) made many people concerned about how to avoid this type of catastrophe (Carey et al., 2014).

This chapter deals with the hazards, risks, and disasters related to volcano—ice interactions. Following an outline of the main interactions that take place during eruptive activity at ice-clad volcanoes, a brief description of the disasters that occurred at Mount St Helens and Nevado del Ruiz is made. Case studies of the recent occurrence of such interactions in Mexico, Colombia, Ecuador, and Chile are then presented, taking into account challenges encountered by scientists and officials when dealing with the eruptive activity in populated areas. In the case of Popocatépetl (Delgado Granados, Julio Miranda, and Cortés Ramos), the difficulty of making decisions based on incomplete information and a large degree of uncertainty is portrayed. The study of Tetelzingo lahar (Carrasco Núñez) is a study of a deposit recognized in the field and interpreted to be a case of important past volcano—ice interactions; the results and identification of the involved process help to evaluate this type of hazard and the possibility of its occurrence in the future. The eruption of Nevado del Huila (Pulgarín Alzate) is a great example of how the volcanological community in Colombia has acquired a large expertise in dealing with volcanoes and especially ice-capped volcanoes. This assertion refers not only to the scientific knowledge developed but also to the expertise in handling difficult decision-making situations. Ecuador has had many volcanic eruptions in the last decade. The experiences collected by the Ecuadorian volcanologists have helped them develop well-oiled procedures for facing eruptive crises. Nonetheless, Cotopaxi volcano, a long-lived, dormant volcano has shown recent signs of unrest (Mothes and Cáceres Correa). The description of the lahar products and the processes that have produced lahar deposits are not only scientifically pertinent but assist the risk management process. Chile is a country with a very large number of active volcanoes in its territory. The responses to recent eruptions of Chaitén and Cordón Caulle volcanoes are examples of how the Chilean scientific community can deal with uncertainty and help to make decisions over short time scales. However, this would not be possible if geological and volcanological studies had not been made beforehand and protocols prepared to face eruptive crises as in the cases of Chaitén (2008) and Cerro Hudson (2011) volcanoes. An important part of the process is to know how the lahar events are produced, as in the case of Villarica and Llaima volcanoes (Moreno Roa).

To know and understand how volcanoes interact with ice and snow, under a range of activity, is just the first part of the hazard assessment and risk reduction procedure. A staged approach describes five steps to be followed to mitigate the effects of volcano–ice interactions. The acquisition of knowledge about volcano behavior and interaction with snow and ice is clearly more than a scientific exercise but part of risk-reduction and disaster-prevention strategies.

17.2 VOLCANO–ICE INTERACTIONS

The recurrent phenomenon produced by the interaction between eruptive activity and glaciers is the generation of lahars (Major and Newhall, 1989; Pierson, 1989), which, depending on their magnitude, may constitute a hazard to the population and infrastructure. The water required for the formation of lahars on snow-capped and ice-clad volcanoes derives from the melting of snow and ice caused by volcanic phenomena as mentioned below.

The interactions cannot be reduced to lahar generation alone; they are complex processes at the interface of snow/ice with magma (either intrusive growth or energetic emplacement of pyroclastic debris) and beneath ice caps. So, in addition to lahars, other hazards may occur as the products of the interaction between eruptive activity and glaciers, as suggested by Keys (1996). Subglacial eruptions occurring under valley glaciers or ice caps can be explosive (Corr and Vaughan, 2008) but involve substantial volumes of water generated by the melting ice. The characteristics of the glaciers, namely, temperature distribution, structure, and hydrology, exert fundamental control on subglacial eruptions. At lower latitudes, glaciers on composite volcanoes generally cover small areas, are steep, thin, and have temperatures close to the melting point. Such temperate ice with its hydrology exerts a fundamental control on subglacial eruptions. Cold glaciers at high latitudes and under dry-continental climatic conditions provide different thermal conditions (Smellie, 2000).

In Iceland, for instance, subglacial rhyolitic eruptions have not been observed, but related deposits formed during glacial and interglacial periods indicate that interactions can be of low-to-high explosivity, and effusive to form sheet-like lava bodies (Tuffen et al., 2001; McGarvie et al., 2007; McGarvie, 2009; Stevenson et al., 2009; Owen et al., 2013). Rhyolitic, andesitic, and basaltic eruptions produce phreatomagmatic deposits, vitriclastic tuffs, lavas (including jointed and pillow lavas), and hyaloclastites. When melt water in-filtrates the lavas, it can produce glassy breccias and cause localized steam explosions (Stevenson et al., 2006). Interaction of lavas with thick ice produced "tuyas" (flat-topped, steep slope volcanic structure; Stevenson et al., 2011) in Iceland during the Pleistocene (McGarvie, 2009). Modern, active volcanism interactions may result in atypical behavior for temperate glaciers as in the case of the glaciers of Mount St Helens where extraordinary-magnitude strain rates due to the dome growth provoked a deformation resulting in glacier squeeze in a way that no slip of the glacier over its bed was observed (Walder et al., 2008).

Climate change is also a factor to consider. Several authors have proposed climate change has a strong influence on glacier shrinkage at volcanoes (i.e., Rivera et al., 2006), and also, climatic fluctuations are recognized as triggers for volcanic eruptions (Rampino et al., 1979; Robock, 2000; Capra, 2006; Nowell et al., 2006; Kutterolf et al., 2013). Loading and unloading volcanoes with mass can have eruptions as a consequence (Sigvaldason, 2002; Sigmundsson et al., 2010). Thus, Tuffen (2010) speculates that volcano—ice interactions may increase through the twenty-first century due to the melting of glaciers as a response to climate change. Gradual disappearance of ice from the summits of volcanoes may unload mass that results in an increase of volcanic eruptions.

17.2.1 Eruptive Activity and Ice Bodies

Effects of volcanic activity on snow and ice involve (1) abrasion and melting generated by pyroclastic flows including pyroclastic surges; (2) surface melting due to lava flows; (3) basal melting of ice and snow by subglacial eruptions or geothermal activity; (4) ejection of water from crater lakes; and (5) tephra deposition (Major and Newhall, 1989).

In general, snow and ice affect the character of volcanic activity via increased amounts of subsurface water often contributing to eruptive clouds and flows, increased fluidization of pyroclastic flows and lahars, and lowered basal temperatures and frictional factors of volcanic flows (Waitt et al., 1983; Brugman, 1990). All these processes are strongly influenced by the style and size of the eruption and the characteristics of the involved glaciers in a tem-poral context. They can occur during the eruption (syn-eruptive) or long after it has occurred (post-eruptive) (Manville et al., 2000). Melting of snow or ice during historical eruptions has most often been caused by pyroclastic density currents (42 percent), followed by basal melting ($\sim$30 percent), surficial lake flows (8 percent), or ejection of water from crater lakes (4 percent) (Major and Newhall, 1989).

The interaction of eruptive activity with ice and snow is complex. Depending on how strong the interaction is, short- and long-term glacial dy-namics can be dramatically altered (Brugman and Meier, 1981; Sturm et al., 1986; Thouret et al., 2007). On the other hand, glacial dynamics may themselves trigger eruptions (Sigmundsson et al., 2010). For example, at Villarrica volcano (Chile), most eruptions of the twentieth century may have been triggered by the increase and decrease of load on the volcanic edifice due to seasonal variations in accumulation of snow, and the entrance of water into the hydrothermal system during the ablation season (Lara, 2004). Seven eruptive cycles have been documented at this volcano toward the end of the austral winter (August—October).

17.2.2 Enhanced Heat Flow

The increase in heat flow associated with the various manifestations of volcanic activity is a factor that can contribute to glacier mass loss (Major and Newhall, 1989). In general, heat conductivity of snow and ice is different and depends mainly on their microstructure (Satyawali et al., 2008). It increases with an increase in grain and bond size as well as grain and pore intercept lengths. The heat conductivity changes up to twice its initial value (Schneebeli and Sokratov, 2004), due to changes in structure and texture, but not due to changes in density. Heat conductivity increases during temperature gradient metamorphism. Metamorphism of ice is a part of the overall heat transfer system (Arons and Colbeck, 1995). Heat transfer from tephra fall is ephemeral unless the tephra's temperature is >400 °C. Welded tephra (generally scoria) are an unusual deposition process. However, Thouret et al. (2007) reported that early scoria fall at Nevado del Ruiz was partially welded. Walder (2000a) mentioned that a thermally driven mechanism exists by which a layer of pyroclasts can incorporate snow and generate a slurry because of unstable fluidization of the pyroclasts due to an intense burst of vapor at the contact between the hot grains and snow. The mechanism involves vapor bubbling, particle convection, and scour of the substrate (Walder, 2000b). The theory applies to monodispersed sediments but does not effectively describe the actual pyroclastic flow material (Walder, 2000a). However, experiments and further theoretical development do allow description of polydispersed tephra (Walder, 2000b).

Heat transfer in subglacial systems in Iceland takes place in two ways: by dike intrusion and propagation or by sill intrusion at the substrate—ice interface (Wilson and Head, 2002). Both mechanisms can produce large volumes of melt water. The large surface area of dikes causes rapid melting, whereas the lateral spreading of sills efficiently transfers the heat. Hot, convecting melt water retained above the sill can produce continuous melting of the surrounding ice. Depending on the ice thickness, CO_2 may play an important role exerting overpressure beneath the ice. If melt water is drained away and CO_2 is put into contact with the atmosphere, the pressure decreases strongly and ice can subside over the sill. If the CO_2 is not released to the atmosphere, together with exsolved H_2O, the pressure is capable of generating explosive events.

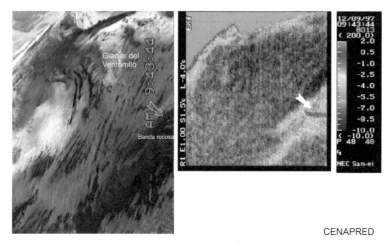

FIGURE 17.1   Visible and thermal infrared image obtained on September 12, 1997. The visible image shows a detail of Ventorrillo glacier and the upper crevasse system and the infrared image shows a thermal anomaly at the right of the glacier (arrows). *Images provided by L. Orozco.*

The appearance of unusual crack patterns or ice thickness changes may be indicative of increased heat flow or deformation under the glacial volcano area (Brugman and Meier, 1981). At Popocatépetl volcano (Mexico), in September 1995, fumaroles were observed on the west side of the glacier (Figure 17.1). This thermal anomaly appeared during the growth of a lava dome inside the crater in mid-August (Delgado Granados et al., 2007). The heat flow related to this lava-dome emplacement event caused an enlargement of the crevasse system as well as low-temperature fumaroles at the tongue of the Ventorrillo glacier, thus provoking ice melting.

17.2.3 Interactions between Tephra and Ice Bodies

17.2.3.1 Tephra Fall

Only a few investigations relate to effects of tephra on glaciers. Numerous studies, however, occur on the effects of debris cover of mountain glaciers (Nakawo and Young, 1981; Bozhinskiy et al., 1986; Mattson et al., 1992; Lundstrom et al., 1993; Adhikary et al., 1997; Takeuchi et al., 2001). Such studies generally show that thin debris covers enhance, but thick debris covers strongly reduce, ablation rates. In both cases, surface debris protects snow and ice surfaces from solar radiation but can warm far above 0 °C. The additional heat from such warm debris can, however, only be efficiently transmitted to the underlying ice by thin debris, whereas for thick debris, the insulating effect predominates. Experiments by Driedger (1981) on effects of ash layers on glacial ablation showed enhanced ablation rates in places with 2–5 mm of ash, whereas ablation was strongly reduced with ash thickness >24 mm. Maximum ablation occurred with 3 mm of ash, and a trace amount of ash caused ablation about 90 percent greater than for ash-free conditions. During the eruption of Ruapehu volcano in 1995−1996, tephra layers of <5 mm caused rapid melting, whereas layers >20 mm thick reduced melting (Manville et al., 2000).

Due to compressing glacier flow, tephra thickness tends to increase toward lower ice margins, causing the reversal of ablation gradients. Reduced ablation also causes an increase in the accumulation area (Benn et al., 2003). Glaciers covered by debris, under equilibrium conditions, show accumulation-area ratios (accumulation area/ablation area) between 0.2 and 0.4, significantly lower than about 0.5−0.7 typically observed on more-or-less clean, mid-latitude glaciers (Benn and Evans, 1998).

At Popocatépetl volcano, tephra deposition on the glacial surface was the most common phenomenon associated with eruptive activity since the beginning of the eruption in 1994 (Julio Miranda et al., 2008). Millimeter-thick tephra layers were deposited on the surface during low-explosivity phases, whereas tephra several centimeters or even up to 1 m thick covered the entire glaciers as a consequence of explosive phases. During the accumulation season, snow covered this tephra deposit. Thick snow layers were buried and preserved by tephra from subsequent explosions, whereas thin snow layers melted under the influence of heat from the deposited tephra. Tephra deposition and subsequent remobilization by various processes resulted in complex interactions with snow and ice. The volume of tephra deposited on the glacier during explosive phases such as occurred in December 2000 explains the "positive" mass balance disturbance recorded by the glacier during the period 2000−2001 after having been negative since 1996 when tephra deposition contributed to greater surface ablation, particularly during periods of low activity. The greatest loss of glacial volume occurred between February 2, 1999, and January 8, 2000, during which period, fewer explosions occurred as compared to the years 1998, 2000, and 2001 (Julio Miranda et al., 2008).

17.2.3.2 Differential Ablation

Tephra covers on glacier surfaces are generally not homogeneous. Thicknesses are variable and depend upon the topography of the glacier surface. Variations in debris thickness result in substantial differential ablation (Benn and Evans, 1998). Differential ablation causes the development of mounds at the glacial surface, which continue to grow until the slope of their sides is steep enough to cause gravity-driven sediment displacements, thus exposing the ice and resuming the melting. The distribution of tephra on the glacier surface of Popocatépetl was irregular over the years (1994−2001; Figure 17.2). The distribution and thickness were determined by the irregular morphology of the glacier surface, explosive intensity, and direction of the prevailing winds. The glacier surface showed major irregularities, especially during the period of maximum ablation, when crevasses increased their dimensions and the surface

FIGURE 17.2   Distribution of tephra over the glaciated area of Popocatépetl volcano in 1997. A heterogeneous distribution of tephra is observed. The greatest accumulation occurred in depressed areas or in areas with less steep slope. *Photograph by H. Delgado Granados.*

showed transverse undulations of the slope (Figure 17.3). This favored the accumulation of tephra in the depressions and less inclined areas, whereas on steeper areas, tephra was thinner. The upper edge of the glacier near the crater was progressively covered by a considerable thickness of tephra. As a result, this part became isolated and experienced little ice loss and, over time, formed an almost flat zone bounded by an escarpment at the front. Eventually, steep-sided unstable blocks of approximately 80,000 m^3 were formed. The possibility of sliding of these blocks became a hazard to mountaineers.

17.2.3.3 Tephra Remobilization

The accumulated volcanic debris on the glacial surface undergoes several cycles of movement and remobilization before final deposition. Water and gravity-driven transport processes are the most important factors responsible for the reworking of volcanic debris. In glacial surfaces covered by debris, differential ablation causes slope increase in some areas such that mass movement of debris initiates debris flows, sliding, or solifluction. Differential ablation processes, tephra reworking, and the development of a surface drainage system on tephra-covered glaciers all result in a distinctive hummocky topography. In places where tephra cover is irregular, differential ablation results in topographic reversal. During a 1996 eruption of Ruapehu volcano, tephra-induced snow avalanches formed where snow was weakened by snow-grain metamorphism, and weakened bonds between grains (Manville et al., 2000). The presence of soluble salts on the tephra further depressed the freezing point and formed additional melt water, which promoted the formation of sediment slurries.

The deposits of several such tephra remobilization processes were observed at Popocatépetl. Remobilization by wind, ice, and snow melting and gravity remobilization occurred repeatedly during eruptive activity of 1994–2001. Remobilization of tephra by melt water was most frequent during

FIGURE 17.3   Morphology of the glacial area of Popocatépetl during the ablation season in July 25, 1996. The surface shows elevated and depressed areas transverse to the slope. A thin tephra cover is observed. *Images provided by Secretaría de Comunicaciones y Transportes.*

the day and occurred at different spatio-temporal scales (Julio Miranda et al., 2008). On a diurnal basis, solar radiation on thin tephra layers caused substantial melting. These melt processes, which were intensified during the ablation season, caused agglomeration of tephra and flows of various sizes. The flows gouged grooves into the ice, into the tephra cover and into the bedrock, thus contributing significantly to the progressive morphological changes of the glacier. Melt water repeatedly caused the saturation of tephra and subsequent flows even during winter. On February 21, 2001 (Figure 17.4),

FIGURE 17.4   Zoom of the glacial area of Popocatépetl in February 21, 2001. Significant ablation on the surface can be seen during the winter, in spite of the tephra cover. Numerous debris flows are observed. Rectangles highlight areas of enhanced differential ablation produced by hot blocks. *Images provided by Secretaría de Comunicaciones y Transportes.*

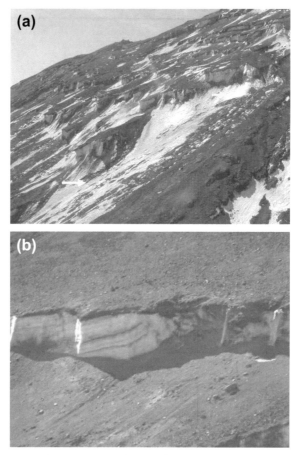

FIGURE 17.5   Zoom of the glacial area of Popocatépetl in April 2004. (a) Lateral view of the ice blocks, the arrows indicating debris fans formed at the base of a block. (b) Zoom to a block front, where a thick tephra layer is observed, as well as the intercalation of tephra and firn. *Photographs by I. Farraz.*

abundant runoff from the remaining glacier was observed despite the tephra cover. Gravitational reworking of tephra began with the fragmentation of the glacier, which developed ice walls exposed to daily ablation, causing the gradual collapse of the top of the blocks covered with tephra. Later, tephra was deposited at the base of the walls, either on top of the immediate block below or on the bedrock and formed volcanic debris fans (Figure 17.5).

17.2.4 Incandescent Material and Ballistic Impacts

Hot material of various grain sizes—often glowing at night—includes ballistic projectiles (i.e., material ejected following a ballistic trajectory; Waitt et al.,

1995; Mastin, 2001; Alatorre Ibargüengoitia and Delgado Granados, 2006). Passive melting occurs when a layer of hot pyroclastic debris is deposited on snow (Pierson, 1989). Melting is favored by an abundance of lithic clasts in the pyroclastic deposit because of their higher heat capacity compared to pumice, as well as the deeper penetration of lithic ballistic missiles in the snow. Static melting is slower than that associated with abrasion and mixture, but may be more sustained and extensive. Ballistic impacts on a glacier cause punctual melting of snow and ice. Glacier melting produced by a ballistic missile is a function of the potential and kinetic energy involved from the moment of ejection of the projectile at the crater, until its impact on the glacier (Delgado Granados et al., 2007). The fall of ballistic projectiles onto a glacier can produce large avalanches as during the eruption of Ruapehu volcano (New Zealand, 1995−1996; Manville et al., 2000). Hot ballistics also caused enough snowmelt to generate minor lahars (Keys, 1996). An additional issue to note about this eruption of Ruapehu volcano is the lake breakout, because the ejection of water from crater lakes also represents a very important hazard.

During its explosive activity in 1994−2001, Popocatépetl ejected incandescent material that fell on glacier-covered slopes of the volcano within ranges of 4 km (Figure 17.6). Most of the glacier area (422.938 m^2) was covered with tephra except the tongues where hot volcanic missiles formed impact craters on the snow and ice. Heat transfer from the hot rocks to the glacier deepened these impact craters later. Small debris flows started at the impact crater. On an area of 54,100 m^2, 185 impact craters with diameters ranging from 1.2 to 7.5 m were counted. As the eruption continued, the tephra cover of the glacier thickened and ice melting due to ballistic impacts decreased.

17.2.5 Pyroclastic Density Currents

17.2.5.1 Pyroclastic Flows

Pyroclastic flows are a type of pyroclastic density current consisting of heterogeneous mixtures of hot volcanic particles and gas flowing due to gravity with a higher density than that of air. They can reach considerable speeds (30−40 m/s; Myers and Brantley, 1995) and are capable of transporting hot debris over large distances (Branney and Kokelaar, 2002). The erosive capacity of pyroclastic flows is determined by size, density, and temperature, as well as the characteristics of the terrain on which they move. On snow, the pyroclastic flow load and grain size are critical parameters, but there is a thickness threshold involved in heat transfer (Walder, 2000a). The formation of pyroclastic flows on volcanoes covered by glaciers is the result of a complex interplay of thermal and mechanical processes (Walder, 2000a). The eruption of November 13, 1985, at Nevado del Ruiz (Colombia) produced a sequence of pyroclastic flows descending across the glacier (Thouret, 1990). The heat of pyroclastic flows and surges and related deposits were responsible for the

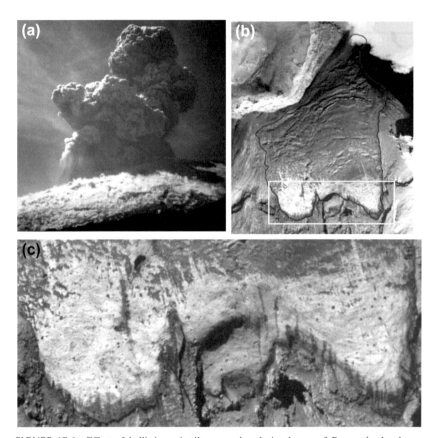

FIGURE 17.6 Effect of ballistic projectiles over the glaciated area of Popocatépetl volcano. (a) Ballistic missiles impacting the glaciated area on January 27, 1999; note that impacts on the ice surface caused steam columns. (b) On February 2, 1999, the glacial area (bounded by the (blue line) with the exception of the front (yellow box) is covered by tephra. (c) Zoom of the glacier front (indicated in (b)). The black spots correspond to impact craters generated by ballistic missiles falling on the glacier. Small debris flows produced by ablation of the glacier are also seen. *Photographs provided by E. Guevara (a) and Secretaría de Comunicaciones y Transportes (b and c).*

melting of snow and ice both on the glacier surface in situ and during transport, removing approximately 16 percent (4.2 km^2) of the volcano's glacier area and a volume of 0.06 km^3 of ice and snow with explosive events, triggering avalanches of ice, snow surface erosion, and avalanches of partially melted snow.

Erosive features resulting from the eruption at Nevado del Ruiz in 1985 had sizes varying from centimeters to meters. Ballistic impacts generated depressions over the glacier of up to 2 m in diameter and depth. Steep ice surfaces were smoothed by abrasion. Pyroclastic flows generated numerous grooves and channels in the snow and firn, and some "levees" made of scoria and lapilli were observed. Toothed "sérac fields" were flattened and in the steeper margins of the glacier (north and west) several gullies, >100 m long and 1−3 m wide, were generated (Thouret, 1990). The interaction of pyroclastic flows with the glacier was probably more vigorous than the effects from pyroclastic surges (Pierson et al., 1990). Ice deformation was observed in the eastern part of the Arenas crater (Thouret, 1990), where the greatest thickness of the pyroclastic deposits (4−6 m) was found: 1- to 3-m-high, plastically-deformed structures induced by ice extrusions were observed through the welded and unwelded pyroclastic deposits. The top of each extrusion was fractured as star-shaped fissures where the dome broke apart as rock avalanches. Rock avalanches can deposit large quantities of debris on glacier surfaces. The effects of increased material loads from volcanic activity on the stability of small/steep glaciers (cf. Röthlisberger, 1987) remains poorly documented.

Pyroclastic flows have played an important role in several eruptions within the last century; some of them have caused damage to infrastructure and human losses. In Japan, Tokachi Dake volcano erupted on May 24, 1926, and the snow cover of the flanks was suddenly melted by pyroclastic flows. The resultant mudflows, with an estimated volume of $5 \times 10^6 \, m^3$, severely damaged several villages (Murai, 1960). During the Crater Peak eruption of September 16−17, 1992 (Spurr volcano in Alaska), pyroclastic flows eroded and incorporated snow and glacier ice and transformed into debris flows that eventually dammed the Chakachatna River (Meyer and Trabant, 1995). However, the pyroclastic flows produced during the eruption of August 18, even though they traveled further down than the others, produced melt water flows but did not generate debris flows.

17.2.5.2 Pyroclastic Surges

During explosive eruptions, pyroclastic flows and surges do occur intercalated in time or gradually transforming from one to the other (Branney and Kokelaar, 2002). At Nevado del Ruiz in 1985, for instance, deposits of both density currents were deposited on the summit ice cap (Calvache, 1990). There, several pyroclastic flows changed to surges, when they flowed from the summit plateau to steep-sided areas (Pierson et al., 1990). The surge deposits were either composed of cohesive fine ash (silt and fine ash) or coarse ash. The surges mainly eroded the snow cover and removed little solid ice, although ice fragments that were 5−20 cm in size were incorporated, and even a 2-m ice block was transported 3.2 km from the crater. Interestingly, some fine-grained surge deposits on the ice cap were composed mainly of eroded and redeposited snow crystals, which may reflect dilute concentrations and low temperatures of the surges (Calvache, 1990).

At Ruapehu volcano in September 2007, ice slurries were produced by wet base surges during very small explosive events (Lube et al., 2009). One of the ice slurries was generated after entrainment of wet base surges into the early spring snow pack over an area of 0.45 km^2. A mixture of clasts, water, and 180,000 m^3 of particulate snow drained over the Whangaehu glacier, where it incorporated 170,000 m^3 of snow and firn ice. High mobility ice slurries can be produced by surges, which are capable of moving large volumes of ice during minor eruptions. The related hazard to this type of large snow-bulking-rate event was reported after the 1995 eruption of Ruapehu volcano (Cronin et al., 1996).

17.2.6 Lahar Generation

Debris flows on ice-clad volcanoes are particularly common due to the abundance of melt water and loose volcanic deposits. These flows initiate when the shear stress exerted by the weight of a debris layer on a steep slope exceeds the yield strength of the sediment. The flow will occur if (1) the yield strength of the sediment is reduced by an increase in pore pressure; (2) the shear stress is increased due to the increase in slope angle due to differential ablation; or (3) the shear stress increases due to the increase in the thickness of the tephra layer. The behavior, morphology, and sedimentological character-istics of the flows strongly depend on the water content (Janda et al., 1981; Benn and Evans, 1998). Reworking occurs at microscale and macroscale, the most obvious being the latter, wherein the movement of water over the surface removes debris, in either a concentrated or diffuse manner.

The eruption of Ruapehu volcano in 1995−1996 had a significant impact on the summit cap and in the glacial valleys both during and after this eruption (Cronin et al., 1996; Manville et al., 2000). Significant thicknesses of tephra were deposited on glaciers and watersheds. The tephra was deposited in the northern and eastern flanks of Ruapehu. On fine-grained tephra deposits (<3 mm) diurnal insolation generated snow melting in contact with the ash. The tephra deposited on the ice surface became moistened. The moisture caused accretion, allowing the formation of lobes. The accretion of fine ash particles reduced surface contact between the tephra and snow, and promoted movement of the tephra. During the winters of 1995 and 1996, the daytime solifluction and surface-tension-accretion processes acted together causing creeping of the tephra deposits. The most intense melting occurred under snow packs covered by thin layers of fine ash triggering small debris flows. These flows caused sieving of the tephra de-posits according to grain size, with fine particles being transported, and coarser particles being left behind. The flows ceased when the slope of the channels decreased, or when the sediment concentration became too high due to the entry of more material or water percolation into permeable snow or regolith. Runoff occurred in relatively impermeable fine-grained tephra layers and over the surfaces of lahar deposits. A few days after deposition of the tephra, runoff formed grooves on the snow surface. Once formed, the grooves began to enlarge, especially in their downslope regions. Runoff was channeled through long furrows, eroding regolith, incorporating old pre-eruptive material and forming small fans.

Pyroclastic flows at Popocatépetl volcano are not frequent, but their interaction with the glacier area has generated lahars (Julio Miranda et al., 2005). Pumice-bearing pyroclastic flows descended over the glaciated area generating lahars on January 22, 2001. The total mass removed from the glacier area during this event was approximately 1.0×10^6 m^3. However, a water volume of approximately 1.6×10^5 m^3 promoted the formation of lahars. The estimated lahar volume was approximately 4×10^5 m^3. During the eruption of Nevado del Ruiz (Colombia) in 1985, 50 percent of the total volume of melt water participated in the formation of lahars (Thouret, 1990).

17.2.7 Jökulhlaups

Volcanic eruptions beneath a glacier are among the most violent expressions of the volcano—ice interactions because of the rapid melting of ice and production of hazardous floods (Björnsson, 2002). In Iceland, volcano—ice interactions produce melt water that may drain or accumulate in subglacial lakes and eventually drain as "jökulhlaups". Subglacial melting may occur in several ways (Wilson and Head, 2002) producing subglacial lakes. These lakes beneath ice-surface depressions drain periodically in outburst floods (Björnsson, 2002). The lakes expand gradually upheaving the glacier when water flowing toward the depression increases the basal water pressure. The water drains out of the lake under the ice dam once the hydraulic seal is broken, and then the depression surface is flattened. The water flows along the ice—bedrock interface through narrow passages mainly controlled by tunnel enlargement caused by ice melting of the walls. Melting is explained in terms of the heat generated by the flowing water and the thermal energy stored in the lake (Nye, 1976; Björnsson, 1992).

Jökulhlaups can be generated at geothermal systems or active volcanoes. At Skafta cauldrons, jökulhlaups have occurred every 2—3 years since 1955 (Björnsson, 1977). During these events, the discharge rises very rapidly and recedes slowly. In the case of jökulhlaups triggered by volcanic eruptions, the volume of melt water is larger as long as the eruptions melt thicker glaciers. At Grímsvötn, jökulhlaups typically increase exponentially to the peak and fall rapidly afterward, and the duration of large floods is generally shorter than the smaller ones, which reach their peak in 2—3 weeks and end 1 week after (Björnsson, 2002). At steep-sided stratovolcanoes covered by ice and snow, jökulhlaups can be fast and dangerous because they may transform into lahars and debris-laden floods as in the case of the 1362 AD eruption of Öraefajökull volcano, where the flood lasted less than a day with peak discharges of $>10^5$ m^3/s (Thorarinsson, 1958). In the case of Katla volcano, its eruptions cut across 400 m of ice cover in 1—2 h, melt large volumes of ice quickly, and break off large blocks of ice from the glacier's margin, generating very large jökulhlaups consisting of water, ice, volcaniclastic particles, and sediment with volumes of 0.7—1.6 km^3 (Gudmundsson et al., 1997; Larsen, 2000).

The jökulhlaups of the Myrdalsjökull cauldron system are considered the most hazardous of Iceland due to the rate of increase of discharge and peak discharges (Björnsson, 2002).

17.3 VOLCANO–ICE INTERACTIONS AS DISASTER GENERATORS: MOUNT ST HELENS AND NEVADO DEL RUIZ

Mount St Helens erupted on May 18, 1980, close to a major industrialized area in the northwest United States. The eruption consisted of a series of complex processes including a sector collapse of the volcano after a bulging of the north flank (Figure 17.7), production of a debris avalanche, a volcanic blast directed toward the north triggered by the collapse, lahars generated by the melting of snow and ice, and pyroclastic flows and ash falls (Lipman and Mullineaux, 1981). Prior to the eruption, thin ash layers covered the 5 km^2 of the glaciated area of Mount St Helens (Brugman and Meier, 1981). In fact, approximately 3-mm layers of ash produce up to 90 percent more ablation of snow than in ash-free conditions (Driedger, 1981). However, the sector collapse suddenly removed 70 percent of the ice volume (0.1 km^3). The resulting melt water caused lahars by three distinctive processes (Janda et al., 1981): In the headwaters of the North Fork of the Toutle River, lahars formed by dewatering of the debris avalanche; on the south and east flanks, lahars generated by the catastrophic ejection of mixtures of debris, ash, and lapilli; and minor lahars formed by melting of debris-laden ice and snow, or by pyroclastic flows/ recently deposited tephra. An additional threat became apparent after the eruption as water ponded in multiple tributary valleys that were blocked by the debris avalanche deposit. During the eruption, lahars damaged 27 bridges, and nearly 200 homes. Blast and lahars destroyed >298 km of highways and roads,

FIGURE 17.7 Aerial view of the bulged north flank of Mount St. Helens before the May 18, 1980, eruption.

and 24 km of railways. Fifty-seven fatalities were caused by the eruption although not all related to the volcano–ice interactions. The impact on wildlife was strong and natural channels of the affected rivers changed significantly (Brantley and Myers, 1997; Tilling et al., 1990).

Nevado del Ruiz volcano (Colombia) erupted on November 13, 1985, causing one of the largest natural disasters of the twentieth century. After one year of precursory activity, the eruption began at approximately 3:00 PM, with minor ash fall reported in towns on the northern sector 1 h later (Naranjo et al., 1986). Around 9:00 PM, the main explosive activity started, producing ash falls 20 km north and 80 km (up to 400 km) northeast of the volcano. The eruption also generated minor pyroclastic surges and flows, and their interaction with the ice and snow over the surface of a glacier with a volume of 3×10^8 m^3 (Figure 17.8) caused extensive surface melting (25–30 percent of mass producing 4.3×10^7 m^3 of melt water; Thouret, 1990) to form three major lahars. A lahar from the northern flank of the volcano reached the towns of Mariquita (11:30 PM) and Honda (2:00 AM on November 14) destroying 20 houses and causing two fatalities. Two lahars, from the eastern and northeastern flanks of the volcano, at Lagunillas and Azufrado valleys coalesced and flooded the town of Armero (11:00 PM) leaving a toll of 25,000 victims there. A lahar on the western flank reached the town of Chinchina (10:30 PM) claiming 1,000 lives. The mean velocities of the lahars were up to 17 m/s, peak discharges up to 48,000 m^3/s, and the total volume of lahars was 9×10^7 m^3 (Pierson et al., 1990). The lessons learned from this eruption are: (1) catastrophic lahars can be produced at ice-clad volcanoes by relatively small eruptions; (2) the consideration of surface area of the snow and ice cap is more important than the total ice volume when assessing laharic hazards; (3) movement of hot rocks over ice or snow is not enough to generate lahars,

FIGURE 17.8   Aerial view of the glaciated area of Nevado del Ruiz volcano on September 27, 1985. *Photograph by E. Parra.*

but their mixture for an efficient and rapid heat transfer; (4) bulking of lahars is produced by entrainment of water and eroded sediment; and (5) lahars can maintain a relatively high velocity when confined in valleys and cause destruction at distances of approximately 100 km from the vent (Pierson et al., 1990).

17.4 VOLCANO–ICE INTERACTIONS IN MEXICO, COLOMBIA, ECUADOR, AND CHILE: DEALING WITH RELATED HAZARDS

17.4.1 Volcano–Ice Interactions at Mexican Volcanoes: Witnessing Volcano–Ice Interaction Hazards

17.4.1.1 The Laharic Event of January 22, 2001, at Popocatépetl Volcano

The development of lahars during January 22, 2001, marked the end of an eruptive sequence that had begun in November 2000. During this period, eruptive events had accelerated ice loss and led to the eventual extinction of Ventorrillo glacier on Popocatépetl (Julio Miranda et al., 2005; Delgado Granados et al., 2007). During November–December 2000, swarms of volcano-tectonic earthquakes of variable magnitudes were detected in the southeast sector of the volcano, and forced the authorities to raise the alert level. This activity was associated with the growth of lava Dome 12, the largest ever recorded up to then. Swarms of volcano-tectonic earthquakes were preceded by harmonic tremor being greater than amplitudes previously observed. Ash emissions continued for a 90-min duration. The activity increased on December 12, with a large number of exhalations (up to 200 per day), many with 5- to 6-km-high ash columns. At night, the glow of the crater and the emission of incandescent fragments could be seen from a distance. Ash fell on nearby villages. On December 15, the harmonic tremor increased to become an astonishing continuous signal, with amplitudes saturating the instruments at all monitoring stations, including distant stations. The seismicity was felt by people at distances of 12–14 km from the vent, and was recorded by the National Seismological Service at distances of over 150 km from the volcano. The harmonic tremor lasted approximately 10 h and was associated with the extrusion of magma at a very high supply rate. A dramatic drop in the level of activity occurred in the morning of December 16, followed 16 h later by a new signal of low-frequency harmonic tremor with increasing amplitude. This signal also reached saturation levels in all seismic stations, and lasted about 9.5 h. Inclinometers detected signals of contemporary large oscillation tremor episodes. The amplitude of the oscillations was in the range of 100 μrad, reaching a peak-to-peak amplitude of 200 μrad.

Between December 13 and 19, 2000, SO_2 measurements made with a correlation spectrometer had values in excess of 50,000 t/day. On December 19, measured values reached >100,000 t/day, which contrasted with the previous weeks' average of 5,000 t/d. In about 25 h of accumulated harmonic tremor, the seismic energy released by the volcano exceeded the released energy over the entire year 2000. The dome grew at an average rate of 18−20 m³/s during episodes of harmonic tremor (Espinasa Pereña, 2013). This growth rate was two orders of magnitude higher than was previously recorded. Episodes of tranquility and large-amplitude harmonic tremor were interpreted with the predictive model of Shimazaki and Nakata (1980), which allowed prediction in advance of the start of the next major episode of dome growth, expected to be on December 18. By December 15, the authorities raised the alert level. Due to the magnitude of the monitored signals and the high emission rate of lava (forming the largest dome observed by then), the Civil Protection authorities declared an increase in the alert level defining a safety radius of 13 km. This included several towns, such as Santiago Xalitzintla and San Pedro Benito Juarez, which were not only the closest towns to the vent but also the most exposed to laharic events due to their location. Precautionary evacuation of inhabitants began on the night of December 15. The state and municipal level authorities decided which settlements had to be evacuated, but some mayors of towns outside the safety radius of 13 km also decided to evacuate people. Nearly 41,000 people left the area; the Civil Protection authorities mobilized about half; approximately 20,000 were self-evacuated. Nearly 14,000 agreed to be transported to shelters where they remained for 10 days. The other evacuees stayed with relatives or friends. More details on the risk management approach and procedures can be found in De la Cruz-Reyna and Tilling (2008).

The total volume of accumulated lava inside the crater of Popocatépetl volcano on December 18 was estimated to be between 15 and 19 million m³. This value exceeded the combined volumes of all previous domes. The vertical growth estimate for Dome 12 was so large that with 20−30 additional hours of growth, the top of the dome would have reached the lowest level of the crater rim, increasing the probability of an overflow of the dome and production of block-and-ash pyroclastic flows. After a three-day period of relative calm, on December 18 a new explosive phase began. Eruptions on December 18 and 19, of relatively low explosivity and duration, ejected a large amount of incandescent fragments over the flanks of the volcano. The media made a live broadcast of these eruptions. After December 19, the activity decreased significantly indicating that the magma supply rate had changed. The first explosion destroyed Dome 12 on December 24, and incandescent fragments reached 2.5 km from the vent, producing an ash column of about 5 km above the crater. On January 13, 2001 Dome 12 was partially destroyed and deflated. Civil Protection authorities were confident that the nature and the size of the event were understood and that the expected range of future activity was established, so the safety radius was reduced to 12 km. No villages are located within this radius, so the evacuees were able to return to their homes. However, on January 22, it became clear that this decision might have been premature. First, a 2.8 magnitude volcano-tectonic earthquake at 14:58 h east of the crater occurred. Seventeen minutes later, an exhalation of water vapor began reaching 1 km in height. An hour later, a large ash emission began, and a few minutes later, the explosivity increased, generating pyroclastic flows descending along the gullies of the volcano to estimated distances of 4−6 km. Pyroclastic flows traveled over the glacier, removing approximately 1 million cubic meters of ice and completely destroying the glacier (Julio Miranda et al., 2005), to produce lahars (Delgado Granados et al., 2007) traveling an additional 15 km and stopping just 1 km from the town of Santiago Xalitzintla (Tanarro et al., 2010). Lahars incorporated blocks up to 0.5 m in diameter. The average width of the flow was approximately 7 m. At 16:40 h, the eruption column had a height of 8 km above the crater. Ash fall was reported in Santiago Xalitzintla. This was a VEI (Volcanic Explosivity Index) = 3−4 eruption. These explosions did not remove >10−20 percent of the volume of Dome 12, so the old crater was almost filled to the brim (Macías and Siebe, 2005).

Further documentation on the volcano−ice interactions can be found in several sources: concerning glacial retreat, in Andrés et al. (2007) (confirming observations by Huggel and Delgado-Granados, 2000; Julio-Miranda and Delgado-Granados, 2003), on the implementation of a simple methodology to estimate lahar volumes and velocity in Muñoz-Salinas et al. (2007), and on rapidly updating channel morphology to perform simulations in Muñoz-Salinas et al. (2008, 2010). All these efforts provide useful tools for the Civil Protection officials, for better protecting the people in the vicinity of the stream channels of the volcano.

17.4.1.2 Volcano−Ice Interactions in the Geologic Record: the Tetelzingo Lahar at Citlaltépetl Volcano

Citlaltépetl (Pico de Orizaba) volcano forms the southernmost end of a nearly N-S-trending volcanic range at the eastern sector of the Trans-Mexican Volcanic Belt. This range has been very active during the Late Pleistocene, producing many large catastrophic collapses from different volcanoes such as Cofre de Perote (Carrasco-Núñez et al., 2010), La Gloria, and Las Cumbres volcanic fields (Carrasco-Núñez et al., 2006).

Citlaltépetl is an ice-capped active volcano (Figure 17.9), with a long and complex eruptive history, typical of a stratovolcano (Carrasco-Núñez, 2000), involving alternating periods of construction and destruction of the volcanic edifice. At least two major catastrophic collapses have been identified producing a large volume of deposits: the Jamapa avalanche and the Teteltzingo avalanche-induced cohesive lahar (Carrasco-Núñez et al., 1993). The latter formed a voluminous debris avalanche that rapidly transformed to a cohesive lahar very close to its source as a result of the sector collapse of hydrothermally-altered rock of the ancestral Espolón de Oro cone (Carrasco-Núñez et al., 1993).

FIGURE 17.9   Aerial view showing the north face of the main glacier of Citlaltépetl (Pico de Orizaba) volcano. *Photograph by G. Carrasco-Núñez.*

FIGURE 17.10   The Tetelzingo avalanche-induced lahar showing the large amount of hydro-thermally altered clay matrix, with a cohesive nature. *Photograph by G. Carrasco-Núñez.*

Interestingly, this deposit was dated at 16,550 + 145 -140 yr BP and 16,365 ± 110 yr BP, which corresponds approximately to the end of the maximum glacier extent in Mexico (Capra et al., 2013). The role of the glacier has been very important in the origin of those catastrophic events, particularly the latter one, that involved the production of large amounts of clay (>3−5 percent) causing a characteristic cohesive nature to the flows (Figure 17.10). The pervasive interaction of intense degassing activity through a central conduit for long periods of time, derived from a very active hydrothermal system within the glacier ice that extended more widely during the Late Pleistocene (Carrasco-Núñez et al., 1993; Capra et al., 2013). It produced a constant supply of pore water and large amounts of hydrothermally-altered clay minerals that are exposed in the upper remnants of former volcanic structures that were destroyed by successive collapsing events. The intense hydrothermal alteration could have been enhanced by an acid-sulfate leaching process, where a slow, steady-state supply of water above the water table allows for small to moderate fluxes of concentrated acidic fluids (Carrasco-Núñez et al., 1993). This has important implications because glaciated volcanoes are sites where hydrothermal alteration is enhanced and large portions of the volcanic edifice become unstable and can easily be removed favoring the generation of avalanche-induced cohesive lahars such as the Teteltzingo lahar derived from the Citlaltépetl volcano.

17.4.2 Volcano—Ice Interactions at Colombian Volcanoes: Eruptions of Nevado del Huila Volcano (2007—2008), Associated Lahars and Volcanic Crisis Management

17.4.2.1 Eruptions of Nevado del Huila Volcano during 2007—2008

Nevado del Huila Volcano (5,364 m asl), in southwest Colombia, has four ice-capped peaks. The only known historical activity of this volcano had been related to hot springs, fumaroles, and low instrumental seismic activity until the eruptions of February 19, 2007, April 18, 2007, and November 20, 2008, took place (Figure 17.11; Pulgarín et al., 2009, 2011).

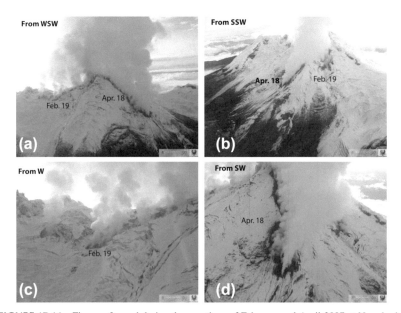

FIGURE 17.11   Fissures formed during the eruptions of February and April 2007 at Nevado del Huila (Colombia). *Photographs belong to the Servicio Geológico Colombiano, provided by B. Pulgarín.*

17.4.2.2 Phreatic Eruption and Lahar of February 19, 2007

This phreatic eruption (Pulgarín et al., 2008; Santacoloma et al., 2009; Londoño and Cardona, 2010) was associated with (1) seismicity; (2) formation of a N−S fissure 2 km long and 80 m wide at the top of the volcano, between Central and La Cresta peaks, through which gas fountains rose up to approximately 2 km; (3) an ash column approximately 4 km in height that covered the upper part of the Central Peak with a 1 m layer, decreasing to the west down to the Páez river; and (4) the formation of a small lahar that moved on the western flank of the edifice along two drainages, La Azufrada (from the longest glacier tongue) and Bellavista gullies, for about 6 km, reaching the Páez river, after which it traveled about 40 km with a maximum depth of 2−3 m causing no damage as most of the sediment load was deposited in the first few kilometers from the source. Near the town of Belalcázar, 48 km down the river, the flow was a small river flood, only 0.5 m above the normal river level.

17.4.2.3 Phreatic Eruption and Lahar on April 18, 2007

This eruption (Pulgarín et al., 2008; Santacoloma et al., 2009; Londoño and Cardona, 2010) resulted from (1) associated seismicity; (2) formation of a new SW−NE fissure crossing the Central Peak, 2.3 km and 80 m wide, and cutting the one formed in February 2007; (3) a small ash emission covering most of both sides of the Central Peak and also dispersed to the west; (4) loss of a portion (estimated to be about 400,000 m^3) of the El Oso glacier tongue front (the longest one on the east); and (5) formation of lahars larger than that formed in February 2007. The lahars descended along both sides of the volcano, to the Páez River in the west and the Símbola River in the east. They affected major road infrastructure and isolated Belalcázar town, the main town near the volcano, but caused no loss of life due to a timely warning and evacuation of the population. The lahar was about 10 m deep, traveled approximately 160 km and reached the Betania Reservoir. It had a volume of about 50,000,000 m^3 and a maximum speed of 80 km/h estimated for the first 50 km. No hot pyroclastic deposits melted snow or ice, but every eruption caused a separate crack in the edifice triggering water flow from the cracks. The volcano's steep sides and the absence of a crater precluded the possibility of water storage in a subglacial basin and/or in a hydrothermal (hot) reservoir. As a result, during the eruption, melt water eroded the rock and was released suddenly, melting part of the snow−ice cap and thus forming lahars.

17.4.2.4 Phreatomagmatic Eruption and Lahar on November 20, 2008

In this eruption (Pulgarín et al., 2008, 2009; Santacoloma et al., 2009; Londoño and Cardona, 2010), the explosion produced a 500-m-wide crater in the glaciated area between the Central and South peaks (western sector) and the immediate generation of a primary lahar, which traveled along the western flank of the volcano until the Páez river ($\sim$5 km) and, from there, >100 km down to the Magdalena river and to the Betania Reservoir (CHB, 1989) ($\sim$160 km from the volcano). Here, the lahar was strongly diluted and the reduced sediment load precluded great damage being done to the reservoir, but affected several small towns and infrastructure along the river. The lahar had average flood heights of approximately 45, 20 and 5 m, for the upper, middle, and lower parts of the Páez River valley, respectively, and a volume between 350 and 400 million cubic meters similar to the one generated by an earthquake in June 1994 in this same valley, in which about 1,100 people were killed (Cardona, 1995), much more than in the volcanogenic one that occurred on November 20, 2008, in which 10 people were killed (Pulgarín et al., 2011). The calculated speeds of the lahar were approximately 100 km/h in the proximal parts and up to 20 km/h in distal zones (Pulgarín et al., 2009). The main impact of the lahar was again on the population of Belalcázar. In this town, the flood reached between 19 and 23 m above the river level and a large sediment deposition occurred, causing a 7- to 11-m elevation of the riverbed (INGEOMINAS-Nasa Kiwe, 2009). As a result of this activity in the Central Peak of Nevado del Huila volcano, a dome began to appear in the new crater. The growth started in November 20, 2008, and lasted until May 2009, resulting in a volume of 35,000,000 m^3 and small incandescent block falls to the W, before another larger dome began to grow in the crater next to the former dome, from October 2009 (Cardona et al., 2009). Even at the end of 2013, the growth continues and has entirely engulfed the first one, producing small incandescent rock falls. It was estimated to be approximately 1000 m in average diameter with a height of 240 m and a volume of 82,000,000 m^3 in May 2010 and 370 m high with an average diameter of 1,300 m and a volume of 150,000,000 m^3 in October 2010.

17.4.2.5 Glacier Changes

Nevado del Huila glaciers suffered large changes. According to Pulgarín et al. (2009, 2010) these are: (1) the formation of two huge fissures approximately 2 km long and 80 m wide during the February and April 2007 explosions; (2) covering of the ice surface with ash and lahar formation by melting snow and ice, leaving behind channels on the glacier surface; (3) abundant fumarolic activity after melting of snow and ice, not only in the large fissure walls but also outside of them; (4) formation of new crack complexes in which ash layers could be seen; (5) tilting and fall of large ice blocks into and out of the resulting fissures from the 2007 eruptions; and (6) generation of small-scale snow—ice avalanches, some of which traveled <1 km down from the glacier. The emplacement of the two contiguous domes between the Central and South peaks caused severe deformation of the glacier. These morphological changes drastically decreased the areal extent and volume of the whole glacier mass. Since December 2009, the glacier area changed to approximately 9.8 km², representing a glacier loss of >1 km² in 2.5 years and continues to shrink.

17.4.2.6 Dealing with the Hazards and Risks

The educational and communication programs and the coordinated work made by all the involved institutions for the management of the crises (INGEO-MINAS, Nasa Kiwe, government, communities, local radio stations, Red Cross, and others), as well as the opportune information on the eruption by INGEOMINAS, strongly helped to safeguard the human lives compared with the seismogenic debris flow that occurred in the same basin in 1994 when about 1,100 people were killed. To better assess these events, further studies and modeling are required (Huggel et al., 2007; Worni et al., 2012).

17.4.3 Volcano–Ice Interactions at Ecuadorian Volcanoes. Variation in Lahar Matrices and Hazards Implications: Case Study Cotopaxi Volcano

17.4.3.1 Primary Lahars and their Generation at Cotopaxi

Cotopaxi volcano is located in the Eastern Cordillera of the Ecuadorian Andes, 60 km S of Quito and NE of Latacunga, a regional capital. The 5,897-m-high active volcano is notable for its extreme relief (Figure 17.12), massive size (22-km diameter) and its glacier-clad slopes. Its recent historical eruptions have all been of andesitic composition; however, rhyolitic eruptions occur about every 2,000 years (Hall and Mothes, 2008). The historical lahar record

FIGURE 17.12   Cotopaxi volcano and main drainage systems toward the north. *Photograph by P. Mothes.*

since 1534 AD includes at least 11 different lahar units intercalated between tephras (Mothes et al., 2004). Two main drainages occur; the northern drainage that flows 325 km to the Pacific Ocean and the SW drainage whose flow ends up in the Atlantic after traversing the Amazon River system.

Most lahars produced at Cotopaxi volcano have been matrix-poor, noncohesive, and clast supported. The 1877 AD lahars, in which gravel and sand sizes predominate, were of this nature. By mixing with stream water they transformed to hyperconcentrated stream flow (HCSF) some 40 km downstream from the vent. In contrast, cohesive lahars are not easily diluted by water and rarely transform to HCSF since the strong internal, interparticle attractive forces mute attenuation. However, a few exceptions occur in the stratigraphic record. These examples serve to remind us that although a snow-clad volcano is known to produce similar types of events, it may indeed have a record of generating other, less-frequently occurring phenomena, which in turn results in broad hazard implications.

The June 26, 1877 AD, lahars at Cotopaxi were formed through erosion and melting of the ice cap by the scoria-rich pyroclastic flows, which had runouts beyond the cone's base. The main components are scoria-rich sands and gravels, but preexisting blocks and debris were also ingested into the flow front. It is still possible today to walk to the lahar's lateral reach and find stringers of the scoria "krumpkoff" heads. These lahars had velocities of 25−30 m/s near the cone and slowed down to 10−15 m/s 40 km downstream on reduced gradients. In both N and S drainages, the lahars arrived at the main cities in little less than an hour. The valley of Latacunga and the city of that name suffered due to the broad reach of the flows and the effect of lahar surges coming down several main "quebradas" (ravines; Figure 17.12). At San Rafael, the transit of the lahar from the volcano was via restricting canyons and only 5 km upstream of this population center, the lahar burst out of its confining channel, spread out, and slowed down. At that time, most of the San Rafael area comprised haciendas, all of which would have been buried. Heading northward from San Rafael, the lahars again became restricted in canyons until arriving at the coastal plain area near the Pacific side of the city of Esmeraldas. Historical accounts say that the northern flowing lahar took 18 h to arrive at Esmeraldas, where it was still transporting parts of houses and bodies of people and animals from the Sierra. The southern flowing lahar had a similar fate and once past Latacunga and Salcedo, the lahar dropped into a deep canyon and made its way down to the Amazon basin lowland via the Pastaza river system. At one point, while passing the outskirts of Baños, at the foot of Tungurahua volcano, the lahar took out a bridge 30 m high above the Pastaza River.

The fate of medium size to small lahars on Cotopaxi is controlled by morphology. First, the actual crater rim is lower on the west side, thus preferentially favoring debauching of pyroclastic flows onto this flank. The second factor is the existence of the broad sloping plain on the north−northeast foot of the cone, which acts as a sediment trap to lahars. This plain is called Limpiopungo and is underlain by a thick package of lavas hundreds of meters thick, which extend 40 km to the north forming the formidable canyon of the Rio Pita, the main route for lahars. Since the lava plateau is so broad and relatively flat, it is an immense depositional zone for most lahars. Since 1534, only four historical lahars have been reported in the Chillos Valley (Mothes et al., 2004). For lahars transiting into the Latacunga valley, there is no lava plateau to cause them to spread out, deposit, and temporarily slow down. All lahars on the W and S flanks are immediately channelized into deep canyons that carry them 25 km to the Latacunga valley, where only then can deposition begin. Probably due to the westward-oriented crater and the lack of a broad distilling plain, the valley of Latacunga has suffered about 11 historical lahar events (Hall and Mothes, 2008).

Lahar deposits of earlier historical eruptions (1534, 1744, and 1768) commonly produced larger volumes, bigger deposits, higher amounts of clasts possibly from bulking, and broader inundation areas. Their generation was mainly from pyroclastic flows running over a glacier surface, which hundreds of years ago was far bigger than that existing today. For example, the Chillos Valley Lahar (CVL), generated 4,500 years BP by a rhyolitic pyroclastic flow and sector collapse on Cotopaxi's N flank, has a bulk volume of about 3 km^3. At its widest, the CVL's flow was 11 km across and about 100 m deep. Along its flow trajectory of 326 km, little transformation occurred in its lithology, due to a cohesive matrix rich in fines. More recently, lahars associated with Cotopaxi's 1742 eruption (VEI = 4) that occurred after 200 years of repose, also have higher fine grain contents. A beige pumice lapilli (marker) layer fell over the volcano's W flank and was followed by the transit in nearby stream channels of a matrix-rich lahar, whose deposit directly overlies this pumice "marker" unit. The matrix-rich 1742 lahar deposit occurring in the Cutuchi Quarry (25 km from the vent) is notable for its ability to have transported large 2- to 3-m diameter clasts in suspension. This characteristic continues down valley to Mulalo (30 km from the vent), implying minimal water entrainment. Nonetheless, at Salcedo, 65 km downstream, less silt has been retained, implying that some mixing with water had occurred, although large clasts are still present. Silt concentrations in these deposits are about 20 percent. Since there is an absence of clay-size particles in the matrix, water miscibility is greater than with a true matrix-rich lahar, such as observed in the Electron or Osceola deposits at Mt Rainier (Scott et al., 1995; Vallance and Scott, 1997). Thus, the 1742 lahar retained "matrix-rich" characteristics for >50 to 60 km downstream.

The generating factor of Cotopaxi's "pseudo" cohesive lahars has probably been pyroclastic flows with a high concentration of fine sand to silt grains. These fine-grained particles may have formed from abrasion while the pyroclastic flow descended 2,000 vertical meters from the crater rim. At Cotopaxi, hydrothermal alteration and clay formation are not observed. Given that matrix-rich lahars are slow to transform downstream to more water-rich flows, the potential impact to essential infrastructure, property, and life is greater due to the solid nature of the flow, greater discharge and higher velocities. It may be difficult to know, during a future Cotopaxi eruption process, whether a fine-grained pyroclastic flow is in transit and a cohesive lahar may form. While water is becoming less available on Cotopaxi due to receding glaciers, the now highly fractured and mushy ice makes the remaining ice cap more easily melted, and thus, future lahar generation is still a likely scenario.

17.4.3.2 Potential to Produce Lahar Flows from Cotopaxi's Glaciers

Cotopaxi is an ice-capped volcano endangering >100,000 people at Latacunga and Los Chillos valleys. Cáceres (unpublished data) described a total glacierized area of 13.9 km^2 for 2006 at Cotopaxi (Figure 17.13). An average melted thickness of 4 m could produce a hypothetical lahar of 60 million m^3, whereas an average melted thickness of 8 m could produce a lahar of 120 million m^3 (Figure 17.13; Jordan et al., 2005; Mothes et al., 2004; Cáceres, 2010). The unrest of Cotopaxi volcano has raised the level of concern of Ecuadorian authorities since 2002. Fumarolic activity at the crater of Cotopaxi has removed ice cover at the volcano's summit (Figure 17.14).

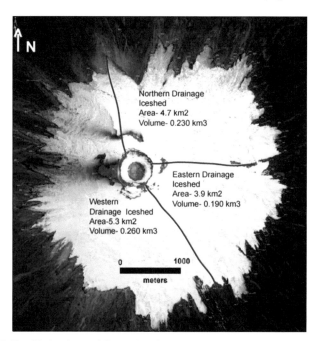

FIGURE 17.13   Glaciated area of Cotopaxi. *Cáceres, 2010; unpublished data. Aerial photograph provided by B. Cáceres.*

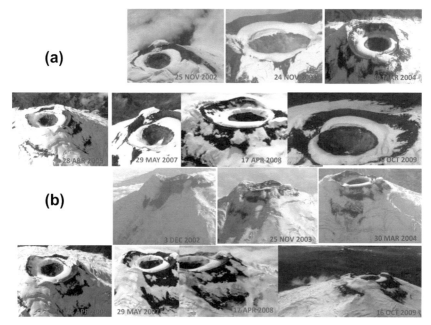

FIGURE 17.14   (a) Decrease in the size of the crater glacier, melting and partial collapse of the crater glacier at the south summit into the crater and formation of melting fans below the crater glacier. (b) Western flank, deglaciated areas extend due to constant fumarolic activity. *Photographs by P. Ramón and S. Vallejo.*

17.4.4 Volcano–Ice Interactions at Chilean Volcanoes. Lahars at Villarica and Llaima Volcanoes, Southern Andes, Chile: Origin and Eruption Styles

17.4.4.1 Villarica and Llaima Volcanoes

Villarrica (2,847 m; 39.4° S) and Llaima (3,125 m; 38.7° S) are the most active stratovolcanoes of the Southern Andes of Chile, and their products are mainly basalts and basaltic andesite. The most frequent eruptions at both within historical times have been mainly Strombolian with a few being Hawaiian in type. The two stratovolcanoes have almost perfect conical morphologies and are covered by large glaciers together with seasonal snow; hence, lahar generation is one of the main hazards associated with their eruptive activity. Moreover, both volcanoes show steep slopes and significant relief at the summit. The lahar deposits show a number of internal structures such as lamination, grading, lenses, a middle layer of subangular to subrounded cobbles, a coarse sand-rich matrix, and some large floating blocks on top of them. Most lahar deposits are clay-poor (less than about 5 percent clay + silt); thus, they are generally induced by sudden water release due to melting of ice and snow during lava eruptions. The solid fraction of lahars, in most cases, consists of sediments from recent till deposits that rest on the volcano's flanks. These materials are loose clastic deposits that are easily merged as debris and may quickly form lahar flows (Moreno and Clavero, 2006).

17.4.4.2 Origin of Lahars at Villarica and Llaima Volcanoes

As observed in historic eruptions, significant lahars are formed only when a high lava effusion rate takes place at the volcanoes' summits. The most important factors are the lateral opening of fissures on the flanks near and/or at the summit of the volcanoes and the resulting high effusion rate (Naranjo and Moreno, 2004). In fact, when a large volume of magma ascends through the main conduits, lava tends to be extruded from fissures up to 1 km long, formed on the flanks and/or at the summit. Then large lava fountaining takes place from the fissure(s) through Strombolian type eruptions. On the other hand, when small magma volumes move upward along the conduits, gentle outpouring of lava is extruded from the main craters with low effusion rates through Hawaiian-type eruptions. In this case, no lahars are generated and only some flooding results, while most of the water evaporates. In the first case, the melted water rushes down the slopes of the volcanoes along the different creeks, incorporating loose debris that yield highly turbulent and high-speed debris flows.

Lava eruptions of Villarrica volcano have taken place through the main crater or through radial flank fissures, depending on the magma volume ascent and the resulting lava effusion rate, as explained before. During the 1984 eruption, that lasted 1.5 months, the lava effusion rate reached only about $20 \, m^3/s$ (Moreno, 1993), and all lava flows poured out gently from the main crater without lahar generation (Figure 17.15). Indeed, only a small flood was observed in the Correntoso drainage (Moreno, 1993; Moreno and Clavero, 2006). However, in the 1971 eruption (at 23:45 h local time), an enigmatic red spot was observed in the ice surface on the NNE midslope of the volcano, at about 1 km below the main crater. This red spot remained for 5 min until unexpectedly a first fissure 1 km long opened from this red spot upward to the summit in seconds and a 400-m-high lava fountain was observed (a mechanism attributed to a dike intrusion and extrusion). When the lava fell on top of the ice, a fan of lava debris was formed instantaneously, covering the ice downwards at a velocity of 360 km/h (Marangunic, 1974). A hollow in the ice started to form in that place and generated a huge column of water vapor. Minutes later, a second fissure 1 km long opened on the SSW opposite flank, and another lava fountain 400 m high was extruded. Both lava fountains had a very high effusion rate, up to $500 \, m^3/s$ (Moreno, 1993), that produced large volume lahars that lasted for 6 h. Along the Correntoso drainage, a discharge of nearly $20 \times 10^6 \, m^3$ of water and debris, through several lahar waves, was observed at a velocity of 60 km/h (Marangunic, 1974). Two lava flows formed toward the NE and SW with a volume up to about $11 \times 10^6 \, m^3$, surprisingly

FIGURE 17.15   Eruption of Villarrica volcano in 1984. Vapor columns rise at the boundaries of the lava flows, which opened a channel through the glacier. *Photograph provided by H. Moreno.*

similar to the lava volume extruded in 1984 that reached nearly 10×10^6 m^3, but in this case, the lava flows discharged mildly within 1.5 months. The trend of the fissure vents on the volcano flanks seemed to be random depending on the internal cone structure as well as on the magma volume injection and its route to the surface through radial dykes. Fissures had opened to the N during the 1908 eruption, toward the NW and SE in 1948, toward the SW in 1963, toward the S in 1964 (lahars swept and destroyed Coñaripe village), and toward the NNE and SSW in 1971. In all cases, large lahars were formed radially on the volcano slopes, depending on the orientation of fissures and associated dike injections (Moreno, 2000; Naranjo and Moreno, 2004).

Lahars on Llaima volcano have origins and features similar to the Villarrica lahars (Moreno and Naranjo, 1991, 2003). In the 1957 eruption, lateral fissures opened toward the N, E, and SE on the volcano slopes with lava fountaining at high rates and large lahars forming associated with voluminous basaltic lava flows. During the eruption on May 17, 1994, a NNE fissure about 500 m long opened from the main crater toward the SSW, from where four small lava fountains up to 200 m high were observed, together with some explosions along the fissure (Moreno and Fuentealba, 1994; Moreno et al., 1994). A lava flow started to advance over the ice but suddenly sank into the glacier, running westward beneath the ice. The lava effusion rate was about 60 m^3/s and an estimated 4.5×10^5 m^3 of lava flow was extruded. Rough estimations suggest that at least $3-4 \times 10^6$ m^3 of ice were melted by the lava flow, but a lahar of only 2.5×10^6 m^3 was formed and ran down the Calbuco drainage, destroying some bridges and houses; thus, most of the water was evaporated. In the 2008−2009 eruption, lava fountaining took place on the S edge of the main crater with an effusion rate of about 20−30 m^3/s that yielded only floods along the Calbuco west drainage, and no damage was reported, even on small wooden bridges.

Based on these observations, the critical lava effusion rate to form small lahars instead of floods, is around 50 m^3/s: significant lahars were produced with an effusion rate of about 60 m^3/s. On the other hand, with high effusion rates of around 500 m^3/s supplied by lava fountaining, large destructive lahars up to 50×10^6 m^3 have been observed. Hence, there is a wide span from 60 to 500 m^3/s of lava fountaining onto the ice cover that can produce lahars. With rates of lava fountaining exceeding 100-m^3/s voluminous lahars can be produced. The eruption style that includes lava fountaining from lateral fissures (typical Hawaiian style), as well as Strombolian explosions with the generation of mushroom columns of gases and pyroclasts that reach up to 9 km high, is not typical of a single-style eruption. A new type of eruption is therefore proposed, which is a combination of Hawaiian-type lava fountaining with the generation of an explosion column that can reach a VEI = 3.

17.5 SPECIFIC ASPECTS OF HAZARD/RISK ASSESSMENT AT ICE-CLAD VOLCANOES

The International Association of Volcanology and Chemistry of the Earth's Interior (IAVCEI) suggests a guideline for professional conduct during volcanic crises (IAVCEI Subcommittee for Crises Protocols, 1999). Every country however has a particular way to handle volcanic hazards. An approach concerning the particular case of volcano−ice interactions is shown in Figure 17.16. Specific aspects to be mentioned are the necessity to estimate glacier volumes and the problem of estimating the probability of hazardous events. In addition, the glacier surface area (snow and firn) and the duration of mechanical and thermal interactions are key issues. Mechanical scouring and heat transfer can be more relevant than ice volume in several cases.

As glaciers on volcanoes rest on rather uniform and steep slopes, slope-related estimates of glacier thickness are recommended (Driedger and Kennard, 1986; Haeberli and Hoelzle, 1995) where direct measurements are difficult. Snow cover alone might also be important, particularly during wintertime (i.e., snow characteristics, texture, and structure as mentioned in Section 17.2.2). The degree of activity of the volcano provides a semi-quantitative indication concerning the probability of future eruptive activity, but changes in snow and ice as induced by climate change and/or volcanic activity can superimpose fast or slow trends with respect to hazards and risks

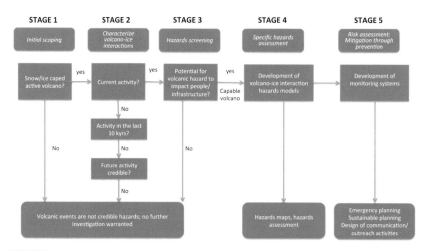

FIGURE 17.16   Staged approach for hazards/risk assessment at ice-clad volcanoes. *Adapted from IAEA (2012).*

related to volcano—ice interactions as mentioned above (Section 17.2). Geological maps and stratigraphic analyses calibrated with radiometric dating of units constitute an essential empirical basis for anticipating possible future events (especially far-reaching flows), preparing corresponding hazard maps and collecting data for parameterizing numerical model calculations. Numerical model simulations can add refined hazard maps together with flow speeds and corresponding warning time within the framework of short-term measures such as early warning systems and evacuation plans. They especially also serve to inform the population, as with other rapid, dangerous processes in nature, vulnerability is not only a function of the physical exposure to hazards but also of the degree of knowledge on how the processes may affect lives and property and on how to behave in an emergency case (Bruendl and Margreth, 2014). Finally, monitoring of ice- and snow-covered volcanoes is very important (Huggel et al., 2007) for the early detection of the related events and the development of warning systems to protect the society as a whole.

ACKNOWLEDGMENTS

The views described in the lines above have been enriched from discussions with several colleagues although we claim responsibility for our opinions. We acknowledge editorial help from Guillermo Ontiveros. Early reviews from Wilfried Haeberli and Colin Whiteman greatly improved the first draft. The final manuscript has strongly benefited from the comments and critical reviews by Carolyn Dridger and Jean-Claude Thouret. Financial support for scientific research from CONACyT (Grants 45433 and 83633), and DGAPA-UNAM (IN113914) is acknowledged.

REFERENCES

Adhikary, S., Seko, K., Nakawo, M., Ageta, Y., Miyazaki, N., 1997. Effect of surface dust on snow melt. Bull. Glacier Res. 15, 85−92.

Alatorre-Ibargüengoitia, M.A., Delgado-Granados, H., 2006. Experimental determination of drag coefficient for volcanic materials: calibration and application to Popocatépetl volcano (Mexico) ballistic projectiles. Geophysical Research Letters 33, L11302. http://dx.doi.org/10.1029/2006GL026195.

Andres, N., Zamorano, J.J., Sanjose, J.J., Atkinson, A., Palacios, D., 2007. Glacier retreat during the recent eruptive period of Popocatépetl volcano, Mexico. Ann. Glaciol. 45 (1), 73−82.

Arons, E.M., Colbeck, S.C., 1995. Geometry of heat and mass transfer in dry snow: a review of theory and experiment. Rev. Geophy. 33 (4), 463−493.

Benn, D.I., Evans, D.J.A., 1998. Glaciers and Glaciation. Arnold, London, 734 pp.

Benn, D.I., Kirkbride, M.P., Lewis, O., Brazier, V., 2003. Glaciated valley landsystems. In: Evans, D.J. (Ed.), Glacial Land Systems. Arnold, London, pp. 372−406.

Björnsson, H., 1977. The cause of Jökulhlaups in the Skaftá river, Vatnajökull. Jökull 27, 71−78.

Björnsson, H., 1992. Jökulhlaups in Iceland: prediction, characteristics and simulation. Ann. Glaciol. 16, 96−106.

Björnsson, H., 2002. Subglacial lakes and jökulhlaups in iceland. Global Planet. Change 35 (3), 255−271.

Bozhinskiy, A.N., Krass, M.S., Popovnin, V.V., 1986. Role of debris cover in the thermal physics of glaciers. Journal of Glaciology 32 (111), 255−266.

Branney, M., Kokelaar, P., 2002. Pyroclastic density currents and the sedimentation of ignimbrites. Geol. Soc. London 27, 136.

Brantley, S.R., Myers, B., 1997. Mount St. Helens—from the 1980 eruption to 1996. US Geological Survey Fact Sheet, 70−97.

Bründl, M., Margreth, S., 2014. Integrative risk management: the example of snow avalanches. In: Haeberli, W., Whiteman, C. (Eds.), Snow and Ice-related Hazards, Risks and Disasters. Elsevier, pp. 263−301.

Brugman, M., 1990. How do fire and ice interact on a volcano? Geosci. Can. 17 (3), 126.

Brugman, M., Meier, M., 1981. Response of glaciers to the eruptions of Mt. St. Helens. In: Lipman, P.W., Mullineaux, D.R. (Eds.), The 1980 Eruptions of Mt. St. Helens, US Geological Survey Professional Paper 1250, pp. 743−756. Washington, USA.

Cáceres, B., 2010. Actualización del Inventario de Tres Casquetes Glaciares del Ecuador (Master's thesis). University of Nice, France, 84 pp.

Calvache, M.L., 1990. Pyroclastic deposits of the November 13, 1985 eruption of Nevado del Ruiz volcano, Colombia. J. Volcanol. Geothermal Res. 41, 67−78.

Capra, L., 2006. Abrupt climatic changes as triggering mechanisms of massive volcanic collapses. J. Volcanol. Geothermal Res. 155 (3), 329−333.

Capra, L., Bernal, J.P., Carrasco-Núñez, G., Roverato, M., 2013. Climatic fluctuations as a significant contributing factor for volcanic collapses. Evidence from Mexico during the Late Pleistocene. Global Planet. Change 100, 194−203.

Cardona, O.D., 1995. El sismo del 6 de junio de 1994: atención de la emergencia y planteamientos para la reconstrucción. In: El desastre y la reconstrucción del Páez. Special Issue of Desastre y Sociedad, Red de estudios sociales en prevención de desastres en América Latina, Cauca y Huila, Colombia, junio 1994−junio 1995, 4(3), pp. 77−104.

Cardona, C., Santacoloma, C., White, R., McCausland, W., Trujillo, N., Narváez, A., Bolaños, R., Manzo, O., 2009. Sismicidad tipo "drumbeat" asociada a la erupción y emplazamiento de un domo en el volcán Nevado del Huila, noviembre de 2008. In: Memorias XII Congreso Colombiano de Geología. Paipa, Colombia, 7−11 September 2009.

Carey, M., McDowell, G., Huggel, C., Jackson, M., Portocarrero, C., Reynolds, J.M., Vicuña,, L., 2014. Integrated approaches to adaptation and disaster risk reduction in dynamic socio-cryospheric systems. In: Haeberli, W., Whiteman, C. (Eds.), Snow and Ice-related Hazards, Risks and Disasters. Elsevier, pp. 219−261.

Carrasco-Núñez, G., 2000. Structure and proximal stratigraphy of Citlaltépetl Volcano (Pico de Orizaba), Mexico. Geol. Soc. Am. Spec. Pap. 334, 247−262.

Carrasco-Núñez, G., Diaz-Castellón, R., Siebert, L., Hubbard, B., Sheridan, M.F., Rodríguez, S.R., 2006. Multiple edifice-collapse events in the Eastern Mexican Volcanic Belt: the role of sloping substrate and implications for hazard assessment. J. Volcanol. Geothermal Res. 158 (1), 151−176.

Carrasco-Núñez, G., Siebert, L., Díaz-Castellón, R., Vázquez-Selem, L., Capra, L., 2010. Evolution and hazards of a long-quiescent compound shield-like volcano: Cofre de Perote, Eastern Trans-Mexican Volcanic Belt. J. Volcanol. Geothermal Res. 197 (1), 209−224.

Carrasco-Núñez, G., Vallance, J.W., Rose, W.I., 1993. A voluminous avalanche-induced lahar from Citlaltépetl volcano, Mexico: implications for hazard assessment. J. Volcanol. Geothermal Res. 59 (1), 35−46.

CHB (Central Hidroeléctrica de Betania), 1989. Plan de Contingencia ante Eventos Hidrológicos y Situaciones Especiales: proyecto Yaguará (Santa Elena). Neiva, 33 pp.

Corr, H.F.J., Vaughan, D.G., 2008. A recent volcanic eruption beneath the West Antarctic ice sheet. Nature Geosci. 1 (2), 122−125. http://dx.doi.org/10.1038/ngeo106.

Cronin, S.J., Neall, V.E., Lecointre, J.A., Palmer, A.S., 1996. Unusual "snow slurry" lahars from Ruapehu volcano, New Zealand, September 1995. Geology 24 (12), 1107−1110.

De la Cruz-Reyna, S., Tilling, R.I., 2008. Scientific and public responses to the ongoing volcanic crisis at Popocatépetl Volcano, Mexico: Importance of an effective hazards-warning system. J. Volcanol. Geothermal Res. 170 (1), 121−134.

Delgado-Granados, H., Julio-Miranda, P., Huggel, C., Ortega del Valle, S., Alatorre-Ibargüengoitia, M.A., 2007. Chronicle of a death foretold: extinction of the small-size tropical glaciers of Popocatépetl volcano (Mexico). Global Planet. Change 56 (1), 13−22.

Driedger, C.L., 1981. Effect of ash thickness on snow ablation. In: Lipman, P.W., Mullineaux, D.R. (Eds.), The 1980 Eruptions of Mt. St. Helens, Washington. US Geological Survey Prof Paper 1250, pp. 757−760.

Driedger, C.L., Kennard, P.M., 1986. Glacier volume estimation on cascade volcanoes: an analysis and comparison with other methods. Ann. Glaciol. 8, 59−64.

Espinasa-Pereña, R. (Ed.), 2013. Historia de la actividad del Volcán Popocatépetl: 17 años de erupciones. Centro Nacional de Prevención de Desastres (CENAPRED). Secretaría de Gobernación, México, p. 69 (e-book).

Gudmundsson, M.T., Sigmundsson, F., Björnsson, H., 1997. Ice−volcano interaction of the 1996 Gjálp subglacial eruption, Vatnajökull, Iceland. Nature 389 (6654), 954−957.

Haeberli, W., Hoelzle, M., 1995. Application of inventory data for estimating characteristics of and regional climate-change effects on mountain glaciers: a pilot study with the European Alps. Ann. Glaciol. 21, 206−212.

Hall, M., Mothes, P., 2008. The rhyolitic−andesitic eruptive history of Cotopaxi volcano, Ecuador. Bull. Volcanol. 70 (6), 675−702.

Huggel, C., Delgado-Granados, H., 2000. Glacier monitoring at Popocatépetl volcano, Mexico: glacier shrinkage and possible causes. In: Hegg, C., Vonder Mühll, D. (Eds.), Beitraege zur Geomorphologie. Proceedings der Fachtagung derSchweizerischen Geomorphologischen

Gesellschaft vom 8.-10. Juli 1999 in Bramois (Kt. Wallis). Birmensdorf, Eidgenoessische Forschungsanstalt WSL, pp. 97−106.

Huggel, C., Ceballos, J.L., Pulgarin, B., Ramirez, J., Thouret, J., 2007. Review and reassessment of hazards owing to volcano−glacier interactions in Colombia. Ann. Glaciol. 45, 128−136.

IAVCEI Subcommittee for Crises Protocols, 1999. Professional conduct of scientists during volcanic crises. Bulletin of Volcanology 60, 323−334.

IIAEA (International Atomic Energy Agency), 2012. Volcanic hazards in Site Evaluation for Nuclear Installations. IAEA Safety Standards for protecting people and the environment, Specific Safety Guide, SSG-21, p. 16.

INGEOMINAS-Nasa Kiwe, 2009. Informe de los escenarios de amenaza por flujos de lodo "avalanchas" en la cuenca del río Páez. INGEOMINAS, Informe interno, Popayán, 71 pp.

Janda, R.J., Scott, K.M., Nolan, K.M., Martinson, H.A., 1981. Lahar movement effects and deposits. In: Lipman, P.W., Mullineaux, D.R. (Eds.), The 1980 Eruptions of Mt. St. Helens, Washington. US Geological Survey Prof Paper, 1250, pp. 461−478.

Jordan, E., Ungerechts, L., Caceres, B., Penafiel, A., Francou, B., 2005. Estimation by photogrammetry of the glacier recession on the Cotopaxi Volcano (Ecuador) between 1956 and 1997. Hydrol. Sci. J. 50 (6), 949−961.

Julio-Miranda, P., Delgado-Granados, H., 2003. Fast hazard evaluation employing digital photogrammetry: popocatépetl glaciers, Mexico. Geofísica Internacional 42 (2), 275−283.

Julio-Miranda, P., Delgado-Granados, H., Huggel, C., Kääb, A., 2008. Impact of the eruptive activity on glacier evolution at Popocatépetl Volcano (Mexico) during 1994−2004. J. Volcanol. Geothermal Res. 170 (1), 86−98.

Julio-Miranda, P., González-Huesca, A.E., Delgado-Granados, H., Kääb, A., 2005. Glacier melting and lahar formation during January 22, 2001 eruption, Popocatépetl Volcano (Mexico). Zeitschrift fur Geomorphology 140, 93−102.

Keys, J.R., 1996. Secondary alpine hazards induced by the 1995−1996 eruption of Ruapehu volcano, New Zealand. In: Proceedings of the International Snow Science Workshop. Canadian Avalanche Association, Banff, pp. 279−284.

Kutterolf, S., Jegen, M., Mitrovica, J.X., Kwasnitschka, T., Freundt, A., Huybers, P.J., 2013. A detection of Milankovitch frequencies in global volcanic activity. Geology 41 (2), 227−230. http://dx.doi.org/10.1130/G33419.1.

Lara, L., 2004. Overview of Villarrica volcano. In: Lara, L., Clavero, J. (Eds.), Villarrica Volcano (39.5°S), Southern Andes, Chile, 61. Servicio Nacional de Geología y Minería, Santiago de Chile, pp. 5−12.

Larsen, G., 2000. Holocene eruptions on the Katla volcanic system, Iceland: notes on characteristics and environmental impact. Jökull 50, 1−28.

Lipman, P.W., Mullineaux, D.W., 1981. The 1980 Eruptions of Mt. St. Helens, Washington. US Department of the Interior, US Geological Survey, 844 pp.

Londoño, J., Cardona, C., 2010. Seismicity associated to the reactivation of Nevado del Huila Volcano, Colombia. In: Proceedings of Cities on Volcanoes 6. Tenerife, Puerto de La Cruz, España, p. 67.

Lube, G., Cronin, S.J., Procter, J.N., 2009. Explaining the extreme mobility of volcanic ice-slurry flows, Ruapehu volcano, New Zealand. Geology 37 (1), 15−18. http://dx.doi.org/10.1130/G25352A.

Lundstrom, S.C., McCafferty, A.E., Coe, J.A., 1993. Photogrammetric analysis of 1984−89 surface altitude change of the partially debris-covered Eliot Glacier, Mount Hood, Oregon, USA. Ann. Glaciol. 17, 167−170.

Macías, J.L., Siebe, C., 2005. Popocatépetl's crater filled to the brim: significance for hazard evaluation. J. Volcanol. Geothermal Res. 141 (3), 327−330.

Major, J.J., Newhall, G.C., 1989. Snow and ice perturbation during historical volcanic eruptions and the formation of lahars and floods. Bull. Volcanol. 52 (1), 1−27.

Manville, V., Hodgson, K.A., Houghton, B.F., White, J.D.L., 2000. Tephra, snow and water: complex sedimentary responses at an active snow-capped stratovolcano, Ruapehu, New Zealand. Bull. Volcanol. 62 (4−5), 278−293.

Marangunic, C., 1974. The lahar provoked by the eruption of the Villarrica Volcano on December of 1971. In: Abstracts International Symposium on Volcanology. IAVCEI, Santiago de Chile, p. 49.

Mastin, L.G., 2001. A simple calculator of ballistic trajectories for blocks ejected during volcanic eruptions. US Geological Survey Open-file Report, 01−45.

Mattson, L.E., Gardner, J.S., Young, G.J., 1992. Ablation on debris covered glaciers: an example from Rakhiot glacier, Punjab, Himalaya, 218. IAHS Publications—publications of the International Association of Hydrological Sciences, 289−296.

McGarvie, D., 2009. Rhyolitic volcano−ice interactions in Iceland. J. Volcanol. Geothermal Res. 185 (4), 367−389.

McGarvie, D.W., Stevenson, J.A., Burgess, R., Tuffen, H., Tindle, A.G., 2007. Volcano−ice interactions at Prestahnukur, Iceland: rhyolite eruption during the last interglacial−glacial transition. Ann. Glaciol. 45, 38−47.

Meyer, D.F., Trabant, D.C., 1995. Lahars from the 1992 eruptions of crater peak, Mount Spurr volcano, Alaska. In: Keith, T.E.C. (Ed.), The 1992 eruptions of Crater Peak vent, Mount Spurr volcano, Alaska, 2139. U.S. Geological Survey Bulletin, pp. 183−198.

Moreno, H., 1993. Volcán Villarrica: Geología y Evaluación del Riesgo Volcánico, regions IX y X, 39°25'S. Unpublished Fondecyt Report, 1247, 1−112.

Moreno, H., 2000. Mapa de Peligros Volcánicos del Volcán Villarrica. Documentos de Trabajo, Servicio Nacional de Geología y Minería, 17. (Escala 1:75.000).

Moreno, H., Clavero, J., 2006. Geología del área del volcán Villarrica, Regiones de la Araucanía y de los Lagos. Servicio Nacional de Geología y Minería, Carta Geológica de Chile, Serie Geología Básica, Mapa M124. [Escala 1:50.000].

Moreno, H., Fuentealba, G., 1994. The May 17−19 1994Llaima volcano eruption, southern Andes 38° 42'S−71° 44'W. Andean Geol. 21 (1), 167−171.

Moreno, H., Naranjo, J.A., 1991. Geology and short-term hazards of Llaima volcano: the most active of the Chilean Andes (38° 45'S). In: International Conference on Active Volcanoes and Risk Mitigation. Napoli, Italy.

Moreno, H., Naranjo, J.A., 2003. Mapa de peligros del volcán Llaima, Región de la Araucanía. Servicio Nacional de Geología y Minería, Carta Geológica de Chile, Serie Geología Ambiental, Mapa M77. [Escala 1:75.000].

Moreno, H., Lescinsky, D., Rivera, A., 1994. The 17 May 1994 eruption of Llaima volcano, southern Chile; an example of ice-lava interaction. Abstracts with Programs, Geological Society of America 26 (7), 117.

Mothes, P., Hall, M.L., Andrade, D., Samaniego, P., Pierson, T.C., Ruiz, A.G., Yepes, H., 2004. Character, stratigraphy and magnitude of historical lahars of Cotopaxi volcano (Ecuador). Acta Vulcanol. 16 (1/2), 85.

Muñoz-Salinas, E., Castillo-Rodríguez, M., Manea, V., Manea, M., Palacios, D., 2010. On the geochronological method versus flow simulation software application for lahar risk mapping: a case study of Popocatépetl volcano, Mexico. Geogr. Ann. A Phys. Geogr. 92 (3), 311−328.

Muñoz-Salinas, E., Manea, V.C., Palacios, D., Castillo-Rodríguez, M., 2007. Estimation of lahar flow velocity on Popocatépetl volcano (Mexico). Geomorphology 92 (1), 91−99.

Muñoz-Salinas, E., Renschler, C.S., Palacios, D., Namikawa, L.M., 2008. Updating channel morphology in digital elevation models: lahar assessment for Tenenepanco-Huiloac Gorge, Popocatépetl volcano, Mexico. Nat. Hazards 45 (2), 309−320.

Murai, I., 1960. On the mudflows of the 1926 eruption of volcano Tokachi-dake, Central Hokkaido, Japan. Bulletin of the Earthquake Research institute 38, 55−70.

Myers and Brantley, 1995. Volcano Hazards Fact Sheet: Hazardous phenomena at volcanoes, US geological survey Open-File Report, 95−231.

Nakawo, M., Young, G.J., 1981. Field experiments to determine the effect of a debris layer on ablation of glacier ice. Ann. Glaciol. 2 (1), 85−91.

Naranjo, J.A., Moreno, H., 2004. Pucón town laharic debris-flow hazards from Villarrica volcano, Southern Andes (39,4°S): causes for different scenarios. In: Lara, L., Clavero, J. (Eds.), Villarrica Volcano, Southern Andes, Chile, 61. Servicio Nacional de Geología y Minería, Boletín N°, p. 72.

Naranjo, J.L., Sigurdsson, H., Carey, S.N., Fritz, W., 1986. Eruption of the Nevado del Ruiz volcano, Colombia, on 13 November 1985: tephra fall and lahars. Science 233 (4767), 961−963.

Nowell, D.A.G., Jones, M.C., Pyle, D.M., 2006. Episodic quaternary volcanism in France and Germany. J. Quaternary Sci. 21 (6), 645−675. http://dx.doi.org/10.1002/jqs.1005, 2006.

Nye, J.F., 1976. Water flow in glaciers: Jökulhlaups, tunnels and veins. J. Glaciol. 17 (76), 181−207.

Owen, J., Tuffen, H., McGarvie, D.W., 2013. Explosive subglacial rhyolitic eruptions in Iceland are fuelled by high magmatic H_2O and closed-system degassing. Geology 41 (2), 251−254. http://dx.doi.org/10.1130/G33647.1.

Parra, E., Cepeda, H., 1990. Volcanic hazard maps of the Nevado del Ruiz volcano, Colombia. J. Volcanol. Geothermal Res. 42 (1), 117−127.

Pierson, T.C., 1989. Hazardous hydrologic consequences of volcanic eruptions and goals for mitigative action: an overview. In: Starosolsky, O., Melder, O.M. (Eds.), Hydrology of Disasters. Proceedings of the Technical Conference in Geneva, WMO. James and James, London, pp. 220−236.

Pierson, T.C., Janda, R.J., Thouret, J.C., Borrero, C.A., 1990. Perturbation and melting of snow and ice by the 13 November 1985 eruption of Nevado del Ruiz, Colombia, and consequent mobilization, flow and deposition of lahars. J. Volcanol. Geothermal Res. 41 (1), 17−66.

Pulgarín, B., Agudelo, A., Calvache, M., Cardona, C., Santacoloma, C., Monsalve, M.L., 2011. Nevado del Huila (Colombia): 2007−2008 eruptions, lahars and crisis management. In: Britkreuz, C., Gursky, H.J. (Eds.), Geo-risk management - a German Latin American approach, C538. Freiberger Forschungshefte, Geowissenchaften, pp. 69−80.

Pulgarín, B., Cardona, C., Agudelo, A., Santacoloma, C., Monsalve, M.L., Calvache, M., Murcia, H., Ibáñez, D., García, J., Murcia, C., Cuellar, M., Ordoñez, M., Medina, E., Balanta, R., Calderón, Y., Leiva, O., 2009. Erupciones Históricas Recientes del Volcán Nevado del Huila, cambios morfológicos y lahares asociados. In: Memorias XII Congreso Colombiano de Geología, 7−11 September 2009 (Paipa, Colombia).

Pulgarín, B., Cardona, C., Santacoloma, C., Agudelo, A., Calvache, M., Monsalve, M.L., 2008. Erupciones del volcán Nevado del Huila y cambios en su masa glaciar: 2007. Boletín Geológico INGEOMINAS 42, 113−132.

Pulgarín, B., Cardona, C., Santacoloma, C., Agudelo, A., Calvache, M., Monsalve, M.L., 2010. Erupciones del volcán Nevado del Huila (Colombia) en febrero y abril de 2007 y cambios en su masa glaciar. In: López, C.D., Ramírez, J. (Eds.), Glaciares, nieves y hielos de América

Latina, cambio climático y amenazas glaciares, Colección Glaciares, Nevados y Medio Ambiente. INGEOMINAS, pp. 279−305.

Rampino, M.R., Self, S., Fairbridge, R.W., 1979. Can rapid climatic change cause volcanic eruptions? Science 206 (4420), 826−829.

Rivera, A., Bown, F., Mella, R., Wendt, J., Casassa, G., Acuña, C., Rignot, E., Clavero, J., Brock, B., 2006. Ice volumetric changes on active volcanoes in southern Chile. Ann. Glaciol. 43 (1), 111−122.

Robock, A., 2000. Volcanic eruptions and climate. Rev. Geophy. 38, 191−219. http://dx.doi.org/10.1029/1998RG000054.

Röthlisberger, H., 1987. Sliding phenomena in a steep section of Balmhorngletscher, Switzerland. J. Geophy. Res. Solid Earth (1978−2012) 92 (B9), 8999−9014.

Santacoloma, C., Cardona, C.E., White, R., Mccausland, W., Trujillo, N., Bolaños, R., Manzo, O., Narváez, A., 2009. Aspectos sísmicos de las erupciones freáticas y freatomagmática del volcán Nevado del Huila−Colombia. In: Memorias XII Congreso Colombiano de Geología, 7−11 de Septiembre de 2009 (Paipa).

Satyawali, P.K., Singh, A.K., Dewali, S.K., Kumar, P., Kumar, V., 2008. Time dependence of snow microstructure and associated effective thermal conductivity. Ann. Glaciol. 49 (1), 43−50.

Schneebeli, M., Sokratov, S.A., 2004. Tomography of temperature gradient metamorphism of snow and associated changes in heat conductivity. Hydrol. Process. 18 (18), 3655−3665.

Scott, K.M., Vallance, J.W., Pringle, P.T., 1995. Sedimentology, behavior, and hazards of debris flows at Mount Rainier, Washington. US Geological Survey Professional Paper, 1547, 56 p.

Shimazaki, K., Nakata, T., 1980. Time-predictable recurrence model for large earthquakes. Geophys. Res. Lett. 7 (4), 279−282.

Sigmundsson, F., Pinel, V., Lund, B., Albino, F., Pagli, C., Geirsson, H., Sturkell, E., 2010. Climate effects on volcanism: influence on magmatic systems of loading and unloading from ice mass variations, with examples from Iceland. Philos. Trans. Roy. Soc. A Math. Phys. Eng. Sci. 368 (1919), 2519−2534. http://dx.doi.org/10.1098/rsta.2010.0042.

Sigvaldason, G.E., 2002. Volcanic and tectonic processes coinciding with glaciation and crustal rebound: an early Holocene rhyolitic eruption in the Dyngjufjöll volcanic centre and the formation of the Askja caldera, north Iceland. Bull. Volcanol. 64, 192−205. http://dx.doi.org/10.1007/s00445-002-0204-7.

Smellie, J.L., 2000. Subglacial eruptions. In: Sigurdsson, H., Houghton, B., Rymer, H., Stix, J., McNutt, S. (Eds.), Encyclopedia of Volcanoes. Elsevier, pp. 403−418.

Stevenson, J.A., Gilbert, J.S., McGarvie, D.W., Smellie, J.L., 2011. Explosive rhyolite tuya formation: classic examples from Kerlingarfjöll, Iceland. Quatrnary Sci. Rev. 30, 192−209.

Stevenson, J.A., McGarvie, D.W., Smellie, J.L., Gilbert, J.S., 2006. Subglacial and ice-contact volcanism at the Öræfajökull stratovolcano, Iceland. Bull. Volcanol. 68 (7−8), 737−752.

Stevenson, J.A., Smellie, J.L., McGarvie, D.W., Gilbert, J.S., Cameron, B.I., 2009. Subglacial intermediate volcanism at Kerlingarfjöll, Iceland: magma−water interactions beneath thick ice. J. Volcanol. Geothermal Res. 185 (4), 337−351.

Sturm, M., Benson, C., MacKeith, P., 1986. Effects of the 1966−68 eruptions of Mount Redoubt on the flow of Drift Glacier, Alaska, U.S.A. J. Glaciol. 32 (112), 355−362.

Takeuchi, Y., Kayastha, R.B., Naito, N., Kadota, T., Izumi, K., 2001. Comparison of meteorological features in the debris-free and debris-covered areas at Khumbu glacier, Nepal Himalayas, in the premonsoon season, 1999. Bull. Glaciol. Res. 18, 15−18.

Tanarro, L.M., Andrés, N., Zamorano, J.J., Palacios, D., Renschler, C.S., 2010. Geomorphological evolution of a fluvial channel after primary lahar deposition: Huiloac Gorge, Popocatépetl volcano (Mexico). Geomorphology 122 (1), 178−190.

Thorarinsson, S., 1958. The Öræfajökull eruption of 1362. Acta Naturalia Islandica 2 (2), 100.

Thouret, J.C., 1990. Effects of the November 13, 1985 eruption on the snow pack and ice cap of Nevado del Ruiz volcano, Colombia. J. Volcanol. Geothermal Res. 41 (1), 177−201.

Thouret, J.C., Ramírez, J., Gibert-Malengreau, B., Vargas, C.A., Naranjo, J.L., Vandemeulebrouck, J., Valla, F., Funk, M., 2007. Volcano glacier interactions on composite cones and lahar generation: Nevado del Ruiz, Colombia, case study. Ann. Glaciol. 45 (1), 115−127.

Tilling, R.I., Topinka, L.J., Swanson, D.A., 1990. Eruptions of Mount St. Helens: Past, Present, and Future. The Climactic Eruption of May 18, 1980, U.S. Geological Survey, General Interest Publication, http://pubs.usgs.gov/gip/msh/contents.html.

Tuffen, H., 2010. How will melting of ice affect volcanic hazards in twenty-first century? Phil. Trans. Roy. Soc. A Math. Phys. Eng. Sci. 368 (1919), 2535−2558.

Tuffen, H., Gilbert, J.S., McGarvie, D.W., 2001. Products of an effusive subglacial rhyolite eruption: Bláhnúkur, Torfajökull, Iceland. Bull. Volcanol. 63 (2−3), 179−190.

Vallance, J.W., Scott, K.M., 1997. The Osceola mudflow from Mount Rainier: sedimentology and hazard implications of a huge clay-rich debris flow. Geol. Soc. Am. Bull. 109 (2), 143−163.

Waitt, R.B., Mastin, L.G., Miller, T.P., 1995. Ballistic showers during crater peak eruptions of Mount Spurr volcano, summer 1992. In: Keith, T.E.C. (Ed.), The 1992 eruptions of Crater Peak Vent, Mount Spurr volcano, Alaska, 2139. US Geological Survey Bulletin, pp. 89−106.

Waitt, R.B., Pierson, T.C., MacLeod, N.S., Janda, R.J., Voight, B., Holcomb, R.T., 1983. Eruption-triggered avalanche, flood and lahar at Mount St. Helens—effects of winter snowpack. Science 221 (4618), 1394−1397.

Walder, J.S., 2000a. Pyroclast/snow interactions and thermally driven slurry formation. Part 1: theory for monodisperse grains beds. Bull. Volcanol. 62, 105−118.

Walder, J.S., 2000b. Pyroclast/snow interactions and thermally driven slurry formation. Part 2: experiments and theoretical extension to polydisperse tephra. Bull. Volcanol. 62 (2), 119−129.

Walder, J.S., Schilling, S.P., Vallance, J.W., LaHusen, R.G., 2008. Effects of lava-dome growth on the crater glacier of Mount St. Helens, Washington. In: Sherrod, D.R., Scott, W.E., Stauffer, P.H. (Eds.), A Volcano Rekindled: The Renewed Eruption of Mount St. Helens, 2004−2006. US Geological Survey Professional Paper, 1750, pp. 257−276.

Wilson, L., Head, J.W., 2002. Heat transfer and melting in subglacial basaltic volcanic eruptions: implications for volcanic deposit morphology and meltwater volumes, vol. 202. Geol. Soc. London, Special Publications, 5−26.

Worni, R., Huggel, C., Stoffel, M., Pulgarín, B., 2012. Challenges of modeling current very large lahars at Nevado del Huila Volcano, Colombia. Bull. Volcanol. 74, 309−324.

Chapter 18

Floating Ice and Ice Pressure Challenge to Ships

Ivana Kubat [1], Captain David Fowler [2] and Mohamed Sayed [1]

[1] *National Research Council of Canada, Coastal and River Engineering, Ottawa, Ontario, Canada,*
[2] *Retired Canadian Coast Guard Captain, McDougall, ON, Canada*

ABSTRACT

The presence of floating ice in northern regions bears significant economic, environmental, and social implications. Navigation is one of the areas where the influence of floating ice is particularly evident. The history of ice threats to navigation and hardships experienced by early mariners are well known. At present, even with all the new technologies, ice can pose serious challenges. A major threat arises when ice around a ship starts to converge. The compression or ice pressure builds up if wind and water currents drive the ice cover against a land boundary. The ice cover compresses and ice accumulates to form ridges. If a ship is caught in such a situation, pressures on the hull will be high. Additionally, the ridges can introduce serious impediments to ship progress. Consequently, ice resistance would dramatically increase, and the ship can become beset. In extreme cases, ships can be damaged, and smaller vessels can be completely lifted onto the ice.

To understand how ice interacts with ships, we must start with ice properties and characteristics of ice covers. Ice ridges, for example, may represent formidable obstacles to ships. The way those ridges form and their properties determine the level of resistance that ships may encounter.

Although the effects of ice pressure are important over all Arctic and northern waters, the focus of this discussion is on the Canadian experience. The particular conditions that lead to ice pressure build-up and the impact on shipping at various specific locations in Canada are surveyed. The discussions conclude with a description of a Captain's experience in dealing with ice pressure on the Great Lakes.

18.1 INTRODUCTION

Seasonal floating ice covers an appreciable part of Northern Oceans, rivers, and lakes. It is important for diverse reasons. Arctic regions contain substantial oil, gas, and mineral resources. Dealing with ice is central to the ongoing and

Snow and Ice-Related Hazards, Risks, and Disasters. http://dx.doi.org/10.1016/B978-0-12-394849-6.00018-4
647

accelerating development of those resources. For example, offshore drilling platforms must be designed to withstand ice forces. Ice is also a major consideration in the prevention and response to any risk of pollution. Transportation of Arctic resources is another area where ice can be both an obstacle and an aid. Although ice is a major threat to shipping, ice roads provide an efficient and relatively safe means of transport. Resource development is only one facet of active Arctic operations. Inuit communities hunting, travel, and supply revolve around the presence and action of floating ice. In addition, tourism is increasingly spreading in many Arctic regions.

In the more southern waters of the Great Lakes, the formation and movement of ice affects a multitude of users. Ice conditions are of concern to mariners, industries, municipalities, property owners, recreational groups, and the general public.

A large commercial shipping industry operates year round, although activity is much lower between December and March when the St Lawrence Seaway is closed. Ferries operate throughout the Great Lakes, but they tend to be small vessels for local traffic. Those that continue to operate in the winter have to suspend operations whenever ice pressure is present. Some areas of commercial fishing exist such as in Lake Erie's Long Point Bay and southern Lake Huron. Many fishing vessels operate year round and often brave the ice to carry out their trade (Figure 18.1). A major fishing industry also operates along the East Coast of Canada, often hampered by severe ice conditions.

The Great Lakes basin is home to a population numbered in tens of millions. Numerous water intakes for municipal drinking water and industrial water supply are located throughout the lakes. Ice conditions, particularly moving ice and ice under pressure, can cause problems with both the amount and quality of water available at the intakes. Icebreaker activity near intakes

FIGURE 18.1   **Great Lakes fishing vessels under icebreaker escort near Goderich, Ontario.**

can also cause difficulties. In some areas, hydroelectric facilities set up ice booms to prevent moving ice from flowing into the intakes. High ice pressure has been known to burst the ice booms resulting in problems for the power utility. When the ice booms are in place on the St Lawrence River, shipping traffic is restricted or stopped.

Waterfront property and facilities may be damaged by adverse ice conditions. Damage to riverfront property is particularly prevalent in the spring when moving ice carries away docks, shoreline protection, and occasionally buildings (Figure 18.2). Severe ice pressure can push ice ashore onto areas of low elevation and has been known to overwhelm entire buildings. Many property owners try to protect their docks and facilities with bubbler systems. Although these are effective in land fast ice conditions, they are useless when the ice is under pressure and moving. In some areas of the Great Lakes, fish farms have extensive exposure to ice. Special coordination is required with icebreakers when conducting ice-clearing operations in spring. For northern communities, ice pile-ups against docks and in harbors during the summer may severely hinder operations.

Snowmobiles, all-terrain vehicles and even road vehicles use the ice for travel when it becomes sufficiently thick. This generally occurs only on areas of land fast ice, such as the North Channel of Lake Huron, the channels of Georgian Bay, the Labrador Coast, and in Arctic regions. Some areas have routes that are marked with small evergreen trees placed in the ice and snow. Ice movement severely disrupts or destroys these routes and causes safety hazards for users. There is sometimes conflict between ice route users and marine traffic.

Using the ice for recreational activities is a common occurrence in some areas. In addition to the highly popular pastimes of snowmobiling and ice fishing, people use the ice for sail boarding, kite boarding, ice boating, Nordic skiing, and skating.

FIGURE 18.2   **Yukon River Ice Jam near Eagle, AK, May 13, 2009.** *Aerial photograph by M. Nolan.*

18.1.1 Ice Properties and Behavior

The structure and mechanical properties of floating ice are rather complex. Both saline sea ice and freshwater river and lake ice have a predominantly columnar structure. Sea ice structure additionally includes pockets of brine. In general, freshwater ice is stronger and more brittle than sea ice. The strength of ice increases at colder temperatures, and decreases during the warming of springtime. Growth of floating ice begins in the fall as air temperatures dip below the freezing point. The balance of heat transfer determines growth rates and maximum equilibrium thickness. For example, seasonal ice in the Arctic reaches a maximum thickness of 2 m. The maximum thickness of freshwater ice over Northern lakes and rivers is much smaller (typically from 0.45 to 0.85 m for the Great Lakes). Seasonal sea ice that survives one or more melt seasons is called multiyear ice. Brine drains from such multiyear ice during the warm periods. The ice thus becomes stronger and grows to larger thicknesses than seasonal ice does. A rich body of literature deals with the growth, structure, and properties of ice, comprehensively documented in Michel (1978), and Weeks (2010).

Most of the knowledge of mechanical properties of ice deals with ice sheets that are <1 m in diameter. In this small-scale condition, ice structure and deformation are mostly uniform. At larger scales, ranging from a few meters to a few kilometers, the deformation and strength of ice are much more complex and not well understood. It is these length scales that affect ships and offshore structures. Over such lengths, the ice cover is non-uniform. It may contain separate floes, cracks, leads, ridges, and accumulations of broken ice blocks (ice rubble). Such heterogeneous ice covers deform in complex modes when driven by wind and water currents. As an ice cover deforms in compression or shear, it breaks into blocks that can accumulate and form ridges and rubble fields. When the environmental forces cause divergence, tensile stresses create cracks or leads of open water. An ice cover can fail in several ways: buckling, downward or upward bending, and crushing, as well as cracking or lead opening. Failure of an ice cover generally encompasses a mixture of those modes. Ships and offshore structures in the ice can experience such complex interaction processes of ice failure.

An important issue on which to focus is what happens when ice around a ship starts to converge or undergoes compression, as this poses a major threat to navigation. Problems generally arise as wind or water currents drive the ice cover against a land boundary. The ice cover compresses, pressures build-up, and ridges form. If a ship is caught in such a situation, pressures on the hull will be high. Moreover, ridges introduce additional impediments to movement. Consequently, ice resistance would dramatically increase, and the ship can become beset. In extreme cases, ships can be damaged, and smaller vessels can be completely lifted onto the ice.

The following discussion addresses the formation and characteristics of ridges since they form an important component of the compression processes. The discussion turns next to the build-up and dissipation of pressure. Examples of besetting events are then surveyed. The chapter concludes with a section on a Captain's experience in dealing with pressure events over the Great Lakes.

18.2 ICE RIDGES

Ice ridges form as an ice cover undergoes compressive or shear failure. Ice extending over hundreds of meters to several kilometers generally deforms over localized zones where ridging takes place. Once the ice breaks or two areas of ice impinge on each other, ice breaks into relatively small rectangular blocks typically about five times the ice thickness. Blocks accumulate to form a sail above water level and a keel below. Pressure or compressive ridges generally follow narrow paths that may meander (Figure 18.3). Shear ridges follow distinctive straight lines (Figure 18.4). Sometimes, under sufficient environmental forcing, accumulations of ice blocks or ice rubble coalesce to form wide fields called "hummock fields."

The frequency of occurrence of ice ridges, their geometry, and strength have long been the subjects of considerable interest for a variety of reasons. Early mariners have always been concerned by the presence of such formidable obstacles. In addition to navigation, the influence of ice ridging is important for the design of offshore structures and buried pipelines. At larger geophysical scales of hundreds of kilometers, ridging is of interest to meteorological and climate studies.

The early observations of ice ridges were recorded by mariners. A classic book by Zubov (1945) included a description of ridging and methods to evaluate the forces due to such formations on ships. More recent investigations

FIGURE 18.3 **Ice ridge in the Gulf of St Lawrence.** *Courtesy S. Prinsenberg.*

FIGURE 18.4   **Shear ridge in Beaufort Sea.** *Courtesy R. Frederking.*

over the past three decades (many cited by Weeks, 2010) dealt with characterization of ridge geometry, morphology, and the frequencies of occurrence over different Northern regions, as well as modeling of the ridging process. A recent article by Strub-Klein and Sudom (2012) provides a comprehensive analysis of available information on the geometry and morphology of ice ridges.

Figure 18.5 sketches a typical cross-section of a first-year ice ridge. A relatively strong "consolidated layer" forms near water level where water freezes within the voids between the blocks of ice that comprises the ridge. Most of the force that a ship must generate to break through the ridge is exerted on the consolidated layer. The sail and keel consist of accumulations of relatively loose ice blocks, and their strength is usually very small compared to the consolidated layer.

Much research, including manual drilling with ice augers and the use of upward looking sonars and electromagnetic induction sensors, was directed at measuring the peak sail heights and keel depths of ridges, as well as the shapes

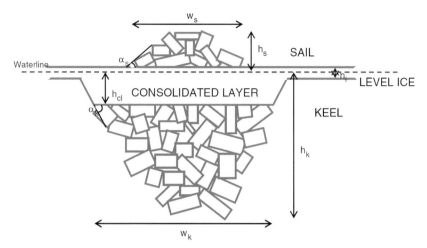

FIGURE 18.5   A sketch of a typical cross-section of a first-year ice ridge. *After Strub-Klein and Sudom (2012).*

of the ridges. The results of numerous studies giving statistics of sail heights and keel depths, and correlations of ridge dimensions and shape parameters, have been compiled by Strub-Klein and Sudom (2012). They cover a wide geographic area including the Bering and Chuckchi Seas, the Beaufort Sea, Svalbard waters, the Barents Sea, the Russian Arctic Ocean, the East Coast of Canada, the Baltic Sea, the Sea of Azov, the Caspian Sea, and Offshore Sakhalin.

Typically, the ratio of keel depth to maximum sail height is approximately 4:1. The mean maximum of ridge sail heights is 2 m, and the peak value is 8 m. For the maxima of keel depths, the mean value is 8 m, and the peak value is 28 m. The mean keel width is 36 m, and the maximum is 202 m.

Efforts to model the ridging process have mainly concerned equating the energy expended by the converging ice cover to the potential energy required to build the sail and keel. The models also consider energy dissipation. The first such model was developed by Parmeter and Coon (1972). Subsequent computer simulations using a Discrete Element approach were reported by Hopkins et al. (1991). The latter modeling approach implies that the maximum sail height and maximum keel depth are limited by the energy that an ice cover can expend, which is limited by the thickness of floating ice. Thus, one can expect a physical limitation on extreme ridge dimensions, a conclusion reached by Melling (2002) and Melling and Riedell (2004). Their analysis was supported by physical arguments, laboratory tests, and computer simulations, and agrees with the recent data analysis of Strub-Klein and Sudom (2012) that the maximum possible keel depth of a first-year ridge is 28 m.

18.3 PRESSURE BUILD-UP AND DISSIPATION

Pressured ice forms in all regions where ice is dynamic and constrained, including those listed above. Although pressured ice (compressive ice) poses difficulties almost everywhere in Northern regions, the focus of this chapter is on Canadian waters. At the scales of interest to shipping operations, local conditions can greatly influence the build-up of ice pressure. This poses a major difficulty in predicting risks to navigation. Pressure build-up depends on environmental driving forces that arise due the action of wind, water currents, and swells. Coastline conditions will determine the potential for the occurrence and severity of pressure episodes in response to the environmental conditions. Local geography and idiosyncrasies of the environment must be considered for each specific region to produce reliable forecasts of pressure risks.

In the Canadian Arctic, local conditions may vary significantly. Microclimates of local areas can be drastically different from the forecasts of global models. For example, ship Captains have observed that wind can change direction at a shoreline, and often follows the shoreline. This contrasts with what happens in southern climates where onshore winds often occur in the afternoon due to heating of the land during the day. Over Frobisher Bay, an easterly wind blowing toward the Incognita Peninsula (the southwest shore of the Bay) may veer to the southeast at the shoreline. The same phenomenon can be observed in Prince Regent Inlet at the northwest side of Baffin Island where westerly winds turn to the southwest and follow the Baffin shore. Another common phenomenon in the Arctic is the katabatic wind, in which air moves rapidly down slopes, often from an icecap. Captains have observed sudden gale-force winds near shores even though there are light winds elsewhere in the vicinity. A funneling effect is often present in areas, such as long fiords. Concerning water currents, local tidal components are usually critical to the build-up of ice pressure, in addition to large-scale water circulation. This is particularly evident at the entrance to narrow channels and inlets where the tidal stream is the strongest.

Indications of ice pressure, such as ridges and block accumulations, are of obvious interest to mariners. Ridging is a pronounced indicator of pressure. Although existing ridges could have formed during past deformation events, if the ice field is immobile, ridged ice indicates the presence of pressure. The severity of the pressure increases with ice thickness. Another good visual indicator of pressure is the state of small broken ice blocks. Block accumulations, with water in the pores, usually freeze quickly in cold temperatures. The resulting pile-ups of frozen ice pose a serious impediment to ships. Progress becomes difficult, and the risk of damage increases.

Dissipation of ice pressure is another process that concerns mariners. Relief can happen very quickly once the driving force has been removed, or can take hours depending on the extent of the ice field. Wind is not the only factor causing ice pressure to dissipate. Swell also plays a role, along with tide, current, and type of ice regime under pressure. Wind can die down, but the pressure can persist. It is therefore difficult to predict dissipation of the pressure.

Efforts to understand and predict ice pressure conditions date back to the encounters of early explorers with ice. A good example is the "Karluk's" encounter with pressured ice during the summer of 1913, which is well documented (McKinlay, 1977). More recent investigations include reports of the Manhattan and the Canadian Coast Guard (CCG) icebreaker the Louis S. St Laurent, and reports describing challenges of navigating in the Arctic, the Gulf of St Lawrence, the Northern Sea Route, and the Gulf of Finland (Bradford, 1971, 1972a,b, 1978; Brigham et al., 1999; Maillet, 1997; Pärn et al., 2007; Stolee, 1970; Swithinbank, 1970a,b; Voelker and Seibold, 1990). Over the past 20 years, interest in pressured ice has grown substantially. Research projects are now underway in Europe and Canada to address the various aspects of forecasting and dealing with pressured ice. Russia also has a long-standing interest and expertise in dealing with pressured ice. For example, Mironov et al. (2012) provided a summary of numerous studies carried out for several decades at the Arctic and Antarctic Research Institute. It described ice phenomena threatening Arctic shipping including a number of examples. The most recent research is described by Kubat et al. (2011, 2012, 2013), Kõuts et al. (2012), Leisti et al. (2011), Lensu et al. (2013a,b), and Lilover et al. (2012).

18.4 REGIONAL CONDITIONS AND INCIDENTS OF BESETTING

A survey of reports of besetting and events of ice pressure was conducted by Kubat et al. (2012). The survey included interviews with Captains and documentation of their comments on pressure ice risks. The reported incidents of besetting obtained through that work are shown in Figure 18.6. Most events took place along the St Lawrence Seaway and on the East Coast of Canada, a reflection of the busy shipping traffic over those regions. Conditions are much more severe over the Arctic. The smaller number of reported incidents there is a result of the relatively low ship traffic. The situation would change if Arctic shipping increased. This section examines the pressured ice risks over specific regions, and presents some of the reported incidents of besetting.

The Beaufort Sea: In the southern Beaufort Sea ice movement against the land boundary is the main factor associated with ice pressure. The edge of the land fast ice, which consists of level undeformed ice, also acts as a boundary that makes pressure build-up possible. In addition, local ice convergence caused by the Mackenzie River outflow and differential movements between the polar and seasonal pack ice areas are recognized as causes of ice pressure.

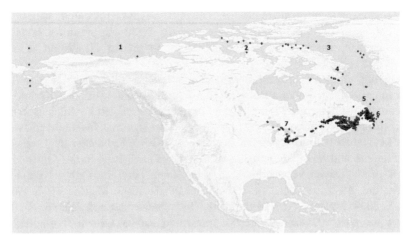

FIGURE 18.6   **Map of reported incidents of ship besetting and damage due to ice pressure, data obtained from Kubat et al., 2012.** 1—Beaufort Sea, 2—Peel Sound, 3—Baffin Bay, 4—Frobisher Bay, 5—Labrador Coast, 6—Newfoundland East Coast, and 7—Great Lakes.

Winter ice zones in the Southern Beaufort Sea (Kovacs and Mellor, 1974) are illustrated in Figure 18.7. Pressures can develop as ice moves southwards against the coastline or the edge of the land fast ice where a shear wall may develop (Figure 18.8).

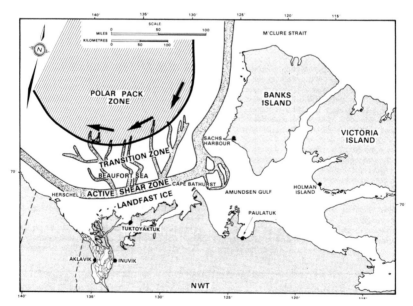

FIGURE 18.7   **Winter ice zones in the Southern Beaufort Sea.** *After Kovacs and Mellor (1974).*

FIGURE 18.8   Shear wall at the edge of land fast ice in the Southern Beaufort Sea. *Courtesy G. Timco.*

The Northwest Passage—Peel Sound and Franklin Strait: The Canadian Archipelago consists of many narrow channels that are covered with ice most of the year. Peel Sound and Franklin Strait present challenges since they are susceptible to multiyear ice invasion from the Queen Elizabeth Islands, and influx of severe ice from the M'Clintock Channel. That area has about 20 percent of multiyear ice coverage through the year, and may decrease to 10 percent in September. Open water is present from August to October. The limited width of the channel, approximately 35 km, can contribute to pressured ice build-up and besetting of vessels as they have less room for maneuvering to avoid pressure zones than in unrestricted areas. Ice conditions in Southern Peel Sound during the mobilization of two multipurpose lower-ice-classed vessels are shown in Figures 18.9 and 18.10. The vessels needed assistance from higher-ice-classed icebreaker Kigoriak during the transit from Southern Peel Sound into Franklin Strait in late July of 2003.

Baffin Bay: in May 1970, a number of pile-up incidents occurred during the Manhattan and Louis S St Laurent voyage to Lancaster Sound. Ice pressure forced both ships together, and rubble formed around the hulls up to 3 m with some ice blocks up to 2 m thick (Figure 18.11). The ice regime consisted of a combination of first-year and old ice. Neither ship could break out for 5 days. In addition to pile-ups against the hull, several ridges also formed in the vicinity of the vessels. During that event, the Louis S St Laurent was forced toward the Manhattan. The starboard side was creased for 51 m fracturing five frames and buckling six frames and the ice between the ships was pushed under their bows and trimmed them 3.2 m by the stern (Stolee, 1970).

FIGURE 18.9　Ice conditions in Peel Sound, July 2003. *Courtesy Captain D. Connelly.*

FIGURE 18.10　Mobilization of ice breaking, anchor handling—tug—supply vessels in July 2003—southern Peel Sound. *Courtesy Captain D. Connelly.*

Frobisher Bay: Typically, Frobisher Bay would be clear of ice or would have low ice concentrations during July. However, ice conditions over the bay were relatively severe during the 2008 and 2012 seasons. During those two years, the prevailing northeast wind caused a delay of ice clearance in the bay before the beginning of the shipping season. The bay was fully covered by combinations of first-year and multiyear ice. In that region, the ice field is largely influenced by the surface currents that are driven by the tidal streams in and out of the bay. This tidal current effectively moves the ice field as long as there is no obstruction such as fast ice, land, or greater influences, such as high winds or wind driven current. In areas where there are geographical obstructions, the flow gets choked and ice pressure therefore increases. This appears

FIGURE 18.11   **Rubble pile-up (3.3 m) at the Manhattan in 1970.** *From J.D. Bradford.*

to be the case in the area of the Frobisher Bay west of Cape Osborn where it narrows. Here, a number of vessels became beset during the summers of 2008 and 2012, and had to be escorted by icebreakers. One of the vessels was severely damaged when her bow was holed. Photographs of ice conditions during those events are shown in Figures 18.12–18.14.

Labrador Coast: Ocean swells play a major role in ice pressure build-up along the Labrador Coast, and to some extent over the southern Arctic. An ocean swell not only contributes to pressured ice conditions, but it can dramatically change the strength and make-up of the ice cover. In such situations, the pressure driven by swells generally persists long after the wind is

FIGURE 18.12   **Ice conditions in the Frobisher Bay in 2008. CCG icebreaker escorting a beset vessel.** *Courtesy Captain J. Vanthiel.*

FIGURE 18.13   **Ice conditions in the Frobisher Bay in 2012.** *Courtesy D. Lambert.*

FIGURE 18.14   **Vessels beset in the Frobisher Bay—July 2012.** *Courtesy D. Lambert.*

over. One example of a pressure build-up took place during the summer of 2007 along the Labrador Coast. The crew of the Umiak 1, a powerful ice-breaking bulk carrier, had a challenge going into Edwards Cove in Voisey's Bay (Figure 18.15). The ice was much worse than the crew had experienced in the depths of winter. The shear zone consisted of a rubble field about eight miles wide that was impenetrable for 10 days. The vessel was in a band of heavily broken, rafted floes, none of which was particularly thick, but were piled together. The transit felt like digging a trench through

FIGURE 18.15   **Umiak 1 beset in heavy ice conditions in June 2007; approach to Edwards Cove.** *Courtesy T. Keane, Fednav.*

the ice according to T. Keane, FedNav. Similar conditions were also experienced in April 2012.

The East Coast of Newfoundland: In 2007, a strong southwesterly wind caused ice to move further south than in usual years. Combinations of factors such as coastline irregularities, large number of shoals, wind, ocean current, and further drift of pack ice caused severe pressure build-up in April and May at the East Coast off Newfoundland and Labrador. The conditions were worsened by the incursions of multiyear ice. As a result, a large number of fishing vessels were trapped in the pressured ice field (Figures 18.16−18.18).

FIGURE 18.16   **CCG icebreaker freeing vessels beset in pressured ice—East Coast of Newfoundland in April 2007.** *Courtesy Captain John Broderick.*

FIGURE 18.17   Ice conditions off East Coast of Newfoundland—April 2007. *Courtesy Captain John Broderick.*

FIGURE 18.18   Vessel damaged in pressured ice—East Coast of Newfoundland 2007. *Courtesy Ron Morrow, CIS.*

Great alarm ensued because of the dangerous conditions, and a large ice-breaking and Search and Rescue operation was conducted by the CCG.

18.5 PRESSURED ICE ON THE GREAT LAKES

The North American Great Lakes are one of the most interesting areas on the planet. They consist of five large lakes, containing 20 percent of the world's freshwater, connected by navigable rivers and canals (Figure 18.19). In the

FIGURE 18.19   **Map of the Great Lakes.**

winter, much of this freshwater freezes creating challenges for those living and working here. The mariner will find that major differences exist between ice operations on the Great Lakes compared to ice operations in sea and ocean areas. The navigational challenges that are present year round in the narrow channels and confined waters of the Great Lake are exacerbated by the ice and cold. The North American continental climate is only slightly moderated by the lakes, unlike coastal areas where the ocean provides considerable moderation. Temperatures of −20° to −30 °C are not unusual. Strong winds are common in winter causing freshwater ice accretion on ships, which is much harder to remove from vessel structures than saltwater ice. Snow, blowing snow, and sea smoke reduce visibility. Commercial shipping is only one of the many groups and organizations in this highly populated area that have an interest in the ice, and some of these interests conflict. However, there are many resources available on the Great Lakes to help all concerned, including good ice reconnaissance, expert weather forecasting, excellent port facilities, and comprehensive icebreaking services.

18.6 FRESHWATER ICE

The freshwater ice occurring in the Great Lakes is different from the salt-water ice found in coastal regions. The mariner will notice that it feels harder and "crunchier" than saltwater ice. The sound of the icebreaking is sharper than saltwater ice, and it tends to break with jagged edges (Figure 18.20). When under pressure, the pieces of ice tend to hold this jagged shape more so than saltwater ice, forming ridges with well-defined profiles. In some conditions, pressure causes large, rectangular pieces, locally called "plate ice," to slide or raft under other pieces, then freeze together resulting in floes of ice that appear the same as the surrounding ice, but are actually two or sometimes three times thicker. During melt, the ice becomes "rotten," turning black and gray with many small melt holes. Rotten freshwater ice bends easily and does not shatter as the ship passes through it, leaving a well-defined vessel track. Rotten ice under pressure tends to disintegrate into a white slush. In some areas, sediment entrapped in the ice contributes to melt.

FIGURE 18.20   **Freshwater ice fracturing.**

18.7 CAUSES OF ICE UNDER PRESSURE IN THE GREAT LAKES

The causes of ice pressure in the Great Lakes are the same as in other areas of the world: wind and current or a combination of these. An additional cause, common in the Great Lakes, is ice pressure created by vessel movement, which is experienced in narrow channels and restricted waterways. Wind-induced pressure is mostly short-lived and predominantly occurs on the eastern shores of the lakes due to the prevailing winds that blow from a direction between southwest and northwest. The area most affected by wind-induced ice pressure is Lake Erie, which lies with its long axis in an east−west direction. As the shallowest of the Great Lakes, Erie is often completely ice covered in the winter. The relatively milder temperatures of this most southerly Great Lake keeps the ice relatively soft and fractured, allowing significant ice movement. A strong southwest wind can blow almost unobstructed along the length of Lake Erie and cause huge ridges in the area from the Welland Canal to Buffalo and the Niagara River (Figures 18.21 and 18.22). A continuous ridge field extending 10 nautical miles west from the port of Buffalo is not unusual. Although Lake Ontario is oriented with its long axis east−west

FIGURE 18.21   **Ice ridges near Port Colborne, Lake Erie, March 2004.**

FIGURE 18.22   Ice ridges near Port Colborne, Lake Erie, March 2004.

similar to Erie, the deeper water of this lake results in very little ice development. Other areas of particular concern for wind generated ice pressure are
Whitefish Bay on the southeast corner of Lake Superior, and the Straits of
Mackinac, which is the passage between the northern ends of Lake Michigan
and Lake Huron.

Ice pressure caused by river current is perhaps the biggest concern on the
Great Lakes and nowhere more so than the connecting waterways between
Lake Huron and Lake Erie. Water flows from Lake Huron, southward down
the St Clair River, into Lake St Clair and then down the Detroit River to Lake
Erie. A northerly wind in Lake Huron can funnel ice into the southern basin,
causing severe ice pressure. A natural ice bridge often forms above the
entrance to the St Clair River, but, if it breaks, the three to five knot current
sends ice hurtling down the river, building up on shoals, islands, and bends in
the river. The ice can pile up right to the bottom of the river, forming a jam
and creating the risk of flooding in this heavily populated and industrialized
area. Ice jams can also occur in the narrow channels of the Detroit River, but
they are generally less common and not as severe as those on the St Clair
River to the north. Ice build-up on the many smaller rivers and creeks flowing
into the Great Lakes often causes jams in the spring, resulting in flooding in
low-lying areas. When these ice jams break free, large patches of ice filled
with debris such as trees, docks, small buildings, and even small boats flow
out into the deeper waters used by ships and can represent a danger to
navigation.

Wind generated currents occur mostly at entrances to harbors. The port of
Goderich on Lake Huron, for instance, usually has a surface current that is
perpendicular to the harbor entrance. Ice movement due to this current creates

very difficult conditions for vessels attempting to transit the immediate entrance to this port. When the current combines with an adverse wind, an ice transit at this port can be too dangerous to attempt.

18.8 ENVIRONMENTAL CONCERNS

Ice and ice movement on the Great Lakes is an environmental concern. Ice coverage has an effect on the regional climatic conditions. When cold wind blows over open water, it picks up the moisture and causes lake-effect snow squalls in the so-called "snow belt" areas in the lee of the Great Lakes. Although these snow squalls are quite localized, the large amount of snow and poor visibility cause travel chaos. Once an area is completely ice covered, snow squall activity ceases. Ice coverage also affects the seasonal water levels of the Great Lakes. During warmer years with less ice coverage, increased evaporation lowers water levels, causing problems for shoreline facilities and shipping. Ice movement can cause shoreline erosion, damaging sensitive wetland areas, and waterfront habitat. Water quality can be affected by ice scouring that mixes sediment into the water column. Wildlife, particularly deer, travel on the ice. Unusual ice movement and vessel tracks which create open water have resulted in deer mortality as the animals are unable to scramble onto solid ice once they fall into the water.

18.9 SHIPPING CONCERNS

The opening of the St Lawrence Seaway in 1959 made ports of the Great Lakes a destination for ships from all over the world. The Seaway extends from Montreal to Lake Erie. When the cold temperatures of winter arrive and ice starts to form, the locks have difficulty operating as ice plays havoc with lock gates. The ice booms deployed at hydroelectric facilities isolate large sections of the river, conflicting with the movement of shipping. Ships have difficulty in maneuvering in ice-filled channels. Buoys marking the channel edges are obscured or sometimes swept away by ice. As the winter season approaches and ice starts to form, ocean-going vessels rush to leave and domestic vessels prepare to scale back or cease operations for the winter. The St Lawrence Seaway, including the Welland Canal, shuts down for the season in late December and remains closed until the third week of March. Traffic continues on the St Lawrence River from Montreal downriver to the Atlantic.

The "Soo" Locks at Sault Ste. Marie, Michigan, allow vessel transit between Lake Superior and Lake Huron. They shut down every year between January 15 and March 25. Shipping traffic continues on Lakes Huron, Michigan, and Erie and their connecting waterways throughout the winter months, but at a comparatively low volume. Domestic bulk petroleum, bulk concentrates such as salt and iron ore, and occasionally manufactured goods such as steel, are shipped in the winter months.

An early freeze-up in the Great Lakes may cause problems for shipping. Numerous vessels attempting to depart the Seaway combined with limited icebreaker availability and problems with infrastructure due to freezing and ice build-up can cause delays. Once the locks in the St Lawrence River and the Welland Canal are closed, icebreakers are often kept busy with vessels voyaging between Lakes Erie, Huron, Michigan, and Superior. As winter conditions worsen, the majority of the Great Lakes fleet will lay up for the winter. Upon closure of the Soo Locks on January 15, vessel traffic decreases significantly.

The shipping industry on the Great Lakes starts to get busy again in mid-March. This is the most likely time of the year for problems with ice. The ice is often at its thickest. Land fast ice starts to break up and move around. Pack ice may become softer and more fractured, and floe size decreases. Natural ice bridges can break. Melting snow raises water levels and creates stronger currents. Ice jams are possible. Just as the natural ice cover starts to deteriorate and move around, shipping traffic increases. Icebreakers can become very busy helping shipping through the late winter ice. Ridges in areas such as Whitefish Bay, the Straits of Mackinac, and eastern Lake Erie can cause major problems. When strong winds occur, severe ice pressure adds to the problems.

Ice coverage on the Great Lakes varies from year to year. The trend has been toward improved ice conditions in the last few decades. The winter of 2013–2014, however, saw a return to very cold conditions with extensive ice coverage, causing problems and concerns for the marine industry.

18.10 DEALING WITH PRESSURED ICE: A SHIP MASTER'S PERSPECTIVE

The best way for ship Masters to deal with ice, especially ice under pressure, is to avoid it. Avoidance can be difficult on the Great Lakes because of the more confined waters compared to coastal and offshore areas. When the connecting waterways, such as the St Mary's, St Clair, and Detroit rivers, are packed with ice, pressure is inevitable and avoidance is impossible (Figure 18.23). In the more open waters of the lakes, altering course to avoid ice is an option. Good tactical information can be obtained from the ice charts, which are produced from satellite and visual observations and are published frequently during ice season. Vessel Traffic Services and government icebreakers provide information and advice. Most commercial vessels on the Great Lakes share ice and route information with each other over very-high frequency radio.

Large commercial vessels on the Great Lakes tend to follow standard routes that are recommended by the Lake Carriers Association. These routes are shown on both Canadian and American nautical charts and usually have separate courses for upbound and downbound vessels. Ice conditions may

FIGURE 18.23   **Icebreaker escort in Detroit River.**

require a departure from the normal shipping routes. Masters navigating in ice-covered waters must avoid the temptation to comfortably follow the accustomed routes and be prepared to go around difficult ice regimes. Mariners must watch the forecasted wind and avoid ice-covered areas with strong onshore winds. Conversely, following shore leads, which are navigable open water areas that have been opened by offshore winds, is a risky undertaking. A shift in wind can close the shore lead like a steel trap.

In 1994, the tug Princess was trapped on the north shore of Lake Erie in severe ice pressure. The pressure was strong enough to completely lift the vessel out of the water and roll it onto its side. As the crew abandoned the ship onto the ice, the Master was injured and was evacuated by helicopter. A United States Coast Guard (USCG) Bay Class icebreaker attempting a rescue was beset and experienced an uplifting of the stern. By the next morning, when the larger and more powerful Canadian icebreaker Samuel Risley arrived on scene, ice pressure had disappeared and both vessels were completely back in the water. The crew of the Princess was able to board it and start its engine. The tug was escorted to a dock in Amherstburg, Ontario, but sank the next night due to damage from the ice.

Vessels approaching port or river entrances in ice can expect a significant worsening of ice conditions. At harbor entrances, local winds and currents often create ice pressure. Navigation near breakwalls can be very tricky as unusual currents are often encountered near breakwalls, which can cause unexpected ice movement. Shear lines develop as the moving ice further offshore encounters stationary ice near the harbor. Vessels can easily swing off course as the bow passes the shear line and one part of the vessel is in moving ice while the other is not. Night navigation or poor visibility makes shear lines much harder to see, adding to the danger of these features. At river entrances, moving ice tends to build up in thickness as the current funnels it into the river. Ships that are proceeding easily through the ice in offshore areas often see a drastic drop in speed and maneuverability when they approach or enter rivers. A vessel stopped by the ice in a river is in serious danger as she will continue to move downriver at the mercy of the ice pack.

Ship Masters must use a combination of caution and boldness when entering areas of moving and pressured ice. If maneuverability becomes a problem when approaching confined waters, the best action may be to turn or back away from the dangerous area. Caution has to be exercised when backing in ice due to the vulnerability of rudders and propellers. When attempting to escape dangerous ice at harbor and river entrances, the best escape direction is often toward the center of the lake, where ice conditions are usually a lot better. If committed to an approach, bold engine and rudder movements may be required to keep the ship moving and under control. Good judgment and lots of experience are vital. If conditions are unknown or questionable, it is often best to stay well offshore, then approach a harbor perpendicular to the shore, if water depths and channels allow. Ports on the southern shore of Lake Erie are a good example of this. Rather than a long approach close to the shore, it is preferable to keep well out in the lake before swinging directly in to the port. Even a small change of wind direction, from slightly offshore to slightly onshore, can be enough to create large areas of pressured ice, with rafting and ridging that will cause even the most capable vessels to become beset.

The CCG and the USCG operate icebreakers in the Great Lakes. The two Coast Guards work harmoniously to strategically deploy available icebreakers in the best locations possible, taking into account the capability of the ice-breakers, the ice conditions, and the traffic. It is common to see Canadian icebreakers working on the American side of the border and vice versa. Ice-breaking for flood control takes priority over vessel assistance. Commercial tugs also provide significant icebreaking assistance to commercial vessels. Tugs operate mainly in ports, breaking ice alongside slips and wharfs, making tracks through the harbor ice, and breaking out turning basins. In particularly bad ice years, large commercial vessels occasionally hire tugs as dedicated escorts.

A common icebreaking technique in the Great Lakes is to prepare and maintain tracks in the ice designed to allow commercial vessels to proceed independently (Figure 18.24). In some areas, such as the St Mary's River, only the ice in the navigable channels is broken. The fast ice on the edge of the channels is left intact. Vessel speed is regulated to minimize collateral damage to the ice from vessel wake. In other areas, such as the parts of the St Clair and Detroit rivers with deep water right up to the shoreline, ice is broken from shore to shore. Vessel traffic can keep ice flowing and help prevent ice jams.

FIGURE 18.24   **Rock Cut showing groomed tracks in the ice.** *Photograph source: USCG.*

The Edgar B. Speer, an American 1,004-ft long bulk carrier became stuck in ice at the Rock Cut, St Mary's River on January 18, 2004. The vessel's beam of 105 ft left very little room between the ship and the edge of the narrow channel. Ice under pressure formed around the ship. It was thought that the current also forced ice under the vessel, effectively grounding her on ice. On January 21, the vessel was freed with assistance from USCG icebreakers and four tugs. Ice problems frequently occur in the Rock Cut with vessels becoming beset in a similar manner to the Edgar B. Speer.

Navigation in the tracks becomes difficult when very cold temperatures are present. As vessels pass through the ice, they grind it up into a thick brash. As water is exposed to the cold air, the brash freezes into a thicker layer than it was prior to passage of the vessel. Over time, the ice in the tracks will become significantly thicker than the fast ice on the sides of the tracks. Icebreakers have the option of creating a new track if there is room, cutting relief tracks on either side of the original track in an attempt to reduce the pressure, or using a combination of natural current and their own propeller wash to flush the brash ice out of the channel.

Any significant ice pressure will move or close the tracks. A combination of warm temperatures and strong winds may create ice pressure. This can occur throughout the ice season but is more prevalent in the spring. Icebreakers become very busy when this happens. On the Great Lakes, when commercial traffic is high, icebreakers will sometimes organize vessels into convoys for escort through difficult areas of ice. In discussion with the Masters of the commercial ships, the icebreaker's Master will assign the vessels positions in the convoy based on their capability. When additional icebreakers are available, they will usually start at the end of the convoy so that they can approach

FIGURE 18.25   Canadian Coast Guard icebreaker Samuel Risley backing alongside the
beset vessel.

from astern of ships that are having difficulty and get them moving again. If
one particular ship is having problems, the second icebreaker may position
itself directly ahead of that ship. In very difficult ice conditions, particularly
when strong ice pressure is present, keeping a convoy intact may not be
possible. Once a commercial ship is stopped by pressured ice, the friction
along the parallel midbody of the ship may be too great for it to start moving
again. The icebreaker will pass down the side of the beset ship and then move
quickly ahead (Figure 18.25). The ice along the side of the beset vessel will
flow into the track of the icebreaker, breaking the friction and pressure, and
allowing the beset vessel to move ahead again. In a convoy situation, when the
next ship ahead stops, all following ships must stop to avoid collision. In
pressured ice, the entire convoy quickly becomes beset. The icebreaker is then
forced to escort ships clear of the ice individually.

Icebreaker techniques for escorting vessels in pressured ice are many and
varied. The icebreaker must allow the escorted vessel to follow much more
closely than would be considered prudent in easier ice conditions because the
track made by the icebreaker closes quickly. This increases the risk of collision.
In severe ice pressure, the escorted vessel may have to be as close as 20 m,
sometimes less, to the stern of the icebreaker. If the icebreaker gets stopped by
heavy ice, collision is likely. If the escorted vessel gets stopped, the icebreaker
can have considerable difficulty backing into the pressured ice to free the beset
vessel. In the many narrow channels occurring in the Great Lakes system, not
only does wind and current cause ice pressure but the channel restrictions also
add to the problem. The hull of the ship forces ice up against the channel sides
and adds further ice pressure. Sometimes, the ice pressure is so severe that it is

not possible to move the escorted vessel. Although the best solution may be to wait until conditions improve, this is not always a safe option. If the entire ice field is moving toward a shoal, then every effort must be made to get the ship moving in a safe direction. Sometimes, two icebreakers working together are better able to get a beset vessel moving in pressured ice. Towing equipment on an icebreaker may be used in extreme situations. One example occurred in the St Clair River when the ocean-going vessel Mallard was beset in moving ice with engine cooling problems and was in danger of running aground. The CCG icebreaker Samuel Risley was able to successfully tow the beset vessel through the heavy ice to a safe dock. Another CCG icebreaker, Griffon, provided ice-breaking support for the towing operation.

Turning a vessel in pressured ice can be very difficult. Vessels underway in ice want to maintain their heading. A turn must be initiated with bold rudder movements and started well in advance of a normal open water turn. In fast ice or pressured ice, the vessel may have to make a series of ahead and astern passes to force their way around a turn. When establishing and grooming tracks in narrow channels, icebreakers need to pay particular attention to the turns, creating entry and exit points, and cutting relief areas for ice to flow away from the bow and stern of the commercial ship. When providing direct icebreaker support to a vessel stuck in a turn, the icebreaker should break the ice around the stern of the beset vessel on the side opposite the direction of turn to give the ice somewhere to move, then proceed ahead of the beset vessel to create a low pressure area on the side of the bow toward the direction of turn. However, sometimes the conditions are so severe that even an icebreaker can get beset in pressured ice (Figures 18.26 and 18.27).

FIGURE 18.26   Canadian Coast Guard icebreaker Griffon beset with tanker near Long Point. Lake Erie. *Photograph source: United States Coast Guard.*

FIGURE 18.27 **Canadian Coast Guard icebreaker Griffon beset with tanker near Long Point. Lake Erie.** *Photograph source: United States Coast Guard.*

18.11 CONCLUSION AND PERSPECTIVE FOR THE FUTURE

The build-up of ice pressure at scales of hundreds of meters is a natural phenomenon that has significant safety, and environmental and economic implications. Even powerful icebreakers can become beset if environmental forces become unfavorable. Mariners often observe that apparently easy ice conditions can quickly change into impenetrable ice fields if wind, or current turn toward a coastline. A beset ship becomes very vulnerable to damage. In addition to direct damage due to ice action, a beset ship can be thrust against hazards or pushed aground. Safety of crews can then be seriously jeopardized, and pollution of the environment can become a serious problem. Aside from concerns over safety and environmental pollution, the excessive resistance of ice and the need for icebreaker escort dramatically increases fuel consumption and delays transit times. The adverse effect on the efficiency and economy of shipping operations is considerable.

Shipping along Arctic corridors is increasing in both frequency of voyages and tonnage. This increase in shipping will likely persist regardless of the climatic conditions. In addition to the shipping of Arctic resources and supply of Northern communities, new routes aimed at reducing transit times are constantly sought and exploited. Tourism is also growing rapidly. The escalating shipping in Arctic regions will lead to more encounters with pressured ice conditions. The potential risks may be further aggravated when the increased traffic includes ships with limited strength and power, which may venture into dangerous zones. Further, some of the mariners venturing into such zones might lack the necessary experience with navigation in ice.

There are research initiatives underway to enhance the knowledge of the processes of ice pressure build-up and to develop technologies that can enhance navigation through high-risk zones. One research project, called SAFEWIN, took place in Europe. Another research effort is underway at the National Research Council of Canada. The research has a number of thrusts: developing accurate forecast models to predict ice pressure and ridging, documentation and analysis of besetting records, developing guidance to characterize the potential of besetting risks for specific regions, and fusion of imagery and environmental data into systems that can support navigation in hazardous ice zones. Training of mariners is also expanding to include navigation through pressured ice zones. The Marine Institute's Center of Marine Simulations in St John's, Newfoundland, Canada, The Transatlantic Ice Academy in Kalmar, Sweden, and Maritime Training Center of Admiral Makarov Academy in Saint Petersburg, Russia include such training in its courses.

REFERENCES

Bradford, J.D., 1971. Sea-ice pressure generation and its effect on navigation in the Gulf of St. Lawrence area. J. Navig. 24 (4), 512−520. http://dx.doi.org/10.1017/S0373463300022359.

Bradford, J.D., 1972a. Sea ice pressures observed on the second "Manhattan" voyage. Arctic 25 (1), 34−39. Publisher: AINA (Arctic Institute of North America).

Bradford, J.D., 1972b. A preliminary report of the observations of sea ice pressure and its effects on merchant vessels under icebreaker escort. In: Proceedings of an International Conference held in Reykjavik, Iceland, May 10−13, 1971, pp. 154−158.

Bradford, J.D., 1978. Icebreaking Capability of CCGS "Labrador" in Western Barrow Strait, October 23−28, 1973. Marine Science Directorate, manuscript Report Series No. 50.

Brigham, L., Grishchenko, V.D., Kamesaki, K., 1999. The natural and societal challenges of the northern sea route. In: The Natural Environment, Ice Navigation and Ship Technology. Springer, pp. 47−120.

Hopkins, M.A., et al., 1991. On the numerical simulation of the sea ice ridging process. JGR 96 (3), 4809−4820.

Kõuts, T., Pavelson, J., Lilover, M-J., 2012. Monitoring of Ice Dynamics Using Bottom-Mounted ADCP in the Central Gulf of Finland in 2010, http://dx.doi.org/10.1109/BALTIC.2012. 6249179. Conference: Baltic International Symposium (BALTIC) 2012 IEEE/OES.

Kovacs, A., Mellor, M., 1974. Sea ice morphology and ice as a geologic agent in the southern Beaufort Sea. In: Symposium on the Coast and Shelf of the Beaufort Sea. Arctic Institute of North America, Arlington, VA, pp. 113−164.

Kubat, I., Babaei, M.H., Sayed, M., 2012. Quantifying Ice Pressure Conditions and Predicting the Risk of Ship Besetting. ICETECH'12, Banff, Alberta, Paper No. ICETECH10-130-R0.

Kubat, I., Babaei, H., Sayed, M., 2013. Analysis of besetting incidents in Frobisher Bay during the 2012 shipping season. In: Proceedings of the 22nd International Conference on Port and Ocean Engineering under Arctic Conditions, 2013. Espoo, Finland. Poac13−164.

Kubat, I., Watson, D., Sayed, M., 2011. Characterization of pressured ice threat to shipping. In: Proceedings of the 21st International Conference on Port and Ocean Engineering under Arctic Conditions, 2011. Montreal, Canada, poac11−136.

Leisti, H., Kaups, K., Lehtiranta, J., Lindfors, M., Suominen, M., Lensu, M., Haapala, J., Riska, K., Kõuts, T., July 2011. Observations of ships in compressive ice. In: The 21st International Conference on Port and Ocean Engineering under Arctic Conditions. Montreal, Canada.

Lensu, M., Haapala, J., Lehtiranta, J., Eriksson, P., Kujala, P., Suominen, M., Mård, A., Vedenpää, L., Kõuts, T., Lilover, M.-J., 2013b. Forecasting of compressive ice conditions. In: Proceedings of the 22st International Conference on Port and Ocean Engineering under Arctic Conditions, 2013. Espoo, Finland, poac13−208.

Lensu, M., Suominen, M., Haapala, J., Külaots, R., Elder, B., 2013a. Measurements of pack ice stresses in the Baltic. In: Proceedings of the 22st International Conference on Port and Ocean Engineering under Arctic Conditions, 2013. Espoo, Finland, poac13−209.

Lilover, M.-J., Kõuts, T., Vahter, K., 2012. Ships in compressive ice—hazard forecast by means of fuzzy logic model. In: The 21st IAHR International Symposium on Ice, Dalian, China, June 2012.

Maillet, A., 1997. The Detection of Ice Pressure in the Gulf of St. Lawrence Using Acoustics. OCEANS '97. MTS/IEEE Conference, Halifax, Nova Scotia, Canada.

McKinlay, W.L., 1977. Karluk, the Great Untold Story of Arctic Exploration. St. Martin's Press, New York.

Mironov, Y., Klyachkin, S., Benzeman, V., Adamovich, N., Gorbunov, Y., Agorov, A., Yulin, A., Panov, V., Frolov, S., 2012. Ice Phenomena Threatening Arctic Shipping. Backbone Publishing Company. ISBN: 13 978-0-984-7864-2-8.

Melling, H., 2002. Sea ice of the northern Canadian Arctic Archipelago. J. Geophys. Res. 107 (C11), 3181. http://dx.doi.org/10.1029/2001JC001102.

Melling, H., Riedel, D.A., 2004. Draft and movement of pack ice in the Beaufort Sea: a time-series presentation April 1990−August 1999. Can. Tech. Rep. Hydrogr. Ocean Sci. 238 (v + 24 pp.).

Michel, B., 1978. Ice Mechanics. Les Presses de L'Universite Laval, ISBN 0-7746-6876-8.

Parmerter, R.R., Coon, M.D., 1972. Model of pressure ridge formation in sea ice. J. Geophys. Res. 77 (33), 6565−6575.

Pärn, O., Haapala, J., Kõuts, T., Jüri, E., Kaj, J., September 2007. On the relationship between sea ice deformation and ship damages in the Gulf of Finland in winter 2003. Est. J. Eng. 13 (3), 201−214.

Stolee, E., 1970. Report on the Voyage in the Canadian Arctic of CCGS Louis S St. Laurent. Queen's Printer for Canada, Ottawa.

Strub-Klein, L., Sudom, D., 2012. A comprehensive analysis of the morphology of first-year sea ice ridges. Cold Reg. Sci. Technol. 82, 94−109. http://dx.doi.org/10.1016/j.coldregions.2012.05.014.

Swithinbank, C.W.M., 1970a. A report of Manhattan's first Arctic voyage in 1969. Polar Rec. 15, 60−61.

Swithinbank, C.W.M., 1970b. Second Arctic voyage of SS Manhattan, 1970. Polar Rec. 15, 355−356. http://dx.doi.org/10.1017/S0032247400061301.

Voelker, R.P., Seibold, F., September 1990. Polar ice-breaker caught in active shear ridge. In: Proceedings of the Second International Conference on Ice Technology. Cambridge University.

Weeks, W.F., 2010. On Sea Ice. University of Alaska Press, Fairbanks, Alaska, ISBN 978-1-60223-079-8, 664 pp.

Zubov, N.N., 1945. Arctic Ice. 1963. US Navy Electronics Laboratory, San Diego, California. Access online, 2012: http://archive.org/details/arcticice00zubo (last visited 18.04.12).

Chapter 19

Retreat Instability of Tidewater Glaciers and Marine Ice Sheets

Andreas Vieli

Department of Geography, University of Zurich, Switzerland

ABSTRACT

The terminus dynamics of tidewater glaciers and marine-based ice sheets are strongly affected by the process of calving and water depth at their termini, resulting in an unstable retreat behavior and accelerated mass loss related to submarine over-deepenings. The governing controls and existing approaches related to this dynamic instability are discussed in this chapter in relation to the Columbia Glacier in Alaska and numerical modeling experiments. The related marine ice sheet instability is further explained, together with the stabilizing influence of floating ice tongues and ice shelves on the grounding line behavior. The chapter concludes with wider implications and future challenges of this dynamic instability as a hazard, such as sea-level rise and accelerated discharge of icebergs.

19.1 INTRODUCTION

Tidewater glaciers are glaciers that terminate in the sea where they discharge icebergs into the ocean through the process of calving (see the example in Figure 19.1). Such ocean-terminating glaciers have long been observed to behave dynamically, fundamentally differently from land-terminating glaciers (Meier and Post, 1987; Post, 1975). Length changes commonly appear asynchronous to climatic forcing and seem to be strongly influenced by basal topography. Based on observations from Alaskan tidewater glaciers, Post (1975) proposed the well-known "tidewater glacier cycle" that relates to a cycle of slow terminus advance followed by rapid unstable retreat through a submarine overdeepening during which climatic conditions may be comparable. The sudden phases of accelerated retreats with enhanced iceberg production pose a hazard not only for shipping routes (Dickson, 1978) but also for fjord ecosystems and importantly contribute to accelerated sea-level rise, as indicated in recent widespread mass loss from tidewater outlet glaciers in Alaska, in the Arctic, in Greenland, and in the Antarctica Peninsula (Gardner et al., 2013; Pritchard et al.,

Snow and Ice-Related Hazards, Risks, and Disasters. http://dx.doi.org/10.1016/B978-0-12-394849-6.00019-6

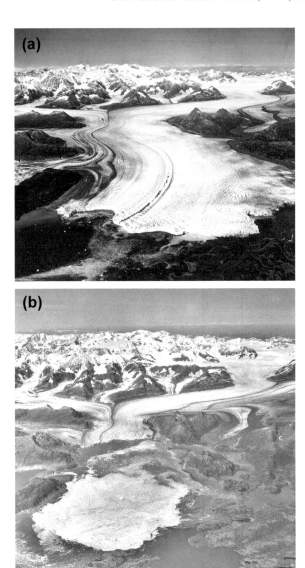

FIGURE 19.1   View of Columbia Glacier in Alaska in (a) 1969 before the onset of rapid retreat and with its terminus resting on a shallow morainal shoal; and in (b) 1996 after it retreated by about 10 km. By 2014, the terminus has retreated behind the major confluence zone. *Photographs: U. S. Geological Survey.*

2009; Cook and Vaughan, 2010). This dynamic instability similarly occurs for the much larger marine-based ice streams of the Antarctic ice sheet (Rignot et al., 2008), but potential rates of mass loss are much larger and the presence of floating ice shelves complicates this mechanism (Joughin et al., 2012).

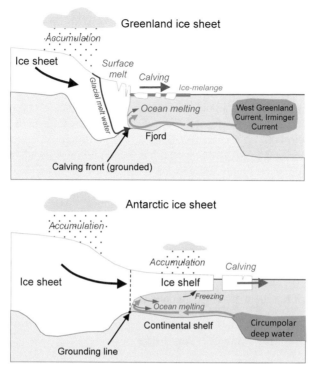

FIGURE 19.2 Sketch of the geometric setting and forcing processes of tidewater glaciers and marine outlet glaciers of the Greenland ice sheet (top) and of marine-based ice streams of the Antarctic ice sheet (bottom). Red refers to processes of mass loss (ablation) and blue to mass gain. A detailed discussion of the forcing mechanisms is given in Section 19.3.

Tidewater glaciers are widespread in latitudes >45° where the equilibrium line altitude (ELA) is close to or virtually below sea level. Thus, the surface area that is exposed to surface ablation is reduced or entirely absent and is replaced by the process of iceberg calving (Figure 19.2). In more temperate and maritime climates, tidewater glacier termini are mostly fully grounded (Van der Veen, 2002) and are widespread, for example, in Alaska, the Canadian Arctic, Svalbard, Novaya Zemlya, South Greenland, and parts of the Antarctic Peninsula. In contrast, in colder climates, such as North Greenland and mainland Antarctica, glaciers and ice streams commonly terminate in floating extensions in the form of ice tongues or ice shelves (Figure 19.2). The location of the transition from grounded to floating ice is referred to as the "grounding line." The term tidewater glacier refers in general to glaciers that terminate in a fjord and are laterally bounded by valley walls; however, the instability discussed in this chapter is also applicable to ice streams that are bounded by stagnant or slowly flowing inland ice.

19.1.1 Mass Budget of Ocean-Terminating Glaciers

The mass budget of such ocean-terminating glaciers differs fundamentally from their land-terminating counterpart in that the ablation process is dominated by the mechanism of iceberg calving, and surface melt plays only a minor role or is completely absent in the very cold climate of Antarctica (Figure 19.2). For the Greenland ice sheet, calving accounts for approximately 50 percent of ablation, and for the Antarctic ice sheet, it is close to 90 percent (Van der Veen, 2002). Submarine melting at the ice ocean contact provides an additional mechanism for ablation which can, in particular for glaciers with a floating terminus, be highly efficient and rates in the order of meters per day have been estimated for glaciers in Greenland and Alaska (Motyka et al., 2011; Enderlin and Howat, 2013) and even in Antarctica, mass loss from ocean melting is comparable to losses from calving (Depoorter et al., 2013). Note that in mass budget estimates for calving, submarine melting is often implicitly included, as until recently, it was difficult to separate the two ablation processes. For example, for the Antarctic ice sheet, of the above-mentioned 90 percent mass loss through calving, about half can be accounted for by submarine melt at the underside of ice shelves.

19.1.2 Tidewater Glacier Dynamics

Tidewater glaciers are, in general, behaving far more dynamically than land-terminating glaciers. They possess, unlike their land-terminating counterparts, non-diminishing ice flux at their termini, that is, in the case of a stable terminus, equal to the rate of ice lost by calving (calving rate). These ocean-terminating glaciers are further characterized by fast flow at their terminus with flow speeds in the order of a few hundred meters to several kilometers per year (Meier and Post, 1987; Joughin et al., 2010). The proximity of the terminus to flotation results in generally low effective pressure, and thus, the fast flow is dominated by basal motion rather than internal ice deformation (Meier and Post, 1987). The flow typically accelerates toward the terminus, causing high along-flow extension and consequently intense crevassing (Vieli et al., 2000). Sudden dynamic changes consisting of rapid retreat, flow acceleration, and surface thinning are common for tidewater glaciers with retreat rates reaching kilometers per year, manifold increases in flow speed and thinning rates of several tens of meters per year. Such rapid dynamic changes have, in the last decade, been widely observed for tidewater glaciers (Carr et al., 2013a; Joughin et al., 2004; Howat et al., 2007; Moon et al., 2012) and, for example, for the Greenland ice sheet make up about half of the mass loss of the ice sheet (dynamic mass loss) with surface melt from atmospheric warming contributing the other half (Van den Broeke et al., 2009). As illustrated in Figure 19.3, high rates of coastal thinning of the Greenland and the Antarctic ice sheet are concentrated at tidewater outlet glaciers or marine-based ice streams

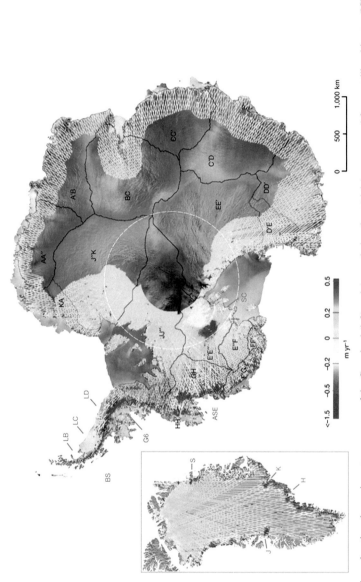

FIGURE 19.3   Surface elevation change in meters per year of the Greenland (left) and Antarctic ice sheet (right) as measured from satellite altimetry (ICESat) over the period 2003–2007. *Copyright Nature Publishing Group. Figure from Pritchard et al., 2009.*

(Pritchard et al., 2009) that are known to be strongly influenced by the shape of the bed (Meier and Post, 1987; Vieli et al., 2001).

In the following sections, the dynamic behavior of ocean-terminating glaciers and in particular the instability related to bed geometry is discussed. In Section 19.2, the case of glaciers with fully grounded calving termini is considered and the dynamic instability related to bed topography for both ocean- and freshwater-terminating glaciers explained. Section 19.3 discusses the influence of external forcing and potential triggering mechanisms for rapid change, such as climatic, oceanic, or sea ice-related forcing. In Section 19.4, the case of marine ice sheets (ice sheets with a bed below sea level) is considered, and the influence of floating ice tongues and ice shelves on this instability is explained. In Section 19.5, a discussion of the wider implications of the tidewater instability is given.

19.2 TIDEWATER RETREAT INSTABILITY AND CALVING

One of the earliest and, to date, most comprehensive studies on the dynamics of a tidewater glacier, with a fully grounded terminus, is from the Columbia Glacier in Alaska (Figure 19.1), and it provides an excellent example to introduce and illustrate the retreat instability of tidewater glaciers.

Until the late 1970s, the terminus of Columbia Glacier was stable at a shallow morainal bank with flow speeds at the terminus of about 1 km/a (Figure 19.4; Meier and Post, 1987; Van der Veen, 1996; Krimmel, 2001). Then, the terminus lost contact with its moraine, and a retreat set in, which accelerated when the terminus reached deeper water in the upstream direction (Figure 19.4). Retreat rates peaked at 2 km/a, and the terminus receded a total of 20 km from 1980 to 2010 and is still ongoing (Figures 19.4 and 19.5; McNabb and Hock, 2014). Along with the retreat and the terminus reaching deeper water, both calving rates and flow speed at the terminus dramatically increased reaching values of up to 12 km/a (Figure 19.4; Pfeffer, 2007a). Further, the phase of rapid retreat seemed to be independent of year-to-year variations in surface mass balance and was not in direct correspondence with climatic forcing (Meier and Post, 1987). With the onset of retreat, the delivery of icebergs to Prince William Sound strongly increased and posed a serious hazard to shipping routes, in particular for oil tankers to and from the pipeline port in Valdez, Alaska (Dickson, 1978). This hazard from icebergs in the late 1970s was the main reason for the start of a detailed investigation by the U. S. Geological Survey on the dynamical behavior of Columbia Glacier and tidewater glaciers in general (Post, 1975).

Other tidewater glaciers in Alaska (e.g., Muir and Tyndall Glacier; McNabb and Hock, 2014; Larsen et al., 2007), in the Arctic (Carr et al., 2013a) and, more recently, ocean-terminating glaciers of the Greenland ice sheet, showed very similar rapid dynamic changes that are potentially related to basal topography. Prominent examples of rapid change in Greenland are Helheim

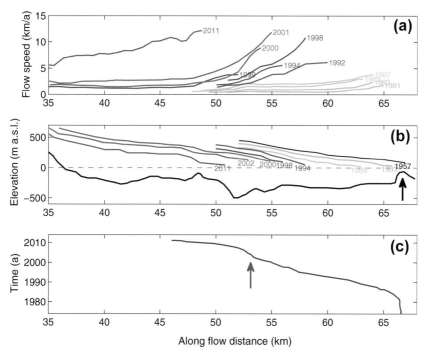

FIGURE 19.4   Observed evolution of along-flow geometry and flow speed of Columbia Glacier, Alaska. (a) Surface flow speed for selected years between 1981 and 2011. (b) Partial surface profiles for selected years between 1957 and 2011 (colored lines) and the basal topography (black line) with the black arrow marking the position of the shallow morainal shoal. (c) Evolution of annual terminus position with time. The colors roughly refer to the time-coded colors of terminus positions in Figure 19.5 and the red arrow marks the position of a significant narrowing of the bed channel as shown in Figure 19.5. *All data are from Krimmel (2001), Pfeffer (2007a), McNabb et al. (2012), and McNabb and Hock (2014).*

and Kangerdluqssuaq Glaciers in East Greenland, which experienced, in the early 2000s, rapid retreat, dramatic thinning, and flow acceleration (Howat et al., 2007; Stearns and Hamilton, 2007). Although some regional retreat trends seem to exist for tidewater outlet glaciers in Greenland (Moon and Joughin, 2008), the large variations in retreat and flow between individual glaciers point beyond the common external forcing to a strong influence from geometric factors (McFadden et al., 2011; Carr et al., 2013b).

Such rapid unstable retreats, as observed for Columbia Glacier, occur in grounded tidewater glaciers mostly related to glacier recession from an initial shallow morainal bank or bedrock sill into a basal overdeepening upstream. Explanations for this retreat instability generally invoke the relationship of increasing calving rates or ice flux at the terminus with increasing water depth (Brown et al., 1982; Van der Veen, 1996), but they conceptually differ in their approach and specifically in their treatment of calving. These differing approaches and calving models are explained in some detail below.

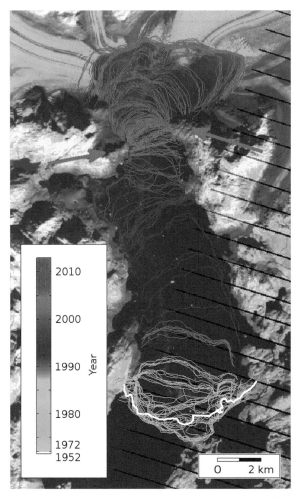

FIGURE 19.5 Mapped front positions (colored) documenting the retreat of Columbia Glacier since 1972. The background LandSat-image is from 2012 and the red arrows mark a distinct narrowing in the fjord where retreat slowed down substantially. *See Figure 19.4(c). Figure from McNabb and Hock (2014), copyright AGU, Wiley.*

19.2.1 Calving Rate/Water Depth Relationship

A first and widely used model of calving was derived from a study of the Columbia Glacier retreat (Brown et al., 1982) and relates calving rate u_c linearly to water depth d at the terminus:

$$u_c = c \cdot d, \tag{19.1}$$

where c is a constant taken as 27.1/a. This empirical relationship was originally based on data from 12 Alaskan glaciers, but since has been extended in similar form with data from glaciers elsewhere in the world (Benn et al., 2007). A similar linear relationship to water depth has been proposed for glaciers calving into freshwater but with a 14 times smaller proportionality constant (Funk and Röthlisberger, 1989; Warren and Kirkbride, 2003). Importantly, in such empirical calving models, the calving rate is directly given by the bed geometry of the glacier and therefore is independent of its dynamics (e.g., flow rate, thinning rate).

Returning to the case of a tidewater glacier, such as the Columbia Glacier in the late 1970s, which was initially stable at a shallow morainal bank and now starts to retreat slightly (e.g., as a result of a climatic perturbation) into deeper water. Due to the reverse bed slope in the upstream direction (Figure 19.4(b)), according to Eqn (19.1), the calving rate increases and leads to enhanced mass loss through calving. The mass deficit results, in turn, in further retreat and accelerated calving as long as the bed deepens upstream. This positive feedback is only stopped when eventually the terminus retreats over the deepest point of the basal depression and reaches shallower water again.

The calving model used here is, however, only empirical and based on data from mostly stable calving termini and with only few terminating in water deeper than 150 m. Indeed, during the rapid retreat phase of Columbia Glacier, calving rates were observed to increase more than expected (Van der Veen, 1996) from the linear relationship derived by Brown et al. (1982).

19.2.2 Flotation Calving Criterion

An alternative approach to calving uses a criterion that determines the position of the terminus at any given time, with the calving rate then being a secondary result of the length change rate and the flow speed at the terminus. The simplest such criteria are based on the observation that grounded termini are in general close to flotation, with the flotation thickness h_f given by the flotation condition

$$h_f = \frac{d \cdot \rho_w}{\rho_i}, \tag{19.2}$$

where ρ_i is the density of ice (910 kg/m^3) and ρ_w the density of ocean water (1,030 kg/m^3). This criterion assumes that any ice that is below a critical height or percentage above the flotation thickness h_f calves away due to buoyancy forces (Van der Veen, 1996; Vieli et al., 2001).

This flotation-criterion approach has also been used in combination with a numerical flow model by Vieli et al. (2001) to explain the dynamical retreat instability over a basal overdeepening as illustrated in the model experiments in Figure 19.6(a) and (b) for a synthetic glacier geometry.

For a glacier that is initially stable at a bedrock high, the ELA is at time zero increased by 50 m to simulate a slightly negative surface mass balance

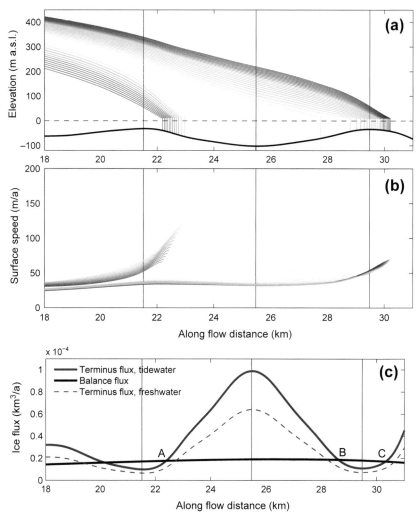

FIGURE 19.6 Modeled retreat of a small tidewater glacier through a basal depression using a flotation criterion for calving (Vieli et al., 2001). The retreat has been triggered by a small reduction in surface mass balance. (a) Modeled response of the glacier surface profiles are shown in intervals of 10 years going from black, through blue and green to red over a total period of 720 years. (b) Evolution of the velocity profiles along the glacier. (c) Mass budget approach from Section 19.2.5 showing required steady-state ice flux at the terminus (thick blue line) for ocean water and any position along the glacier according to Eqn (19.3) (Schoof, 2007) and for freshwater (blue dashed line). The thick black line shows the balance flux for a given climate. The gray shaded areas mark positions of theoretically unstable calving termini. The letters A, B, and C mark the crossing points between terminus flux and balance flux and therefore mark potential terminus positions.

and initiate surface thinning that in turn leads to a slow retreat (Figure 19.6(a) and (b)). Although the retreat and terminus thinning is slow, eventually the calving front retreats over the bedrock high and reaches deeper water. Ice thickness and flow at the terminus start to increase in deeper water, resulting in increased speed and ice flux (Figure 19.6(b)) and thus provoking a mass deficit and surface drawdown in the terminus region. This thinning leads, when applying the floatation criterion, to further retreat into deeper water and therefore enhanced ice flux and further thinning and retreat. This provides an unstable thinning-retreat feedback in the situation of a reverse bed slope as indicated by the gray shading in Figure 19.6.

Eventually, the terminus retreats over the deepest point and reaches shallower water depths that decrease in upstream direction. This reduces ice flow, thinning, and calving, and therefore retreat slows down and the terminus is able to restabilize (Figure 19.6). In the given example, glacier size, water depth, and consequently rates of retreat and calving are in general relatively small. However, the unstable behavior would qualitatively be similar for larger and faster glaciers.

In contrast to the previous empirical water-depth calving model, in this case, it is the glacier dynamics and more specifically the surface thinning that determines the calving rate and amount of mass lost through calving. The thinning and its propagation upstream is itself controlled by the amount of flow acceleration from the terminus and calving simply acts to remove the ice delivered to the terminus. The flow and consequently the ice flux at the terminus are strongly influenced by the flotation thickness (Eqn (19.2)), which is essentially given by the water depth at the terminus. Based on theoretical considerations, both ice flow and longitudinal stretching at a vertical ice cliff at or near flotation are non-linearly increasing with water depth (see Section 19.2.5) and thus are driving this retreat instability related to reverse bed slopes.

Although the two approaches described under Sections 19.2.1 and 19.2.2 are fundamentally different in how they treat calving, the implications for the dynamics of grounded tidewater glaciers are qualitatively the same. Enhanced mass loss through calving increases with water depth linearly or non-linearly, thus producing unstable retreat when retreating into basal overdeepenings.

19.2.3 Other Calving Models

Although the two calving models discussed above reproduce the basic calving behavior of tidewater glaciers, they are crude and have significant deficiencies. For example, the flotation criterion does not allow for floating ice tongues that are often observed in colder climates such as Northern Greenland or Antarctica. Even Columbia Glacier formed a temporary floating ice tongue in 2007, when it retreated rapidly into the widening bay upstream of the channel narrowing (Figure 19.5, red arrows) and started to discharge large tabular-shaped icebergs (Walter et al., 2010).

A more universal criterion determines the position at which full depth crevassing is estimated based on longitudinal stretching rates (Benn et al., 2007). This model is more physical, allows for both floating or grounded calving termini (Nick et al., 2010), and has been successfully applied to simulate changes of Greenland outlet glaciers (Nick et al., 2012, 2013; Vieli and Nick, 2011).

New empirical relationships relating the calving rate linearly to the horizontal ice-flow spreading rate (Levermann et al., 2012; Alley et al., 2007b) are consistent with observed stable positions of current ice shelves, but they are empirical and applications have so far been limited to large-scale floating ice shelves.

Recent new developments use statistical approaches (Bassis, 2011; Colgan et al., 2012) or fracture and damage mechanics (Bassis and Jacobs, 2013) in combination with numerical flow models to simulate the process of calving. However, these approaches require more sophisticated flow models and have not yet been applied and tested extensively on real-world tidewater glaciers. Nevertheless, a recent application of Monte Carlo simulations of calving to Columbia Glacier produced promising results (Colgan et al., 2012).

Importantly, most of the recent new approaches to calving are still strongly influenced by bed geometry and cause enhanced mass loss in deeper water. They therefore produce a similar dynamic instability, as explained above, which may in the presence of a floating tongue, be somewhat reduced (Nick et al., 2013). A good introduction and overview of different calving models is given in Benn et al. (2007).

19.2.4 Basal Motion and Water Pressure Feedback

Retreat instability related to basal overdeepenings is further enhanced by a positive feedback between basal motion and water pressure at the base of the glacier.

In contrast to land-terminating glaciers, the terminus of tidewater glaciers is typically close to flotation, and basal water pressure is therefore high and consequently effective pressure (given by the difference between ice overburden pressure and basal water pressure) is close to zero. Basal motion is in general inversely proportional to effective pressure (Fowler, 2010). Therefore, for tidewater glacier termini approaching flotation, basal motion is increasing rapidly toward the terminus (Meier and Post, 1987; Vieli et al., 2000). A small reduction in the ice thickness of the terminus, or deepening in water depth during retreat, further reduces the effective pressure leading to strongly enhanced basal motion and in turn to stretching, mass drawdown, thinning, and further retreat. This effect has been investigated in detail by Pfeffer (2007a) and provides an additional positive feedback for unstable retreat in the situation of basal overdeepenings.

19.2.5 Concept of Steady-State Flux, and Mass Budget

An alternative consideration of the retreat instability is provided by the concept of the steady-state mass budget of a tidewater glacier, as used for marine ice sheets (Schoof, 2007). The stability of the terminus is evaluated on the basis of steady-state mass fluxes and illustrated on the geometric config-uration used in the numerical modeling example above (Figure 19.6).

The basic principle is that at the location of the terminus the balance flux q_b (given by the integrated surface mass balance at this position) has to be balanced by the ice flux due to ice flow at that position. For a given surface mass balance, the balance flux at the terminus can be calculated for any po-sition along the domain (see black line in Figure 19.6(c)). The ice flux at the calving terminus at or near flotation (termed terminus flux q_t, blue line in Figure 19.6(c)) is dictated by ice-flow mechanics and the water depth and can be approximated by the boundary layer theory developed for grounding line flux of marine ice sheets (Schoof, 2007):

$$q_t = \left(\frac{A(\rho_i g)^{n+1}(1 - \rho_i/\rho_w)^n}{4^n C} \right)^{\frac{1}{m+1}} \cdot \left[h_f \right]^{\frac{m+n+3}{m+1}}, \tag{19.3}$$

where A is the rate factor and a measure of ice softness, g is the gravitational acceleration, C is the basal sliding coefficient using a Weertmann-type sliding relation, and n and m are the exponents of the flow law and basal sliding relation and about 3 and 1/3 to 1, respectively. The flotation thickness h_f is given by the water depth through the flotation condition (Eqn (19.2)). The terminus flux q_t is essentially given by ice flow, the flotation condition, and the difference in hydrostatic pressure between the ice and ocean at a vertical ice cliff. The latter results in an equation for the longitudinal stretching rate $\partial u/\partial x$ at the calving front approximated by

$$\frac{\partial u}{\partial x} = A \left[\frac{g \rho_i}{4} \left(1 - \frac{\rho_i}{\rho_w} \right) \right]^n h_f^n. \tag{19.4}$$

This frontal boundary condition indicates that in deeper water with increased frontal thickness h_f, stretching rates and thus crevassing are enhanced. Further, it enhances flow acceleration toward the terminus when retreating into deeper water (see, e.g., Figure 19.4(a)).

According to Eqn (19.3), the terminus ice flux q_t (blue line in Figure 19.6(c)) required in steady state at the calving terminus increases non-linearly (to the power of 3.5–4.7) with flotation thickness h_f (and there-fore water depth). For maintaining the terminus at a stable position, both the balance flux q_b (black line in Figure 19.6(c)) and the terminus flux q_t (blue line) have to balance each other, which is the case in the crossing points of the two flux lines in Figure 19.6(c) (at points A, B, and C). Terminus positions for which the flux decreases along the flow correspond to areas of reverse bed slopes and are intrinsically unstable (position B and gray shaded areas in Figure 19.6; Schoof, 2007). If the terminus retreated slightly from position B, the water depth and consequently the terminus flux q_t would increase and be higher than the long-term balance flux q_b. The glacier would then be in a mass deficit, would retreat further, and not return to its initial position. It would stabilize again upstream at the crossing point A, where the terminus flux and consequently water depth decrease enough to match the balance flux.

If the terminus would be slightly advanced from position B, the water depth and consequently the terminus flux q_t would decrease below the long-term balance flux. The glacier would then have a mass surplus and advance further over the basal high to a new stable position at crossing point C.

For this steady-state concept, an instability in terminus position is caused, as before by basal topography and the reverse bed slope, specifically. Further, it is based on quantities solely at the glacier terminus and therefore requires only minimal data, but it also does not consider the transient behavior of the terminus (e.g., rates of retreat or thinning). However, this concept does allow a simple analysis of terminus stability and an evaluation of the related thresholds in ice flux for instability.

For example, if the terminus is located at point C (Figure 19.6(c)) and the long-term surface mass balance decreases due to atmospheric warming, the balance flux line (black) shifts downward, and thus, the stable terminus position (point C) moves along the terminus flux line (blue) upstream. If the surface mass balance would be suppressed enough, the line would lie below the terminus flux line and would no longer cross it. Consequently, the glacier would not reach a nearby stable position and would retreat through the basal depression up to a crossing point (near position A) where the bed shallows upstream again.

Alternatively, for a surface mass balance that is constant with time, the terminus flux curve could move up and down and cause the terminus to cross a threshold and retreat. According to Eqn. (19.3), long-term sea-level rise and enhanced basal sliding are potential processes changing this terminus flux. Note, however, that Eqn (19.3) is only an approximation of an idealized case, and in reality, other forcing mechanisms may influence the terminus flux and trigger retreat (see Section 19.3).

Note that a similar steady-state stability concept as above could be applied to the water-depth model, as calving rate and terminus flux are in steady-state equivalent.

19.2.6 Hysteresis and Irreversibility

The strong modulating effect from basal topography on terminal ice flux and therefore terminus stability allows in the case of basal overdeepenings potentially more than one stable state of position of a glacier terminus, as illustrated in the conceptual model above (Section 19.2.5) through the two stable crossing

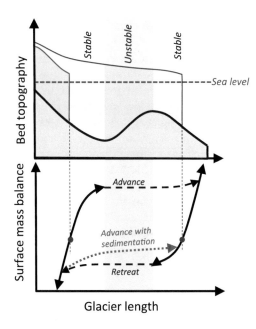

FIGURE 19.7   Schematic illustration of the hysteresis behavior of tidewater glaciers in relation to an overdeepening of the bed. The dashed arrows mark rapid advance or retreat where the terminus jumps from one stable branch to the next. The red dotted arrow indicates a potential advance that involves a shallowing of the bed through high sedimentation at the terminus. The gray shading indicates terminus positions that are unstable and the two blue lines shows two possible stable glacier geometries to the same climate conditions (surface mass balance).

points A and C in Figure 19.6(c). This means, for a given basal overdeepening, that the possible stable terminus positions describe a hysteresis between glacier length and climate (Nick and Oerlemans, 2006; Schoof, 2007), as illustrated in the schematic in Figure 19.7. The two stable branches are located on down-sloping beds in the along-flow direction and are separated by areas of reverse bed slopes. On which branch the terminus is positioned depends on the history of the terminus position, for example, whether it was already located on the same branch or not. A switch from the advanced to the retreated branch in Figure 19.7 is possible when the lower threshold of stability in ice flux is crossed (e.g., surface mass balance negative enough). On the other hand, an advance from the retreated to the advanced branch needs to overcome the high peak in terminus flux due to the basal overdeepening that usually requires balance fluxes and thereby surface mass balance to increase manifold. As a consequence, unstable retreat through such basal depressions is in the short term irreversible (Meier and Post, 1987) as they would require unrealistically high shifts to positive mass balance, unless other processes enabling advance are involved.

19.2.7 Tidewater Glacier Advance

One such potential advance mechanism, especially in colder climates, is through the formation and advance of a floating ice tongue or ice shelf, as observed, for example, during the readvance of Helheim Glacier in Greenland in 2006 (Howat et al., 2007). Such an ice tongue provides an additional backstress to the grounded ice upstream and therefore acts to reduce the ice flux at the grounding line (Section 19.4.1).

In warmer climates where floating ice tongues are absent, the process of sedimentation and plowing of a morainal bank beneath the terminus would reduce the water depth and ease readvance (see also red dashed line in Figure 19.7). This process is limited, however, by the availability and delivery of sediment to the calving margin, and thus, although sedimentation rates can be high for tidewater glaciers, it requires time scales of several decades to centuries. Examples of such an advance mechanism through sedimentation are the advancing Hubbard and Taku Glaciers in Alaska (Motyka et al., 2006; Post and Motyka, 1995). The latter built up a terminal moraine that almost completely protected the terminus from the influence of ocean water and reduced calving rates practically to zero, resulting in a 7 km re-advance through a fjord over the last 120 years. Note that such a readvance of tidewater glaciers does not necessarily require atmospheric cooling.

19.2.8 Tidewater Glacier Cycle

Building on this readvance mechanism, Post (1975) and Meier and Post (1987) suggested the well-known "tidewater glacier cycle" that is characterized by (1) slow advance over several decades to centuries and (2) rapid unstable retreat through basal overdeepening in fjords. Advance is thereby strongly controlled by the rate of sedimentation, whereas rapid retreat is a result of the water depth-related instability. As advance is slow and retreat fast, most tidewater glaciers should be advancing under a constant climate. Glacier change therefore appears asynchronous and independent of climate, and is in this respect often cited in the literature. Although Meier and Post (1987) mentioned that the relationship between climate and length change is not direct, this does not mean they are entirely independent (Pfeffer, 2007b). Changes in climate likely act as an initial trigger for rapid retreat and once on the way retreat continues, irrespective of a reversal to colder climate. This is consistent with observations from Greenland indicating strong seasonal (summer) retreat often preceding phases of large-scale retreat (Carr et al., 2013b). Atmospheric forcing has also been suggested as a partial trigger for the near synchronous major retreat and speed-up of outlet glaciers in Greenland (Moon and Joughin, 2008; Howat et al., 2008). Such atmospheric forcing is, however, not the sole triggering factor, and over the last 10 years, it became clear that changes in oceanic and sea-ice conditions strongly influence calving (Straneo et al., 2013) and may initiate unstable tidewater glacier retreat (Section 19.3).

19.2.9 Inland Propagation and Response

In the modeling experiment above and in general agreement with observations, the thinning and acceleration related to tidewater glacier retreat originate from the terminus and then propagate upstream (Nick et al., 2009). This inland propagation speed increases proportionally to flow speed and surface slope (Cuffey and Paterson, 2010). For fast flowing tidewater glaciers, this means that thinning can rapidly propagate inland and access mass from further upstream relatively quickly and discharge it to the ocean. For ice sheets, this implies that the terminus at the ocean boundary and the inland ice sheet are tightly coupled through the fast flowing outlet glaciers and ice streams (Payne et al., 2004), and that the involved response time scales are much shorter than for purely land-based ice sheets.

19.2.10 Effect of Channel Width

In the explanations above, the dynamics of tidewater glaciers have been considered for a one-dimensional case along a flowline and variations in channel width were not taken into account. In reality, the width of glacial fjords varies, which in addition to basal topography influences the dynamical behavior and stability of tidewater glacier termini. A glacier channel that widens upstream has a similar effect to a reverse bed slope in that once the glacier retreats from a channel narrowing into a wider section, the flux at the terminus increases due to mass conservation and reduced lateral stresses with the result that mass loss increases (Jamieson et al., 2012). Conversely, a channel that narrows in the upstream direction reduces ice flux at the terminus during retreat, and thus has a stabilizing effect. This influence of width on retreat behavior is supported by numerical modeling studies (Enderlin et al., 2013a; Jamieson et al., 2012) and observational records of tidewater glacier retreat (Warren and Glasser, 1992; Carr et al., 2014).

19.2.11 Glaciers Calving into Freshwater

The dynamic instability in the situation of a basal depression described above applies similarly to glaciers that terminate in freshwater (lakes). This is exemplified by rapid retreat of several lake-terminating outlet glaciers of the Patagonian ice field, such as Glaciar Upsala and Glaciar O'Higgins (Sakakibara et al., 2013; Casassa et al., 1997). For the same bed geometry, compared to ocean-terminating glaciers, the rates of calving, retreat, and mass loss of freshwater-terminating glaciers are substantially reduced, which is essentially a consequence of the reduced density of freshwater ($1,000 \text{ kg/m}^3$) compared to ocean water ($1,030 \text{ kg/m}^3$). This difference in water density impacts in several ways.

First, the flotation thickness for a given water depth is reduced for freshwater resulting in a freeboard height above the water level that is 25 percent below the freeboard in the case of ocean water. Thus, assuming a grounded terminus at or near flotation, in freshwater (1) longitudinal stretching at the calving terminus is reduced by a factor two (Eqn (19.4)) and (2) terminus ice

FIGURE 19.8   Retreating ice tongue of Rhonegletscher in the Swiss Alps in September 2009 terminating in a newly formed lake. The lowest part of the ice tongue is already floating and about to detach from the main part as indicated by the large water-filled crevasses. The ice tongue is about 300 m before it reaches the lake and the two lakes on the left and right sides are connected beneath the ice. *Photograph: Andreas Vieli.*

flux is reduced by 20−30 percent (Eqn (19.3)), which implies decreased crevassing and calving rates for the same water depth.

Second, the reduction in flotation thickness also impacts on the effective pressure at the bed and thus reduces basal motion for a given ice thickness and basal water pressure. Third, an indirect effect, subaqueous melt is strongly reduced in freshwater due to a lack of density contrast between glacier melt water and ambient lake water inhibiting buoyancy-driven convection at the ice−water interface (Section 19.3.2).

In the context of the current general recession of glaciers in mountainous areas of the world as a result of atmospheric warming, the emergence of new terminal lakes may be a common feature (Linsbauer et al., 2012). These lakes may, besides posing a hazard for glacial lake outburst floods or tsunami waves from calving (Clague and O'Connor, 2014), accelerate retreat and mass loss through the water depth-dependent instability mechanism described above. Examples of such accelerated retreat are widespread and occur throughout the world, for example, the Rhonegletscher in the Swiss Alps (Figure 19.8).

19.2.12 Long-term Evolution

The unstable retreat behavior related to submarine basal overdeepenings and width variations of glaciers are also well reflected in the paleorecord. Large morainal banks in fjords, especially where fjords narrow, are typical and are indicative of periods with prolonged stable terminus positions (Hunter et al., 1996). In deep fjord areas, submarine moraines are generally absent, indicating rapid retreat. Similarly, grounding zone wedges deposited at the grounding line of paleo ice streams off the Antarctic coast seem in places to be related to the narrowing in ice stream troughs (Jamieson et al., 2012).

The long-term evolution of basal overdeepenings is the result of multiple cycles of erosion and sediment deposition from glaciers and feeds back to the dynamics of tidewater glaciers as exemplified by the advance through sedimentation and bulldozing in a fjord (Powell, 1991). Strong variations in iceberg discharge from tidewater glaciers are also evident in the sedimentary record within these fjords in the form of pulses of ice-rafted debris (IRD) and variations in fjord-temperature proxies enabling the reconstruction of calving activity as, for example, in the case of Helheim Glacier in Greenland since 1890 (Andresen et al., 2011). Extreme examples in this respect are the short periodic pulses of strongly enhanced iceberg discharge from the Laurentide ice sheet during the Quaternary period and again evident as IRD in ocean sediment cores in the North Atlantic (Heinrich, 1988). Besides internal thermomechanical feedbacks of the Laurentide ice sheet (Alley and MacAyeal, 1994; Abe-Ouchi et al., 2013), these so-called Heinrich events may have been further promoted by rapid retreat into the marine overdeepening of the Hudson Bay.

19.3 TRIGGERING AND FORCING MECHANISMS

The profound influence of basal topography on terminus dynamics implies a highly non-linear and threshold-like response to external forcing. This means that relatively small and short-term variations in climate may be able to push the glacier into a state of unstable retreat. Indeed, prolonged periods of summer calving due to the processes discussed further below, often mark the beginning of large-scale retreat as exemplified by tidewater outlet glaciers in Greenland and the Arctic. Besides surface mass balance, other forcing factors, discussed below, may act as triggers for such rapid retreat of glaciers that are in a threshold position regarding bed topography. Good reviews of recent tidewater glacier change and related forcing factors in the Arctic and in Greenland are given in Carr et al. (2013a) and Straneo et al. (2013).

19.3.1 Atmospheric Forcing

Changes in atmospheric conditions and hence surface mass balance (ablation and accumulation) have in the short-term relatively little direct effect on the terminus behavior. Surface ablation rates are in the order of a few meters per year and act over limited areas, which means that the mass loss in the terminus region through this process is relatively small compared to the total ice flux delivered to the calving front. Further, it takes a long time for longer-term mass changes occurring upstream (accumulation anomalies) to propagate to the terminus and impact on terminus dynamics through thickness change (Vieli and Nick, 2011). Over long time periods (decades to centuries), surface mass balance change may, however, be able to tip a glacier over the threshold of stability and trigger rapid retreat as illustrated in the modeling example above (see Section 19.2.2 in Figure 19.6) and proposed for the tidewater glacier cycle. Enhanced melt-water production due to atmospheric warming may further enhance crevasse formation through the process of hydro-fracturing as suggested for seasonal variations of glaciers in Greenland (Sohn et al., 1998; Weertman, 1973; Van der Veen, 1998).

The recent observations of seasonal terminus fluctuations and the tendency for regionally synchronous dynamic changes of tidewater glaciers in Greenland (Moon et al., 2012; Howat et al., 2008) indicate that other forcing factors may act more rapidly and, thereby, directly affect the terminus behavior and potentially trigger rapid retreat. These forcing factors and related mechanisms are discussed below in more detail.

19.3.2 Ocean Forcing

Melt at the ice–ocean interface is a process that has in recent years been recognized to have a strong influence on the changes of calving termini, both for floating ice tongues and for vertical, grounded calving fronts. For tidewater glaciers in Greenland and Alaska, melt rates in the order of meters per day have been estimated (Motyka et al., 2003, 2011; Enderlin and Howat, 2013; Rignot et al., 2010). Recent acceleration and retreat of Greenland tidewater outlet glaciers have further been linked to the arrival of warmer subsurface ocean water (Straneo and Heimbach, 2013; Rignot et al., 2012, Figure 19.2). Beneath Antarctic ice shelves, melt rates are smaller and in the order of a few meters to tens of meters per year (Joughin et al., 2012; Rignot and Jacobs, 2002). Because this process is acting over relatively large areas, mass loss through ocean melt is comparable to calving (Depoorter et al., 2013).

For vertical grounded calving fronts, the available area for submarine melt is limited, and total mass loss through calving is dominant; the exception being the tidewater glaciers that are exposed to warm fjord waters, such as LeConte Glacier in Alaska. There, fjord temperatures in September are $>7\,°C$, producing average submarine melt rates over the calving face of 10–17/md and therefore dominating mass loss over iceberg calving (Motyka et al., 2003, 2013).

Ocean melting is, however, commonly very localized and more enhanced at a depth where the warmest waters are typically reaching the ice face (Figure 19.2). This means that even in cases of colder fjord temperatures with significantly smaller average submarine melt rates than calving rates, locally, ocean melt can erode, undercut, and therefore oversteepen the ice cliff below the water line (Figure 19.2). This increases the longitudinal stresses at the terminus and therefore promotes enhanced crevassing and calving (O'Leary and Christoffersen, 2013). Thus, ocean melt is effective through modifying the ice-cliff geometry. Although this process seems, for larger tidewater glaciers, to be concentrated at a depth where vertical convection occurs (Sciascia et al., 2012; Xu et al., 2012), for smaller glaciers with low calving rates, erosion of a

FIGURE 19.9   Calving front of Hansbreen in 1999, a small grounded tidewater glacier in southern Svalbard. At the water line, a notch is eroded out due to the melt from surface waters and wave action. *Photograph: Andreas Vieli.*

notch at the water line through waves and wind action has been identified as an important control for calving (Figure 19.9; Vieli et al., 2001). This is especially true for glaciers terminating in lakes (Kirkbride and Warren, 1997; Kääb and Haeberli, 2001).

The onset of retreat, acceleration, and thinning of Jakobshavn Isbrae in Greenland in 1997 coincided with the arrival of warm subsurface waters of the West Greenland and Irminger Current, suggesting a direct and rapid influence of submarine melt on ice dynamics (Holland et al., 2008). This oceanic influence is further supported by order-of-magnitude higher rates of melt compared to surface ablation. Evidence for warm subsurface ocean water leading to dynamic mass loss of outlet glaciers is apparent in other areas of Greenland (Straneo and Heimbach, 2013; Straneo et al., 2011) and also from paleo-reconstruction from fjord sediments (Lloyd et al., 2011; Andresen et al., 2011), but in general, data of subsurface ocean temperature and salinity are rare, especially in fjords and in the vicinity of calving termini (Straneo et al., 2013). Further, temporal variability in fjord water conditions is high, observations mostly short term, and fjord circulation responsible for transporting ocean waters to the calving termini is not well understood (Straneo et al., 2011).

Despite ocean temperatures around Greenland of only a few degrees above the melting point of ice, due to the high heat capacity of ocean water, plenty of energy is available for melt, and the key issue is how to bring the energy into ice contact at the calving face. Plumes driven by the buoyant freshwater from ice melting help thereby to entrain relatively warm ambient fjord waters required for submarine melting (Straneo and Heimbach, 2013). These buoyant plumes are further enhanced by subglacial discharge that is released at depth at the calving terminus (Figure 19.2), and are evident from upwelling of turbid

FIGURE 19.10   Photograph of the grounded calving front of Store Glacier, a tidewater outlet glacier in West Greenland, taken on June 25, 2008, from a time-lapse camera *(Courtesy: Jason Box, GEUS)*. Note the area of upwelling of a muddy buoyancy plume reaching the surface at the calving front in the middle of the picture. The dark-grey color of the upwelling water indicates a high suspended sediment concentration originating at least partly from subglacial discharge.

waters often observed in front of tidewater termini (Figure 19.10). The subglacial discharge released at the termini has emerged as a crucial factor for controlling submarine melt and provides a link to atmospheric forcing (Motyka et al., 2013). Enhanced surface melt due to atmospheric warming increases the subglacial discharge and therefore buoyancy-driven submarine melting.

Submarine melting has been suggested as the trigger for the recent widespread rapid dynamic mass loss from tidewater glaciers of the Greenland ice sheet due to an anomalous inflow of subtropical waters from the North Atlantic (Irminger and West Greenland Current; Figure 19.2), which in turn is driven by changes in atmospheric circulation and a long-term increase in heat content of the upper ocean in the North Atlantic (Straneo and Heimbach, 2013). A detailed discussion of the influence of the North Atlantic on the ice dynamics of the Greenland ice sheet is given in a review by Straneo and Heimbach (2013) and for ocean forcing on ice sheet dynamics in general by Joughin et al. (2012).

19.3.3 Sea Ice and Ice Mélange

Another factor suggested as a potential trigger of tidewater glacier change is a reduction in the winter ice-melange cover in front of calving termini. This mixture of icebergs and sea ice that freezes together during the cold season prevents the already heavily crevassed and fractured ice near the calving terminus from detaching and breaking off and can reduce or even fully suppress calving (Amundson et al., 2010). An extension of the period of ice melange-free fjords due to an atmospheric warming will increase the length of the season of summer calving as observed on outlet glaciers in Greenland (Howat et al., 2010; Carr et al., 2013b) and elsewhere in the Arctic (Carr et al., 2014). For Jakobshavn Isbrae, the onset of seasonal retreat of the floating ice tongue coincides with the breakup of the ice mélange and occurs clearly before air temperatures rise above zero degrees (Joughin et al., 2008).

In summary, atmospheric warming, reduced sea-ice mélange, and enhanced ocean melt may act as triggers for initiating retreat of tidewater glaciers, and, depending on the terminus type or climatic/oceanic regime, their relative importance may vary (Nick et al., 2013). Further, these forcing mechanisms are not independent of each other, and thus, the larger scale-coupled atmosphere-ocean circulation system needs to be considered for understanding and predicting the influence of the oceanic melt of tidewater glacier dynamics (Straneo and Heimbach, 2013).

19.4 MARINE ICE SHEETS AND ICE SHELVES

19.4.1 Marine Ice Sheet Instability

For ice sheets, such as the West Antarctic ice sheet (WAIS), that rest on a bed well below sea level and that deepens upstream (Figure 19.2), it has long been postulated that they are, in the absence of a buttressing ice shelf, dynamically inherently unstable (Weertman, 1974; Mercer, 1978; Thomas, 1979). With a sea-level equivalent of the West Antarctic ice sheet of over 3 m, and the recent accelerated dynamic mass loss of major marine ice streams such as Pine Island Glacier (PIG; Shepherd et al., 2002; Rignot et al., 2008), this has raised concerns regarding their potential contribution to future sea-level rise and provoked intense research on the stability of such marine ice sheets (Joughin and Alley, 2011).

If one ignores the effect of a floating ice shelf in front of a marine ice stream, in principle, the same flux instability occurs as already described in Section 19.2.5 for the grounded tidewater glaciers. Essentially, the ice flux required for stability increases non-linearly with water depth (Eqn (19.3)) thus, leading to unstable retreat in the situation of a reversed bed slope. This instability has been theoretically and numerically demonstrated for simplified cases of marine ice sheets (Schoof, 2007).

In very cold polar climates, such as in Antarctica, the marine ice streams typically flow into laterally confined ice shelves or floating ice tongues that exert a stabilizing effect on the grounding line. These floating extensions of ice streams experience resistance to flow from their sides through lateral shear or friction. This lateral drag is transferred upstream to the grounding line and thus is "buttressing" the grounded ice upstream (Goldberg et al., 2009; Weertman, 1974; Dupont and Alley, 2006). This buttressing effect potentially allows stability even in the case of a reverse bed slope, as is currently the case for many larger ice streams in West Antarctica (Gudmundsson et al., 2012). With a thinning or collapse of an ice shelf, this buttressing effect is reduced or removed thus exposing marine ice sheets to the flux instability (described in Section 19.2.5). Similarly, the deposition and buildup of sediment at the grounding line of marine ice streams acts to reduce water depth and thus has been suggested to potentially slow down or even stabilize retreat of the grounding line (Alley et al., 2007).

19.4.2 Ice Shelves and Stability

The profound influence of ice shelves or ice tongues on the stability of marine-terminating ice masses requires a more detailed discussion of their controls and the potential deterioration processes. In general, ice shelves and floating ice tongues are observed in climates where the mean annual air temperature (MAAT) is below $-5\,°C$ (Van der Veen, 2002). This temperature criterion is likely related to the process of hydrofracturing (Section 19.3.1). Below an MAAT of $-5\,°\,C$, none to very limited amounts of surface melt occur. Above this threshold, enough surface melt water is available to pond up on the very flat surface of ice shelves. In zones of extensional crevasses, typical of ice shelves, this ponding water promotes a further deepening of crevasses through the process of hydrofracturing and allows crevasses to penetrate through the full shelf thickness (Weertman, 1973; Van der Veen, 1998). Before the collapse of the Larsen B ice shelf in 2002, such extensive water ponding had been observed (Figure 19.11; Scambos et al., 2000). Further, the distribution of ice shelves in the Antarctic Peninsula is consistent with such a thermal limit, and

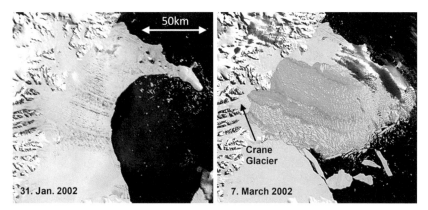

FIGURE 19.11 Satellite images (moderate-resolution imaging spectroradiometer (MODIS)) of the Larsen B ice shelf in the Antarctic Peninsula shortly before (left) and after (right) the collapse of the ice shelf in 2002. Before the collapse, melt ponds at the ice shelf surface are widespread. The light blue areas in the right-hand image mark small fragments of icebergs that broke up during the collapse of the ice shelf. *Courtesy: National Snow and Ice Data Center (NSIDC), University of Colorado, Boulder; MODIS images from the National Aeronautics and Space Administration (NASA) Terra Satellite.*

the disintegration of several ice shelves (Larsen A, Wordie, Wilkins, Larsen B) over the recent 50 years follows the general southward shift of the −5 °C isotherm as a result of the strong atmospheric warming trend (Cook and Vaughan, 2010; Vaughan and Doake, 1996).

With the collapse of the Larsen B ice shelf in 2002 (Figure 19.11), its former tributaries such as Crane Glacier started to speed-up by a factor of 5 within a few years and thinned dramatically (several 10 m/a) in order to dynamically adjust to the loss of buttressing downstream (Figure 19.12; Hulbe et al., 2008; Rignot et al., 2004; Rignot, 2006; Scambos et al., 2004). In Greenland, large floating ice tongues mostly occur in the cold North, with the prominent exception of Jakobshavn Isbrae. Until 1997, Jakobshavn had a 10-km-long floating ice tongue that started to retreat and disintegrate rapidly and by 2004 was mostly gone (Csatho et al., 2008). With the retreat of the floating terminus, the ice stream sped up from the terminus by more than a factor of two and led to strong thinning that is propagating inland (Joughin et al., 2004, 2008). Although the process of hydrofracturing may also have contributed (Sohn et al., 1998), thinning from enhanced submarine melt at the underside of the ice tongue, and reduced periods of winter sea ice, have been proposed as the dominant causes for the ice-tongue disintegration (Holland et al., 2008; Joughin et al., 2008; Vieli and Nick, 2011).

The acceleration of PIG, its grounding line retreat, and enhanced mass loss have been mainly attributed to the thinning, and consequent reduction in buttressing, of its ice shelf (Rignot, 1998; Payne et al., 2004; Shepherd et al., 2004), which in turn is caused by enhanced submarine melting at its underside, in particular near the grounding line. The onset of the thinning of PIG has been

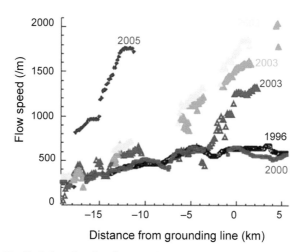

FIGURE 19.12 Evolution of surface flow speed along Crane Glacier, a former tributary of the Larsen B ice shelf, which disintegrated in February 2002 (Figure 19.11). Negative distance values on the x-axis refer to upstream direction. *Figure adapted from Rignot (2006).*

linked to warmer water reaching the cavity beneath the ice shelf (Jenkins et al., 2010). Similarly, strong coastal thinning of other marine ice streams around Antarctica has been linked to thinning of ice shelves that are exposed to warmer subsurface waters (Pritchard et al., 2012).

In contrast to grounded termini, calving from ice shelves, rather than collapse, is in general rather episodic and in the form of large tabular icebergs (Bassis and Jacobs, 2013) that detach periodically at intervals of several years to decades. Thus, a single calving event, such as the recent detachment of a 30-km-wide tabular iceberg of Pine Island ice shelf in autumn 2013, is generally seen as not directly related to climatic forcing but part of a periodic cycle of advance and calving.

A recent large-scale analysis of a 50-year record of terminus-position change of 175 outlet glaciers along 5,400 km coast of the East Antarctic ice sheet revealed that, despite large fluctuations between individual glaciers, clear long-term trends of synchronous glacier advance and retreat are observed (Miles et al., 2013). These changes correspond to trends in air temperature, sea-ice extent, and the larger-scale atmospheric circulation, but surface melt on these glaciers is very limited and most of these possess laterally unconfined floating ice tongues with minimal buttressing effect. Thus, the relationship between upstream dynamics of these outlet glaciers and terminus change is not very direct or clear.

19.5 WIDER IMPLICATIONS AS HAZARDS

The hazardous implications of tidewater glacier and ice sheet instability discussed above, including icebergs, and a potentially rapid contribution to sea-level rise, will now be considered.

19.5.1 Enhanced Iceberg Discharge

With the sudden increase of iceberg discharge to the ocean, the effect of the bed-related retreat instability goes well beyond glacier-related change. Enhanced iceberg production poses a serious threat to shipping routes, as exemplified by the case of the rapidly retreating Columbia Glacier in late the 1970s (Dickson, 1978), and current offshore oil exploration. Glacier terminus type strongly affects the size of icebergs and thus how far they can reach and act as a hazard. Icebergs that calve from grounded termini mostly break up into fragments that are relatively small and affect mostly the local fjords or coasts. Floating ice tongues or ice shelves such as in Antarctica or Northern Greenland, however, often produce much larger tabular icebergs of kilometers to tens of kilometers in diameter and can, through ocean currents and wind drift, act as hazards hundreds of kilometers away from their point of origin. The most famous example perhaps is the fatal collision of the Titanic passenger ship in 1912 with an iceberg originating from the Greenland ice sheet. This event initiated the continuous monitoring of icebergs in the North Atlantic by the U S Coast Guard in 1914 (International Ice Patrol). Another example, from Antarctica, is the large iceberg B-09B that detached in 1987 from the Ross Ice Shelf in West Antarctica, drifted several thousand kilometers westwards along the Antarctic coast and eventually collided in 2010 with the large 80-km-long floating ice tongue of the Mertz Glacier in East Antarctica. As a result, the Mertz Glacier tongue broke off, which reduced the size and strength of the nearby coastal polynya (Kusahara et al., 2011). This seems to reduce the local formation of sea ice and of Antarctic Bottom Waters, and therefore affects the thermohaline circulation (Kusahara et al., 2011), and also the carbon uptake of these water masses (Shadwick et al., 2013) and the habitat of local penguin colonies. It is important to note that, even for stable glaciers, iceberg discharge forms an essential part of the mass balance of ocean-terminating glaciers, and the calving instability or collapse of an ice shelf acts to temporarily and locally enhance such iceberg production.

The break-off and rolling over of large icebergs produce shallow water waves with long wave lengths that can locally cause tsunamis. They pose a hazard to coastal settlements in the fjords of Greenland that are distant from the actual calving fronts. Waves of several meters have been recorded from calving events at Jakobshavn Isbrae even 50 km away (Amundson et al., 2008), and their action is also evident in the vegetation-free coastal banks along the fjords. In Illulissat in West Greenland, such wave events from calving are known as "kaneling" and have violently stirred the water in the fishing harbor in the past (Mikkelsen and Ingerslev, 2002; Reeh, 1985).

Impulse waves from calving into freshwater lakes in mountainous regions are known to trigger further hazardous processes (Haeberli, 1977) such as the breaching of morainal dams with devastating glacier outburst floods (Clague and O'Connor, 2014).

For waves from iceberg over-rolling, theoretical considerations estimate tsunami wave heights in deep water of about one-hundredth of the thickness of a capsizing iceberg, thus resulting in potential waves of several meters height that are further amplified when reaching shallow waters at the shoreline (MacAyeal et al., 2011). In 1995, such an iceberg tsunami caused substantial damage to the fishing harbor of Uummannaq in West Greenland (see "Tsunami Greenland 1995" in youtube.com http://www.youtube.com/watch?v=_2NvwlnKVtU). With the growing interest in tourism and exploration of natural resources in Greenlandic fjords, enhanced iceberg production may become a more apparent hazard in the future.

Locally, such enhanced iceberg production also delivers large amounts of freshwater into the proximal fjords and may thereby affect the composition and circulation of the fjord waters and ecology. Further, the presence of large numbers of icebergs may affect the fishing and hunting practices of the local population.

Enhanced freshwater delivery from the Greenland ice sheet in the form of icebergs may potentially have an impact on thermohaline circulation in the North Atlantic and thereby affect the climate system. Experiments with coupled atmosphere—ocean circulation models indicate a potential weakening of this Atlantic Meridional Overturning Circulation (AMOC) for extreme scenarios of dynamic mass loss and surface melt from the Greenland ice sheet, but the effect is an order of magnitude smaller than the uncertainty of the model simulations and insufficient to cause an abrupt slowdown of the AMOC. Currently, no clear weakening trend of the AMOC is detectable in observations (IPCC, 2013). In addition, these coupled climate models still struggle to transport the freshwater from the Greenland coast to the areas of Atlantic deep-water formation (Straneo and Heimbach, 2013).

19.5.2 New Landscapes

The rapid retreat of tidewater glaciers, as observed for Columbia Glacier, occurs over relatively short time scales of decades and opens up new landscapes and fjords that were previously fully covered by ice (Pfeffer, 2007b). This provides opportunities for colonization by vegetation and fauna, and may also lead to further natural hazards. The freshly emerging fjord walls are typically oversteepened, lack support from ice pressure, and, thus, are fully exposed to gravitational slope processes such as rockfall and landsliding that, in turn, may cause tsunami waves in fjords as observed in West Greenland (Pedersen et al., 2002).

19.5.3 Contribution to Sea Level and Related Challenges

The high potential for rapid dynamic mass loss of ocean-terminating glaciers and ice streams has direct implications for sea-level rise. In 2009, about half of the mass loss of the Greenland ice sheet was due to the strong acceleration of tidewater outlet glaciers (Van den Broeke et al., 2009) and amounts to about 2 mm of sea-level rise since 2000 (Shepherd et al., 2012). Similarly, enhanced ice-stream discharge accounts for almost all the mass loss of the Antarctic ice sheet of 4 mm of sea-level rise since 2000 (Shepherd et al., 2012). It remains to note, however, that although these dynamic changes are related to ocean-terminating glaciers and ice streams, not all of them are necessarily directly due to a bed-related instability and they may partly be a direct result of warmer ocean waters eroding these glacier termini or ice shelves. Further, a considerable part of tidewater glacier ice is already below the sea level and does not further contribute to sea-level rise when melted (Haeberli and Linsbauer, 2013). Future predictions of the contribution of ocean-terminating glaciers to sea-level rise are still uncertain for the reasons discussed below and summarized in Vieli and Nick (2011), Straneo et al. (2013) and Allison et al. (2014).

19.5.3.1 Calving Models

Significant advances in the development of calving models have been made in recent years, but so far, most applications are focused on theoretical cases and these calving approaches have not yet been implemented in larger-scale ice sheet models as they lack thorough validation on real-world examples. The process of calving is complex and the link of climatic and oceanic forcing to calving is still not well understood (Straneo et al., 2013) as it relies heavily on highly simplified parametrizations that require calibration for each individual glacier (Nick et al., 2013). The inclusion of newly emerging calving models into larger-scale ice sheet models is further hampered by numerical or technical issues such as the continuous tracking of the calving front, grid dependency or insufficient spatial resolution. Because of the strong threshold behavior and potential for highly non-linear retreat, predictions of rates of mass loss and retreat are sensitive to the details of calving, flow models and basal geometry (Enderlin et al., 2013a, 2013b).

Regarding future sea-level contribution, a longer-term perspective for tidewater glacier retreat of decades to centuries is needed, and aiming to predict the exact timing of rapid recession of each individual glacier is probably unrealistic. This also requires the evaluation of tidewater glacier models over multidecadal time scales, for which data constraints are relatively rare (Straneo et al., 2013). Although observations of tidewater glacier change, such as terminus positions, surface thinning and flow acceleration, are, through the method of remote sensing becoming more readily available, such time series are mostly limited to the last two decades. Moreover, data on oceanic forcing are even today very rare and for the past rely on reconstructions from proxies with limited resolution.

19.5.3.2 Bed Topography and Model Predictions

Because of the strong dependency of calving rates and tidewater glacier dynamics on basal topography, accurate knowledge of fjord and channel geometry is key for predictions of tidewater glacier change. Typically, the channels of such glaciers are narrow and deep and airborne radio-echo-sounding often struggles to detect the bed due to multiple reflections or attenuation from the heavily crevassed ice or high water content. Inversions for bed topography still suffer from the limitations of their forward models. The majority of bed topographies of the world's tidewater glaciers are today still unknown and any model-based predictions of future contribution to sea level from corresponding glaciers therefore remain uncertain. However, tidewater- or freshwater-terminating glaciers store a significant volume of the world's glacier ice, but their dynamical behavior is not consistent with approaches based upon surface mass balance for land-based glaciers as they can potentially lose mass at much higher rates.

For the Greenland ice sheet, detailed basal topography is known for only a handful of major marine-terminating outlet glaciers, and, thus, current predictions of dynamic mass loss are limited to these glaciers (Nick et al., 2013), or rely on semi-empirical upscaling methods (Price et al., 2011; Pfeffer et al., 2008). The lack of accurate data on bed topography and bathymetry of tidewater glaciers and the deficiencies of numerical models in process representation at the ocean boundary are hindering more accurate predictions. Although major tidewater outlet glaciers currently terminate in deep fjords, these fjords are mostly shallowing inland. Thus, these glaciers will, in the long term, eventually reach much shallower waters that act to stabilize them. The exceptions are a few large outlet glaciers, such as Jakobshavn Isbrae or the Northeast-Greenland ice stream. For the Greenland ice sheet, surface melt may therefore, in the long term, dominate over dynamic mass loss.

For the West Antarctic ice sheet and parts of the East Antarctic ice sheet, bed topography is rather different, with deep, wide basins and troughs that extend far into the ice sheet. Thus, dynamic contributions from unstable grounding line retreat, has the potential to further increase in the future (Joughin and Alley, 2011). Due to the larger spatial scale of marine ice streams in Antarctica, recent efforts from airborne radio-echo-sounding to map their beds were highly successful. Together with significant advances in large-scale ice sheet models that include more realistic treatment of grounding lines, uncertainties in the predictions of the future behavior of the West Antarctic ice sheet have been significantly reduced in recent years. For example numerical models are able to produce past rapid disintegration of the marine sector of the West Antarctic ice sheet (Pollard and DeConto, 2009). Rapid mass loss from PIG equivalent to 3.5–5 mm of eustatic sea-level rise during 2011–2031 is predicted with grounding line retreat in the long-term being irreversible (Favier et al., 2014). The initialization of such numerical models, the potential instability from bed topography, and the coupling to ocean forcing remain, however, a major challenge.

REFERENCES

Abe-Ouchi, A., Saito, F., Kawamura, K., Raymo, M.E., Okuno, J., Takahashi, K., Blatter, H., 2013. Insolation-driven 100,000-year glacial cycles and hysteresis of ice-sheet volume. Nature 500, 190–194.

Alley, R.B., MacAyeal, D.R., 1994. Ice-rafted debris associated with binge/purge oscillations of the Laurentide Ice Sheet. Paleoceanography 9, 503–511.

Alley, R., Anandakrishnan, S., Dupont, T.K., Parzek, B.R., Pollard, D., 2007a. Effect of sedimentation on ice-sheet grounding-line stability. Science 315 (5820), 1838–1841.

Alley, R.B., Horgan, H.J., Joughin, I., Cuffey, K.M., Dupont, T.K., Parizek, B.R., Anandakrishnan, S., Bassis, J., 2007b. A simple law for ice-shelf calving. Science 322 (5906), 1344.

Allison, I., Colgan, W., King, M., Paul, F., 2014. Ice sheets, glaciers and sea level rise. In: Haeberli, W., Whiteman, C. (Eds.), Snow and Ice-Related Hazards, Risks and Disasters. Elsevier, pp. 713–747.

Amundson, J., Truffer, M., Luethi, M., Fahnestock, M., West, M., Motyka, R.J., 2008. Glacier, fjord, and seismic response to recent large calving events, Jakobshavn Ibrae, Greenland. J. Geophys. Res. 35, L22501.

Amundson, J.M., Fahnestock, M., Truffer, M., Brown, J., Luethi, M.P., Motyka, R.J., 2010. Ice melange dynamics and implications for terminus stability, Jakobshavn Isbrae, Greenland. J. Geophys. Res. 115, F01005.

Andresen, C.S., Straneo, F., Ribergaard, M.H., Bjørk, A.A., Andersen, T.J., Kuijpers, A., Nørgaard-Pedersen, N., Kjær, K.H., Schjøth, F., Weckström, K., Ahlstrøm, A.K., 2011. Rapid response of Helheim Glacier in Greenland to climate variability over the past century. Nat. Geosci. 5, 37.

Bassis, J., 2011. The statistical physics of iceberg calving and the emergence of universal calving laws. J. Glaciol. 57 (201), 3–17.

Bassis, J., Jacobs, S., 2013. Diverse calving patterns linked to glacier geometry. Nat. Geosci. 6, 833–836.

Benn, D.I., Warren, C.R., Mottram, R.H., 2007. Calving processes and the dynamics of calving glaciers. Earth Sci. Rev. 82, 143–179.

Brown, C.S., Meier, M.F., Post, A., 1982. Calving Speed of Alaska Tidewater Glaciers, with Application to Columbia Glacier. Tech. rep., US Geological survey professional paper 1258-C.

Carr, R.J., Stokes, C.R.S., Vieli, A., 2013a. Recent progress in understanding marine-terminating Arctic outlet glacier response to climatic and oceanic forcing: twenty years of rapid change. Prog. Phys. Geogr. 37 (4), 436.

Carr, R.J., Stokes, C., Vieli, A., 2013b. Influence of sea ice decline, atmospheric warming, and glacier width on marine-terminating outlet glacier behavior in northwest Greenland at seasonal to interannual timescales. J. Geophys. Res. 118.

Carr, R.J., Stokes, C., Vieli, A., 2014. Recent retreat of major outlet glaciers on Novaya Zemlya, Russian Arctic, influenced by fjord geometry and sea-ice conditions. J. Glaciol. 60 (219), 155–170.

Casassa, G., Brecher, H., Rivera, A., Aniya, M., 1997. A century-long recession record of Glaciar O'Higgins, Chilean Patagonia. Ann. Glaciol. 24 (63), 106–110.

Clague, J.J., O'Connor, J.E., 2014. Glacier-related outburst floods. In: Haeberli, W., Whiteman, C. (Eds.), Snow and Ice-Related Hazards, Risks and Disasters. Elsevier, pp. 487–519.

Colgan, W., Pfeffer, W.T., Rajaram, H., Balog, J., 2012. Monte Carlo ice flow modeling projects a new stable configuration for Columbia Glacier, Alaska, c. 2020. Cryosphere 6, 1395–1409.

Cook, A., Vaughan, D., 2010. Overview of areal changes of the ice shelves on the Antarctic Peninsula over the past 50 years. Cryosphere 4, 77–98.

Csatho, B., Schenk, T., van der Veen, C.J., Krabill, W.B., 2008. Intermittent thinning of Jakobshavn Isbrae, west Greenland, since Little Ice Age. J. Glaciol. 54 (184), 131–144.

Cuffey, K.M., Paterson, W.S.B., 2010. The Physics of Glaciers, fourth ed. Butterworth-Heinemann, Elsevier, Oxford.

Depoorter, M.A., Bamber, J.L., Griggs, J.A., Ligtenberg, S.R.M., van den Broeke, M.R., Moholdt, G., 2013. Calving fluxes and basal melt rates of Antarctic ice shelves. Nature 502, 89–92.

Dickson, D., 1978. Glacier retreat threatens Alaskan oil tanker route. Nature 88–89.

Dupont, T.K., Alley, R.B., 2006. Role of small ice shelves in sea-level rise. Geophys. Res. Lett. 33, L09, 503.

Enderlin, E.M., Howat, I.M., 2013. Submarine melt rate estimates for floating termini of Greenland outlet glaciers (2000–2010). Geophys. Res. Lett. 59 (213), 67–75.

Enderlin, E.M., Howat, I.M., Vieli, A., 2013a. High sensitivity of tidewater outlet glacier dynamics to shape. Cryosphere 7, 1007–1015.

Enderlin, E.M., Howat, I.M., Vieli, A., 2013b. The sensitivity of flowline models of tidewater glaciers to parameter uncertainty. Cryosphere 7, 1579–1590.

Favier, L., Durand, G., Cornford, S.L., Gudmundsson, G.H., Gagliardini, O., Gillet-Chaulet, F., Zwinger, T., Payne, A.J., Brocq, A.M.L., 2014. Retreat of Pine Island Glacier controlled by marine ice-sheet instability. Nat. Clim. Change 4, 117–121.

Fowler, A.C., 2010. Weertman, Lliboutry and the development of sliding. J. Glaciol. 56 (200), 965–972.

Funk, M., Röthlisberger, H., 1989. Forecasting the effects of a planned reservoir that will partially flood the tongue of Unteraargletscher in Switzerland. Ann. Glaciol. 13, 76–80.

Gardner, A.S., et al., 2013. A reconciled estimate of glacier contributions to sea level rise: 2003 to 2009. Science 340, 852–857.

Goldberg, D., Holland, D.M., Schoof, C., 2009. Grounding line movement and buttressing in marine ice sheets. J. Geophys. Res. 114, F04026.

Gudmundsson, G.H., Krug, J., Durand, G., Favier, L., Gagliardini, O., 2012. The stability of grounding lines on retrograde slopes. Cryosphere 6, 1497–1505.

Haeberli, W., 1977. Experience with glacier calving and air bubbling in high Alpine water reservoirs. J. Glaciol. 19 (81), 589–594.

Haeberli, W., Linsbauer, A., 2013. Global glacier volumes and sea level - small but systematic effects of ice below the surface of the ocean and of new local lakes on land. Cryosphere 7, 817–821.

Heinrich, H., 1988. Origin and consequences of cyclic ice rafting in the Northeast Atlantic Ocean during the past 130,000 years. Quat. Res. 29 (2), 142–152.

Holland, D.M., Thomas, T.R.H., deYoung, B., Ribergaard, M.H., Lyberth, B., 2008. Acceleration of Jakobshavn Isbrae triggered by warm subsurface ocean waters. Nat. Geosci. 1, 659–664.

Howat, I.M., Box, J.E., Ahn, Y., Herrington, A., McFadden, E.M., 2010. Seasonal variability in the dynamics of marine terminating outlet glaciers in Greenland. J. Glaciol. 56 (198), 601–613.

Howat, I., Joughin, I., Fahnestock, M., Smith, B., Scambos, T., 2008. Synchronous retreat and acceleration of southeast Greenland outlet glaciers 2000–06: ice dynamics and coupling to climate. J. Glaciol. 54, 646–660.

Howat, I.H., Joughin, I., Scambos, T.A., 2007. Rapid changes of ice discharge from Greenland outlet glaciers. Science 315, 1559–1561.

Hulbe, C.L., Scambos, T.A., Youngberg, T., Lamb, A.K., 2008. Patterns of glacier response to disintegration of the Larsen B ice shelf, Antarctic Peninsula. Global Planet. Change 63, 1–8.

Hunter, L., Powell, R.D., Lawson, D.E., 1996. Morainal-bank sediment budgets and their influence on the stability of tidewater termini of valley glaciers entering Glacier Bay, Alaska, USA. Ann. Glaciol. 22, 211–216.

IPCC, 2013. In: Stocker, T.F., Qin, D., Plattner, G.K., Tignor, M., Allen, S., Boschung, J., Nauels, A., Xia, Y., Bex, V., Midgley, P. (Eds.), Climate Change 2013: The Physical Science Basis. Contribution of Working Group I to the Fifth Assessment Report of the Intergovernmental Panel on Climate Change. Cambridge University Press, Cambridge, United Kingdom and New York, NY, USA, p. 1535.

Jamieson, S.S.R., Vieli, A., Livingstone, S.J., O'Cofaigh, C., Stoke, C., Hillenbrand, C.D., Dowdeswell, J.A., 2012. Ice-stream stability on a reverse bed slope. Nat. Geosci. 5, 799–802.

Jenkins, A., Dutrieux, P., Jacobs, A.S., McPhail, S.D., Perret, J.R., Webb, A.T., White, D., 2010. Observations beneath Pine Island Glacier in West Antarctica and implications for its retreat. Nat. Geosci. 3, 468–472.

Joughin, I., Alley, R.B., 2011. Stability of the West Antarctic ice sheet in a warming world. Nat. Geosci. 4, 506–513.

Joughin, I., Alley, R.B., Holland, D.M., 2012. Ice-sheet response to oceanic forcing. Science 338, 1172.

Joughin, I., Abdalati, W., Fahnestock, M., 2004. Large fluctuations in speed on Greenland's Jakobhavn Isbrae glacier. Nature 432, 608–610.

Joughin, I., Howat, I., Fahnestock, M., Smith, B., Krabill, W., Alley, R.B., Stern, H., Truffer, M., 2008. Continued evolution of Jakobshavn isbrae following its rapid speedup. J. Geophys. Res. 113, F04, 006.

Joughin, I., Smith, B., Howat, I.M., Scambos, T., Moon, T., 2010. Greenland flow variability from ice sheet wide velocity mapping. J. Glaciol. 56 (197), 415−430.

Kääb, A., Haeberli, W., 2001. Evolution of a high-mountain lake in the Swiss Alps. Arct. Antarct. Alp. Res. 33 (4), 385−390.

Kirkbride, M.P., Warren, C., 1997. Calving processes at a grounded ice cliff. Ann. Glaciol. 24, 116−121.

Krimmel, R.M., 2001. Photogrammetric dataset, 1957-2000, and bathymetric measurements for Columbia Glacier, Alaska. U.S. Geol. Surv. Water Resour. Invest. Rep., 01−4089.

Kusahara, K., Hasumi, H., Williams, G.D., 2011. Impact of the Mertz Glacier Tongue calving on dense water formation and export. Nat. Commun. 2 (159).

Larsen, C.F., Motyka, R.J., Arendt, A.A., Echelmeyer, K.A., Geissler, P.E., 2007. Glacier changes in southeast Alaska and northwest British Columbia and contribution to sea level rise. J. Geophys. Res. 112 (F01007), 106−110.

Levermann, A., Albrecht, T., Winkelmann, R., 2012. Kinematic first-order calving law implies potential for abrupt ice-shelf retreat. Cryosphere 6 (2), 273−286.

Linsbauer, A., Paul, F., Haeberli, W., 2012. Modeling glacier thickness distribution and bed topography over entire mountain ranges with GlabTop: application of a fast and robust approach climate. J. Geophys. Res. 117 (F03007), 235−238.

Lloyd, J., Moros, M., Perner, K., Telford, R.J., Kuijpers, A., Jansen, E., McCarthy, D., 2011. A 100-year record of ocean temperature control on the stability of Jakobshavn Isbrae, West Greenland. Geology 39 (9), 867−870.

MacAyeal, D., Abbot, D.S., Sergienko, O.V., 2011. Iceberg-capsize tsunami genesis. Ann. Glaciol. 52 (58), 51−56.

McFadden, E.M., Howat, I., Joughin, I., Smith, B.E., Ahn, Y., 2011. Changes in the dynamics of marine terminating outlet glaciers in west Greenland (2000-2009). J. Geophys. Res. 336116 (F02022).

McNabb, R., Hock, R., 2014. Alaska tidewater glacier terminus positions. J. Geophys. Res. 119.

McNabb, R., Hock, R., O'Neel, S., Rasmussen, L.A., Ahn, Y., Braun, M., Conway, H., Herreid, S., Joughin, I., Pfeffer, W.T., Smith, B.E., Truffer, M., 2012. Using surface velocities to calculate ice thickness and bed topography: a case study at Columbia Glacier, Alaska, USA. J. Glaciol. 58 (212), 1151−1164.

Meier, M.F., Post, A., 1987. Fast tidewater glaciers. J. Geophys. Res. 92 (B9), 9051−9058.

Mercer, J.H., 1978. West Antarctic ice sheet and CO_2 greenhouse effect: a threat to disaster. Nature 271, 321−325.

Mikkelsen, N., Ingerslev, T., 2002. Nomination of the Ilulissat Icefjord for the Inclusion in the World Heritage List. GEUS, Copenhagen.

Miles, B.W.J., Stokes, C.R., Vieli, A., Cox, N.J., 2013. Rapid, climate-driven changes in outlet glaciers on the Pacific coast of East Antarctica. Nature 500, 563−567.

Moon, T., Joughin, I., 2008. Changes in ice front position on Greenland's outlet glaciers from 1992 to 2007. J. Geophys. Res. 113, F02,022.

Moon, T., Joughin, I., Smith, B., Howat, I., 2012. 21st-century evolution of Greenland outlet glacier velocities. Science 336, 576−578.

Motyka, R.J., Dryer, W.P., Amundson, J., Truffer, M., Fahnestock, M., 2013. Rapid submarine melting driven by subglacial discharge, LeConte glacier, Alaska. Geophys. Res. Lett. 40, 1−6.

Motyka, R., Hunter, L., Echelmeyer, K., Conner, C., 2003. Submarine melting at the terminus of a temperate tidewater glacier, LeCont Glacier, Alaska, U.S.A. Ann. Glaciol. 36 (1), 57−65.

Motyka, R.J., Truffer, M., Fahnestock, M.A., Mortesen, J., Rysgaard, S., Howat, I., 2011. Submarine melting of the 1985 Jakobshavn Isbrae floating tongue and the triggering of the current retreat. J. Geophys. Res. 116, F01007. http://dx.doi.org/10.1029/2009JF001632.

Motyka, R.J., Truffer, M., Kuriger, E.M., Bucki, A.K., 2006. Rapid erosion of soft sediments by tidewater glacier advance: Taku Glacier, Alaska, USA. Geophys. Res. Lett. 33, L24504.

Nick, F.M., Oerlemans, J., 2006. Dynamics of tidewater glaciers: comparison of three models. J. Glaciol. 52 (177), 183−190.

Nick, F.M., van der Veen, C.J., Vieli, A., B, D.I., 2010. A physically based calving model applied to marine outlet glaciers and implications for their dynamics. J. Glaciol. 56 (199), 781−794.

Nick, F.M., Luckmann, A., Vieli, A., an der Veen, C.J., van As, D., van de Wal, R., Pattyn, F., Hubbard, A.L., 2012. The response of Petermann Glacier, Greenland, to large calving events, and its future stability in the context of atmospheric and oceanic warming. J. Glaciol. 58 (208), 229−239.

Nick, F.M., Vieli, A., Andersen, M.L., Joughin, I., Payne, A.J., Edwards, T., Pattyn, F., van de Wal, R., 2013. Future sea-level rise from Greenland's main outlet glaciers in a warming climate. Nature 497 (208), 235−238.

Nick, F.M., Vieli, A., Howat, I., Joughin, I., 2009. Large-scale changes in Greenland outlet glacier dynamics triggered at the terminus. Nat. Geosci. 2, 110−114.

O'Leary, M., Christoffersen, P., 2013. Calving on tidewater glaciers amplified by submarine frontal melting. Cryosphere 7, 119−128.

Payne, A.J., Vieli, A., Shepherd, A.P., Wingham, D.J., Rignot, E., 2004. Recent dramatic thinning of largest West Antarctic ice stream triggered by oceans. Geophys. Res. Lett. 31, L23, 401.

Pedersen, S.A.S., Larsen, L.M., Dahl-Jensen, T., Jensen, H.F., Pedersen, G.K., Nielsen, T., Pedersen, A., von Platen-Hallermund, F., Weng, W., 2002. Tsunami-generating rock fall and landslide on the south coast of Nuussuaq, central West Greenland. Geol. Greenl. Surv. Bull. 191, 73−83.

Pfeffer, W.T., 2007a. A simple mechanism for irreversible tidewater glacier retreat. J. Geophys. Res. 112, F03S25.

Pfeffer, W.T., 2007b. The Opening of a New Landscape: Columbia Glacier at Mid Retreat. American Geophysical Union, New York. ISBN: 978-0-8-8790-729-1.

Pfeffer, W.T., Harper, J., O'Neel, S., 2008. Kinematic constraints on glacier contributions to 21st-century sea-level rise. Science 321 (5894), 1340−1343.

Pollard, D., DeConto, R.M., 2009. Modelling West Antarctic ice sheet growth and collapse through the past five million years. Nature 458, 329−333.

Post, A., 1975. Preliminary Hydrography and Historical Terminal Changes of Columbia Glacier. US Geological Survey Hydrologic Investigations Atlas HA-559.

Post, A., Motyka, R., 1995. Taku and LeConte Glaciers, Alaska: calving speed control of late Holocene asynchronous advances and retreats. Phys. Geogr. 16, 59−82.

Powell, R.D., 1991. Grounding line systems as second-order controls on fluctuations of tidewater termini of temperate glaciers. In: Glacial Marine Sedimentation; Palaeo Significance. GSA special paper 261.

Price, S.F., Payne, A.J., Howat, I.M., Smith, B., 2011. Committed sea-level rise for the next century from Greenland ice sheet dynamics during the past decade. Proc. Natl. Acad. Sci. 108 (22), 8978−8983.

Pritchard, H.D., Arthern, R.J., Vaughan, D.G., Edwards, L.A., 2009. Extensive dynamic thinning on the margins of the Greenland and Antarctic ice sheets. Nature 461, 971−975. http://dx.doi.org/10.1038/nature08471.

Pritchard, H.D., Ligtenberg, S.R.M., Fricker, H.A., Vaughan, D.G., van den Broeke, M.R., Padman, L., 2012. Antarctic ice-sheet loss driven by basal melting of ice shelves. Nature 484, 503−505.

Reeh, N., 1985. Long calving waves. In: Proceedings, 8th International Conference on Port and Ocean Engineering under Arctic Conditions, Narssarssuaq, 7-14 September 3, pp. 1310−1327.

Rignot, E., 1998. Fast recession of a West Antarctic glacier. Science 281, 549−551.

Rignot, E., 2006. Changes in ice dynamics and mass balance of the Antarctic ice sheet. Proc. R Soc. London, Ser. A 364 (1844), 1637−1655.

Rignot, E., Jacobs, S.S., 2002. Rapid bottom melting widespread near Antarctic ice sheet grounding lines. Science 296, 5575.

Rignot, E., Bamber, J.L., van den Broeke, M.R., Li, Y., van den Berg, W.J., van Meijgaard, E., 2008. Recent Antarctic ice mass loss from radar interferometry and regional climate modelling. Nat. Geosci. 1, 106−110.

Rignot, E., Casassa, G., Gognineni, P., Krabill, W., Thomas, R., 2004. Accelerated ice discharge from the Antarctic Peninsula following the collapse of Larsen B ice shelf. Geophys. Res. Lett. 31, l18401. http://dx.doi.org/10.1029/2004GL020697.

Rignot, E., Fenty, I., Menemenlis, D., Xu, Y., 2012. Spreading of warm ocean waters around Greenland as a possible cause for glacier acceleration. Ann. Glaciol. 53 (60), 257−266.

Rignot, E., Koppes, M., Velinconga, I., 2010. Rapid submarine melting of the calving faces of West Greenland glaciers. Nat. Geosci. 3, 187−191.

Sakakibara, D., Sugiyama, S., Sawagaki, T., Marinsek, S., Skvarca, P., 2013. Rapid retreat, acceleration and thinning of Glaciar Upsala, Southern Patagonia Icefield, initiated in 2008. Ann. Glaciol. 54 (63), 67−75.

Scambos, T.A., Bohlander, J.A., Shuman, C.A., Skvarca, P., 2004. Glacier acceleration and thinning after ice shelf collapse in the Larsen B embayment, Antarctica. Geophys. Res. Lett. 31, l18402. http://dx.doi.org/10.1029/2004GL020670.

Scambos, T.A., Hulbe, C., Fahnestock, M., Bohlander, J., 2000. The link between climate warming and break-up of ice shelves in the Antarctic Peninsula. J. Glaciol. 46 (154), 517−530.

Schoof, C., 2007. Ice sheet grounding line dynamics: steady states, stability and hysteresis. J. Geophys. Res. 112, F02, 528.

Sciascia, R., Straneo, F., Cenedese, C., Heimbach, P., 2012. Seasonal variability of submarine melt rate and circulation in an East Greenland fjord. J. Geophys. Res. 118, 2492−2502.

Shadwick, E.H., Rintoul, S.R., Tilbrook, B., Williams, G.D., Young, N., Fraser, A.D., Marchant, H., -Tamura, T., 2013. Glacier calving reduced dense water formation and enhanced carbon uptake. Geophys. Res. Lett. 40, 904−909.

Shepherd, A., et al., 2012. A reconciled estimate of ice-sheet mass balance. Science 338, 1183−1189.

Shepherd, A., Wingham, D.J., Mansley, J.A.D., 2002. Inland thinning of the Amundsen sea sector, West Antarctica. Geophys. Res. Lett. 29 (10).

Shepherd, A., Wingham, D., Rignot, E., 2004. Warm ocean is eroding West Antarctic ice sheet. Geophys. Res. Lett. 31, L23, 402.

Sohn, H.G., Jezek, K.C., Van der Veen, C.J., 1998. Jakobshavn glacier, West Greenland: 30 years of spaceborne observations. Geophys. Res. Lett. 25, 2699−2702.

Stearns, L.A., Hamilton, G.S., 2007. Rapid volume loss from two East Greenland outlet glaciers quantified using repeat stereo satellite imagery. Geophys. Res. Lett. 34, L05, 503.

Straneo, F., Heimbach, P., 2013. North Atlantic warming and the retreat of Greenland's outlet glaciers. Cryosphere 504, 36–43.

Straneo, F., Curry, R.G., Sutherland, D.A., Hamilton, G.S., Cenedese, C., Vage, K., Stearns, L.A., 2011. Impact of fjord dynamics and glacial runoff on the circulation near Helheim Glacier. Nat. Geosci. 4, 322–327.

Straneo, F., Heimbach, P., Sergienko, O., Hamilton, G., Catania, G., Griffies, S., Hallberg, R., Jenkins, A., Joughin, I., Motyka, R., Pfeffer, W.T., Price, S.F., Rignot, E., Scambos, T., Truffer, M., Vieli, A., 2013. Challenges to understanding the dynamic response of Greenland's marine terminating glaciers to oceanic and atmospheric forcing. Bull. Am. Meteorol. Soc., 1131–1144.

Thomas, R.H., 1979. The dynamics of marine ice sheets. J. Glaciol. 24, 167–177.

Van den Broeke, M., Bamber, J., Etterna, J., Rignot, E., Schrama, E., van de Berg, W.J., van Meijgaard, E., Velinconga, I., Wouters, B., 2009. Partitioning recent Greenland mass loss. Science 326, 984–986.

Van der Veen, C.J., 1996. Tidewater calving. J. Glaciol. 42 (141), 375–385.

Van der Veen, C.J., 1998. Fracture mechanics approach to penetration of surface crevasses. Cold Reg. Sci. Technol. 27, 31–47.

Van der Veen, C.J., 2002. Calving glaciers. Prog. Phys. Geogr. 26, 96–122.

Vaughan, D.G., Doake, C.S.M., 1996. Recent atmospheric warming and retreat if ice shelves on the Antarctic Peninsula. Nature 379, 328–331.

Vieli, A., Nick, F.M., 2011. Understanding and modelling rapid dynamic changes of tidewater outlet glaciers: issues and implications. Surv. Geophys. 27 (3).

Vieli, A., Funk, M., Blatter, H., 2000. Tidewater glaciers: frontal flow acceleration and basal sliding. Ann. Glaciol. 31, 217–221.

Vieli, A., Funk, M., Blatter, H., 2001. Flow dynamics of tidewater glaciers: a numerical modelling approach. J. Glaciol. 47 (159), 595–606.

Walter, F., O'Neal, S., McNamara, D., Pfeffer, W.T., Bassis, J.N., Fricker, H.A., 2010. Iceberg calving during transition from grounded to floating ice: Columbia Glacier, Alaska. Geophys. Res. Lett. 37, L15501.

Warren, C.R., Glasser, N.F., 1992. Contrasting response of south Greenland glaciers to recent climatic change. Arct. Alp. Res. 24 (2), 124–132.

Warren, C.R., Kirkbride, M.P., 2003. Calving speed and climatic sensitivity of New Zealand lake-terminating glaciers. Ann. Glaciol. 36, 173–178.

Weertman, J., 1973. Can a Water-Filled Crevasse Reach the Bottom Surface of a Glacier? IAHS Publication 95, 185–188.

Weertman, J., 1974. Stability of the junction of an ice sheet and an ice shelf. J. Glaciol. 13 (67), 3–11.

Xu, Y., Rignot, E., Menemenlis, D., Koppes, M., 2012. Numerical experiments on subaqueous melting of Greenland tidewater glaciers in response to ocean warming and enhanced subglacial discharge. Ann. Glaciol. 53 (60), 229–234.

Chapter 20

Ice Sheets, Glaciers, and Sea Level

Ian Allison [1], William Colgan [2], Matt King [3] and Frank Paul [4]

[1] *Antarctic Climate and Ecosystems Cooperative Research Centre, Hobart, Australia,*
[2] *Geological Survey of Denmark and Greenland, Copenhagen, Denmark,* [3] *School of Geography and Environmental Studies, University of Tasmania, Hobart, Australia,* [4] *Department of Geography, University of Zurich, Switzerland*

ABSTRACT

Within the past 125,000 years, variations in Earth's climate have resulted in global sea levels fluctuating from 130 to 140 m lower than present day to 6 to 9 m higher. Presently, global mean sea level is rising at its fastest rate in the past 6,000 years ($\sim$ 3 mm/year). In this chapter, we discuss both the causes and implications of sea-level rise from the perspective of a cryospheric hazard. We also survey the best estimates of sea-level rise and cryospheric mass change from a variety of monitoring techniques. The transfer of terrestrial ice into the sea has contributed about 50 percent of the sea-level rise since 1993, and probably exceeded the combined sea-level changes due to thermal expansion, changes in terrestrial water storage, and changes in ocean basin size since 2003. This cryospheric contribution to sea-level rise is approximately equally split between the combined ice sheets of Greenland and Antarctica, and the global population of about 200,000 glaciers. The societal effects of sea-level rise will be highly varied throughout the world, with some locations experiencing relative sea-level drop, whereas others experience a relative sea-level rise several times the global mean. Perhaps counter-intuitively, the sea-level rise due to terrestrial ice loss will be most substantial in areas furthest from the source of melting ice. Although this cryospheric hazard will unfold over a much longer time scale than many of the other hazards discussed in this volume, the ramifications of sea-level rise will likely be more widespread and profound. Some implications discussed here include coastal inundation, increased coastal flood frequency and groundwater salinization.

Snow and Ice-Related Hazards, Risks, and Disasters. http://dx.doi.org/10.1016/B978-0-12-394849-6.00020-2

20.1 CONTEMPORARY SEA-LEVEL RISE IN A GEOLOGIC PERSPECTIVE

Future warming of Earth will inevitably lead to sea-level rise due to both the thermal expansion of ocean water and the direct impact of melting terrestrial ice (i.e., ice sheets, glaciers, and permafrost) on global ocean volume. Averaged over the globe, sea level rises by 1 mm for each 362.5×10^9 tonnes of ice stored above sea level, or approximately 400 km^3 of ice, that melts or flows into the ocean (Cogley, 2012). The amount of ice stored in the Greenland Ice Sheet is equal to 7.4 m of sea-level equivalent (SLE), after allowing for that portion of the ice that is below present sea level (Bamber et al., 2013), whereas the Antarctic Ice Sheets contain 58.3 m of SLE (Fretwell et al., 2013). Estimates of the total number of glaciers on Earth vary considerably because of different subdivisions of contiguous ice masses and regional variation in the minimum size of what is classified as a glacier. The current best estimate, including those perennial ice masses peripheral to the Greenland Ice Sheet and located on Antarctic islands (excluding the Antarctic Peninsula), is around 200,000, covering a total area of about $730,000 \text{ km}^2$ (Pfeffer et al., 2014). These store an estimated 0.4 m of SLE (Huss and Farinotti, 2012; Vaughan et al., 2013). However, when considering water from ice that is below sea level (Loriaux and Casassa, 2013), retained in lakes of the subglacial topography (Haeberli and Linsbauer, 2013), or stored in endorheic (closed) basins (e.g., Neckel et al., 2014), the actual contribution to sea-level rise will be a few percent less than this. Terrestrial permafrost globally contains an estimated 0.1 m of SLE (Zhang et al., 2008).

Over the past 800,000 years, Earth's climate has varied with ice-age cycles of around 100,000 years duration. Sea-level changes of more than 100 m, primarily driven by growth and decay of ice sheets in the Northern Hemisphere, have accompanied these climate oscillations (Figure 20.1). Sea level during the last interglacial, the warm period around 125,000 years ago before the most recent ice age, was 6–9 m higher than at present (e.g., Dutton and Lambeck, 2012; Kopp et al., 2013). During this interglacial, global air temperatures are believed to have been around 2 °C higher than at present. During the peak of the last ice age, about 21,000 years ago, a large amount of water was locked in terrestrial ice sheets, and large continental shelf areas were exposed as sea level fell nearly 130–140 m below its present level (e.g., Lambeck et al., 2002; Stanford et al., 2010; Austermann et al., 2013). As global air temperatures increased after the last ice age, sea level rose by as much as 2 m/century until around 6,000 years ago, reclaiming continental shelves as ocean area. Since 6,000 years ago, the average rate of sea-level rise has been less than 0.1 m/century (Figure 20.1, inset).

Over the last 2,000 years, until the late nineteenth century, sea level changed very little, remaining stable to within about 0.2 m (Lambeck et al., 2004). During the twentieth century, however, sea level rose at a faster rate

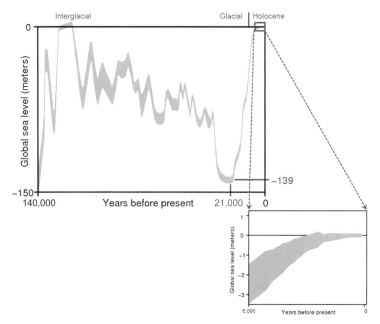

FIGURE 20.1   History of global sea level over the last glacial–interglacial cycle (and inset, over the last 6,000 years) relative to present day. Thickness of the line represents the uncertainty. *From Antarctic Climate and Ecosystems CRC (2008), after Lambeck and Chappell (2001), Lambeck et al. (2002, 2004).*

than has been inferred for any time in the past 6,000 years. This contemporary sea-level rise is primarily due to a combination of the thermal expansion of water in a warming ocean, and the transfer of glacier and ice sheet mass into the ocean via melting and runoff, as well as iceberg calving.

Human activities can also directly influence sea level through changes in terrestrial water storage. Humans add to ocean mass through the depletion of terrestrial groundwater aquifers, and reduce ocean mass through increased terrestrial water storage behind dams. Since the mid-1950s, when a period of large-scale dam building commenced, the combination of these two counteracting activities resulted in a net negative contribution to sea-level change, but since the 1990s this has reversed and the net effect is now a small positive contribution (Church et al., 2013). Although there is a limit to future dam building, a sustained or increased rate of groundwater pumping will result in a more positive sea-level rise contribution from changes in terrestrial water storage in the next few decades (Chao et al., 2008; Cazenave and Llovel, 2010). At the global scale, the sea-level change resulting from changes in ocean basin volume (approximately 0.3 mm/year), such as through isostatic adjustment and/or sedimentation, is smaller than that due to enhanced cryospheric contribution and thermal expansion (Tamisiea, 2011).

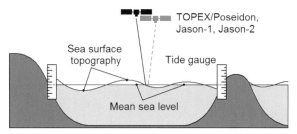

FIGURE 20.2   Shore-based measurements of tides and sea level have been widely made since the nineteenth century. Since 1993, radar altimeters on satellites have measured sea surface height over most of the global oceans.

Past sea levels, reconstructed from tide-gauge records (Figure 20.2) back to 1860 (Church and White, 2011), show a global-average rise of 1.7 ± 0.2 mm/year during the twentieth century. Since 1961, tide-gauge reconstruction suggests an average rise of 1.9 ± 0.4 mm/year. Since 1993, satellite altimeters have provided sea-level data for latitudes up to 66° (N or S), producing sea-level rise estimates that are consistent with tide-gauge reconstructions within the error limits of both techniques (Figure 20.3).

20.2 RECENT GLACIER AND ICE SHEET CONTRIBUTION TO SEA-LEVEL RISE

20.2.1 Recent Sea-Level Rise and Its Sources

Satellite altimeter observations of sea level have been available since 1993, resulting in measurements of sea-level rise that are more spatially comprehensive and more accurate than relying on tide-gauge reconstructions alone. From 1993 to 2009, tide-gauge reconstructions suggest a sea-level rise of 2.8 ± 0.8 mm/year, whereas satellite altimeter observations suggest a sea-level rise of 3.2 ± 0.4 mm/year (Milne et al., 2009; Church and White, 2011; Meyssignac and Cazenave, 2012). This apparent increase in the rate of sea-level rise from the twentieth century mean is an anticipated consequence of observed global warming (IPCC, 2007, 2013).

For the periods 1972 to 2008, and 1993 to 2008 the sum of the individual components of sea-level change equals the total estimated sea-level rise, within the error limits of the various terms (Church et al., 2011). The 95 percent confidence error limits of both the sum of the components and the observed change, however, amount to about 30 percent of the sea-level rise signal for the period. Since 1972, the thermal expansion of sea water has contributed about 45 percent of total sea-level rise, whereas glacier mass loss has contributed about 40 percent, with most of the remaining sea-level rise assumed to be sourced from the ice sheets. For periods prior to 1972, the sea-level budget is not fully closed, indicating errors in some of the estimated

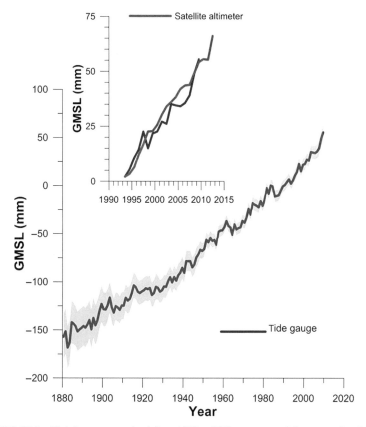

FIGURE 20.3   Global-average sea level from 1880 to 2012 reconstructed from coastal and island tide-gauge data (blue, one standard deviation uncertainty shaded). Inset compares the tide-gauge data with combined TOPEX/Poseidon, Jason-1 and Jason-2 satellite altimeter data since 1993 (red). GMSL, Global mean sea level. *Updated from Church and White (2011) courtesy of J. Church and N. White.*

contributions. The causes of some sea-level fluctuations in the earlier part of the instrumental record cannot be fully explained. Further details are given by Gregory et al. (2013).

Recent corrections to biases in ship-based measurement of subsurface ocean temperature since 1961, and the widespread use of robotic buoys to measure subsurface ocean temperature and salinity to 2,000-m depth (except in ice-covered oceans), have improved estimates of the increase in ocean heat content and resulting thermal expansion of ocean water (Domingues et al., 2008). For the period since 1972, when a reasonably good distribution of ocean temperature data are available, warming of the upper 700 m of the ocean is estimated to have resulted in a sea-level rise of 0.63 mm/year. Over the full ocean depth, including the impact of smaller changes in the abyssal ocean, the thermal expansion is equivalent to 0.80 mm/year (Church et al., 2011). Spatial sampling of the abyssal ocean is particularly sparse and hence the associated thermal expansion estimate is uncertain. More recently, during the post-1993 satellite altimetry period, upper ocean thermal expansion is equivalent to 0.71 mm/year of sea-level rise, and over the full ocean depth the expansion is equivalent to 0.88 mm/year of sea-level rise. Despite this slight acceleration in the sea-level rise associated with thermal expansion, in the last decade the sea-level rise contribution from terrestrial ice has overtaken that due to thermal expansion.

20.2.2 Glacier Contributions

The rate of global glacier wastage increased in the 1980s (e.g., Kaser et al., 2006; Cogley, 2009; Zemp et al., 2009), and around that time a study by Meier (1984) highlighted the significant sea-level rise contribution of meltwater from glaciers. Since then, glacier changes have become an iconic symbol of the impact of climate change, particularly since the link between increases in air temperature and ice melt is easily understood. Glacier area or volume changes do not necessarily occur in a stepwise monotonic or linear fashion; they can be nonlinear and rapid (e.g., terminus retreat from year to year) and cumulatively massive (e.g., the glacier volume loss since the Little Ice Age (Figure 20.4) which, in Europe, lasted from about 1350 to about 1850 (IPCC, 2007)).

The mass balance (or mass budget) of a glacier, typically determined on an annual or seasonal (winter and summer balance) basis, is the balance between mass input, generally mostly as snowfall, and mass lost through melt with runoff, and in some places, iceberg calving. A glacier with a "balanced" mass budget loses the same mass of ice and snow each year as it gains, resulting in a net balance of zero. The mass budget of a glacier is positive (or negative) under conditions of net mass gain (or loss), with any mass imbalance contributing to sea-level fall (or rise) after some delay (e.g., Andermann et al., 2012) and excepting those cases, discussed above, where the glacier melt is

FIGURE 20.4 Trimlines on the Aletsch Gletscher, Switzerland, indicating the previous much greater thickness of this glacier. The trim line marks the boundary between well-vegetated terrain that has remained ice-free for a considerable time and poorly vegetated terrain that until relatively recently (typically since the Little Ice Age) was beneath glacier ice. *Photograph: J. Alean.*

retained on land. Determining volume and mass changes of glaciers, along with finding appropriate ways of translating the sparse sample of direct mass budget measurements (about 100 out of roughly 200,000 glaciers each year) to the global scale, has become a key activity of glacier research (e.g., Raper and Braithwaite, 2006; Bahr et al., 2009; Cogley, 2009; Hock et al., 2009).

A long history of glacier observations has been recorded (e.g., WGMS, 2008). Advancing glaciers during the Little Ice Age were responsible for disastrous flooding and loss of life in the European Alps and elsewhere (e.g., Tufnell, 1984), and hence the scientific exploration of glaciers was already underway by the seventeenth century (e.g., Haeberli, 2008). At the same time, non-scientific exploration of the mountains in the Alps and Norway resulted in numerous landscape paintings and other evidence that allowed very detailed reconstruction of the fluctuations of selected glaciers over the past 400 years (e.g., Zumbühl et al., 2008; Nussbaumer et al., 2011; Zemp et al., 2011).

Internationally coordinated annual measurements of glacier-front variations have been made since the International Glacier Commission was founded in 1894 by F.A. Forel. The current sample of glaciers monitored for change in length or terminus position (~ 600) is about six times larger than those glaciers for which mass budget measurements are made. The sample monitored for length change also includes more larger glaciers, and can thus be considered as being more representative on a regional to global scale (WGMS, 2008). Recent studies have used long-term time-series measurements of glacier-length changes to independently reconstruct hemispheric temperature variability (Leclercq and Oerlemans, 2012), and to estimate glacier volume changes since the beginning of the nineteenth century (Oerlemans et al., 2007; Leclercq et al., 2011). The latter type of studies help to place observed contemporary rates of glacier change into a longer perspective. Other studies have used mass balance models, forced by reconstructed air temperature and precipitation time series from global reanalysis data, to determine global glacier volume changes, and hence their sea-level contribution over the past 60–110 years (e.g., Hirabayashi et al., 2010; Radić and Hock, 2010; Marzeion et al., 2012).

Direct mass balance measurements, which have now been available for about 60 years (Zemp et al., 2009), provide a better quantification of the sea-level contribution from glaciers over recent decades (e.g., Mernild et al., 2013). In essence, annual global glacier mass change can be obtained by multiplying total glacier area with a mean value of climatic surface mass balance. Until recently, however, neither the total glacier area nor the representativeness of locally measured surface mass balances at the global (or even mountain range) scale was accurately known.

The emergence of new technologies and data, such as from differencing digital elevation models (DEMs), laser scanning and altimetry, and gravity measurements from space, provides improved spatial coverage and representativeness of glacier changes (e.g., Arendt et al., 2006; Berthier et al., 2010; Moholdt et al., 2010; Bolch et al., 2013; Jacob et al., 2012). Although spaceborne gravimetry provides a direct measure of mass changes, this technique is only applicable to heavily and compactly glacierized regions (e.g., Gardner et al., 2013). The other methods require knowledge of snow and firn density in the near surface layers in order to convert volume to mass change (e.g., Huss, 2013), but they reveal glacier-wide changes and thus reduce the errors associated with the spatial extrapolation of conventional measurements of surface mass balance (Huss, 2012). Estimating volume change from the difference between DEMs for different periods has the further advantage of being applicable to large regions such as entire mountain ranges (e.g., Gardelle et al., 2013). It can reveal differences in the changes of individual glaciers, indicating how well the measured glaciers represent the overall mean over a given period (Paul and Haeberli, 2008).

Recently, mostly using spaceborne sensors, a near-complete global glacier inventory has been compiled (the "Randolph Glacier Inventory" or RGI; Pfeffer et al., in 2014). The RGI is a data set of glacier outlines in a vector format, with some attribute information added (e.g., glacier area, date of the outline, region), and has enabled a much more precise determination of total global glacier volume (e.g., Huss and Farinotti, 2012; Grinsted, 2013). The RGI also improves the accuracy of global up-scaling of available mass balance measurements (e.g., Gardner et al., 2013). Bringing all data sets and methods together, the scientific community can now determine the contribution of glaciers to sea-level changes from the eighteenth century to the year 2100 within reasonable error bounds.

The reconstructed rate of sea-level rise from 1700 until the end of the Little Ice Age (around 1850 in most regions) is about zero (Church et al., 2013). From 1850 until the 1950s, glacier shrinkage contributed to rising sea level at a rate estimated as near-constant at 0.6–0.8 mm/year (summarized in Vaughan et al., 2013 and in Figure 20.5). The glacier contribution slowed to 0.3 mm/year until the 1980s before increasing again to 0.6 mm/year. The different studies that have reconstructed the past glacier contribution to sea-level change differ slightly in absolute number, but show the same general trends within their uncertainty ranges (Figure 20.5). For the period 2003 to 2009, Gardner et al. (2013) reconciled disparate estimates of glacier wastage to provide a revised estimate of the global glacier mass budget as -259 ± 28 Gt/year. This is equivalent to $+0.71 \pm 0.08$ mm/year SLE, which accounts for 29 ± 13 percent of the observed sea-level rise, and is similar in magnitude to the combined sea-level contribution from both ice sheets over the same period.

20.2.3 Ice Sheet Contributions

In contrast to glaciers, our knowledge about past changes of the ice sheets is largely based on remote sensing observations and numerical modelling. Over the last two decades, three different and independent satellite-based methods

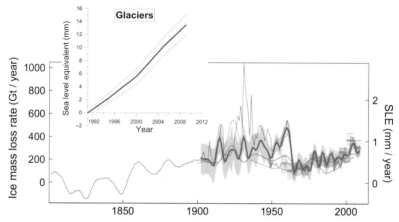

FIGURE 20.5   The rate of global sea-level contribution from glaciers from 1801 to 2010, after IPCC AR5 (Vaughan et al., 2013). The various sources are: *updated from Leclercq et al. (2011)*, arithmetic mean, including Greenland peripheral glaciers (GL) (red line); *updated from Leclercq et al. (2011)*, area-weighted mean, including Greenland peripheral glaciers (GL) (orange line); Marzeion et al. (2012) (thick purple line and error shading); Marzeion et al. (2012), including GL (thin purple line); Hirabayashi et al. (2013) (green line); *updated from Cogley (2009)* (thick blue line and error shading); *updated from Cogley (2009)*, including GL [thin blue line]. The inset shows the IPCC AR5 estimation of the cumulative contribution of glaciers to sea-level rise since the beginning of 1992. SLE, sea-level equivalent.

have been available to estimate changes in the ice mass of Greenland and Antarctica (Figure 20.6). The first approach, known as the mass budget or input—output method (Figure 20.6(1)), estimates the difference between the net surface mass balance over the ice sheet, usually derived from regional climate models (e.g., Lenaerts et al., 2012), and ice discharged from the ice sheet. Ice discharge is calculated near the grounding zone of the ice sheet, the region where the ice enters the ocean and contributes to sea-level rise, even though it is floating and may still be attached to the grounded ice sheet. The ice discharge flux is estimated using ice thickness data combined with an estimated depth-averaged ice velocity. Surface velocity can be directly observed via radar (e.g., Rignot et al., 2011a) or optical methods (e.g., Howat et al., 2011), but needs to be corrected for the shear profile within an ice column in order to convert surface velocity into depth-averaged velocity. Most satellite approaches are limited to measurement of the horizontal component of velocity. Ice thickness measurements are obtained from airborne radar measurements along the grounding zone (e.g., Fretwell et al., 2013) or, where these are not available, from satellite altimeter surface height measurements of the ice where it starts to float and is assumed to be in hydrostatic equilibrium. The input—output method can be applied to the ice sheet as a whole, or to individual drainage basins (Rignot et al., 2011b).

The second approach, known as the volumetric method, relies on repeated airborne or spaceborne altimetry observations (Figure 20.6(2)). Observed

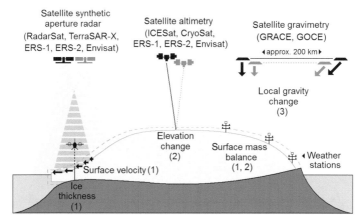

FIGURE 20.6 Remote sensing approaches to assessing ice-sheet mass balance: (1) the input–output method, (2) volumetric change, and (3) local gravimetry change. Numbers in parentheses denote the data requirements of each method, while names denote satellites frequently used in ice sheet mass balance assessments. Surface mass balance over the ice sheet is typically obtained from regional climate models that are verified against in situ point measurements.

changes in ice sheet surface elevation over time are subsequently converted to mass changes using an assumed density of that volume of the ice that has changed. This takes into account numerous factors, including surface mass balance and mean surface temperature that affect the near-surface snow and firn density (e.g., Sørensen et al., 2011).

Finally, the third approach, known as the gravimetry method (Figure 20.6-3), uses space-based observations of temporal variations in the Earth's gravity field to assess changes in regional ice mass (e.g., Velicogna, 2009; Chen et al., 2011; Schrama and Wouters, 2011). Although this last method does not require knowledge of surface mass balance or effective density, it is sensitive to models used to isolate the cryospheric mass change signal from mass change signals due to land and ocean hydrology, atmospheric mass exchange, and solid earth processes. In particular, a gravity signal associated with the ongoing readjustment of the Earth's crust in response to loss of ice mass following the last ice age, known as glacial isostatic adjustment (GIA), is especially nontrivial to accurately account for in glacierized regions (e.g., Ivins et al., 2000; Peltier, 2004; Whitehouse et al., 2012).

The mass change estimates produced by these different methods give a broadly consistent picture of the spatial and temporal variability of mass loss from the Greenland and Antarctic ice sheets as a whole. Substantial variations remain when comparing individual drainage basins, or in the relative contribution of East and West Antarctica to the total. Mass loss over both ice sheets may be characterized as occurring around the lower elevation ice sheet periphery, concentrated at large marine-terminating glaciers, in contrast to slight mass gain in the higher-elevation ice sheet interior. Large differences in the absolute values from initial studies using the different methods have now been substantially reduced with improved analysis and longer time series (Shepherd et al., 2012; Hanna et al., 2013).

Ice sheet mass balance estimates were largely reconciled by the Ice-sheet Mass Balance Intercomparison Experiment (IMBIE) (Shepherd et al., 2012). This effort applied the three mass balance methods described above to common geographical regions and time intervals. Although the uncertainty associated with any single mass balance method was sometimes large, the overall certainty was considerably improved using a combination of methods.

In the satellite observation era, IMBIE estimated a change in Greenland mass of -142 ± 49 Gt/year (0.39 ± 0.14 mm/year of SLE) over the period 1992 to 2011 (Shepherd et al., 2012). The IPCC Fifth Assessment Report (AR5; IPCC, 2013) similarly estimated the Greenland mass budget from an average of 18 recent studies for the same period as -125 ± 25 Gt/year (0.34 ± 0.07 mm/year of SLE; Vaughan et al., 2013) (Figure 20.7). The AR5 also concluded that the rate of mass loss from the Greenland Ice Sheet has accelerated since 1992. While the average loss over the 1992 to 2001 period was 34 (-6 to 74) Gt/year (0.09 (-0.02 to 0.20) mm/year SLE), the loss rate increased to 215 (157 to 274) Gt/year (0.59 (0.43 to 0.76) mm/year SLE) during the 2002 to 2011 period.

Less accurate reconstructed mass balance estimates suggest that since 1840 the Greenland Ice Sheet has contributed 25 ± 10 mm of sea-level rise (Box and Colgan, 2013). Since 1972, the average rate of the Greenland Ice Sheet

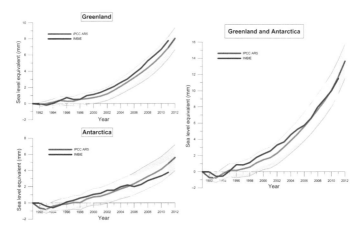

FIGURE 20.7   Cumulative sea-level rise since the beginning of 1992 due to ice loss from the Greenland Ice Sheet, the Antarctic Ice Sheet and both ice sheets combined. The cumulative plots, and their error estimates, are from two recent assessments: IPCC AR5 (Vaughan et al., 2013) and IMBIE (Shepherd et al., 2012). These assessments use the same primary data sources, but quite different ways of selecting and averaging the data. (Compare with the inset to Figure 20.5 which shows the cumulative loss from glaciers).

contribution to sea-level rise has likely been 0.0–0.3 mm/year (Rignot et al., 2008; Box and Colgan, 2013). While Wouters et al. (2013) highlight that short-term variability may substantially bias acceleration estimates when records shorter than one or two decades are analyzed, since 1840 there has been an apparent acceleration in Greenland's sea-level rise contribution at a rate of 0.16 mm/year per century, peaking in the most recent decades (Box and Colgan, 2013). Discharge rates of outlet glaciers vary dramatically over small spatial and temporal scales (Howat et al., 2007; Bjørk et al., 2012; Kjær et al., 2012; Khan et al., 2013) and long-term trends are difficult to quantify. The overall increase in Greenland's contribution to sea-level rise has been punc-tuated by brief periods of net sea-level drawdown, during periods of positive mass balance conditions, the most recent of which was between about 1972 and 1998.

In comparison to Greenland, a paucity of in situ observations and an order-of-magnitude larger ice sheet area make it relatively difficult to estimate Antarctica's sea-level contribution prior to the satellite era. Compounding these problems is the substantial and poorly known GIA signal in Antarctica. Unlike in Greenland, where GIA models are well constrained by ice history data, poor GIA constraints introduce a relatively large GIA-associated un-certainty in gravimetry-derived estimates of Antarctic mass change. This was partially addressed by recent developments in GIA modelling, with two new models used by IMBIE (Shepherd et al., 2012; Whitehouse et al., 2012; Ivins et al., 2013), reducing the East Antarctic ice loss estimated from satellite gravimetry in comparison to previous estimates. Combining various Antarctic mass balance estimates, IMBIE estimated an average Antarctic mass loss between 1992 and 2011 of 71 ± 53 Gt/year (0.20 ± 0.15 mm/year SLE; Shepherd et al., 2012). The IPCC AR5 estimate of Antarctic ice loss for this period, an average of 10 recent studies, was 88 ± 35 Gt/year (0.24 ± 0.10 mm/year SLE) (Vaughan et al., 2013) (Figure 20.7).

Ambiguity remains as to whether East Antarctica as a whole has been close to balance or in substantial positive mass balance since the 1990s (Zwally and Giovinetto, 2011; King et al., 2012; Shepherd et al., 2012; Ivins et al., 2013). Estimation of this is complicated by relatively large snowfall variability (e.g., Boening et al., 2012). Given the relatively poor knowledge of trends and variability in Antarctic mass balance prior to the satellite era, and subsequent time-series variability during the satellite era, it is difficult to infer a statisti-cally significant acceleration in ice sheet mass loss during satellite observa-tions (e.g., Velicogna, 2009; Wouters et al., 2013; Williams et al., 2014).

Like glaciers, the ice sheets lose mass through both melt and runoff as well as iceberg calving. The dominant mechanisms of mass loss, however, vary between the two ice sheets. During the 2000 to 2008 period, about 50 percent of the increased ice loss from Greenland was due to increased melt and runoff, and about 50 percent due to increased iceberg discharge into the sea (van den Broeke et al., 2009). Greenland's low elevation peripheral mass loss was partly offset by a broad region of slight mass gain in the ice sheet interior, which is believed to be due to enhanced snowfall. The enhanced surface melt and runoff from the Greenland Ice Sheet is in response to higher temperatures, and can be reproduced by physically based regional climate models (Ettema et al., 2009; Fettweis et al., 2011). Substantial uncertainty, however, is associated with the parameterization of meltwater that is retained and refreezes within the firn, and hence does not contribute to sea-level rise (Harper et al., 2012). In contrast to Greenland, Antarctica experiences relatively little surface melt. Thus, the primary mechanism of Antarctic ice loss has been due to increased iceberg calving as a result of outlet glacier acceleration (Figure 20.8). Ice shelves and floating ice tongues occupy a substantial portion of the Antarctic coastline, and most ice is lost from Antarctica either by iceberg calving or as melt from the base of these (Depoorter et al., 2013; Rignot et al., 2013). However enhanced submarine melting at ice—ocean interfaces has also contributed to Antarctic ice loss (Holland et al., 2008; Pritchard et al., 2012).

The physical basis underlying the recent acceleration of some marine-terminating glaciers of both ice sheets is less clear. Over the past decade or so, at least five distinct mechanisms have been proposed to explain outlet glacier acceleration.

Firstly, observational evidence indicates a link between increased surface meltwater production and enhanced basal sliding velocity (Figure 20.9) on a variety of time scales (Zwally et al., 2002; Shepherd et al., 2009), although the importance of this mechanism is increasingly discounted (e.g., Schoof, 2010; Tedstone et al., 2013; Sole et al., 2013; Vaughan et al., 2013). Secondly, the acceleration of major glaciers has been postulated to be a short-term dynamic adjustment to force perturbations due to anomalously high thinning of the

FIGURE 20.8   The (a) grounding line and (b) floating tongue of Antarctic outlet glaciers. The grounding line of the Priestley Glacier (a) is marked by a change in surface slope where summer meltwater pools. Upstream of the grounding line the ice is sliding over bedrock and the surface is heavily crevassed. Downstream, the ice, which is several hundred meters thick, moves forward floating on ocean water in a fjord. The ice above sea level contributes to sea level as soon as it becomes afloat. The floating Drygalski Ice Tongue (b) pushes into the Ross Sea, spreading and thinning under its own weight. This ice is already effectively part of the ocean. Icebergs periodically calve from the front of the tongue where the ice is typically 150—200 m thick. *Photographs: I. Allison.*

FIGURE 20.9   Summer meltwater runoff exiting the Greenland Ice Sheet via a proglacial river flowing beside Russell Glacier, before reaching the ocean near Kangerlussuaq, Greenland. Note the people on the riverbank for scale. *Photograph: W. Colgan.*

terminus (Thomas, 2004). Thirdly, several studies have suggested that outlet glacier acceleration can stem from a loss of terminus back stress, such as the loss of an ice shelf or sea ice buttressing, and a subsequent increase in effective driving stress (Scambos et al., 2004; Howat et al., 2008). Fourthly, outlet glacier acceleration may be due to decreased effective basal pressure (ice pressure minus water pressure) due to the thinning of outlet glaciers in response to surface melt, with no change in subglacial water pressure (Pfeffer, 2007).

Finally, recent thermo-mechanical modelling has suggested that glaciers and ice sheets are susceptible to relatively rapid changes in ice temperature due to slight changes in surface meltwater, and thus latent heat, input to the glacier hydrology system (Phillips et al., 2010). Direct observational evidence of such cryo-hydrologic warming is limited (Bader and Small, 1955; Jarvis and Clark, 1974). Since ice rheology (or "effective viscosity") is highly sensitive to ice temperature, the warming of ice has been postulated to potentially result in enhanced ice velocities (van der Veen et al., 2011; Phillips et al., 2013).

Although observational and/or theoretical support can testify to the importance of each of these five mechanisms in various settings, no single governing mechanism has emerged to explain the widespread and sustained acceleration of marine-terminating outlet glaciers over recent decades. Refined diagnostic modelling of recent ice sheet as well as glacier mass loss is ongoing in the cryospheric sciences.

20.3 FUTURE GLACIER AND ICE SHEET CONTRIBUTION TO SEA-LEVEL RISE

Most glaciers are currently out of balance with climate and will continue to decrease in volume and contribute to sea-level rise even if temperatures stabilize (e.g., Bahr et al., 2009; Price et al., 2011; Mernild et al., 2013). Under further warming, changes to the current imbalance will not be monotonic, and during any period individual glaciers may oppose the global trend. In Greenland, the current ice sheet averaged surface mass balance is positive, but this mass gain is more than balanced by ice discharge at the coast. Both snowfall accumulation and surface melt and runoff are anticipated to increase with continued warming, but the melt and runoff rate will increase far more rapidly than precipitation. Analogous to what happened during the decay of the Laurentide Ice Sheet at the end of the last ice age (Carlson et al., 2009), the Greenland Ice Sheet may potentially reach a threshold of inevitable decline, when surface mass balance becomes negative over most of the ice sheet and there is very little net annual mass input. Over long time scales, negative mass balance is unsustainable for any ice mass. An increase in air temperature of 1.6 °C above preindustrial conditions (with a 95 percent confidence interval of 0.8–3.2 °C) is likely sufficient to initiate such irreversible wastage (Robinson et al., 2012). Other estimates of this threshold air temperature increase range from 2 to 5 °C (Gregory and Huybrechts, 2006; Huybrechts et al., 2011). Exceeding this threshold temperature increase for centuries would cause the Greenland Ice Sheet to almost completely disappear. In Antarctica, the surface temperature of the ice sheet proper is generally well below the melting point, and increased surface melt will not be a significant cause of mass loss, even after several centuries of warming. Glaciers on the lower latitude Antarctic Peninsula, however, will undergo enhanced melt and runoff in a warmer world, as could the low-elevation ice shelves elsewhere, and this could result in ice shelf breakup and consequent increased mass loss of grounded ice (e.g., Banwell et al., 2013).

Projections of future sea level are derived from a variety of models. Future atmospheric and ocean temperatures are simulated by coupled atmosphere–ocean climate models, driven by different scenarios of future levels of atmospheric CO_2 and other greenhouse gases. Uncertainty in projections from these climate models arises because of unknown future greenhouse gas levels and because of differences between the parameterizations of physical processes implemented in different models.

Over 90 percent of the net energy increase in the Earth system since 1971 is stored in the ocean, mostly in the upper hundreds of meters (e.g., Domingues et al., 2008). This energy is redistributed horizontally and with depth by ocean circulation, eddies, and turbulent mixing. Projections of future increases in ocean temperature from climate models are used to calculate thermal expansion of the ocean and its global distribution (e.g., Yin et al., 2010).

Climate model estimates of atmospheric temperature and precipitation change are also used to derive regional changes of the surface mass balance of glaciers for different scenarios, and hence to project the change in area and volume of the glaciers. Surface mass balance varies strongly with elevation and also regionally, so glacier locations and the distribution of their area as a function of elevation (hypsometry) are needed to derive the total change to the balance between mass input and output. Although the RGI (Pfeffer et al., 2014) provides an improved source of these data, and has been used to project global sea-level contribution from glaciers over the next century using glacier process-based models (e.g., Marzeion et al., 2012; Giesen and Oerlemans, 2013; Radić et al., 2013), considerable complexity occurs in the response of glaciers, even at a regional level, with long-term evolution and thus their sea-level contribution strongly depending on glacier-specific attributes (e.g., thickness, location, climate, hypsometry). Glaciers are likely to disappear in some mountain ranges (e.g., Zemp et al., 2006; Huss, 2012; Radić et al., 2013), or shrink to very small (<0.5 km^2) topographically protected (e.g., shaded) cirques (DeBeer and Sharp, 2009). Under future widespread glacier wastage, stream flow derived from glacier melt is expected to initially increase, with regionally variable timing of peak stream flow (e.g., Casassa et al., 2009; Bliss et al., 2014). Glacier-derived stream flow will then decrease, once glacier extent has sufficiently declined to the point that further increases in specific runoff (i.e., runoff per unit ice-covered area) no longer modulate stream flow (e.g., Flowers et al., 2005; Huss, 2011).

Recent model projections of the global sea-level contribution from glaciers by 2100 (Marzeion et al., 2012; Giesen and Oerlemans, 2013; Radić et al., 2013) range from 0.07 to 0.17 m for the low greenhouse gas concentration scenario RCP2.6 (a "Representative Concentration Pathway" giving a total radiative forcing from anthropogenic greenhouse gases of 2.6 W/m^2 by 2100; as defined by the IPCC Fifth Assessment Report) to 0.12−0.26 m for the high concentration RCP8.5 (8.5 W/m^2 radiative forcing by 2100) (see also IPCC, 2013). These estimates, however, do not consider glacier ice below sea level (that does not contribute to sea-level rise) and the potential capture on land of meltwater in newly developing lakes or endorheic basins (see Section 20.1). Hence, the actual sea-level rise may be marginally smaller (a few percent) than these estimates.

For the ice sheets of Antarctica and Greenland, coupled climate model projections are used to estimate future changes to surface mass budget. In addition to these surface mass budget changes, however, changes in ice flow and discharge into the ocean play a major role in the overall ice sheet mass budget. Considerable uncertainty exists in projecting changes in iceberg discharge, because the processes that underlie observed outlet glacier acceleration, as discussed above, are poorly constrained in diagnostic models. Like climate models, the ice sheet models that attempt to project both surface mass balance and iceberg discharge at ice sheet scale, are predicated on a deterministic approach. Thus, when a generalized first-principles understanding is not available for a given process, either due to an inability to generalize a locally understood process to other areas or invoke causative rather than first-principles understanding, it is exceedingly difficult to parameterize the process in any model. It can take several years, if not decades, for a novel process, one recently introduced to the scientific literature, to achieve a generalized first-principles understanding suitable for introduction into a community model capable of prognostic simulations.

The IPCC Fourth Assessment Report (AR4; IPCC, 2007) did not specifically model future "dynamic response" of ice sheet outlet glaciers to increased temperatures because of the difficulty in including these processes in prognostic models. Instead, AR4 assigned an empirically estimated additional 0.2 m by the year 2100, on top of the model projections of surface mass balance changes to the ice sheets, to give a likely maximum sea-level rise estimate of 0.79 m by the end of the twenty-first century. But AR4 did not exclude possibly even higher (uncapped) sea-level rise beyond this.

Two large multinational collaborations were initiated to address the uncertainty in the IPCC AR4 sea level projections arising from the unknown future dynamic response of the ice sheets. The ice2sea programme (ice2sea Consortium, 2013), a collaboration of 26 institutions funded by the European Union, investigated ice sheet and glacier processes and developed better physically based models to project future ice sheet change; while the SeaRISE project (Sea-level Response to Ice Sheet Evolution; Bindschadler et al., 2013) compared projections from a number of existing ice sheet models.

Since 2007, better understanding of ice sheet dynamic processes (e.g., Schoof, 2007, 2011), and improved models, which can better reproduce the observed retreat and speed-up of outlet glaciers within the Greenland Ice Sheet (Nick et al., 2009; Vieli and Nick, 2011), have narrowed the uncertainty. Nick et al. (2013) used a glacier flow model with full dynamic treatment of marine termini to project the response of four of the biggest outlet glaciers of Greenland for different future-warming scenarios. The increased sea-level contribution from these glaciers is 80 percent dynamic in origin. Church et al. (2013) extrapolate these results to the entire ice sheet to estimate a likely upper limit of the Greenland Ice Sheet dynamic component of sea level increase by 2100 as 85 mm for RCP8.5, and 63 mm for other RCP scenarios. Combining these with projected surface mass budget changes they estimate a total Greenland contribution to sea level rise by 2100 of 0.07 (0.02 to 0.13) m for RCP2.6, and 0.12 (0.05 to 0.23) m for RCP8.5.

Antarctic surface temperatures are so cold that even with the upper projections of warming over the twenty-first century, little increase in melt will occur. Instead, increased snowfall over the interior will dominate the surface mass budget, and the surface mass budget component of Antarctic change will lead to reduced rate of sea-level increase. Church et al. (2013) used a range of sensitivities of Antarctic precipitation to increasing air temperature, together with the ratio of future warming in the Antarctic to the global average, to estimate surface mass budget changes to the Antarctic Ice Sheet. They determine that Antarctica will contribute a relative sea-level reduction by 2100 of 0.00−0.02 m for RCP2.6, and a relative reduction of 0.01−0.07 m for RCP8.5. To assess the dynamic changes to the Antarctic Ice Sheet, high resolution models incorporating grounding-line migration (e.g., Pattyn et al., 2013) have been applied to glaciers that are experiencing significant present retreat and acceleration in the Pine Island Bay region of Antarctica (Favier et al., 2014). Grounding-line retreat was simulated, but the effect on sea level by 2100 was relatively small because retreat occurred late in the century, tended to be confined to bedrock troughs of limited spatial extent, and much of the lost ice was already floating and displacing ocean water (Payne et al., in press). Based on these and other studies, Church et al. (2013) estimate a total Antarctic contribution to sea-level rise by 2100 of only 0.05 (−0.05 to 0.16) m for RCP2.6, and 0.03 (−0.08 to 0.15) m for RCP8.5.

Summing the results from improved process-based models of the two ice sheets (Agosta et al., 2013; Ligtenberg et al., 2013; Nick et al., 2013) and of global glacier changes (Giesen and Oerlemans, 2013), the ice2sea programme estimated an overall range for the cryospheric contribution to sea-level rise by 2100, for the mid-range A1B emission scenario of IPCC AR4, of between 0.035 and 0.368 m. A common result of the majority of ice2sea projections is that sea level will continue rising after 2100, with sea-level rise in the period 2100 to 2200 greatly exceeding that from present to 2100.

The ten ice sheet models of the Greenland and Antarctic ice sheets compared by the SeaRISE project (Bindschadler et al., 2013) were of varying complexity and grid resolution, but used standard forcing and boundary conditions. The sensitivity of the ice sheets was investigated to prescribed changes of surface mass balance, sub-ice-shelf melting and basal sliding. Although there was a large range in projected sea-level contributions, Greenland was more sensitive to changes in atmospheric temperature and precipitation, whereas Antarctica was most sensitive to increased ice-shelf basal melting. For RCP8.5, the average sea-level contributions were 0.223 m (with a 0.62 m range) from Greenland, and 0.081 m (0.14 m range) from Antarctica.

Semi-empirical models (e.g., Rahmstorf, 2007; Vermeer and Rahmstorf, 2009; Moore et al., 2013) are based on scaling relations between observed sea-level rise and another parameter, such as global-average temperature, and assume that this relation holds into the future. They generally project higher sea-level rise, up to 1.6 m total, over the next century. In another approach, Hansen (2007) assumed a decadal doubling time for cryospheric sea-level rise and suggested a sea-level rise contribution of 5 m by 2100 from all terrestrial ice (Hansen, 2007). However, Pfeffer et al. (2008) argue that kinematic restraints on glacier acceleration would limit a maximum sea-level rise contribution by 2100 to between 0.8 and 2.0 m, although sea-level rise would continue well beyond 2100. These projections, which extrapolate observed mass loss rates, do not acknowledge that the geometry and sea-level rise contribution of tidewater glaciers can eventually stabilize following the onset of acceleration and retreat (Colgan et al., 2012; Nick et al., 2013). Only statistical models, rather than deterministic physical-process models, suggest sea-level rise projections that exceed the AR4 cap of 0.8 m by 2100, and should be cautiously interpreted.

Although each of the last three IPCC assessments have cited a wide range for projected sea-level rise over the twenty-first century, they have all had a similar midpoint and range. The Third Assessment Report (IPCC, 2001) projected a rise of 0.09–0.88 m between 1990 and 2100 for the full range of emission scenarios. The AR4 projected a range of 0.18–0.79 m at 2090–2099 relative to 1980–1999 although, as noted above, did not exclude higher values (IPCC, 2007). And the AR5 projection for sea-level rise by 2100 is a range of 0.28–0.97 m for the four different RCPs (IPCC, 2013). The lower limit of projected sea-level rise over the twenty-first century has slowly increased with successive assessments. The AR5 assessment however discounted a sea-level rise of greater than 1 m over the twenty-first century unless there is a collapse of the marine West Antarctic Ice Sheet. This is considered unlikely, but cannot be completely ruled out.

Figure 20.10 summarizes the relative rate of contribution from the different sources to both historical and projected future sea-level rise as assessed by AR5 (IPCC, 2013). (The AR5 projections are given as the mean value over 2081–2100 ("2090" in Figure 20.10) relative to 1986–2005 ("1996" in Figure 20.10).) The total rate of sea-level rise has generally increased with time since 1901, and is projected to increase even further by the end of the century. The contribution from ice has been at 50 percent or a

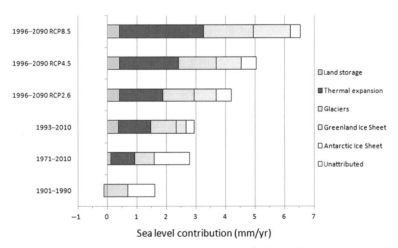

FIGURE 20.10   The average rate of global sea-level rise (mm/yr) and its component contributions, including those from glaciers, the Greenland Ice Sheet and the Antarctic Ice Sheet, as assessed by IPCC AR5 for past periods and for future projections for different Representative Concentration Pathways (RCPs). Prior to modern satellite technology of the 1990s it was not possible to estimate some of the sources, and these are shown as unattributed. This figure does not illustrate the considerable errors associated particularly with the future projections. RCP scenarios were used by IPCC for the Fifth Assessment Report and are defined as the total radiative forcing by 2100 (in Watts per square meter) due to anthropogenic greenhouse gases in the atmosphere. GIS, Greenland Ice Sheet; AIS, Antarctic Ice Sheet. *Data from Church et al. (2013).*

little more of the total contribution, which includes thermal expansion and land storage, and is projected to remain at this percentage over the twenty-first century.

The cryospheric contributions to sea-level rise, particularly from the ice sheets, will continue and will increase well beyond 2100. The total AR5 projection of sea-level rise by 2200 is in the range 0.35—2.03 m, and by 2500 it is in the range 0.50—6.63 m (Church et al., 2013). The relative contribution of ice sheet melt to these projections increases with time, and Church et al. (2013) also note that the ice sheet contributions are probably underestimated because of low confidence in the ability of ice-sheet models to reliably simulate dynamic ice-discharge from Greenland and Antarctica. At 500 years in the future, Bindschadler et al. (2013) project a mean to sea-level rise from Greenland of up to 0.37 m, and a mean contribution from Antarctica of up to 3.35 m.

20.4 IMPLICATIONS OF SEA-LEVEL RISE

As discussed in the introductory chapter to this volume (Haeberli and Whiteman, 2014), the social impacts of changing terrestrial ice volume will not be limited to sea-level rise. Improved understanding of regional variations in climate change and subsequent terrestrial ice and sea-level responses are required for planning locally relevant adaptation and mitigation strategies. Although sea-level rise occurs on a longer time scale than most of the other hazards discussed in this volume, it is a major threat to low lying coastal regions worldwide, including the most densely populated and agriculturally productive regions on Earth. Rising sea level directly impacts coastal regions through inundation and shoreline erosion. Both occur through the combined action of tides, storm surges ("storm tides"), and wave action, underlain by transient sea-level rise forcing; and both result in an increasing frequency of extreme flood events, rather than a monotonic increase in average flood events.

Sea-level rise will have a major impact on human societies due to the significant concentration of communities and infrastructure in coastal regions. The local impacts of sea-level rise are determined by local changes in relative sea level, which accounts for local land uplift or subsidence, atmospheric pressure and gravitational changes, in addition to changes in ocean volume. Many of the Asian mega-deltas, with a total population of over 250 million people, are undergoing land subsidence and are particularly vulnerable to relative sea-level change (Woodroffe et al., 2006). Although coastal subsidence is characteristically geologic in origin, human activity of pumping both groundwater and fossil fuels from near-shore and shallow-coastal areas can enhance local subsidence rates over and above background rates. In Tokyo and Bangkok, for example, groundwater pumping has resulted in bedrock subsidence of 5 and 2 m, respectively, within the past century (Nicholls and

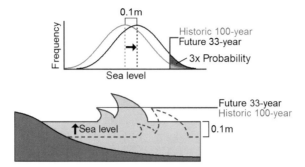

FIGURE 20.11   The influence of sea-level rise on storm frequency. Top: Idealized probability density functions denoting a shift from historic (grey) to future (black) sea-level frequency distributions, resulting in the historic 100-year sea-level event becoming the future 33-year sea-level event. Bottom: Schematic showing the same shift in the historic 100-year sea-level event becoming three times more likely as the future 33-year sea-level event.

Cazenave, 2010). For city planners this effectively corresponds to an equivalent increase in local sea level.

Slight increases in mean local sea level can translate into substantial increases in the frequency of local ocean-flooding events. As a general rule of thumb, a 0.1-m rise in sea-level increases the frequency of coastal flooding by about a factor of three (e.g., Church et al., 2006, Figure 20.11). Thus, a mean sea-level rise of 0.5 m will increase flood frequency by a factor of 3^5, or roughly 250 times, above pre-sea-level rise frequencies. Under such a hypothetical scenario, an event which on average only happened once every 100 years pre-sea-level rise, will happen several times a year after 0.5 m of sea-level rise. For example, in some parts of northern Europe, historical 100-year-event flooding may occur as frequently as every 10 years by about 2070 (Lehner et al., 2006).

More precisely, the regional distribution of increased ocean-flooding depends not only on the regional pattern of sea-level rise, but also storm tides and any future changes that may occur to these. Using tide-gauge information on local storm tides, Hunter (2012) has estimated the multiplication factor for flooding frequency for a sea-level rise of 0.5 m at a number of coastal sites (Figure 20.12). Information from tide gauges and storm-tide models can be used to estimate the likelihood of future flooding, and the planning measures that need to be allowed for adaptation (Hunter, 2010). However, a critical element of projecting changes in recurrence frequency and the likelihood of ocean flooding is also projecting the associated uncertainty in future sea-level rise.

The degree to which coastlines respond to sea-level change is a complex function of geology, geomorphology, geometry, and wave and storm energy. Considering a range of sea-level scenarios for the twenty-first century (0.2—0.8 m sea-level rise) for sandy beaches alone, global-scale modelling suggests land loss of 6,000—17,000 km^2 by 2100 (e.g., Hinkel et al., 2013).

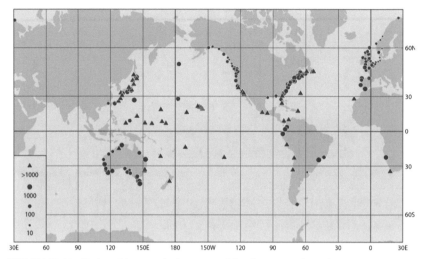

FIGURE 20.12   Projected increase in frequency of flooding events from the sea at coastal and island locations for a sea-level rise of 0.5 m. *Updated from Hunter (2012) by J. Hunter.*

A 1-m sea-level rise is projected to directly displace 8 million people in the South Asia–Pacific region alone by 2100 (Wetzel et al., 2012). In the absence of adaptation, 0.2–4.6 percent of the global population will be flooded annually by 2100, resulting in expected annual losses of 0.3–9.3 percent of global gross domestic product (Hinkel et al., 2014). Such projections of the human consequence of sea-level rise are dependent on not only climate projections, but also demographic and economic projections, as well as understanding drivers and responses of transient changes in local population density.

Globally, sea-level induced migration and infrastructure adaption will likely cost hundreds of billions of dollars by the year 2100. For example, by the year 2100, the projected value of global assets within the 100-year oceanic floodplain will likely be between USD 17 and 210 trillion. Protecting these assets by building dikes would require a global dike maintenance budget of between USD 12 and 71 billion per year by 2100 (Hinkel et al., 2014). The committed sea-level rise associated with maintaining present-day global mean air temperature consigns 6 percent of all UNESCO world heritage sites (40 of 720 UNESCO sites) to inundation within the next two millennia (Marzeion and Levermann, 2014). Under upper limit projections of mean global air temperature change, up to 21 percent of UNESCO world heritage sites may be committed to inundation. The social impacts of sea-level rise and coastal erosion or inundation are therefore not only costly, but culturally profound.

A further impact of sea-level rise is groundwater salinization, causing problems for municipal water supply and agriculture (Gornitz, 1991). This occurs when saline water invades a fresh water table in response to a relative increase in saline water pressure resulting from an increase in sea level.

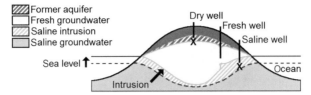

FIGURE 20.13   Schematic overview of how rising sea level can facilitate saline intrusions into coastal fresh water aquifers, potentially causing wells to go dry or saline. Groundwater withdrawal can further enhance the salinization of groundwater.

The salinization process depends on regional factors other than sea-level rise, such as local rainfall, storm tides, and ocean wave climatology (Ferguson and Gleeson, 2012). But, in unconfined aquifer systems that are head-controlled (e.g., head is maintained by surface water features), conventional hydro-geologic parameterizations suggest a 0.5 m increase in sea level can result in a saline intrusion impact extending 5 km inland from the coast (Werner and Simmons, 2009). In most geologic settings, this inland distance may be considered an upper bound estimate of saline intrusion resulting from sea-level rise. Salinization is an especially important concern for small, low-lying islands where people live close to the coast without appreciable surface freshwater sources. In these groundwater dependent communities, salinization can lead to the loss of both drinking water and agricultural irrigation sources (Figure 20.13). In low-lying wetlands, biodiversity adapted to relatively fresh water may be stressed beyond recovery by the intrusion of saline water. In Florida, for example, biodiversity conservation proposals include selecting the "most defensible core sites" within the mangrove ecosystem for heavily engineered management against saline intrusion (Ross et al., 2008). In the absence of sea-level rise, enhanced groundwater withdrawal is the major driver of salinization, as it lowers the effective water pressure of fresh water tables, permitting saline water tables to encroach. Under a wide range of geologic settings and population densities, coastal aquifers will more likely be vulnerable to groundwater withdrawal than projected sea-level rise (Ferguson and Gleeson, 2012).

Sea-level rise does not occur uniformly over the global ocean, but has strong regional variability. This is substantially amplified over decadal to multidecadal timescales through non-secular large-scale climate processes that can produce rates of sea-level change an order of magnitude greater than the global average (Zhang and Church, 2012). For example, while the Gulf of Alaska experienced a relative sea-level decrease of approximately 10 mm/year over the 1993 to 2008 period, the island nations east of the Philippines experienced a relative sea-level increase of approximately 20 mm/year (or six times the global mean) over the same period (Cazenave and Llovel, 2010). Thus, although sea-level rise is often referred to in "global" language, un-derstanding each of the origins of local sea-level rise and how they may

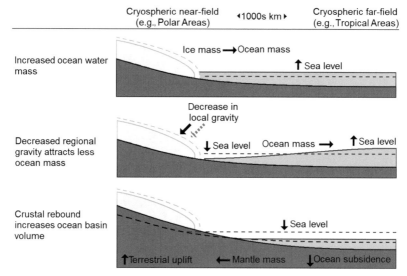

FIGURE 20.14 Schematic overview of three cryospheric processes that influence both absolute and relative sea level. At any coastal location, the net change in sea level due to cryospheric processes is the site-specific sum of all three mechanisms.

change in the future is substantially more important for the implementation of community adaptations.

Perhaps ironically, the sea-level rise due to glacier and ice-sheet mass loss will be felt hardest in areas furthest from cryospheric source areas for two reasons. Firstly, the Arctic and Antarctica are losing sufficient ice mass to modify their respective regional gravitational fields (Figure 20.14). As a consequence, less ocean water mass is gravitationally attracted to these cryospheric regions, and instead ocean water mass is redistributed towards equatorial regions of Earth. This redistribution of existing ocean water can cause a relative sea-level drawdown in excess of 1.5 mm/year in near-field cryospheric regions, and a corresponding amplification of sea-level rise in far-field equatorial regions (Riva et al., 2010). This redistribution of existing ocean mass from cryospheric to non-cryospheric regions may be equivalent to up to 30 percent of the glacier and ice sheet contribution to sea-level rise (Cazenave and Llovel, 2010). Secondly, cryospheric regions losing substantial ice mass experience elastic and viscoelastic rebound as the Earth's crust is partly unloaded of its overburden pressure (Figure 20.14). Bedrock uplift rates in excess of 10 mm/year are typical of the Greenland periphery over the 2007 to 2012 period (Bevis et al., 2012). In cryospheric regions, such as the Gulf of Alaska, these local uplift rates can outpace local net sea-level change, resulting in relative sea-level decrease (Larsen et al., 2005; Cazenave and Llovel, 2010). Glacial isostatic rebound is believed to very modestly increase ocean volume, resulting in an estimated global sea-level decrease of 0.3 mm/year (Tamisiea, 2011).

In polar regions, the presence of melting sea ice can also suppress the thermal expansion of sea water by constraining near-surface water temperatures to the freezing point. By contrast, in tropical regions, the tremendous latent heat sink provided by melting sea ice is unavailable, and excess thermal energy directly contributes to the thermal expansion of sea water (Cazenave and Nerem, 2004). Such nuances in the spatial fingerprints of different sea-level rise mechanisms result in a highly heterogeneous distribution of sea-level rise across Earth. For the purpose of planning for local sea-level associated hazards, "sea-level change" may be a more appropriate term than "sea-level rise" to convey the diverse range of scenarios that will confront human populations around the world over the next several decades.

20.5 CONCLUDING REMARKS

Sea-level rise in response to global climate change is slow but relentless, and thus quite different in nature from most of the other hazards discussed in this volume. But the impact of sea-level rise on coastal regions through increased frequency of high storm-surge events, are sudden and episodic hazards. Millions of people and substantial infrastructure concentrated in the coastal zones of continents, coupled with projections of accelerated socioeconomic growth in coastal zones, indicate that global sea-level rise will be a major challenge with severe implications for human civilization. Since 1992, melting glaciers and ice sheets have added to ocean mass, contributing about 50 percent of the total sea-level rise. Glaciers and ice sheets are expected to continue to contribute this proportion of total global sea-level rise for the rest of the twenty-first century. Many ice masses worldwide are out of equilibrium with the present climate, and will continue to shrink and add to sea level even if global temperature stabilizes.

Projections of future sea-level rise have large uncertainties, not only because of incomplete understanding of the distribution of ocean thermal expansion and processes of ice sheet dynamics, but also because of uncertainties in future greenhouse gas emissions and the climatic response to these emissions. Over the last three IPCC Assessments however, the midpoint projection of sea-level rise by 2100 has remained around 0.4−0.6 m, with a lower probable limit of about 0.2 m. Most recently the IPCC (2013) concluded that only a collapse of marine-based sectors of the Antarctic Ice Sheet could cause global mean sea level to rise substantially above 0.8 m in the twenty-first century, and even then the additional sea-level rise from this would initially, during the twenty-first century, not exceed several tenths of a meter. With continuing warming however, sea level will continue to rise well beyond 2100, with the ice sheets in particular capable of adding several meters to sea level over a number of centuries.

REFERENCES

Agosta, C., Favier, V., Krinner, G., Gallée, H., Fettweis, X., Genthon, C., 2013. High-resolution modeling of the Antarctic mass balance, application for the 20th, 21st and 22nd centuries. Clim Dyn 41, 3247−3260. http://dx.doi.org/10.1007/s00382-013-1903-9.

Andermann, C., Longuevergne, L., Bonnet, S., Crave, A., Davy, P., Gloaguen, R., 2012. Impact of transient groundwater storage on the discharge of Himalayan rivers. Nat Geosci 5 (2), 127−132.

Antarctic Climate and Ecosystems CRC, 2008. Position Analysis: Climate Change, Sea-Level Rise and Extreme Events: Impacts and Adaptation Issues. PA01−0809011, 20 pp. ISSN: 1835−7911.

Arendt, A., Echelmeyer, K., Harrison, W., Lingle, C., Zirnheld, S., Valentine, V., Ritchie, B., Druckenmiller, M., 2006. Updated estimates of glacier volume changes in the western Chugach Mountains, Alaska, and a comparison of regional extrapolation methods. J Geophys Res 111, F03019.

Austermann, J., Mitrovica, J.X., Latychev, K., Milne, G.A., 2013. Barbados-based estimate of ice volume at Last Glacial Maximum affected by subducted plate. Nat Geosci 6 (7), 553−557. http://dx.doi.org/10.1038/ngeo1859.

Bader, H., Small, F., 1955. Sewage Disposal at Ice Cap Installations. In: Snow Ice and Permafrost Research Establishment Report, vol. 21, 1−4.

Bahr, D.B., Dyurgerov, M., Meier, M.F., 2009. Sea-level rise from glaciers and ice caps: a lower bound. Geophys Res Lett 36 (4), L03501.

Bamber, J.L., Griggs, J., Hurkmans, R., Dowdeswell, J., Gogineni, S., Howat, I., et al., 2013. A new bed elevation dataset for Greenland. Cryosphere 7, 499−510.

Banwell, A.F., MacAyeal, D.R., Sergienko, O.V., 2013. Break-up of the Larsen B Ice Shelf triggered by chain-reaction drainage of supraglacial lakes. Geophys Res Lett 40 (22), 5872−5876. http://dx.doi.org/10.1002/2013gl057694.

Berthier, E., Schiefer, E., Clarke, G., Menounos, B., Rémy, F., 2010. Contribution of Alaskan glaciers to sea-level rise derived from satellite imagery. Nat Geosci 3, 92−95. http://dx.doi.org/10.1038/ngeo737.

Bevis, M., Wahr, J., Khan, S., Madsen, F., Brown, A., Willis, M., et al., 2012. Bedrock displacements in Greenland manifest ice mass variations, climate cycles and climate change. Proc Natl Acad Sci USA 109 (30), 11944−11948. http://dx.doi.org/10.1073/pnas.1204664109.

Bindschadler, R.A., Nowicki, S., Abe-Ouchi, A., Aschwanden, A., Choi, H., Fastook, J., et al., 2013. Ice-sheet model sensitivities to environmental forcing and their use in projecting future sea-level (The SeaRISE Project). J Glaciol 59 (224), 194−225.

Bjørk, A.A., Kjær, K.H., Korsgaard, N.J., Khan, S.A., Kjeldsen, K.K., Andresen, C.S., Box, J.E., Larsen, N.K., Funder, S., 2012. An aerial view of 80 years of climate-related glacier fluctuations in southeast Greenland. Nat Geosci 5 (6), 427−432.

Bliss, A., Hock, R., Radić, V., 2014. Global response of glacier runoff to twenty-first century climate change. J Geophys Res Earth Surf 119, 717−730. http://dx.doi.org/10.1002/2013JF002931.

Boening, C., Lebsock, M., Landerer, F., Stephens, G., 2012. Snowfall-driven mass change on the East Antarctic ice sheet. Geophys Res Lett 39 (21), L21501. http://dx.doi.org/10.1029/2012gl053316.

Bolch, T., Sandberg Sørensen, L., Simonsen, S.B., Mölg, N., Machguth, H., Rastner, P., Paul, F., 2013. Mass loss of Greenland's glaciers and ice caps 2003-2008 revealed from ICESat data. Geophys Res Lett 40, 875−881.

Box, J., Colgan, W., 2013. Greenland Ice Sheet mass balance reconstruction. Part III: marine ice loss and total mass balance (1840−2010). J Clim 26, 6990−7002.

Carlson, A.E., Anslow, F.S., Obbink, E.A., LeGrande, A.N., Ullman, D.J., Licciardi, J.M., 2009. Surface-melt driven Laurentide Ice Sheet retreat during the early Holocene. Geophys Res Lett 36, L24502. http://dx.doi.org/10.1029/2009GL040948.

Casassa, G., Lopez, P., Pouyaud, B., Escobar, F., 2009. Detection of changes in glacial run-off in alpine basins: examples from North America, the Alps, central Asia and the Andes. Hydrol Process 23 (1), 31−41.

Cazenave, A., Llovel, W., 2010. Contemporary sea level rise. Ann Rev Mar Sci 2, 145−173. http://dx.doi.org/10.1146/annurev-marine-120308-081105.

Cazenave, A., Nerem, R.S., 2004. Present-day sea level change: observations and causes. Rev Geophys 42 (3).

Chao, B.F., Yao, Y.H., Li, Y.S., 2008. Impact of artificial reservoir water impoundment on global sea level. Science 320 (5873), 212−214. http://dx.doi.org/10.1126/science.1154580.

Chen, J.L., Wilson, C.R., Tapley, B., 2011. Interannual variability of Greenland ice losses from satellite gravimetry. J Geophys Res 116, B07406.

Church, J.A., White, N.J., 2011. Sea-level rise from the late 19th to the early 21st century. Surv Geophys 32, 585−602. http://dx.doi.org/10.1007/s10712-011-9119-1.

Church, J.A., Clark, P.U., Cazenave, A., Gregory, J.M., Jevrejeva, S., Levermann, A., et al., 2013. Sea level change. Chapter 13 in: climate change 2013: the physical science basis. In: Stocker, T.F., Qin, D., Plattner, G.-K., Tignor, M., Allen, S.K., Boschung, J., Nauels, A., Xia, Y., Bex, V., Midgley, P.M. (Eds.), Contribution of Working Group I to the Fifth Assessment Report of the Intergovernmental Panel on Climate Change. Cambridge University Press, Cambridge, United Kingdom and New York, NY, USA.

Church, J.A., Hunter, J.R., McInnes, K.L., White, N.J., 2006. Sea-level rise around the Australian coastline and the changing frequency of extreme sea-level events. Aust Meteorol Mag 55, 253−260.

Church, J.A., White, N.J., Konikow, L.F., Domingues, C.M., Cogley, J.G., Rignot, E., Gregory, J.M., van den Broeke, M.R., Monaghan, A.J., Velicogna, I., 2011. Revisiting the Earth's sea-level and energy budgets from 1961 to 2008. Geophys Res Lett 38, L18601. http://dx.doi.org/10.1029/2011GL048794.

Cogley, J.G., 2009. Geodetic and direct mass-balance measurements: comparison and joint analysis. Ann Glaciol 50 (50), 96−100.

Cogley, J.G., 2012. Area of the ocean. Mar Geod 35, 379−388.

Colgan, W., Pfeffer, W., Rajaram, H., Abdalati, W., Balog, J., 2012. Monte Carlo ice flow modeling projects a new stable configuration for Columbia Glacier, Alaska, c. 2020. Cryosphere 6, 1395−1409.

DeBeer, C., Sharp, M., 2009. Topographic influences on recent changes of very small glaciers in the Monashee Mountains, British Columbia, Canada. J Glaciol 55, 691−700.

Depoorter, M.A., Bamber, J.L., Griggs, J.A., Lenaerts, J., Ligtenberg, S., van den Broeke, M., Moholdt, G., 2013. Calving fluxes and melt rates of Antarctic ice shelves. Nature 502, 89−92.

Domingues, C.M., Church, J.A., White, N.J., Gleckler, P.J., Wijffels, S.E., Barker, P.M., Dunn, J.R., 2008. Improved estimates of upper-ocean warming and multi-decadal sea-level rise. Nature 453, 1090−1093. http://dx.doi.org/10.1038/nature07080.

Dutton, A., Lambeck, K., 2012. Ice volume and sea level during the last interglacial. Science 337, 216−219.

Ettema, J., van den Broeke, M., van Meijaard, M., van de Berg, W., Bamber, J., Box, J., Bales, R., 2009. Higher surface mass balance of the Greenland Ice Sheet revealed by high-resolution climate modeling. Geophys Res Lett 36, L12501. http://dx.doi.org/10.1029/2009GL038110.

Favier, L., Durand, G., Cornford, S.L., Gudmundsson, G.H., Gagliardini, O., Gillet-Chaulet, F., Zwinger, T., Payne, A.J., Le Brocq, A.M., 2014. Retreat of Pine Island Glacier controlled by marine ice-sheet instability. Nat Clim Change 4, 117−121. http://dx.doi.org/10.1038/nclimate2094.

Ferguson, G., Gleeson, T., 2012. Vulnerability of coastal aquifers to groundwater use and climate change. Nat Clim Change 2 (5), 342−345. http://dx.doi.org/10.1038/NCLIMATE1413.

Fettweis, X., Tedesco, M., van den Broeke, M., Ettema, J., 2011. Melting trends over the Greenland Ice Sheet (1958−2009) from spaceborne microwave data and regional climate models. Cryosphere 5, 359−375.

Flowers, G., Marshall, S., Björnsson, H., Clarke, G., 2005. Sensitivity of Vatnajökull ice cap hydrology and dynamics to climate warming over the next 2 centuries. J Geophys Res 110, F02011. http://dx.doi.org/10.1029/2004JF000200.

Fretwell, P.T., Pritchard, H.D., Vaughan, D.G., Bamber, J.L., Barrand, N.E., Bell, R., et al., 2013. Bedmap2: improved ice bed, surface and thickness datasets for Antarctica. Cryosphere 7, 375−393.

Gardelle, J., Berthier, E., Arnaud, Y., Kääb, A., 2013. Region-wide glacier mass balances over the Pamir − Karakoram − Himalaya during 1999−2011. Cryosphere 7, 1263−1286.

Gardner, A., Moholdt, G., Cogley, J., Wouters, B., Arendt, A., Wahr, J., Berthier, E., Hock, R., Pfeffer, W., Kaser, G., Ligtenberg, S., Bolch, T., Sharp, M., Hagen, J., van den Broeke, M., Paul, F., 2013. A reconciled estimate of glacier contributions to sea level rise: 2003 to 2009. Science 340. http://dx.doi.org/10.1126/science.1234532.

Giesen, R.H., Oerlemans, J., 2013. Climate-model induced differences in the 21st century global and regional glacier contributions to sea-level rise. Clim Dyn 41 (11−12), 3283−3300. http://dx.doi.org/10.1007/s00382-013-1743-7.

Gornitz, V., 1991. Global coastal hazards from future sea level rise. Palaeogeogr Palaeoclimatol Palaeoecol 89, 379−398.

Gregory, J.M., Huybrechts, P., 2006. Ice-sheet contributions to future sea level change. Philos Trans Math Phys Eng Sci 364 (1844), 1709−1731.

Gregory, J.M., White, N.J., Church, J.A., Bierkens, M., Box, J.E., van den Broeke, M.R., et al., 2013. Twentieth-century global-mean sea level rise: is the whole greater than the sum of the parts? J Clim 26 (13), 4476−4499. http://dx.doi.org/10.1175/JCLI-D-12-00319.1.

Grinsted, A., 2013. An estimate of global glacier volume. Cryosphere 7, 141−151. http://dx.doi.org/10.5194/tc-7-141-2013.

Haeberli, W., 2008. Changing views of changing glaciers. In: Orlove, B., Wiegandt, E., Luckman, B.H. (Eds.), Darkening Peaks: Glacier Retreat, Science, and Society. Berkeley, USA, pp. 23−32.

Haeberli, W., Linsbauer, A., 2013. Global glacier volumes and sea level − small but systematic effects of ice below the surface of the ocean and of new local lakes on land. Brief communication. Cryosphere 7, 817−821.

Haeberli, W., Whiteman, C., 2014. Snow and ice-related hazards, risks and disasters − a general framework. In: Haeberli, W., Whiteman, C. (Eds.), Snow and Ice-Related Hazards, Risks and Disasters. Elsevier, pp. 1−34.

Hanna, E.H., Navarro, F.J., Pattyn, F., Domingues, C.M., Fettweis, X., Ivins, E.R., Nicholls, R.J., Ritz, C., Smith, B., Tulaczyk, S., Whitehouse, P.L., Zwally, H.J., 2013. Ice-sheet mass balance and climate change. Nature 498, 51−59. http://dx.doi.org/10.1038/nature12238.

Hansen, J.E., 2007. Scientific reticence and sea level rise. Environ Res Lett 2, 024002. http://dx. doi.org/10.1088/1748-9326/2/2/024002.

Harper, J., Humphrey, N., Pfeffer, W., Brown, J., Fettweis, X., 2012. Greenland ice-sheet contribution to sea-level rise buffered by meltwater storage in firn. Nature 491, 240−243.

Hinkel, J., Lincke, D., Vafeidis, A., Perrette, M., Nicholls, R., Tol, R., Marzeion, B., Fettweis, X., Ionescu, C., Levermann, A., 2014. Coastal flood damage and adaptation costs under 21st century sea-level rise. Proc Natl Acad Sci USA 111 (9), 3292−3297. http://dx.doi.org/10. 1073/pnas.1222469111.

Hinkel, J., Nicholls, R.J., Tol, R.S.J., Wang, Z.B., Hamilton, J.M., Boot, G., Vafeidis, A.T., McFadden, L., Ganopolski, A., Klein, R.J.T., 2013. A global analysis of erosion of sandy beaches and sea-level rise: an application of DIVA. Global Planet Change 111, 150−158. http://dx.doi.org/10.1016/j.gloplacha.2013.09.002.

Hirabayashi, Y., Doell, P., Kanae, S., 2010. Global-scale modeling of glacier mass balances for water resources assessments: glacier mass changes between 1948 and 2006. J Hydrol 390 (3−4), 245−256.

Hirabayashi, Y., Zhang, Y., Watanabe, S., Koirala, S., Kanae, S., 2013. Projection of glacier mass changes under a high-emission climate scenario using the global glacier model HYOGA2. Hydrol Res Lett 7, 6−11.

Hock, R., de Woul, M., Radić, V., Dyurgerov, M., 2009. Mountain glaciers and ice caps around Antarctica make a large sea-level rise contribution. Geophys Res Lett 36, L07501.

Holland, P.R., Jenkins, A., Holland, D.M., 2008. The response of ice shelf basal melting to variations in ocean temperature. J Clim 21 (11), 2558−2572.

Howat, I.M., Ahn, Y., Joughin, I., van den Broeke, M.R., Lenaerts, J.T.M., Smith, B., 2011. Mass balance of Greenland's three largest outlet glaciers, 2000−2010. Geophys Res Lett 38, L12501. http://dx.doi.org/10.1029/2011gl047565.

Howat, I., Joughin, I., Fahnestock, M., Smith, B., Scambos, T., 2008. Synchronous retreat and acceleration of southeast Greenland outlet glaciers 2000−06: ice dynamics and coupling to climate. J Glaciol 54, 646−660.

Howat, I.M., Joughin, I., Scambos, T.A., 2007. Rapid changes in ice discharge from Greenland outlet glaciers. Science 315 (5818), 1559−1561.

Hunter, J., 2010. Estimating sea-level extremes under conditions of uncertain sea-level rise. Clim Change 99, 331−350. http://dx.doi.org/10.1007/s10584-009-9671-6.

Hunter, J., 2012. A simple technique for estimating an allowance for uncertain sea-level rise. Clim Change 113, 239−252. http://dx.doi.org/10.1007/s10584-011-0332-1.

Huss, M., 2011. Present and future contribution of glacier storage change to runoff from macroscale drainage basins in Europe. Water Resour Res 47, W07511.

Huss, M., 2012. Extrapolating glacier mass balance to the mountain-range scale: the European Alps 1900−2100. Cryosphere 6, 713−727.

Huss, M., 2013. Density assumptions for converting geodetic glacier volume change to mass change. Cryosphere 7, 877−887.

Huss, M., Farinotti, D., 2012. Distributed ice thickness and volume of all glaciers around the globe. J Geophys Res 117 (F4), F04010. http://dx.doi.org/10.1029/2012JF002523.

Huybrechts, P., Goelzer, H., Janssens, I., Driesschaert, E., Fichefet, T., Goosse, H., Loutre, M.F., 2011. Response of the Greenland and Antarctic ice sheets to multi-millennial greenhouse warming in the earth system model of intermediate complexity LOVECLIM. Surv Geophys 32, 397−416.

ice2sea Consortium, 2013. From Ice to High Seas: Sea-level Rise and European Coastlines. United Kingdom, Cambridge, 50 pp.

IPCC, 2001. Climate change 2001: the scientific basis. In: Houghton, J.T., Ding, Y., Griggs, D.J., Noguer, M., van der Linden, P.J., Dai, X., Maskell, K., Johnson, C.A. (Eds.), Contribution of Working Group I to the Third Assessment Report of the Intergovernmental Panel on Climate Change. Cambridge University Press, Cambridge, United Kingdom and New York, NY, USA, 881 p.

IPCC, 2007. Climate change 2007: the physical science basis. In: Solomon, S., Qin, D., Manning, M., Chen, Z., Marquis, M., Averyt, K.B., Tignor, M., Miller, H.L. (Eds.), Contributions of Working Group I to the Fourth Assessment Report of the Intergovernmental Panel on Climate Change. Cambridge University Press, Cambridge, United Kingdom and New York, NY, USA, 996 p.

IPCC, 2013. Climate change 2013: the physical science basis. In: Stocker, T.F., Qin, D., Plattner, G.-K., Tignor, M., Allen, S.K., Boschung, J., Nauels, A., Xia, Y., Bex, V., Midgley, P.M. (Eds.), Contribution of Working Group I to the Fifth Assessment Report of the Intergovernmental Panel on Climate Change. Cambridge University Press, Cambridge, United Kingdom and New York, NY, USA, 1535 p.

Ivins, E.R., James, T.S., Wahr, J., Schrama, E.J.O., Landerer, F., Simon, K., 2013. Antarctic contribution to sea-level rise observed by GRACE with improved GIA correction. J Geophys Res 118. http://dx.doi.org/10.1002/jgrb.50208.

Ivins, E.R., Raymond, C.A., James, T.S., 2000. The influence of 5000 year-old and younger glacial mass variability on present-day crustal rebound in the Antarctic Peninsula. Earth Planets Space 52 (11), 1023−1029.

Jacob, T., Wahr, J., Pfeffer, W.T., Swenson, S., 2012. Recent contributions of glaciers and ice caps to sea level rise. Nature 482, 514−518.

Jarvis, G., Clark, G., 1974. Thermal effects of crevassing on Steele glacier, Yukon Territory, Canada. J Glaciol 13, 243−254.

Kaser, G., Cogley, J.G., Dyurgerov, M.B., Meier, M.F., Ohmura, A., 2006. Mass balance of glaciers and ice caps: consensus estimates for 1961-2004. Geophys Res Lett 33, L19501.

Khan, S.A., Kjær, K.H., Korsgaard, N.J., Wahr, J., Joughin, I.R., Timm, L.H., Bamber, J.L., van den Broeke, M.R., Stearns, L.A., Hamilton, G.S., Csatho, B.M., Nielsen, K., Hurkmans, R., Babonis, G., 2013. Recurring dynamically induced thinning during 1985 to 2010 on Upernavik Isstrøm, West Greenland. J Geophys Res Earth Surf 118 (1), 111−121.

King, M.A., Bingham, R.J., Moore, P., Whitehouse, P.L., Bentley, M.J., Milne, G.A., 2012. Lower satellite-gravimetry estimates of Antarctic sea-level contribution. Nature 491, 586−589. http://dx.doi.org/10.1038/nature11621.

Kjær, K.H., Khan, S.A., Korsgaard, N.J., Wahr, J., Bamber, J.L., Hurkmans, R., van den Broeke, M.R., Timm, L.H., Kjeldsen, K.K., Bjørk, A.A., Larsen, N.K., Jørgensen, L.T., Færch-Jensen, A., Willerslev, E., 2012. Aerial photographs reveal late-20th-century dynamic ice loss in Northwestern Greenland. Science 337 (6094), 569−573.

Kopp, R.E., Simons, F.J., Mitrovica, J.X., Maloof, A.C., Oppenheimer, M., 2013. A probabilistic assessment of sea level variations within the last interglacial stage. Geophys J Int 193, 711−716.

Lambeck, K., Chappell, J., 2001. Sea level change through the last glacial cycle. Science 292, 679−686.

Lambeck, K.L., Anzidei, M., Antonioli, F., Benini, A., Esposito, A., 2004. Sea level in Roman time in the Central Mediterranean and implications for recent change. Earth Planet Sci Lett 224, 563−575.

Lambeck, K., Yokoyama, Y., Purcell, T., 2002. Into and out of the last glacial maximum: sea-level change during oxygen isotope stages 3 and 2. Quat Sci Rev 21 (1), 343−360.

Larsen, C.F., Motyka, R.J., Freymueller, J.T., Echelmeyer, K.A., Ivins, E.R., 2005. Rapid visco-elastic uplift in southeast Alaska caused by post-little ice age glacial retreat. Earth Planet Sci Lett 237 (3−4), 548−560.

Leclercq, P.W., Oerlemans, J., 2012. Global and hemispheric temperature reconstruction from glacier length fluctuations. Clim Dyn 38, 1065−1079. http://dx.doi.org/10.1007/s00382-011-1145-7.

Leclercq, P.W., Oerlemans, J., Cogley, J.G., 2011. Estimating the glacier contribution to sea-level rise for the period 1800-2005. Surv Geophys 32 (4−5), 519−535.

Lehner, B., Doll, P., Alcamo, J., Henrichs, T., Kaspar, F., 2006. Estimating the impact of global change on flood and drought risks in Europe: a continental integrated analysis. Clim Change 75, 273−299.

Lenaerts, J.T.M., van den Broeke, M.R., van de Berg, W.J., van Meijgaard, E., Kuipers Munneke, P., 2012. A new, high resolution surface mass balance map of Antarctica (1979−2010) based on regional climate modeling. Geophys Res Lett 39 (1−5), L04501.

Ligtenberg, S.R.M., van de Berg, W.J., van den Broeke, M.R., Rae, J.G.L., van Meijgaard, E., 2013. Future surface mass balance of the Antarctic ice sheet and its influence on sea level change,simulated by a regional atmospheric climate model. Clim Dyn 41 (3−4), 867−884. http://dx.doi.org/10.1007/s00382-013-1749-1.

Loriaux, T., Casassa, G., 2013. Evolution of glacial lakes from the Northern Patagonian Icefield and terrestrial water storage in a sea-level rise context. Global Planet Change 102, 33−40.

Marzeion, B., Jarosch, A.H., Hofer, M., 2012. Past and future sea-level change from the surface mass balance of glaciers. Cryosphere 6, 1295−1322.

Marzeion, B., Levermann, A., 2014. Loss of cultural world heritage and currently inhabited places to sea-level rise. Environ Res Lett 9. http://dx.doi.org/10.1088/1748-9326/9/3/034001.

Meier, M.F., 1984. Contribution of small glaciers to global sea level. Science 226 (4681), 1418−1421.

Mernild, S.H., Lipscomb, W.H., Bahr, D.B., Radić, V., Zemp, M., 2013. Global glacier changes: a revised assessment of committed mass losses and sampling uncertainties. Cryosphere 7, 1565−1577.

Meyssignac, B., Cazenave, A., 2012. Sea level: a review of present-day and recent-past changes and variability. J Geodyn 58, 96−109.

Milne, G.A., Gehrels, W.R., Hughes, C.W., Tamisiea, M.E., 2009. Identifying the causes of sea-level change. Nat Geosci 2, 471−478. http://dx.doi.org/10.1038/ngeo544.

Moholdt, G., Nuth, C., Hagen, J.O., Kohler, J., 2010. Recent elevation changes of Svalbard glaciers derived from ICESat laser altimetry. Remote Sens Environ 114, 2756−2767.

Moore, J.C., Grinsted, A., Zwinger, T., Jevrejeva, S., 2013. Semiempirical and process-based global sea level projections. Rev Geophys 51 (3), 484−522. http://dx.doi.org/10.1002/rog.20015.

Neckel, N., Kropacek, J., Bolch, T., Hochschild, V., 2014. Glacier mass changes on the Tibetan Plateau 2003−2009 derived from ICESat laser altimetry measurements. Environ Res Lett 9. http://dx.doi.org/10.1088/1748-9326/9/1/014009.

Nicholls, R., Cazenave, A., 2010. Sea-level rise and its impacts on coastal zones. Science 328, 1517. http://dx.doi.org/10.1126/science.1185782.

Nick, F., Vieli, A., Andersen, M., Joughin, I., Payne, A., Edwards, T., Pattyn, F., van de Wal, R., 2013. Future sea-level rise from Greenland's main outlet glaciers in a warming climate. Nature 497, 235−238.

Nick, F.M., Vieli, A., Howat, I.M., Joughin, I., 2009. Large-scale changes in Greenland outlet glacier dynamics triggered at the terminus. Nat Geosci 2, 110−114.

Nussbaumer, S.U., Nesje, A., Zumbühl, H.J., 2011. Historical glacier fluctuations of Jostedalsbreen and Folgefonna (southern Norway) reassessed by new pictorial and written evidence. Holocene 21 (3), 455−471.

Oerlemans, J., Dyurgerov, M., van de Wal, R.S.W., 2007. Reconstructing the glacier contribution to sea-level rise back to 1850. Cryosphere 1, 59−65.

Paul, F., Haeberli, W., 2008. Spatial variability of glacier elevation changes in the Swiss Alps obtained from two digital elevation models. Geophys Res Lett 35 (5), L21502.

Pattyn, F., Perichon, L., Durand, G., Favier, L., Gagliardini, O., Hindmarsh, R., et al., 2013. Grounding-line migration in plan-view marine ice models: results of the ice2sea MISMIP3d intercomparison. J Glaciol 59, 410−422.

Payne, A.J., Cornford, S., Martin, D., Agosta, C., van den Broeke, M.R., Edwards, T., Gladstone, R., Hellmer, H.H., Krinner, G., Brocq, A.M., Ligtenberg, S.R.M., Lipscomb, W.H., Ng, E.G., Shannon, S.R., Timmermann, R., Vaughan, D.G. Impact of uncertain climate forcing on projections of the West Antarctic ice sheet over the 21st and 22nd centuries. Earth Planet Sci Lett, in press.

Peltier, W.R., 2004. Global glacial isostasy and the surface of the ice-age Earth: the ICE-5G (VM2) model and GRACE. Ann Rev Earth Planet Sci 32, 111−149.

Pfeffer, W.T., 2007. A simple mechanism for irreversible tidewater glacier retreat. J Geophys Res 112, F03S25. http://dx.doi.org/10.1029/2006JF000590.

Pfeffer, W., Arendt, A., Bliss, A., Bolch, T., Cogley, J., Gardner, A., Hagen, J., Hock, R., Kaser, G., Kienholz, C., Miles, E., Moholdt, G., Mölg, N., Paul, F., Radić, V., Rastner, P., Raup, B., Rich, M.J., 2014. Sharp and the Randolph Consortium. The Randolph Glacier Inventory: a globally complete inventory of glaciers. J Glaciol 60 (221), 537−551.

Pfeffer, W.T., Harper, J.T., O'Neel, S., 2008. Kinematic constraints on glacier contributions to 21st-Century sea-level rise. Science 321 (5894), 1340−1343.

Phillips, T., Rajaram, H., Colgan, W., Steffen, K., Abdalati, W., 2013. Evaluation of cryo-hydrologic warming as an explanation for increased ice velocities in the wet snow zone, Sermeq Avannarleq, West Greenland. J Geophys Res 118, 1−16. http://dx.doi.org/10.1002/jgrf.20079.

Phillips, T., Rajaram, H., Steffen, K., 2010. Cryo-hydrologic warming: a potential mechanism for rapid thermal response of ice sheets. Geophys Res Lett 37, L20503. http://dx.doi.org/10.1029/2010GL044397.

Price, S., Payne, A., Howat, I., Smith, B., 2011. Committed sea-level rise for the next century from Greenland Ice Sheet dynamics during the past decade. Proc Natl Acad Sci USA 108, 8978−8983.

Pritchard, H.D., Ligtenberg, S.R.M., Fricker, H.A., Vaughan, D.G., van den Broeke, M.R., Padman, L., 2012. Antarctic ice loss driven by ice-shelf melt. Nature 484, 502−505.

Radić, V., Hock, R., 2010. Regional and global volumes of glaciers derived from statistical upscaling of glacier inventory data. J Geophys Res 115, F01010. http://dx.doi.org/10.1029/2009JF001373.

Radić, V., Bliss, A., Beedlow, A.C., Hock, R., Miles, E., Cogley, J.G., 2013. Regional and global projections of the 21st century glacier mass changes in response to climate scenarios from GCMs. Clim Dyn 42 (1−2), 37−58. http://dx.doi.org/10.1007/s00382-013-1719-7.

Rahmstorf, S., 2007. A semi-empirical approach to projecting future sea-level rise. Science 315 (5810), 368−370.

Raper, S.C.B., Braithwaite, R.J., 2006. Low sea level rise projections from mountain glaciers and icecaps under global warming. Nature 439. http://dx.doi.org/10.1038/nature04448.

Rignot, E., Box, J.E., Burgess, E., Hanna, E., 2008. Mass balance of the Greenland Ice Sheet from 1958 to 2007. Geophys Res Lett 35, L20502.

Rignot, E., Jacobs, S., Mouginot, J., Scheuchl, B., 2013. Ice shelf melting around Antarctica. Science 341 (6143), 266−270. http://dx.doi.org/10.1126/science.1235798.

Rignot, E., Mouginot, J., Scheuchl, B., 2011a. Ice flow of the Antarctic ice sheet. Science 333, 1427−1430.

Rignot, E., Velicogna, I., van den Broeke, M.R., Monaghan, A., Lenaerts, J., 2011b. Acceleration of the contribution of the Greenland and Antarctic ice sheets to sea level rise. Geophys Res Lett 38, L05503.

Riva, R., Bamber, J., Lavallée, D., Wouters, B., 2010. Sea-level fingerprint of continental water and ice mass change from GRACE. Geophys Res Lett 37, L19605. http://dx.doi.org/10.1029/2010GL044770.

Robinson, A., Calov, R., Ganopolski, A., 2012. Multistability and critical thresholds of the Greenland Ice Sheet. Nat Clim Change 2, 429−432.

Ross, M., O'Brien, J., Ford, R., Zhang, K., Morkills, A., 2008. Disturbance and the rising tide: the challenge of biodiversity management on low-island ecosystems. Front Ecol Environ 7, 471−478.

Scambos, T., Bohlander, J., Shuman, C., Skvarca, P., 2004. Glacier acceleration and thinning after ice shelf collapse in the Larsen B embayment, Antarctica. Geophys Res Lett 31, L18402. http://dx.doi.org/10.1029/2004GL020670.

Schoof, C., 2007. Marine ice-sheet dynamics. Part 1. The case of rapid sliding. J Fluid Mech 573, 27−55.

Schoof, C., 2010. Ice-sheet acceleration driven by melt supply variability. Nature 468, 803−806. http://dx.doi.org/10.1038/nature09618.

Schoof, C., 2011. Marine ice sheet dynamics. Part 2. A stokes flow contact problem. J Fluid Mech 679, 122−155.

Schrama, E.J.O., Wouters, B., 2011. Revisiting Greenland Ice Sheet mass loss observed by GRACE. J Geophys Res 116, B02407. http://dx.doi.org/10.1029/2009JB006847.

Shepherd, A., Hubbard, A., Nienow, P., King, M., McMillan, M., Joughin, I., 2009. Greenland Ice Sheet motion coupled with daily melting in late summer. Geophys Res Lett 36, L01501. http://dx.doi.org/10.1029/2008GL035758.

Shepherd, A., Ivins, E., Geruo, A., Barletta, V., Bentley, M., Bettadpur, S., et al., 2012. A reconciled estimate of ice-sheet mass balance. Science 338, 1183−1189.

Sole, A., Nienow, P., Bartholomew, I., Mair, D., Cowton, T., Tedstone, A., King, M.A., 2013. Winter motion mediates dynamic response of the Greenland Ice Sheet to warmer summers. Geophys Res Lett 40 (15), 3940−3944. http://dx.doi.org/10.1002/grl.50764.

Sørensen, L.S., Simonsen, S.B., Nielsen, K., Lucas-Picher, P., Spada, G., Adalgeirsdottir, G., Forsberg, R., Hvidberg, C.S., 2011. Mass balance of the Greenland Ice Sheet (2003−2008) from ICESat data - the impact of interpolation, sampling and firn density. Cryosphere 5, 173−186.

Stanford, J.D., Hemingway, R., Rohling, E.J., Challenor, P.G., Medina-Elizalde, M., Lester, A.J., 2010. Sea-level probability for the last deglaciation: a statistical analysis of far-field records. Global Planet Change 79 (3), 193−203. http://dx.doi.org/10.1016/j.gloplacha.2010.11.002.

Tamisiea, M.E., 2011. Ongoing glacial isostatic contributions to observations of sea level change. Geophys J Int 186 (3), 1036−1044. http://dx.doi.org/10.1111/j.1365-246X.2011.05116.x.

Tedstone, A.J., Nienow, P.W., Sole, A.J., Mair, D.W.F., Cowton, T.R., Bartholomew, I.D., King, M.A., 2013. Greenland Ice Sheet motion insensitive to exceptional meltwater forcing. Proc Natl Acad Sci USA 110 (49), 19719−19724. http://dx.doi.org/10.1073/pnas.1315843110.

Thomas, R., 2004. Force-perturbation analysis of recent thinning and acceleration of Jakobshavn Isbrae, Greenland. J Glaciol 50, 57−66.

Tufnell, L., 1984. Glacier Hazards. Longman, London, 97 pp.

Vaughan, D.G., Comiso, J.C., Allison, I., Carrasco, J., Kaser, G., Kwok, R., Mote, P., Murray, T., Paul, F., Ren, J., Rignot, E., Solomina, O., Steffen, K., Zhang, T., 2013. Observations: cryosphere. Chapter 4 in: climate change 2013: the physical science basis. In: Stocker, T.F., Qin, D., Plattner, G.-K., Tignor, M., Allen, S.K., Boschung, J., Nauels, A., Xia, Y., Bex, V., Midgley, P.M. (Eds.), Contribution of Working Group I to the Fifth Assessment Report of the Intergovernmental Panel on Climate Change. Cambridge University Press, Cambridge, United Kingdom and New York, NY, USA.

van den Broeke, M., Bamber, J., Ettema, J., Rignot, E., Schrama, E., van de Berg, W., et al., 2009. Partitioning recent Greenland mass loss. Science 326. http://dx.doi.org/10.1126/science.1178176.

van der Veen, C., Plummer, J., Stearns, L., 2011. Controls on the recent speed-up of Jakobshavn Isbræ, West Greenland. J Glaciol 57, 770−782.

Velicogna, I., 2009. Increasing rates of ice mass loss from the Greenland and Antarctic ice sheets revealed by GRACE. Geophys Res Lett 36, L19503.

Vermeer, M., Rahmstorf, S., 2009. Global sea level linked to global temperature. Proc Natl Acad Sci USA 106 (51), 21,527−21,532.

Vieli, A., Nick, F.M., 2011. Understanding and modelling rapid dynamic changes of tidewater outlet glaciers: issues and implications. Surv Geophys 32, 437−458.

Werner, A., Simmons, C., 2009. Impact of sea-level rise on sea water intrusion in coastal aquifers. Ground Water 47, 197−204.

Wetzel, F., Kisslin, W., Beissmann, H., Penn, D., 2012. Future climate change driven sea-level rise: secondary consequences from human displacement for island biodiversity. Glob Change Biol 18, 2707−2719.

WGMS, 2008. Global Glacier Changes: Facts and Figures. UNEP, World Glacier Monitoring Service, Zurich.

Whitehouse, P.L., Bentley, M.J., Milne, G.A., King, M.A., Thomas, I.D., 2012. A new glacial isostatic adjustment model for Antarctica: calibrated and tested using observations of relative sea-level change and present-day uplift rates. Geophys J Int 190, 1464−1482.

Williams, S.D.P., Moore, P., King, M.A., Whitehouse, P.L., 2014. Revisiting GRACE Antarctic ice mass trends and accelerations considering autocorrelation. Earth Planet Sci Lett 385, 12−21. http://dx.doi.org/10.1016/j.epsl.2013.10.016.

Woodroffe, C.D., Nicholls, R.J., Saito, Y., Chen, Z., Goobred, S.L., 2006. Landscape variability and the response of Asian megadeltas to environmental change. In: Harvey, N. (Ed.), Global Change Implications for Coasts in the Asia-Pacific Region. Springer, London.

Wouters, B., Bamber, J.L., van den Broeke, M.R., Lenaerts, J.T.M., Sasgen, I., 2013. Limits in detecting acceleration of ice sheet mass loss due to climate variability. Nat Geosci 6 (8), 613−616. http://dx.doi.org/10.1038/ngeo1874.

Yin, J.J., Griffies, S.M., Stouffer, R.J., 2010. Spatial variability of sea level rise in twenty-first century projections. J Clim 23, 4585−4607.

Zemp, M., Haeberli, W., Hoelzle, M., Paul, F., 2006. Alpine glaciers to disappear within decades? Geophys Res Lett 33, L13504. http://dx.doi.org/10.1029/2006GL026319.

Zemp, M., Hoelzle, M., Haeberli, W., 2009. Six decades of glacier mass balance observations − a review of the worldwide monitoring network. Ann Glaciol 50, 101−111.

Zemp, M., Zumbühl, H.J., Nussbaumer, S.U., Masiokas, M.H., Espizua, L.E., Pitte, P., 2011. Extending glacier monitoring into the Little Ice Age and beyond. PAGES News 19 (2), 67−69.

Zhang, X., Church, J.A., 2012. Sea level trends, interannual and decadal variability in the Pacific Ocean. Geophys Res Lett 39 (21). http://dx.doi.org/10.1029/2012gl053240.

Zhang, T., Barry, R.G., Knowles, K., Heginbottom, J.A., Brown, J., 2008. Statistics and characteristics of permafrost and ground-ice distribution in the Northern Hemisphere. Polar Geogr 31, 47–68.

Zumbühl, H.J., Steiner, D., Nussbaumer, S.U., 2008. 19th century glacier representations and fluctuations in the central and western European Alps: an interdisciplinary approach. Global Planet Change 60 (1–2), 42–57.

Zwally, H.J., Abdalati, W., Herring, T., Larson, K., Saba, J., Steffen, K., 2002. Surface melt-induced acceleration of Greenland Ice-Sheet flow. Science 297, 218–222.

Zwally, H.J., Giovinetto, M.B., 2011. Overview and assessment of Antarctic ice-sheet mass balance estimates: 1992–2009. Surv Geophys 32, 351–376.

Index

Printed in the United States
By Bookmasters